NATHANAEL-ISRAEL ISRAEL, PhD

I0055361

Turbulent Origin of the Universe

OTHER BOOKS BY NATHANAEL-ISRAEL ISRAEL

Get them at your local bookstore, or online (e.g. on Amazon, Science180.com/books)

Turbulent Origin of Chemical Particles
Why You Don't Have to Embrace Evolution, Big Bang, or Deny God to Scientifically Prove the Formation of All Chemical Particles

Turbulent Origin of Life
Why You Don't Have to Embrace Evolutionism or Check Your Brain at the Door in the Name of Faith or Science to Accurately Decrypt the Origin of Life Using the Historic Formula of the Universe Formation

Reconciling Science and Creation Accurately
What Science Accurately Teaches about Creation and God's Existence that Atheists, Freethinkers, and even Most Christians Ignore … And How to Demonstrate it Without Taking Sides Between Rationality and Faith

Origin of the Spiritual World
Top Secrets about the Origin of Everything in the Universe that Some Elites Have Hidden from You for Thousands of Years

From Science to Bible's Conclusions
How Decoding the Universe-Origin by Properly Revisiting Scientific Data—That Top Scientists Collected but Wrongly Analyzed—Bizarrely led to the 3500 Years Old Biblical Account of Creation

How God Created Baby Universe
What Children Must Scientifically Learn Early about the Universe Formation to Avoid Dangerously Abandoning God Later in Life Just Like Most College Students Who Embrace Evolution and Big Bang That Deny Biblical Creation

How Baby Universe Was Born
How to Scientifically Talk to Children about the Universe Formation and They will Know Forever How to Correctly Test the Intersection of Science and Faith

Science180 Accurate Scientific Proof of God
Can We Scientifically Explain the Formation of the Universe Through Natural Processes Without Evoking Evolution and Big Bang?

More books written by Nathanael-Israel Israel can be found at Israel120.com/books

NATHANAEL-ISRAEL ISRAEL, PhD

Founder of Science180 & Father of Science180 Cosmology
Founder of Science180 Academy
www.Israel120.com

Turbulent Origin of the Universe

There is Only One Scientific, Simple, Safe, Trustworthy, Unexpensive, Brave, Practical, Nonconformist, Universal, Verifiable Formula that Accurately Decodes the Universe Formation … But You Are Not Using It

$$T = \frac{D}{Ve} + \frac{2\,R\,\pi}{Vo}$$

Science180

Augusta, Georgia
United States of America
www.Science180Publishing.com

Copyright © 2025 by Nathanael-Israel Israel
Visit the author's website at Israel120.com

Turbulent Origin of the Universe
There is Only One Scientific, Simple, Safe, Trustworthy, Unexpensive, Brave, Practical,
Nonconformist, Universal, Verifiable Formula that Accurately Decodes the Universe
Formation ... But You Are Not Using It

First edition: October 2025

Published by Science180
Augusta, Georgia (USA)
www.Science180Publishing.com

Book Cover and Illustrations by Nathanael-Israel Israel

ISBN: 979-8-9932150-1-3

Library of Congress Control Number: 2025920944

More books by the same author can be found at Israel120.com and Science180.com
For information about special discounts available for bulk purchases, please visit
Science180.com/discount for more details.

Science180 can bring authors including Dr. Nathanael-Israel Israel to your live or
recorded events. For more information or to book an event, please visit
Science180.com/speaking

For any questions, please visit Science180.com/contact

To publish your book(s) with Science180 Publishing, go to Science180Publishing.com

To interview the author of this book, visit Israel120.com/interview

To donate, please visit Israel120.com/donate or Science180.com/donate.

Published and Printed in the United States of America.

Content

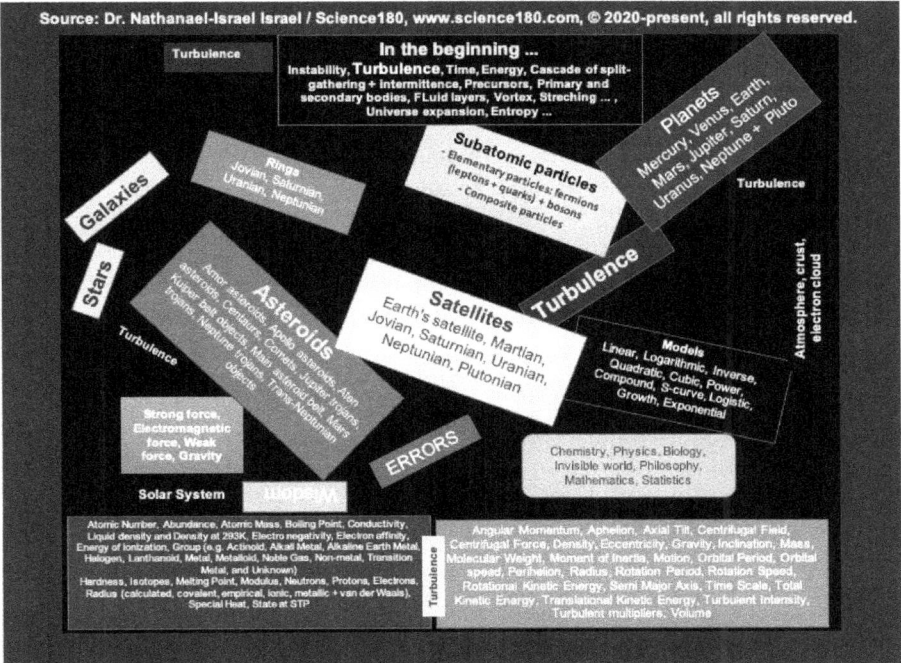

In the beginning ...
Instability, **Turbulence**, Time, Energy, Cascade of split-gathering + intermittence, Precursors, Primary and secondary bodies, FLuid layers, Vortex, Streching ... , Universe expansion, Entropy ...

Turbulence

Planets
Mercury, Venus, Earth, Mars, Jupiter, Saturn, Uranus, Neptune + Pluto

Turbulence

Subatomic particles
- Elementary particles: fermions (leptons + quarks) + bosons
- Composite particles

Galaxies

Rings
Jovian, Saturnian, Uranian, Neptunian

Stars

Turbulence

Satellites
Earth's satellite, Martian, Jovian, Saturnian, Uranian, Neptunian, Plutonian

Turbulence

Atmosphere, crust, electron cloud

Asteroids
Amor asteroids, Apollo asteroids, Aten asteroids, Centaurs, Comets, Jupiter Trojans, Kuiper belt objects, Main asteroid belt, Mars Trojans, Neptune Trojans, Trans-Neptunian objects

Models
Linear, Logarithmic, Inverse, Quadratic, Cubic, Power, Compound, S-curve, Logistic, Growth, Exponential

Strong force, Electromagnetic force, Weak force, Gravity

ERRORS

Chemistry, Physics, Biology, Invisible world, Philosophy, Mathematics, Statistics

Solar System

Wisdom

Atomic Number, Abundance, Atomic Mass, Boiling Point, Conductivity, Liquid density and Density at 293K, Electro negativity, Electron affinity, Energy of Ionization, Group (e.g. Actinoid, Alkali Metal, Alkaline Earth Metal, Halogen, Lanthanoid, Metal, Metalloid, Noble Gas, Non-metal, Transition Metal, and Unknown) Hardness, Isotopes, Melting Point, Modulus, Neutrons, Protons, Electrons, Radius (calculated, covalent, empirical, ionic, metallic + van der Waals), Special Heat, State at STP

Turbulence

Angular Momentum, Aphelion, Axial Tilt, Centrifugal Field, Centrifugal Force, Density, Eccentricity, Gravity, Inclination, Mass, Molecular Weight, Moment of Inertia, Motion, Orbital Period, Orbital speed, Perihelion, Radius, Rotation Period, Rotation Speed, Rotational Kinetic Energy, Semi Major Axis, Time Scale, Total Kinetic Energy, Translational Kinetic Energy, Turbulent Intensity, Turbulent multipliers, Volume

Due to the nature of turbulence, I did not feel comfortable doing a conventional table of content. For events in turbulence are not in a linear sequence as a secular table of content would tend to lay it. However, if you want to see the chronological order that I adopted to write the book, please go to the end of the book.

CHAPTER 1

MY JOURNEY OF WRITING THIS AMAZING BOOK THAT WILL SURELY HELP YOU

1.1. What you will get out of this book, and why and how I wrote it

What is the origin of the universe and how can we know it for sure using science, not by just believing in a myth or some religions that simply rely on faith without rigorous factual proofs? Why do some people think that the world was formed after billions of years processes, while others believe that it was created by God in 6 days? How can we know for sure who is right? And where are the scientific data they rely on to calculate or to believe in those timelines? What formula did they use and how do they know they are accurate? How did they know their math and stories are 100% correct? Those from Missouri (the "show me state" in the US where I got my PhD) would just say: "show me the data first, the formula, the proof, and even the witnesses if possible." If God created the universe, did He really do it in 6 days or are the creation days equivalent to billions of years today? How can a human being scientifically settle the disagreement between the billions of years story, the 6 days of creation narrative, and all other stories or myths about the origin of the universe? How can we scientifically calculate for sure how long it took for the universe to be formed? Wouldn't it be nice if someone could properly analyze the scientific data and the creation stories across the globe to settle their disagreements and help people know for sure the real process that birthed the world?

If you have ever asked any of those questions and are seeking the correct and scientifically-verifiable answer, then this book is for you. How or why am I confidently saying that? This is because, by using the scientific data and methodology, I unearthed (for the first time in history, and I really mean for the first time) deep secrets about the origin of the universe that I am very sure their ramifications will definitely change the world and impact many scientific fields. I don't know you, I don't know where you came from, nor what your profession is, but I am going to guess why you picked up this book and want to read it: you are a great person seeking the truth and want to understand the real story (backed by reliable scientific facts) behind the formation of the universe. By reading this book your life will be changed and you will have a fresh, original perspective on the beginning and formation of the universe using scientific data never tapped into before, because they were encrypted but are now decoded. By the time you finish reading this book, you will fully articulate and comprehend the scientific process and the duration of the formation of universe. Using facts, you will certainly know whether the universe was formed within 6 days as Jews and Christians believe or whether it was formed by a gradual billions of years process like the proponents of the Big Bang theory and evolutionism advocate. This book will also help you to comprehend why some people deny creation but embrace the billions of years theory, while others believe in a supreme being who created everything. In addition to scientifically knowing which story of the formation of the universe is correct or not, you will also understand whether there is any undeniable scientific proof of the existence of God or not, whether the difference in people's perception about the beginning is in the way they were raised, educated, brainwashed, or lack to properly, scientifically analyze the facts on the universe, or something else.

By the time you finish reading this mind blowing masterpiece, you will:

- Bridge the gaps between many existing theories and know how to reassess them in the new light accurately shined on the origin of the universe
- Discover how to bridge various fields of knowledge (including in areas outside your home field of expertise) to arrive at the best understanding of the universe-origin that will transform your career
- Discover programs strategically designed to enlighten you, guide you to navigate and filter the massive data collected on the universe and its content so you know how to answer the world's most challenging origin questions, remove any scientific and philosophical cataracts that may be blocking you, and help bring you many steps closer to your best life today and forever
- Disrupt all religious and scientific chains of repetitive nonsenses about the universe-origin, and turn your attention toward unconventional ideas leading to greater innovation and prosperity
- Draw unconventional inspirations from other disciplines to get the legitimate and irrefutable underlying logic of the occurrence of natural phenomena related to the universe genesis

Nathanael-Israel Israel: Member of the American Association for the Advancement of Science

CHAPTER 1: INTRODUCTION

- Educate yourself and others to better resonate with your target market or customers who are craving something original that breaks wrong explanations of the origin of the universe
- Finally satisfy your audience's burning desire for freedom from beliefs and scientific theories about the universe-origin that suffocate them and bind their mind, faith, unbelief, heart, and education
- Get specific in-depth knowledge, up-to-the-minute information, ideas, and insights about the universe-origin, so that you expand your market, cut useless costs, stop wasting time on inadequate projects, and start focusing on profitable solutions
- Harness the power of the existing scientific data to challenge the status quo of existing theories on the origin of the universe. Visit www.Science180.com/scientific to learn more…
- Learn at your convenience anywhere across the globe, from the only center in the whole world that provides an accurate, condensed, and comprehensive teaching on the origin of the universe and the things in it
- Learn authentic information not from someone who can just read a PowerPoint, but from the true go-to expert (when it comes to critical cosmological problems) who will help you get much closer to the better life you want to live
- Learn how to unearth the secrets of the universe formation and land on high-return discoveries in unexpected places by establishing unusual connections between previously disparate pieces of scientific knowledge, unconnected theories in diverse disciplines, and isolated observations in many disciplines that no one ever thought to be connected
- Open the way to a future of technology, innovation, discoveries, and breakthroughs based on the real decoding of the foundation of the universe and its content
- Properly identify and formulate original hypotheses and relationships worthy of further future investigations using the groundbreaking framework that is guiding and assisting great scientists
- Stand as the lightning bolt that electrifies those who are still struggling to understand the universe-origin
- Synthesize and make sense of prior empirical or scientific data on the universe-origin within the acclaimed Science180 Cosmology's framework
- Ultimately boost your confidence in detecting, confronting, and avoiding wrong theories by knowing the facts and processes involved in the formation of the universe

Science180: Where Universe-Origin is Accurately Decoded, Full Stop

- Undeniably learn how to revolutionize the way scientists and nonscientists use scientific data to demonstrate the formation of the universe and how to use that insight to boost your profit and improve lives permanently
- Use the proven universe-origin formula to excite yourself, and your audience, to positively raise their eyebrows, to keep them on their toes and to always be hungry for more food for their thoughts
- Unlock the secrets of the turbulence that occurred during the formation of the universe and pioneer new ways to study the universe through the glimpse of what I, Dr. Nathanael-Israel Israel, called the "split-gathering of precursors" as you discover, understand, and enjoy deep secrets of the formation of the universe with Science180's perspective.

But first, allow me to tell you a brief story about how I wrote this book and how my journey prepared me to view the scientific data collected on the universe from a different perspective than that of the physicists and other theorists on whom most people rely to tell the story of the universe.

As a holder of a doctorate degree (PhD) in plant, insect, and microbial sciences from a renowned university in the USA, where I graduated as the Doctoral Marshal, meaning the first of my class of hundreds of students, I, during my entire life (including my schooling years and the first few years of my professional life), never thought about writing a book on the origin of living things (which is more related to what I have studied) and much less a book on the origin of the universe (which seems beyond my original field), until something happened in 2013, which led to finding myself progressively addressing some challenging, unsolved questions in science including those tackled in this book.

Considering my past, those who know me will not be surprised at me writing a book on advanced knowledge in physics. Indeed, natural sciences (including physics, mathematics, chemistry, and biology) have always been some of my favorite domains of interest even before I reached high school. At high school, I majored in mathematics, physics, and biology, and I self-taught myself 11th grade and 12th grade courses leading to the obtention of my high school diploma in mathematics, physics, and biology. In high school, I used to be the one that my physics, mathematics, and biology teachers frequently sent to the blackboard to explain to my classmates the answer of our examinations and homework using my exam papers, which always got an A+ grade. Despite my shyness, I was accustomed to doing my best to help my classmates to advance their understanding in science. Foreseeing some life challenges coming on my way based on some financial struggles I was already going through, I abandoned the general education curriculum (which would have directly landed me at a university) to instead attend a vocational school after passing a test and winning a national scholarship, which covered my tuition fees, housing, and a $10 USD per month (pocket money). It was at that school in West Africa that I obtained my high school diploma by self-teaching, for the courses at that vocational school were based on tropical agriculture and crop production, which were not the curriculum of the general education system I was attending.

My "success" story of obtaining those high school diplomas in those days inspired many people and some students were interested in me teaching them the courses of the general educational system after the regular class hours. Therefore, at one point, even before I started my education at university, I created an evening school program attended by more than 20 students, who wanted to follow my path to get their high school education in physics, mathematics, and biology in parallel to the training we were getting in agriculture, which is perceived as the engine or foundation of the economy of most African countries. I used to teach math and physics to those students after my school hours. Many of the students I trained passed their high school diploma and some of them are even university professors today. Looking back, I felt like physics and mathematics have always been in my DNA, and if it was not life's challenges that made me to strategically leave the secular general educational system temporally to join a vocational school to seek a diploma in tropical agriculture, I could have never taken the path of life sciences that led to my Bachelor's Degree in Agronomy, a Master of Science Degree in Agricultural Engineering and Natural Resources Management in Benin (West Africa), where I graduated first of my class before moving to the USA where I got my Ph.D. in Plant, Insect, and Microbial Sciences, again as first of my class.

Despite my academic records, I never really felt like my calling was in pure secular research. Before I finished my Ph.D. in the USA, I told my PhD advisor, the late Professor Gene Stevens (I wish he was still alive to see my accomplishments) that I did not think my mission in life was to teach or to do conventional academic research as it has been done. In fact, although I like research and am good at it, which can be proven by some awards and honors I received and scientific articles I published in peer reviewed journals with high impact fact, I was not really pleased with some limitations put on scientists in some academic environments. Because I always seek to think freely and not lay my life just for a financial gain, I felt like, after my PhD, I might better enjoy life doing research in nonacademic fields. A few months before I graduated from my PhD, I got a job (at a 500 Fortune company in the USA) earning about $100,000 USD per year. But it did not take too long after getting that first job before I realized that my calling was somewhere else outside the big multinational company where I was paid to do "research" like an illiterate or an idiot who, like a restless robot commanded to work without considering my full potential and needs, must produce data to justify some products for sale, regardless of the harms they could cause to customers and the planet. Welcome to the corporate world of research done for profit, not really to help customers!

At one point, after that biotechnology research scientist experience ended for reasons I explained in my incoming autobiography, I funded my first owned business, Later, I created an online news company, Global Diaspora News (www.GlobalDiasporaNews.com) that sources and disseminates knowledge across the globe with an emphasis placed on the international diaspora and their stakeholders.

However, despite the great feedback, service to the world, and experience I got from those businesses, they did not satisfy my thirst and hunger for success, nor did they get me close to the multi-revised version of my modest and downgraded American dream I was initially told could be reached with a good education, nor did they get me near the fiction and movie-like fantasy expectation that some people tried to sow in me before I headed to the Land of Uncle Sam, which I wished I had the space and time to describe under the reign of President Donald Trump (the 45th and 47th US President), during which unconventional shocking waves were regularly sent to the whole world from Washington DC, which also happens to be the first city I found myself after leaving the "Quartier Latin de l'Afrique", Benin (West Africa), thinking that I was going to the "paradise" on Earth, not knowing that, about 20 years later, I would be deeply writing a book in 2020 during which human beings finally got their first predator, a coronavirus virus, which pandemic ravaged the world. As the world was going through unprecedent pain in that memorable 2020, I was also pushing to finish the writing of the first draft of this book! In other words, although I finished writing the bulk of this book in 2020, I have to wait until the right time, 2025, before I published it and eight others I wrote on various aspects of the origin of the universe, of life, and of chemicals, targeting different audiences and age groups. To set the record straight, the first ideas and inspiration about this book came to me during the second administration of President Barack Hussein Obama (born 1961, 44th president of the US, 2009–2017). Then, I discovered turbulence and wrote the first draft of this book during the first administration of President Donald John Trump (born 1946, 45th president of the US, 2017-2021). Then, I reviewed this book, split the first draft, and expanded it into 9 books during the administration of President Joseph Robinette Biden, Jr. (born 1942, 46th president of the US, 2021-2025). Finally it was in 2025, the first year of the second administration of President Donald John Trump (47th President of the US) that I published this book and 8 others (see details at www.Science180.com/books).

Due to my thirst for knowledge and my desire to have the big picture of things in the world, I delighted in broadening my knowledge, skills, and expertise in various domains. I then became a member of several professional associations including the American Association for the Advancement of Science; American Society for Microbiology; American Society of Agricultural and Biological Engineers; American Society of Agronomy; American Society of Biochemistry and Molecular Biology; Crop Science Society of America; Soil Science Society of America; American Chemical Society; French Writers Association; etc. Throughout the years, I increased my ability to view scientific problems from various angles and I gained deep insight into various things in life, some of which led to my digging into the origin of the universe.

As evidences accumulated during my search of the origin of the world, I felt like I had to tell the story that I discovered without initially laying hypothesis as conventionally done in science. Time and space did not allow me to tell the entire story of the motivation to write this book here, but to make a long story short, I just did not get up one day and decide to write this book. It has been my life journey, and need to solve problems I faced in life that indirectly led me into this research and by the time I resolved to write this book, I have found myself with thousands of pages I wrote, and from which I had to pick extracts to tell the story I endeavored to share with the world in a language that I hope can be understood not only by the scientists, but also by the profane and/or those who may be perceived as not having enough scientific background, yet who are also interested in the origin of the world.

Unlike most scientific studies which are funded by governmental or private grants, this research was 100% funded by myself, and I had to rely on what I have to cover the fees involved in the writing, publication, and promotion of my books about the origin of the universe, life, chemicals, etc. Consequently, I was not limited by guidelines from anyone or any institution which might have forced me to follow a certain methodology of research and publication with strict deadlines and other restricting obligations. This freedom of thinking and acting allowed me to freely explore the topics I addressed in this book and others as I deemed relevant without worrying about anybody wrongly judging my methods, speed, scope of the work, flexibility, strategies, plans, visions, deadlines, objectives, and many other aspects of this research and beyond. This means that this book is my pure reflection on the topics I addressed, and it was not controlled by anybody which could have restrained me. I could have wished to work with others while investigating the problem, but by the time I finished this book, I realized that, if I had involved people and institutions into the research, I could not have been able to achieve results presented here with the same originality. In other words, although I did not initially plan to write this book in secret, I can now confess that, waiting for it to be ready before presenting it to the world seems to be the best way to get the task done although I paid a big price. As you read my autobiography and other books I wrote on the origin of the universe and the things and beings in it, you may better understand the history, efforts, strategies, and methodology of my writing of this book.

Despite what I said above, some people may still be interested in knowing more about why did I hide the discovery and waited until completing the entire theory before revealing it to the world? Indeed, the idea of hiding the preliminary data was first made known to me a little before April 2013 and emphasized to me on December 2nd, 2016 (meaning 4 years before the writing of the first draft of this page that you are reading now). Looking back, I could see that if I had even published the data in 2013 as they were coming to me, many things would have been left out, while criticism and competition could have sunk my enthusiasm in completing the work. Before I made the decision to hide my findings until I fully have the complete story, and my strategy all set to make a significant impact, I was feeling like if I was not careful, patient, and endurant, some people or systems could have forced me to start publishing preliminary data, which could have caused some problems. These efforts by others could have also confused me or caused others interested in the topic to wander in many directions, which may satisfy their prejudicial hunger or thirst, but which would contradict the truth and complicate my efforts. I also felt like if I blew the research to the world early, some people may get jealous even before I reached the end, and others may block or prevent me from finishing the work and present it to the whole world. For the ignorance of some people is what is feeding the so-called wealth or prosperity of others, which usually do not want the captives to be freed but to keep working for them. In contrast, some people may start honoring me for my modest contribution to the understanding of the origin of the universe, which could mislead many things and even have a detrimental effect on me for taking the glory of something I do not deserve. Regardless of their positivity or negativity, comments from others could have drifted me away from the target for I may have risked to turn my attention toward them and/or try to respond to some of them, therefore annoying myself with things which would have consumed my time and/or diluted my originality to think through everything before going live. Therefore, I decided that, even if I have to suffer as much as needed, I must finish the project before publicizing it. I was also hoping that this strategy could allow me to explore as many angles as I perceived are relevant without crumbling or dispersing my efforts or missing the big picture. Honestly, there were times I wished I could talk to some friends, but I did not have many (near me) and I could not trust anyone to keep the secret (for years) until its due time. Hence from 2013 until I wrote this book in 2020, the only person who was aware of it did not "fortunately" understand much of it (else I may have been more concerned about that person blowing it to other people who may leak part of it on social medias and/or to family members before I even finished the introduction).

By the time I was ready to reveal the work to the whole world, I realized that some institutions and even nations could try to classify or conceal the findings if they could get a hold of it beforehand, and consequently keep the secrets for themselves, while the whole world needs to know the truth that people have sought to know for thousands of years. I also sensed that if I can reveal part of the secrets to some businesses, they could make a lot of money from the technology that could be invented based on my work, but those companies may bury the work forever. After weighing the options associated with publishing the results, I felt like the best strategy is to ensure that the work is quickly introduced to the whole world in a widely publicized way not only to preserve some of my copyrights, but also to allow every nation to be informed of the discovery. To prevent secular publishing companies from slowing me down, leaking, and diluting the results, while some of them may try to twist my discoveries to falsely claim the ownership of my hard work, I realized that I should not submit my books to any traditional publishers, but I needed to create my own publishing company to publish my books and articles. This decision was also based on my previous experiences with book publishing with other companies for other books I wrote.

Instead of seeking how to be rich from my discovery as some may do, my goal was (among others) to ensure that the book is published in its original form without delay and with the possibility that it can reach as many people as possible at a cheap and affordable price. At one point, I even contemplated making the entire work free and allowing people to directly download it from the website www.Science180.com, which will publish future updates, but I felt like doing so may cause the world not to be interested in it, for I have also come to realize that people tend to mistreat free things for which they did not put much effort into. It was toward that end that I funded Science180, a publishing company in the USA, to carry the publication not only of my books, but also those written by others. Knowing what I addressed in this book and others, I expect them to be translated into many languages very soon.

Considering the amount of data that I dealt with and the process that led to the writing of this book, I know that, although errors can still exist in it, it presents the proper pathway to finally understand how the universe was formed. I expect a positive, scientific awakening or revolution to follow the publication of this book and the others in which I tackled different domains of knowledge, leading to a historic technological and innovation revolution across the globe. Unlike all other books written about the origin of the universe, the current book is a radical one, which is not written using a conventional methodology, but using a unique strategic way of thinking.

By the time I finished analyzing the amount of data existing in the universe, I realized that, not only human beings have many scientific data, but they also have several philosophical (including religious) viewpoints that are difficult to reconcile in one book. I also apprehended that everybody is not equipped to handle the same number of scientific proofs, and some people sometimes cling to their religious views even in the presence of facts that disprove them. Another thing I learned is that most people do not like reading things which directly oppose their conviction just as Muslims do not read the Bible, nor do the Christians the Koran, nor do Evolutionists like to read Creationist books, and vice versa. Because they seem to see literatures written about religions or ideologies they do not believe in as unauthentic and unworthy to be read, most people are imprisoned by what they think they know.

Therefore, because all sorts of human beings have been questioning the origin of the universe, I needed to ensure I address each of them by providing enough evidences or food for their thought, while being aware that no mere human being can ever fully satisfy or convince everyone. In 2020, to ensure I appeal to my audience according to their education, knowledge, philosophy, and religion, I initially decided to write 3 versions of this book:

- A scientific or secular version
- A public version (that the general public or nonscientists can understand)
- A philosophical version (that I also called the religious or prophetic version)

Splitting the work into 3 main books also initially allowed me to fully express my thoughts and findings without diluting them or being too diplomatic in my presentation of the evidences or being fearful that people will not read them. For instance, I felt like it is not strategic to talk about religion in the scientific version or to talk too much about science in the public and religion versions. I do not need to emphasis that some scientists do not want to read some religious books just as some religious people do not want to read some scientific books, while the non-scientists cannot read scientific books. Hence, I thrive to help people feel comfortable reading the version(s) that pleases and fits their needs, and if they can read all these versions, they can get the full picture of my understanding of the origin of the universe. Because the current version of the book is the scientific one, some people who do not have a scientific background (at least that of a middle school student) may find some scientific chapters difficult to read. The public version of the book did not focus on raw scientific data which can annoy, scare, or confuse the profane, but emphasizes key information that presents the story of the origin of the universe in a language that the non-scientists can understand. In my book called *"Reconciling Science and Creation Accurately"*, I detailed religious data that support the scientific evidences. Instead of trying to unite all religions, I clearly expressed my religious viewpoints as I dealt with this critical topic of the genesis of the universe, which cannot be completely understood without referring to some spiritual facts, which science can likely never be able to explain using conventional methods. Yet, as I was working on the first version of this scientific version, it became too long, being more than 2000 pages. Therefore, because the chapters I have on chemical elements alone were more than 300 pages, I decided to isolate them to make another book solely devoted to chemistry. Hence, I did not detail all the chemical aspects of the formation of the universe in this book, but in the book devoted to the origin of chemical particles. Furthermore, because the chapter I wrote on the origin of living things became too long for this book, I removed it and instead wrote a completely different book on the origin of life: *"Turbulent Origin of Life"*. I also called that book the biological version of my book on the origin of the universe. Likewise, as I was working on the religious book, I came across key information in the lost and rejected scriptures called the pseudepigrapha. Therefore, to avoid offending some people who may not accept that some of those books are authentic, I had to split the initial religious book into two: a Biblical version and a pseudepigraphic version.

By the time I finished all of these books, my experience with my children also inspired me to write 2 children's versions: a religious children's version that believing children and their parents will enjoy, and a secular children's version targeting children of unbelievers.

Before I proceed with recounting my understanding of the formation of the universe, I would like to say that, if people were about to write a storybook, they can likely proceed in 5 ways:

1. They are the main actor of the story or the main actor of the story told them how the events of the story happened and they then recount it to others.

2. Without much data, they just made up a story by assuming things which do not consider the massive data available in all domains of knowledge, including those they don't know but which exist and those they know but refuse to use because they do not believe in or like them.

3. They try to build a story solely based on what they know in their domain of expertise or belief only, while they ignore all other domains of knowledge, not because they do not like them, but because they do not want to even address them because they are beyond their expertise and/or they do not have the resources to address them, yet they strongly think or believe that their partial version of the story is 100% correct and that any other version is incorrect.

4. They do not try to tell a story by researching things, but after hearing things here and there, they just recount what they think could have happened and/or what people are talking about.

5. They learned pieces of information related to events which have occurred and/or could have occurred or which currently exist after the events of the story happened and then, after deeply studying the characteristics of the things they learned about, they try to piece together the events of the story by reverse engineering them based on the data, not just based on what anybody has told them or what they have read in any book or what they imagined.

I would like to mention that my story (that you are about to read) was compiled based on the method #5 of storytelling I just mentioned. Nevertheless, seeing the speed and the thought process with which I wrote this book and all the supplemental data attached to it, I sometimes felt like I was thinking through things under the influence of something or someone. I will later tell you what I mean. For now, hold on and let us get into some details about how everything could have begun with what I called the mother of all turbulences.

In the end, by the time I published this book in 2025, I wrote 9 books on the origin or the universe, as presented below:

1. *"Turbulent Origin of the Universe":* The scientific version (which is the current version you are reading)

2. *"Turbulent Origin of Chemical Particles":* The chemical version (which focuses on the origin of chemical particles)

3. *"Turbulent Origin of Life":* The biological version (which deals with the origin of life)

4. *"From Science to Bible's Conclusions":* The public version (which is the summary of my findings in a language fitting the general public)

5. *"Reconciling Science and Creation Accurately":* The philosophical version (which I also called the prophetic or Biblical version, and in which I handled creation from the Biblical perceptive as compared to that of other religions)

6. *"Origin of the Spiritual World":* The pseudepigraphic version (which contains information hidden in the lost and rejected scriptures called the Pseudepigrapha, written before the Common Era)

7. *"How God Created Baby Universe":* The religious children's version addressing both science and faith (which diluted the main points for children ages 7-12, and who belief in God)

8. *"How Baby Universe Was Born":* The secular children's version addressing science only (which diluted the main points for children ages 7-12, who do NOT believe in God)

9. *"Science180 Accurate Scientific Proof of God":* God's existence book (a book to scientifically test the existence of God)

On the next pages, I presented more details on these books, so you can better put this book you are reading in the context of my global vision of writing on the origin of the universe and its content.

Throughout my writing, wherever you see "universe-origin", please know that I meant "origin of the universe" or "the origin of the universe". Likewise, wherever you see "life-origin", please understand that I meant "origin of life" or "the origin of life". In the same manner, wherever I mention "chemicals-origin", please know that I am referring to "origin of chemicals" or "the origin of chemicals".

Nathanael-Israel Israel: Member of the American Association for the Advancement of Science

'Science180 Academy' Success Strategy:
SCIENCE180 BOOKS THAT WILL HELP YOU!

I, Nathanael-Israel Israel, broke down my discovery about the formation of the universe into many books so that you, the readers, can pick the ones that correspond to your needs and interests without disappointing you or wasting your precious time. These books come in many versions (e.g. scientific version, public version, chemical version, biological version, biblical or prophetic version, pseudepigraphic version, and a children's version) targeting people according to their expertise, educational background, and interests as briefed below:

1. **"TURBULENT ORIGIN OF THE UNIVERSE"** (This is the scientific version of my book tailored to scientists and anyone interested in the detailed scientific demonstration of the universe formation). In this book I used the "mother of all turbulences" to scientifically demonstrate the formation of the universe so that scientists can understand and reorient the course of their research, teaching, and publishing, and accept the truth to better live today and forever. Get *"Turbulent Origin of the Universe"* today to begin an incredible journey of accurately decoding the universe and change your life forever! Learn more at Science180.com/scientific

2. **"RECONCILING SCIENCE AND CREATION ACCURATELY"** (this is the book that I called the "Biblical or prophetic version of my book on the universe-origin, and it targets Christians and anyone interested in knowing the Biblical perspective of the creation of the universe). This important book accurately demonstrates the marvelous creation and formation of the universe by God in six consecutive 24-hour-days, and answers many questions about the universe creation, so that after acknowledging Him (who deserves all the glory now and forever), human beings can choose life and avoid the terrible judgment awaiting the unbelievers in the world to come. Get this thoughtful book now to figure out what happened at the beginning, what is coming up, and why it is time to urgently rethink everything you have been told about the universe-origin so you don't eventually regret! Don't say I did not tell you! Learn more at Science180.com/biblical

3. **"TURBULENT ORIGIN OF CHEMICAL PARTICLES"** (Called the "chemical version" of my book on the universe-origin, this elegant book targets chemists, biochemists, and anyone interested in chemistry). With *"Turbulent Origin of Chemical Particles"*, the accurate decrypting and understanding of the formation of chemicals has never been profitable and easy. Hence this great book is THE ultimate how-to guide for great people wanting to correctly decode the origin of the chemicals and positively transform their lives. Get this celebrated book today. Learn more at Science180.com/chemical

4. **"ORIGIN OF THE SPIRITUAL WORLD"** (This book is what I called the pseudepigraphic or hidden version of my books on the universe-origin, and it is meant for believers who want to tap into a higher level of scriptural secrets that most people may not believe). This book draws the attention of the world toward the pseudepigrapha (a collection of hidden and rejected books, yet filled with deep secrets still valuable today) and explaining how, since thousands of years, God has already revealed deep details about the supernatural origin of the universe, but people (including those who believe or claim to believe in Him) have just refused to literally accept God's mysterious story of creation, which can never be understood by just sticking with conventional science. If you believe in God, have some origin-related questions which answers you cannot find anywhere, not even in the Bible, and if you want to tap into historically neglected revelations to answer fundamental universe and life questions, then be sure to get a copy of *"Origin of the Spiritual World"* today. Learn more at Science180.com/pseudepigrapha

5. **"FROM SCIENCE TO BIBLE'S CONCLUSIONS"** (I called this book the "public version" of my book on the origin of the universe and it is tailored for the general public, and it is a great summary of the scientific version from a perspective that laypeople will fully understand). In this book, I, Nathanael-Israel Israel, broke down the complicated (scientific, philosophical including religious) data about the origin of the universe in a simple language that the general public can fully understand, and know in order to live happily forever. Quickly grab and read this scientifically verifiable, bestselling book to finally get the accurate, jaw-dropping answer that has been rationally shaking both believers, skeptics, and all freethinkers. Don't wait! Learn more at Science180.com/public

6. **"TURBULENT ORIGIN OF LIFE"** (This is the biological or life version of my book on the origin of the universe). It is meant to suit both scientists, nonscientists, and all kids of laypeople, and it decodes the origin of all forms of life so human beings can understand and better live. As of 2025, this book is my only book devoted to the origin of all forms of life, and it will help you to grasp in a simple language what is needed to fully understand the formation of all forms of life. Whether you are a scientist or a layperson, a believer, or a skeptic, you cannot afford to ignore the greater, better, faster, simpler, cheaper, easier, and accurate formula unlocked in this important book that successfully decoded the origin of life. Get *"Turbulence Origin of Life"* today and change lives. Don't wait. Learn more at Science180.com/life

7. **"HOW BABY UNIVERSE WAS BORN"** (How was the universe formed? Did God really form it like some people believe, or did it come out of some long processes? How can we scientifically prove and break down this difficult mystery in a language that children will fully understand and like?) Get the answers as you read this book that I called the "children version" of my book on the origin of the universe and life. Accurately explaining the complex formation of the universe and of life to children can be very hard in our modern world, but by getting *"How Baby Universe was Born"*, you will know the proven formula to help children to easily understand their huge universe-origin and life-origin questions with confidence, humor, and joy. They will surely belly laugh and thank you for it! It is time to buy this pragmatic book and offer it to the children in your life today. Learn more at Science180.com/children

8. **"HOW GOD CREATED BABY UNIVERSE"**. The most difficult part of writing scientific things to children is how to break down complex technical concepts into simple words that they and even anyone who can read and clearly understand (without losing the accurate details and facts). When the topic to address is about the origin of the universe, the task is even more challenging for most people, but not for Nathanael-Israel Israel. As long as you can read, you will find this amazing book extremely helpful to grasp all complicated concepts needed to properly crack the origin of the universe in a language that even children ages 7-12 and anyone who did not go very far in school can fully comprehend.

9. **"SCIENCE180 ACCURATE SCIENTIFIC PROOF OF GOD"** (Whether you are a believer, an unbeliever, a freethinker, administrator, politician, curriculum designer, curriculum specialist, education policymaker, librarian, school board member, parent, researcher, student, teacher, clergy, or a layperson, as long as you are really seeking to scientifically understand the rational proof of the existence of God, *"Science180 Accurate Scientific Proof of God"* is the much-admired book written for great people just like you). As long as you are interested in the first and the only scientific book that talks to anti-creationists, evolutionists, big bang proponents, atheists, and all other freethinkers and rationalists about the universe formation and they bigly beg to know more about God, the creator, that they mistakenly deny; then this book is for you. As long as you are really seeking to scientifically understand the rational proof of the existence of God, *"Science180 Accurate Scientific Proof of God"* is the much-admired book written for great people just like you. Grab it today and start reading it. Don't wait any longer! Learn more at Science180.com/godproof

If you want to have the entire big picture of my discovery of the origin of the universe, life, and chemicals, and to enlighten your life and career, then plan to get all or some of these books that best suit your needs and interests. For more details, visit Science180.com/books

1.2. What is turbulence and why I talked about the mother of all turbulences?

After 7 years of advanced research about the origin of the universe and its content, and while I was looking for scientific terms and theories to accommodate and explain the trends I was seeing in the data, I can confidently say that the formation and the physics of things in the universe are related to a unique and gigantic turbulence. Before I start dealing with and breaking down turbulence, as best I can, I felt like I should first define what it is and what is known about it. Turbulence is considered as "the last great unsolved problem of classical physics". No clear definition exists for turbulence, a domain of research which did not register much major advances after more than 500 years of its study. Indeed, some of the initial details (e.g. helicoidal structures and "disordered" movement) in turbulence (even the term turbulence itself) were first discovered (at least qualitatively) by Leonardo da Vinci (born 1452 – died 1519). It was in the 20th centuries that scientific advancements allowed to better see some structures present in turbulence. Since then, progress has been made, yet still a lot needs to be understood.

CHAPTER 1: INTRODUCTION

If you have ever seen the movement of cream poured into coffee or the movement of clouds in the atmosphere, you have already experienced turbulence. Fluid (e.g. air) flows around a vehicle or an airplane are also very turbulent. Turbulence is usually defined by those researching in that field as a state of fluid motion characterized by seemingly random and chaotic multi-dimensional vorticity. In other words, vortices or eddies are always present in a turbulent fluid. In general, when most people and even the experts in the field think about turbulence and its features, they usually refer to it as chaos, a state of disorder, randomness, trouble, or a disordered motion of a "crowd." However, when I looked at turbulence data as I explained in this book, I did not see chaos, but a high level of order that people so far have failed to perceive. Therefore, after my discoveries in turbulence (as you will see in my books), I felt like I should craft my own definition of turbulence and I did so on October 28, 2018. Indeed, while scientists define turbulence as a state of disorder and chaos, I have come to realize that in contrast to all those nonsenses and testimonies of ignorance, turbulence is a complicated phenomenon that can lead to an ordered state of matter that did not always allow scientists to comprehend it because they were approaching it with a limited and corrupted mindset, which does not even know that every kind of matter in the universe is a product of the original turbulence, which occurred at the beginning of the universe. I saw order everywhere in turbulence, but the problem is that the order in turbulence is so complex and filled with details that a mere human being using the current limited computational machine cannot easily understand it. In other words, turbulence contains too many data and grabbing the full picture of its information is impossible to those who like to think linearly. To see some details in turbulence, scientists will have to magnify their data and the scale of their experiments as well as the power of the equipment they use to visualize these data. Unfortunately, scientists are limited in what they can see and when they see things, they cannot properly analyze them because they think turbulence is a disorder. They also tend to focus on bigger things, while neglecting the smaller ones, which should help them to have a complete understanding of how things were formed. This understanding did not come to me overnight but after a lot of research and writing about the massive data collected on things in the universe in many domains of knowledge. I also realized that the initial turbulence in the universe occurred at many scales and locations in space. Although many things in the universe today are no longer going through the kind of strong turbulence at the beginning of the universe, they are still "feeling" or bearing the consequences of that turbulence, which birthed them. To put it in another way, many kinds of turbulences have been occurring in the universe, but they all can be linked to the initial turbulence that I called the "Mother of all Turbulences", the strongest and largest turbulence of all time. Before that turbulence, no other turbulence has ever existed and since it occurred so the universe could be formed, no other turbulence as strong has occurred yet although many strong turbulences are still happening in the universe on many scales. Other names that can be used to describe the "Mother of all Turbulences"

Science180: Where Universe-Origin is Accurately Decoded, Full Stop

are the:

- Major Turbulence
- Universal Turbulence
- Turbulence at the Beginning
- Genesis Turbulence
- Father of all Turbulences
- Turbulence #1
- Initial Turbulence

Therefore, the theory of the origin of the universe that I elaborated in this book using the "Mother of all Turbulences" can be called "Turbulence 1", or "Theory of Turbulence 1", or "Turbulence 1 Theory".

One thing is to ask how the universe was formed, but another one is to know what needs to be demonstrated to answer that question using a sound methodology that also addresses what is present in the universe, while counting an enjoyable and logical story, and yet on target. Therefore, before I move to the next chapter, I felt like I needed to say something about what was really known for sure about turbulence. Although I wrote another book completely devoted to turbulence for the experts in that field, I felt like at this point, I should briefly mention some key areas of interest and research in that field. By doing so, the readers can have a glimpse at why and how I addressed some challenging problems in this book.

Indeed the scientific community has done a lot of research to try to explain things in the universe. However, some mysteries remain unexplained, leaving scientists scratching their heads and wondering if they could ever crack the unsolved, most difficult problems in the world. Turbulence is one of the top unsolved problems in science and one that many of the top scientists which have ever lived, including many Nobel Prize winners in Physics, thought, and taught that God cannot even explain. For more than 5 centuries, scientists, particularly physicists had a tough time explaining the phenomena of turbulence in fluids, and mathematicians have nightmares about it as well (Gupta, 2015). Nobel laureate, Richard Feynman, declared turbulence as "the most important unsolved problem in classical physics." When asked what he would ask God if given the opportunity, Quantum physicist Werner Heisenberg said *"I'll ask him two questions. Why relativity? And why turbulence? I really believe he will have an answer to the first."*

I tried to understand what is really known about the so-called chaos in turbulence, but I could not find much things set in stone until I finally came across an honest and frank lecture on turbulence by Professor William George, a dual citizen of the US and Sweden, a professor of turbulence and one of the scientists whose work I have read because of frankness. In fact, when I was reading some of his articles on the state of knowledge and perspectives on turbulence, I often sensed his honesty regarding how very little has been done since scientists started working on turbulence. To try to review what is known in

Nathanael-Israel Israel: Member of the American Association for the Advancement of Science

turbulence for sure, the following statement of Professor William George can be useful (George, 2013b):

"When I [Professor William K. George] started working in turbulence in the 1970's, I really thought we would be much farther ahead in our understanding and in our ability to predict them numerically. The fact is that, in my opinion, we have made almost no real progress in the past 30 years. And the reason I believe is that almost all of the funding and effort in that field of research has gone into attempts to do experiments and numerical analysis that could have been recognized at the outset to simply not measure up. Some of these efforts have been truly heroic in terms of the effort and ingenuity needed to perform them. But the reason they contributed little is that they began with a very poor understanding of the problem. And in particular, they failed to recognize what was necessary to truly make a contribution to our understanding – as opposed to simply using a new experimental technique or proving that a bigger computation could be done. I hope to show that these efforts were doomed from the outset: there simply was not enough computational power to compute an interesting flow. Or the experiments were far too small. I'm pretty sure the same is true today. Sometimes the best experiment or simulation is the one that was NOT done. The people who fund our work really don't like paying for something twice. So, if you do something that really was not quite ready to be done, you preclude someone (often yourself) from doing it correctly. Few really understand, and most believe things that are incorrect or unjustifiable. While some experts in the field of turbulence trust some existing theories are a gospel, I [Professor William K. George] is very careful in doing so... Now not everyone who teaches courses on this subject (and especially those who write books about it) will tell you this, but the truth is we really don't know a whole lot for sure about turbulence. And worse, we even disagree about what we think we know! There are indeed some things some researchers think they understand pretty well, but most of them are based on assumptions and logical constructions about how some researchers believe turbulence behaves within certain limits. However, even these ideas have never really been tested in controlled laboratory experiments, because no one has ever had the large-scale facilities required to do so. The acceptance of most ideas in turbulence is perhaps more due to the time elapsed since they were proposed and found to be in reasonable agreement with a limited data base, than that they have been subjected to experimental tests over the range of their assumed validity. Therefore, it might be wise to view most 'established' laws and theories of turbulence as more like religious creeds than matters of fact."

Turbulence is a field of research where no one knows for sure the right equations to lay or the methods to solve them without creating more equations, while less variables exist to tackle them. On a larger scale such as that of fluid dynamics as a whole, the top experts are not even sure if the questions and equations used are even well posed. For instance, the Navier-Stokes equations (named after the French mathematician, Henri Navier, and the British physicist, Sir George Gabriel Stokes), which has been dominating fluid dynamics since the 19th century, is not even proven yet. A $1 million USD prize that Clay Mathematics Institute announced to give to the person who will prove that the equation was even well laid (Navier–Stokes existence and smoothness problem) is still on. Today, turbulence is usually expressed or predicted using what is called the "Reynolds number", named after the British physicist, Osborne Reynolds (1842–1912), a dimensionless quantity in fluid mechanics used to forecast flow patterns of fluid flows. When the Reynolds number of a fluid is low, it is believed to be laminar (meaning the fluid layers are moved like smooth sheets); but when the Reynolds number of a fluid is high, turbulence occurs in the fluid.

I was presenting the viewpoints of some experts in the field not just to expose that little is known about turbulence, but to also show that, while I was trying to scientifically explain the formation of the universe through turbulence, I did not have much proven existing methodologies of research to confidently rely on, but I had to creatively invent some formulas and methods myself based on the data I was dealing with and how I perceived they can be better presented. By the time I finished writing this book, I understood that little is known in turbulence because it points to the origin of the universe and everything in it, and mastering it may be beyond some of the far-reaching limits of the human spectrum of knowledge.

As of today, to better understand turbulence and the anatomy of turbulent motions, some scientists think that they should decompose turbulence into scales and mathematically describe the following (Eggers and Villermaux, 2008):

- the different scales of motion;
- the equations that describe these scales;
- how these scales get their energy;
- how these scales transfer energy to each other;
- how structures (e.g. vortices) at different scales interact with (or evolve from) each other
- etc.

Seen from the angle of jet physics involving fluid breakup, turbulence presents other challenges. Indeed, in fluid dynamics, a jet is a "stream of matter having a more or less columnar shape". Jets occur on the scale of the universe as well as on subatomic length scales" (Eggers and Villermaux. 2008). According to that author, almost all classical physics comes into play in jet dynamics, and some of the most recurrent questions are:

- *"Will the jet break, and if so, how long will it take?*
- *How sensitive is the jet to background turbulence or the presence of a dense, viscous outer medium in relative motion?*
- *What is the impact of viscosity on the stages of the separation between the fluid fragments?*
- *After breakup, how disperse in size will the fragments be?*
- *If they are dispersed, how can the size distribution be made narrower, or broader?"*

These questions are some of the most difficult ones in turbulence not only as it applies to physics and mathematics, but also to other scientific domains in general. Their successful resolution will have enormous implications in many disciplines. Therefore, as I dealt with the origin of the universe, I also addressed the diversity of bodies in the universe, the equations which can describe them without forgetting how they get their energy and how they interact with one another on small and large scales. I also detailed fluid breakup mechanisms involved in the fragmentation and gathering together of the precursors of bodies in the universe.

As I was working and reflecting on the data available on the universe, I realized that they are the astronomical expression of what some scientists are looking for in the lab and complaining not to be able to collect or comprehend because of the limited size of their lab experiments and the limits imposed on human scientific spectra of understanding. In other words, it appeared to me that most physicists trying to understand turbulence do not know that the vast amount of data available on celestial bodies are about turbulence. To put it another way, I perceived that the lack of advancement in the field of turbulence and other fields related to the origin of the universe is not due to a lack of data, but to the fact that people did not know the meaning of the data already collected and the thought process they needed to get the most out of them.

Therefore, to explain the meaning of the scientific data available on the universe, I had to design my own methodology of studying turbulence, for no clear guideline existed that I could follow. Toward that end, I understood that in the Solar System for instance, explaining energy transfer is like elucidating how the planetary systems and the Sun got their energy from the precursor of the Solar System or how a planet and its satellites got their energy from the precursor of their planetary system, and so on and so forth. Based on what I found, coherent structures found in turbulent fluids seem to mainly get their energy from their precursors, not just from the interactions with one another. I came to realize that trying to explain turbulence by focusing only on statistical analysis is like going into a town, collecting data on human beings or on various wild animals, plants, fungi, and other types of organisms, then do a statistical analysis on those data, and claim, based on the correlations and coefficients of determination of these regressions between the variables, that one understands the genesis, the functioning, mechanism, energy transfer of the entire ecosystem. In other words, the existence of great correlations or other types of relationships between variables should not be sufficient to understand the origin, development, and underlying mechanisms of an ecosystem. For each organism in an ecosystem has its unicity, which is shaped by the environment and other factors. Some ecologists better understand this relationship between living organisms and their environment. Even within the same species, the behavior of its individuals varies a lot not only with time and space, but also according to many other factors. Unfortunately, many physicists still do not understand that physical bodies, small or big, are unique and have been shaped by their environment and that it is incorrect to try to use statistics to formulate and average equations and withdraw conclusions about the origin and mechanisms controlling the universe. For instance, although most turbulence researchers know that initial conditions greatly affect the outcome of a turbulent flow, they still fail to understand that each physical event including turbulence is unique, and likely not identical to no other, and consequently should be treated accordingly. Furthermore, even structures occurring inside a turbulent flow are also unique and therefore, averaging parameters about turbulent structures (even the largest coherent structures) in an attempt to understand laws underlying them is like averaging statistical parameters or variables over many species in an ecosystem. Of course, certain coherent structures in turbulence have similarities. However, the physical similarities and differences between these structures cannot be enough to fully understand their origin, functioning, and fate. That is why not only most turbulence experiments are hard to repeat or reproduce, and their interpretations after centuries of research still did not yield sound conclusions yet.

The anatomy of a body is different from its physiology. Anatomy describes the bodily structures of a living organism as revealed by external observations and internal ones after dissection and separation of constituent parts, etc. In contrast, physiology mainly deals with the functioning of these parts of that body indeed. Yet, both anatomy and physiology are required to understand the constitution and functioning of a body. Consequently, to study turbulence, its "anatomy" and "physiology" are needed. Furthermore, just as anatomy and physiology cannot be enough to predict the behavior of an organism in a society and how it interacts with others of the same and different kinds, it may be impossible to focus on the morphology and interactions between coherent structures in turbulence to fully understand turbulence.

For instance, based on the statistical similarities and differences between species, some people claim to understand the origin of species including which ones descend from others. On the other hand, others do not even try to explain how life was created or formed, but they stick to stories revealed in philosophical and/or religious books. Some people strongly believe that God created the universe and everything in it, but others do not even believe in the existence of God, and much less in Him creating anything or everything. In most secular schools, some people perceive evolutionism as a predominant foundation of most biological theories in a way that those who dare to embrace creationism and other forms of theories are not usually welcomed. Similarly, some creationists attack those who talk about evolutionism (and vice versa) as if nothing good in biology can help explain some creationist stories. In the end, the evolutionist and the creationists tend not to listen to one another properly to understand where they may be wrong or how to better help one another. I did not seek to address or litigate all of those problems in this book (*"Turbulent Origin of the Universe"*), but I provided details in other books including *"Reconciling Science and Creation Accurately"*. For unless human beings can better seek and understand the foundation of the universe, they can never use statistics or other theories to explain matter in living and nonliving things, and the interactions between them. Our mind needs to be enlightened so we can receive mysterious secrets about the origin and fate of the universe and its constituents.

Nathanael-Israel Israel: Member of the American Association for the Advancement of Science

CHAPTER 1: INTRODUCTION

"Turbulent origin of the universe" is the first and only scientific book that accurately explains everything you need to unconventionally, easily, affordably, and enjoyably decode the universe formation. In this book, filled with great diagrams and digestible scientific facts, you will discover, learn, or get:

- The all-in-one, proven & uncomplicated scientific formula that accurately decoded the formation of the universe, and that explained the birthdate of the stars, planets, satellites, asteroids, and all other celestial bodies in the universe, so you can position yourself to stay on top of your competitors, avoid repeating crucial mistakes that many people have ignorantly made at their own perils
- Extraordinary, unprecedented, accurate insights into the first factors (e.g. early universe physics) that defined the history and formation of the universe so you can tap into deep scientific secrets you ignore, and set yourself apart from others
- The new physics that will revolutionize science forever and land you into a zone of original ideas that improve lives nonstop regardless of your expertise
- The 4 simple things without which it is impossible for anyone to ever understand the formation of the universe, think accurately, work differently, achieve, or perform better for superior results
- The verified key to move the cosmological mountains of misunderstanding, so you can confidently free your mind from doubts, improve your health, and prevent you from any danger connected with sticking with wrong assumptions
- Save time and money, and enjoy your life once you remove errors holding your true understanding of the universe-origin captive
- Historic scientific proof of whether a planet was formed in 2.82 days, whether a satellite was formed in 3.32 days, and whether a star was formed in 3.69 days after the beginning of the universe; so you can creatively produce and address a broader work spectrum by learning how to effectively communicate with and establish unusual connections between otherwise disconnected and disparate scientific data
- The scientific formula that successfully tested the existence of God in a way that shocked believers, skeptics, and all other freethinkers
- Why the scientific community has failed to sufficiently explain the origin of the universe; and understand how existing theories have missed and undefined central ideas, and imposed limits on the vision of scientists
- Specific in-depth knowledge, up-to-the-minute information, and ideas so you can expand your market, cut useless costs, stop wasting time on inadequate projects, and start focusing on the profitable solutions (Science180.com/scientific)

Science180: Where Universe-Origin is Accurately Decoded, Full Stop

- How Science180 Academy can strategically enlighten you, guide you to navigate and filter the massive data collected on the universe, so you can answer the world's most challenging questions, remove any scientific and philosophical cataracts that may be blocking you, and bring you many steps closer to your best life
- How to better resonate with your target market that is craving something original that breaks wrong explanations of the universe-origin

As you continue reading *"Turbulent Origin of the Universe"* today to get an incredible journey of accurately decoding the universe and change your life forever!

1.3. Take home message

Having a PhD in plant, insect, and microbial sciences in the USA, I never thought I would be writing books on the origin of the universe until some experiences led me into it. After years of investigation of the scientific data, I realized that the universe was formed after processes which can be explained using turbulence, a scientific field where very little is known despite hundreds of years of investigation of the matter. I therefore had to design my own method to approach the data. To accommodate the needs of my readers, I split my findings into 9 books published in 2025, and many more are coming. I encourage you to read those that interest you and also share them with others.

Dr. Nathanael-Israel Israel, the author of this book, is told by people that he is the "#1 Universe-Origin, Life-Origin, and Chemicals-Origin Expert". He is the founder of Science180 (www.Science180.com). To learn more about how he may help you, visit Israel120.com.

CHAPTER 2

PEOPLE WENT CRAZY FOR THIS STORY ABOUT WHAT NEEDS TO BE CONSIDERED TO BREAK THE CODE OF THE UNIVERSE-ORIGIN .... CAN THIS KNOCK OUT THE BIGGEST NAME IN COSMOLOGY?

2.1. Matters in the universe

I cannot just start talking about the formation of the universe without first explaining what the universe is, and/or what is in it as of today, and which bodies I decided to focus my investigation on. Therefore, before I start detailing the way the universe was formed, I would like to briefly address the diversity and characteristics of the bodies and systems of bodies in the universe today. Furthermore, to test the soundness of my findings, people will likely refer to data currently available on bodies in the universe. In other words, although it is hard or impossible to find things in the current universe that could look exactly like what existed in the beginning, a lot of data are available in various domains of knowledge which can help address some problems referring to the origin.

Although it could be expected that I present a full review of literature on the bodies in the universe and everything about them at this point, space and time did not allow me to do so. However, on the website associated with my work on the origin of the universe (www.Science180.com), I provided additional information. Here, I just aimed at presenting some big trends about the bodies in the universe.

Things and beings in the universe can be classified into 2 groups:
- physical world and
- spiritual or invisible world

Although the physical can be defined as things that can be seen, it is important to mention that even in the physical world, there are things that everybody does not see the same, but that are believed to exist and to be real. Furthermore, although certain things can be seen with the naked eye, others require microscopes or telescopes before being seen. Many others could still not be seen by any means even after using the most advanced scientific equipment. Moreover, things that some people cannot see with the naked eye can be clearly seen by others. The existence of these kinds of matter still classified as physical, yet never seen, is sometimes "proven" using techniques targeting the manifestation of their presence. Many other matters postulated to be physical were never seen nor their presence ever felt or manifested by any experiment. This implies that the border between the physical and the spiritual world is not easy to define and consequently, some people who may refuse to be labeled as spiritual or religious accept to believe in things usually labeled as scientific hypothesis, some of which are plainly like some religious dogma they blame some religious people to accept in the name of faith!

Most people in the universe believe in the existence of some sort of spiritual world that many cannot see, although testimonies abound about some people being able to see into invisible things beyond the limit of the human spectra. In my book *"Reconciling Science and Creation Accurately"*, I talked about how some people (e.g. ministers, clerics, preachers, magicians, prophets, traditional healers, and other spiritual leaders) can see and even converse with some spirits, spiritual things, and spiritual beings, which are beyond the scope of this book. Without going into detail about accounts and testimonies of prophets and magicians who, according to their rank, can access the spiritual world to some extent, it is at least scientifically proven that some wild animals can see and sense things beyond the human spectra and much more than other organisms. For instance, some carnivores are said to clearly see in the darkness, while most human beings and herbivores cannot. Some animals can perceive, sense, or smell their prey from miles away, while others can only sense things near them. Although human beings have invented ways to see certain things previously invisible, many things are still imperceptible. What could I say about the role that telescopes, microscopes, radiations, and other advanced equipment have played in the human appreciation of the world? Despite what we think we know today, a lot is still unknown and, considering the limits of the spectra of everything in the universe, I think that it will never be possible for mere human beings to fully know everything. For what some people know and can see, others do not. Although I also understand that people perceive the world differently, my goal here is not to judge people and their belief or to see things in a dichotomic way such as:

- a positive world versus a negative one,
- a clean one versus and unclean one,
- a pure one versus an impure one,
- a good thing versus an evil one.

Instead of trying to be exhaustive in reviewing all the bodies in the universe and their characteristics, my goal of the following segments is to allow the readers to have a glimpse at some of the things I will be addressing in this book, knowing that more details can be found elsewhere, including in my other books, on my websites (e.g. www.Science180.com and www.Israel120.com), through seminars and other materials.

The matters in the universe are hierarchically clustered in structures ranging from subatomic particles to the largest scale such as that of clusters of clusters … of galaxies. Some subatomic particles are clustered into atoms, which at their turn are clustered into molecules and chemical compounds, which at their turn are clustered into minerals and rocks, which are parts of celestial bodies. Celestial bodies can be classified as planets, asteroids, satellites, stars, and the clusters of bodies they form such as planetary systems, stellar systems, galaxies, and clusters of galaxies and so on and so forth. A stellar system contains at least a star. Some stars are isolated while others are arranged into globular stars and others are organized into galaxies. At their turn, galaxies are organized into clusters of galaxies, some of which are arranged into superclusters, which, on much higher scales, can be organized into large-scale galactic filaments and so on and so forth until the largest structure of the universe (that no human being can fully see or understand) may be reached. In other books, I talked about how many levels of clusters I think exist in the universe.

Galaxies are known to contain thousands or even millions of stars. Just like the Sun, most stars are believed to have their own stellar system that consists of a star orbited by planetary systems and/or asteroid systems. Some asteroids also have their own system, meaning they are orbited by their own satellite. In general, planetary systems consist of a primary planet orbited by a satellite (if any) or a system of satellites (if more than 1). Although as of 2020, no satellite in the Solar System is known to be orbited by its own satellite(s), meaning satellites of a satellite, it is better not to rule out the possibility of some satellites (maybe in other stellar systems) having their own satellites. Different portions in this book are devoted to each of the types of celestial bodies and particles I mentioned in this section.

Among the many galaxies and clusters of galaxies I addressed in this book, the Milky Way Galaxy is the most studied in the universe. My investigation of planets, asteroids, and satellites was based on those in the Solar System. Consequently, I studied more than 461 bodies in the Solar System including the Sun and more than 460 bodies orbiting it:

- 9 planets
- 241 asteroids
- 210 satellites (known as of 2020)
- Rings systems
- Atmosphere and crust of the celestial bodies

The names of the planets in the Solar System are in order from the Sun:

- Mercury
- Venus
- Earth
- Mars
- Jupiter
- Saturn
- Uranus
- Neptune
- Pluto (although some scientists no longer consider it as a planet since 2005, the general public still does).

While some of these planets are terrestrial (i.e. Mercury, Venus, Earth, Mars, and Pluto), meaning they have a hard crust, others are giant gases (i.e. Jupiter and Saturn), while others are giant ice (i.e. Uranus and Neptune).

The 241 asteroids are distributed as follow:

- 11 Amor asteroids
- 56 Apollo asteroids
- 14 Aten asteroids
- 8 Centaurs
- 24 Comets
- 13 Jupiter trojans
- 4 Kuiper belt objects
- 61 Main belt asteroids
- 7 Mars trojans
- 9 Neptune trojans
- 34 Trans-Neptunian objects

Indeed, originally, asteroids used to be defined as small celestial bodies (including those called minor planets or planetoids) found between the Sun and the orbit of Jupiter. At one point, some scientists limited the definition of asteroids to those found only between the orbit of Mars and that of Jupiter. As more bodies are discovered in the Solar System, the definition was expanded. Some asteroids are located near the orbit of the Earth and therefore they are called "Near Earth Objects" or "Near Earth Asteroids". Certain asteroids that move near Earth usually cross Earth's orbit, causing some scientists to believe that one day, the "Near Earth Asteroids" may end up hitting the Earth, therefore stoking fear into some people who think that asteroids can really hit us very soon. However, things have probably been flying by the Earth since the beginning, but it is just now that the scientists have the technology to "see" some of them. Kind of like how I used to walk over the bushes in my yard for years but did not see snakes there until I started clearing it out. Anyway, to try to avoid this type of incident, a lot of research has been done to catalog and study the movement of these asteroids. In general, celestial bodies which diameter is less than 1 meter are called meteoroids. Millions of asteroids were found in the Solar System and the list keeps growing.

Now, I will define the types of asteroids I listed above (NASA, 2014; Wikipedia, 2016a, b, c, d, e, f, g, h, i, j, k, l). Indeed, named after the asteroid 1221 Amor, Amor asteroids are near-Earth asteroids that approach the orbit of the Earth from beyond, but do not cross it. However, they cross the orbit of Mars. Named after 1862 Apollo, Apollo asteroids are near-Earth asteroids that cross the orbit of Earth. Their semi-major axis (i.e. average distance from the Sun) is generally greater than that of the Earth (1 AU), but their perihelion (closest distance from the Sun) is less than the Earth's aphelion (farthest distance from the Sun), 1.017 AU. By the way, also called the astronomical unit, 1 AU is the distance from the Earth to the Sun. Named after 2062 Aten (the first asteroid in their group), Aten asteroids are near-Earth asteroids which semi-major axis is less than 1 AU and aphelion is greater than 0.983 AU. Because the aphelion of most known Aten asteroids is higher than 1 AU, their orbits are not always entirely contained within the Earth's orbit.

Concerning the main belt asteroids, millions of asteroids are found in a belt located between the orbit of Mars and Jupiter. Millions of them have a diameter less than 1 km. The biggest of them is called Ceres.

Centaurs are minor planets usually characterized by a semi major axis between that of the giant planets (Jupiter, Saturn, Neptune, and Uranus) and an orbit sometimes crossing that of those planets.

Most comets are known for their ability to outgas (like the exhaust of an airplane) when they pass near the Sun and they sometimes have a tail. The orbit of most comets is believed to lie at a semi major axis equal to or greater than that of Jupiter. Nevertheless, new comets are found at the periphery of the asteroid belt, therefore pointing out that between the main asteroid belt and a little after Jupiter, something predisposes some asteroids to display a dust like tail as a comet. At the edge of the asteroid belt, scientists discovered a group of asteroids called main comet belt that acts as a comet sometimes when they are near their perihelion. The orbit of most comets is believed to be very elliptic. Some people believe that, as comets are near their perihelion, their orbital velocity increases and their internal matter interacts with the space they pass through in a way that the resulting reaction causes a luminescence that can be visually seen even with the naked eye.

Finally, trojans are groups of celestial bodies (minor planets, satellites, asteroids, etc.) that share the same orbit with a planet or a satellite. They remain at the same position (called Lagrangian points) and do not interfere with the movement of the primary body near them. There are many types of trojans, but usually, they can be divided into two main groups:

- the leading trojans which are ahead of a planetary system on the same orbit they share;
- the trailing trojans which are behind a planetary system with which they

share the same orbit.

Astronomers were able to observe Mars trojans, Jupiter trojans, Uranus trojans, Neptune trojans, and even Earth trojans and Venus trojans. As of 2016, data on trojans was available for seven Mars trojans, 13 Neptune trojans, one trojan of Earth, and more than 6000 Jupiter trojans. Some asteroids have their own trojans. For instance, Ceres and Vesta have some trojans that at least temporarily share their orbits. Some Jupiter trojans are very large. For instance, the diameter of the largest Jupiter trojans is estimated at about 200 km. Some trojans have their own satellite. For instance, the Mars trojans called 5261 Eureka has its own satellite. The existence of these satellites around trojans and asteroids suggested to me that the law that explains the formation of the satellites may not be exclusively linked to the size of the primary bodies. As of 2016, no trojan has been found around Saturn (the planet that has the largest ring in the Solar System). One of the things that struck me in 2016 when I was reading about trojans was that, despite their small size as compared to that of the planets they lead or trail, they have almost the same orbital speed as these planets, therefore confirming some of my 2013 inspirational thoughts that the orbital speed of celestial bodies does not depend on their size, but their position.

The 210 satellites are classified as:

- The Moon (Earth's satellite)
- 2 Martian satellites
- 79 Jovian satellites
- 82 Saturnian satellites
- 27 Uranian satellites
- 14 Neptunian satellites
- 5 Plutonian satellites

Throughout this book, unless specified otherwise, when I talk about satellites, I mean natural satellites or natural moons of planets, not the artificial or man-made ones. Sometimes, to avoid confusion when talking about the Moon, the only natural satellite of the Earth as of 2020, I say Moon (Earth's satellite).

As far as subatomic particles and atoms are concerned, I explored their similitudes and differences according to their subdivisions:

- elementary particles: fermions and bosons; the fermions being the leptons (electron, muon, tau, electron neutrino, muon neutrino, and tau neutrino); and the quarks (up quark, down quark, charm quark, strange quark, top quark, and bottom quark); while the bosons are the scalar bosons (e.g. Higgs bosons) and the Gauge bosons (e.g. photon, W bosons, Z bosons, and the gluons).
- composite particles: hadrons (e.g. baryon, meson) and others hadrons (e.g. atomic nuclei, atoms, superatoms, exotic atoms, and molecules), and the
- quasiparticles (e.g. phonons, excitons, plasmon, polaritons, polarons,

and magnons).

I also studied minerals, rocks, and lava. Particles and celestial bodies in the universe are characterized by different variables. For instance, celestial bodies are described using several variables such as:

- Density
- Discovery year
- Eccentricity
- Escape velocity
- Gravity
- Inclination
- Mass
- Motion
- Orbital angular momentum
- Orbital moment of inertia
- Orbital period (sidereal, tropical, and synodic)
- Orbital velocity (max, mean, and min)
- Perihelion
- Radius (equatorial and polar)
- Random multipliers (turbulent multipliers)
- Ring system
- Rotation period
- Rotation speed at equator
- Rotational angular momentum
- Rotational kinetic energy
- Rotational moment of inertia
- Satellite number
- Semi major axis
- Tilt (axial)
- Timescale
- Total kinetic energy
- Translational kinetic energy
- Turbulent intensity
- Turbulent kinetic energy
- Visual magnitude
- Volume
- Hundreds of other variables and ratios available on www.Science180.com

To download a fresh copy of the complete list of the variables used to describe celestial bodies, to help you understand the full picture of the variables I studied

before cracking the code of the universe, visit
www.Science180.com/UniverseVariablesList.

In contrast, atoms, molecules, minerals, and rocks are described using different variables such as:

- Atomic number
- Abundance
- Conductivity
- Density
- Electronegativity (e.g. Pauling electronegativity)
- Electron affinity
- Energy of ionization
- Hardness
- Number of neutrons
- Number of protons, etc.

To get the complete list of the variables I studied to decode the origin of chemical particles, visit www.Science180.com/ChemicalVariablesList

For most of the variables I mentioned above, some of the analysis I did refer to the following:

- Mean
- Minimum
- Maximum
- Range
- Standard deviation
- Variance
- Skewness
- Kurtosis
- Sum
- Coefficient of variation
- Turbulence intensity
- Graphs (e.g. histograms, pie charts, scatter charts)
- Probability functions
- Probability distribution
- Analysis of variance (ANOVA)
- Regression (compound, cubic, exponential, growth, inverse, linear, logarithmic, logistic, power, quadratic, and S) and multiple regressions
- Derivatives of functions (first, second, third, fourth, and fifth derivatives)
- Etc.

During my research, I carefully studied each of all the variables I mentioned above (and in this book, a chapter is devoted to most of them individually) hoping that, by the time you finish this book, the story helps you understand the events that could have taken place and shaped the diversity of things in nature. During my study of these variables, I kept the big picture in my mind and as any specific idea was coming to me along the way, I always noted how it could fit into the big picture, and how it could or could not affect other variables I was exploring. In other words, I never finished writing about a variable without addressing and rehearsing in my mind how I think it would contribute to solving the main problem of how the universe began. As ideas were coming into my mind, I never judged them nor believed that any of them hold the ultimate key to unlock the secrets behind the origin of the universe. Although sometimes, some trends were compelling, I did not yield to any pressure or interest to start focusing on them as the solution. By the time I "finished" studying the variables, I wrote thousands of pages, made hundreds of graphs and tables. Time and space did not allow me to present here all of the findings which shaped my perspective about the origin of the universe, but as you check out the website of this book (www.Science180.com), and the articles and books I wrote, you will find more advanced details about how I reached the conclusion I presented in this book. Consequently, instead of starting to judge this book based on the limited data you may have or that I presented here, which is just a tiny portion of the vast amount of information I relied on, and which I cannot exhaustively present here, it may be better to also review each of the equations, inequations, and other complex things that I took into consideration before I dared to state what I wrote.

While I was investigating the bodies, I realized that although most systems of bodies in the universe seem to be organized hierarchically and that some systems of secondary bodies seem to orbit larger ones called their primary bodies, the relationship between them cannot be properly understood or explained by just viewing these bodies as simple fractals or other geometric figures which simplify complexity.

When addressing the interactions between particles and celestial bodies in the universe, people tend to use forces, some of which are termed fundamental forces in nature. As of 2020, four fundamental forces are believed to control all interactions in nature:

- strong force also called strong nuclear interaction
- electromagnetic force
- weak force also called weak nuclear interaction
- gravity

While the electromagnetic force and the weak force have been unified into the electroweak force, gravity still stands alone and is not much understood although the classical theory of Isaac Newton and the relativity theory of Albert Einstein claimed to address it. Although I addressed this issue later in this book, more details can also be found in my book on the origin of chemical particles, where I also reviewed what is known about those forces and the interpretation that I think better fits them considering my understanding of the origin of the universe.

Not only did I study these variables in detail since 2013, but I also sought scientific relationships between them since 2015, meaning years before I finalized this book. In 2016 for instance, I spent almost six months trying to unravel the meaning of thousands of regressions between the variables. I understood that most of the variables I mentioned above are not linearly connected, and a regression between two variables which is very significant for certain types of bodies is not always significant for other types of bodies, therefore making a general conclusion nearly impossible if the analysis is based on the regressions only. In other words, I realized that to succeed in properly modeling the data, what is needed was not just a sophisticated regression or modeling between the variables, but a deep intimate knowledge and understanding of the meaning of the data and the events that took place during the formation of the universe. To put this strategy into different words, I realized that, if I could have a deep understanding of the main events that occurred during the formation of the bodies in the Solar System for instance, I could better narrow down the variables which are key to properly elaborate the mathematical aspect of the modeling. Without a clear understanding of the data, modeling would mean nothing even if they are backed up by strong coefficients of determination. Additionally, by 2016, I felt like if I started modeling the data quickly, the trends I would start getting may force me to look at the data in certain directions, which may be just a mathematical demonstration rather than the true physical and holistic direction toward explaining reality. Therefore, although sometimes I was tempted to model the data, even after I finally discovered that they are coded by a turbulence, I refused to engage in advanced mathematical modeling until I break down and unravel the layers of facts and events surrounding the complexity of the things in the universe. As I explained later, one of the key strategies to better understand the data led to my discovery of the turbulence zones without which I think it could have been impossible to properly address the data statistically in a meaningful way. Therefore, I waited until I clearly knew the events that could have taken place at some key positions of the turbulence before I modeled the data. Having not used such a strategy, many theorists ended up doing countless regressions with powerful coefficients of determination, but unfortunately, those models are just pure mathematical demonstrations of fictions pertaining to no physical reality. Consequently, I did not plan to bother the readers of this book with complicated equations, but I will instead tell a story simply based on my twelve years (2013-2025) investigation and understanding of the origin of the universe.

2.2. Existing theories to explain the universe

As most scientific investigations would proceed, I could have wished to provide a detailed review and references of all the existing theories on the origin of the universe at this point, but I refused to do so at this point (but later as I see fit) because, frankly, I did not find any good one which I really like and which would not mislead or confuse the readers. Another way of expressing this is that I declined to present a conventional review of literature here not only because I do not think it is important to spent too much time and resources on so many wrong theories written, but I also know that doing so will lead many readers astray, confused and promote things I know are incorrect. Each person also has his or her own view and thought about the origin. Therefore, although I quote some existing works that I think are relevant, I focused on what could have happened, while I would like to encourage the readers to be very careful when reading things on the origin of the universe and to think properly for themselves, for such research or thoughts are very consequential. What people think or believe to be the origin (that we cannot see) of the world can highly affect how they perceive life and everything else in the world, including their relationship with others. I have seen many people I used to respect so much making mistakes in the interpretation of scientific data, yet they talk as if their imagination is a gospel.

Moreover, throughout my writing, I declined to use certain terminologies and keywords commonly found in existing theories because they are filled with errors, confusion, and nonsense, which will highly harden the comprehension of my story. Subsequently, I tried to use simple words and expressions found in common vocabulary so that my explanation of the formation of the universe is not easily misunderstood or misinterpreted because of puzzling terminologies in existing theories. Toward that end, I also refrained from finger-pointing and criticizing existing theories each time I explain my viewpoints which could contrast them. I wrote a different book on errors in modern science, where I addressed things that I think are wrong and that need to be fixed in modern science. Because I even did not have enough space in this book to completely go over all the facts I found, I thought it was also better to focus on the telling of my story than criticizing all the time what is wrong in existing philosophies or models. Likewise, I would also highly caution those who will attempt to translate my work into other languages to be very careful and make efforts to stick with the simple terms that I used in my writing rather than embracing the complex and wrong scientific jargons which address things completely different than what I meant with my modest vocabulary.

Finally, considering what I know by deeply studying the available data in science and in other domains of knowledge, I also came to understand that, when people start talking about things they don't know, they assume false things which they can defend until their "death", unless something prevents them from doing so or allows them to find their own mistakes. In other words, this world could have been better if people can understand and/or focus on their calling, mission, or areas of expertise (that they can grow as they learn more things outside of it) rather than crafting and forcing wrong theories on others. Toward that end, when I was not confident with any explanation, I did not bother to elaborate on it. Similarly, I encourage you not to rush to your conclusions before fully reading mine and properly analyzing the supplemental data I provided on the aforementioned website.

Early in my investigation in 2013, I realized that it will be impossible to explain the origin of the universe by just sticking with or merely combining some linear reasoning found in separate fields of knowledge. It appeared to me that the system of equations needed to properly explain the origin of the universe must consider many facts in the existing fields of knowledge such as chemistry, physics, biology, philosophy, mathematics, the spiritual and invisible world, etc.

Throughout the writing of my book, each time I had a different or an improved understanding of the data, I always surrendered my previous understanding and interpretation I had analyzed to the facts of any new information or data I came up with along the years. Although I had various pieces of interpretations and perspectives in each of the chapters I wrote during my investigation, as I was putting together the final book, I always visited my old writings and surrendered them to the new fact-oriented understanding I was having of the big picture as time went by. By doing so, I not only took into consideration my early thoughts about the data, but I also reconciled my viewpoints and old pre-conceptions with the new options which were appearing with new insights into the data. In other words, until the final version of the book was "complete", I kept looking at my old writings with a fresh and increasingly enlightened eye based on the new data or understanding which I trained myself to be happy about, while not letting my previous progress neglect my need to always have room for improvement. Sometimes, I showed the history of some of my thoughts, hoping that it may also help you know how I came to my conclusions. Therefore, the theory I laid in this book was what, as of 2025, I perceived as the best explanation or "solution" of the systems of equations I laid based on the data I analyzed. Should new data arise and I find time to revisit this book, I am confident that I would still find room for improvement. Likewise, if I will ever come to realize that anything I said in this book is terribly wrong, I will not shy from coming forward to correct it, at least by informing the public about it, be it in the form of another book, or article (online or print), or other options available to me at that time. Therefore, to be in the loop, please remember to be in touch, at least via the website www.Science180.com. I hope you will enjoy what I discovered and if you have any question as you read this book, do not hesitate to contact me.

2.3. Take home message

From the plethora of data available in the universe that may need explanation, the findings you will come across in this book about the origin and formation of the universe are based on my study of hundreds of variables collected on:

- More than 461 bodies in the Solar System including the Sun, 9 planets (Mercury, Venus, Earth, Mars, Jupiter, Saturn, Uranus, Neptune, and Pluto), 241 asteroids, 210 satellites, ring systems, atmosphere, and crust of the celestial bodies
- Stars in the universe and their clusters into galaxies, globular clusters,

clusters of galaxies on various scales
- Chemical elements and subatomic particles known as of 2025, etc.

Some of the variables that I studied on celestial bodies include:
- Angular momentum
- Axial tilt
- Density
- Eccentricity
- Escape velocity
- Gravity
- Kinetic energy (i.e. translational kinetic energy, rotational kinetic energy, and total kinetic energy)
- Mass
- Orbital inclination
- Orbital period, and orbital speed
- Radius
- Rotation period, and rotation speed
- Semi major axis
- Timescale
- Volume and many more variables available at www.Science180.com/UniverseVariablesList

For most of the aforementioned variables, I will do some descriptive and analytical statistics, regression analysis, and other mathematical works before finalizing my conclusion about the formation of the universe. I handled details about chemical particles and life in other books.

'Science180 Academy' Success Strategy:
SCIENCE180 ACADEMY OVERVIEW

Science180 Academy is a training, speaking, consulting, and mentoring program designed to groom and empower people of all backgrounds in the truth about the origin of the universe, life, and chemicals. According to their background and interest, trainees are taught different levels of scientific facts to grasp a deeper understanding of the origin of the universe, how to properly think to unearth mysteries hidden in the massive scientific data collected across the globe, but which is unfortunately less analyzed. If you want to be enlightened and equipped so you can cause positive changes in your respective field of expertise, then Science180 Academy program is for you.

Science180 Academy does not confer college credit, grant degrees, or grade its attendants, participants, or students. It is not an accredited university or college, but is the one-stop-destination for universe-origin, life-origin, and chemicals-origin experts. It is where scientists and laypeople get all their origin-related questions properly answered. It is the only place where the accurate interpretation of the universe-origin, life-origin, and chemicals-origin holistically data matters a lot.

Science180 Academy brings together Dr. Nathanael-Israel Israel (the Founder of Science180) and other experts to deliver outstanding value, insight, and lessons to assist you to accurately understand the true origin of the universe, chemicals, and life, so you can tap into that knowledge to improve lives perpetually. Nathanael-Israel's goal is to give you practicable and undeniable proofs of the formation of the universe so you can be fired up to become the best version of you, and to cause positive changes to your initiatives that will profit you today and forever. For Nathanael-Israel, decoding the origin of the universe and everything in it is not a job, but his life mission, and helping others to fully understand that is his mission. Visit Science180Academy.com today to start.

If you are still wondering if Science180 Academy is for you, let me also inform you that some of Science180's clients and prospects have a profound technical knowledge and background in science, while others don't. Some are creationists (e.g. Science180 creationism, Young Earth creationists, Old Earth creationists, Intelligent design proponents), others are anti-creationists. Some are believers, others are freethinkers (including atheists, humanists, rationalists, agnostics, nontheists, nonreligious, skeptics, nonbelievers, religiously unaffiliated, spiritual-not-religious, ex-believers, and doubters).

Nathanael-Israel Israel: Known as the #1 International Authority that Truly Unlocked the Secrets of the Turbulence that Shaped the Universe

Regardless of their background, belief, or disbelief, Science180 works with each of these people to figure out their needs, priorities, and the products and services that best fit them. Science180 improves their knowledge, experience, performance, and answer their questions (related to the universe-origin, life-origin, and chemicals-origin) by crafting a personalized program that perfectly matches their interests, needs, and things that are dear and meaningful to them whether it is to:

- Become the leader that captures the heart of your followers, prospects, and customers craving for an unconventional explanation of the origin of the universe, life, and chemicals
- Benefit from continual updates and assistance during your journey to decode the universe, and clear your way for the universe-origin related freedom, power, technology, innovation, and breakthroughs of the future.
- Challenge the cosmological status quo and embrace the real change that will disrupt the cages that were holding you
- Connect with practical tips about how to decode the origin of the universe, life, and chemicals and protect yourself from wrong theories in the literature and the media
- Disrupt all chains of repetitive nonsenses about the universe-origin, life-origin, and turn your attention toward unconventional ideas leading to greater innovation and prosperity
- Empower and align yourself with Science180, the historic company that has done what no other organization has ever done: accurately decode the origin of the cosmos and its content
- Empower yourself to leave unforgettable marks and to stand tall as a symbol of freedom, power, creativity, and originality in your field of expertise
- Fearlessly push the boundaries of the human abilities to properly understand what is perceived as un-understandable, mysterious, supernatural, unimaginable, impossible, and unthinkable that holds you back
- Free yourself from boring explanations of the origin of the universe, life, and chemicals and embrace the proven theory that opens doors to unparallel opportunities
- Get inside secrets about how to locate flaws in origin-related theories so you can save time, money, and other resources to improve lives

- Have a reliable access to the world's authority on origin-related matters and get your origin questions professionally answered with the truth step-by-step
- If you are a teacher, discover tools to better engage with students and introduce them to new ideas; and if you are a physicist, understand the mother of all turbulences that shaped everything in the universe
- Learn how to set on fire all false universe-origin theories and life-origin theories that are enslaving humankind
- Protect yourself and loved ones by keeping all of you secured and empowered with the true knowledge of the origin of the universe
- Satisfy your burning desire for freedom from beliefs and scientific theories about the universe-origin and life-origin that suffocate you and bind your mind, faith, unbelief, heart, and education
- Scientifically test and know whether there is a God that created the universe or not, and which God it is
- Stand as the lightning bolt that electrifies your colleagues who are still struggling to understand the universe-origin
- Ultimately boost your confidence in detecting, confronting, and avoiding wrong theories by knowing the facts and processes involved in the formation of the universe

To register or to learn more, visit Science180Academy.com today.

CHAPTER 3

YOU THINK YOU KNOW HOW THE UNIVERSE STARTED? WAS THERE A BIG BANG, GOD, A SINGULARITY, SOMETHING ELSE OR NOTHING AT ALL? WAIT UNTIL YOU FINISH READING THIS CHAPTER TO PROPERLY ASSESS HOW MUCH YOU REALLY KNOW ABOUT THE MOTHER OF ALL MATTERS IN THE UNIVERSE …

Before I started mathematically demonstrating the origin of the universe, I introduced (from chapter 3 to chapter 12) the machinery that explains the processes hidden behind the vast amount of data I studied. Later in this book, I proved them using the data I analyzed. I used this approach to also distance my writing from most of the existing theories which tend to put more emphasis on mathematical modeling without explaining the machinery which supports their data. If I had chosen to first present all the proofs of everything I said in this chapter and the few that follow first, I could have needed to address most of the things from chapter 12 until the end of this book beforehand, which may have bored some people. Instead, I chose to start with concepts which I realized summarized the content of the mathematical demonstrations I did later in this book. This strategy can also help the readers to have a glimpse at the mathematics to come that some may not be interested in cracking. Now, let us see what I have to say as an opening statement concerning the beginning of the world.

3.1. Sudden appearance of a mysterious initial matter in the universe

In the beginning, a certain kind of matter, which I call the "turbulent prima materia", or the "turbulent original particle", or the "turbulent original substance", or the "foundational matter", nothing quite similar to any matter known today, meaning its characteristics are different from that of any matter known today, mysteriously appeared in the universe "out of nothing" ..., and through very complex, dynamic, and turbulent processes, that turbulent prima materia was progressively molded into all types of matter, bodies, systems or clusters of matters and of bodies known in the universe today. I labeled the first matter or original substance in the universe using the term "turbulent prima materia" instead of "prima materia" not only to distance myself from the existing theories on "prima materia", which is not well explained and therefore does not reflect the reality as I perceive it after spending years studying the origin of the universe, but also to emphasize the role that turbulence has played in the formation of the universe.

When the origin of the universe is mentioned, many critical questions are usually asked such as:

- What was the initial matter and where did it come from and why did it appear?
- How did I know about that matter?
- Who sent it and where did it come from?
- Did a beginning even ever exist?
- Why was there a beginning at all?
- Have things always existed without a need of a beginning?
- Who made the things in the beginning?
- Were things created or formed or progressively fashioned?
- Where are we coming from and where are we going?
- Why are we even here?
- How were living and nonliving things in the world formed?
- Are we living in an illusion?
- How can we precisely know from A to Z how everything in the world came into existence?

All theories I have read concerning the origin of the world have started with some initial assumptions. Although many books claim to have explanations about the beginning of the universe, no existing scientific book has provided a detailed account of what could have happened in and at the beginning. Of course, since the beginning of time, efforts have been made to try to answer some of the above questions that most human beings have inquired at one point during their lifetime. However, the variety of scientific and philosophical (including religious) viewpoints in the world seem to complicate the answer and make it nearly impossible to elaborate a theory on the origin of the universe without touching some spiritual things that most people have never seen or could see. Besides some religious people who stick to the genesis story of their reference religious book(s), those who define themselves as atheists and/or people who do not believe in any religion also have or tend to have their own ways of explaining their perspective on the origin of the universe. Despite the divergences, the perceptions or the narrations occasionally overlap and some people even agree with viewpoints which may sound contradictory, biased, or unfounded. But how can we think others are biased or wrong in their belief or viewpoint if we cannot have an undeniable and convincing truth to impartially reject them while confirming ours? And how can people reject their belief about the origin they cannot see, while they do not even know what is currently happening around them? To make a long story short, despite the amount of truth or light which would be shed on the origin of the universe, there will always be things that people will never accept, not because these things are necessarily wrong, but because most human beings tend to stick to certain kinds of dogma they put above every demonstration, and that they will never get rid of, unless they yield to their faith or pre-established assumptions, hypotheses, or prejudices usually found in religions and other forms of ideologies. Therefore, in *"Turbulent Origin of the Universe"*, I did not seek to explain my religious or philosophical viewpoints nor try to convince anyone of them, but to focus on undeniable scientific facts, which are relevant and unavoidable for any realistic attempt to explain the origin of the universe without mentioning or annoying anyone with religious doctrines and dogmas, which may cause some people to throw this book away without even reading it. Subsequently, some answers to the questions I asked above are better approached in other books:

- *"Reconciling Science and Creation Accurately"*
- *"Origin of the Spiritual World"*
- *"From Science to Bible's Conclusions"*
- *"How God Created Baby Universe"*
- *"Science180 Accurate Scientific Proof of God".*

Most theories on the origin of the universe agree that at one point, an initial matter appeared before going through some processes to yield the current world. While some people believe that the original matter was created by God in a few days, others think that it was formed by chance and/or by other procedures which they claim lasted millions or billions of years. Here, I do not seek to state who is right or wrong, but by the time you finish reading this book, you can better understand my position based on scientific facts.

When I talked about the beginning, I meant the beginning of all kinds of matter including the one I framed as the original matter or the precursor of all matters, or the mother of all matters. For, after carefully looking at everything in the universe, it appeared to me that their origin can be traced back to a kind of initial matter that went through some transformations according to several variables, constraints, opportunities, and complex phenomena or reactions at various locations. Although I could have timed the length of each event that I think could have taken place, I refrained from doing so this early, but to go over the entire thought process that caused me to come to the timeline I explained toward the end of this book. In other words, I indeed also scientifically demonstrated in this book how long it could have taken from the beginning to the end of the formation of this world and some ages or stages it could have gone through until today.

Due to the complexity of what could have happened at the beginning of the world and the simultaneous reactions which could have occurred at various locations in the early universe, I felt it difficult to chronologically and exhaustively describe the processes or steps that could have taken place in a linear form from the beginning of the world to its current state. To put it in another way, I would like to advise you not to confuse my efforts to tell a story, chapter by chapter, as the exact chronological succession of events that occurred from the beginning of the world until now. To emphasize this warning, the progression of events from one chapter to another must not be confused with the chronology account of events which happened during the formation of the universe. Although I tried to begin the narration with what I perceived as some of the foremost events, it is important to keep in mind that the formation of the world was not a linear succession of events one after another, from simpler forms of matter or events to more complex or dynamic ones.

Just as during the development of a living organism, an egg can go through steps of differentiation to become different types of cells, tissues, organs, apparels, or systems, which are all connected, so also the "turbulent prima materia" went through complex processes to birth all kinds of matters and systems of bodies known in the world today. In other words, the original matter in the universe could have been like an "undetermined" particle, which had the potential of becoming anything just like at the beginning of the development of a living thing, the first cells have the potential to become anything but are later specialized into particular cells because of some unique changes in the expression of their DNA, etc. Similarly, the initial matter in the universe had a programming, which was affected to yield different types of matter depending on the conditions it was subdued to. The difficult task is to explain the machinery of the processes involved in the shaping and distribution of that "turbulent prima materia" into the bodies present in the universe today. Before I explain (in the next chapters) how the initial matter was shaped into the diversified world we know today, and that I predict toward the end of this book as something which could also end one day, I will first elucidate what I perceived could have been the characteristics of the initial matter which appeared at the beginning.

3.2. Characteristics of the turbulent prima materia, the first matter in the universe

Seeing the immensity of the universe and the inability of human beings to fully apprehend it using even the most advanced telescopes and other tools to explore the world, it seemed to me that the initial matter in the universe could have occupied a very huge portion of space which can be labeled as a very deep and

Nathanael-Israel Israel: Acknowledged as The World's Most Accurate Universe-Origin Scientist

wide amount of matter. Because we don't even know the limits of the universe, and I wondered if any scientific equipment can even ever allow scientists to know with certainty these limits, I did not bother to estimate the size of the space occupied by the initial matter. In other words, although I perceived that the universe has some limits, and I provided some details on that toward the end of this book, I was satisfied to express and accept the size of the space occurred by the initial matter as very deep and wide, but not a point as some theories claimed.

As the mother or father of all matters, the "turbulent prima materia" was formless, leaving the process which shaped it after its appearance to be the reason of the forms of matter we see in the world today. From this point forward, for consistency reasons, I preferred using the term "mother" instead of "father" to describe the parent of a body. I also favored the term "daughter" to refer to bodies born from a mother body. I could have used the term "son" or "child" or "parent" to address the relationship between the bodies, but I chose to stick with the term "mother" and "daughter" bodies, not because I tried to elevate any gender above the other, but just to be consistent with the commonly used language. Moreover, although people seem to like the term "mother nature" to address nature, I have not read anything yet about "father nature". Yet, most religious people tend to refer to God as a "Father" not a "Mother". I explained this mystery in *"Reconciling Science and Creation Accurately"* and in *"Origin of the Spiritual World"*.

Although it could sound easy to label the initial matter as having a specific form or as made of some chemical elements known today, the work I have done suggested that giving a shape or a form to the initial matter, and then use it to explain other matters in the universe is one of the reasons the previous scientific theories have failed to explain the formation of the universe. Considering the initial matter as formless also put more pressure and burden on me to properly explain how it was transformed into the current types of matters and bodies in the universe including those that can be seen using the most advanced microscopes and telescopes, without forgetting those that can be seen with a naked eye. In other words, to the best of my ability, I tried to demonstrate the formation of every matter and bodies, not only the smallest ones but also the gigantic ones without neglecting those we cannot even see.

As of today, matter is generally found in 4 states: solid, liquid, gas, and plasma. Considering the similarities and differences of these states of matter, and the processes that the "turbulent prima materia" could have gone through before birthing the matters or bodies we know as of today, I perceived that the status of the "turbulent prima materia" was none of the 4 states of matter known today. In other words, the initial state of the original matter has had to go through complex changes and processes so that each of the current states could be formed. As of today, plasmas seem to be less complex than gases, liquids, and solids. The day scientists will know more about plasmas, they will find them as complex as, if not more complex than, the other types of states of matter. For now, I don't want to engage too much into those details. Anyway, I felt like it can be possible that the state of matter which is the closest to the state of the original matter could be a kind of plasma, but, again, not a plasma as we know today. Although it is impossible to properly describe the turbulent prima materia, at one point, it could have looked like a flash of lightning, a flame of a burning fire, a hot fire, or like the magma or lava of a volcano. I used all these terms to try to help people visualize the initial matter because when I see some bodies in the universe such as the Sun, the magma or lava that comes out the interior of the Earth for instance, I felt like they witness of what the initial matter could have been made of. Here, I do not mean that the turbulent prima materia was a magma or lava or a light, for each of these types of matter were formed after the original was molded, and later, I will explain how that happened.

Because I don't think that any particle in the early universe was left unchanged by the processes which shaped the world, I don't think that anything in the universe today is an exact photocopy of the initial matter at the beginning. To put it another way, the first state of the early universe was different than the current state of the world and consequently, their physics is different and could not be properly approached using the modern methods in physics, which are based on current forms of matter and laws. The laws of physics must have changed as the universe was being formed and like I scientifically demonstrated (not assumed) later in this book, the laws of nature are not the same everywhere in the universe nor the same throughout the steps or stages of changes that the universe went through from its beginning until today. Consequently, I had to use an unconventional reasoning to address certain challenging and mysterious things rather than forcing myself to demonstrate them using the modern scientific methodology of research, which does not always apply to everything in the current world, and much less to things at the beginning or during the formation of the universe.

3.3. The "original mysterious scattering" of the "turbulent prima materia" leading to the beginning of its spreading

The "original mysterious scattering" that I am going to describe now can also be called:

- sudden burst asunder of the "turbulent prima materia", or the
- breaking open or breaking apart of the "turbulent prima materia"

One of the difficult questions I faced while working on this theory was whether or not the turbulent prima materia was at one point together before being pushed away, separated, or split. Or were the things in the universe formed at the same place they are today without migrating from one place to another? After a careful analysis of the available data (e.g. spatial distribution of bodies in the universe according to their system, the distance between them, their speed, size, inclination, eccentricity, etc.), it appeared to me that the formation of the systems of matters or systems of bodies in the universe began at a place different from where they are today, but during their formation, they were relocated at the same time particles inside of them were also moved and formed. The migration of the precursors of bodies that occurred and that I explained in this book is completely different from the existing migration theories which have no clue about the mother of all turbulences.

Soon after its appearance, the turbulent prima materia could have been broken open by a violent event which can be expressed as a burst asunder. Something caused the turbulent prima materia to break apart suddenly, which could have been accompanied by a huge noise as that of an "explosion" due to factors including its internal pressure. I put the word "explosion" in quote because it would not have been a mere explosion as we know today. When I talk about internal pressure of the turbulent prima materia, it should not be viewed exactly as what can occur to chemicals based on current states of matter today. For any explosion or pressure in the world today involves chemicals which were not present yet at the beginning I am narrating here. Therefore, instead of using the word "explosion", I prefer using the term the "original mysterious scattering".

The original mysterious scattering divided the turbulent prima materia into pieces or blocks of matter. It could also have been possible that the turbulent prima materia broke apart because its surface tension was not strong enough to hold it together forever. The breaking apart of the turbulent prima materia exposed its content and "propelled" the broken pieces into many directions. In other words, after the first stage of the formation of the universe, which was the sudden appearance of a bulk of initial matter, the second stage could have been the sudden breakup of that bulk of matter into fragments or pieces of bodies of different sizes. Each piece of the initial matter that was burst asunder can be called a "broken piece of the turbulent prima materia". I also called these pieces the "major daughters of the turbulent prima materia" for they were the immediate daughters of the turbulent prima materia, and although some of them could have been small (because of intermittence which I explained later), some of them could have been very large and deep. In other words, some of the broken pieces of the turbulent prima materia were very huge and bigger than what is called galaxies (although they were not galaxies yet); and the word "pieces" may not be the best way to address them. I could have even used the term broken "mega pieces" of the turbulent prima materia, but because they descended from the turbulent prima materia, I was content with labeling them as pieces of the turbulent prima materia.

By this stage, energy (that I will detail in another chapter very soon) was already born. The energy communicated to each of the "broken pieces of the turbulent prima materia" could have propelled them into motion and the bodies which were from them depended on the changes that their precursors could have gone through.

The "broken pieces of the initial matter" started distancing themselves from the position they were before their mothers were broken. Their movement initiated the expansion of the early universe which, even up until today never ceased expanding. In other words, the expansion of the universe known today is a consequence of the force which propelled the broken pieces of the turbulent prima materia into motion and spread them over the early universe. However, as I explained in the following sections, the "original mysterious scattering" was not just a simple "explosion" from a center to a periphery, but it also inculcated or communicated specific motions to the broken pieces of the turbulent prima materia and their daughter bodies afterwards.

Today, because of the resistance of their crust, some celestial bodies such as terrestrial planets, satellites, and asteroids may not be able to burst asunder despite the high pressure in their interior. But when an escape is found in the crust, the internal pressure can release lava, which leads to volcanoes. In contrast, because they lack a crust, stars "explode" all the time at their surface, releasing some of their content (e.g. photons) into space. Although it can be possible in some places in the universe, I do not mean that stars today break apart into many pieces, but I mean that their surface explodes all the time. For instance, the surface of the Sun explodes all the time, but the Sun itself could no longer break apart because its precursor went through changes which "tied" its constituents together and prevented them from collapsing or breaking into pieces.

Some "major daughters of the turbulent prima materia" went through additional "explosions" to yield smaller daughters. Under some circumstances, there could have been a cascade of "explosions" until a certain point where the "multi-broken pieces of the turbulent prima materia" could no longer explode. By this time, changes were already happening in the bulk of matter which were breaking apart. Those changes were not just the fruit of an explosion, but also that of a turbulent instability that was occurring as things were moving. The size of the broken bulk of matter varies a lot. Some of them would later become clusters of galaxies. Some of the precursors of galaxies also exploded and yielded smaller clusters of stars, some of which seem to orbit a center of mass, which does not always have a celestial body. Hence, unlike stellar systems which have a star at their "center", some galaxies have no celestial body at their center. Others are satellite galaxies seemingly orbiting a larger galaxy. All those clusters of galaxies and clusters of other celestial bodies in the universe are the product of how their precursors broke up and exploded during the cascade of events that shaped all the bodies in the universe. The daughter bodies born from each "explosive" event could have tended to form systems of bodies. For instance, bodies in each cluster of stars, cluster of galaxies, or stellar system could have descended from the same parent. In the end, the organization of the universe into clusters of galaxies of various shapes and sizes resulted from the cascade of fragmentation and gathering together that the turbulent prima materia had to go through to birth the bodies in the universe. Bodies in each defined cluster of galaxies or cluster of stars or bodies can be traced back to a common precursor.

As I worked on the distribution of the celestial bodies, I also realized that the fragmentation of the precursors was not by means of a mere random explosion. Even though I could have detailed here how the broken mega pieces of the precursors were molded into clusters of galaxies or stars, I reserved that process for another chapter later. For, I need to present other foundational data first. Because more data exists on the planetary systems (planets and their satellites), I decided to first reveal my findings about them before elaborating on what could have happened for the galaxies and stars. In the chapter on the 9 system-additive variables and the turbulent multipliers, I will also explain some laws related to how the mother precursors of the bodies were broken into primary bodies and secondary bodies, following specific laws involving fluid flow, turbulence, shearing, diffusion, and other complex phenomena related to fluid mechanics and dynamics. Under the influence of the force that scattered them, the precursors of some bodies also diffused through space. On top of the scattering force or the energy communicated to the precursors of bodies, their internal composition (e.g. density, viscosity) could have also affected their diffusion and consequently the distance from the original position of their mother. In the chapter on semi major axis and its increment, I will revisit this concept. Fig. 1 is a sketch that summarizes some of what I have been saying so far.

TURBULENT ORIGIN OF THE UNIVERSE

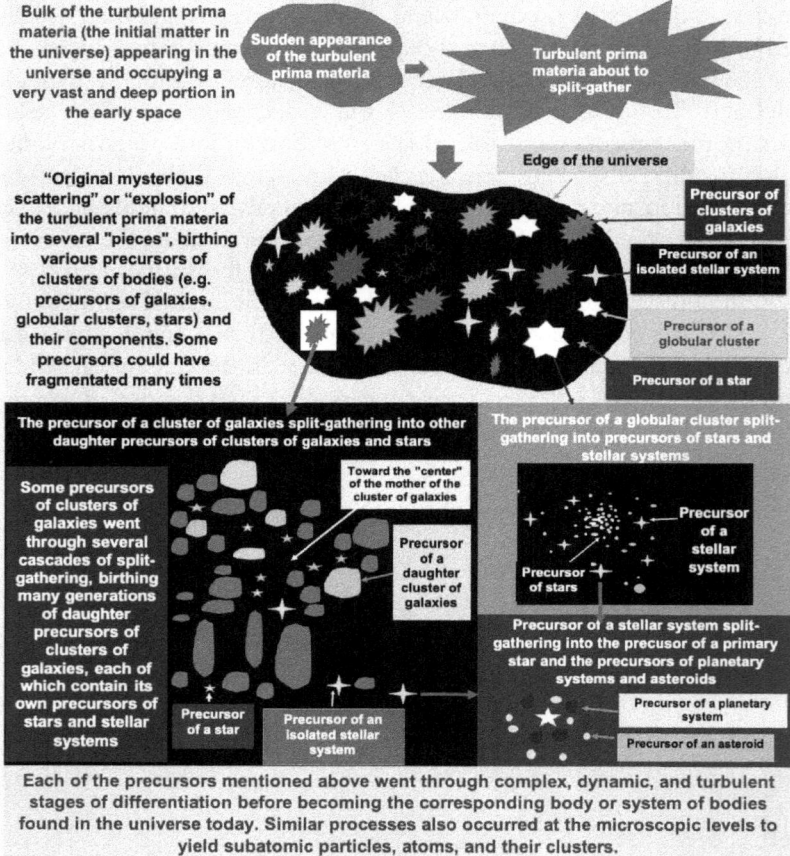

Fig. 1: Breakup of the turbulent prima materia into pieces of major daughters, which started spreading across the universe and became the precursors of clusters of galaxies, stellar systems, stars, and other bodies and particles found in the universe

To get a prettier and color version of this graph, visit www.Science180.com/TurbulentPrimaMaterialBreakup

To summarize, the universe did not start from a single point as some theories state, but from a very deep and massive amount of initial matter. And if the expansion of the universe can be reversed and the movement of all bodies and systems of bodies in the universe properly traced, it can be possible to estimate the direction of the location or region in space where the massive turbulent prima materia could have started its journey. Because of the expansion of the universe and the inability of current scientific equipment to properly measure the size, speed, and other characteristics of all remote celestial bodies, I did not spend

Nathanael-Israel Israel: Acknowledged as The World's Most Accurate Universe-Origin Scientist

much time trying to investigate the initial location of the turbulent prima materia. However, knowing that every particle and celestial body in the universe is moving, it seemed to me that the current location of things in the universe is not where they have always been since the beginning. For instance, although the Earth moves around the Sun and the movement of the Sun in the sky seems to be the same, the location of the Sun and that of the entire Solar System is constantly changing. Similarly, although it is difficult or impossible to properly study the movement of all galaxies and clusters of stars in the universe, they are moving and changing places. Therefore, the position of the turbulent prima materia just at the beginning of the universe should be different from that of the current universe for the whole universe is moving. By the end of this book, I will explain why I think that the universe could have started at a position in the direction located above the north pole, not the south, west, or east.

3.4. Fate of the "broken pieces" of the turbulent prima materia

As I was working on the variables used to describe particles and the celestial bodies in the universe, I was always thinking about what could have happened so that the trends I was seeing in the data could exist. Toward that end, I was writing down all the ideas or options of events which were passing through my mind. It was as if I got the equation of a derivative function and then tried to write the equation of the integral associated with it. By doing this intellectual gymnastic for all the variables I studied throughout the years, I narrowed down many options of the possible proceedings which could have happened. Some options of possible events did not fit the big picture. Based on the data I analyzed and the partial insights each of them gave me, I realized that the broken pieces of the turbulent prima materia that I addressed earlier went through changes and processes including their breakup or fragmentation, squeezing or compression, and other stages of differentiation which are the foundations of the story in this book:

- Instability of fluids
- Split of fluid bodies and formation or precursors of bodies
- Birth, split, and transfer of energy
- Initiation of movement (e.g. revolution, rotation)
- Fluid flow
- Formation of fluid layers
- Acquisition and transfer of momentum
- Initiation and development of turbulence
- Fluid breakup
- Fluid mixing
- Organization and separation of fluids into layers, fluid ligaments, and other jet features and physics associated with fluid expansion, stretching,

necking, destabilization, pinch off, breakup, relaxation, recoiling, propulsion of fragmentated bodies of fluids, tilting, or overturning

- Positioning and spacing of bodies
- Sizing of bodies
- Gathering together or collection of fluid layers
- Spiraling or spinning of layers of fluids to birth a kind of spherical bulks of fluids
- Formation, squeezing, and loosening of vorticial structures
- Birth and strengthening of various forces (e.g. forces that compressed matter to increase its density or loosen it to decrease density)
- Wrapping, tilting, stretching, squeezing of vorticial structures
- Coalescence of fluid bodies
- Etc.

The aforementioned events or physical phenomena did not occur in the order I listed them, and they did not come to my understanding in one day. But after laying out many equations (through which I will walk you in this book), I felt like the events which could have explained the formation of the universe should turn around those processes I just mentioned. As you read the chapters to come, you will better understand what I mean. But for now, let me give a metaphoric example.

While exploring some natural phenomena related to living organisms, I noticed some similarities with the processes of the formation of the universe. In other words, many of the processes I used to explain the formation of the universe are also found in some biological systems, reactions, and pathways. For instance, in cell biology, the first or initial cells during the development of an organism are undetermined, but as they go through stages of differentiation, biochemical modifications including cellular division, gene expression, transcriptomics and epigenetics modifications intervene to transfer the initially undetermined cells into determined cells, which are not always able to reverse back to other kinds of cells including the undetermined cells. Similarly, not only was the turbulent prima materia undetermined, but it was also transformed into specific daughter bodies most of which are unable to reverse back to their precursors.

Unlike living organisms which bodies usually start with one egg or one cell, which goes through many mitoses before its differentiation starts, in the case of the universe, things started with a huge bulk of a single kind of matter (turbulent prima materia), which filled the early universe. Just as undetermined cells started splitting into different parts with different functions, so also did the bulk of the turbulent prima materia go through fragmentations simultaneously associated with modifications of the characteristics of their daughter bodies. Although some chemical elements can be transformed into others, in general, the mechanisms of transforming particles of one kind into another do not usually occur naturally at

Nathanael-Israel Israel: Acknowledged as The World's Most Accurate Universe-Origin Scientist

a large scale today as it was in the beginning. For the energy and processes involved in shaping things in the universe have been costly and it is not easy to reverse natural laws and things. Even at the level of human beings who have a will, it is not easy to change human nature. Else using their free will, human beings could have been more easily changing their mind and nature to adapt to new requirements or to become whatever or whosoever they want to be. In other words, natural laws are very strongly forged and are not easily breakable using part of the energy which formed them.

3.5. Take home message

In the beginning of the formation of the universe, the "turbulent prima materia" (which is a kind of matter different from any matter known today) mysteriously appeared in the universe out of nothing…, and through very complex, dynamic, and turbulent processes, that turbulent prima materia was progressively molded into all types of matter, bodies, systems, or clusters of matters, and of bodies known in the universe today. The initial matter in the universe was formless and could have occupied a very huge portion of space which can be labeled as a very deep and wide amount of matter. The state of the matter in the "turbulent prima materia" was none of the 4 common states of matter known today (solid, liquid, gas, and plasma). The initial state of the original matter went through complex changes and processes so that each of the current states of matter could be formed accordingly. The formation of the systems of matter or systems of bodies in the universe began at a place different from where they are today. During their formation, celestial bodies were relocated at the same time particles inside of them were being formed and moved. Quickly after its appearance, the turbulent prima materia could have been broken open into giant pieces by a violent event accompanied by a huge noise as that of an "explosion". The original mysterious scattering is a term I coined to explain how the turbulent prima materia was divided into blocks of matter of various scales. This event exposed and propelled the broken pieces of the initial matter into many directions. These broken blocks of the initial matter started distancing themselves from the position they were before their mothers were broken. After a series of burn asunder, the turbulent explosion continued until it could not anymore. In other words, cascades of turbulent "explosions" occurred until a certain point where the multi-broken pieces of the turbulent prima materia could no longer explosively split. Meanwhile, turbulent changes were already happening in the bulk of matter that were breaking apart. To sum up, the original mysterious scattering of the turbulent prima materia led to the beginning of the spreading of bodies in the universe and the turbulent modifications of the clusters formed during the turbulent cascades of "explosive" events, which played a crucial role in the origin and fate of the universe. Some of the aforementioned broken pieces of the turbulent prima materia quickly became the precursors of the systems of bodies (clusters of galaxies, galaxies, and all other celestial bodies and particles found in the universe). Some precursors of galaxies exploded and yielded smaller clusters of precursors of stars of various configurations. The broken pieces of the turbulent prima materia went through changes and processes including their breakup or fragmentation, squeezing or compression, and other stages of differentiation and in the end, they yielded all of the bodies in the universe. Now, I will explain how the fragmentation of the initial matter took place.

'Science180 Academy' Success Strategy:
SCIENCE180 ACADEMY PROGRAMS

Owned by Science180, Science180 Academy is a training, speaking, consulting, and mentoring program specialized in everything universe-origin, life-origin, chemicals-origin, and anything at the intersection of reason and faith, or science and religion.

Science180 Academy deals with different subjects according to the needs of its members or target groups. When people register to Science180 Academy, they must choose the program(s) they want to focus on so their training can be properly personalized accordingly. This is similar to how people register to a university, and take classes in a specific department matching their needs!

Science180's breakthroughs are so complex and dense that it is not realistic or good to try to explain all in just one academy, else people will be overwhelmed, disinterested, and confused by the plethora of data to handle. In other words, Science180 Academy offers a wide range of origin-related training in various domains strategically designed to allow people to choose the most suitable for their needs so that, regardless of their background or field of expertise, people can equip themselves, align their mindset, improve lives today and forever using the accurate explanation of the origin of the universe, of life, and of chemicals. Science180 Academy curriculum is based on 12 years of deep unconventional research that culminated with the publication of many much-admired books on the formation of the universe and its content (Science180.com/books). The content of each Science180 Academy is strategically crafted by Dr. Nathanael-Israel Israel (who is acknowledged as the internationally-acclaimed world's authority in origin-related issues) to suit both scientists and nonscientists, religious and nonreligious people, leaders as well as followers, so they can fully decode the proofs of the formation of the universe, of life, and of chemicals they have been wanting to demonstrate or grasp.

The current programs of Science180 Academy are:

1. SCIENCE180 ACADEMY OF COSMOLOGY (Designed for all scientists who want to scientifically study cosmology, the science of the origin and fate of the universe)

2. SCIENCE180 ACADEMY OF TURBULENCE (This is a perfect fit for scientists and other experts interested in studying abiotic turbulence). Examples of these people include:
- o   Automobile manufacturers
- o   Chemical producers
- o   Commercial space businesses manufacturers
- o   Space agencies
- o   Aircraft manufacturers
- o   Technology companies
- o   Turbulence companies, experts, and scientists

3. SCIENCE180 ACADEMY OF LIFE SCIENCES (Tailored to those who want to study biotic turbulence):
- o   Life science specialists, associations, or organizations
- o   Healthcare companies
- o   Biotechnological companies

4. SCIENCE180 ACADEMY OF CHEMISTRY (Designed for chemists, biochemists, scientists, and other educated people who want to understand the origin of chemical particles)

5. SCIENCE180 ACADEMY FOR LAYPEOPLE OR THE GENERAL PUBLIC (Very fit for any layperson or "less" educated people who wants to learn (in a simple language) deep insights that even those who went to university for years were unable to decrypt by themselves, so these laypeople can be equipped to eliminate all forms of scientific and religious universe-origin prejudices)

6. SCIENCE180 ACADEMY FOR CHILDREN (This Academy breaks down origin key topics into language that children can fully understand). This is the only Science180 Academy that your whole family will like and enjoy together, and which will set children on the path of success by accurately showing them early in life the formation of the universe, and how to detect errors in theories or stories that would misguide them as they grow up. Therefore, you need to add this great, efficient, trustworthy, and cost-effective "Science180 Academy for Children" to the strategic journey of children toward their best tomorrow.

7. SCIENCE180 ACADEMY OF THE PSEUDEPIGRAPHA AND SPIRITUAL WORLD (Only one ancient blueprint has the reliable power to help you to accurately decrypt the spiritual origin and history of everything in the universe. If you are a believer and want to delve into the prophetic, angelic, and higher order of knowledge based on the spiritual world, then this Science180 Academy is for you. This program is suitable for those who took at least "Science180 Academy of Creationism". For you to enjoy the courses in this Academy, you need to have at least learned about or attended Science180 Academy of Creationism. If not, you may waste your time trying to grasp simple and supernatural things that cannot be scientifically proven in this Academy, but in Science180 Academy of Creationism).

8. SCIENCE180 ACADEMY OF CREATIONISM (Science180 Creationism is a scientific theory spearheaded by the groundbreaking discoveries of Nathanael-Israel Israel, that scientifically explained the origin of the universe, life, and chemicals using turbulence, and that mathematically reconciled science and the Biblical account of creation for the first time in history. Science180 is different from all existing creationist theories known before 2025.

Science180 Creationism reconciled science with the Biblical account of creation, including scientifically proving that the Earth was formed on Day 3, while the Moon and the Sun were formed on Day 4 of creation!). As you attend "Science180 Academy of Creationism", you will receive accurate answers to all your questions concerning the creation of the universe). The target audience of "Science180 Academy of Creationism" include:
- o   Christians of all ages and all educational backgrounds
- o   Christian or Bible colleges, universities, and schools
- o   School boards of education and textbook initiative leaders
- o   Churches, church associations, and Christian ministries
- o   Anti-creationists wanting to explore the Biblical creation narrative

9. SCIENCE180 ACADEMY FOR FREETHINKERS & ALL ANTI-CREATIONISTS (This Science180 Academy is designed for evolutionists, anti-creationists, and all other types of unbelievers seeking to rationally explore and understand alternative arguments for creation or formation or origin of the universe, life, and chemicals from a fresh, scientific perspective).

10. SCIENCE180 ACADEMY OF LEADERSHIP-(Also called "Science180 Academy for Leaders", this program will enlighten leaders of organizations on how to solve their people problems, process problems, and profit problems related to the origin of the universe, of life, and chemicals according to their domain of expertise). With "Science180 Academy of Leadership", leaders will gain new insights so they can cast new visions and avoid focusing on screwed-up processes, products, and services related to universe-origin initiatives that need to be fixed, faced, or dealt with.

Science180 Academy of Leadership will also equip leaders to address process problems related to inefficiency, gaps, missed opportunities, wasted time and efforts, too many steps, bureaucracy, useless layers between organization and customers concerning the innovation, research methodology, research, product development, strategic planning, workforce diversity in alignment with the historic Science180 breakthroughs so that they can sell more often at full price, avoid regrets in the end, open new markets focusing on real solutions, expand their products and services lines, cut useless costs and research, stop wasting time on useless products that will yield nothing, start focusing on the real money-making problems, reach and convert more prospects into profitable and loyal customers, speed up time to market, avoid spending resources on unprofitable projects but on profitable ones, take their organization to a higher level, open new groundbreaking doors, boost their margins of profits, beat their competitors, and make big profits, conserve more cash and spend it on more valuable things, and outpace their competition as long as their products or services are related to the origin of the universe, life, and chemicals. Perfect fits for Science180 Academy of Leadership include leaders of:
- International scientific organizations
- Academies of sciences
- Universities, colleges, and schools
- Automobile manufacturers
- Government space agencies
- Commercial space businesses, Private aerospace companies, aircraft manufacturers
- Technology companies, chemical producers, healthcare companies, biotechnologies companies
- Any other organization that can benefit from the insight into the origin of the universe, life, and chemicals

11. SCIENCE180 ACADEMY FOR GOVERNMENTAL AGENCIES (Do you want to know how and why most nations and governments are wasting millions of dollars on universe-origin and life-origin researches they don't need … and how to avoid it? Indeed, for most developed nations, and even for some under developed countries, universe-origin projects can cost billions of US dollars and other expensive things that cannot be afforded without sacrificing crucial priorities. Even in developed countries, the impact and the return of investment of the space researches are subject of intense political and economic debates.

What if your nation or institution can reduce wasteful spending on universe-origin research and life-origin research, as well as your dependency of wrong theories on the origin of the universe and life? "Science180 Academy for governmental agencies" will show you how to use the latest scientific breakthrough to better understand the origin of the universe without wasting money on what is already known or what we think we don't know, but that most scientists ignore. Having spent years accurately decoding the origin of the universe, of life, and of chemicals, Dr. Nathanael-Israel Israel delivers science-backed insight to properly understand all the processes connected to the universe formation—so you don't waste more money and time on trying to research the beginning of the cosmos, but to focus on reducing budget of spatial agencies, focus on real science, cutting-edge research, and things that inevitably lead to discovery and innovation). Perfect matches for "Science180 Academy for governmental agencies" include leaders of:

- o Department of Defense including its military branches (Army, Marine Corps, Navy, Air Force, Space Force, Coast Guard, and National Guard)
- o Department of Energy
- o Aeronautics / Space agencies (e.g. NASA, ESA)
- o National Science Foundation (e.g. NSF)
- o Other key governmental agencies

12. OTHER SCIENCE180 ACADEMY: If you did not relate with any of the Science180 Academies mentioned above, but you are still interested in learning something specific about the origin of the universe, life, and chemicals that better fits your needs, please visit Science180Academy.com to contact us so we can discuss that with you.

CHAPTER 4

EVERYONE STOPPED FOR THIS SECRET OF THE THEORIES OF FRAGMENTATION OR BREAKUP OF MATTER IN THE EARLY UNIVERSE

4.1. Fragmentation of matter theories

Because from this point onward, I will talk a lot about processes of fluid fragmentation, I felt like before I go any deeper into anything else, I needed to briefly review what is known about that subject. Indeed, scientists have been interested in understanding how things fragment in nature. They found that fragmentation phenomena are ubiquitous and can be grouped into three classes that I will review below:

- Aggregation
- Maximum entropy principle and random breakups
- Sequential cascades of breakups

4.1.1. Aggregation

Aggregation is defined as a process by which an "*ensemble of initially small elementary particles form clusters of increasing average size as they collide and merge, and the evolution is towards ever larger sizes*" (Eggers and Villermaux. 2008). This model has been used to try to explain the growth of aerosols in clouds (Seinfeld and Pandis, 1998). Talking about collision, some scientists think that a fragmentation following a drop collision is a phenomenon potentially crucial to understanding the structure of the Solar System (Stern et al., 2006). Yet, a deep review of fragmentations following a drop collision concluded that such scenarios are scarce in nature (Eggers and Villermaux. 2008). That study implied that the Solar System and most systems in the universe were unlikely produced by a collision impact. In other words, I think that fragmentations following a drop collision are scarce probably because they are not the main event that have driven the formation of most bodies in the universe.

I was accustomed to use the term aggregation to explain the process by which precursors of celestial bodies were clustered into entities originally separated. However, the more I went deeper into the understanding of the processes involved, the more I realized that collision was not what really happened. As I was profoundly reflecting on how aggregation is defined in the previous paragraph and which embraces the notion of collusion, I felt like, to avoid confusion and complication in my writing and to distance my work from some errors in some existing theories, I decided to use a different word. Therefore, instead of saying "aggregate", I chose to use the terms "gather together", "assemble together" or "collect". Consequently, instead of using the term "aggregation", I adopted the terms "gathering together", "assembling together", "collection" to express the process by which originally separated, dispersed, or isolated precursors of bodies were collected or gathered together into more unified entities or systems of entities under the influence of turbulence. By the time you finish reading this book, you will better understand why I avoided using terms which imply the notion of collision or merging for they can mislead. Similarly, nowhere else in my writing, should you see anything called "accretion". For the processes by which matter was clustered into more condensed bodies or systems of bodies were not by means of what is called accretion as postulated in some theories such as the Big Bang theory. As I later explained in this book, the gathering together of the precursors of the bodies into their daughter bodies involved, among many things, the formation of new connections between their constituents and their topological transformation.

As I clarified along the way, I came to comprehend that the split and the gathering together of the precursors of bodies or precursors of systems of bodies into their daughter bodies were not separated or isolated in time. For both processes could have started almost simultaneously although the gathering together may have continued even after the split or its duration could have been longer than that of the split. In other words, as the precursors of the bodies were splitting, their constitutive or internal components were also being gathered together. Therefore, to translate the joint process of the splitting and gathering together of the precursors of the bodies, I invented the term "split-gathering together" or "split-gathering" in short.

4.1.2. Maximum entropy principle and random breakups

The maximum entropy principle and random breakups approach consider the random splitting of an initial volume into various disjoint elements, all in one step (Eggers and Villermaux. 2008). This is an ensemble theory commonly used in statistical physics, with no explicit reference to dynamics. Similar ideas were developed for the kinetic theory of gases (Mayer and Mayer, 1966) and the physics of polymers (Stockmayer, 1943). The idea is to visualize a given volume $V_0 = d_0^3$

as a set of $K = (d_0/d_m)^3$ elementary bricks of volume $v_m = d_m^3$, whose linear sizes are for instance linked by the requirement $We = d_0/d_m$ (Eggers and Villermaux. 2008). Like I explained throughout this document, the fragmentation of the precursors of bodies in the universe was not random and did not happen all in one step. The chapter on turbulent multipliers explains why the breakup of matter was not random, but was defined by specific laws I will present later.

4.1.3. Sequential cascades of breakups

The process of aggregation is inverse to that of sequential cascade of breakups, which implies size reduction. Indeed, the sequential cascade of breakups is a process whereby a mother body gives rise to daughter bodies, which themselves break into smaller bodies, producing bodies of even smaller sizes, and so on and so forth. Originally imagined to be a progression in which large eddies broke down into even smaller eddies, the belief of a cascade of breakups in physical space was the foundation of the famous Lewis F. Richardson's poem, "*Big Whorls Have Little Whorls*":

> "Big whorls have little whorls
> That feed on their velocity,
> And little whorls have lesser whorls
> And so on to viscosity."

Although advances in flow visualization allowed to better see vortical motions in such a way that some scientists questioned the cascade idea, it is generally admitted that energy transfer (e.g. in fluids) usually ends up at the dissipative scales, either by direct transfer from large scales to small ones, or by a progression of transfers, or some combination of both (George, 2013a**)**.

In 1941, the Russian mathematician, Kolmogorov (Kolmogorov, 1941) was the first person to model sequential cascades of breakups. He was inspired by ore grinding, a process where the size of brittle solid particles is repeatedly reduced. Kolmogorov later applied that model to turbulence (Kolmogorov, 1962; Monin and Yaglom, 1975), and despite its errors, his turbulence model has been applauded by many in that field of research, not because it is perfect, but because no better replacement has been found in more than 60 years.

The model of sequential cascades of breakups was later extended to liquid atomization. For instance, spray formation is a sequential process where mother drops give rise to daughter drops, which themselves break into smaller drops, producing ever smaller sizes. There are many variants of cascade process (Konno and al., 1983; Novikov and Dommermuth, 1997; Martinez-Bazan, 1999; Gorokhovski and Saveliev, 2003). In each variant of cascade process, a drop of initial volume V_0 breaks, after n steps of the cascade, a family of drops of volume V_n related to V_0 by a random multiplier (α_i) smaller than unity is believed to emerge.

4.1.4. Comparison of three scenarios of fragmentation

When the aforementioned three classes of fragmentation models were compared to experimentally measured and statistically converged drop size distributions, it was found that "m*aximum entropy principle and aggregation scenarios were far from the truth: nature does not aggregate nor split liquid volumes at random*" (Eggers and Villermaux, 2008). The inability of the aggregation scenario and of the random breakup theory to explain things in nature suggested that natural things were not formed by a random process nor by a simple aggregation. Similarly, the size of celestial bodies in the universe would not have been defined by chance, or by a random event, or by a simple process. Nevertheless, sequential cascades of breakups seemed to be closer to reality found in nature, for it explained the fragmentation data better than the other two models. Although the conclusion suggested that no perfect model exists yet to properly explain fragmentation of bodies in nature and more less at the beginning of the universe, I felt like, despite its imperfection, sequential cascades of breakups better describe the fragmentation of the precursors of celestial bodies.

As I got more insight into the study of the variables which describe celestial bodies, I recognized that, although the sequential cascade of breakups scenario could help prove that everything in the universe could have descended from a common origin, it could have been coupled by a process of gathering the broken pieces of precursors into unique entities or systems of entities. In other words, I felt like the formation of the universe balanced a kind of sequential cascade of breakups with a process of gathering the constituents of the daughter bodies following the breakups. To put it another way, no clear model of fragmentation existed yet that I could have used to easily explain how precursors of bodies in the universe were split-gathered together into the bodies and systems of bodies known today. It took me years of investigating the data collected on subatomic particles, atoms, molecules, minerals, rocks, celestial bodies, and clusters of galaxies before I finally understood the process by which their precursors could have been split-gathered into the daughter bodies or clusters of matter known today. Before diving into the split-gathering of the precursors of bodies, I first reviewed what is known about fluid breakup today and how it can apply to the story of the formation of the universe.

4.2. Fluid breakup as of today, and how it applies to the formation of the universe

Unlike what most scientists would have done, I did not investigate or review the literature about fluid breakup when I started working on the subject. For about 4 years, I analyzed a lot of data on celestial bodies and tried modeling them before realizing that I needed to check existing theories on fluid dynamics. In fact, it took me years of investigation before I understood that the process of the formation of the universe involved fluid breakup. I understood that if I had started my investigation with the review of existing theories on fluid breakup, I would have limited myself and missed the point of the split-gathering of the precursors of the bodies in the universe.

The breakup of fluids or fluid thread breakup is the process by which a single mass of fluid breaks into several smaller fluid masses. The first correct analysis of fluid thread breakup is said to be determined qualitatively by Young (Young, 1805), and mathematically by Laplace between 1804 and 1805 (Laplace, 1805). They attributed the driver of thread breakup to surface tension properties and they also deduced the importance of mean curvature in the creation of excess pressure in the fluid thread. They showed that surface tension can behave in two ways: an elastic mechanism that can support a hanging droplet and a pressure mechanism due to capillary pressure that promotes thread breakup. More than 200 years later, an extensive review of jet physics pointed out that the breakup of jets is most often driven by surface tension (Eggers and Villermaux. 2008).

In this paragraph, I summarized what Wikipedia said about fluid breakup (Wikipedia, 2019a). Indeed, fluid breakup is characterized by the elongation of the fluid mass forming thin, thread-like regions between larger nodules of fluid. The thread-like regions continue to thin until they break, forming individual droplets of fluid. Thread breakup occurs where a fluid in a vacuum forms a free surface with surface energy. If more surface area is present than the minimum required to contain the volume of fluid, the system has an excess of surface energy. A system not at the minimum energy state will attempt to rearrange so as to move toward a lower energy state, leading to the breakup of the fluid into smaller masses to minimize the system surface energy by reducing its surface area. The exact outcome of the thread breakup process is dependent on the surface tension, viscosity, density, and diameter of the thread undergoing breakup. The breakup process in a fluid thread or jet usually begins with the development of small perturbations (e.g. vibrations of the fluid container or non-uniformity in the shear stress) on the free surface of the fluid. The wavelength of the perturbation is the critical parameter determining whether a given fluid thread will breakup into smaller fluid masses. The viscosity of a fluid can also decrease how rapidly a given perturbation will grow or decay with time. Changes in the internal pressure of a fluid thread can be induced by capillary pressure as the free surface of the thread deforms. Capillary pressure can also depend on the curvature of the interface at a given location at the surface. This implies that fluid pressure depends on two radii of curvature that shape the surface of the fluid. Within the thinned area of a fluid thread undergoing breakup, the first radius of curvature is smaller than the radius of curvature in the thickened area, leading to a pressure gradient that would tend to force liquid from the thinned to thickened areas. However, the second radius of curvature remains important to the breakup process. For some perturbation wavelengths, the effect of the second radius of curvature can overcome the pressure effect of the first radius of curvature, inducing a larger pressure in the thickened regions than the thinned regions. This would push fluid back toward the thinned regions and tend to return the thread to its original, undisturbed shape. However, for other perturbation wavelengths, the capillary pressure induced by the second radius of curvature will reinforce that of the first radius of curvature. This will drive fluid from the thinned to the thickened regions and further promote thread breakup.

Considering the process of breakup that I reviewed above and the trends in the data I analyzed, I perceived that the fragmentation and gathering together of the precursors of matters in the universe are 2 phenomena which must be simultaneously taken into consideration to fully explain the formation of bodies. For one did not happen without the other. In other words, regardless of their size, the precursors of the bodies and systems of bodies in the universe were not gathered together without being previously fragmented and loose. I do not mean that the gathering together of bodies must always happen after a fragmentation. It may seem impossible or unreasonable for things or bodies in the universe to seem gathered or assembled or collected if, at one point, their components were not initially fragmented into separate entities. There are levels of fragmentation and of gathering together, and things were not fragmentated and collected the same way. As I presented in the next chapters, the diversity of these phenomena also contributed to the variation of size, density, mass, speed, and many other characteristics of bodies in the universe. As apostrophized early in this section, the surface tension of the precursors of bodies could have played a major role in the breakup of bodies in the universe. Properties that affect surface tension could have also affected the breakup of the precursors of celestial bodies. Consequently, surface tension properties could have contributed to driving the breakup or the fragmentation of the precursors of celestial bodies, hence its importance in explaining the formation of things in the universe.

Because under certain conditions, a single mass of fluid is able to break into many smaller fluid masses, similarly, considering the data I have analyzed, it appeared to me that the precursors of the celestial bodies could have gone through special "cascades" of breakups that yielded the bodies in the universe. Some of those breakups could have started with the elongation of fluid masses of fluid flows of some precursors of bodies.

As the precursors of bodies were moving under the influence of turbulence (that I detailed in the incoming chapters), some could have been amassed into larger clusters of matter that were still going through changes. To explain astronomical phenomena using expressions more adequate to lab settings, I would say that some precursors could have been like gigantic "nodules" of fluids and smaller "nodules". Some of those "fluid nodules" could have been elongated. Under some circumstances, this elongation could have led to the formation of thin, thread-like regions between larger "nodules of fluid". These thread-like regions could have continued to thin until they broke from the ligament that originally attached them to their mother precursor, forming individual fluid masses with various sizes and characteristics. Here, the size of the fluid threads could have been as small as possible, but also as astronomically big as possible.

For anything was possible and I perceived that the processes which formed the smaller kinds of matter also formed the big ones, but with different calibration or scaling. The size and location of the precursors of bodies could have been some of the features that defined how the machinery which molded them shaped their destiny. In other words, the characteristics of the environments where the precursors of bodies landed during the fragmentation and clustering of the initial matter are one of the factors that fixed what they became.

As fluid threads of the precursors of bodies were being formed, some could have been broken up because of factors including their free surface energy, which was also going through changes. Today, a surface tension is one of the main parameters used to explain fluid breakup, but during the formation of the bodies in the universe, some kind of dynamic surface tension also existed according to the characteristics of the stages of the bodies which were being formed. The precursors which surface area were more than the minimum required to contain the volume of their fluids could have had a surplus of surface energy. The precursors which were not at their minimum energy state would have attempted to rearrange in order to diminish their energy state by reducing their surface area. This rearrangement could have led to the breakup of their fluids into smaller masses. While all those processes were going on, other characteristics acquired by the precursors of the bodies affected the ways their constitutive bodies were rearranged. Some of those factors that I will explain later include viscosity, density, spatial localization, and composition of the thread. Each of those factors could have affected how rapidly a perturbated fluid could have grown or decayed over time. For instance, during the formation of the bodies in the universe, the surface of the precursors could have also been perturbated because of their movement through space, and also because of other factors such as the turbulence that was occurring inside of them. According to the perturbation that acted upon them, some precursors could have produced a thread that contributed to fragmentating them into other bodies. In contrast, the thread of precursors which perturbation was weak might have stayed in a state that did not allow them to be fragmented. I later detailed each of those topics in my writing.

4.3. Take home message

As I studied the bodies in the universe, I realized that they were formed during an event which broke up bulks of original matter into smaller blocks or pieces of matter. Hence, I reviewed the process of breakup of matter to see if anything was already done that can help solve the problem that I was facing regarding how to explain the organization of matters in the universe. Unfortunately, no clear model of fragmentation existed yet that I could have used to easily explain how the precursors of bodies in the universe were split-gathered together into the bodies and systems of bodies in the universe before I finally understood the process by which their precursors could have been split-gathered into the daughter bodies or clusters of matter known today. The split-gathering of the precursors of matters in the universe are 2 phenomena which must be simultaneously considered to fully explain the formation of bodies. For, regardless of their size, the precursors of the bodies and systems of bodies in the universe were not gathered together without being previously split from something; and nothing split without being reorganized in a certain way somehow. As the precursors of the celestial bodies were splitting, their constitutive or internal components were also being gathered together. Therefore, to translate the joint process of the splitting and the gathering together of the precursors of the bodies, I coined the term "split-gathering together" or "split-gathering" in short. The data I analyzed showed that the breakup of matter in the early universe and its organization into systems of bodies can be explained using the physics of fluid breakup.

CHAPTER 5

'I THOUGHT IT WAS SAFE': WHY SMART PEOPLE ARE ABANDONING EXISTING THEORIES REGARDING THE BIRTH OF TIME, ENERGY, AND EXPANSION OF THE UNIVERSE THAT COMPLICATE THE UNDERSTANDING OF REALITY?

Time can be defined as the progression or duration of events measurable by the movements (e.g. revolution and rotation) of the celestial bodies (e.g. Sun and Earth) from the past, present, and future. Before the beginning of physical matter, no celestial body was formed yet. Consequently, no physical body (known as of today) had existed yet so that time could be measured as done today. Therefore, before the beginning of the universe, no time known to men existed but an era which some people perceive as perpetuity or eternity, a kind of time not definable using a physical change of things or movements as known today. To put it in another way, time was born at the moment the bulk of the turbulent prima materia started moving and differentiating into things, which properties started changing. Unfortunately, time is one of the most misunderstood concepts in science. I devoted a completely different book to it.

As things started moving and gathering together, energy was born and transferred to them. Although energy exists in many forms (e.g. nuclear, molecular, electric), which I later addressed in this book, below, I focused on the kinetic energy (which can be associated with motion). As I was pondering on the definition of kinetic energy, some questions came to my mind:

- How can energy move things if itself was not created or formed by something which was moving or which can communicate a movement?
- Why were things launched into motion during the formation of the

universe?

- How was energy trapped in celestial bodies and even in atomic and subatomic particles?
- Why are things moving in the universe instead of being static?

As I went deeper in my investigation, I realized that the definition of kinetic energy by itself hides a secret about the origin of the universe. The fact that a kinetic energy is associated with a matter in motion suggests that at one point during the formation of the universe, something could have "pushed" the turbulent prima materia into motion so energy could be communicated to it and to its daughter bodies. For, from the smallest things to the biggest ones, and from the living to the nonliving things, everything in the universe is moving, though some movements are not easily seen with the naked eye or with telescopes or microscopes. In other words, everything in nature contains energy, and, to some extent, everything is under the influence of its energy. Each body (small or big) in the universe is a store of energy which, under certain conditions can be released to do work or produce heat, but also to manifest in other types of things or beings.

Although their movement cannot be always seen or felt, most constituents of ordinary matter are moving. Whatever you may be holding in your hand right now (be it the hard copy of my book or the electronic device you may be reading it from), or the chair you are sitting on, or the ground you are standing on, everything contains something which is moving. Although some people may not know it, even the Earth on which you are on right now, is moving.

I could not forget that day in 2013 when, as we were talking while I was driving to go meet my website developer, someone who was riding with me could not understand or believe me when I explained to her that everything in the universe has constituents that are moving. I told her that, inside atoms, electrons are moving, but she did not think that to be true. I had to strategically stop that debate, for I did not want to hurt her feelings or make her angry at her belief and determination that things are static. That should have not surprised me, for, even as of today, some people (including those who went to school) still believe that the Earth is flat! Anyway, the fact that things are moving in nature suggested to me that, to properly explain the origin of matters, I must not only address how they came into existence but also how they get their energy. Without an energetic input, most movements are impossible, implying that at the beginning of the formation of the universe, there could have existed a mighty source of energy and/or a form of energy which was communicated to the initial matter. Otherwise, where could the universe find the energy required to ignite movements and form the various things it contains? Therefore, the story of the origin of the universe would be incomplete if the process which initiated or defined the movement of matter and clusters of bodies is not properly explained.

Along the years, it appeared to me that something happened to the turbulent prima materia for it to start moving so that its daughter bodies could contain things, which are moving although we cannot always see them with the naked eye. When I was retrospectively trying to come up with series of initial events that could explain the data (scientific and philosophical) available on the distribution of kinetic energy, mass, radius, motion (e.g. rotation, revolution), speed (e.g. orbital and rotational), and position of the bodies in the universe, I understood that, as the bulk of the turbulent prima materia was broken into mega pieces under the influence of a violent "explosion", energy was also communicated to these pieces. As the cascade of "explosions" or of fragmentations of precursors of bodies went down hierarchically, energy and movement were also transferred accordingly. In fact, I did not entertain the thought of an early "explosion" in the universe until the time I was writing this section on 30 June 2020, when Covid-19 was hitting the world very hard. In other words, during the 7 previous years of the writing of this book, I never believed or thought that I would be referring to an "explosive" event early during the formation of the universe.

Considering the data that I dealt with, I perceived that, the turbulent prima materia could have already been rich in energy and could have been moved from the top (i.e. north pole) of the universe toward the bottom (i.e. south pole) of the early universe. Just as most celestial bodies in the Solar System rotate generally east to west, I felt like things in the whole universe may be moving north to south, and that movement could still be going on, but we cannot feel it, because of the lack of reference. In another book, I talked more about the shape, fate, and poles of the universe.

When I studied the functionalities and characteristics of things in nature, I felt like, the appearance of energy could have been so quick that separating the birth of the turbulent prima materia and the appearance of energy in it could not make any difference. For, in the beginning, things happened very fast. In other terms, the time between the appearance of the first matter in the universe and the communication of energy to it could have been very short to a human mind seeking to understand it chronologically. It can also be accepted that the appearance of matter and energy in the universe could have been simultaneous. For instance, kinetic energy implies a mass in motion of course; but a motion cannot start without the existence of a mass first, although both mass and motion can also appear at the same time. Therefore, for the sake of trying to explain things to a human mind who tends to see things linearly, it is all right to separate the appearance of the turbulent prima materia from its acquisition of energy, which implies a movement communicated to its massive body undergoing processes of split-gathering into daughter bodies.

Anyway, by or at the time that the bulk of the turbulent prima materia "exploded", energy occurred in the universe and the processes that followed would have determined how the energy in the universe by that time would be divided into clusters of bodies on different scales. As cascades of "explosions" followed the initial one, they could have affected (e.g. increased or transformed) the energy in the universe due to the release of their internal pressure. In other words, as the turbulent prima materia was destabilized, the movement communicated to its daughter clusters of bodies could have been one of the initial factors responsible for the birth and distribution of energy in the universe. To put it another way, energy could have been communicated to the precursors of bodies by the processes that destabilized the turbulent prima materia and ignited its movement and separation into small components embedded into bigger ones, and so on so forth as the cascade of breakup continued until it "dissipated" into systems that most human beings fail to see.

The kinetic energy sequestered in bodies was like an impartation from their mother precursors. As the precursors of some clusters of matter were moving during the genesis of bodies, some of their energy was used to tilt their orbital plane and rotational axis (as they were being established), while part of the energy was used to lock their constituents into compartments at distances that depended on environmental conditions in the early universe. At the chemical level, a significant portion of the energy used to form the particles is contained inside their constitutive particles. As of today, when chemical compounds are scratched or damaged to some extent, some of their energy can be leaked or released. This partially explains why rubbing two materials against each other can cause them to heat up and even catch fire. Do I need to recall here that it is said that fire was discovered in the antiquity as people were rubbing 2 stones against one another? Also, by rubbing two hands against each other for a while, heat can be felt. That heat is part of the energy being released from the elements that are being "damaged" or "destabilized" during the rubbing. In other words, chemical particles (e.g. subatomic particles, atoms, or molecules) and celestial bodies (e.g. stellar systems and galaxies) are just stores of the energy that was available in the early universe. Because scientific laws suggest that energy cannot be created or destroyed, it appeared to me that the energy in the clusters of bodies in the universe could have been born at an early stage of the formation of the universe. In other words, every type of matter could have been produced by the transformation that the turbulent prima materia went through during the turbulence that shaped things in the universe. The energy which was communicated to a mother precursor of bodies as it was being molded was trapped into its daughter bodies. Part of the energy found in the bodies in the universe today could have been affected by their interactions with other bodies in their vicinity.

One of the difficult tasks I addressed in this book is how energy was transferred or split among the bodies in the universe. By the time I finished writing my books about the origin of the universe, I realized that, to some extent, the law of conservation of energy and conservation of momentum in the universe as expressed in some existing theories is a joke. For, other phenomena including spiritual and mystic activities have been occurring in some bodies in the universe and changing their amount of energy. The energy contained in matter could be higher than what scientific instruments can measure. For there are forms of energy which are not physical or chemical and yet real. Energy is not just a function of mass and speed only. As I explained in other chapters later, the energy contained in chemical elements and in celestial bodies does not depend on their mass and speed only, but also on other characteristics related to how they were formed. For the way energy was stored inside bodies and the amount of it that can be released if the storage process can be "reversed" depends on many factors.

As I close this section on energy, if I can speak a language that can appear profane, I would like to say that the original matter could have been a kind of primal fire or flame from which all other bodies were born just like sparks of flames can fly outward from a major flame and birth smaller flames of fires. Another way to express this is that, at certain stages of the development or differentiation of the precursors of bodies, some of them could have been like a flaming fire or wind. This can also explain why most matters can burn or melt when a significant amount of fire or heat is applied to them. For fire was present in some of the stages of differentiation that some, if not all, precursors of chemical particles had to pass through. While some chemical elements like gold are not always affected by fire but purified by it, others can be denaturized, but not damaged enough to always yield other chemical elements. The heat in the precursors of the particles in the universe could have been very much higher than that used in scientific laboratories and could have played a role in their differentiation into different daughter bodies. Later, I will provide more details on the storage of energy in the bodies.

Take home message

Before the beginning of the universe, no time known to man existed, but an era that some people perceive as perpetuity or eternity, a kind of time not definable using a physical change of things or movements as they are known today. Time (as known to humans) was born at the moment the bulk of the turbulent prima materia started moving and differentiating into things, which properties started changing. At one point during the formation of the universe, the turbulent prima materia was "pushed" into motion so energy could be communicated to it. As of today, everything in nature is moving and/or is made of particles which are moving. Without an energetic input, most movements are impossible, implying that at the beginning of the formation of the universe, there could have existed a mighty source of energy and/or a form of energy which was communicated to the initial matter. As I studied in the universe, it appeared to me that, as the bulk of the turbulent prima materia was broken into mega pieces under the influence of a violent "explosion", energy was also communicated to these pieces. As the cascade of "explosions" or cascade of fragmentations of the precursors of bodies went down hierarchically, energy and movement were also transferred accordingly. The appearance of energy could have been so quick that separating the birth of the turbulent prima materia and the appearance of its energy could not make any difference. As the precursors of some clusters of matter were moving during the genesis of their bodies, some of their energy was used to tilt their orbital plane and rotational axis, while part of their energy was used to lock their constituents into compartments at distances that depended on environmental conditions in the early universe. At the chemical level, a significant portion of the energy used to form the particles was contained inside their constitutive particles.

Nathanael-Israel Israel: Member of the American Society of Biochemistry and Molecular Biology

- Help children to easily sort out their origin-related questions using strategies that get them to tap into deep secrets that even highly educated people ignore
- Clearly explain to children how to mathematically know without a doubt whether God created the universe as the Bible says or billions of years evolution processes formed it

Accurately explaining the complex formation of the universe and of life to children can be very hard in our modern world, but by getting *"How God Created Baby Universe"*, you will know the proven formula to help children to easily understand their huge universe-origin and life-origin questions with confidence, humor, and joy. They will surely laugh aloud while reading this book and thank you for it! It is time to buy this pragmatic book to help the children in your life today.

Member of the American Association for the Advancement of Science, American Chemical Society, and the American Society for Microbiology, **Dr. Nathanael-Israel Israel is** a Beninese-American scientist and international consultant, who shows the world how to scientifically decode the formation of the universe, of life, and who is known as the creator of the Chemicals Turbulent Origin Formula™, the inventor of the Life Turbulent Origin Formula™, and the discoverer of the Universe Creation Formula™. Learn more at Israel120.com.

Another Book by Nathanael-Israel Israel:
HOW BABY UNIVERSE WAS BORN

If you don't believe in God or you hate God, or you don't think there is anything or anyone called God, but you want your children to understand how the universe was formed from a scientifically-proven perspective that considers the facts, then this book is for your children.

Dr. Nathanael-Israel Israel is the founder of Science180, the American organization that helps people enter the realm of true knowledge about the universe formation. In other words, he is known as the first human being to ever use modern science to give people the state-of-the-art decoding experience of the origin of the universe and of life.

CHAPTER 6

WHAT IS PREVENTING SCIENTISTS FROM UNDERSTANDING THE BEGINNING AND DEVELOPMENT OF THE INSTABILITY AND TURBULENCE IN THE UNIVERSE?

In simple terms, I defined instability as a phenomenon or process that destabilizes things. For instance, when a fluid is hardly shaken, it can be destabilized. Similarly, as the "broken pieces of the turbulent prima materia" started moving and stretching under the influence of the "original mysterious scattering", the whole universe was destabilized and a major turbulence broke out.

As I looked at various bodies in nature, I found that when things and particularly fluids are heavily destabilized, their constituents rearrange differently into "pockets" or compartments which composition and size can depend on the intensity and nature of the instability. For instance, when cream is poured into a coffee, an instable mixture can be seen. Similarly, when water or any other liquid is put in a bowl and then thrown into the air, depending on the intensity used and the way it is thrown, that liquid not only can leave the bowl but as it starts moving in the air, it can be scattered into clusters of fluid of different sizes, speeds, shapes, etc. You can even do this experiment yourself by taking water in a small bowl, go outside, throw it into the air, and observe. As you do so, proceed safely! In other words, there is a way to destabilize or scatter a fluid and cause it to yield daughter clusters of fluid, which can gain a rotational or orbital movement.

When I was growing up as a child, my friends and I used to play a game by spinning a dried shell of a snail from which "all" the living materials were removed. Once the shell is "cleaned", there is a way to cut its edge and shape it like a cone. Then, we used our middle finger and thumb to rotate that conical shell by launching it into motion on the ground. After moving through the air and landing on the ground, the conic shell starts spinning or rotating while also orbiting around a center of its elliptic trajectory. The trajectory, speed, position, duration of the motion of the conical shell usually depended on how we launched it, the intensity of the force we applied to it, and many other factors. The joy in playing that game is to watch the conical shell move in a certain way and to see how long it could do so before dying off and falling to the ground. What I was trying to convey here is that, there is a way to simultaneously launch things into a simultaneous orbital and rotational movement using a single force.

After studying the types of matter in chemistry, biology, physics, from the smallest to the biggest, and after carefully reflecting on the thousands of pages I wrote about their characteristics, it appeared to me that the origin and movement of matters and celestial bodies can be traced back to a major instability that occurred during the early stage of the formation of the universe. I perceived that the instability of the initial matter in the universe triggered the formation of precursors of bodies on different scales and explains their clustering in the early universe. Even as of today, scientific evidences suggest that the universe is expanding. At the same time, some natural forces (that I will explain later in this book and also in my book titled "*Turbulent Origin of Chemical Particles*") aid in maintaining things together. Some of those forces are called fundamental forces in nature. For instance, people often use gravity to explain why things tend to fall back on Earth when they are thrown into the air with a certain speed. These facts suggested that, for the universe to keep expanding while things in it still exist and generally "maintain" their identity, the initial instability at the beginning could have been able to ignite a kind of fragmentation of the initial matter, while also allowing it to start gathering according to processes which, in the end, allowed the formation of a complex and dynamic universe filled with moving things. Here, I did not seek yet to explain what destabilized or caused the instability of the turbulent prima materia. In other chapters later and also in the philosophical versions of this book, I delved into the mathematical demonstration of how I discovered and came to my conclusion.

Under the initial instability, the state of the turbulent prima materia started changing. At one point, the "state of matter" was plasma-like or fluid-like. This fluid-like matter could have moved in a manner compatible with the direction of the force which started the instability of their mother. As the instable fluids were moving, they started splitting and collecting themselves into precursors of bodies of all sizes: from small to large scales. These precursors of bodies became the precursors of other bodies and so on and so forth until the smallest bodies or clusters of bodies were formed. As the matter which constituted the initial universe was moving and mixing with one another, vorticial structures were born. The movement of some fluid was like a flow. The instability of these flows played a huge role in the organization, size, location, and structure of subatomic particles, atoms, molecules, planetary systems, stellar systems, clusters of galaxies, and the universe as a whole. The variation of the instability in the fluids in the precursors of bodies was responsible for the difference in certain characteristics (e.g. speed, density, and viscosity) of the precursors of bodies and defined how they were assembled into different bodies on the microscopic, macroscopic, and astronomical scales. From the smallest types of matter (including but not limited to subatomic particles) to galactic systems, passing by planetary systems, stellar systems, bodies (small and big) in the universe were assembled into different groups based on how their precursors were moved around by the events that "disturbed" the turbulent prima materia, the initial matter in the universe.

It is well documented in fluid mechanics that when fluids move side by side, the interface of their interactions can give rise to instabilities. This instability disturbs the fluids and births certain structures. Although I could not properly state the type of instability which occurred during the genesis of the universe, it may be useful to mention that many types of fluid instability have been studied in nature. One of the first studied flow instabilities is called the "Kelvin–Helmholtz instability" which is postulated to occur when a velocity difference exists across the interface between two fluids.

Similarly, considering the trends of the data I studied (details coming up in the next chapters), I perceived that, as the precursors of matters started flowing, fluid layers were formed. Stacked on top of the others, these fluid layers generally moved at different speeds. In the chapter on orbital speed, I provided more details on the speed of the fluid layers. Due to the variation of their speed and according to their size and other characteristics, a turbulence started in them and soon after, patterns or re-arrangements of their matter emerged. In other words, the instability of the fluids grew into turbulence. The intensity of the turbulence is not always the same from one layer to the other and from one precursor of matter to the other. Not only did turbulence contribute to splitting the precursors of the systems of bodies into different bodies, but it also contributed to imparting on them different characteristics according to their location and the scale of their turbulence. Later in this book, I detailed the characteristics of turbulence.

The pattern of the turbulence at the origin of the universe had structures within it at all imaginable scales, from the invisible or the tinniest imaginable scales to the biggest conceivable ones. Each of those scales of turbulence went through different developmental stages and led to the diversity of the bodies and systems of bodies in the universe as of today. During the cascade of breakups which occurred as the precursors were being split-gathered together, the energy that the mother precursors communicated to their daughters affected the instability and the turbulence of the precursor of the latter. As I explained later, many factors controlled the development of the instability, turbulence, and the destiny of the bodies born from each precursor. For instance, the kinetic energy that the precursor of some daughter bodies received from their mothers allowed them to overcome the eventual resistance that the viscous force of their constituents could have had. Therefore, the fluids in the daughter bodies could have been subject to different intensities of turbulence, leading to the formation of different bodies accordingly. In case a small amount of energy was

Nathanael-Israel Israel: Acclaimed at the Standout Scientific Authority Who Accurately Decoded the universe-origin communicated to a daughter precursor, the level of its turbulence could have been relatively smaller.

At this point, I felt like it may be useful to summarize what I have been saying so far. Indeed, all the bodies in the universe came from a bulk of initial matter called turbulent prima materia, which, after appearing out of nothing, was broken up or fragmented by a turbulent "explosion" (that I called the original mysterious scattering), which destabilized it and pushed its daughter bodies into different directions. The turbulent prima materia was molded into different types of matters and bodies or systems of bodies in the universe. It may be impossible for a mere human being to fully know the nature of the original matter for it had gone through many changes to yield the matter in the universe today. However, soon after its appearance or as it appeared, the bulk of the turbulent prima materia covered a deep domain in space and was filled with energy, which was imparted onto the bodies as the bulk of matter in the universe was scattered. After the fragmentation of the turbulent prima materia, the volume of the space occupied by its daughter bodies could have increased in the universe as they travelled through space and occupied different positions as they were being simultaneously reshaped. The broken pieces of the bulk of turbulent prima materia birthed many daughter bodies which became the precursors of the bodies known today. As they were moving and being shaped, some of these daughter bodies were also fragmented and gathered together into other great-daughter bodies and so forth, leading to a hierarchical organization of the universe into clusters of bodies on various scales. The nature of the descendant daughter bodies depended on many factors including their location. The energy richness of the primordial matter can explain why all matters in the universe contain energy and can be seen as a "modified version of assembled energetic particles". During the formation of the universe, turbulence occurred on different scales, with different intensities, at different positions, and led to the formation of things in the universe. From the smallest microscopic scales to the largest astronomical ones, the scales of the turbulence that I studied in my books can be classified as:

- Invisible or spiritual scale that no naked eye or scientific equipment will ever see
- Sub-subatomic scale: scales smaller than that of the current elementary particles
- Subatomic scale: scale of the currently known subatomic particles
- Atomic scale
- Molecular scale
- Mineral scale
- Rock scale
- Asteroid scale
- Satellite system scale
- Planetary system scale

- Stellar system scale
- Globular cluster scale
- Galactic scale
- Clusters of galaxies scale
- Scale of the whole universe: Mother of all scales of turbulence

One of the tasks I handled in the remainder of this book is to explain how the turbulence on each of those scales shaped the universe known as of today.

Take home message

As the "broken pieces of the turbulent prima materia" started moving and stretching under the influence of the "original mysterious scattering", the whole universe was destabilized and a major turbulence broke out. The origin and movement of matters and celestial bodies as of today can be traced back to the major instability that occurred during the early stage of the formation of the universe. That instability could have triggered the formation of the precursors of bodies on different scales and could partially explain how the scattering and clustering of the precursors of bodies or matters started in the early universe. The initial instability changed the state of the turbulent prima materia. As the matters constituting the initial universe were moving and mixing with one another, vortical structures were born. As the instable fluids in the early universe were moving, they started splitting and collecting themselves into precursors of bodies of all sizes: from small to large scales. The movement of the fluids was like a flow which instability played a huge role in the organization, size, location, and structure of subatomic particles, atoms, molecules, planetary systems, stellar systems, clusters of galaxies, and the universe as a whole. From the smallest types of matter (including but not limited to subatomic particles) to the galactic systems passing by the planetary systems, stellar systems, bodies (small and big) in the universe were assembled into different groups based on how their precursors were moved around by the events that "disturbed" the turbulent prima materia, the initial matter in the universe. As the precursors of matters started flowing, fluid layers were formed, and, stacked on top of the others, they moved at different speeds. A turbulence started in them and soon after, patterns or re-arrangements of their matter emerged. Turbulence occurred on different scales, with different intensities, at different positions, and led to the formation of things in the universe.

Nathanael-Israel Israel: Acclaimed at the Standout Scientific Authority Who Accurately Decoded the universe-origin

CHAPTER 7

DID YOU KNOW THAT THE PRECURSORS OF CELESTIAL BODIES BROKE AWAY OR SEPARATED FROM THEIR SIBLINGS ACCORDING TO SPECIFIC RULES HIDDEN IN WHAT I TERMED "TURBULENT SPLIT-GATHERING"?

7.1. Cascade of turbulent split-gathering of the precursors of bodies

Although their turbulence intensity varied, the processes involved in the breakup and gathering together of the precursors of bodies in the universe were turbulent. Therefore, instead of simply using the term "breakup" and "gathering together" to illustrate them, I decided to use "turbulent breakup" and "turbulent gathering together" to emphasize the role of turbulence during the formation of the universe. Therefore, throughout this book, even if I neglect to mention the adjective "turbulent" to qualify them, all the splitting, fragmentation, collection together, and gathering together of precursors of bodies during the formation of the universe were turbulent. Using the term "turbulent breakup" or "turbulent gathering together" also allowed me to distance the type of breakup I addressed in my books from all other kinds of fluid breakup currently known.

The precursors of matter and of systems of matter can be classified in many ways. The original mysterious scattering launched the organization of the daughters of the turbulent prima materia into various precursors of bodies at different scales. The fragmentation of some precursors led to the formation of new and smaller ones. The study of fluid drop satellites and the review of jet physics pertaining to drop breakup (e.g. water drop breakup) suggested to me that, on smaller scales, some precursors of bodies could have come together and merged to yield bigger bodies as they were mixing with others, near them. The mergence I am talking about here should not be confused with the coalescence described in some theories.

As soon the turbulent prima materia was destabilized and started moving and scattering through space, its nature immediately started changing and, according to the characteristics of their environment, new natures were imparted into the clusters of bodies which were being formed and which, at their turn, went through other changes before yielding their own daughter bodies and so on and so forth until the fragmentation "stopped". The transformation that the precursors of matters went through was not just a simple metamorphosis but deep changes beyond a mere modification of shape, or form. Deeper structural changes occurred without forgetting the modifications of the ways the types of matter and their constituents interacted with others. As I explained later in the chapter related to timescale, those processes were very quick and their duration was affected by the turbulence intensity and other environmental factors.

Considering the nomenclature and characteristics (e.g. size and constitution) of the particles and celestial bodies in the universe, I decided to classify the precursors of bodies into 3 major groups:

- Precursors of celestial bodies and of their clusters
- Precursors of chemical particles and of their clusters (e.g. elementary and composite particles) such as subatomic particles, atoms, and clusters of atoms (e.g. molecules, minerals, and rocks)
- Precursors of spiritual or nonphysical things and beings

Throughout this book, I used the terms precursors of matters and precursors of systems of bodies because some matters or clusters of matter are small (e.g. atoms and subatomic particles), while others are big (e.g. planets and satellites), and others are even bigger (e.g. planetary systems, stellar systems, and galaxies). I did not feel comfortable using the term precursor of matter to address the precursors of galaxies or the precursors of stellar systems, for although they are made of matter, they are much larger than some other clusters of matter. However, to be precise, although they may be viewed as very small, atoms are not just simple types of matter, but they are also clusters of matter which can be classified into nucleons (found in the nucleus) and electrons (which are generally organized in shells or orbitals around the nucleus). Therefore, when I use the term "cluster of matter", I mean any system of matter, whether they are as small as atoms or as big as galaxies. Below, I explored each of the main precursors.

The precursors of the celestial bodies and of their clusters can be divided into the following:

- Precursors of galaxies and of clusters of galaxies
- Precursors of stellar systems and of clusters of non-galactical stars (meaning stars which are not organized as a galaxy)
- Precursors of stars and of the planetary systems and asteroid systems orbiting them

- Precursors of planets and the satellites and rings orbiting them
- Precursors of atmosphere around celestial bodies
- Precursors of all other known and unknown types of celestial bodies.

From this point onward, when I use the term "precursors of systems of bodies" or "precursors of clusters of bodies" I mean any of the precursors I just mentioned. But when I say "initial matter" or "original matter", I do not mean any of the precursors I just mentioned, but the turbulent prima materia (the first kind of matter at the beginning of the universe). In other words, the precursors of the systems of bodies occurred at a later stage of the development and differentiation of the turbulent prima materia.

To be sure we are on the same page, I would like to emphasize that at the very early stage of the formation of the precursors of bodies in the universe, no definite particle, chemical element, or celestial body was fully formed yet but their precursors had to go through turbulent changes which defined the destiny of their daughters. Because of the variation of the impact that the machinery that shaped the universe had on each precursor, the precursors of bodies became different things. In other words, because of the variation of the changes they went through, the precursors of bodies were differentiated into specific bodies accordingly. Some bodies ended up being similar because their precursors went through similar processes. In contrast, others looked different because their precursors were impacted differently by the machinery that molded things. Due to the cascades of breakups, some smaller precursors of bodies were embedded inside bigger ones on various scales.

To demonstrate how the precursors were transformed into the current bodies, I had to sometimes rely on modern physics (although it has its limitations). At one point, I needed to select key physical laws that could help me explain what could have happened during the genesis of the universe. I also needed to know the state of matter of the precursors of the bodies. In those days, it appeared to me that, as the precursors of bodies were changing, so also did the state of their constitutive matters. The state of the matter in the precursors in the early universe were not a mere plasma, or gas, or liquid or solid known today, but a specific fluid-like or plasma-like. I could have said gas-like instead of fluid-like, but I decided to say fluid-like because a main part of the physics I used, and which I explained later in this book, to address the formation of bodies in the universe, is based on the physics of fluids, which of course applied to both gases and liquids. I also used physics applied to solids to address some advanced stages of the formation of the universe. Considering the characteristics and nomenclature of the current bodies in the universe, I classified the breakup of the precursors accordingly to fit what is already known.

7.2. Generations of turbulent split-gathering of matter and their clusters

The sequential cascades of split-gathering that I addressed here are the continuation of the description I started in the chapters on the fragmentation of the turbulent prima materia. I could have presented this section earlier, but I felt like it was better to first address how instability and turbulence shaped the clusters of fluids as they were being split. Hence, I squeezed the section on instability and turbulence before continuing with the breakup of the precursors.

As a reminder of how I came up with the concept of generations of split-gathering, I would have to say that it all started on the evening on Friday July 3rd, 2020. That evening, I was very exhausted and unlike other Fridays of the last 7 years, I had to close my computer by 7 PM. As I was laying on my couch, the idea of the generations of fluid breakup just came into my mind. I then quickly got up to make a note. By that day, I was well aware of the sequential cascades of breakups that could have happened in the universe, but I never thought about using the words "generation of breakup" to describe them.

When applied to living things, the term "generation" refers to organisms belonging to the same range of age and having similar characteristics. For instance, in a family, parents usually belong to the same generation, while children usually belong to another one, grandchildren belong to another one, and so on and so forth. This definition may exclude some polygamist families where some men in their old ages have children separated by more than 50 years, meaning belonging to different generations. In the same village or town, people who are not from the same family can also be classified into generations according to their age range. Because it takes some time for an organism to develop before being able to reproduce, children and their parents do not have the same age; and knowing the reproductive age of parents, the maximum age of their biological children can be estimated. Moreover, for living organisms, the parents and children can be all alive at some point, but in the context of the formation of nonliving things (as I later explained in this book), the parents (mother precursors) must "die" so that their natural children (daughter bodies) can live.

In the context of the formation of bodies in the universe, I defined a generation of split-gathering of precursors of matters or precursors of systems of matters as their division into groups of precursors which reflect not necessarily their date of birth, but mostly the stage, hierarchical position inside other systems of bodies or along the chains of breakups, and the nature of most of the daughter matters or bodies. The generations of split-gathering can also be called generation of breakup, generation of breakup and gathering together, provided it is understood that precursors of bodies were gathered together as their mothers were being broken up or reorganized.

Although the generations of breakup of precursors of bodies can be perceived as a classification of the mother-to-daughter relationships, I would not define it as a measure of the time that these bodies were born. Instead, clusters of bodies of the same generation of breakup could have more likely been born after the turbulent primary materia went through a similar number of cascades of breakup. Even so, as I explained later, due to phenomena such as intermittency (the uneven distribution of size, energy, and its dissipation in physical spaces), some smaller bodies or clusters of bodies were formed as bigger clusters were being split without their precursors going through many cascades of breakups. For instance, the precursors of galaxies belong to the same generation of breakup, while the precursors of stars belong to another generation, and the precursors of planets belong to another generation yet, while the precursors of satellites belong to another one. In other words, the galaxies, stars, planets, satellites, atoms, subatomic particles, etc., were not products of the same generation of breakup. The precursors of atoms and subatomic particles (e.g. electrons, neutrons, and protons) belong also to different generations of breakups just as the precursors of a planet and that of a galaxy belong to different generations.

Therefore, to facilitate the classification of the "cascades" of breakups between generations of breakups, precursors of bodies or precursors of clusters of bodies differ mostly by their characteristics (e.g. size and stage of differentiation or fragmentation of the turbulent prima materia they descended from). Like I explained later in the chapters on timescale, it did not take much time for certain precursors to be born. Consequently, what I labeled as "cascades" of breakups could have occurred very quick and in some instances were spontaneous or simultaneous in such a way that it could be better to view the "cascades" of breakup as a succession of fragmentations, which could have varied more in space than in time.

As some bigger bodies were forming, smaller ones inside of them were also going through their own process of fragmentation and gathering together. Under other circumstances, some mother precursors must have literally gone through some fragmentations before their daughter bodies could have gone through their own. For instance, unlike human beings whose children have a developmental growth younger than their mother's, meaning that mothers birth children, who must grow up to become adults before birthing their own children (therefore giving grandchildren to their mothers), the split-gathering of matters and clusters of bodies was different. All daughter bodies in the universe could not have waited for their mothers to finish their development before going through their own. When the mothers of some celestial bodies were going through their development stages, some of their daughters could have already birthed their own children, which could even be growing and taking good care of their own children before their grandmas finished their formation or development. For instance, as I explained in the chapters on timescale, some satellites were formed before their primary planet and the star in their stellar system. Part of this dynamic and complex process is due to the fact that the same turbulent prima materia was used to mold both the children and their mother and grandma. One of the main differences is the amount of the original material involved in the formation of each cluster and the environment where the formation occurred.

Furthermore, the diversity of the precursors of bodies belonging to each generation of breakups can explain the diversity of bodies in the universe according to their types. For instance, the diversity of stars in the universe is a consequence of various precursors of stars born at different locations in the universe following similar stages of breakup of their mother precursors. Likewise, on the scale of stellar systems, the plurality of planets and asteroids in the Solar System for instance is due to the fact that many precursors of planets and asteroids were split-gathered during similar stages of breakup of such precursors. Considering the organization, characteristics, and the current classification of the clusters of bodies and clusters of matter in the universe, I concluded that about 9 generations of breakup and generations of gathering together of precursors could have occurred during the formation of the universe.

1ˢᵗ generation of turbulent split-gathering, leading to the cascades of precursors of clusters of galaxies and clusters of globular clusters

The 1ˢᵗ generation of split-gathering could have split the bulk of the turbulent prima materia into the precursors of the largest clusters of galaxies, which at their turn were split-gathered into smaller clusters of galaxies and so on and so forth until the precursors of the smallest galaxies were reached and could no longer be split-gathered into any other precursor of galaxies. In other words, the bulk of the turbulent prima materia went through a cascade of split-gathering to yield different sizes of clusters of galaxies embedding other smaller clusters until the precursor of the "youngest" galaxy was reached or formed. Considering how quick everything was formed, I don't think that the word young or old galaxies make since, but I used them to try to express a chronology. The hierarchical breakup of the precursors of the clusters of galaxies along the cascade of split-gathering under the influence of the original mysterious scattering explains why as of today, the universe looks like a filament of clusters of galaxies organized as a network. In the chapter on galaxies, I provided more details.

As of today, galaxies are believed to be organized into galaxy clusters, which at their turn are organized into superclusters, which at their turn are organized into clusters of superclusters, which at their turn can be organized into clusters of clusters of superclusters and so on so forth until all higher levels of clusters of galaxies are included in the largest system called the universe.

Here, I would like to introduce a new term I called "level of clustering". Applied to galaxies, I defined the levels of clustering of galaxies, not as the number of galaxies, but the rank of the hierarchical clustering or inclusion of clusters of galaxies. For instance, using the example at the beginning of this paragraph, if I stop the hierarchization of the clusters of galaxies with the cluster of clusters of superclusters of galaxies, the levels of the clusters can be defined as:

- Level 1: galaxies

- Level 2: clusters of galaxies
- Level 3: superclusters of galaxies
- Level 4: cluster of clusters of superclusters of galaxies

In other words, the higher someone can go into understanding the full picture or organization of the entire universe, the larger the size of the clusters of galaxies he or she would find until the largest cluster of everything (meaning the entire scale of the whole universe) is attained. However, I honestly don't know if scientific equipment can ever discover all of the limits of the universe and have such a full picture of everything.

In the philosophical versions of this book (e.g. in *"Reconciling Science and Creation Accurately"* and in *"Origin of the Spiritual World"*), I provided more details regarding why some religious stories regarding "heavens" can imply that the universe may contain no more than 10 levels of hierarchical clusters. Anyway, for now, let's leave this philosophical exercise for my other books and get back on track to finish the 1st generation of split-gathering.

Along the aforenamed cascades of split-gathering of the precursors of the clusters of galaxies, as a consequence of the intermittence (on an astronomical scale) that I explained in the next chapter, precursors of globular clusters were also formed between precursors of galaxies and precursors of clusters of galaxies. In simple terms, a globular cluster is an organization or clustering of stars that are not part of a galaxy. In other words, unlike stars that are found and often well organized in galaxies, stars in globular clusters are not arranged as if they are orbiting a galactic center, but they are spread over a certain domain of space. As of today, many globular clusters are found between galaxies because their precursors were formed during the process which split and shaped the precursors of the galaxies. In the chapter on galaxies and stars, I detailed the globular clusters. The precursor of the Milky Way Galaxy was one of those birthed during the cascade of split-gathering of the precursors of galaxies.

2nd generation of turbulent split-gathering leading to the precursors of stellar systems

The 2nd generation of split-gathering caused the precursors of galaxies and of globular clusters to birth the precursors of stellar systems. The precursors of some stellar systems were clustered into groups, while others were isolated. Similarly, during this stage, the precursors of some isolated stars were also born. By this stage, the precursor of the Milky Way Galaxy birthed the precursors of many stellar systems including the precursor of the Solar System.

3rd generation of turbulent split-gathering, causing the precursors of stellar systems to form the precursors of primary stars and the precursors of the bodies orbiting them (e.g. planetary systems, planets without satellites, asteroid systems, and asteroids without satellites)

The 3rd generation of split-gathering is the split of the precursor of the stellar systems into the precursors of primary stars and the precursors of the secondary bodies that would later orbit them. The precursors of the secondary bodies of the stars were the precursors of:

- planetary systems
- planets that would not have a satellite
- asteroid systems
- asteroids that would not have a satellite
- all other matters found between and in primary precursors in stellar systems (as I stated above)

The precursors of the primary stars later went through changes to yield the primary stars. Because the precursors of some stars were birthed in environments that did not favor them being orbited by secondary bodies, some stars were formed without planets or asteroids orbiting them. In other words, according to their location, size, and environmental conditions during their split-gathering, some stars could have been formed without a planet, asteroid, and any other bodies orbiting them.

Before finalizing their formation, the precursors of asteroids and planets that would not have any satellite orbiting them did not go through additional split-gathering into other celestial bodies. In contrast, as I explained in the next section, the precursors of the planetary systems and asteroid systems had to go through another split-gathering before yielding their daughter bodies.

4th generation of turbulent split-gathering, causing the precursors of planetary systems to birth the precursors of their primary planets and precursors of their satellites, while the precursors of asteroid systems birthed the precursor of their primary asteroids and precursors of their satellites

The 4th generation of split-gathering divided the precursors of the planetary systems and the precursors of asteroid systems into the precursors of their primary bodies and the precursors of the system of secondary bodies orbiting them. This means that the 4th generation of breakup is about the fragmentation of the precursor a planetary system into the precursor of a primary planet and the precursor of the secondary bodies orbiting that primary planet. It is also about the split-gathering of the precursor of an asteroid system into the precursor of its primary asteroid and the precursor of the satellite(s) of that asteroid. Once it split from the precursor of its planetary system, the precursor of the primary planet was collected into a unique body that underwent other changes to get its final characteristics. The precursor of the primary asteroid also did the same. However, unlike the precursor of the primary planets and of the primary asteroids which were directly collected without splitting into other celestial bodies, the precursors of the satellites had to go through another split-gathering before yielding their satellites. When only one satellite was formed around a planet, the precursor of that satellite was just wrapped into a single celestial body, which completed its formation after undergoing some changes.

5th generation of turbulent split-gathering, causing the precursors of satellite systems to form the precursors of individual satellites and of rings

The 5th generation of breakup was the process which split-gathered the precursors of all secondary bodies of planetary systems and secondary bodies of asteroid systems into their individual bodies. Indeed, satellites are not the only secondary bodies in planetary systems or in asteroid systems. Rings (that I will discuss later) are also secondary bodies. In general, the precursors of the secondary bodies in planetary systems and asteroid systems were split-gathered into the precursors of individual satellites and of rings (if any). For instance, after splitting from the precursors of their primary planet, the precursors of the satellite systems went through turbulent processes that divided them into the precursors of individual satellites. The precursors of the satellites and rings of asteroids did the same. Later in this book, I explained how the precursors of the rings and of the satellites birthed their daughter bodies.

6th generation of turbulent split-gathering leading to the precursors of minerals, mineraloids, and rocks

As the cascade of breakup of precursors was progressing from the upper level of clusters of galaxies to the lower levels such as that of the satellites and asteroids (which can be considered as the lowest level of organization of celestial bodies), internal clustering occurred inside the precursors of celestial bodies (e.g. asteroids, planets, satellites, and stars), which were unable to break into any other celestial body. If I can use the term "unsplitable" to describe the ability of the precursors of asteroids, planets, satellites, and stars to hold together without breaking into any other celestial bodies, I would define the 6th generation of split-gathering as the internal clustering of precursors of "unsplitable" celestial bodies into precursors of minerals, mineraloids, and rocks. For instance, in the precursors of the crust of the terrestrial planets (e.g. Earth), precursors of rocks were assembled and each of them birthed specific daughter bodies like rocks, minerals, and mineraloids. Although some rocks and minerals could have been born or formed many years after the formation of the universe, the organization of minerals and rocks in the crust the Earth for instance implies that their precursors went through dynamic and turbulent rearrangement, clustering, and mixing which affected their purity and other characteristics. Other chapters in this book are devoted to rocks and minerals.

7th generation of turbulent split-gathering, leading to the precursors of atoms and their clustering into molecules and chemical compounds

As the precursors of minerals, mineraloids, and rocks were being split-gathered, precursors of composite particles such as atoms, molecules, and chemical compounds were formed. The precursors of some atoms and molecules were split-gathered differently to form specific atoms, some of which are not ordinary or conventional atoms, but exotic ones, having characteristics not seen with ordinary atoms on Earth. In other stellar systems beyond the Solar System, because of their environment and the processes that formed them, some atoms can have unfamiliar and strange characteristics not seem on Earth. Because the inventory of atoms in the Solar System is not exhaustive yet, we are very far from knowing the diversity of the precursors of atoms which could have existed in the universe. The plethora of isotopes on Earth alone illustrates the many combinations or arrangements of the precursors of atoms that could have existed in the early universe. In my book on the origin of chemical elements, I provided meticulous details.

8th generation of turbulent split-gathering, leading to the formation of the precursors of subatomic particles

As the precursors I mentioned above were splitting and collecting themselves into clusters, the microscopic matter inside of them were also splitting into precursors of subatomic particles and precursors of other kinds of particles smaller than the currently known subatomic particles. If the vast number of subatomic particles in nature could be revealed, we would be shocked at how much little science has scratched the surface of the gigantic chemical iceberg. Therefore, I expect the details of the precursors of subatomic particles to be very limited. This also calls for tolerance toward those who mistakenly invented hypothetical particles not seen yet and which will probably never be discovered. As more particles are found and additional details understood, some particles like electrons (which are still considered as one of the smallest elementary particles) would be proven to consist of other particles.

The 8th generation of turbulent split-gathering is about the breakup and gathering together of all discovered and discoverable subatomic particles. Because of their small size, I think that particles which precursors belong to the 8th generation of turbulent split-gathering would cause many problems to the scientific community. Although more of these kinds of particles will be discovered, scientific equipment, existing and probably uncorrectable mistakes in some existing theories and dogma locked into some scientific strongholds would probably limit their quick and real understanding. In other terms, these particles will cause so many problems and confusion to the scientific world and will be a field of contradictions and controversies due to the huge variation of their characteristics and the inability of scientists to have a consensus on their nature, functions, origin, and changes. The confusion of some theories in quantum physics is already illustrative!

9th generation of turbulent split-gathering, leading to the precursors of the smallest particles that will never be scientifically discovered

Some particles are smaller than the currently known elementary particles which exist in nature and, due to their size and the limits of scientific inquiries, some of them will never be detected by physical equipment. The 9th generation of turbulent split-gathering is about the precursors of all kinds of particles that will never be known by science. The characteristics of the precursors of the particles under the influence of the 9th generation of turbulent split-gathering could have been close to those of the turbulent prima materia, the initial matter in the universe. These particles may belong to the spiritual or invisible spectra beyond that which is accessible by the most advanced and sophisticated science yet to come. The precursors of some types of matter found in the spiritual world can also belong to the 9th generation of turbulent split-gathering.

I had to address these "invisible particles" of the 9th generation of split-gathering to agree with certain things I believe in, but that I cannot demonstrate here yet. I wish I can address at this time the kinds of matter involved in some spiritual details (e.g. the prophetic, magic, miracles, and witchcraft) that some people have experienced and testified to have seen, touched, but which a mere human being cannot explain by modern physics. For although scientists will keep searching for the meaning of things in nature, I know that science will never discover certain details of matter, for they are hidden and can be unlocked only by codes which the modern scientific methodology has been rejecting for centuries and would likely continue to discard.

Before I close this segment on generations of breakup and gathering, I would like to emphasize that, because of intermittence (e.g. size intermittence) and the way clusters of bodies were broken and gathered into systems, some precursors birthed during any of the aforementioned 9 generations of split-gathering could have been formed in generations previous to that in which I put them. In other words, the generations of split-gathering are not inflexible, linear, and obligatory steps or stages through which the precursors of all of the bodies or clusters of bodies must have gone through. For instance, the precursors of some stars could have been formed before the precursors of some galaxy clusters were split-gathered together. Similarly, breakups leading to the formation of some subatomic particles could have happened before the ones leading to the formation of some minerals and rocks. Likewise, the precursors of some particles (e.g. those of the 8th and 9th generation of split-gathering) could have completed their formation even before some stars completed theirs. Therefore, the best way to visualize the daughter bodies of the 9 generations of split-gathering is as products of events occurring almost at the "same" time but at different scales and intensities of turbulence. As I exposed in the chapters on the 9 system-additive variables and the turbulent multipliers, the precursors of the stellar systems, planetary systems, asteroid systems, and satellite systems did not split-gather into the precursors of their daughter bodies by chance, but according to precise laws I will explain very soon. For now, let's conclude this chapter and then take a look at the intermittence of the bodies and systems of bodies in the universe.

7.3. Take home message

Following the destabilization of the turbulent prima materia, new natures were imparted into the clusters of bodies, which were being formed and which, at their turn, went through other changes before yielding their own daughter bodies and so on and so forth until the fragmentation "stopped". The transformation that the precursors of matters went through was not just a simple metamorphosis, but deeper structural changes occurred without forgetting the modifications of the ways the types of matter and their constituents interacted with others. Some of the precursors of bodies formed were the precursors of galaxies and clusters of galaxies, precursors of stellar systems, precursors of stars and of the planetary systems and asteroid systems, precursors of planets and their satellites, precursors of atmosphere and rings around celestial bodies, precursors of all other types of celestial bodies and precursors of chemical particles found in and between the celestial bodies.

I defined a generation of split-gathering of the precursors of matters as their division into groups of precursors which reflect not necessarily their date of birth, but mostly the stage, hierarchical position inside other systems of bodies or along the chains of breakups, and the nature of most of the daughter matters or bodies. The generations of split-gathering can also be called generation of breakup, generation of breakup and gathering together, provided it is understood that precursors of bodies were gathered together as their mothers were being broken up or reorganized. I determined that about 9 generations of split-gathering of precursors could have occurred during the formation of the universe:

- 1st generation of turbulent split-gathering, leading to the cascades of precursors of clusters of galaxies and clusters of globular clusters
- 2nd generation of turbulent split-gathering leading to the precursors of stellar systems
- 3rd generation of turbulent split-gathering causing the precursors of stellar systems to form the precursors of primary stars and the precursors of the bodies orbiting them (e.g. planetary systems, planets without satellites, asteroid systems, and asteroids without satellites)

- 4th generation of turbulent split-gathering, causing the precursors of planetary systems to birth the precursors of their primary planets and the precursors of their satellites, while the precursors of asteroid systems birthed the precursor of their primary asteroids and the precursors of their satellites

- 5th generation of turbulent split-gathering, causing the precursors of satellite systems to form the precursors of individual satellites and the precursors of rings

- 6th generation of split-gathering leading to the precursors of minerals, mineraloids, and rocks

- 7th generation of turbulent split-gathering, leading to the precursors of atoms and their clustering into molecules and chemical compounds

- 8th generation of turbulent split-gathering, leading to the formation of the precursors of subatomic particles

- 9th generation of turbulent split-gathering, leading to the precursors of the smallest particles that will never be scientifically discovered.

'Science180 Academy' Success Strategy
SCIENCE180 SEMINARS

People whose awareness is raised by Science180 usually ask me to go deeper or they wonder "what's else?". That is one of the reasons Science180 trains them through strategic work sessions (during seminars or training sessions) that transfer customizable skills and solutions to them. Science180 Seminars are client-centered and tailored to strongly engage the clients so they maximize the discovery of and the tapping into new opportunities, and exponentially outperform their expectations. Science180 offers customizable seminars that can be labeled as a colloquy, conference, consultation, discussion, forum, keynote speech, lecture, lesson, meeting, symposium, summit, study group, tutorial, workshop or working section accordingly on any topic related to:

- Universe-origin for scientists and mathematicians, philosophers, laypeople, and the general public
- Universe-origin or universe creation for believers
- Life-origin for life scientists, for all other scientists, and for believers
- Chemical-origin for scientists
- Universe-origin seminars for children
- Universe and life-origin for pseudepigraphic believers

As you contact us with your needs, we can customize your program accordingly. Learn more at Science180Seminars.com.

CHAPTER 8

IF TURBULENCE IS STILL AN UNSOLVED PROBLEM IN SCIENCE, WHAT CAN WE LEARN FROM THE GENERATIONS OF TURBULENT INTERMITTENCE THAT SHAPED THE UNIVERSE?

A signal is generally said to be intermittent if long periods with events of low magnitude usually separate rare events of large magnitude. Although intermittence can be addressed for at least several of the variables that I studied (e.g. radius, mass, volume, kinetic energy, angular momentum, and moment of inertia), I chose to focus this chapter on the intermittence of the radius or size of the bodies. Radius or size intermittency can be defined as the frequent presence of smaller bodies, smaller clusters of matter, or smaller systems of bodies between bigger or larger ones. For instance, larger satellites are usually separated by smaller ones. Large planets are separated by small asteroids. Likewise, small clusters of stars and even isolated stars are usually found between galaxies and clusters of galaxies.

In general, the size intermittence in the universe could be due to how the turbulent prima materia and its daughters at each generation of split-gathering were broken up and gathered together into clusters of bodies. I illustrated this in the chapter on the radius of the bodies. Intermittence could have been caused by many factors including the fact that the cleavage lines, cleavage points, interface of separation, or the interface of breakup of the precursors of the bodies and their clusters was not neat or sharp, therefore leading to the formation of smaller bodies or systems of bodies between bigger or larger ones. For instance, just like when you break a piece of corn bread into half, some pieces fall off, so also to some extent, when the precursors of bodies were being broken during the formation of the universe, pieces of bodies came off particularly around the breaking points.

Although I realized that the explanation of size intermittency would help explain the distribution of the size of the bodies in the universe, until July 4th, 2020, the day after I discovered the generations of turbulent split-gathering, I never thought about classifying the bodies according to their generation of intermittence. The intermittence idea came to my mind as I was reclining in a chair in my backyard on Saturday July 4th, 2020, when the Americans were celebrating their Independence Day. In those days, as I had set a 120 days deadline to write this book, I used to take Saturdays off and recline in the backyard of my house, listening to some educational videos such as those related to secrets of successful people, leadership, visual intelligence, how to finish complex tasks, how to trust and cooperate with others, etc. Most of the time I reclined while watching my children play in the backyard, a lot of ideas usually came to me. That Saturday, it appeared to me that, just as the turbulent generations of split-gathering, size intermittence also had different generations according to the scales of the bodies.

As I was trying to gaze at that idea, I felt like smaller galaxies or clusters of galaxies can be located between larger ones. Similarly, smaller stars can be found between larger ones just as smaller clusters of stars are also found between larger clusters of stars. Likewise, smaller planetary systems and asteroid systems are located between bigger ones. At the satellite levels, smaller satellites can also be found between bigger ones. As I was reflecting on the generations of split-gathering, I also felt like each of them could have been associated with a generation of intermittence. To put it into other words, on all scales, smaller bodies or smaller systems of bodies are usually found between bigger ones. Sometimes, no small body is found between big ones and, as I explained later, viscosity and other characteristics of the fluids in the precursors can help explain that. Just like I classified the turbulent split-gathering into 9 generations, I also managed to group below the turbulent intermittence into 9 generations accordingly to the generation of split-gathering.

1st generation of turbulent intermittence, leading to the presence of smaller clusters of galaxies between larger ones and smaller clusters of globular clusters between larger ones

Because of this intermittence, smaller superclusters of galaxies are found between bigger ones just as smaller clusters of non-galactic stars can be found between bigger ones. This generation of intermittency explains why between large galaxy clusters, smaller galaxies are found. Because the precursors of the galaxies had different sizes, the galaxies which were born from them also ended up having various sizes, which distribution is a footprint of the processes that split-gathered their precursors.

2nd generation of turbulent intermittence: intermittence of stellar systems

All stellar systems do not have the same size. Some stellar systems are much bigger than the Solar System, while others are smaller than it. The 2nd generation of intermittence can explain the presence of smaller stellar systems between larger ones.

3rd generation of turbulent intermittence: intermittence of stars

This intermittence explains the presence of smaller stars between bigger stars and even between stellar systems of all sizes.

4th generation of turbulent intermittence: intermittence of planetary systems and asteroids system

This intermittence caused:

- smaller planetary systems to be present between larger ones,
- smaller planets between larger ones,
- smaller asteroid systems or asteroids to be present between larger ones,
- asteroids to be found between planets.

For instance, the small Martian planetary system is found between the big Jovian planetary system and the Earth-Moon system (which is larger than the Martian planetary system). Similarly, bigger planets are separated by smaller bodies. In the Solar System for instance, planets are separate by smaller bodies, not always smaller planets, but asteroids and smaller rocks. For instance, although Mars is located between Jupiter and Earth, the bigger planets are not always separated by smaller ones, but by asteroids which are bodies smaller than them.

5th generation of turbulent intermittence: intermittence of satellites

The 5th generation of intermittence explains why smaller satellites are found between larger ones. Indeed, in most satellite systems (meaning clusters of satellites around a primary planet), the satellites do not always have the same size. Usually, smaller satellites separate bigger ones. The 5th generation of size intermittence explains how the discrepancy in the size of the precursors of the satellites was imparted into the satellites. In the chapter on the radius of celestial bodies, I provided plenty of details about size intermittence.

6th generation of intermittence: intermittence of minerals, mineraloids, and rocks

The 6th generation of intermittence explains why smaller amounts or deposits of minerals, rocks, and mineraloids exist between larger ones. Rocks have different sizes and between bigger rocks, smaller ones exist, pointing to the distribution of their precursors during the formation of bodies in the universe. Some rocks composed of complex compounds are separated by simpler rocks because of the

6th generation of intermittence. Likewise, minerals are not equally distributed. Although some minerals are pure and composed of the same chemical elements, particles collected or dug from mines, ores, and wells suggest that smaller cluster of minerals exist between bigger ones.

7th generation of turbulent intermittence: intermittence of atoms, molecules, and chemical compounds

Atoms and molecules come in different sizes. Some atoms and molecules are very big while others are small. Likewise, bigger atoms and bigger molecules are mixed with smaller ones. For instance, soil analyses suggest that, in most soil samples, many types of atoms and molecules can be found and mixed. Even in most pure samples, impurities are usually found, meaning that the precursors of the atoms and minerals were separated by the precursors of other kinds of atoms and molecules, not always of the same size. The 7th generation of intermittence explains the presence of small atoms, molecules, and chemical compounds between larger ones.

8th generation of turbulent intermittence: intermittence of subatomic particles

This generation of intermittence is also responsible for the presence of smaller subatomic particles between larger ones. In fact, within a given atom or composite particle, the subatomic particles are not usually of the same size. Even in the nucleus of atoms, the particles do not have the same size. For instance, the size of the protons, neutrons, and gluons is different. Although all electrons are believed to have the same size, based on the trends in the data I analyzed (see my book *"Turbulent Origin of Chemical Particles"*), I showed that some electrons are bigger than others. In other words, there is intermittence in the size distribution of electrons as well as most of the other subatomic particles. All this was caused by how their precursors were split-gathered together during the formation of the subatomic particles.

9th generation of turbulent intermittence, leading to the precursors of invisible bodies (which will never be scientifically discovered) between and within larger ones

Although I could not see them with the naked eye, the size of some invisible particles that science will never discover could be intermittent. Nevertheless, the closer the invisible particles may be to the turbulent prima materia, the more similar their size would be. The detail of the 9th generation of intermittence can be appreciated only by those who can "see in the spirit", meaning beyond what the most sophisticated scientific tools cannot see yet or never. The philosophical versions I wrote for this book illustrate what I mean.

Before I close this section on generations of intermittence, I would like to mention that, throughout this book, to honor the law of intermittence which abounds in nature, I did not try to force the chapters, sections, or paragraphs to have the same length as some grammatical rules require. I spent more or less time on some topics based on the amount of information I had on them. When I was not inspired to write something, I did not force myself to do so just for the sake of balancing the number of words across chapters, sections, and paragraphs as required by some linguistic rulebooks. In other words, I purposely left some chapters very long, while others are very short. Similarly, some paragraphs or descriptions of some topics are longer than others. If I had tried to adjust the length of the chapters, segments, and paragraphs, I would have written useless and inapplicable things just to please some grammarians or stylists.

Take home message

Intermittence is a propriety that reflects how long periods with events of low magnitude usually separate rare events of large magnitude in a process or system. For instance, the size intermittency is about the frequent presence of smaller bodies, smaller clusters of matter, or smaller systems of bodies between bigger or larger bodies. That is why in a system of celestial bodies, larger bodies are usually separated by smaller ones. For instance, large planets are separated by small asteroids. Similarly, small clusters of stars and even isolated stars are usually found between galaxies and clusters of galaxies. The size intermittence of celestial bodies in the universe could be due to how the turbulent prima materia and its daughters at each generation of split-gathering were broken up and gathered together into daughter bodies or clusters of bodies. Intermittence was also caused by the fact that the cleavage lines, cleavage points, interface of separation, or the interface of breakup of the precursors of the bodies and their clusters was not neat or sharp, therefore leading to the formation of smaller bodies or smaller systems of bodies between bigger or larger ones. I grouped the turbulent intermittence into 9 generations accordingly to the generation of split-gathering:

- 1st generation of turbulent intermittence, leading to the presence of smaller clusters of galaxies between larger ones and smaller clusters of globular clusters between larger ones
- 2nd generation of turbulent intermittence: intermittence of stellar systems
- 3rd generation of turbulent intermittence: intermittence of stars
- 4th generation of turbulent intermittence: intermittence of planetary systems and asteroids system
- 5th generation of turbulent intermittence: intermittence of satellites
- 6th generation of intermittence: intermittence of minerals, mineraloids, and rocks
- 7th generation of turbulent intermittence: intermittence of atoms, molecules, and chemical compounds
- 8th generation of turbulent intermittence: intermittence of subatomic

particles

- 9th generation of turbulent intermittence, leading to the precursors of invisible bodies (which will never be scientifically discovered) between and within larger ones.

97

Science180: Efficient, Trustworthy, and Cost-Effective Company to Add to Your Strategic Journey Toward Your Best Tomorrow

CHAPTER 9

DISCOVER THE SECRETS BEHIND WHY PRIMARY BODIES IN THE UNIVERSE ARE SURROUNDED BY SECONDARY BODIES LIKE A RICH BROTHER–THAT RECEIVED A LARGE INHERITANCE FROM HIS PARENT–IS SURROUNDED BY HIS SIBLINGS

9.1. Split and organization of the precursors of a system of bodies into the precursors of a primary body and of secondary bodies

Looking at the data on the clusters of matter in the universe, I noticed that they mostly consist of a kind of "central" body, which can also be called the primary body, "around" which secondary bodies are positioned. By "central" bodies, I do not mean bodies located at the exact center of the system they "dominate", but I mean that their position is roughly near what can be labeled as the "center" of their system. Some may call it the barycenter. The primary body in each system of bodies is generally "orbited" by others bodies. For instance, just as electrons were believed to orbit the nucleus and/or moved in a shell or orbital around the nucleus, so also satellites orbit their primary planets. At their turn, planets and asteroids orbit their primary star, which in the Solar System is the Sun. The Sun itself (if not the Solar System) at its turn is postulated to be moving around the "center" of mass of the Milky Way Galaxy, the galaxy that the Sun belongs to. Because of their immense size, very remote location, and the inability of current scientific equipment to properly measure them, all galaxies are not "proven" to be orbiting other primary body bigger than them. Even if they do, their movement may not be easily perceived from the Earth. Nevertheless, when I looked at some pictures of galaxy clusters, I felt like some galaxies could be orbiting the "center" of the galaxy cluster they belong to. Finally, primary bodies are not elementary bodies, for they generally consist of other bodies which, when carefully analyzed, seem like entities of matter squeezed, wrapped, and organized in different manners appealing to a phenomenon similar to what could have "ejected" away or split the precursors of secondary bodies from the precursor of the primary bodies. In the chapter of the "9 system-additive variables", I will illustrate what I meant and why, among many characteristics, primary bodies are usually bigger than the secondary bodies in their system.

By studying the variation of the characteristics of the secondary bodies in each system of bodies and by comparing them across many systems, I perceived that the secondary bodies in each system were born during a turbulent flow of the fluids of their precursors. When I tried to "reverse-engineer" what could have happened for the systems of matter to have many similarities, I realized that, early in its genesis, the precursor of each system of bodies was split-gathered together into the precursors of up to 2 types of daughter bodies:

- the precursor of the primary body or precursor of the system of primary bodies and sometimes
- the precursor of the secondary body or precursor of the system of secondary bodies (if many)

In other words, when a mother precursor was broken up or fragmentated, it yielded the precursor of a primary daughter and the precursor of a secondary daughter. In some instance, more than one primary body and more than one secondary body were formed. In other cases, no secondary body was formed at all. For instance, the precursor of the Solar System was fragmentated and gathered together into 2 main precursors:

- the precursor of the primary body in the Solar System which was the precursor of the Sun and the
- precursor of the systems of secondary bodies in the Solar System which was the precursor of the bodies orbiting the Sun

As I did some mathematics with some of the variables I studied, I realized that the proportion of the mother precursors that landed into the primary bodies and the secondary bodies was not by chance, but followed specific laws that I addressed in the chapters on the 9 system-additive-variables and the turbulent multipliers. In Fig. 2, I sketched the breakup of a mother precursor.

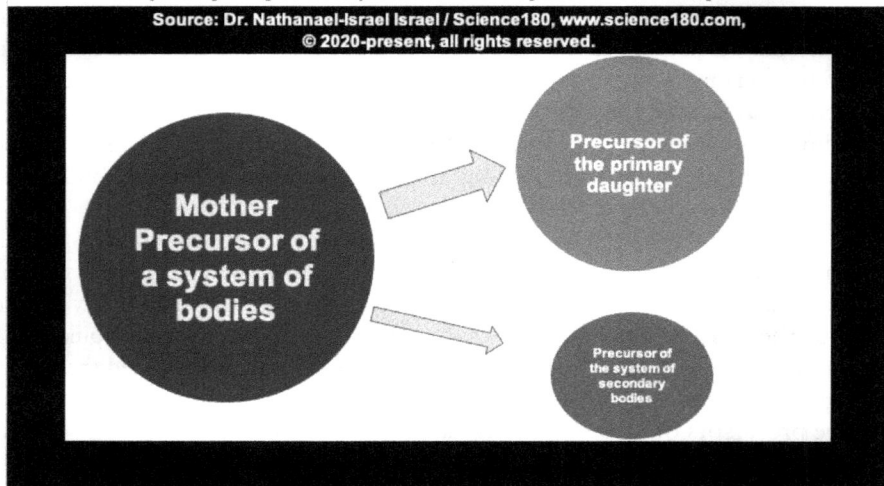

Fig. 2: Fragmentation of the mother precursor of a system of bodies into the precursor of the primary body and the precursor of the system of secondary bodies

It is important to underline that, at the early stage of the development and differentiation of the precursors, no defined shape was acquired yet. As the precursors of the primary bodies were being molded into primary bodies, they did not move much farther away from the "position" of their mother precursors as the precursor of the secondary bodies did. Another way of saying this is that, in each system, after splitting from the precursor of their secondary bodies, the precursors of the primary bodies stayed near the "position" of their mother precursors, while the precursors of the secondary bodies were shifted farther away by a turbulent flow, which characteristics and development would determine the nature of the daughter secondary bodies. The mechanism that split-gathered the precursors of each system of bodies into the precursor of a primary body and the precursor of secondary bodies was also responsible for the destiny, scaling of movement, and other characteristics of the daughter bodies. The birth of the precursors of secondary bodies depended on many factors such as the size, location, energy, and other initial conditions of the mother precursors. Hence, the mother of some secondary precursors yielded many daughters, while others yielded just one.

To illustrate the same concept at the atomic level, the precursor of some atoms was split-gathered into the precursor of the nucleus and the precursor of the electrons, which would later "orbit" the nucleus. Another way of saying this is that the precursor of the system of secondary bodies of each atom was split-gathered into the precursors of electrons around atoms. Similarly, as illustrated below, the precursor of a planetary system was split-gathered into the precursor of a planet and the precursor of its satellites (Fig. 3) through various processes that people have failed to re-engineer or realize.

Fig. 3: Layout (as of today) of the primary planet and satellites born from the precursor of a planetary system

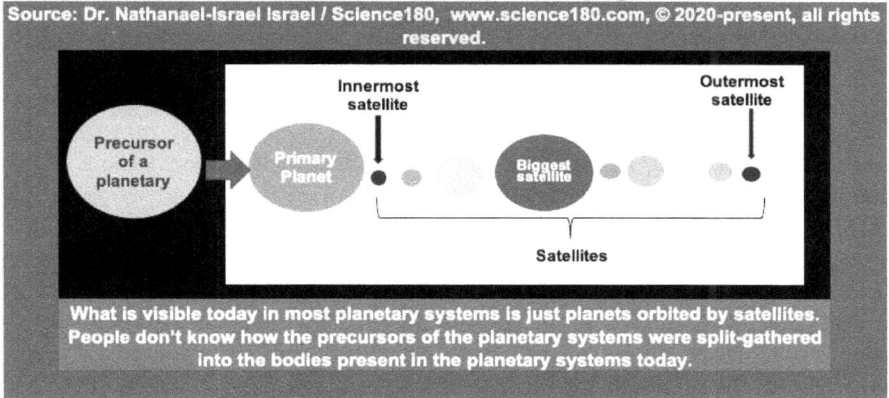

What is visible today in most planetary systems is just planets orbited by satellites. People don't know how the precursors of the planetary systems were split-gathered into the bodies present in the planetary systems today.

Although I could go on and on to illustrate the precursors of all the systems of bodies known today, I stopped those examples here, leaving the details for other chapters to come. At this point, I would like to state that just as some particles and celestial bodies are not systems of bodies, so also were their precursors. For instance, although many galaxies are found in the universe, many stars are known to belong to no galaxy, but they are isolated and sometimes in clusters (e.g. globular clusters) different from galaxies. Similarly, some smaller bodies are isolated instead of being associated with others in a system of bodies. In the Solar System for instance, some planets (e.g. Mercury and Venus) do not have a satellite. In other words, the precursors of Mercury and Venus were not split into a primary body orbited by a secondary body, but they were just molded into one body each: Mercury and Venus. The precursor of the system of secondary bodies in each planetary system is what was split-gathered into the satellites around each planet. Likewise, although some asteroids are known to have their own satellites, most of them do not have a satellite. At the scale of microscopic particles, while atoms are known to have a nucleus orbited by at least an electron, many other types of particles found in nature are not orbited by others. For instance, electrons and other types of leptons (e.g. muon) can also be found in nature not around a nucleus, but in "free" and "isolated" forms.

As of today, most secondary bodies (e.g. satellites) orbit their primary body (e.g. planet) while other secondary bodies (like electrons) are said to be in a shell around their primary body (e.g. nucleus of an atom). However, when the bodies in the universe were being formed, it took some processes before their movement (e.g. rotational and orbital) could have been appointed by turbulent events they went through and which shaped and defined many of their characteristics. In other words, secondary bodies did not just start orbiting the primary body in their system overnight or by chance, but complex mechanisms were deployed so that, as the mother precursors were being molded into their daughter bodies, the changes that the daughters went through caused them to acquire specific movements in the end. To put it another way, the processes of the formation of the bodies in the universe conferred specific movements and positions to them as daughters inheriting traits from their parents. In the next chapter and in that on orbital speed, I will detail how movements were acquired and maintained until today.

In general, precursors tend to cluster around their mother or where their mother had been, just as children gravitate around their parent, which, at their turn, orbit or gravitate around their own parents, and so on and so forth. This explains why satellites orbit their primary planet, which, at its turn, orbits a primary star, which at its turn can orbit the center of its galaxy. For instance, the Moon orbits the Earth, and the Earth orbits the Sun, and the Sun orbits the center of the Milky Way Galaxy. However, because the way they broke up from their mother precursors, some daughter precursors are not able to properly orbit their mothers, but their trajectory can still be traced back to their origin. It is as children weaned or independent from their parents, yet, their roots can still be traced back to the parents. Just as some children stay in the neighborhood of their parents during their entire lifespan, some celestial bodies and matter also stayed in the vicinities of their mothers. Some bodies are still orbiting the remainder of the body of their mother precursor, while others (having a non-circular trajectory) escaped the position of their mother. In the next chapters, I will explain what is the remainder of the mother precursor of each system of bodies and how bodies escaped their mother precursors. Furthermore, until today, the internal constituents of some daughter bodies (e.g. stars, planets, and satellites) are still moving just as the celestial bodies in the universe are not static. This explains why even inside microscopic bodies (e.g. atoms), some subatomic particles (e.g. electrons) are believed to still be moving almost just as planets moving around their primary stars.

For now, bear with me as I am first eagerly laying some foundations of the mathematics that I will do later; otherwise, the incoming mathematical demonstrations may confuse some people and make them to lose track of the real story supporting them. For I have realized that one failure of the existing theories on the origin of the universe is that, people tend to quickly jump into mathematically and statistically modeling things and showing strong coefficient of determinations, while unfortunately they fail to properly interpret what the data in those models mean and much less what is the real story that can explain the machinery of everything. To avoid following that path of the useless modeling, which has filled the scientific and philosophical world with countless theories lacking meaningful interpretations of reality, I decided to first present my perspective of the formation of the universe before diving into the mathematics which led me to the conclusion I have been presenting so far.

As I addressed in my book on the origin of life, it is not by chance that most cells of organisms have a nucleus which is a primary body usually surrounded by many organelles, which are like secondary bodies. In that book, I also explained how the structure of each organelle respects turbulence laws, most of which I presented in this book. In general, a closer look at the organization of livings things points to clusters of primary bodies surrounded by secondary bodies.

9.2. Inclusion levels of primary and secondary bodies in the system of bodies

When I started doing some mathematical demonstrations on the variables I collected on the bodies in the universe, I quickly noticed that the trend of the regressions I got when I mixed the planets, asteroids, and satellites together was different than that I got when I considered planets and asteroids separately from satellites. As I carefully investigated the data (as you will see in the chapters to come), I realized that planets and asteroids are different sets of primary bodies in the Solar System, whereas satellites are also different sets of secondary bodies in the planetary systems. The mathematics you will see in the incoming chapters also suggested to me that the way the fluids in the precursors of the bodies flowed from the mother to their daughters affected the relationship between the variables from one type of body to the others, and from one system of bodies to the others.

As a reminder, in a system of bodies, a primary body is the "central" body which is usually orbited by secondary bodies. Although a satellite is a secondary body in a planetary system, it is important to note that, in the Solar System for instance, a satellite is not a direct secondary body of the Sun. In other words, planets and asteroids are direct secondary bodies of the Sun, while satellites are direct secondary bodies of the planets and asteroids. Using this logic, I initially considered the first level of primary body in a system as the main primary body, whereas the first level secondary bodies are those directly orbiting the primary body. On the scale of the Solar System, the Sun is the first level primary body. Because they directly orbit the Sun, planets and asteroids in the Solar System are first level secondary bodies of the Sun or of the Solar System. On the scale of the Milky Way Galaxy, stars at the galactic center can be perceived as first level of primary bodies, whereas stars at the edge or in the galactic arm are first level of secondary bodies. On the galactic scale, stellar systems orbiting the center of the galaxy can be generally perceived as the first level of secondary bodies. To sum up, based on my first glimpse of the data in the Solar System, I initially considered the Sun as a level 1 primary body, the planets and asteroids as a level 1 secondary bodies and the satellites as a level 2 secondary bodies. I also noticed that all of the secondary bodies in a system of bodies are not the same. Some are big and others are small. For instance, in the Solar System, Jupiter is the largest secondary body. In the Jovian planetary system, Ganymede (the largest Jovian satellite) is the largest satellite or the largest secondary body. Furthermore, among all the satellites in the Solar System, Ganymede is the largest one. Therefore, relative to the Solar System, Jupiter is the largest level 1 secondary body in the Solar System, whereas Ganymede is the largest level 2 secondary body in the Solar System.

To avoid using relative reference, I felt like it would be better to classify the primary bodies and secondary bodies according to what I called "inclusion levels" of the primary bodies and secondary bodies, which I defined as a hierarchical classification of the level of inclusion into others based on their movement (usually revolution) according to their hierarchical position in the cascade of fragmentation or split-gathering of bodies in the universe. I then found that, as the scale of the system of bodies changes (e.g. from planetary system to stellar system to galactical system), the references did the same. To avoid talking about relative references, I decided to classify the levels of inclusion on the scale of the whole universe. In other words, instead of considering the Solar System as the largest scale of reference when classifying the levels of primary and secondary bodies, I considered the whole universe as the largest scale. Based on the work I did on the generations of split-gathering and generations of intermittence, I designed 9 inclusion levels of primary and secondary bodies:

- Inclusion level 1: organization of the invisible world
- Inclusion level 2: microscopic scales such as that of atoms (e.g. electron and other leptons orbing the nucleus of atoms)
- Inclusion level 3: satellite system scales (e.g. satellites orbiting planets and asteroids)
- Inclusion level 4: planetary system scales (e.g. planets and asteroids orbiting a star)
- Inclusion level 5: stellar system scales (e.g. stars orbiting a galaxy)
- Inclusion level 6: galactic system scales (e.g. galaxies orbiting a cluster of galaxies)
- Inclusion level 7: scales of clusters of galaxies (e.g. clusters of galaxies orbiting a supercluster)
- Inclusion level 8: all other higher order scales including sets of superclusters
- Inclusion level 9: scale of the whole universe (all largest clusters of galaxies embedded in the global system formed by the universe)

Below I expounded on that classification.

Inclusion level 1: organization of the invisible world

In the invisible world, there may exist some associations between particles or beings involving secondary bodies orbiting or placed around primary bodies. Particles mostly born from the 9th generation of split-gathering (meaning smaller particles that will never be discovered by scientific equipment) belong to hidden systems. In the nucleus of atoms and in the space around the nucleus for instance, some particles may exist but, due to their small size, they will never be discovered. I intentionally use the term "never" instead of "may" because, seeing some details about the universe and the scientific methods used to explore them, I believe that

Nathanael-Israel Israel: Founder of Science180, the accurate universe-origin, life-origin, and chemicals-origin decoder

human beings will never comprehend 100% all of the makeup of the universe if they continue to rely on scientific means only. Either they could not comprehend the subatomic particles beyond certain limits or they could not reach the limits of the universe to access the diversity of all of its content. I classified the particles or bodies belonging to the invisible world that I described above as inclusion level 1 bodies.

Inclusion level 2: microscopic scales such as that of atoms (e.g. electron and other leptons orbing the nucleus of atoms)

Inside any type of celestial bodies (e.g. planets, satellites, asteroids, and stars), atoms are usually the main ordinary systems of particles, which are organized as primary bodies "orbited" by secondary bodies. Although some molecules and other chemical compounds can be organized like clusters of atoms around which other atoms are arranged, these compounds do not seem to be like primary bodies orbited by secondary bodies. Therefore, I used the configuration of electrons (leptons) around the nucleus of atoms to define the 2nd level of primary bodies and secondary bodies. In other words, I considered the particles in the nucleus of atoms as level 2 primary bodies and the electrons as level 2 secondary bodies.

As I explained in my book *"Turbulent Origin of Chemical Particles")*, the nucleus of most atoms contains many kinds of particles such as the nucleons (e.g. protons and neutrons), which are the most known by the general public and other particles like bosons (e.g. Higgs bosons, W bosons, gluons), which are less known by the public. The number of these particles varies according to the type of atoms. Although the nucleus of the protium (the ordinary hydrogen) is said to have only one proton, the nucleus of most atoms is not just a single or simple primary body, but a system of primary bodies. For instance, each cluster of neutrons, protons, and bosons are systems of particles. Another way of explaining this is that, unlike systems of celestial bodies (e.g. stellar systems, planetary system, asteroid asteroids) for which the primary body is a single body, for most atoms, the primary bodies (which can be perceived as the nucleus) is a system of bodies. Because no one has been able to observe subatomic particles yet, it is difficult to say whether the types of primary bodies inside the nucleus of atoms are separated or mixed entities. In other words, the cluster of protons may not be necessarily separated from that of the neutrons and from that of the bosons. Instead, all protons, neutrons, and bosons may just be mixed all together. No matter how the particles in the nucleus are organized, they are complex and diversified.

While many subatomic particles are found in atoms and other composite particles, others are not, but are isolated. Some of those isolated subatomic particles can be sometimes found in association with others in atoms. For instance, protons (a particle usually found in the nucleus) can also be found in isolation and not orbited by any electron or any other lepton. Similarly, electrons and other types of leptons usually found around the nucleus of atoms can be found isolated and not in orbit or in orbitals around a nucleus. For instance, some free electrons exist in nature just as electronic current, which is the basis of certain forms of electricity. In other words, some subatomic particles are primary bodies not orbited by a secondary body, while others are secondary bodies not orbiting a primary body. I classified those subatomic particles as the level 2 primary bodies. All the clustering of atoms from minerals to rocks passing by all kinds of chemical compounds can be classified in inclusion level 2. To learn more about this chemistry, refer to my book *"Turbulent Origin of Chemical Particles"*.

Inclusion level 3: satellite system scales (e.g. satellites orbiting planets and asteroids)

Inclusion level 3 is about planetary systems orbited by satellites and also about asteroids systems where primary asteroids are orbited by satellites.

Inclusion level 4: planetary system scales (e.g. planets and asteroids orbiting a star)

Inclusion level 4 is about planets and asteroids orbiting a star. It can also be referred to as when a planetary system or an asteroid system is considered with respect to the primary star in a stellar system. Just as the Sun is orbited by planets and asteroids and together, they form the Solar System, many other stars in the universe could probably have their own stellar systems. Some stars do not have many planetary systems but are orbited just by a single planetary system or even just by a planet or a single asteroid. Some stars may not have any other celestial body orbiting them at all.

Inclusion level 5: stellar system scales (e.g. stars orbiting a galaxy)

When stars or stellar systems are viewed as bodies or systems of bodies orbiting the galactic center, inclusion level 5 comes into play.

Inclusion level 6: galactic system scales (e.g. galaxies orbiting a cluster of galaxies)

Inclusion level 6 is about galaxies orbiting or moving around a cluster of galaxies.

Inclusion level 7: scales of clusters of galaxies (e.g. clusters of galaxies orbiting a supercluster)

When clusters of galaxies are studied with respect to superclusters of galaxies, inclusion level 7 is concerned.

Inclusion level 8: all other higher order scales including sets of superclusters of galaxies

Looking at some pictures of galaxies, it seemed to me that some clusters of galaxies may even be moving around other clusters as if some galaxies or clusters

Nathanael-Israel Israel: Founder of Science180, the accurate universe-origin, life-origin, and chemicals-origin decoder

of galaxies are primary systems of bodies while those orbiting or moving around them are systems of secondary bodies. Even if those movements exist, because of their scale and their distance from the Earth, it is difficult, and even impossible, for current scientific equipment to properly measure the orbital movement of galaxies and know which ones are primary or secondary. Because I cannot rule out the motion of superclusters of galaxies around other superclusters, I prefer to address the inclusion level 8 not as a matter of secondary superclusters orbiting primary ones, but just as a matter of belonging. In other words, inclusion level 8 primary and secondary bodies are any organization of the universe into clusters larger than that of superclusters of galaxies orbiting others. Another way of saying this is that any clustering of bodies expressing how large galaxy superclusters include large ones, and so on and so forth until the largest possible supercluster of galaxies, is related to inclusion level 8. Because the universe is still not fully explored yet, many sublevels of inclusion level 8 will be found.

Inclusion level 9: scale of the whole universe (all largest clusters of galaxies embedded in the global system forming by the universe)

The inclusion level 9 is about the global scale of the whole universe containing sets of various inclusion levels of superclusters of galaxies and intermittent bodies or systems of bodies as explained in the chapter on generations of intermittence.

Although it may sound imperfect, this classification of the inclusion levels of primary bodies and secondary bodies is important because it will help to properly model and better predict the "evolution" or the dynamical equations of the variables related to the bodies according to their positions in the cascade or generations of split-gathering. Another way of explaining this is that a proper classification of the generations of split-gathering, generations of intermittence, and the levels of the primary and secondary bodies or systems of bodies is very important to accurately lay the foundations of the physical scenarios of the cascades of turbulence and how their statistics change during the stages of development or differentiation of the bodies in the universe. Even if the number of the inclusion levels can change with time because of the advancement of science, the logic of the inclusion levels will still stand. The following graph (Fig. 4) recapitulates the inclusion levels.

Fig. 4: Hierarchy of the systems of bodies in the universe according to their inclusion levels

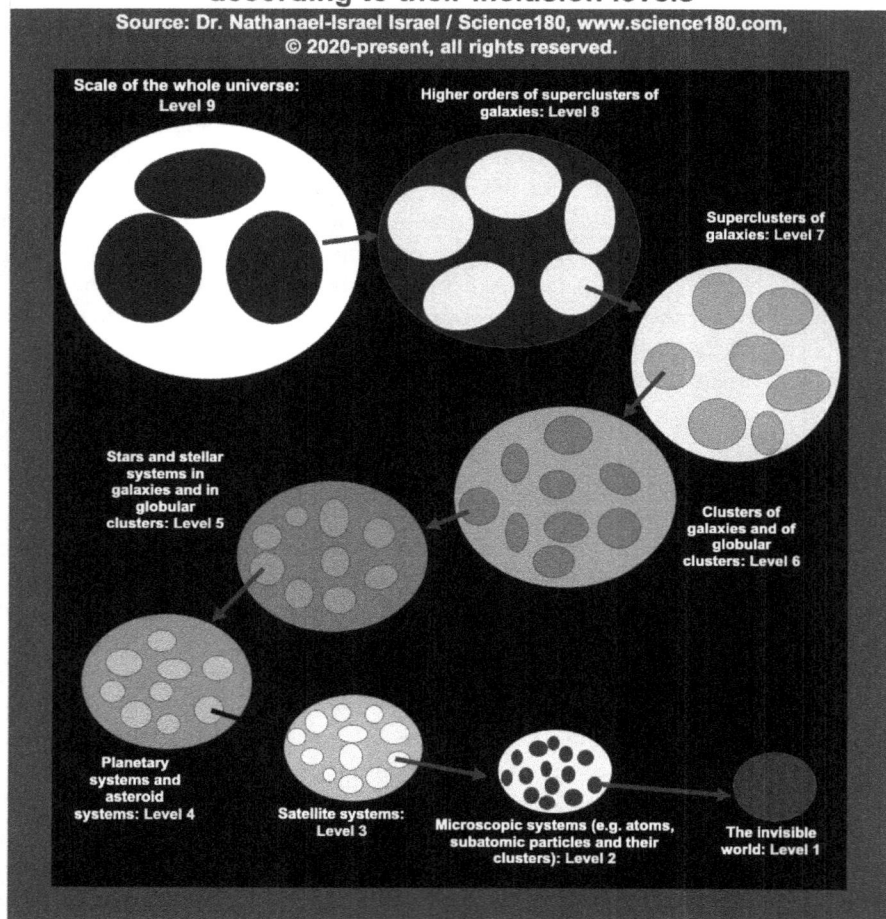

To get a prettier and color version of this graph, visit
www.Science180.com/InclusionLevels

9.3. Take home message

In general, most matters or systems of matter in the universe are organized as primary bodies and secondary bodies. For instance, just as electrons are believed to orbit the nucleus and/or move in a kind of shell or orbital around the nucleus, "similarly", satellites orbit their primary planets. At their turn, planets and asteroids orbit their primary star, which in the Solar System is the Sun. The Sun itself (if not the Solar System) at its turn is moving around the galactic center of the Milky Way Galaxy, the galaxy that the Sun belongs to. Even galaxies are clustered into systems and so on and so forth. In short, I realized that the

Nathanael-Israel Israel: Founder of Science180, the accurate universe-origin, life-origin, and chemicals-origin decoder

precursors of bodies and of systems of bodies were early split-gathered together usually into the precursors of 2 types of daughter bodies:

- the precursor of the primary body or system of primary bodies and sometimes
- the precursor of the secondary body or system of secondary bodies (if any), and in some cases no secondary body was formed at all.

In other words, the breakup or fragmentation of a mother precursor yielded the precursor of a primary daughter and sometimes the precursor of one or more secondary daughter(s). In some instance, more than one primary body and more than one secondary body could have been formed. In other cases, no secondary body was formed at all. For example, the precursor of the Solar System was split-gathered into:

- the precursor of the Sun and the
- precursor of the bodies orbiting the Sun

The split-gathering of precursors of bodies into primary bodies and secondary bodies along with the cascade of split-gathering led to the nesting of systems of bodies according to what I called inclusion levels of primary bodies and secondary bodies in the systems of bodies:

- Inclusion level 1: organization of the invisible world that most people cannot see
- Inclusion level 2: microscopic scales such as that of atoms (e.g. electrons and other leptons orbing the nucleus of atoms)
- Inclusion level 3: satellite system scales (e.g. satellites orbiting planets and asteroids)
- Inclusion level 4: planetary system scales (e.g. planets and asteroids orbiting a star)
- Inclusion level 5: stellar system scales (e.g. stars orbiting a galaxy)
- Inclusion level 6: galactic system scales (e.g. galaxies orbiting a galaxy cluster)
- Inclusion level 7: scales of galaxy clusters (e.g. galaxy clusters orbiting a galaxy supercluster)
- Inclusion level 8: all other higher order scales including sets of galaxy superclusters
- Inclusion level 9: scale of the whole universe (all largest clusters of galaxies embedded in the global system formed by the universe).

'Science180 Academy' Success Strategy
SCIENCE180 CONSULTING

Because Science180's trainings, seminars, or strategic work sessions (through which it transfers skills and training solutions) are great, some customers want to go even deeper on a long-term, sustainable basis. That is where Science180 Consulting, one-on-one consulting, and mentoring (that some people may prefer calling coaching programs) comes in. That is where Science180 can truly change people's behavior on a long term basis according to their specific needs. With Science180 Consulting, you will discover and understand the deep secrets of the formation of the universe, life, and chemicals around you. Hear Dr. Nathanael-Israel Israel's personal selection and teaching on key topics that will help you break the code of the universe formation and functioning. All strategically designed to enlighten you, guide you to navigate and filter the massive data collected on the universe and its content so you know how to answer the world's most challenging origin questions, remove any scientific and philosophical cataracts that may be blocking you, and help bring you many steps closer to your best life today and forever. Science180 Consulting will train you, transfer unconventional skills to you and change your behavior so you go deeper. To get started today or to learn more, go to Science180Consulting.com.

Nathanael-Israel Israel: Founder of Science180, the accurate universe-origin, life-origin, and chemicals-origin decoder

CHAPTER 10

THE 12 UNDENIABLE SECRETS ABOUT THE FLUID LAYERS OF THE PRECURSORS OF BODIES "THEY" DON'T TEACH YOU AT ANY UNIVERSITY

10.1. Formation and shearing of layers of fluids in the precursors of bodies

As the precursors of bodies were being split-gathered, their fluids were flowing because, among other things, they were not a solid that could have moved altogether in a block. During their flow, fluid layers were formed and moved at various speeds according to their location. Considering the characteristics of the bodies in the universe as of today, I felt like they inherited many proprieties from the fluid layers of their mother precursors. In other words, the characteristics of the fluid layers affected the destiny of their daughter bodies. Another way of saying this is that the developmental stages or differentiation processes that the fluid layers of the precursors had gone through had defined the characteristics of their daughter bodies. As I explained later, the characteristics of the layers (e.g. size, speed, position, and depth) were not random, but were consistent with the laws of turbulence that I discovered.

Considering the organization and speed of celestial bodies (see chapter 15 on orbital speed), I understood that their precursors could have shear flowed. The shear flow of the fluids in the precursors of bodies consisted of fluid layers sliding over one another, with each layer moving faster than the one beneath it. Indeed, in the segments below and in the chapters on orbital speed and rotational speed, I explained how the movement (including shearing) of the fluids created a shear stress and shear strain, which acted on the fluids to give or transfer to their layers and to their daughter bodies different speed gradient, momentum, kinetic energy, and other characteristics. In the next graph, I sketched the fluid layers in the precursor of a system of bodies. Then, I addressed some characteristics of the

111

fluid layers of the precursors of bodies.

Considering the physics of fluid ligament, pinching off, undulation, necking, and break up (see chapter on ligament and vortex), it is highly possible that the precursors of some bodies could not have been initially organized into layers, but were just formed as the tube or filament or ligament of their precursors broke up. In other words, although I illustrated the precursors of the bodies as layers stacked by one on top of the others, it is very important to keep in mind that some precursors were not individual layers. I provided more details on that in the chapter referring to the necking, pinching off, undulation, and breakup of the precursors of the bodies.

Finally, although it may be easy to say that a shear flow (in which fluid components sheared pass one another) dominated during the formation of the universe, it is also important to mention that extensional flows (in which fluid components flow away from or towards one other) could have also occurred during the formation of the universe. Indeed, the "original mysterious scattering" which split the turbulent prima material could have contributed to causing some fluids to flow extensionally away from their mother precursors. This extensional flow of fluids seemed to have been imparted to some daughter bodies along the cascade of breakup.

10.2. Intermittence of fluid layers

By studying the size and the distribution of the bodies in the universe, I realized that, during the formation of the universe, the fluid layers of the precursors of bodies did not have the same size, thickness, or depth. Some layers were tiny, while others were very large and deep. Between bigger or larger layers, smaller (or tiny) ones were sometimes found. As I explained later, characteristics such as viscosity of the fluids could have also affected the intermittence of the size of the layers. For instance, fluid layers that were very viscous could have been less intermittent. The presence of smaller layers between larger ones could have been a root of the size intermittence of the daughter bodies. During the separation of the fluid layers, the "cleavage" points, lines, junctions, separation, or breakup interface were not smooth all of the time, but they led to the formation of even tinnier layers or clusters of matter, which, after going through some changes, led to the formation of tinnier bodies between larger ones. In other words, the intermittence of the daughter bodies was caused by the intermittence of the layers of fluids of the precursors, the characteristics of the fluids, and the way the layers of fluids were turbulently split-gathered. In this segment, I addressed intermittence by referring to the size of the layers and their daughter bodies, but in fact, intermittence also occurred with many other variables (e.g. energy, mass, and volume) that I addressed in other chapters (e.g. see chapter on the 9 system-additive variables). The next graph (Fig. 5) illustrates an intermittence of fluid layers. Based on the data I analyzed, the intermittence is usually higher in the most turbulent zone, but smaller in the outer zones. Details on the turbulent zones are coming up very soon.

Fig. 5: Intermittence of Layers of fluids in the precursor of a system of bodies having 5 turbulence zones

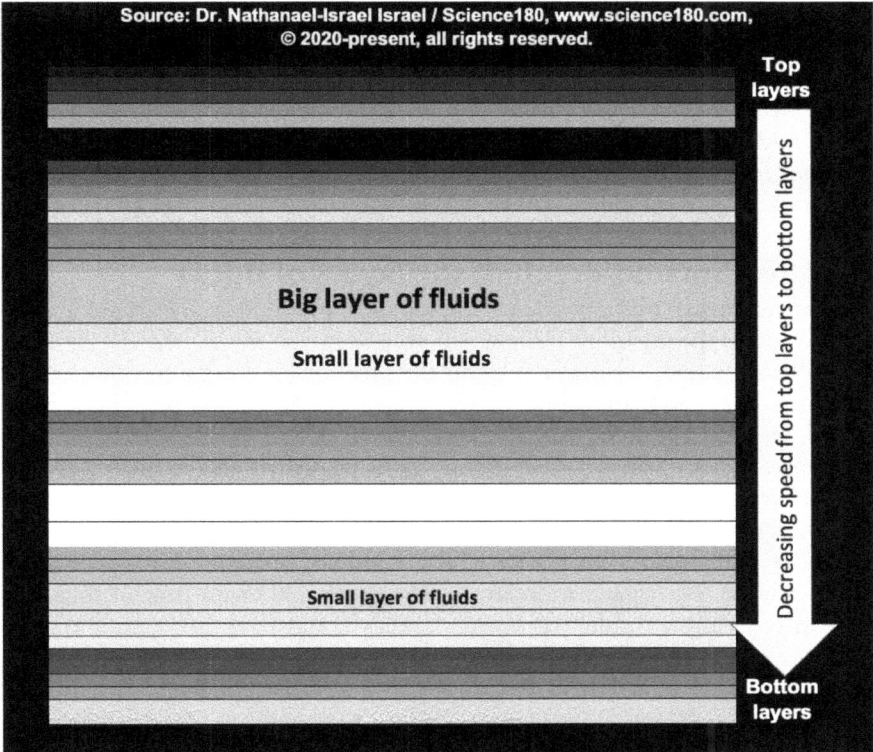

Top layers

Decreasing speed from top layers to bottom layers

Big layer of fluids

Small layer of fluids

Small layer of fluids

Bottom layers

10.3. Mixing of fluids in the layers

As the fluid layers were flowing, under the influence of the turbulence they were going through, structures were formed inside of them. Some structures were coherently distinguishable, while others were not. The flowing fluid layers were not completely isolated from one another. Some mixed with others just as inside each layer, structures (some of which were vorticial) were mixing and rearranging. Some small layers were "swallowed" by bigger ones, meaning that some small layers and structures ended up being incorporated into bigger ones. While some layers were stacked on top of others, others were mixing as their fluids were flowing. In other words, some layers coalesced or merged to form bigger ones, while others were completely separated from others. Another way of saying this is that, some big or deep layers consisted of smaller layers, while others are tiny and did not mix with others.

Because of the intensity of the turbulence in the fluids during the formation of the universe, the movement and path of the fluid layers were not generally smooth, but very turbulent. In other words, the fluid layers of the precursors of bodies did not always move smoothly passing adjacent layers without mixing to some extent. The degree of the mixing of the fluids in the layers affected the size, purity, composition, and other characteristics of some daughter bodies. For instance, bigger daughter bodies could have been born from larger or deeper layers mixed to a higher extent, while smaller bodies could have descended from thinner layers that were not mixed with many others. Thicker and very turbulent layers could have favored the formation of larger bodies than thinner layers could have had. In the segment on the topological changes of the fluids, I expounded on the mixing of the fluids.

10.4. Position and depth of the layers of fluids

In the precursors of primary bodies, the top or upper fluid layers could have been located around the position of the north pole of the bodies as of today, while the bottom layers could have been around the position of the south pole. In other words, the north and south pole of the primary bodies could have been mostly formed respectively from the fluids in the top and bottom layers of their precursors. In contrast, as for the secondary bodies, the fluids in the top layers of the bulk of their precursor were molded into the innermost secondary bodies, while the bottom layers were shaped into the outermost secondary bodies. The precursors of the secondary bodies which could not break up into other systems of bodies were also organized in layers. The north and south pole of those fluid layers in the "unbreakable" secondary bodies could have been predominantly formed from the fluids in their top and lower layers respectively. Consequently, in celestial bodies which are not as solid as the terrestrial planets, the movement of the fluids or plasma at the north and south pole may not always be the same. Similarly, the vortices, cyclones, and other turbulent structures formed at the poles of some celestial bodies should not be the same. Depending on the location of the bodies, movements at the south pole may differ from that at the north pole. In the chapter on vortex and ligament, I better elaborated on this trend and also addressed the mysterious cyclones found in the poles of Jupiter and Saturn.

10.5. Stratification of the crust and atmosphere of celestial bodies due to the fluid layers of their precursors

The presence of fluid layers in the precursors can also explain the strata found in the crust and atmosphere of some bodies today. Regardless of the density of the crust of celestial bodies, strata or residue of layers can still be seen. Although the constituents of the fluid layers were mixed, compressed, and shaped during their turbulent split-gathering to form their daughter bodies, the footprint of the layers in the precursors can still be seen in the stratification of the daughter bodies. Another way of saying this is that, the changes that the fluid layers in the precursors had gone through during the formation of their daughter bodies were not enough to erase the traces or footmark of the potential location of some of those layers.

In the segments related to the gathering together of fluid layers into well-shaped (e.g. spherical) bodies, I sketched how the fluid layers in the precursors were translated into strata in the daughter bodies. Additionally, after the formation of the universe, some deposits (e.g. sedimentation) also affected or caused the stratification of some upper layers of celestial bodies. For instance, some strata of rocks or minerals in the upper crust of the Earth could have been formed after the deposition of materials which are not related to the fluid layers of the precursor of the Earth.

10.6. Speed of the fluid layers later translated into orbital and rotational speed of the daughter bodies

The fluid layers did not move at the same speed. While some layers moved fast, others were slow. The top layers had a higher speed than the bottom layers. The decrease of the speed of the fluid layers in the precursors can also be illustrated by how the speed of the flow in a river generally decreases from the top layer to the bottom one. In the chapter on orbital speed, I explained the reason of that trend. Later, as some layers were wrapped into spherical bodies, their speed was converted into the orbital speed and rotational speed of their daughter bodies. In other words, although the entire bulk of the fluid layers in the precursors was moving as a whole, inside of it, their constitutive layers had different speeds. The speed of the fluid layers before their split is one of the key variables that defined the orbital speed of their daughter bodies. That is why the orbital speed of the bodies in any system does not depend on their size but mainly on the position of the layers of their precursor inside the stack of layers of all bodies belonging to their systems before their splitting.

Because the fluid layers in the precursors were organized one on top of the others, the layers on top separated from the bulk of layers before those at the bottom. In other words, as the bulk of the fluid was flowing, the time it took for the layers to be separated and launched into their separate path was positively correlated with the depth at which they were located. Consequently, in each system of bodies, the distance separating the position that a layer starts moving away from its mother precursor and the position at which it was separated from the last stack of layers it belonged to before going onto its own trajectory increased from the top layers to the bottom ones. Nevertheless, after the bodies were formed and positioned on their orbit, the distance separating them could still be expanding as the universe itself is said to be expanding. In the chapters on semi major axis and timeline, I provided ample details.

10.7. Direction of the flow of the fluid layers can explain the counterclockwise movement of some bodies in the Solar System as seen from the north pole

The way the precursors of bodies were molded into complex forms defined the characteristics (e.g. speed, size, and location) of their daughter bodies. For instance, the direction and turbulence of the fluid flows communicated to their daughter bodies movements in specific directions. In general, the fluid layers of the precursors flowed into the direction of the force which was pushing them away from the initial position of their mother layer. The types of movement (e.g. rotational and orbital) that some bodies exhibit today were progressively born out of the movement induced by the instability and turbulence in their precursors during the formation of the universe.

The fact that the fluid layers flowed into the direction communicated by their mother precursors can explain why for instance prograde planets and asteroids in the Solar System orbit the Sun in the same direction just as prograde satellites orbit their primary planets in the same direction. Later, I will explain why prograde and retrograde secondary bodies do not orbit their primary body in the same direction. Considering the direction of the revolution of planets and asteroids in the Solar System, it seemed to me that the direction into which the precursor of the bodies orbiting the Sun was launched away from the precursor of the Sun was a counterclockwise direction, when viewed from the north hemisphere. As I explained later, the movement of and the way that the vortical structures of the precursors of bodies were separated, flipped over, tilted, and stretched some daughter bodies, hence the motion of some bodies is prograde and that of others is retrograde. This also impacted the orbital inclination, eccentricity, and axial tilt of some bodies. I expounded on these things later in this book.

10.8. Topological and structural changes in the fluids of the precursors of bodies

It would have been incomplete to talk about the events which shaped the beginning of the universe without addressing how the instability or turbulence of the fluids affected their topology. To properly explain fluid instability and how it affected the distribution of the precursors of matters, careful attention must be given to the topological changes in the flowing fluids. Therefore, this section on topological changes is very important and is one of the foundations of my story of the formation of the universe. For, as explained below, these topological changes of the fluids of the precursors contributed to the formation of diverse types of matter and systems of bodies in the universe.

Indeed, it is well established in physics that instability in a fluid changes its topology. For instance, the addition or pouring of cream into a cup of coffee destabilizes the coffee and, the intensity of the instability affects the shape of the fluid "filaments" that are formed. Although the nature of the matter in the precursors of the bodies in the universe was different from that of the matters known today, the instability of the turbulent prima materia affected the topology of the fluids in the layers of the precursors of bodies.

The late Erik Mollo-Christensen, an expert in flow instability, had found that flow instability is everywhere in nature and is the first step in the chain of events which generate turbulence. Flow instability can lead to the formation of waves and turbulence. Turbulence is believed to usually occur when the instability of a flow is beyond a certain critical value (Mollo-Christensen, 1969). Even today, when an instability such as that of Kelvin–Helmholtz occurs, the interface separating two initially parallel streams having different speeds is naturally unstable and as the instability grows, several types of structures can be formed. Similarly, the interface of the fluid layers of the precursors of bodies could have birthed the formation of very unstable structures, which mixed and interacted with others. Similarly, flow instability and the turbulence that accompanied it was one of the first events during the formation of the matters in the universe. However, seeing the characteristics of the bodies in the universe and the time it could have taken to form them, the instability of the precursors could have been very violent and very fast. More details are in the chapter on timeline.

As the precursors of bodies were being split-gathered under the influence of the turbulence they were going through, the identity of their surface and interior changed. As new structures were being formed in the interior of the layers, new forms were also acquired externally. The internal and external changes continued until they reached a point that could no longer allow major changes. That point was about the time the formation of the bodies could have been declared "finished". In other words, the topological changes that the precursors of bodies went through were not indefinite and their duration varied according to the characteristics (e.g. size, location, and speed) of the layers. As the clusters of bodies of different scales were interacting with one another, their degree of freedom decreased until these clusters could no longer rearrange much, therefore locking themselves into an "equilibrium" stage defined by their characteristics (e.g. position, movement, composition, and speed).

While some of those changes are too small to be seen today by the naked eye and the bodies born from them could not be seen either, other changes were so immensely big that the current scientific equipment could not yet or will never fully apprehend their size. Another way of saying this is that, from the invisible scales (smaller than the microscopic scales) to the astronomical ones (that of higher orders of galaxy superclusters), topological changes occurred in the early universe. The variations of the topological changes were responsible for or are some of the foundations of the diversity of bodies and systems of bodies in the universe. In other words, the scale of the topological changes in the fluids of the precursors defined the scale or size of the bodies or systems of bodies in the universe today. I am not saying that the topological changes are the only factors that defined the properties of the bodies in the universe, but without their disparity, things in the universe could have been different and less diverse.

While some changes were similar, others were different. If the intensity of the instability was different at the beginning of the universe, a completely different world would have been formed. Just by visualizing the topological changes occurring in an instable fluid as time passes by, it seems to me that the structural reorganization of matters in an instable fluid depends also on the duration of the instability or the time it would take for the degree of freedom of the fluids to be significantly reduced. For instance, a certain volume of coffee mixed with a certain volume of cream can yield different structural mixtures depending on the intensity with which the mixture is stirred and shaken. Likewise, the precursors of the bodies in the universe could have birthed different kinds of daughter bodies if the turbulence they went through was different. Another way of saying this is that the phenomena that destabilized the early universe and transferred different intensities of turbulence into the fluid layers of the precursors of bodies were responsible for many characteristics of the bodies in the universe today.

While the instability of the precursors could have moved their fluids in different directions, causing some splashing, sloshing, and other types of movements, the trends in the data I analyzed suggested that the way the fluids were rearranged followed specific turbulence patterns. When some clusters of fluid layers were moving and hitting one another, some of their components could have splashed and increased the mixing of their constituents. All those movements and mixtures of fluids affected the turbulence that the precursors and their daughter underwent. I later provided more details on the mixing of the fluids of the precursors of bodies.

As the precursors of bodies were moving through space, they went through stages of "differentiation" including their expansion, split, shrinking, compression, or squeezing. It has been proven that, when a fluid undergoes topological transitions (both breakup and merging), the identity of its surface changes (Eggers and Villermaux, 2008). The same author also reported that the rate at which perturbations grow in space depends on the jet speed. This suggests to me that the rate at which the perturbation of the precursors of the bodies grew in space during the formation of the universe could have depended on the characteristics (e.g. speed, size, and energy) of their jet. These characteristics of the precursors could have determined the size and the distribution of their descendants or daughter bodies. For instance, if the precursor of the Solar System was launched with a different speed and at a different angle and direction, the Sun and the other bodies in the Solar System could have been different. Similarly, if the size of that precursor was different, the bodies in the Solar System would have been different. By the time I walk you through the variables that I studied, you will better understand what I mean.

To sum it up, during the formation of the bodies in the universe, the precursors underwent several topological transitions, which contributed to changing their identity, properties, shape, and many other features. The daughter bodies were molded according to the "impartation" of proprieties they received from their mother precursors. In other words, the bodies were formed after some laws inherited from or dictated by their mothers. Unlike nonliving things, children of human beings do not always act like their mothers or parents. In addition to what they receive from their parents, children and even adult human beings make their own choices, which can change the course of their lives. In contrast, the story of the genesis or the destiny of matters was mostly defined by their mother precursors, the environments they were formed, and the changes their precursors went through. For instance, if the intensity of the perturbation that broke and separated the fluids in the precursors of bodies was different than what it was, the universe would have been different. In other words, the intensity of the perturbation of the mother precursors influenced the characteristics (e.g. size) of their daughter precursors. The instability that the daughter precursors went through also affected their shape, composition, and positioning. To some extent, if the turbulence that occurred during the formation of the universe was calibrated differently, a different world could have existed, meaning that the precursors of the universe would have been fragmentated into other types of bodies which may not match what they are today. This also implies that, in different stellar systems, other types of bodies or matters different from what is present in the Solar System could be found. Therefore, it can be very dangerous for scientists to model the universe or its constituents just by basing their models on experimental data collected only on Earth or only in the Solar System.

10.9. Turbulence zones: formation and delimitation

As I amply demonstrated in the incoming chapters, the fluid layers in the precursors of bodies in the most developed turbulence can be divided into regions that I called the turbulence zones. Due to its importance and the amount of information involved, I devoted a completely different book to turbulence zones. For the sake of space, I will pinpoint here just some main traits. Indeed, the fluid layers that were at the top and at the bottom of the precursor of the secondary bodies were not as turbulent as those which were somewhere "between" them. By "between", I do not mean a geometric or spatial middle, but a position which was neither the top one nor the bottom ones. Because primary bodies are generally one body, I did not divide the fluid layers of their precursors into turbulence zones. However, because the secondary bodies are usually many, I split them into turbulence zones so I could better study them.

Considering the similarities and differences in the tendencies of the variables that I have analyzed and the characteristics of the bodies from the innermost to the outermost ones in each system and across many systems of bodies (e.g. planetary systems), I realized that, in most cases, the turbulence that occurred in the fluids of the precursors during the formation of the universe could be traced back to the precursors of 7 turbulence zones that I called:

- Turbulence Zone 0

- Turbulence Zone 1
- Turbulence Zone 2
- Turbulence Zone 3
- Turbulence Zone 4
- Turbulence Zone 5
- Turbulence Zone 6

I could have labeled them Zone 1 to Zone 7, but for a few years, I worked on 5 turbulence zones (Zone 1 to Zone 5) before I realized, while working on the ring systems, that 2 more zones were needed to explain the full picture of the turbulence in the precursors of the bodies in the universe. Hence, I called those additional two zones Zone 0 and Zone 6 instead of having to rewrite everything using the new labels (from Zone 1 to Zone 7).

Indeed, the turbulence Zone 0 is the innermost one. In planetary systems, Zone 0 hosts the innermost rings and is located between the primary planet and the innermost satellite, whereas in stellar systems, it is located between the primary star and the innermost planet or innermost asteroids (whichever is the closest to the star). In the chapter on rings, I expounded more on turbulence Zone 0.

Turbulence Zone 1 is immediately downstream from Zone 0. It contains bodies bigger than those found in Zone 0. In the planetary systems for instance, Zone 1 is the region where the innermost and fastest satellites are found, whereas in stellar systems, Zone 1 is the region where the fastest and innermost planets are found. The turbulence in Zone 1 can be said to be "laminar" or "smooth" flowing. But this does not mean that no turbulence existed in Zone 0.

Turbulence Zone 2 is a transition zone from the "laminar" turbulence to the fully developed turbulence, which will be reached in Zone 3. In other words, Zone 2 is not as turbulent as Zone 3, but is more turbulent than Zone 1. That is why Zone 2 can also be called the "transition zone into the highest and most developed turbulence". For downstream of Zone 2, Zone 3 (which is the most turbulent zone) is found.

Turbulence Zone 3 is the zone where the highest turbulence is found. It is the zone or region where the fluids of the precursor of bodies were most highly turbulent. The biggest secondary bodies in a fluid flow are usually found in its Zone 3. In the stellar systems for instance, the largest planets are found in Zone 3. In the planetary systems, the largest satellites are usually found in Zone 3. One of the key characteristics of the outermost satellite in Zone 3 for instance is that it has the smallest rotational angular speed (when all the bodies in Zones 1, 2, and 3 are concerned).

Turbulence Zone 4 is also a transition zone, but unlike Zone 2, Zone 4 is a transition zone out of turbulence. In other words, located downstream of Zone 3, Zone 4 is the zone where the fluid flow of the precursors could have transitioned out of the developed turbulence in Zone 3. The data I studied suggested that the characteristics of the bodies in Zone 4 and Zone 2 are more similar to one another than to those in any of the other zones. Viewed from the perspective of the stack or pile of fluid layers before their split, the fluid layers of the precursors of the bodies in Zone 2 could have been located on top of those in Zone 3, which is located on top of those in Zone 4. In other words, the fluid layers of the bodies in Zone 3 could have been located between the fluid layers in Zone 2 (on top) and those in Zone 4 (at the bottom).

Turbulence Zone 5 is the outermost zone where the remote celestial bodies in each system are found. It is the outer zone located downstream of Zone 4. The flow of the fluids of the precursors in Zone 5 could have been the most laminar one and the speed of bodies in Zone 5 are the slowest. Bodies in Zone 5 (and sometimes in Zone 4) usually have the highest orbital inclination and eccentricity.

Turbulence Zone 6 is located downstream of the outermost bodies and does not contain "any" celestial bodies. It is the transition zone where the influence of the system can be felt, but which does not contain any well-defined celestial body. Particles under the influence of the system of bodies can be found until a point where no influence can be felt.

Because all turbulences are not fully developed, all turbulent flows do not have all 7 zones. Therefore, the largest bodies are not always found in Zone 3. For instance, the largest satellites in some planetary systems are not found in Zone 3. The presence of all the turbulence zones depends on the initial conditions (e.g. size, intensity, and location) of the concerned fluids. The number and size of the turbulence zones depended on the size of the layers and the development of the turbulence of their fluids. While some precursors had all 7 zones, others did not. Before they were split, the layers of the fluids in the top or upper turbulence zones were the fastest while those in the lower turbulence zone was the slowest. Similarly, the daughter bodies of the fluids in the top turbulence zones were closer to their primary bodies than those on the bottom turbulence zones. Usually, when I studied the satellites, I focused on 5 turbulence zones that, in short, I called Zones 1, 2, 3, 4, and 5.

The delimitation of the turbulence zones is key to understanding how I approached turbulence. As you will read in the chapters to come, the development of turbulence involves many events including fluid flow, fragmentation, squeezing, ligament formation, breakup, pinching off, recoiling, etc., all of which dynamically occurred during the formation of the universe as everything was changing together until a kind of "equilibrium", which limited the degree of freedom of the movement of some constituents of matter, was reached. Even after that "equilibrium" was attained, things in nature have still been under the influence of many changes. For, since the beginning of the turbulence that shaped things in the universe, nothing has been static in nature. Everything has been changing to some extent although some changes may not always be seen or felt. A careful and strategic study of the relationship between the variables collected on the bodies (small and big) in the universe allowed me to better understand how the precursors of the bodies evolved under the influence of turbulence. The characteristics of the turbulence zones are an imprint or footprint of what happened to the matter that birthed the bodies in the universe.

After splitting from their mother, the precursors of the turbulence zones went through a turbulent development based on the conditions they inherited from their mother. The precursors of some turbulence zones also split into different bodies or systems of bodies. For instance, the precursor of the Solar System could have been organized into 7 zones, with Zone 3 being around the region of the giant planetary systems, which were split to yield different precursors of planetary systems, which were later split into different planets and satellites, and the precursor of each satellite system had to go through its own split. At different levels in the universe, precursors of bodies were split and reorganized differently before splitting again and so on and so forth until their turbulence "die off" somehow and no further split occurred except those that occurred at the atomic and subatomic levels.

I was writing this book before I discovered the notion of the fluid layers and how it is key to better understand the formation of the universe. Beforehand, I used to perceive that the precursors of the bodies could have come off their mother in the reverse order of their distance from the semi major axis. For instance, I used to think that the outermost bodies could have come off before the innermost bodies. But I realized in 2020 that this way of perceiving the genesis of the bodies is wrong.

10.10. Separation of fluid layers as they took different paths

At one point, the bulk of fluid layers in the precursors were moving together as if the layers were bound together by a "yoke". At one point, the fluid layers took different paths. According to their environmental conditions (including the turbulence, speed, and shearing), the fluids in the layers were differently organized in clusters, which could no longer stay together, and consequently ended up taking different tracks. Accordingly, the fluid layers and their daughter bodies were separated from one another. The separation of some fluid layers or of their daughter bodies occurred when the change in their speed or direction was too large for them to continue adhering to one another.

By the time the precursors of the celestial bodies were separated, some particles could have been already formed. But considering the difference in the abundance of chemical elements in the atmosphere of the celestial bodies in the Solar System (see details in my book *"Turbulent Origin of Chemical Particles"*), I showed that all the chemical particles could not have been formed and completely differentiated before the separation of the precursors of the celestial bodies they belonged to. The continuation of the turbulence that each precursor of bodies or precursor of systems of bodies had gone through after splitting from their mother precursors determined the chemical composition of their daughter bodies. Consequently, the similarities and differences between the chemical composition of the bodies in the universe reflect the processes that their precursors went through. The precursors of the bodies that are near one another or that came from similar or related mother precursors ended up being more similar. Later, I will better explain how the precursors of the bodies were broken up and what caused some to yield satellites, planets, or planetary systems, while others became stars, stellar systems, and their clusters including galaxies.

10.11. Gathering together of the fluids in the layers into unique bodies or systems of bodies

Even before the fluid layers were separated, their constituents and themselves were being collected into structures (e.g. vortical structures) and after their separation, some of them were gathered together into unique bodies. The internal organization of the fluids as they were moving under the influence of turbulence contributed to causing them to collect. Looking at the pictures of hurricanes and galaxies (that I detailed in the chapter on galaxies), I felt like the spiraling of fluids of the precursors of some massive bodies could have occurred many times. In other words, some spirals could have been wrapped around many times. Similarly, the data on celestial bodies suggested that the fluid layers of some precursors of bodies could have been wrapped around or swirled many times, while others could have not. It also sounded like bodies that are at the outskirt or edge of the system of bodies could have been wrapped up differently and even more loosely than the innermost ones. Images of some spiraled galaxies suggested that their outer stars are not always wrapped tightly, nor pulled or pushed toward the center of their galaxies as some inner ones, therefore making me think that the force which "attracted" the fluids and wrapped them around the primary body or system of primary bodies had some limits. In other words, something could have prevented some fluid bodies of the precursors to be compacted into a single and tight body. That is why some celestial bodies have no other celestial body orbiting them, while others have many. Some bodies are even a spiral of systems of bodies. Many things I said in this section apply to both celestial bodies and even microscopic particles. In Fig. 6, I sketched how fluid layers split-gathered into separated layers, which, after being wound up, birthed unique daughters.

Because of the fluid layers of their precursors, celestial bodies and even particles are surrounded by layers of fluids or layers of particles, which must be crossed before reaching the surface of their primary body. That is why the Earth for instance is surrounded by layers of gas, including the ozone in the atmosphere, while the outer portion of the Earth is made of layers forming the crust.

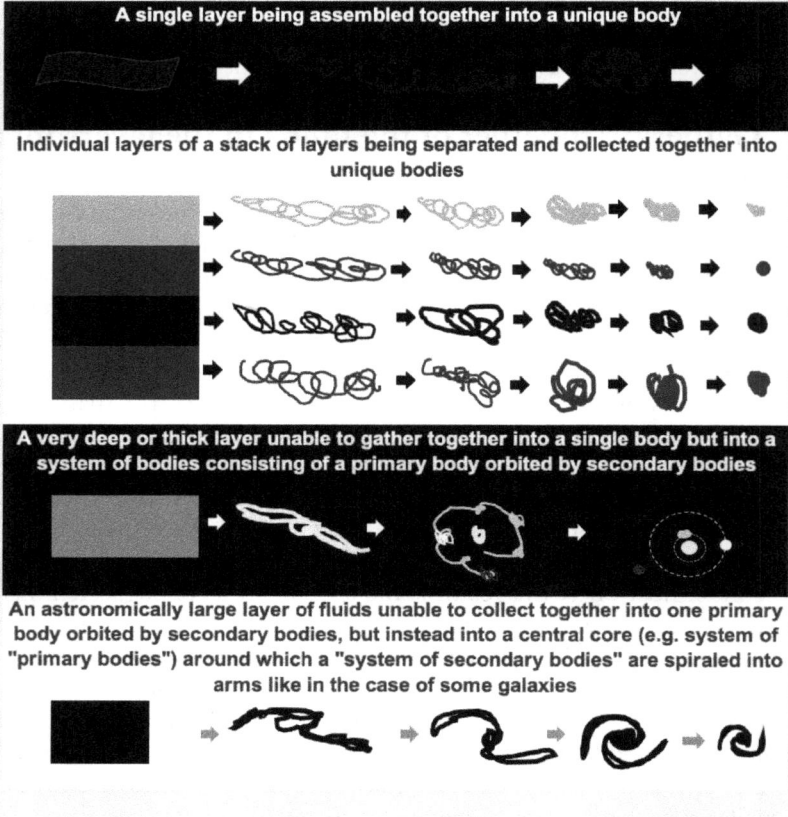

Fig. 6: Gathering together of fluid layers of the precursors of bodies into individual bodies, systems of bodies, and astronomical spiraling structures
Source: Dr. Nathanael-Israel Israel / Science180, www.science180.com,
© 2020-present, all rights reserved.

Fig. 7 illustrates how a stack of fluid layers could have birthed a set of spherical bodies.

Fig. 7: Layers of fluids in the precursors gathered together into spherical bodies

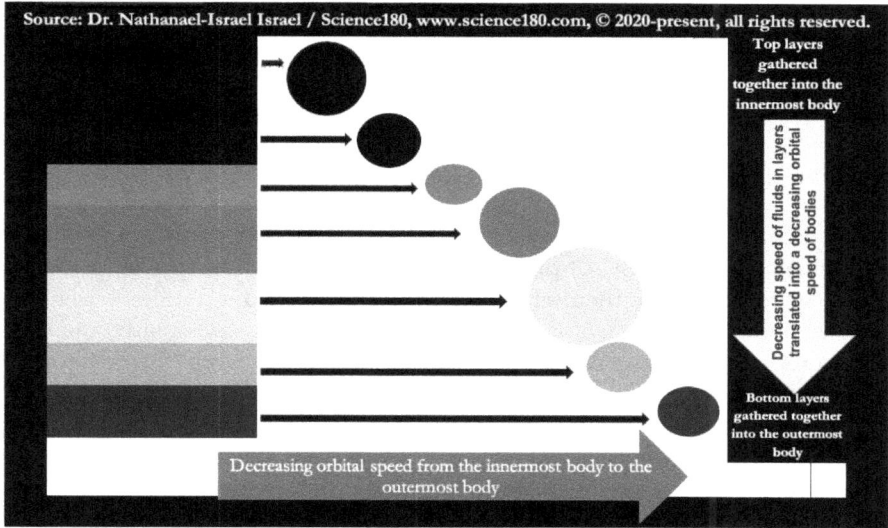

Top layers gathered together into the innermost body

Decreasing speed of fluids in layers translated into a decreasing orbital speed of bodies

Bottom layers gathered together into the outermost body

Decreasing orbital speed from the innermost body to the outermost body

As illustrated in Fig. 8, under some circumstances, the fluid layers of some precursors of bodies could have been gathered together in such a way that the top layers were clustered around the north pole whereas the bottom layers were clustered around the south pole. In other conditions, other factors could have mixed the fluids and prevented them from following the same position in the celestial bodies as in the fluids of their precursors.

Fig. 8: Gathering together of the layers of fluids in the precursor of a body into a spherical body

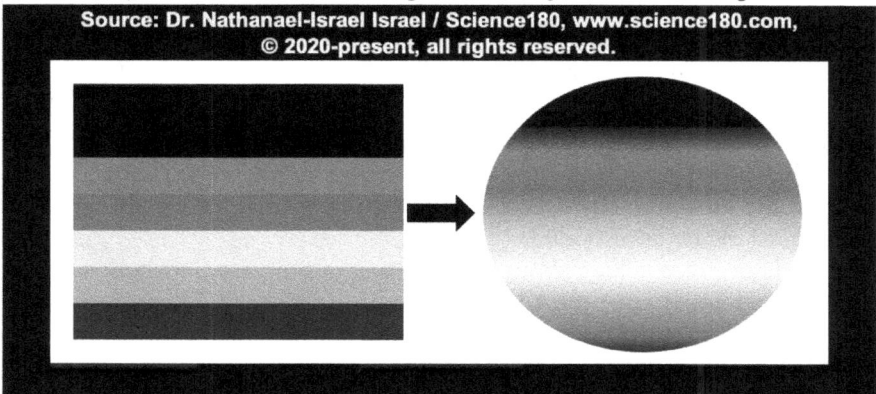

As some vortices or eddies were being "rolled" to form spherical bodies, they

Science180: Nonconformist, Rule-Breaker, and Accurate Demonstrator of the Universe-Origin were squeezed. The energy in the precursors also affected the squeezing intensity or the applied pressure. As the fluid layers were being wrapped, they applied a squeezing force to their daughter bodies. At the microscopic levels as well as the astronomical levels, squeezing of the precursors of the bodies occurred. In the chapter on vortices, I revisited how the squeezing of the vortices on a small and large scale could have also affected the density and other characteristics of the bodies. As I detailed in the chapters on rotational speed, vortex, gravity, and the fundamental forces in nature, the processes that collected the bodies before and after the layers of their precursors were separated played a crucial role in the establishment of gravity, shape, rotational movement, and the ellipticity (flattening and oblateness) of the celestial bodies. For instance, as some celestial bodies were collecting themselves into their "final shape", others were like globular bodies rapidly "whirled" round on their axis. Consequently, the north and south poles of most celestial bodies were flattened, while the equator was raised up, therefore giving some celestial bodies (e.g. planets) an ablate shape. I will later provide evidences as to why most of the larger celestial bodies are oblate.

In the giant gas planets (e.g. Jupiter and Saturn) or in stars (e.g. the Sun), which outer surfaces are not solidified, vorticial structures carrying the footprint of the turbulence that occurred during their formation, still exist. As I detailed in the chapters on vortices and ligament, the cyclones found on Jupiter and at the pole of Saturn are examples. For instance, in the case of terrestrial planets (those with a solid crust), the vorticial structures could have been solidified into rocks and then piled up to yield the crust. For instance, the data I studied suggested that, as the fluids of the precursors were collecting themselves, smaller clusters of fluids inside of larger ones were assembled together into harder bodies such as chemical compounds, minerals, and rocks. The section on generations of split-gathering and inclusion levels also gave a glimpse into such processes.

Considering the organization of the subatomic particles, atoms, molecules, chemical compounds, minerals, rocks, crust of terrestrial planets, atmosphere of celestial bodies, the plasma of the Sun and the other variables I studied in my books on the origin of the universe, I showed the gathering together of the bodies (small and big) in the universe involved processes associated with the following phenomena: amalgamation, bending, braiding, clumping, kinking, knobbing, lumping, netting, pulling, pushing, shearing, spiraling, twisting, winding, and many other processes I already discussed in previous chapters. As I get into the details in the incoming chapters, I will explain how these processes took place according to the types of bodies.

I could not finish talking about the gathering of precursors without saying the role that cold could have played in the change of state of some precursors of bodies. Indeed, as some precursors were being molded during the turbulence of the genesis of the universe, their constitutive particles, and the interactions between them were changing until they were locked into systems, which defined their motion in harmony with their environment. One of the factors that contributed to reducing the interactions between particles was the temperature of their environment. For instance, when an originally hot precursor landed in an environment which temperature was below certain values (e.g. freezing point), that precursor can freeze at least at its surface, therefore limiting the interactions between some of its constituents. In contrast, when a precursor is very big and extremely hot, its internal energy and heat can overcome the cold of its surrounding environment in a way that it cannot freeze or solidify. Instead, such huge precursors can stay very hot and maintain some of their initial heat and configuration of their constituents. Although these kinds of bodies could still lose energy and heat over time, they could still keep their fluid characteristics and heat. Their size and internal energy could have defined how long some precursors could have stayed fluids before becoming solid. The surface of some precursors could never be solid and could emit light like stars. In other words, some stars are like the "remnant" of precursor bodies that were too big to allow the cold in their environment to freeze them or force them to become a solid. It is possible that some stars can die (even today) if for instance their energy can no longer be sufficient to overcome the cold of their environment and/or if the constituents of their surface can undergo some modifications that can solidify them and limit the release of photons and other particles of light. Here, the death of a star is not its disappearance, but a state where its surface stops emitting light because its matters were transferred into matters that could not give light.

Depending on their size, spatial localization, speed, and the characteristics of its constituents, a same precursor can lead to the formation of different daughter bodies. For instance, as of today, a liquid can be converted into ice by lowering its temperature below a certain critical value. However, depending on the duration of the temperature lowering, a same water can lead to the formation of various types of ice with different structures or different states of matter. For instance, amorphous water is formed by quickly lowering the temperature of water. Although it is a kind of ice, amorphous water does not have regular repeating crystal structure like normal ice. Similarly, the outcome of some precursors was highly affected by the duration of the process they went through. In the next chapter, I will review some types of precursors of bodies.

10.12. Take home message

As the fluids in the precursors of bodies were flowing during the formation of the universe, they became instable and that instability grew into turbulence. According to the size and scale of their fluids, the precursors of the clusters of matters in the universe went through various intensities of instability, some of which were very turbulent. As the precursors of matters started flowing, fluid layers were formed, and stacked on top of the others according to the system they belong to. In each stack of fluids, the fluid layers moved at different speeds, which contributed to producing turbulence on different scales, with different intensities, at different positions, and led to the formation of various precursors of bodies in the universe. From the invisible or the smallest microscopic and even submicroscopic scales to the largest astronomical ones, processes similar to what I just explained occurred. Each scale of turbulence went through different developmental stages and led to the diversity of the bodies and systems of bodies in the universe today. The variation of the scales of the turbulence explains the diversity of the bodies (spiritual bodies, subatomic particles, atoms, molecules, minerals, rocks, celestial bodies, and their clusters) in the universe. The fluids in the precursors were characterized by:

- The formation and shearing of fluid layers
- Intermittence and mixing of the fluid layers
- The top layers led to the formation of innermost secondary bodies, whereas the bottom ones birthed the outermost secondary bodies
- The stratification of the fluid layers of the precursors is responsible for the stratification of the crust and atmosphere of celestial bodies
- The speed of the fluid layers was translated into the orbital and rotational speed of the daughter bodies
- The direction of the fluid flow in the layers can explain the counterclockwise movement (as seen from the north pole) of some bodies in the Solar System
- Topological and structural changes of the fluids of the precursors of bodies account for some of the variations in the composition of the bodies

The fluid layers in the precursors of bodies can be divided into 7 regions that I called turbulence zones:

- Turbulence Zone 0: the innermost zone where the innermost rings are located in the planetary systems, meaning between the primary planet and the innermost satellite.
- Turbulence Zone 1: immediately downstream of Zone 0 and encompassed bodies bigger than those found in Zone 0. For instance, Zone 1 of planetary systems is where the innermost and fastest satellites are found.
- Turbulence Zone 2: a transition zone from the "laminar" turbulence of Zone 1 to the fully developed turbulence which is reached in Zone 3.
- Turbulence Zone 3: the zone where the highest turbulence leading to the biggest secondary bodies in a fluid flow are usually found. A key characteristic of Zone 3 of the satellite systems is that, the rotation period increased from the innermost satellite in Zone 1 until the outermost satellites in Zone 3, meaning that the outermost satellite in Zone 3 has the highest rotation period among all of the bodies in Zones 1, 2, and 3, and after Zone 3, the rotation period dropped, therefore serving like an inflection point for the rotation period.
- Turbulence Zone 4: a transition zone but, unlike Zone 2, Zone 4 is a

transition zone out of turbulence.

- Turbulence Zone 5: the outermost zone where the remote and slowest celestial bodies in each system are found.
- Turbulence Zone 6: located downstream of the outermost bodies and does not contain "any" celestial bodies.

In general, I mostly dealt with 5 zones (Zones 1, 2, 3, 4, and 5). According to their position, turbulence intensity, speed, and shearing rate, the fluids in the layers were differently organized into clusters, and, at one point, because they could no longer stay together, they took different paths according to their position in the stack of layers. Then, the separated fluid layers which took the same trajectory were gathered together into unique bodies or systems of bodies. During the process, the structures (e.g. vortices or eddies) formed in these layers were squeezed, while the layers as a whole were "rolled" to form spherical bodies. The processes that collected the bodies before and after the layers of their precursors were separated played a crucial role in the establishment of gravity, shape, rotational movement, and the ellipticity (flattening and oblateness) of the celestial bodies. The gathering together of the bodies involved processes associated with the following phenomena: amalgamation, bending, braiding, clumping, kinking, knotting, lumping, netting, pulling, pushing, shearing, spiraling, twisting, and winding.

CHAPTER 11

THE REMARKABLE SCIENTIFIC ADVANCEMENT THAT WILL CHANGE FOREVER THE RATIONAL EXPLANATION OF THE PRECURSORS OF BODIES AND SPEED IN THE UNIVERSE

Although the specific examples I used in this chapter were based on particles and bodies in the Solar System, to some extent, some of them can also be applied to other stellar systems in the universe. Likewise, because it is the most studied stellar system, the Solar System is the focus of most of the examples and mathematics I did in the other chapters in this book. During the aforementioned cascade of split-gathering which led to the formation of the Milky Way Galaxy, the precursor of the Solar System was born. It soon began moving in space and quickly, was broken into many precursors of bodies including the precursor of the Sun and the precursor of the other bodies in the Solar System. The precursor of the Sun was close to the location where the precursor of the Solar System was before splitting. In other words, the precursor of the Solar System split-gathered to birth the precursor of the Sun and the precursor of all bodies orbiting the Sun (i.e. the precursors of the planetary systems, asteroids systems, and their satellites and rings).

11.1. Precursors of subatomic particles, atoms, and molecules

At the atomic levels, the precursors of atoms were split-gathered into the precursors of the nucleus and precursors of the secondary bodies (e.g. electrons) "orbiting" them or found in shells or orbitals around them. Similar split-gathering processes occurred on subatomic particles. Due to the lack of reliable experimental data on the size of all subatomic particles and atoms, and the fact that no one has ever seen any of these particles, I relied on my understanding of the data I analyzed to address the precursors of microscopic particles. The complex, sometimes controversial, and contradictory data in quantum physics also point at how scientists are still struggling to characterize atoms and subatomic particles and much less, how they do not properly understand their origin.

The precursors of electrons around the nucleus could have also been organized into tiny layers according to the scale of the turbulence which occurred at their level. Current knowledge in classical and quantum physics does not provide sufficient information to properly model the size of all electrons according to their positions around the nucleus.

Because some atoms do not have many electrons, the turbulence of the fluids in the precursors of electrons around nuclei may not have had fluids in all turbulence zones. However, for bigger atoms, fluid layers of the precursors of their electrons could have existed in all turbulence zones. Unfortunately, very little is known about the characteristics of the innermost electrons of the heaviest atoms. For instance, when I was studying the ionization energy of atoms (see my book *"Turbulent Origin of Chemical Particles"* for more details), I realized that most literatures did not go beyond the 30th ionization energy, yet some atoms have more than 100 electrons, suggesting that most ionization energy data were not collected on electrons closer to the nucleus than that responsible for the 30th ionization energy. Among all the variables I studied on electrons, ionization energy was the one that gave me a better insight into potential behavior of layers of the precursors of electrons around atomic nuclei according to their potential position. In other words, data do not exist to model the distribution of electrons and much more the way their precursors could have been organized into layers or turbulence zones.

Unlike celestial bodies where layers of precursors of secondary bodies were mostly converted into secondary bodies orbiting their primary bodies, it is the case for electrons, the fluid layers of their precursors could not have always been fully wrapped around or gathered together to form electrons. Another way of expressing this is that the fluid layers of the precursors of some, if not, of all electrons could have failed to wind up, spiral, or cluster into one body which can orbit the nucleus, therefore they were scattered into a shell or orbital around the nucleus. On a macroscopic scale, this trend is also found with the precursors of some celestial bodies, such as that of the main belt asteroids, which, because it failed to collect into one body, was split into many asteroids, which cover the space of the main belt located between Mars and Jupiter. The inability of the fluid layers of the precursors of electrons to collect into a single body could explain why electrons are postulated not as a single body orbiting the nucleus, but as matters spread over an electronic shell or cloud. However, as I explained later in the chapter on radius and rings, when some precursors of secondary bodies were gathered into their daughter bodies, some of their fluids were left out, and consequently they became rings surrounding some bodies such as satellites.

Similarly, it is possible that even in the electronic shell, some electrons may be surrounded by smaller particles, which belonged to their mother precursors and which were not compressed or collected with the electrons themselves. In other words, even in an electronic shell, the size and particles of the constitutive bodies may differ according to their position with respect to the nucleus of their atoms.

The precursors of atoms could have gone through processes like those of the precursors of the systems of celestial bodies, but this time at a microscopic scale. Because as of 2025, about 118 chemical elements, accounting for more than one thousand isotopes where discovered (see my book *"Turbulent Origin of Chemical Particles"*), the turbulence which shaped these atoms could have been different, hence different atoms were obtained.

11.2. Precursors of planetary systems and asteroids systems

As the precursor of the bodies orbiting the Sun was escaping the precursor of the Sun, its bulk of fluids started flowing together before being decomposed into the precursors of various bodies. All of the bodies orbiting the Sun today had their own precursors, which, at the early stage of their movement, were initially fluid layers. Not only were those layers moving at different speeds, but their size was also different. On the scale of the Solar System, the largest layers birthed the largest planetary systems, while the smallest layers birthed the smallest asteroids. In other words, the size of the planetary systems and asteroid systems as of today was defined by the size of the fluid layers of their precursors.

The layers of the precursors of the planetary systems were not just stacked one on top of the other, but they were separated by tiny layers which were the foundation of some asteroids. For instance, between the fluid layers of the precursor of Mercury and the precursor of Venus, tiny fluid layers existed and were the precursors of the small asteroids found in the space between Mercury and Venus. Similarly, between the fluid layers of the precursor of Venus and the precursor of the Earth-Moon system, there were tiny fluid layers, which were the foundation of the precursor of the asteroids located between Venus and the Earth-Moon system. Likewise, between the layer of fluids of the precursor of the Earth-Moon system and the precursor of Martian planetary system, there were smaller layers, which became the precursor of the asteroids between Mars and the Earth. In general, many asteroids found between the planets in the Solar System descended from fluid layers found between the fluid layers of the planetary systems. Once the fluid layers were separated, they yielded different precursors such as the precursors of the planetary systems and asteroid systems.

The precursors of the planetary systems and of the planets which have no satellite were the precursors of: Mercury, Venus, Earth-Moon system, Martian Planetary System, Jovian Planetary System, Saturnian Planetary System, Uranian Planetary System, Neptunian Planetary System, and Plutonian Planetary System. The precursors of the asteroids in the Solar System were the precursors of the

Amor asteroids, Apollo asteroids, Aten asteroids, centaurs, comets, Kuiper belt objects, main belt asteroids, Trans-Neptunian objects, and trojans (e.g. Mars trojans, Jupiter trojans, and Neptune trojans).

Besides the aforementioned tiny fluid layers of the precursors of the asteroids found between the layers of the precursors of the planetary systems, smaller amounts of fluids came off from the edges of the layers of the planetary systems as they were splitting from one another. These bodies of fluids which came off the fluid layers became the precursors of other small bodies including asteroids, dusts, and microscopic particles found in the interplanetary spaces. In other words, the precursors of certain bodies in the Solar System were not originally defined as fluid layers, but were formed after some fluid "came off" during the separation of larger precursors or larger fluid layers. In other words, as I later explained in this book, while some bodies were being formed, the prevailing turbulence shook their precursors, sometimes "expelling" or ejecting fluid fragments from the bigger bulk of bodies being shaped, therefore, yielding other smaller bodies between larger ones. This is not true on the scale of the Solar System only, but also on that of other stellar systems, galactic systems and even their clusters. For the intermittence in the fluid layers was found on all scales of bodies: from the subatomic levels to the galactic ones passing through the stellar systems, planetary systems, and satellites systems, etc. The way smaller precursors of bodies were formed between bigger ones is the root of the intermittence found in the universe. In Fig. 9, I illustrated what the fluid layers in the precursors of the bodies orbiting the Sun could have looked like before their separation.

As I explained in other chapters later, although in this graph I placed the Uranian and Neptunian planetary systems in Zone 4, they can also be placed in Zone 3.

The precursors of the planetary systems in the Solar System split-gathered to yield the precursors of the primary planets and the precursor of all the bodies orbiting them (satellites and rings). Likewise, the precursors of the asteroid systems were split-gathered into the precursors of primary asteroids and the precursors of their satellites, rings, and dusts found around those asteroids. As the precursors of the bodies orbiting the Sun (i.e. planetary systems, asteroid systems) were being birthed, a fluid ligament was initially formed to attach them to the precursor of the Sun. That ligament later detached, and, after undergoing some changes, it split into its daughter bodies. The fluid bodies in that ligament continued their movement in space, splitting into several bodies (e.g. precursors of planets and asteroids) during a very turbulent event that respected the laws I explained in the incoming chapters. After the fluid bodies were separated, many other ligaments could have been formed on a scale smaller than that of the Solar System (e.g. planetary system level). Before breaking into pieces, some fluid ligaments stretched until reaching a point beyond which they could no longer expand. The rotation of the vorticial structures in some fluid bodies also drew or attracted more nearby fluids into some precursors. As the fluid bodies were rotating, their content was squeezed. The collection and the squeezing could have continued until these bodies acquired their shape. Depending on their localization and the events that defined them, the precursors of some celestial bodies in the Solar System became a liquid bulk, others became a gas bulk, some a solid bulk, and others gained some complex structures that contemporary science is struggling to fully explain. In the chapters on ligaments and vortex, I explained why some celestial bodies in the Solar System ended up being solid, while others became giant gas and others giant ice, whereas the largest body in the Solar System is a star.

Fig. 9: Fluid layers (according to the turbulence zones) of the precursors of the bodies orbiting the Sun

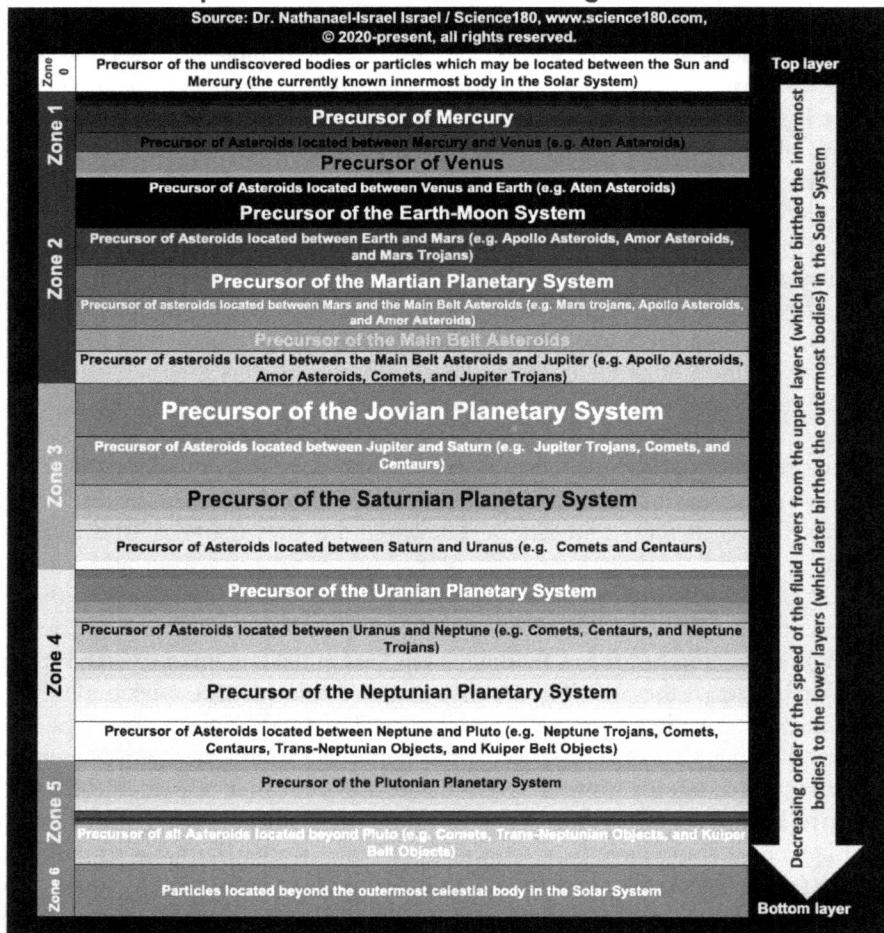

Zone		Top layer
Zone 0	Precursor of the undiscovered bodies or particles which may be located between the Sun and Mercury (the currently known innermost body in the Solar System)	
Zone 1	**Precursor of Mercury**	
	Precursor of Asteroids located between Mercury and Venus (e.g. Aten Asteroids)	
	Precursor of Venus	
	Precursor of Asteroids located between Venus and Earth (e.g. Aten Asteroids)	
Zone 2	**Precursor of the Earth-Moon System**	
	Precursor of Asteroids located between Earth and Mars (e.g. Apollo Asteroids, Amor Asteroids, and Mars Trojans)	
	Precursor of the Martian Planetary System	
	Precursor of asteroids located between Mars and the Main Belt Asteroids (e.g. Mars trojans, Apollo Asteroids, and Amor Asteroids)	
	Precursor of the Main Belt Asteroids	
	Precursor of asteroids located between the Main Belt Asteroids and Jupiter (e.g. Apollo Asteroids, Amor Asteroids, Comets, and Jupiter Trojans)	
Zone 3	**Precursor of the Jovian Planetary System**	
	Precursor of Asteroids located between Jupiter and Saturn (e.g. Jupiter Trojans, Comets, and Centaurs)	
	Precursor of the Saturnian Planetary System	
	Precursor of Asteroids located between Saturn and Uranus (e.g. Comets and Centaurs)	
Zone 4	**Precursor of the Uranian Planetary System**	
	Precursor of Asteroids located between Uranus and Neptune (e.g. Comets, Centaurs, and Neptune Trojans)	
	Precursor of the Neptunian Planetary System	
	Precursor of Asteroids located between Neptune and Pluto (e.g. Neptune Trojans, Comets, Centaurs, Trans-Neptunian Objects, and Kuiper Belt Objects)	
Zone 5	**Precursor of the Plutonian Planetary System**	
	Precursor of all Asteroids located beyond Pluto (e.g. Comets, Trans-Neptunian Objects, and Kuiper Belt Objects)	
Zone 6	Particles located beyond the outermost celestial body in the Solar System	Bottom layer

Decreasing order of the speed of the fluid layers from the upper layers (which later birthed the innermost bodies) to the lower layers (which later birthed the outermost bodies) in the Solar System

Get a beautiful color version of this graph at
www.Science180.com/FluidLayersPrecursorsBodiesOrbitingSun

After descending from the precursors of their planetary systems, the precursor of each planet went through changes until becoming a fully formed primary planet. Likewise, the precursor of each satellite became a fully shaped satellite. As of 2025, although no satellite of a satellite was discovered or confirmed in the Solar System yet, I cannot exclude that such bodies can exist in the Solar System or in other stellar systems. For, the precursors of some satellites could have birthed satellites and satellites of satellites, and so on and so forth. Before providing more specific information on the formation of the planets and the main belt asteroids in the Solar System, I did Fig. 10 to illustrate their position and orbit.

Fig. 10: Sketch of the planets and main belt asteroids in the Solar System

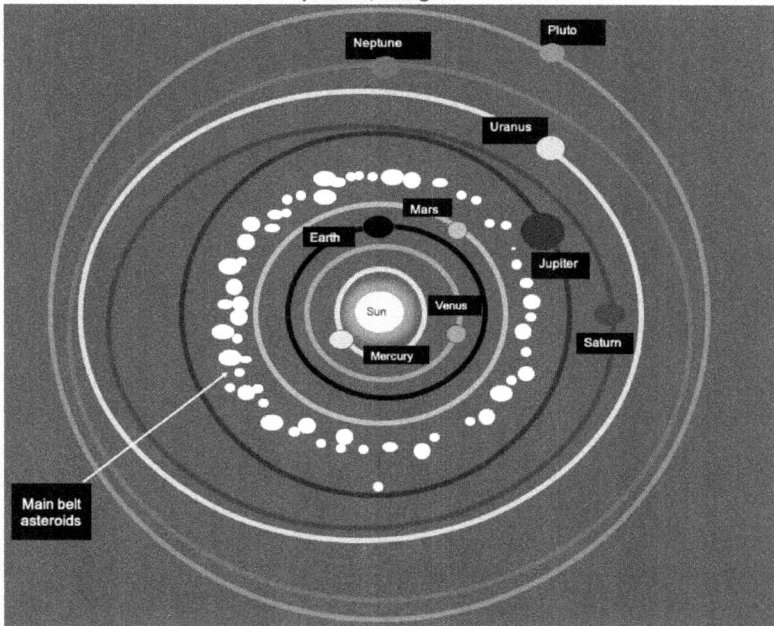

Now, I will elaborate on the precursor of each planet and planetary system in the Solar System.

11.2.1. Precursors of Mercury and Venus

Due to factors I explained later, the precursor of Mercury and the precursor of Venus were assembled into individual planets, none of which, as of 2025, are known to be orbited by a satellite. Because these planets are close to the Sun, and

Science180: The All-In-One Proven & Uncomplicated Universe-Origin Formula because the Solar radiation also challenges the detection of the innermost bodies in the Solar System, it is possible that Mercury and Venus may have some satellites, which are too small to be detected or they don't have a satellite at all. In any case, recent discoveries at least suggest that the orbits of these planets are embedded in a trail of dusts similar to rings. In the chapter on rings of celestial bodies, I better elaborated on these facts.

During the formation of most planets in the Solar System, precursors of satellites were also formed as a product of the split-gathering. While the precursors of some planetary systems birthed only one satellite (e.g. precursor of the Earth-Moon system), the precursor of the other planetary systems in the Solar System birthed at least 2 satellites. In the chapter on the satellites in the Solar System, I detailed how the number of satellites varies according to the turbulence zones and other factors.

11.2.2. Precursor of the Earth-Moon system

After the precursor of the Earth-Moon system split from the precursor of the bodies orbiting the Sun, it was split-gathered into the precursor of the Earth and the precursor of the Moon. Although the giant planets are the ones usually thought to have a ring system, recent evidences suggest that the orbit of the Earth may be embedded in a trail of dust or particles that look like rings. The precursor of that trail of dust around the orbit of Earth could have been formed as the precursor of the Earth-Moon system was splitting into its daughter bodies. After splitting, each of these 2 precursors went through a "maturation" process, meaning all the processes which gathered their fluid layers into a spherical shape. As illustrated in the following graph, the precursor of the Earth yielded the Earth.

After going through some changes and traveling thousands of miles away from the original position of the precursor of the Earth-Moon system, the precursor of the Moon became the Moon. Because of the intensity of the turbulence of the precursor of the Earth-Moon system, the precursor of the Moon yielded a big body which characteristics match that of the satellites found in Zone 3. Later chapters shed some light onto those characteristics.

The formation of the Earth as well as that of celestial bodies having a crust involved many developments that allowed the fluid layers of the precursor to form microscopic particles (e.g. subatomic particles, atoms, and molecules), which were gathered together into minerals and rocks as explained in the chapter on the generations of split-gathering and intermittence. The processes by which the rocks, minerals, and mineraloids were "bound" together or "stuck" together involved processes I introduced in the previous chapter and also in others later. My book *"Turbulent Origin of Chemical Particles"* explained how, at the chemical level, fluid layers were transformed into various chemical particles. The particles were hardened and stuck together on microscopic scales as well as astronomical scales which controlled how the Earth itself got its form. The sticking together of the particles involved the cohesion between them and the formation of bonds which precursors were born since the time fluid elements started gathering together. While some particles were bound tightly, others were loose. I expounded on these processes in other chapters and in my book *"Turbulent Origin of Chemical Particles"*.

11.2.3. Precursor of the Martian Planetary System

The precursor of the Martian planetary system went through processes that split-gathered it into the precursor of Mars and the precursor of the Martian satellites. Indeed, after the fluid layers of the precursor of the Martian satellites split from the fluid layers of the precursor of Mars, they moved away before splitting into 2 bodies. The layers which were on top became the fluid layers of the innermost Martian satellite and the layers at the bottom became the precursor of the outermost Martian satellite. As addressed in other chapters (e.g. chapters on radius, orbital speed, semi major axis, and 9 system-additive variables), the precursor of the Martian satellites did not go through a strong turbulence. This also contributed to forming the Martian satellites very small to Mars, for their precursors did not travel too far before being gathered together.

11.2.4. Precursor of the Jovian Planetary System

Unlike the precursor of the Earth-Moon system (which birthed just one satellite), and unlike the precursor of the Martian satellites (which birthed just 2 satellites), the precursor of the Jovian planetary system birthed more than 70 satellites. Indeed, as the precursor of the Jovian planetary system was being split-gathered, it birthed the precursor of Jupiter and the precursor of the bodies orbiting Jupiter. The precursor of Jupiter went through changes and then became Jupiter, the largest planet in the Solar System. The precursor of the secondary bodies in the Jovian planetary system went through another turbulence, which split-gathered it into the precursor of the Jovian satellites and Jovian rings. In the chapter on rings, I described the characteristics and processes involved in the formation of these rings. After splitting from their mother precursor, the fluid layers of the precursor of the Jovian satellites produced many satellites. Under some conditions, the fluids in many layers were collected into a single satellite, while in other conditions, a single layer would have yielded more than one satellite, hence the semi major axis of some satellites is the same. As I explained later, the semi major axis of the Jovian satellites known as of today gave a clue into their position in the stack of fluid layers of the precursor of all the Jovian satellites.

In fact, as I better explained in the chapter on rings, not all of the fluids in each layer were gathered into satellites. Some fluids in the layers were not gathered, but were left out and became rings surrounding the satellites. Some fluid layers yielded more than one satellite and others merged before yielding a satellite. Later, I explained why the precursor of some planets like Jupiter was unable to collect itself into a terrestrial planet but into a gas giant.

11.2.5. Precursor of the Saturnian Planetary System

The precursor of the Saturnian planetary system was split-gathered into the precursor of Saturn, the precursor of the Saturnian satellites, and the precursor of the Saturnian rings. Later, I explained how the intensity of the turbulence in the Saturnian planetary system affected the density, size, and the predominance of rings in that system. The size of the precursor of Saturn could have been one of the factors which prevented it from becoming a terrestrial planet but a giant gas.

11.2.6. Precursor of the Uranian Planetary System

After going through some processes, the precursor of the Uranian planetary system was split-gathered into the precursor of Uranus, and the precursor of the Uranian satellites and Uranian rings. As I explained in the chapters on vortex and ligament, one of the key characteristics of the precursor of the Uranian planetary system is that it was tilted as it split from the remainder of the precursor of the bodies orbiting the Sun.

11.2.7. Precursor of the Neptunian Planetary System

The precursor of the Neptunian planetary system was divided into the precursor of Neptune and the precursor of the Neptunian satellites and Neptunian rings. Due to some characteristics that shaped the precursor of the Neptunian planetary system, the precursor of Neptune was unable to become a terrestrial planet or a giant gas, but a giant ice.

11.2.8. Precursor of the Plutonian Planetary System

Finally, despite being smaller than the precursor of Mercury and Venus, the precursor of the Plutonian planetary system was split gathered into the precursor of Pluto and the precursor of the Plutonian satellites. The precursor of the Plutonian satellites was split-gathered into individual bodies yielding the precursor of Charon (the largest Plutonian satellite) and the precursors of 4 more Plutonian satellites. As I explained in the chapter on rotational angular speed, orbital speed, and radius, the precursor of Charon could have affected the movement of the precursor of Pluto, and vice versa. By the time the precursor of Pluto and of Charon finished their formation, Pluto and Charon formed a binary system characterized by a movement that some qualified as chaotic.

11.2.9. Precursors of asteroids

According to the classification available for asteroids at the time of the publication of this book, the precursors of the asteroids could have been the precursors of the:

- Amor asteroids
- Apollo asteroids

- Aten asteroids
- Centaurs
- Comets
- Kuiper belt objects
- Main asteroid belt
- Trans-Neptunian objects
- Trojans (e.g. Mars trojans, Jupiter trojans, and Neptune trojans)

The precursors of the asteroid systems went through similar processes as the precursors of planetary systems to birth their primary asteroids and their satellites accordingly. To birth the asteroids, the precursors of the asteroids went through processes similar to those of the planets which have no satellite, except that the scales of the size, turbulence intensity, and the daughter bodies were different. Fig. 11 illustrates the position of some Jovian trojans and Mars trojans.

Fig. 11: Sketch of the Jovian trojans and Mars trojans in the Solar System

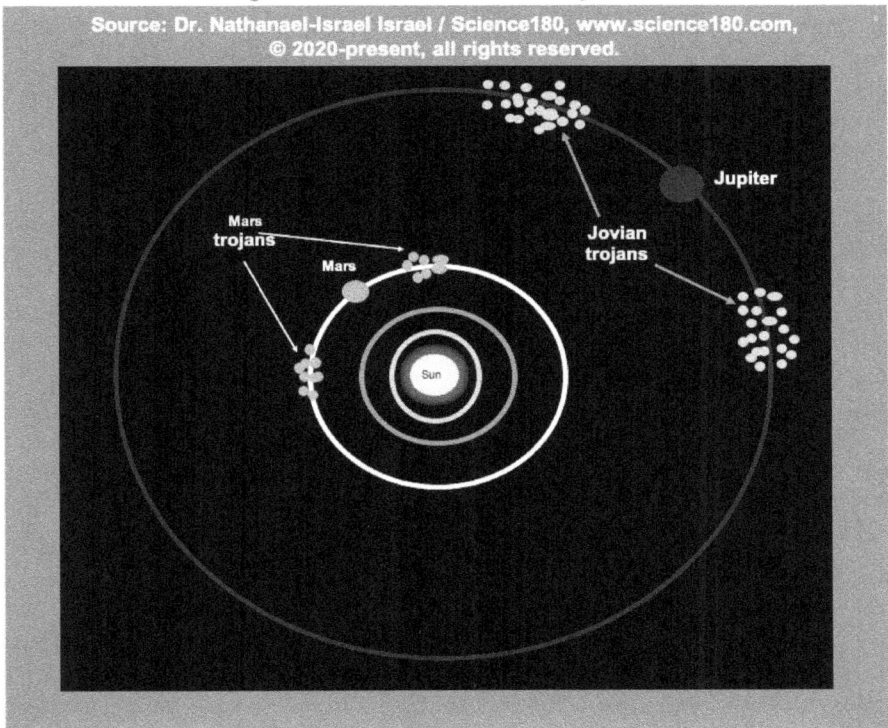

Looking at the position of the trojans of the planets, I felt like the precursors of the planets and their trojans could have been part of a common system of layers or nearby layers before separating from one another. In fact, considering the trends I found for how fluid layers split-gathered and how the position of the celestial bodies gave a clue about their positions in the stack of fluid layers that their precursors once belonged to, I felt like the precursors of the trojans could have been leading, trailing, and surrounding the fluid layers of the precursor of their primary planets.

When the precursor of the trojans separated from the precursor of the planetary system that they led or trailed, it could have been "ejected" with so much energy which split it instead of allowing it to collect into one body. A recoiling force may also have been involved. It is also possible that the precursors of the trojans were not associated with or incorporated into the precursor of the planetary systems because the precursor of the forces which gathered together fluids in the precursor of the planetary system was unable to attract and gather the precursor of the trojans and mix them with the precursor of the planet. In other words, the fluid layers of the precursor of some trojans could have been the leading and trailing portion of the leftover of the fluid layer of the precursor of the planetary systems. Another way of explaining this complex process is that, all the fluid bodies which were toward the periphery of the precursor of each planetary system were not gatherable into the precursor of that planetary system, hence they were left alone and went through their own split-gathering to yield small asteroids sharing the "same" orbit or trajectory as the planetary system they led and trailed. Because the precursors of trojans were not ejected from the precursor of the planetary system, but were positioned in front and in the back of the later, their daughter bodies were not able to orbit the primary planet like satellites do. As I explained in the chapter on orbital speed, the orbital speed of the trojans is in the same order of magnitude as the orbital speed of the planet they lead or trail. Hence, in the end, trojans trail or lead primary planets and share the same orbit with them and orbit the same star just as the planet they lead or trail does. In Fig. 12, I illustrated how the fluid layers of the planetary systems and of the trojans trailing and leading them may have been positioned. The graph also shows how the precursor of a planetary system was then split-gathered into a primary planet orbited by satellites, while the precursors of the trojans were split-gathered into clusters of trojans positioned in front or behind the position or orbit of the primary planetary system.

11.2.10. Precursors of rings

The origin of the rings in the Solar System has been a hot topic in a lot of research. I wish I could expound on the precursor of the ring systems here, but I do not have anything else to add to what I already expressed about them when I was dealing with the precursors of the planetary systems and of their planets and satellites. I could not have even created a different title for the precursors of rings, but I felt like they needed one and, to respect the law of intermittence that is so abundant with celestial bodies, I did not force myself to lengthen this section. In other words, just as some celestial bodies and systems of bodies are smaller than others, bear with me that this section on the precursors of rings is small yet deserves to have its own segment. In the chapter on rings, I explained how they were formed.

Fig. 12: Split-gathering of the fluid layers of the precursor of a planetary system and its trojans

Source: Dr. Nathanael-Israel Israel / Science180, www.science180.com, © 2020-present, all rights reserved.

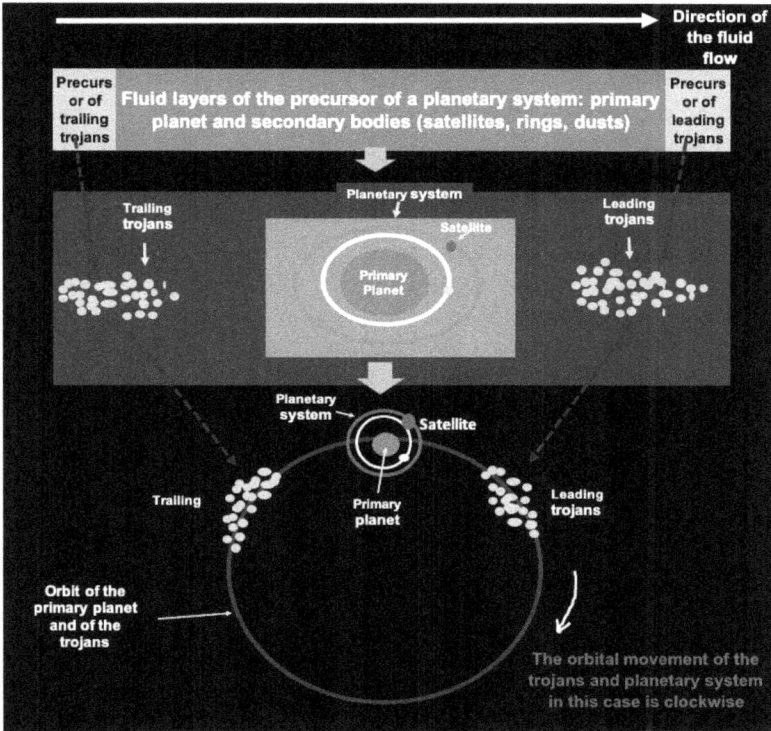

11.3. Precursors of the atmosphere, medium, or space between bodies and systems of bodies in the universe

Before I address the origin of the medium or space between the bodies in the universe, I felt like I should first review some of those media. In the chapter on the formation of galaxies, stars, and globular clusters, I revisited the voids or empty space in the universe. Indeed, interplanetary dust clouds and zodiacal clouds are found in the spaces between planets in the same stellar system. Also called cosmic dusts, these clouds are believed to be freely-moving small particles. Their mass is aid to be negligible and similar to that of an asteroid with a radius

15 km with a density of 2.5 g/cm^3 (Pavlov, 1999).

Considering the types of bodies in the universe, I classified the space between the clusters of bodies in the universe as:

- Intergalactic space (space between galaxies and their clusters)
- Interstellar space (space between stellar systems and stars)
- Interplanetary space (space between planetary systems of the same stellar system)
- Inter asteroid space (space between asteroids of the same stellar system)
- Inter satellite space (space between satellites in the same planetary system)
- Inter mineral space (space between minerals)
- Inter rock space (space between rocks)
- Inter molecular space (space between molecules and chemical compounds)
- Inter atomic space (space between atoms)
- Inter subatomic particles space (space between subatomic particles)
- Invisible world space (space between spiritual things and beings invisible to the naked eye and even to the most advanced scientific equipment).

In addition to celestial bodies, dusts were also formed in space by the time the universe was fashioned. Also called extraterrestrial dusts or space dusts, cosmic dusts are dusts (which size is in the order of micrometer) present in outer space. Sometimes, they fall on the Earth. According to their astronomical location, cosmic dusts are categorized into:

- intergalactic dusts
- interstellar dusts
- interplanetary dusts (e.g. zodiacal clouds)
- circumplanetary dusts (e.g. planetary rings)

Because of the remote position and the limit in the accuracy of scientific equipment, certain things in space thought to be dust can be big bodies that could not be properly seen yet from the Earth. For instance, when it comes to galaxies, what scientists think is dust could be planets, satellites, and even stars that scientific equipment cannot resolve yet. Therefore, I personally believe that the physics of cosmic dusts is filled with errors or assumptions, which will be revealed in the future. As I carefully studied celestial bodies and their environment, I understood like dusts or tiny particles exist between and around almost most bodies in the universe, not just around astronomical celestial bodies. In other words, in addition to the well-defined celestial bodies which dominate them, some of those spaces are filled with tiny particles, dusts, or clouds, which I classified into:

- Intergalactic dusts
- Interstellar dusts
- Interplanetary dusts

- Inter asteroid dusts
- Inter satellite dusts
- Inter molecular dust
- Inter atomic dusts
- sub-subatomic dusts
- Atomic dusts
- Subatomic particle dusts
- Other types of cosmic dusts

I used the term sub-subatomic dusts because I believe that some particles tinier than the known subatomic particles may be filling some space between atoms, molecules, and even subatomic particles. Sub-subatomic particles may not be like dusts, but could be much tinier. Moreover, some particles found in space today could not have been there since the beginning of the universe, but were formed after some particles reacted with others.

Some people believe that the spaces between celestial bodies are filled with matter and radiation called "medium" (e.g. interplanetary medium which is believed to be the material which fills the Solar System, and through which the bodies in the Solar System move). In light of this, before I move to the next segment, it is important to mention that one of the errors that some ancient theories of the origin of the universe is the perception of the matter in the space between celestial bodies. Some theorists have referred to what they called "ether", a substance some people believe to fill space and which they claimed to explain gravity.

In the next paragraphs, I presented the precursors of the aforementioned media or spaces. Indeed, as the layers of fluids which were originally stacked one on top of the other were separating, a space appeared between them. In other words, after their separation, 2 originally consecutive layers of fluids became surrounded by space. Above the space between them could have been the layers which were on top and below the space between them was the layer of fluid which was below. Some fluid bodies of the precursors could have leaked into these layers, therefore causing some spaces to be filled with particles, while others can be empty. The number of particles between bodies in space could have been defined by the amount of fluid which were leaked into them or which landed into them. Some particles in those spaces were not produced by a simple leakage, but were literally formed there during the split-gathering of the precursors of bodies.

By the time the precursors of bodies were gathered together, the space that appeared around them became their atmosphere and/or the space where their secondary daughter bodies were born. In other words, the genesis of the atmosphere of the celestial bodies can be traced back to the separation of the fluid layers of the celestial bodies they surround. Similarly, as the layers of fluids of the precursors of the secondary bodies were going through their own split-gathering into their daughter secondary bodies, additional space was created between these layers. Part of these new spaces became the precursors of the atmosphere of each secondary body.

For example, as the precursor of the Solar System was being split-gathered into the precursor of the Sun and the precursor of the bodies orbiting the Sun, a space was formed around the precursor of the Sun and that space was the precursor of the atmosphere of the Sun. Then, as the layers of the precursor of the bodies orbiting the Sun started split-gathering into the precursors of asteroids and planetary systems, space was formed around and between them. For the asteroids and planets which have no satellites, the immediate space which was formed around their layers became the precursors of their atmosphere. As the precursor of a planetary system split-gathered into the precursor of a primary planet and the precursor of the system of satellites, the space around the precursor of the primary planet as it was being wound-up into the planet became the precursor of the atmosphere of the planet. In contrast, the precursor of the system of satellites had to split before the precursors of the atmosphere of each satellite could be formed around the individual satellites. Because of their radius which is many kilometers wide, some satellites, asteroids, planets, and stars have bigger atmospheres.

Unlike celestial bodies that have gigantic atmospheres around them according to their size, atoms and subatomic particles do not have much space around them. Yet, when they were being formed "empty voids" were formed between their constitutive particles. In the case of atoms surrounded by electrons, the space around the nucleus where the electrons are located is called "electron cloud". That space is not empty as some people think. The radius of atoms and subatomic particles is usually very small, millions fold smaller than a millimeter (see my book *"Turbulent Origin of Chemical Particles"*); consequently, the space around the nucleus" or their primary bodies is also very small. Likewise, the cloud of matter between electrons is extremely small. Although these subatomic particles cannot be seen, spaces exist between them. For instance, the space between chemical elements affects the length of their bonds. Just as I explained for celestial bodies, when the tiny layers of the constitutive particles of atoms and subatomic particles were being split-gathered, a tiny space was also formed between the daughter bodies. For each particle, the size of that space defined the size of what could be called "atmosphere", but which in the case of some microscopic particles is the space between the electrons and the nucleons. Just as the atmosphere around a celestial body is under the influence of the celestial body yet all the space between two celestial bodies is not always under the influence of neither of them, so also between two atoms or subatomic particles, a space near them can be under their respective influence, but a certain space between them is not controlled by either of them. The length of the space between atoms and subatomic particles affect their "bonding" to one another and their reactivity with others.

Regardless of the size of the bodies, the space exists between them, and at the global scale the "empty" space between bodies in the universe was initiated since the split-gathering of the fluid layers of the precursors of these bodies and then amplified by the expansion of the universe. The atmosphere of celestial bodies is just a tiny portion of the sky. In my book *"Turbulent Origin of Chemical Particles"*, I devoted many chapters to the composition and origin of the atmosphere of celestial bodies. In that book, I also delved into details concerning the space between microscopic particles. For their composition is not random but was affected by specific rules. Because of the differences in the environment where the precursors of the atmosphere or clouds around particles were formed, different particles were molded. The amount of fluid bodies that landed into the precursors of those atmospheres or spaces defined the abundance of particles found in them.

In the case of the Earth, the precursor of the space formed around the precursor of the Earth-Moon system was the precursor of the immediate sky we can see from the Earth. On the scale of the Solar System, and that of the stars and other celestial bodies we can see in the sky, the space formed between their bodies altogether is the visible sky. In the case of the Earth, the precursor of the atmosphere birthed all the particles in it. Some particles could have leaked or volatized from the Earth after its formation to station in the atmosphere. Likewise, other particles in the atmosphere were formed after some reactions involving solar radiations. In Fig. 13, I sketched how an atmosphere or space filed with particles (atmospheric particles) was formed after the fluid layers of the precursors of bodies separated.

Fig. 13: Precursors of the atmosphere, the space between bodies, and the particles which occupy them

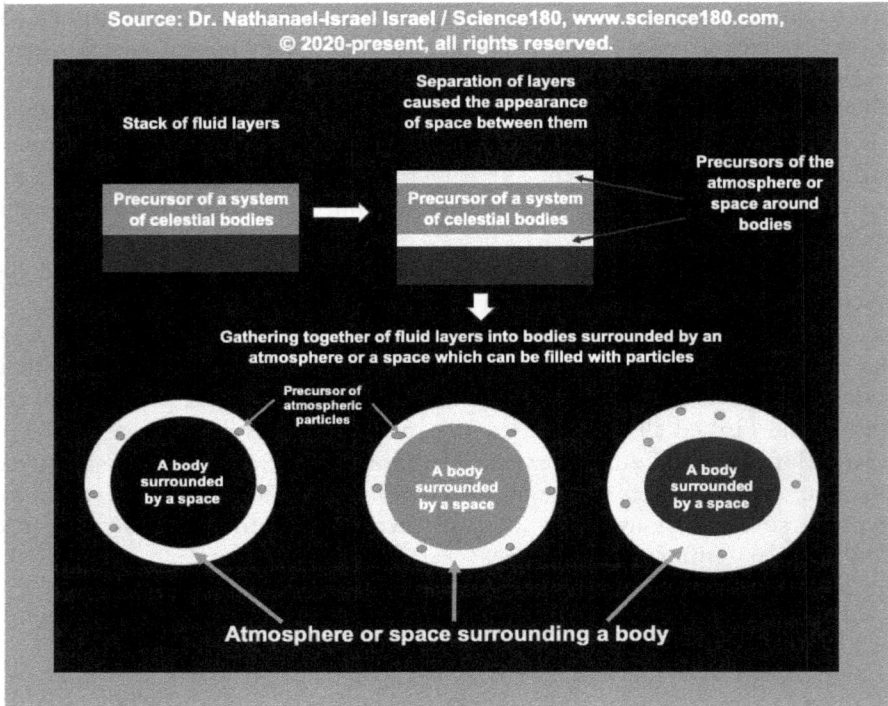

11.4. Precursors of spiritual things and beings

Although the current version of this book is not about religion, but very scientific facts, I felt like before resting the foundations of the precursors of matter, it would not hurt to just mention a few things about what I think could have happened to the precursors of spiritual things and beings. Because some people may not read the philosophical versions of this book (where I handled religious aspects of the formation of the universe), yet they believe in spiritual things related to God, demons, magic, witchcraft, occultism, I felt like it is important to talk here a little bit about some generic information about the formation of the spiritual world. Indeed, all religions talk about extraterrestrial beings also qualified as spiritual beings, holy angels, demons, above which God is placed although they do not perceive God the same way.

Invisible things and/or spiritual beings present in the universe, and which are well accepted by faith by religious people, also had their precursors. Considering the stories and revelations in many religious books, other beings exist in nature that are unknown to most human beings. For example, holy angels, God, and demons (that some call bad angels) are types of beings that are not common to a mere human being. Although most humans deal with demons every day, they do not know much about God Himself and His holy angels. Even believers do not know much about holy angels. Some angels may have manifested themselves in forms of UFOs (Unidentified Flying Objects). Although some UFOs can be fake or fictive, some UFOs are real beings which manifest themselves in ways strange to what ordinary human beings are accustomed to. Some UFOs can be military spacecrafts, while others can be fictive. More recently, people have been worrying about UFOs, some of which are believed to be spiritual beings. Some people even think that UFOs will destroy the earth or human civilizations someday. Many people have even testified having seen UFOs physically and images of those encounters abound on the internet. In contrast, some people doubt their existence. Considering that some holy angels and demons can manifest as UFOs, our world is already invaded by UFOs. Many people believe that eventually this world will end someday with the culmination of angelic activities. In other words, one day, human civilization could be invaded by other beings like extraterrestrial beings.

The data I dealt with suggested that the turbulent prima materia did not go through much modification before its daughter bodies produced the matter used to form spiritual beings and spiritual things. In other words, spiritual beings and spiritual things could consist of matter that is very close to the initial matter in the universe. This can also explain why spiritual beings are able to move very fast and reach thousands and even millions of miles away in a glimpse of an eye. Moreover, the ability of a spiritual being to be transformed into many types of bodies can be related to the ability of the turbulent prima materia to become anything. In the philosophical versions of this book, I elaborated on how angels can change themselves into different shapes and mimic other beings.

Spiritual beings like holy angels, demons, witches, spirits involved in magic, UFOs, and others (which existence for now cannot be properly proven physically without using faith, for most of them are invisible), had also been made from a precursor of matter, which I labeled as the "precursor of spiritual beings and spiritual things" except God. I excluded God from this debate because I do not want to dare to address His origin. In none of my books, did I ever attempt to explain the origin of the spiritual entity called God and which is differently perceived by the plethora of religions across the globe. Although I did not seek to discuss here which religion is wrong or right about the perception of God, which some people think does not even exist, I personally believe that no mere human being can ever explain the origin of everything and one of the things or beings which origin no one needs to try to explain is God. In other words, although I used scientific data to explain the origin of the universe, I never tried and I will never try to explain the origin of God by any means, for I personally believe that God has no beginning and no end. Saying and trying to explain the beginning of everything is a contradiction, an impossibility, and even if it is possible, it is an infinite task which can never be done even using all the computational power known in modern science, which is not even able to properly compute some turbulence modeling on the scale of a laboratory yet! Nonetheless, because I had a burden to also answer some spiritual questions on the origin of the universe and its content, I would recommend that you consult the philosophical versions of this book (as I already introduced) if you want to know more about how I explained the spiritual origin of the physical world.

Talking about the spiritual world, Christians believe that human beings consist of 3 parts: flesh of the physical bodies, the soul, and the spirit. The soul and the spirit are spiritual entities. Before I close this section of the precursor of the spiritual world, I would like to say something about the fragmentation of the energy of the soul. I have another book devoted exclusively to spirits and souls. Just like the precursors of celestial bodies had a certain amount of energy in them, which was fragmentated to yield different precursors of daughter bodies, which in the end birthed different bodies, so also the fetus or a living organism for instance has an initial energy which is split over the cells of that organism. As the organism grows, the energy of its soul is more "diluted" over newly-formed cells in such a way that the immunity is negatively affected. That is why the immunity of most organisms decreases as they age. For instance, most babies have a higher immunity than adults. This is because the energy of their soul is stronger and is not as depleted or "diluted" as that of an adult. As children grow, they use part of their soul energy to meet different needs, make choices, and satisfy their desires. As they use energy to accomplish activities in life, children deplete their energy reserve and, in the end, by the time they reach adulthood, they move more slowly and accomplish less. In contrast, in the case of nonliving things like celestial bodies, the energy of their precursors is used to make different daughter bodies according to processes embedded in them that I explained in another book related to "heredity" of nonliving things. The same process partially explains why, in each system of bodies, the innermost bodies are the fastest, while the outermost bodies are the slowest. Just as the biggest bodies are located in Zone 3 (which is neither the innermost zone nor the outermost zone), with human beings, the biggest decisions and things in life are accomplished during a period of their lifespan between 25-45 years of age, which is the turbulence Zone 3 of most human being lifespan. When human beings miss accomplishing things around that period of age, they have more difficulty doing much during the rest of their life. As people age, when the energy in their soul decreases to a certain level, they can naturally die, which is a kind of split-gathering separating the spiritual parts (soul and spirit) from the physical body (flesh). At that time, the dead flesh will go through some biogeochemical cycles to return its constituents (chemical particles) to the soil or Earth, while the soul and spirit will enter another realm that most religions talk about but disagree about. While some people believe that the soul and spirit may spend eternity in either hell or heaven, others do not even believe in anything after death or after the life on Earth. This disagreement also concurs with the laws of split-gathering, for daughter bodies are usually smaller than their mother precursor and difficult to comprehend. In other words, details of the daughter bodies of the current human beings should be smaller in size than the physical bodies known as of today. To put it another way, the life that human beings are currently living is balancing or summing the processes that will split-gather their daughter bodies (spirit, soul, and flesh) at the time of "death" and/or afterward. The choices that people make in their lifetime lead their life into certain directions and beyond certain limits, people lock themselves into "irreversible" systems that define the rest of their life in this world and in that to come, which most people do not agree with. Hence, details about the soul and spirit cannot be fully understood by a mere human being because, these spiritual things belong to another realm smaller but eternally bigger than that which the spectrum of human beings can fully perceive today. Hence, for now, most spiritual things are meant to be believed, and when human beings will fully enter the spiritual world after their death and/or at the end of their life on Earth, they can better see the huge details they failed to grasp while they were living in their physical bodies on Earth. Like I explained throughout this book, most mother bodies change and even die as their daughter bodies were being born. Hence, a mother body cannot fully know the destiny of its daughters while it has not split yet and if it split, it must cease to exist as a mother. Hence, human beings cannot know everything about their spiritual components (soul and spirit) using their physical senses while they are still alive in the flesh. They must either believe what is said about the spiritual world or die before knowing more details about realties of the current world, and at that time, it may be too late for some people to change anything in their life before death. Hence, some religions encourage people to make better choices now, for after death, people will know a lot of things they ignore today, but that knowledge will not be useful, for their destiny after death will be shaped by their life today, just as how it is impossible for most daughter bodies to reverse themselves into their mother bodies and then split-gather their reversed mother bodies into another type of daughter bodies. Regretting a mistake is not sufficient to fix it. For the forces and energy which are required to split-gather things are not in the power of the daughter bodies to reverse themselves into their mother and then, transform their mother into other daughter bodies different than what they are before the "reverse engineering". This is also why many people fail to even know who they are, for human beings are made of things that we cannot fully know as long as we are still in our current bodies and seeing things through the lens of what we believe or what we think we know. Likewise, we cannot know the Earth if we are still on Earth, and we cannot know this world if we are still in it. But after we leave here, we will know better and those who refuse to accept this truth today will be shocked too late. As of today, just as celestial bodies and even chemical particles are surrounded by huge voids (see the chapter on galaxies and stars), so also human beings surround their lives with empty things and choices which do not always profit them. Some build very huge homes they do not even inhabit, while others are living with materials they barely use or can afford, and some are still lacking the minimum they need. These things are mysteries that cannot be fully proven scientifically, but which can be predicted if we can have the right philosophical ideology of faith! I provided ample details in other books including:

- *"Reconciling Science and Creation Accurately"*
- *"Origin of the Spiritual World"*
- *"From Science to Bible's Conclusions"*
- *"How God Created Baby Universe"*
- *"Science180 Accurate Scientific Proof of God"*

11.5. Positioning of the celestial bodies at a specific distance

As the precursors of bodies were being assembled and their movement set, they were dynamically led onto their orbit where they will spend the rest of their "lifespan". This implies that, when the bodies were being formed, their position was changing in space and at the end of their formation, the trajectory of their revolution about their primary body was set. For instance, the chapter on semi major axis detailed the positioning of celestial bodies with respect to their primary bodies. Just as most celestial bodies in the Solar System are moving in certain directions, I also think that as a whole, the universe may also be moving from top toward the bottom, meaning a movement from above the north pole toward the south pole.

In the chapter on timescale, I explained the time it took for the bodies to leave their mother precursors and reach their position. The time it took to position the bodies at their position in orbit could have been affected by the longitudinal stretch and detachment of the ligaments of the precursors of bodies from their mother. Indeed, stretching can dramatically alter the development of the instability in the precursors of bodies and how they break. For instance, a thread of honey can be elongated by gravity to a very small radius before breaking (if it ever does). Other kinds of fluids (e.g. inviscid fluids) can be elongated to long ligaments provided the stretch is strong enough. When a pipette dipped into still water is rapidly removed vertically, a liquid ligament can be produced from the elongation of the bridge between the liquid and the pipette end. When the motion is fast enough, the ligament can stretch while remaining smooth. As soon as the stretching stops, because the ligament is no longer connected to the liquid bulk, it can break into drops.

As the fluids in the precursors of bodies were flowing under the influence of turbulence, the instability they had gone through could have changed with time. As the precursors of the celestial bodies were being fragmented, there came a time when their ligaments were connected to the bulk of their mother body they were separating from. During that time, the fluid ligament could have stretched until it reached a stopping point. At that time, the connection of the ligament to the initial bulk body stopped, meaning that the ligament broke from its mother. In some cases, this breakup was followed by the breakup into pieces of the fluids in the ligament itself eventually. In other words, after the ligament broke from its mother, its fluid layers flowed before separating from one another according to their position.

The size and the environmental conditions of the fluid ligament of some precursors affected the final state and shape of their daughter bodies. For instance, some bodies such as the satellite Prometheus are elongated because the ligament of their precursors could have been long and was not able to change after splitting from their mother precursors. The freezing or solidification of some ligaments in cold conditions can also explain why some bodies (e.g. Saturnian satellites like Pan, Atlas, and Iapetus) have long ridges around their equator that make them look like flying saucers. Many main belt asteroids also have elongated shape. The way the precursors of the bodies were collected after splitting from their mother precursors could also explain some shapes. That is also why the shape of many satellites is not circular but has many sides.

The forces that shaped the bodies can also be related to the interactions they have with the environment and with one another. For instance, the interactions between the precursors of celestial bodies and their particles as they were moving and were being gathered together were the foundations of the fundamental forces in nature. In the chapters on gravity and vortex, I elaborated on how the precursors of the bodies interacted with one another in a process which evolved until the "completion" of their formation.

I cannot finish talking about the position of bodies in the universe without addressing the expansion of the universe itself. Indeed, the initial push or launch of the precursors of matter was partially responsible for the so-called "expansion of the universe" that cosmologists have been trying to explain. For, even after their formation, celestial bodies still have a residual tendency to move away from the position of their mother precursors. As bodies distanced themselves from the position of their primary bodies, they could have encountered colder environments which sucked some of their energy. As bodies lose energy, their expansion away from their initial position could also slow down. Hence, a decrease of the expansion of the universe that some scientists have been trying to demonstrate. Depending on their position in space, some bodies or systems can seem to be expanding faster than others. On Earth for instance, directions are determined based on the four cardinal points: North, South, East, and West. However, as seen from Earth, the position of celestial bodies in space cannot be fully defined using just the cardinal points. Therefore, it may even be misleading to try to define the position of bodies in the universe or the expansion of the universe itself by just referring to cardinal points. Another difficulty is that it is impossible to properly define spatial position of bodies in space by referring to pictures or visualization of the sky. For it is hard to determine a 3D configuration of something just by using a 2D picture. Using data on albedo or the brightness of bodies to define their distance and size can be misleading. Hence, I did not focus on some of those data in my investigation.

11.6. Precursors of speeds

To what I already said about the speed of bodies, I would like to add that, orbital and rotational speed were not formed spontaneously, but they were produced by the processes that formed the bodies. This implies that each type of speed also had its own precursor. In many chapters of this book, I talked about orbital speed and rotational speed of the celestial bodies. Even for the subatomic particles, atoms, and molecules, things have been moving although the precision of the measurements can be hard to reach. However, the rotation and revolution of the celestial bodies and the speed associated with them were established as the bodies were being split-gathered and then put into orbit. Before each celestial body was placed into its orbit, there was a time when nothing was called rotation and revolution yet. Between the time the precursor of a celestial body was born and the time it reached its position in orbit, the speed it had was different from the speed its daughter got once in orbit. Similarly, before the shape of a body was completed, the rotational speed of its precursor was different from that of the "final" body. In other words, between the birth of the precursor of a celestial body and its reach of its trajectory or orbit, the speed that animated it can be called the "pre-orbital" speed. Similarly, the rotational speed a celestial body got before acquiring its final shape can be labeled the "pre-rotational" speed. Both the "pre-orbital" speed and the "pre-rotational" speed went through changes and were also affected by the processes which molded the celestial bodies. To put it another way, the "pre-orbital" speed and the "pre-rotational" speed were the precursors of the orbital speed and of the rotational speed respectively. These two precursors of speeds were themselves born out of the movement which animated the early stages of the precursors of these bodies. In other words, as the precursors of the celestial bodies started moving under the influence of the push they got from their mothers, the movement of their fluids progressively birthed the precursor of the orbital speed and the precursor of the rotational speed, which matured into the orbital speed and rotational speed as the fluids underwent processes which shaped the characteristics of their daughter bodies and placed them in their "final" orbit.

As the fluid layers of the precursors of bodies were flowing, some fluid lines were curled to form vortices. The curls are like movement of a line or region of fluids in 3 different axes (x, y, and z). Vorticity is defined as the curl of the velocity of a flow at a point while circulation relates to a surface. Both variables are used to define a vortex. The balance of the rotation of the vortices in the fluid layers of the precursor of the celestial bodies could have contributed to yielding the rotation of these celestial bodies. The more viscous the fluids of a precursor of a body were, the more they could have resisted rotation. Later, I explained how viscosity of the precursors of bodies can help explain the slow rotation of some planets such as Mercury and Venus. The way the vortices were stretched, twisted, and tilted affected the orbital inclination of their daughter bodies.

At this point, I would like to talk about why I think that the speed of clusters of bodies bound together could be smaller than that of their individual bodies before their binding. In fact, on June 29, 2018, as I was reading the news and came across an article about how the speed of photons bound together was smaller than the speed of individual photons, I felt like the clustering of particles can slow them as a system of bodies. Indeed, for a long time, scientists have been forming molecules by mixing atoms. However, efforts to stick photons together to make bigger and heavier photons have been vain until a few years ago. For scientists, individual photons that make up light do not interact. Part of the problem can be due to the high speed of photons that constitute light. In 2018, a group of researchers at the Massachusetts Institute of Technology (MIT), Harvard University in the USA, said that they managed to cause photons to stick together and form heavier and larger triplets of photons. Once they managed to get the photons together, they noticed that "the heavier photon molecules" were considerably less nimble, and rather than moving at their regular speed of about 300,000 kilometers per second (the speed of light), they were moving 100,000 times slower than normal noninteracting photons" (Liang et al., 2018). Some of the members of this same group of scientists mentioned in the reference above were able to put together two photons, suggesting that, under certain conditions, it is possible to form bigger bodies from smaller ones, even from very tiny particles like photons that scientists have assumed for so long to be massless and incapable of interacting with other particles. This data suggested to me that, just as the clustering of photons into heavier compounds slowed down their speed, so also the speed of clusters of precursors of bodies could have been reduced as individual precursors were being gathered together to form clusters of bigger ones, or as smaller entities were being formed or shaped inside the bigger ones. In other words, the speed of the precursors of celestial bodies and their matter could have been much higher than the speed of the bodies (small or big) that can be observed in nature today.

11.7. Take home message

During the cascade of split-gathering that led to the formation of the Milky Way Galaxy, the precursor of the Solar System was born. It soon started moving in space and was quickly broken into many precursors of bodies including the precursor of the Sun and the precursor of the bodies orbiting the Sun. As the precursors of the secondary bodies in the Solar System were splitting from the precursor of the Sun, the bulk of its fluids started flowing together before being organized into fluid layers, which became the precursors of many bodies. The fluid layers of the precursors of the planetary systems were not just stacked one on top of the other, but they were separated by tiny layers which were the foundational layers of some asteroids. The precursors of the planetary systems and of the planets which have no satellite were the precursors of Mercury, Venus, Earth-Moon system, Martian planetary system, Jovian planetary system, Saturnian planetary system, Uranian planetary system, Neptunian planetary system, and the

Plutonian planetary system. The precursors of the asteroids in the Solar System were the precursors of the Amor asteroids, Apollo asteroids, Aten asteroids, centaurs, comets, Kuiper belt objects, main belt asteroids, Trans-Neptunian objects, trojans (e.g. Mars trojans, Jupiter trojans, and Neptune trojans), etc.

The precursors of the planetary systems in the Solar System split-gathered to yield the precursors of the primary planets and the precursor of all the bodies orbiting them (satellites, rings, and dusts, etc.). For instance, according to its position in the stack of fluids of the precursor of all the bodies orbiting the Sun, after the precursor of the Earth-Moon system split from that stack of fluid, it was split-gathered into the precursor of the Earth and the precursor of the Moon. After going through some changes and traveling thousands of miles away from the original position of the precursor of the Earth-Moon system, the precursor of the Moon became the Moon. As the bodies of fluids were being gathered together, the space which appeared between them became their atmosphere and/or the space when the secondary daughter bodies of the fluid layers were born. As the precursors of bodies were being assembled and their movement set, they were dynamically led into their orbit where they will spend the rest of their "lifespan". Speed (e.g. orbital and rotational) were given to them according to the position of their precursors in the stack of fluid layers of their mother and the intensity of turbulence which shook them.

At the atomic levels, the precursors of atoms were split-gathered into the precursors of the nucleus and the precursor of the secondary bodies (e.g. electrons) "orbiting" them or found in shells or orbitals around the nucleus. Likewise, looking at some subatomic particles, similar processes of split-gathering could have occurred. The precursors of electrons around the nucleus could have also been organized into tiny layers according to the scale of the turbulence which occurred at their level. Because some atoms do not have many electrons, the turbulence of the fluids in the precursors of electrons around nuclei may not have had fluids in all turbulence zones. The precursors of the atoms could have gone through processes similar to those of the precursors of the systems of celestial bodies, but this time at a microscopic scale. Special matters formed from the turbulent prima materia were also used to form spiritual entities. Spiritual bodies also had their precursor, but I detailed that in other books.

CHAPTER 12

DISCOVER WHAT THE "9 SYSTEM-ADDITIVE VARIABLES" ARE SAYING ABOUT THE 99% VERSUS 1% RULE, WHICH IS ALMOST SIMILAR TO HOW A HANDFUL OF WEALTHY PEOPLE AMASSED ALMOST 99% OF THE WORLD'S MONEY–DID THEY REALLY DO THAT?

After laying the foundations of the story about the formation of the universe, I will now start the mathematical demonstrations that I know you will enjoy!

12.1. Definition of the 9 system-additive variables

For each of the following variables, the total value in a system can be calculated by adding the value of each of the bodies in that system:

1. Mass
2. Orbital angular momentum
3. Orbital moment of inertia
4. Rotational angular momentum
5. Rotational kinetic energy
6. Rotational moment of inertia
7. Total kinetic energy
8. Translational kinetic energy
9. Volume

I found that the kinetic energy of a planetary system can be estimated by adding the kinetic energy of the primary planet and the satellites in that planetary system. Similarly, the kinetic energy in the Solar System can be estimated by adding the kinetic energy of all the bodies in the Solar System. The same estimation can be done for any of these previously mentioned 9 variables.

Moreover, when I was studying these variables characterizing, I noticed that, as long as their relative values (%) of the primary bodies with respect to the total in their system, or the relative values of the bodies in turbulence Zone 3 with respect to the total in each system of secondary bodies were concerned, the trends of the 9 variables are similar. For instance, some similarities exist between how the total kinetic energy, the translational kinetic energy, and the rotational kinetic energy of the precursors of the systems of bodies was split between the primary bodies and their secondary bodies. In general, more than 99.9% of each of these kinds of kinetic energy of the precursors of the systems of bodies went into the precursors of the primary bodies, while less than 0.1% went into the precursor of the secondary bodies. Likewise, as the precursors of the secondary split in bodies, more than 99% of its energy went into the precursors of the bodies in turbulence Zone 3, while the rest went into the bodies in the other turbulence zones. Interestingly, the trend I summarized above for the primary bodies and the secondary bodies also is found with the 8 other variables. Consequently, as I started writing about each of these 9 variables, I noticed that the sentences and numbers I was using to describe each of them were almost the same, except that they related to different variables. Therefore, instead of studying each of these variables individually, and devoting a specific chapter to each, I decided to combine and address their trends as a whole. Combining all of these variables in a table and writing about them altogether not only can reduce the number of pages you would have to read but also can offer a comparative insight into their trends. Toward that end, to avoid listing all of these 9 variables each time I talk about each, I decided to label them as the "9 system-additive variables". I used the term system-additive because their value in a system of bodies can be known by adding the values of all the individual bodies in that system. I do not mean that the other variables I studied besides these 9 are unimportant, but I just needed a way to label these 9 variables so I can better summarize my writing and also avoid having you read similar descriptions 9 times. Such a qualificative could not be used for instance for the radius or semi major axis of the bodies. For instance, the radius of a system of bodies cannot be approached by adding the radius of each constitutive body. Similarly, the semi major axis of a system of bodies is not the sum of the semi major of each of its bodies.

12.2. During the split-gathering of the precursor of the Solar System, 99.9% of the value of its 9 system-additive variables went into the precursor of the Sun and less than 0.1% went into the precursors of the bodies orbiting the Sun

Using the data available on the celestial bodies in the Solar System today, I calculated the sum of the each of the 9 system-additive variables of the bodies in the Solar System, that of the Sun and that of all the bodies orbiting the Sun. The graph below (Fig. 14) illustrates the relative value of the Sun with respect to all of the totals available on all the bodies in the Solar System.

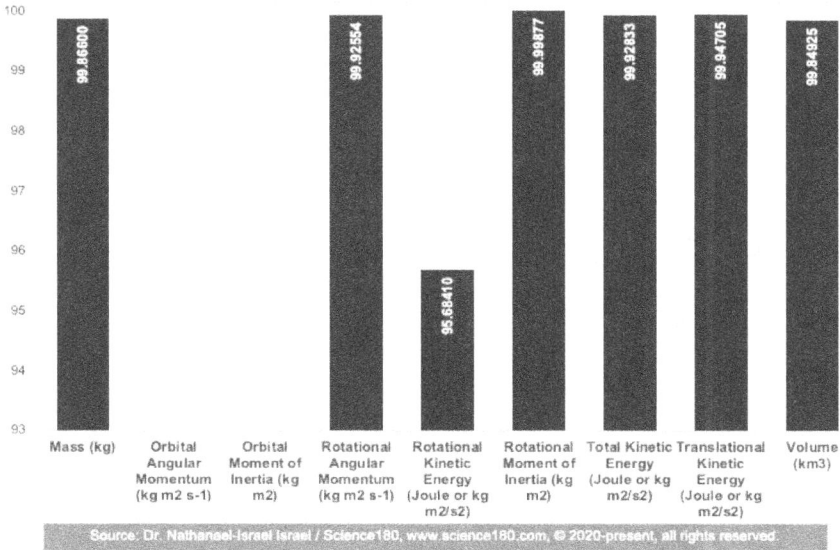

Fig. 14: Ratio "Sun / All bodies in the Solar System" (%) to the 9 "System-Additive" Variables

The contribution of the Sun to the orbital angular momentum and the orbital moment of inertia of the bodies in the Solar System is not mentioned because the semi major of the Sun with respect to the Sun could not be calculated. In the following sections, I will address the orbital angular momentum and the orbital moment of inertia of the bodies orbiting the Sun.

The data suggests that, as the precursor of the Solar System was being fragmentated into its daughter bodies, 99.9% of the value of each of the 9 system-additive variables was gathered into the precursor of the Sun and the rest (less than 0.01% was pushed into the precursor of the secondary bodies and systems of secondary bodies in the Solar System. After going through some turbulent changes, the precursor of the Sun "finalized" its formation, became the Sun, which kept 99.9% of the value of any of the 9 system-additive variables of the bodies in the Solar System. In the next paragraphs, I illustrated what could have happened to the 3 kinds of kinetic energy I studied. In other words, out of the 9 system-additive variables, I focused my illustration by referring most of the time to the laws that can explain the distribution and variation of the kinetic energy in the precursors of the systems of bodies and in their daughter.

Indeed, considering the trends of the 9 system-additive variables for different types of systems of bodies, it appeared to me that, the general rule of the split-

Science180: The Undeniable Scientific Challenge to All Metaphorical, Figurative, Loose, Liberal, or Vague Explanations of the Universe-Origin

gathering of kinetic energy in the precursors of the systems of bodies is that, more than 99% of the kinetic energy of the precursor of the system of bodies was pushed into the precursor of the primary body and less than 0.1% was pushed into the precursors of the systems of secondary bodies. As the precursor of the system of secondary bodies started flowing and mixing, different daughter precursors of secondary bodies were born and the kinetic energy in them depended on various factors. The precursor of some of those daughter secondary bodies went through their own split-gathering by generally following the 99.9% and less than 0.1% rule of the amount of the kinetic energy they pushed into their daughter primary bodies and secondary bodies. As the precursor of the Solar System was being split-gathered into its daughter bodies, 99.9% of that total kinetic energy was gathered into the precursor of the Sun and the rest (less than 0.1%) was pushed into the precursor of the secondary bodies and systems of secondary bodies in the Solar System. After going through some changes, the precursor of the Sun "finalized" its formation, became the Sun, which in the end kept the 99.9% of the energy in the Solar System.

Similarly, out of the total translational kinetic energy of the celestial bodies I studied in the Solar System, more than 99% of it was accumulated in the Sun. Indeed, as the precursor of the Solar System was split-gathered into bodies, 99.9% of its translational energy went into the precursor of the Sun while less than 0.1% of it was pushed into the precursor of all the secondary bodies and system of secondary bodies around the Sun. After going through some processes, the precursor of the Sun became the Sun which as of today contains more than 99.9% of the translational kinetic energy in the Solar System. The precursor of the secondary bodies in the Solar System started flowing and later split its translational kinetic energy over the various kinds as I explained in the following section.

Furthermore, I found that the Sun alone accumulated 95.7% of the total rotational kinetic energy of the bodies in the Solar System. The remaining 4.3% of the rotational kinetic energy in the Solar System is found in the planetary systems and asteroids. Consequently, I also deducted that during the formation of the universe, the precursor of the Solar System could have pushed around 95.7% of the rotational kinetic energy into the precursor of the Sun and the remaining 4.3% of its rotational kinetic energy could have been pushed into the precursors of the planetary systems and the asteroids.

12.3. During the split-gathering of the precursor of the bodies orbiting the Sun, more than 99.9% of the 9 system-additive variables went into the precursors of the planetary systems and less than 0.1% went into the precursors of the asteroids

As the precursor of the systems of secondary bodies in the Solar System started flowing away from the precursor of the Sun, the value of the 9 system-additive variables in it was split differently according to the bodies which were being born. The precursor of the secondary bodies in the Solar System birthed the:

- Precursors of the planetary systems and the
- Precursors of the asteroids and their systems.

Although I was aware of the existence of particles between the planetary systems and between the asteroids, I did not consider them in the mathematics related to the 9 system-additive variables because these particles are negligible. I did not mean that their existence is not important, but that the value of each of the 9 system-additive variables of those interplanetary particles and particles found between the asteroids is negligible.

In general, about 99.9% of the value of any of the 9 system-additive variables of the precursors of the secondary bodies in the Solar System were accumulated or pushed into the precursors of the planetary systems. This implies that less than 0.1% of the 9 system-additive variables of the precursors of the secondary bodies in the Solar system was sent into the precursors of the asteroids and systems of asteroids. Consequently, after going through some dynamic changes, and the formation of the bodies in the Solar System was "completed", about 99.9% of the value of any of the 9 system-additive variables of the bodies orbiting the Sun is found in the planetary systems, while about 0.1% is found in asteroids. For instance, as the precursor of the system of secondary bodies in the Solar System started flowing away from the precursor of the Sun, the total kinetic energy in it was split differently according to the bodies which were being born. The total kinetic energy in the precursor of the secondary bodies was used to form the:

- Precursors of the planetary systems and the
- Precursors of the asteroids and their systems.

About 99.9% of the total kinetic energy of the precursors of the secondary bodies in the Solar System accumulated into the precursors of the planetary systems. This implies that less than 0.1% of the total kinetic energy of the precursors of the secondary bodies in the Solar System was pushed into the precursors of the asteroids and systems of asteroids. Consequently, after going through some dynamic changes, the formation of all the bodies in the Solar System "completed, about 99.9% of the total kinetic energy orbiting the Sun is found in the planetary systems, while about 0.1% is found in asteroids.

Indeed, as the precursor of the secondary bodies in the Solar System started flowing and after splitting into various kinds of bodies, it pushed 99.9% of its translational kinetic energy into the precursors of planetary systems. In the end, the planetary systems account for 99.9% of the translational kinetic energy in the secondary bodies in the Solar System.

Similarly, because about 99.6% of the total rotational kinetic energy in the precursors of the secondary bodies orbiting the Sun were accumulated by the precursors of the Jovian and Neptunian planetary systems alone, these 2 planetary

159

Science180: The Undeniable Scientific Challenge to All Metaphorical, Figurative, Loose, Liberal, or Vague Explanations of the Universe-Origin

systems contain about that percentage of the rotational kinetic energy available in the bodies orbiting the Sun as of today.

To sum it up, as the precursor of the secondary bodies or systems of secondary bodies of the precursor of the Solar System were being split-gathered, more than 99.9% of their kinetic energy went into the precursors of the planetary systems and less than 0.1% went into the precursors of the asteroids. The percentages (%) of the 9 system-additive variables that the planetary systems and the asteroids have are illustrated in the Fig. 15.

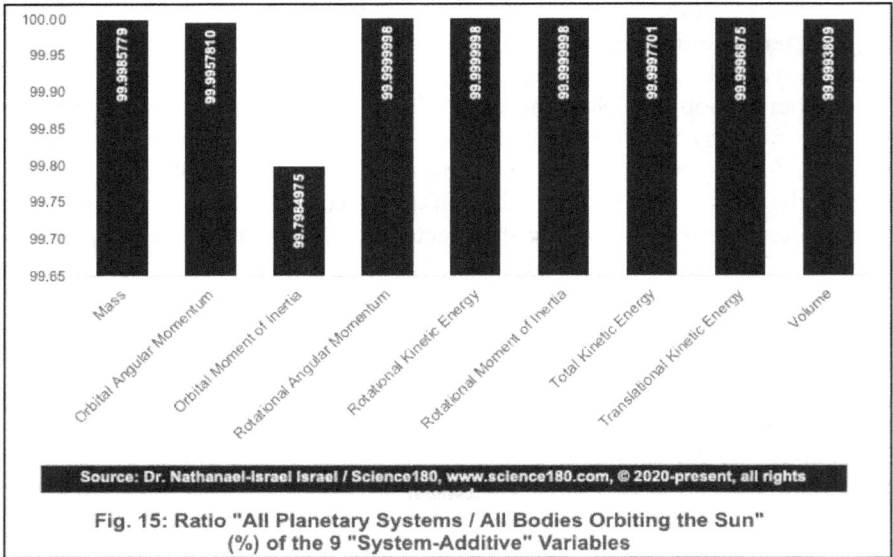

Fig. 15: Ratio "All Planetary Systems / All Bodies Orbiting the Sun" (%) of the 9 "System-Additive" Variables

Below, I detailed the sum of each of the 9 system-additive variables according to the types of asteroids and planetary systems (Fig. 16-24). My goal is to present how the precursor of the secondary bodies was split among the asteroids and the planetary systems. The Jovian and Saturnian planetary systems are the 2 most dominant ones.

Uranian Planetary System
3.2540%

Mercury Planetary System
0.0124%

Venus Planetary System
0.1824%

Earth Planetary System
0.2266%

Martian Planetary System
0.0240%

Saturnian Planetary System
21.3053%

Neptunian Planetary System
3.8390%

Plutonian Planetary System
0.0005%

Apollo asteroids
0.0000%

Aten asteroids
0.0000%

Amor asteroids
0.0000%

Jovian Planetary System
71.1543%

Centaurs
0.0000%

Mars trojans
0.0000%

Comets
0.0000%

Kuiper belt objects
0.0001%

Jupiter trojans
0.0000%

Main asteroid belt
0.0001%

Trans-Neptunian objects
0.0012%

Neptune trojans
0.0000%

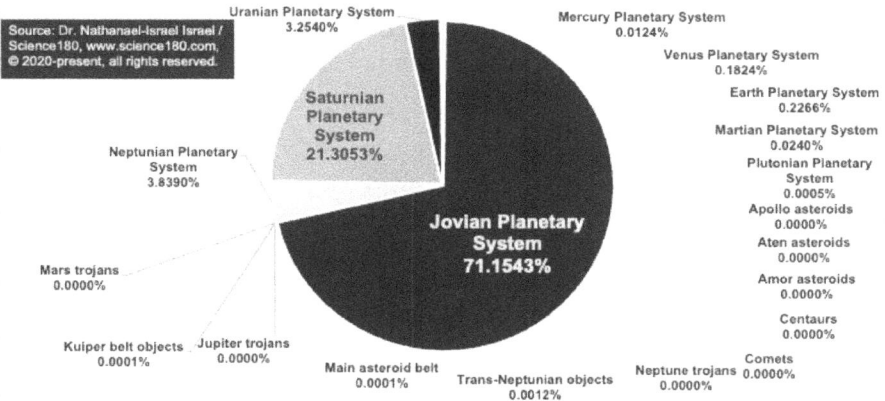

Fig. 16: Relative Mass of the Systems of Bodies (Asteroids and Planetary Systems) Orbiting the Sun

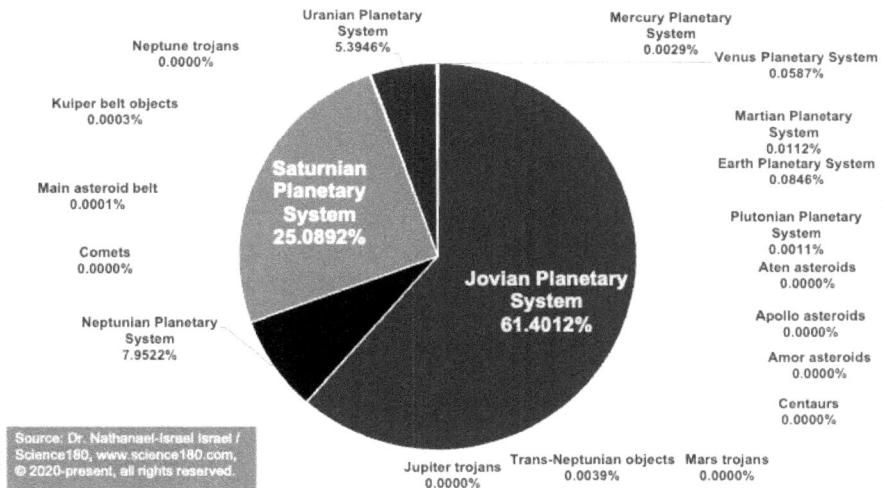

Uranian Planetary System
5.3946%

Neptune trojans
0.0000%

Mercury Planetary System
0.0029%

Venus Planetary System
0.0587%

Kuiper belt objects
0.0003%

Martian Planetary System
0.0112%

Earth Planetary System
0.0846%

Main asteroid belt
0.0001%

Saturnian Planetary System
25.0892%

Comets
0.0000%

Plutonian Planetary System
0.0011%

Aten asteroids
0.0000%

Neptunian Planetary System
7.9522%

Jovian Planetary System
61.4012%

Apollo asteroids
0.0000%

Amor asteroids
0.0000%

Centaurs
0.0000%

Jupiter trojans
0.0000%

Trans-Neptunian objects
0.0039%

Mars trojans
0.0000%

Fig. 17: Relative Orbital Angular Momentum of the Systems of Bodies (Asteroids and Planetary Systems) Orbiting the Sun

When the asteroids and planetary systems in the Solar System were analyzed, the Neptunian planetary system was the one with the highest orbital moment of inertia, accumulating 40.5% of the orbital moment of inertia of all the bodies orbiting the Sun (Fig. 18). It is followed by the Saturnian and Jovian planetary systems which orbital moment of inertia is about 22% that of all the bodies orbiting the Sun. As far as the bodies orbiting the Sun were concerned, I noticed that, unlike mass, volume, energy, and radius which maximum values were found in the Jovian planetary system, orbital moment of inertia was not found there.

Science180: The Undeniable Scientific Challenge to All Metaphorical, Figurative, Loose, Liberal, or Vague Explanations of the Universe-Origin

This suggests that orbital moment of inertia is distributed differently regardless of size and energy. The high orbital moment of inertia of the Neptunian planetary system is also related to how its satellites are so highly spread.

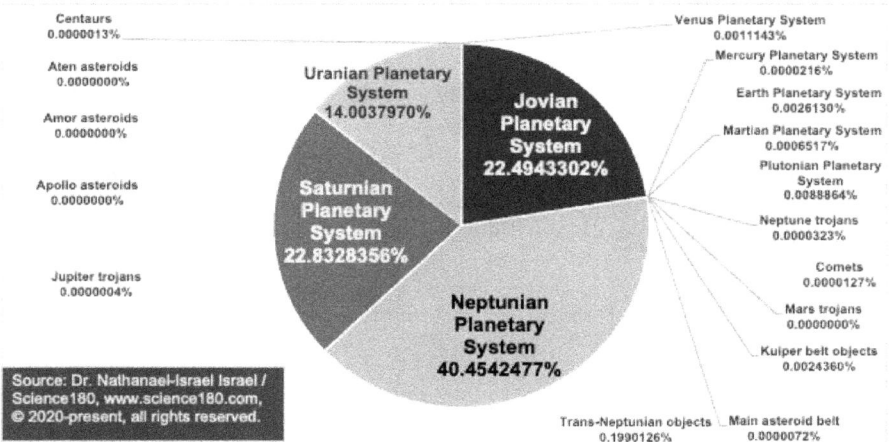

Fig. 18: Relative Orbital Moment of Inertia of the Systems of Bodies (Asteroids and Planetary Systems) Orbiting the Sun

More than 99.9% of the total rotational angular momentum of the bodies orbiting the Sun is found in the planetary systems (Fig. 19). The Jovian planetary system alone has 82.9% of the total rotational angular momentum concentrated in asteroids and the planetary systems of the Solar System. Together with the Saturnian planetary system, they account for 99.4% of the rotational angular moment of all the bodies orbiting the Sun.

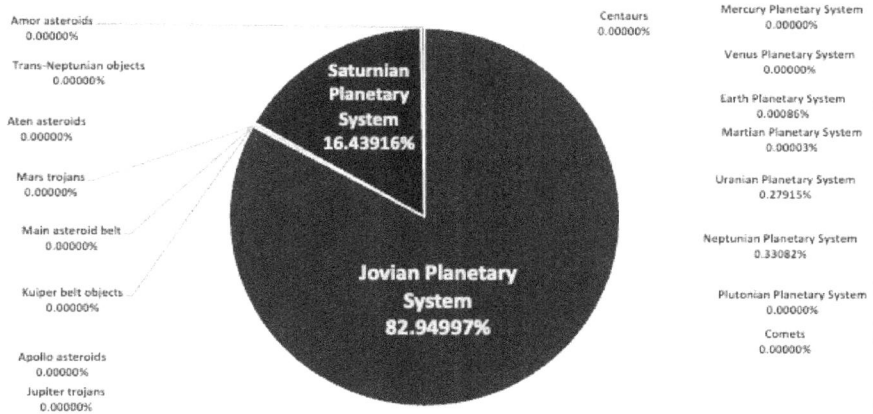

**Fig. 19: Relative Rotational Angular Momentum of the
Systems of Bodies (Asteroids and Planetary Systems)
Orbiting the Sun**

The Jovian and Neptunian planetary systems alone account for 99.6% of the total rotational kinetic energy in the bodies orbiting the Sun (Fig. 20).

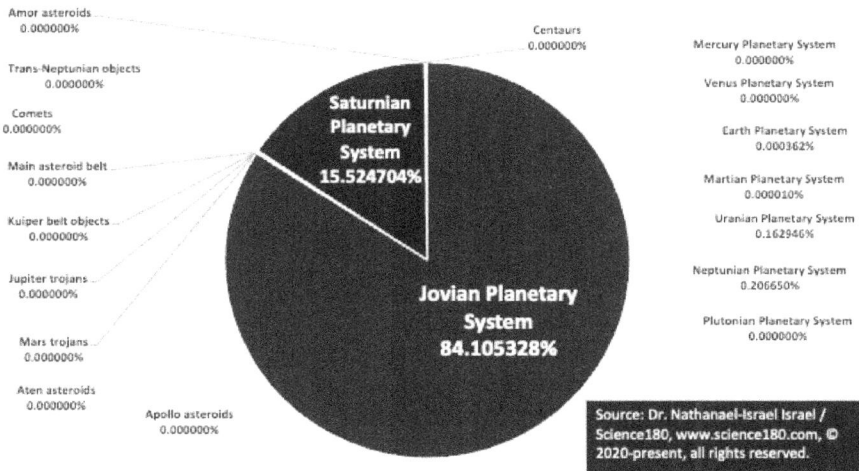

**Fig. 20: Relative Rotational Kinetic Energy of the Systems of
Bodies (Asteroids and Planetary Systems) Orbiting the Sun**

Planetary systems occupy 99.9% of the total rotational moment of inertia of the bodies orbiting the Sun.

Science180: The Undeniable Scientific Challenge to All Metaphorical,
Figurative, Loose, Liberal, or Vague Explanations of the Universe-Origin

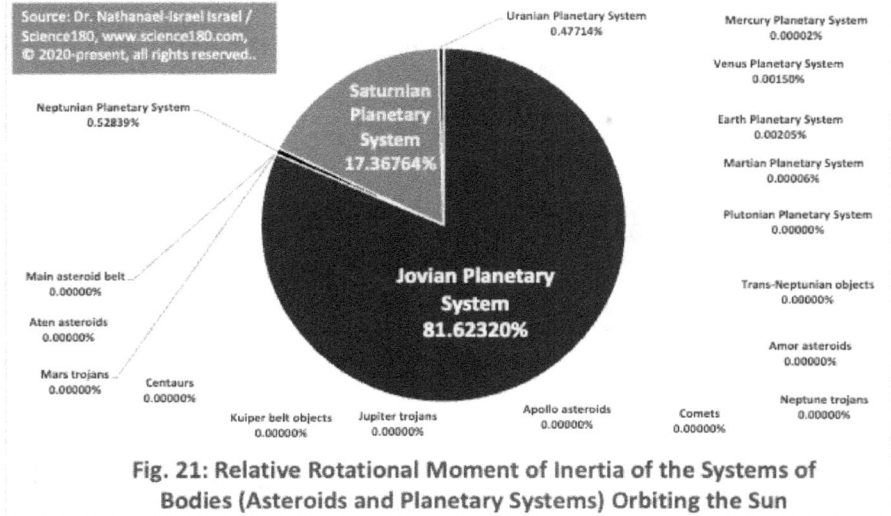

Fig. 21: Relative Rotational Moment of Inertia of the Systems of Bodies (Asteroids and Planetary Systems) Orbiting the Sun

The amount of total kinetic energy sequestered by the precursors of the planetary systems varies according to their types (Fig. 21). As illustrated below, more than 97% of the total kinetic energy orbiting the Sun is sequestered in the 4 giant planetary systems (Fig. 22). Together, the Jovian and the Saturnian planetary systems account for 96.3% of the total kinetic energy of the bodies orbiting the Sun. Because these 2 planetary systems fit the definition of turbulence Zone 3, meaning the most turbulent zone, it is fair to state that, most of the total kinetic energy in the precursors of the secondary bodies in the Solar System were sequestered in what could have been the Zone 3 of the fluid flow of the precursor the secondary bodies in the Solar System. The Jovian planetary system alone accounts for 82.3% of the total kinetic energy of the bodies orbiting the Sun. Besides Venus and the Earth-Moon planetary system which each account for almost 1% of the total kinetic energy orbiting the Sun, any of the remaining planetary systems account for less than 1% of the total kinetic energy of the bodies orbiting the Sun. By the end of the formation of the Solar System, the relative total energy of the systems of bodies is distributed as shown in Fig. 22.

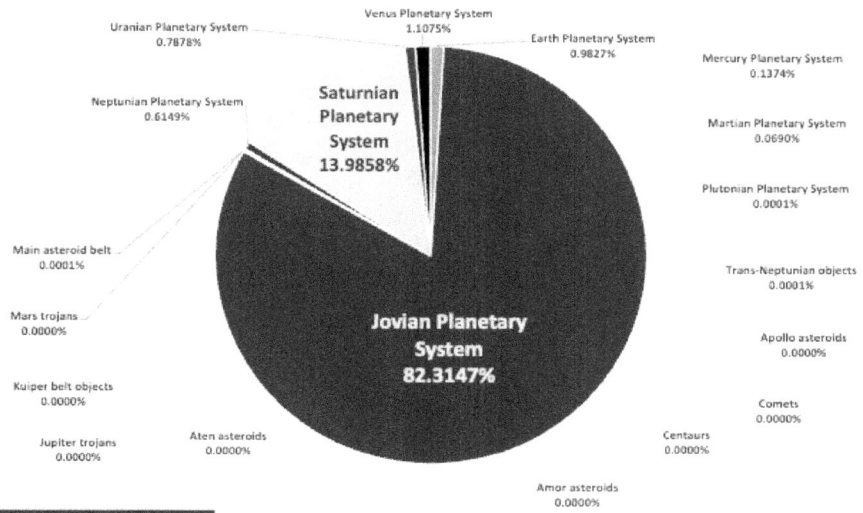

Source: Dr. Nathanael-Israel Israel / Science180, www.science180.com, © 2020-present, all rights reserved.

Fig. 22: Relative Total Kinetic Energy of the Systems of Bodies (Asteroids and Planetary Systems) Orbiting the Sun

Likewise, the Jovian and Saturnian planetary systems alone account for about 95.1% of the total translational kinetic energy of all the bodies orbiting the Sun, implying that the precursors of the most turbulent planetary systems accumulated the bulk of the translational kinetic energy of the precursors of all the secondary bodies. Consequently, after their formation was "completed", the Jovian and Saturnian planetary systems concentrate about 95.1% of the translational energy of the bodies orbiting the Sun. The Jovian planetary system alone accounted for 81.7% of the translational kinetic energy orbiting the Sun (Fig. 23).

Science180: The Undeniable Scientific Challenge to All Metaphorical, Figurative, Loose, Liberal, or Vague Explanations of the Universe-Origin

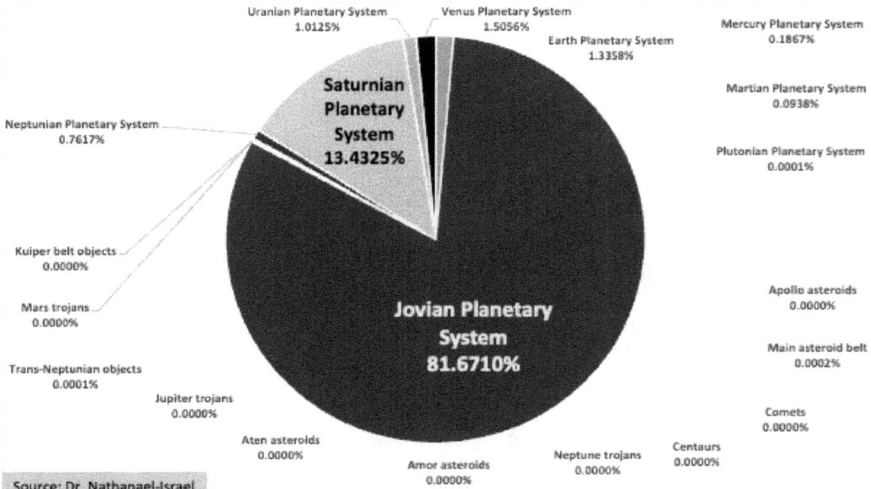

Fig. 23: Relative Translational Kinetic Energy of the Systems of Bodies (Asteroids & Planetary Systems) Orbiting the Sun

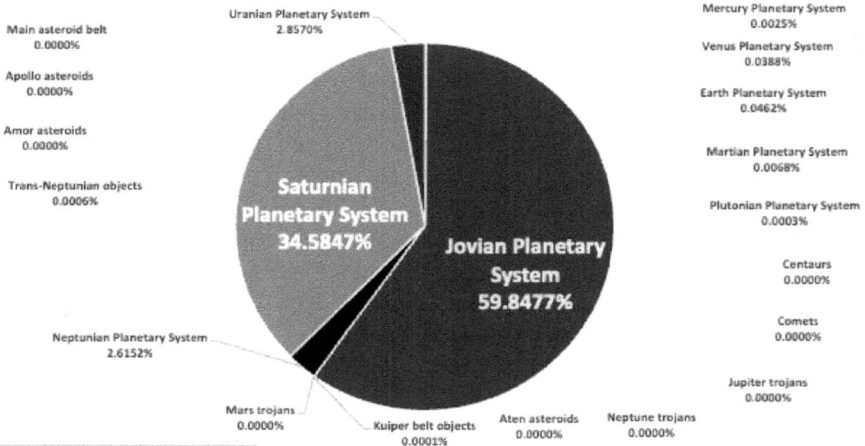

Fig. 24: Relative Volume of the Systems of Bodies (Asteroids and Planetary Systems) Orbiting the Sun

Below, I focused the analysis of the planetary systems and their constituents: planets and satellites.

12.4. More than 99% of the 9 system-additive variables of the planetary systems is accumulated in the 4 giant planetary systems; and the Jovian and Saturnian planetary systems together usually account for more than 95%

The amount of each of the 9 system-additive variables sequestered by the precursors of the planetary systems varies according to the types of planetary systems. More than 96% of the 9 system-additive variables of the bodies orbiting the Sun is sequestered in the 4 giant planetary systems. Indeed, besides the total kinetic energy and the translational kinetic energy for which the 4 giant planetary systems account for 97.7% and 96.9% respectively, for the other 7 system-additive variables (i.e. mass, orbital angular momentum, orbital moment of inertia, rotational angular momentum, rotational kinetic energy, rotational moment of inertia, and volume), the 4 giant planetary systems account for more than 99% of the total value found in the planetary systems.

Usually, the Jovian and Saturnian planetary systems account for 95% of the 9 system-additive variables of the bodies orbiting the Sun. In fact, besides the orbital angular momentum and the orbital moment of inertia which are more split over the 4 giant planets, the Jovian and Saturnian satellites generally account for more than 90% of the total value of any of the other 7 system-additive variables in the planetary systems. In other words, the Jovian and Saturnian planetary systems alone contribute by the following toward the total value of all the planetary systems in the Solar System:

- **92.46%** of the mass
- 86.49% of the orbital angular momentum
- 45.42% of the orbital moment of inertia
- **99.39%** of the rotational angular momentum
- **99.63%** of the rotational kinetic energy (%)
- **98.99%** of the rotational moment of inertia
- **96.30%** of the total kinetic energy
- **95.10%** of the translational kinetic energy
- **94.43%** of the volume

Based on the data I presented in this section so far, because the 4 giant planetary systems fit the definition of turbulence Zone 3, the most turbulent zone of the flow of the fluids that was the precursor of all secondary bodies in the Solar System, it is fair to state that, most of the 9 system-additive variables in the precursors of the secondary bodies in the Solar System were sequestered in the most turbulent zone, Zone 3. Even if I narrowed down that Zone 3 to the Jovian and Saturnian planetary systems only, it can still be fair to state that most of the 9 system-additive variables in the precursors of the secondary bodies in the Solar System were sequestered in the most turbulent zone, Zone 3.

For instance, the Jovian and Saturnian planetary systems account for 96.3% of the total kinetic energy of the bodies orbiting the Sun. The Jovian planetary system alone accounts for 82.3% of the total kinetic energy of the bodies orbiting the Sun. Besides, Venus and the Earth-Moon planetary systems which each account for almost 1% of the total kinetic energy orbiting the Sun, any of the remaining planetary systems account for less than 1% of the total kinetic energy of the bodies orbiting the Sun.

The following graph shows the percentage or relative value of the 9 system-additive variables of the asteroids and planetary systems with respect to the value of all the bodies orbiting the Sun. In other words, when each of the 9 system-additive variables is considered, the relative value presented in Fig. 25 is the percentage that each planetary system contributes toward the total value of all the planetary systems.

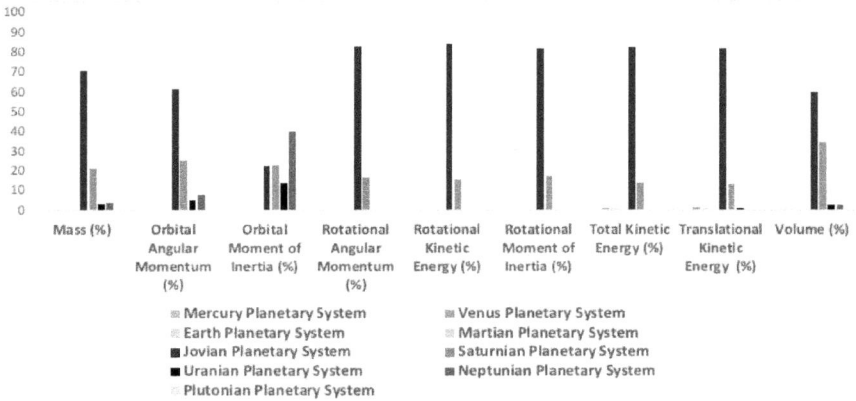

Fig. 25: Relative value (%) of each of the 9 "system-additive" variables of each planetary system with respect to the total value in all the planetary systems in the Solar System

For more clarity, I did the following graphs (Fig. 26-34) to illustrate each of the 9 system-additive variables.

Nathanael-Israel Israel: Known as the World's Most Accurate Universe-Origin Mathematician

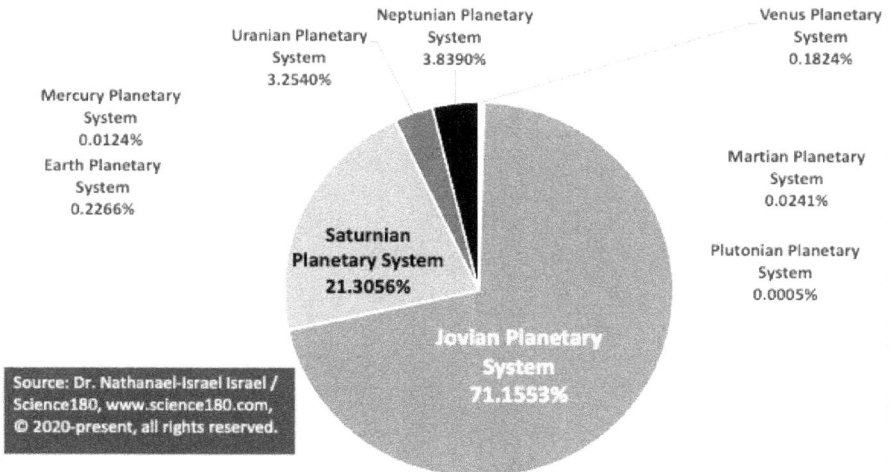

Fig. 26: Relative Mass of the Planetary Systems in the Solar System

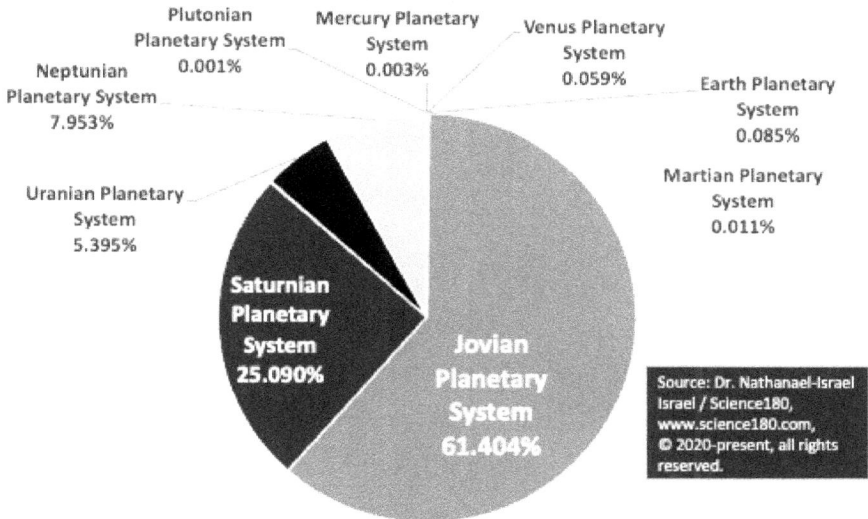

Fig. 27: Relative Orbital Angular Momentum of the Planetary Systems in the Solar System

The Neptunian planetary system accounts for 40.5% of the total orbital moment of inertia of all the planetary systems in the Solar System (Fig. 28). That is almost double the orbital moment of inertia of the Jovian planetary system or the Saturnian planetary system. Despite the difference in their size and speed, the Jovian and Saturnian planetary systems have almost the same orbital moment of

Science180: The Undeniable Scientific Challenge to All Metaphorical, Figurative, Loose, Liberal, or Vague Explanations of the Universe-Origin

inertia. Together, the 4-giant planetary systems account for 99.9% of the total orbital moment of inertia in the planetary systems of the Solar System. Orbital moment of inertia is one of the variables that seems to be more evenly distributed among the giant satellites than any other variable (e.g. mass, volume, and kinetic energy) I studied, suggesting that orbital moment of inertia may have something special to add to the complexity of the distribution and functioning of the bodies in the Solar System and in other systems as well.

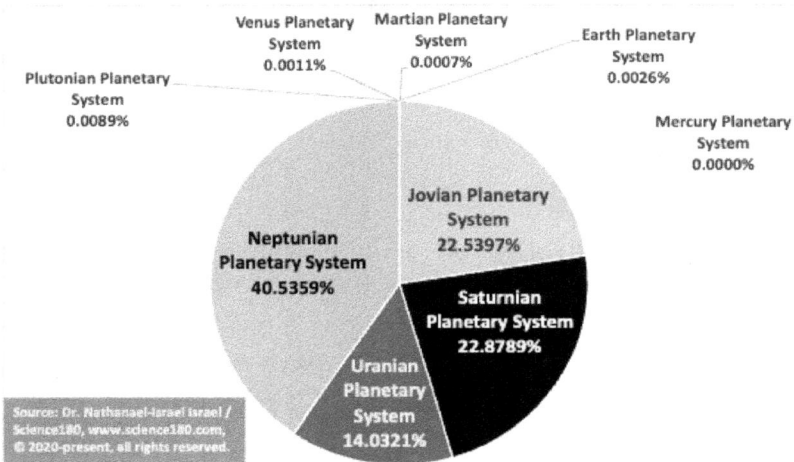

Fig. 28: Relative Orbital Moment of Inertia of the Planetary Systems in the Solar System

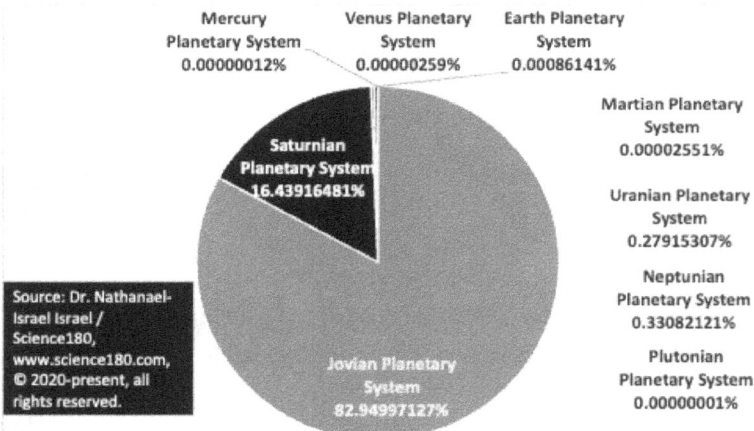

Fig. 29: Relative Rotational Angular Momentum of the Planetary Systems in the Solar System

The Jovian and Saturnian planetary systems account respectively for 84.1% and 15.5% of the total rotational kinetic energy of all planetary systems in the Solar System (Fig. 30). These two planetary systems alone are 99.7% the total rotational kinetic energy of all the planetary systems in the Solar System.

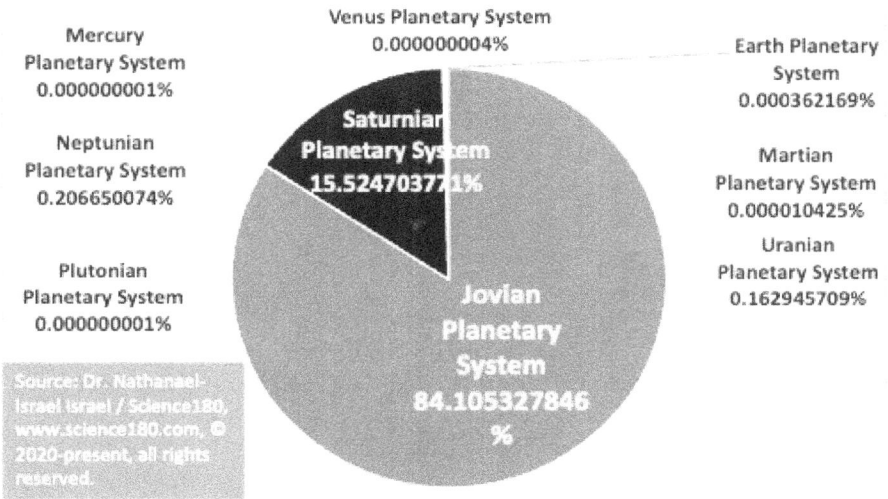

Mercury Planetary System 0.000000001%

Venus Planetary System 0.000000004%

Earth Planetary System 0.000362169%

Neptunian Planetary System 0.206650074%

Saturnian Planetary System 15.5247037?1%

Martian Planetary System 0.000010425%

Uranian Planetary System 0.162945709%

Plutonian Planetary System 0.000000001%

Jovian Planetary System 84.105327846%

Fig. 30: Relative Rotational Kinetic Energy of the Planetary Systems in the Solar System

Together, the Jovian and Saturnian planetary systems account for 99% of the total rotational moment of inertia in the planetary systems of the Solar System. The Jovian planetary system alone accounts for 81.6% of the total rotational moment of inertia of the asteroids and planetary systems in the Solar System (Fig. 31). This contrasts the value of orbital moment of inertia, which maximum value was found in the Neptunian planetary system. The rotational moment of inertia of the Neptunian planetary system is just 0.5% that of all the bodies orbiting the Sun. Yet, for orbital moment of inertia, the Neptunian planetary system accounts for 40.5% of the total rotational moment of inertia of the bodies orbiting the Sun. The 2 biggest planetary systems (Jovian and Saturnian) account for 98.9% of the total rotational moment of inertia of all the bodies orbiting the Sun. The rotational moment of inertia of the Jovian planetary system is 4.7 times that of the Saturnian planetary system.

The high rotational moment of inertia of the Jovian planetary system suggests that the bodies in the precursors of the Jovian planetary system would have resisted acceleration and that more torque could have been needed to change their rotation. This fact holds a key to the constitution of the precursor of the Jovian planetary system. That precursors could have been very dense and viscous, hence a higher rotational moment of inertia. In other words, because the precursor of the Jovian planetary system could have been very viscous or dense, it could have required a higher torque to change its rotation, which in return was translated into a higher rotational moment of inertia of the bodies in that planetary system. This may partially explain why the range of the semi major axis of the Jovian satellites is very small.

Talking about rotational moment of inertia, I would like to recall that, inertia is defined as the resistance of an object to change its position and state of motion. It is not a force, but a tendency to resist. Depending on the state of interest, 3 types of inertia are usually considered:

- Inertia of rest (inability of a body to change by itself its state of rest)
- Inertia of motion (inability of a body to change by itself its state of motion)
- Inertia of direction (inability of a body to change by itself the direction of its motion).

Just as how it is believed that mass determines the force needed for a specific acceleration, rotational inertia is believed to determine the torque needed for a specific angular acceleration about a rotational axis. Rotational inertia is a property which expresses how difficult it is to change the rotational speed of an object around a rotational axis. That is why the moment of inertia of an object is viewed as its resistance to acceleration. Therefore, a larger rotational moment of inertia usually implies that more torque is required to change the body's rotation rate.

Inertia and the moment of inertia associated with it have played a key role in the study of motion. For instance, moment of inertia played a critical role in the famous 3 laws of motion published by Isaac Newton in 1687. For example, Isaac Newton's first law of motion states that "*an object at rest stays at rest and an object in motion stays in motion with the same speed and in the same direction unless acted upon by an unbalanced force.*" In other words, for him, objects tend to "*keep on doing what they're doing.*" His second law related the force acting on an object to its mass and acceleration. This law is mathematically expressed by the famous equation:

$$F = m * a$$

with F is a force, "m" is the mass, and "a" is the acceleration.

Finally, his third law of motion states that every action is opposed by an equal reaction.

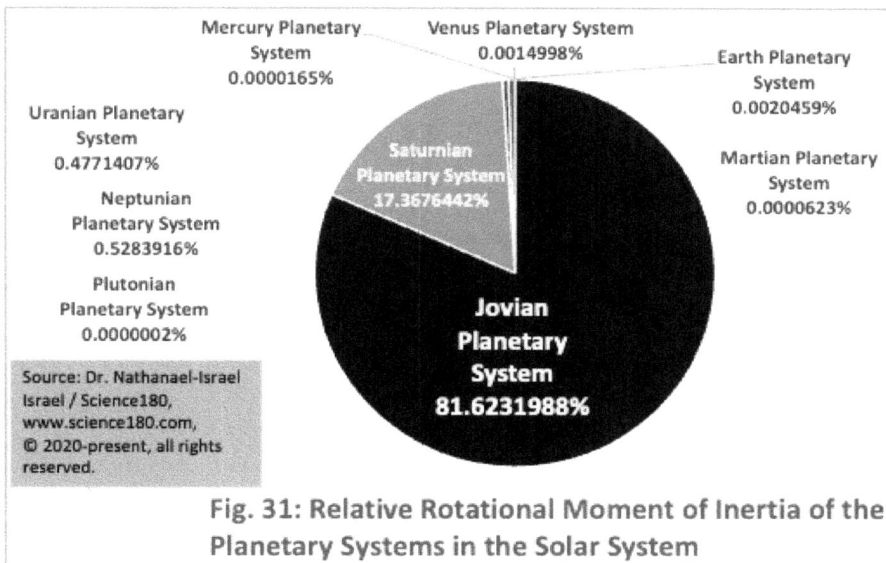

Fig. 31: Relative Rotational Moment of Inertia of the Planetary Systems in the Solar System

As shown in Fig. 32, as the total kinetic energy in the precursor of the planetary systems in the Solar System was being split-gathered into the precursor of the planetary systems, the precursor of the Jovian and Saturnian planetary systems accumulated respectively 82.3% and 14% of it. Consequently, by the time the formation of the Solar System was completed, the Jovian and Saturnian planetary systems gathered 96.3% the total kinetic energy of all the planetary systems in the Solar System. Although Saturn is almost as big as Jupiter, its total kinetic energy is much smaller than that of Jupiter because, due the circumstances of its formation, Jupiter is denser and has a higher orbital and rotational speed than Saturn.

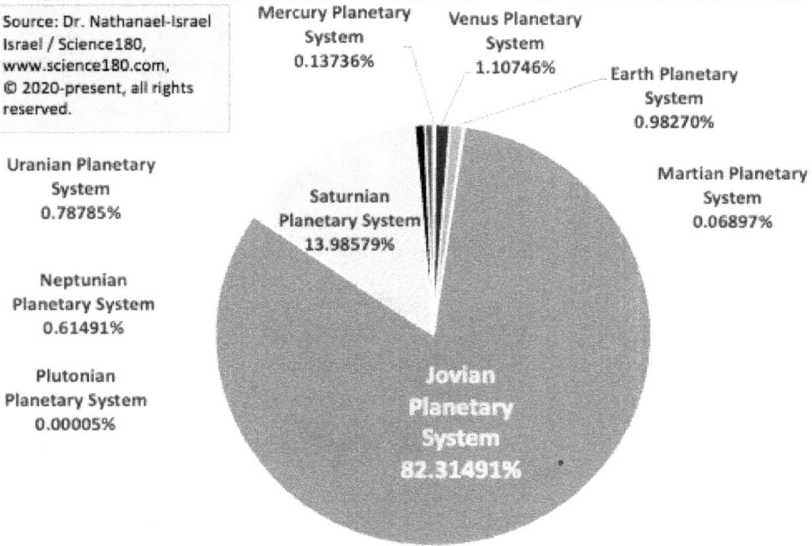

Mercury Planetary System 0.13736%

Venus Planetary System 1.10746%

Earth Planetary System 0.98270%

Uranian Planetary System 0.78785%

Saturnian Planetary System 13.98579%

Martian Planetary System 0.06897%

Neptunian Planetary System 0.61491%

Plutonian Planetary System 0.00005%

Jovian Planetary System 82.31491%

Fig. 32: Relative Total Kinetic Energy of the Planetary Systems in the Solar System

The Jovian and Saturnian planetary systems account respectively for 81.7% and 13.4% of the translational kinetic energy in the planetary systems in the Solar System, meaning that together, these two planetary systems account for more than 95% of the translational kinetic energy in all the planetary systems in the Solar System (Fig. 33).

Mercury Planetary System 0.1867%

Venus Planetary System 1.5056%

Earth Planetary System 1.3359%

Saturnian Planetary System 13.4325%

Martian Planetary System 0.0938%

Uranian Planetary System 1.0125%

Neptunian Planetary System 0.7617%

Jovian Planetary System 81.6712%

Plutonian Planetary System 0.0001%

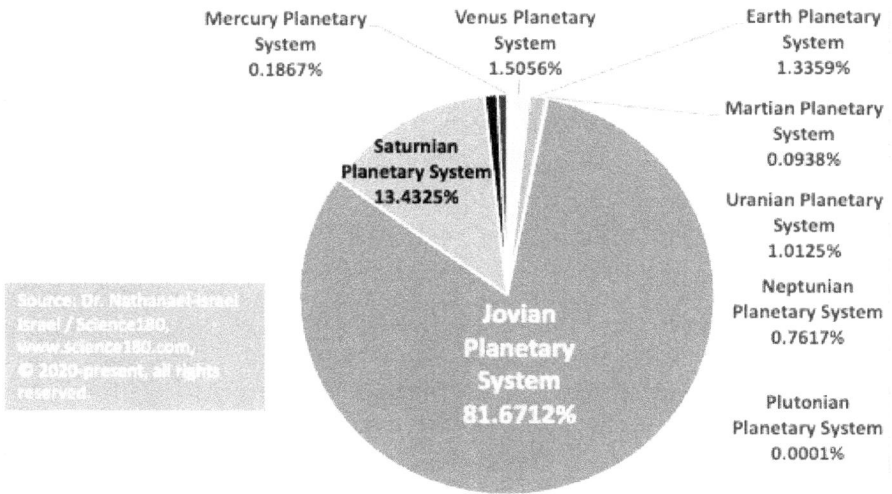

Fig. 33: Relative Translational Kinetic Energy of the Planetary Systems in the Solar System

Similarly, most of the volume of the planetary systems is found in the Jovian and Saturnian systems (Fig. 34).

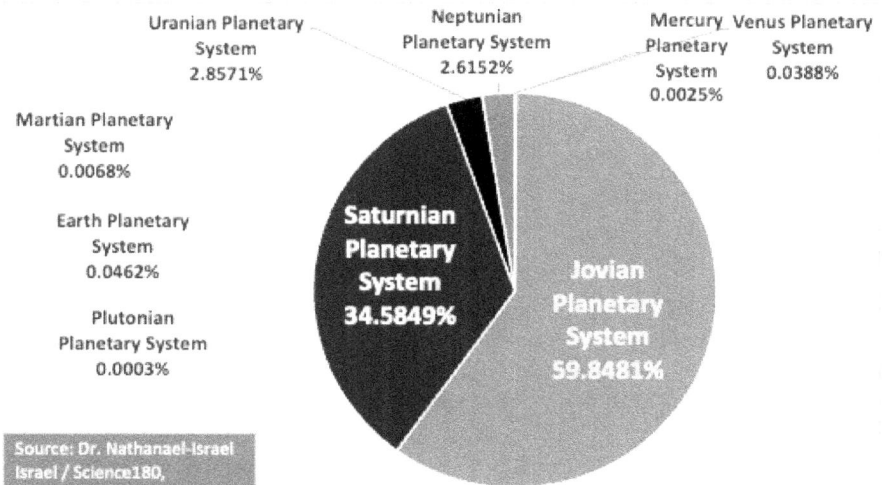

Uranian Planetary System 2.8571%

Neptunian Planetary System 2.6152%

Mercury Planetary System 0.0025%

Venus Planetary System 0.0388%

Martian Planetary System 0.0068%

Earth Planetary System 0.0462%

Plutonian Planetary System 0.0003%

Saturnian Planetary System 34.5849%

Jovian Planetary System 59.8481%

Fig. 34: Relative Volume of the Planetary Systems in the Solar System

Science180: The Undeniable Scientific Challenge to All Metaphorical, Figurative, Loose, Liberal, or Vague Explanations of the Universe-Origin

12.5. During the split-gathering of the precursors of the planetary systems, more than 99% of each of the 9 system-additive variables went into the precursors of the primary planets and less than 1% went into the precursors of their satellites

As each of the 9 system-additive variables of the precursor of each planetary system was being split-gathered, the precursor of the primary planet sequestered on average more than 99% of the total, while the precursor of the satellites gathered together less than 0.1% (Fig. 35). In fact, the percentage for the planets varies between 88.21% and 100% for each of the 9 system-additive variables of the planetary systems. The smallest percentage was found with the precursor of the Plutonian planetary system which sometimes sent less than 99% of each of the 9 system-additive variables into the precursor of Pluto, its primary planet. In fact, for the precursor of the Plutonian planetary, it is for only 4 out of the 9 system-additive variables that 99.9% of the value of the precursor of the planetary system was sent into the precursor of Pluto. For 3 other variables, about 97% of the value of the Plutonian planetary system was sent into the precursor of Pluto and finally, for 2 variables, about 89% was sent into the precursor of Pluto. In other words, for the Plutonian planetary system, the precursor of the primary planet, Pluto, received the following percentage of the value of the precursors of the Plutonian planetary system:

- 89.16% of the mass
- **99.99**% of the orbital angular momentum
- **99.99**% of the orbital moment of inertia
- 96.93% of the rotational angular momentum
- 96.93% of the rotational kinetic energy
- 96.93% of the rotational moment of inertia
- **99.97**% of the total kinetic energy
- **99.97**% of the translational kinetic energy
- 88.21% of the volume

Due to some initial conditions (e.g. position, energy, and size), the precursors of some planets were not able to split-gather part of the matter into the precursor of the satellites capable of orbiting them. Consequently, some of those precursors of planets ended up lacking a satellite around their planets. That was the case for Mercury and Venus, which have no satellite. For a better graphical view, I presented the contribution of the planets to the 9 system-additive variables of their planetary system in the following graph.

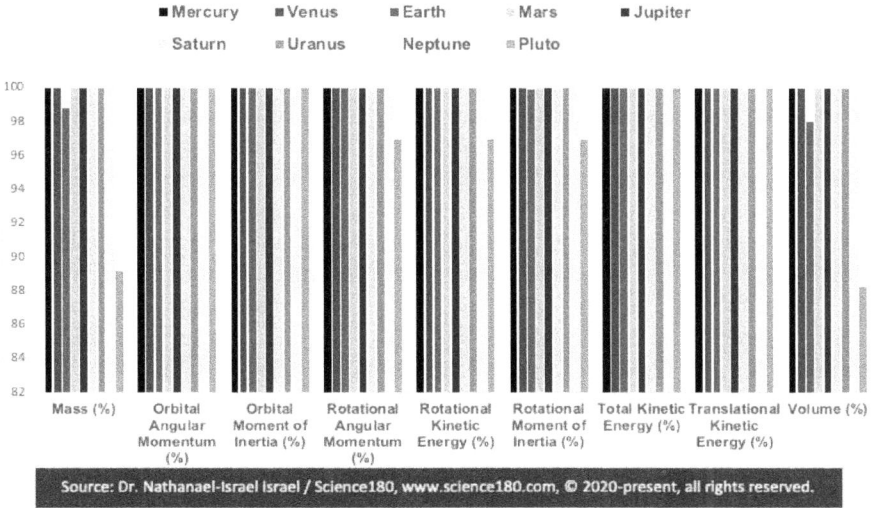

Fig. 35: Contribution of the Primary Planets to the 9 "system-additive" variables of their Planetary Systems

12.6. The precursors of Jupiter and Saturn collected most of the 9 system-additive variables found in planets in the Solar System

The precursors of Jupiter and Saturn could have accumulated more than 90% of the total value of most of the 9 system-additive variables (Fig. 36). In fact, except for the orbital moment of inertia for which Jupiter and Saturn accumulate just 45.42% of the total value found in the 9 planets in the Solar System, while Neptune alone accumulated 40.5% of the total value, for any of the other 8 system-additive variables, Jupiter and Saturn together accumulated more than 95% of the total value found in the planets. This trend is similar to that found for the planetary systems. For instance, by the time the formation of the planets was completed, Jupiter and Saturn respectively sequestered 82.3% and 14% of the total kinetic energy of the planets in the Solar System and together, they make up 96.3% of the total kinetic energy of all the planets in the Solar System. Likewise, Jupiter and Saturn account for 84.1% and 15.5% of the total rotational kinetic energy of the planets in the Solar System and together, they also make up 99.6% of the rotational kinetic energy of all planets in the Solar System.

Moreover, Jupiter has 82.9% of the rotational angular momentum of the planets in the Solar System, followed by Saturn which contributes 16.4% to the total. When it comes to orbital moment of inertia, the 2 biggest planets in the Solar System do not account for the biggest share anymore. Indeed, Neptune has the highest orbital moment of inertia in the Solar System, accounting for 40.5% of the total orbital moment of inertia of the planets in the Solar System. It is

177

followed by Saturn and Jupiter which each has almost 22% of the total orbital moment of inertia of the planets in the Solar System. The four giant planets account for 99.9% of the total orbital moment of inertia of the planets in the Solar System. The smallest orbital moment of inertia of the planets in the Solar System was recorded with Mercury because of its smallest semi major axis despite it having the highest orbital speed among the planets. In contrast, the highest orbital moment of inertia of the planets in the Solar System was found with Neptune, and this is because the semi major axis of Neptune played a significant role in the high value of its orbital moment of inertia. In contrast, Jupiter and Saturn have almost a similar orbital moment of inertia because what one of them gained with its high mass is lost by its smaller semi major axis and vice versa. Indeed, the mass of Jupiter is higher than that of Saturn, but its semi major axis is smaller than that of Saturn. In the balance, the product of their mass and semi major axis which contributed to their orbital moment of inertia is almost the same. The next graph recapitulated the aforementioned trends.

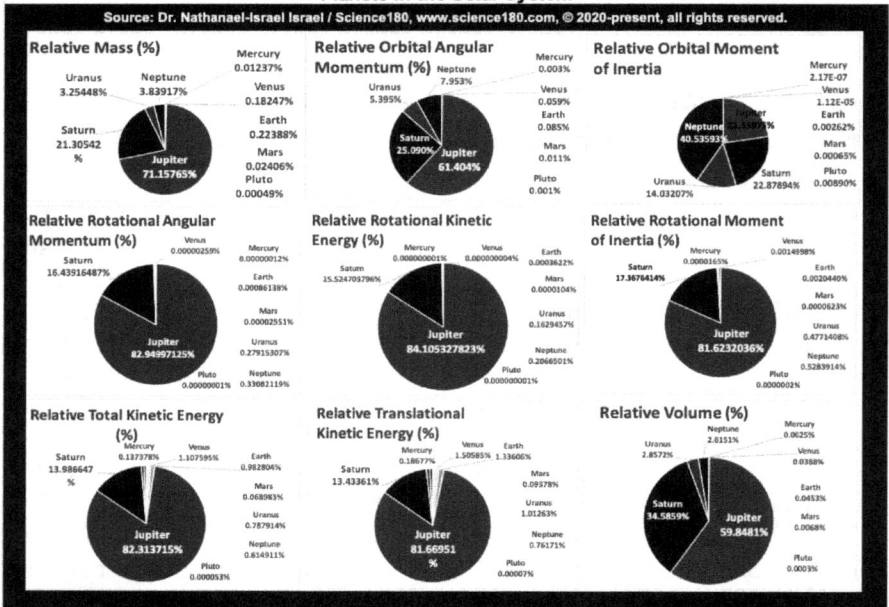

Fig. 36: Relative value (%) of the 9 "system-additive" variables of the Planets with respect to all Planets in the Solar System

Nathanael-Israel Israel: Known as the World's Most Accurate Universe-Origin Mathematician

12.7. A positive relationship exists between some of the 9 system-additive variables of the precursors of the planetary systems and that in the precursors of the primary planets and the precursors of their satellites

For the precursors of the planetary systems which were able to yield a precursor of a planet and a precursor of satellite(s), a positive relationship existed between the 9 system-additive variables of the precursors of the planetary system and that of the precursors of the satellite(s) and the precursor of its primary planet (Fig. 37). This implies that the amount of the 9 system-additive variables gathered together inside the precursor of the planets and precursors of their satellites depended on the amount of those 9 system-additive variables present in the precursors of their planetary system.

For instance, after the formation of the planetary systems, some of the 9 variables (e.g. total kinetic energy, rotational kinetic energy, and translational kinetic energy) in the planetary systems had a positive impact on that in the planets ($r^2=1$) and that in their satellites ($r2>0.99$). Similarly, a positive relationship exists between these 3 variables in the primary planets and that in their satellites. Below, I present some details.

Fig. 37: Regression between the Total Kinetic Energy, the Translational Kinetic Energy, and the Rotational Kinetic Energy of the Planetary Systems, their Primary Planet, and their Satellites

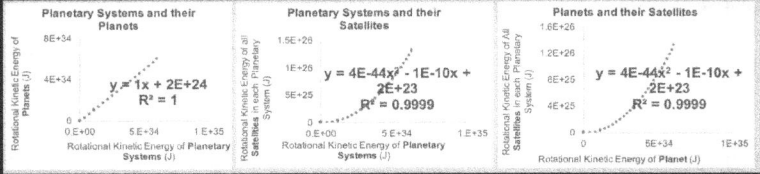

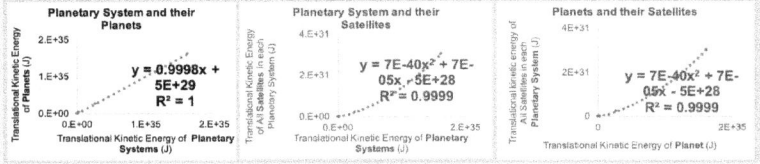

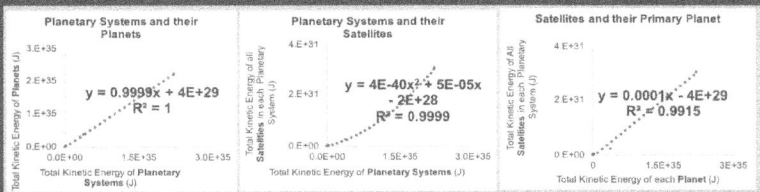

The rotational kinetic energy in the precursors of the planetary systems affected that in the precursors of their planets and satellites

The rotational kinetic energy of the planetary system positively affected that in their primary planet ($r2=1$) and that in their satellites ($r2>0.99$).

The translational kinetic energy in the precursors of the planetary systems affected that in the precursors of their planets and satellites

The translational kinetic energy in the planetary systems has a positive impact on that in the planets ($r2=1$) and that in their satellites ($r2>0.99$). The translational kinetic energy of the planets is more than 99.9% that of their planetary systems, suggesting that, during the formation of the planetary systems, the precursors of each system sequestrated more than 99.99% of its energy into the primary planet (Fig. 38).

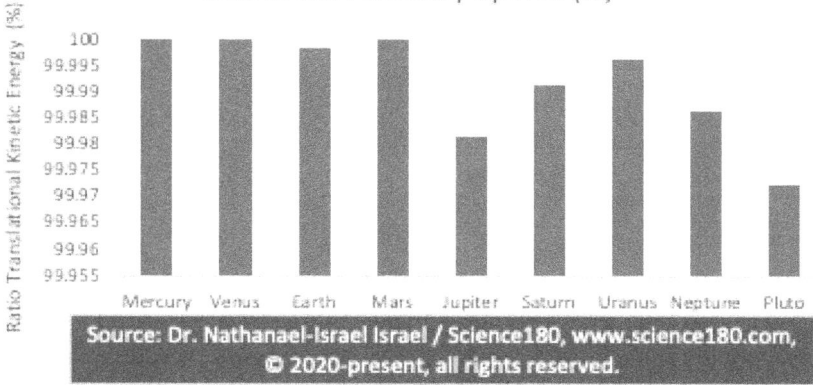

Fig. 38: Ratio Translational Kinetic Energy of Planet to that of the Planetary System (%)

The translational kinetic energy of the satellites depended on their primary planet, the turbulence zone, and the interaction between these 2 variables ($p<0.000$). The energy in the Jovian satellites is 13.3 times that of the Saturnian satellites. In general, more than 99% of the translational kinetic energy of the satellites is usually concentrated in turbulence Zone 3 (Fig. 39).

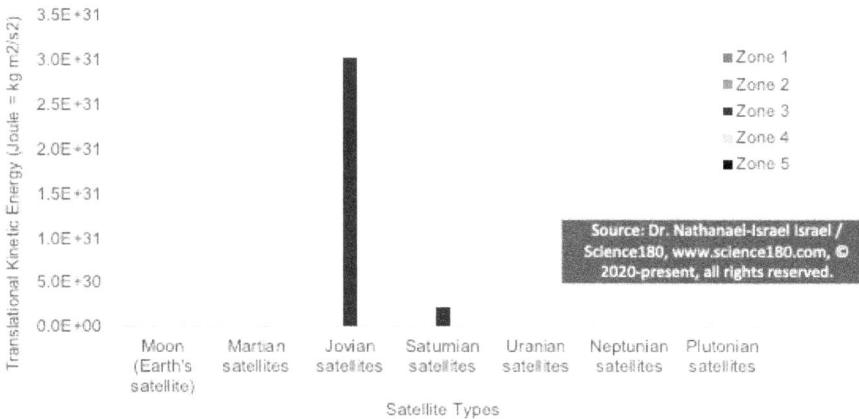

Fig. 39: Translational Kinetic Energy of Satellites in the Solar System

The total kinetic energy in the precursors of the planetary systems affected that in the precursors of their planets and satellites

As presented in the previous regressions, when the energy of the precursor of each planetary system was being split, 99.9% of it was sequestered into the

Science180: The Undeniable Scientific Challenge to All Metaphorical, Figurative, Loose, Liberal, or Vague Explanations of the Universe-Origin

precursor of the primary planets and less than 0.1% was pushed into the precursors of the satellites. Due to some initial conditions (e.g. position and size), the precursors of some planets were not able to split-gather part of their matter into the precursor of a satellite capable of orbiting them. Consequently, the precursors of some planets ended up not having a satellite around their planets. That was the case for Mercury and Venus. For the precursors of the planetary systems that were able to yield a precursor of both planet and satellite(s), a positive relationship existed between the energy of the precursors of the planetary systems and the energy in the precursor of their satellite(s) and the precursor of their primary planet. Therefore, after the formation of the planetary systems, the total kinetic energy in the planetary systems had a positive impact on that in the planets ($r2=1$) and that in their satellites ($r2>0.99$). Similarly, a positive relationship exists between the total energy in the primary planets and that in their satellites.

12.8. During the split-gathering of the precursors of the satellites, more than 99% of the 9 system-additive variables were accumulated in their most turbulent zone, usually Zone 3

To facilitate the comparison of the data, Fig. 40 presents the data in forms of percentage of each turbulence with respect to the value of the satellites in all the turbulence zones. I called these ratios "turbulent multipliers" and I devoted a separate chapter for them. The graph below shows that most of the percentages of any of the 9 system-additive variables are found in Zone 3, the most turbulent zone.

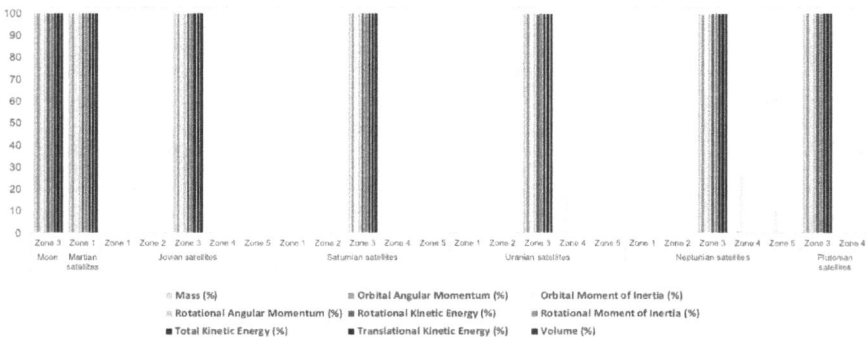

Fig. 40: Sum of the turbulent multipliers (%) of each of the 9 "system-additive" variables of all the satellites in each turbulence zone according to their types

Based on the data above and others I studied, it appeared to me that as the fluids in the precursors of the satellites started flowing away from the precursor of their primary planets, the value of the 9 system-additive variables was split differently over the turbulence zones. In general, more than 99% of the 9 system-additive variables of the precursor of the satellites was concentrated in turbulence Zone 3. Indeed, while considering each of the 9 system-additive variables and all of the 6 planetary systems that have satellites in Zone 3, the average percent of Zone 3 is 99%, the standard deviation is 5.53% and the coefficient of variation is 5.59%. Because none of the Martian satellites belong to Zone 3, the 9 system-additive variables of their satellites were found in Zone 1. Because the Moon, the only satellite of the Earth, fits the classification of turbulence Zone 3, 100% of the value of the 9 system-additive variables of the Moon belongs to Zone 3.

Out of the 54 percentages available for the Zone 3 (meaning when considering all of the 9 variables and the 6 planetary systems that have Zone 3), only 2 values are less than 99%. In other words, 52 of the 54 percentages available for the 9 system-additive variables for Zone 3 are higher than 99%. In fact, the average percentage for these 52 data are for Zone 3, meaning for 96.3% of the data available in Zone 3, is 96.3%, the standard deviation is 0.17%, and the coefficient of variation is 0.17%.

The only 2 percentages that are less than 99% are about the orbital moment of inertia in Zone 3 of the:

- Uranian satellites (81.92%) and
- Neptunian satellites (63.15%).

While among all the 9 system-additive variables, the smallest percentage in Zone 3 was recorded with the orbital moment of inertia of the Uranian satellites and the Neptunian satellites, these same satellites recorded the highest value recoded in Zones 4 and 5 for the same variable, orbital moment of inertia. For instance, the highest percentage ever recorded in Zone 4 for any of these 9 system-additive variables was for the orbital moment of inertia of:

- Uranian satellites (15.95%) and
- Neptunian satellites (26.61%).

This means that, while considering the orbital moment of inertia in Zones 3 and 4, the following percentage was observed for the following satellites:

- Uranian satellites (97.87%) and
- Neptunian satellites (89.761%)

Furthermore, while the orbital moment of inertia in Zone 5 is negligible for the Uranian satellites (2.13%), for the Neptunian satellites, it is 10.24%. In other words, Zones 3, 4, and 5 of the Neptunian satellites accumulated 99.9% of the total orbital moment of inertia in the Neptunian satellites.

Below, I elaborated on the 3 kinds of kinetic energy and on the rotational angular momentum.

More than 99% of the total kinetic energy of the satellites is accumulated in the most turbulent zone, usually Zone 3

Indeed, more than 99% of the total kinetic energy of the secondary bodies was concentrated in Zone 3, the most turbulent zone (Fig. 41). As the precursors of the satellites started flowing, its total kinetic energy was split differently over the turbulence zones. In general, more than 99% of the total kinetic energy of the precursor of the satellites was concentrated in turbulence Zone 3. Because some satellite systems do not have Zone 3, the energy of their satellites was found in their most turbulent zone. For instance, the energy of the precursor of the Martian satellites was not even able to form bodies in Zone 3, hence the Martian satellites are only found in Zone 1.

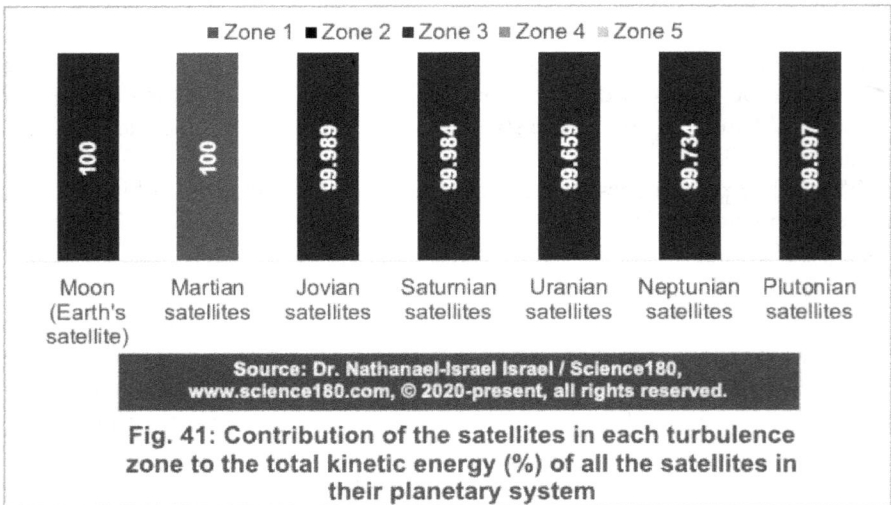

Fig. 41: Contribution of the satellites in each turbulence zone to the total kinetic energy (%) of all the satellites in their planetary system

As I addressed in other chapters (e.g. chapter on semi major axis), the Neptunian satellites are farther away from Neptune than the satellites of any other planet. Before the days of "Turbulent Universal Science" that I pioneered, the cause of such fact has never been explained properly. When I was trying to explain these trends, I noticed, among many other things, that the total kinetic energy in the Neptunian satellites in Zones 4 and 5 is higher than that of the satellites in the same zones for any other planets in the Solar System. The high energy of the precursors of bodies in Zones 4 and 5 could have propelled the Neptunian satellites in Zones 4 and 5 to go very far, hence they are the satellites with the highest semi major axis and increment of semi major axis in the Solar System. To put it another way, the Neptunian satellites in Zones 4 and 5 were propelled with more energy than the satellites in Zones 4 and 5 of the other satellite systems. As I explained later, the high rotational angular speed (translated into a higher energy with which the precursors of the Neptunian satellites were propelled) could contribute to explaining their high semi major axis. I will better expound on this.

More than 99% of the rotational kinetic energy of the satellites is accumulated in the most turbulent zone, usually Zone 3

As the precursors of the satellites started flowing, they accumulated more than 99% of their rotational kinetic energy into the satellites in the most turbulent zone, usually Zone 3 (Fig. 42). In each planetary system, more than 99% of the rotational kinetic energy of the satellites is usually concentrated in the most turbulent zone, usually Zone 3. In general, the mean rotational kinetic energy of the satellites in Zone 1 is usually less than that in Zone 2, which is often less than that in Zone 3. The mean rotational kinetic energy in Zone 4 although usually less than that in Zone 3, is often more than that in Zone 5. In other words, the mean rotational kinetic energy of the satellites in Zone 5 is usually less than that in any other zone. This trend also explains why the Jovian and Saturnian planetary systems accumulated about 99.6% of the rotational kinetic energy in the secondary bodies orbiting the Sun, for the precursors of these 2 planetary systems were located in what was the most turbulent zone, Zone 3, of the fluid flow of the bodies orbiting the Sun.

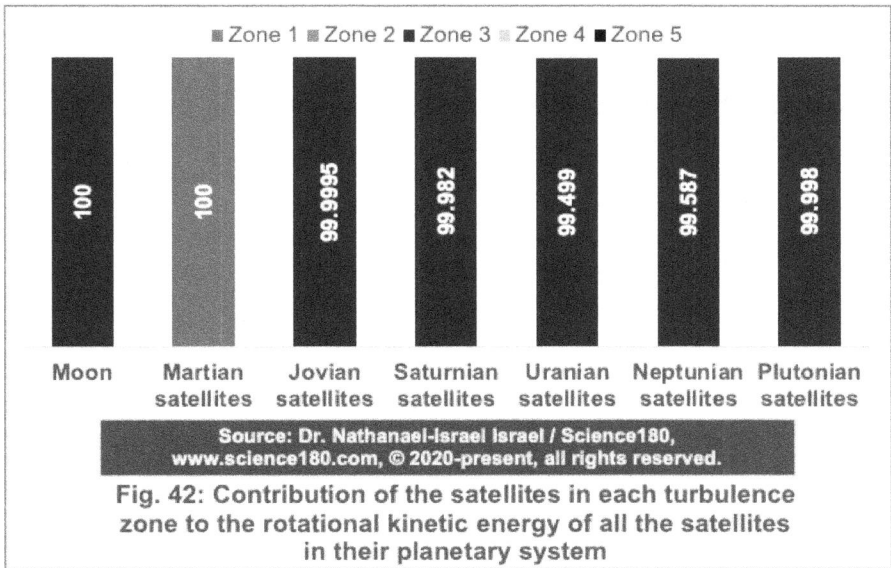

Fig. 42: Contribution of the satellites in each turbulence zone to the rotational kinetic energy of all the satellites in their planetary system

As of 2020, in the Solar System, the rotational kinetic energy of the satellites is $1/5.13E+08$th that of all the planetary systems. The Jovian satellites alone account for 96% of the total rotational kinetic energy found in the satellites in the Solar System (Fig. 43). Saturnian satellites account for just 2.8%, meaning that the rotational kinetic energy of the Jovian satellites is 34.4 times that of the Saturnian satellites and 427 times that of the Moon. The total rotational kinetic energy of the Martian satellites is lower than that of the satellites of any other planet in the Solar System.

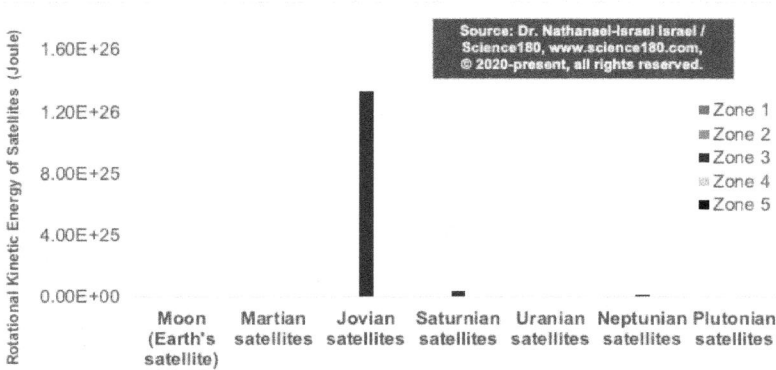

Fig. 43: Rotational Kinetic Energy of Satellites according to their Turbulence Zones

The rotational kinetic energy in the satellites as of 2020 is presented in the table below.

More than 99% of the translational kinetic energy of the satellites is accumulated in the most turbulent zone, usually Zone 3

When the precursors of the secondary bodies of each planetary system started flowing and breaking up into each secondary body, they gathered more than 99% of their energy into bodies in the most turbulent zone, usually Zone 3. In other words, just as 99% of the translational kinetic energy orbiting the Sun is found in the 4 giant planetary systems (which are the most turbulent zones), so also about 99% of the energy of the satellites was gathered into the satellites in the most turbulent zone (Fig. 44). In each system, since the bodies were positioned at different places, with different masses and speed, their translational kinetic energy ended up being also different. Nevertheless, the fingerprint of the law that split and gathered the translational kinetic energy into those bodies can be detected as explained above. Additionally, since energy is said to be conserved, and never created or lost, the energy balance in the bodies of the Solar System is a witness of the process that broke up and assembled the energy in the precursor of the Solar System into different compartments know today as bodies in the Solar System.

As of today, more than 99% of the translational kinetic energy in a system is generally found in the primary body and over 99% of the translational kinetic energy of the bodies orbiting a primary body is found in the most turbulent zone,

Zone 3. The body with the highest translational kinetic energy is not always the biggest body and the ratio of the translational kinetic energy of a primary body to that of the biggest body orbiting it is not a constant. The next figure illustrated the percentage of the translational kinetic energy of the turbulence zones.

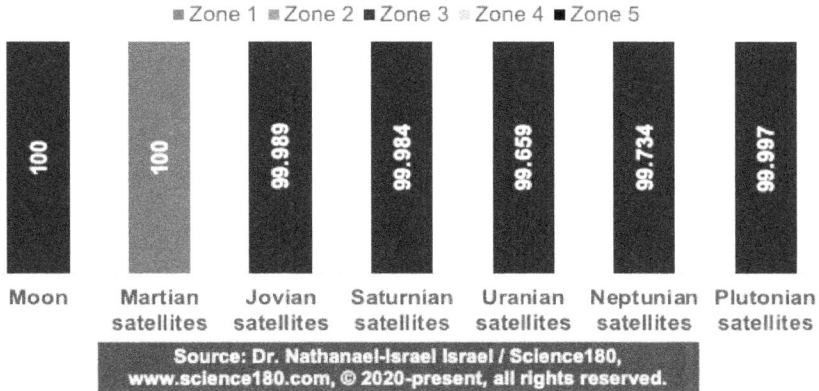

Fig. 44: Contribution of the satellites in each turbulence zone to the translational kinetic energy of all the satellites in their planetary system

More than 99% of the rotational angular momentum of the satellites in each planetary system is accumulated in the most turbulent zone, usually Zone 3

The total rotational angular momentum of the satellites in the Solar System is $1/6.23E+07$th that of all the planetary systems in the Solar System. The Jovian satellites alone account for 84.1% of the rotational angular momentum of all the satellites in the Solar System. Indeed, the total rotational angular momentum of the Jovian satellites is 6.8 times to $8.33E+10$ times that of the satellites of any other planetary system in the Solar System. For instance, the rotational angular momentum of the Jovian satellites is 47.1 times that of the Moon and 6.8 times that of the Saturnian satellites. The rotational angular momentum of the Neptunian satellites is 10 times that of the Uranian satellites. Besides Mars which satellites are in Zone 1 only, more than 99.9% of the rotational angular momentum of the satellites in any planetary system is usually concentrated in less than 5 satellites in turbulence Zone 3.

The higher value of the rotational angular momentum of the Neptunian satellites as compared to that of the Uranian satellites may help explain why the semi major axis of the Neptunian satellites is so high. For the high rotational angular momentum of the precursors of the Neptunian satellites was translated into a rotational energy which contributed to expanding the semi major axis of the satellites which came out of it.

12.9. Most of the 9 system-additive variables of the precursors of the satellites was concentrated in a few satellites

As the precursor of the satellites in each planetary system split to yield the satellites, most of the values of each of the 9 system-additive variables were concentrated in a few satellites. Indeed, less than 5 satellites account for more than 99% of the value of the 9 system-additive variables of the satellites in each planetary system. Indeed, in each planetary system, less than 5 satellites account for more than 99% of the total kinetic energy of its satellites. For instance, Phobos, the biggest Martian satellite, alone accounts for 94.2% of the total kinetic energy of the Martian satellites. Of the 79 Jovian satellites known in 2020, four (meaning 5% of the Jovian satellites) account for more than 99.99% of the total kinetic energy of the Jovian satellites. The biggest Saturnian satellite, Titan, alone accumulated 91.2% of the total kinetic energy of all 82 Saturnian satellites. Four of the 27 Uranian satellites accumulated 97.7% of the total kinetic energy of the Uranian satellites. The total kinetic energy of the biggest Neptunian satellite, Triton, is 99% that of all 14 Neptunian satellites. Likewise, the biggest Plutonian satellite (Charon) accounts for 99.9% of the total kinetic energy of the Plutonian satellites.

In the same way, in each planetary system, less than 5 satellites account for more than 99% of the translational kinetic energy and the rotational kinetic energy of its satellites. For example, Phobos, the biggest Martian satellite, alone accounts for 99.7% of the rotational kinetic energy of the Martian satellites. Out of the 67 Jovian satellites known as of 2017, four (meaning 5% of the Jovian satellites) account for more than 99.9% of the rotational kinetic energy of the Jovian satellites. The biggest Saturnian satellite, Titan, alone accumulated 95.5% of the total rotational kinetic energy of all 62 Saturnian satellites known as of 2017. The 4 biggest Saturnian satellites account for more than 99% of the total rotational kinetic energy of the Saturnian satellites. Four of the 27 Uranian satellites accumulated 98.2% of the total rotational kinetic energy of the Uranian satellites. The rotational kinetic energy of the biggest Neptunian satellite, Triton, is 99.4% that of all 14 Neptunian satellites. Finally, the biggest Plutonian satellite (Charon) accounts for 99.9% of the rotational kinetic energy of the Plutonian satellites.

As far as the rotational angular momentum is concerned, Phobos, the biggest Martian satellite, accounts for 98.8% of the total rotational angular momentum of the 2 Martian satellites. Out of the 67 Jovian satellites known as of 2017, four (meaning 5% of the Jovian satellites) account for more than 99.9% of the total rotational angular momentum of the Jovian satellites. Io, the innermost Jovian satellite in Zone 3, has the highest rotational angular momentum, accounting for 43.9% of the total rotational angular momentum of the Jovian satellites.

Ganymede (the biggest Jovian satellite) came next as it has 37.7% of the total rotational angular momentum of the Jovian satellites. The biggest Saturnian satellite, Titan, alone accumulated 99% of the total rotational angular momentum of all the 62 Saturnian satellites known as of 2019. Out of the 27 Uranian satellites, 4 accumulated 99.6% of the total rotational angular momentum of all Uranian satellites. Titania, the biggest Uranian satellite has 38% of the total rotational angular momentum found in the Uranian satellites. More than 99% of the rotational angular momentum of the Neptunian satellites is found in Triton, the biggest Neptunian satellite. Similarly, the biggest Plutonian satellite (Charon) accounts for more than 99% of the rotational angular momentum of the Plutonian satellites.

Before I address other aspects of the split-gathering of the precursors of the bodies into their daughter bodies, I would like to emphasize a key observation I made concerning the problem of the conservation of angular momentum. Indeed, momentum is a physics term that refers to the quantity of motion that an object has. In other words, momentum is a "mass in motion". Because all objects have a mass, an object on the move has a momentum. For instance, in sports, a team that is on the move (i.e. winning) is said to have the momentum of the game. The conservation of angular momentum and of energy are 2 of the key conservation laws in physics. Even at the microscopic scale where classical physics seems to fail and quantum mechanics prevails, the conservation laws are said to exist due to so-called "inherent symmetries present in nature". According to the law of conservation of angular momentum, when no external torque acts on an object, no change of angular momentum will occur. It is believed that, when no external torque is applied to an object spinning in a closed system, its angular momentum will not change. This law of the conservation of angular momentum is usually applied to explain the spin of an ice skater, which is one of the most famous accounts used to explain conservation of angular momentum. Indeed, the net torque of an ice skater executing a spin is believed to be very close to zero, because: 1) there is relatively little friction between their skates and the ice, and 2) the friction is exerted very close to the pivot point. Therefore, when ice skaters are spinning on the tip of their skate with their arms extended, their angular momentum is said to be conserved because the net torque on them is said to be negligibly small. Their rate of spin increases greatly when they pull in their arms, decreasing their moment of inertia. The work they do to pull in their arms results in an increase in rotational kinetic energy.

According to the theories on the formation of the Solar System and its constituents, angular momentum is one of the most challenging unresolved issues. This is because one of the main theories of the formation of the Solar System, the solar nebula model, implies a conservation of angular momentum among the celestial bodies, whereby in reality, angular momentum is not conserved all the time. Despite efforts to use migration and collision of bodies at the origin of the Solar System to justify the distribution of angular momentum, several factors still exist and point to major errors in the nebula model of the formation of the Solar System. The scientific community is well aware of this issue and all theories about the formation of the Solar System point at this problem that no one has fully explained yet before the days of "Science180 Cosmology" that I spearheaded.

One of the reasons that angular momentum of bodies in the Solar System is not constant or conserved is that during their genesis, their precursors descended from a specific mother, and underwent some turbulence that applied a non-null torque to them and these precursors were moving in environments where they lost energy and took different trajectory paths, which affected several physical variables. I better elaborated on this in the chapters on rotational angular speed and semi major axis. Because previous theories of the formation of the universe did not properly appreciate what happened at the beginning, by trying to guess, their authors ended up thinking that angular momentum must be conserved. If they had understood how speed, rotation, and mass were acquired by the bodies, the authors of these theories could have realized that angular momentum cannot be conserved particularly for all bodies in a system.

The fact that despite the lack of conservation of the rotational angular momentum, the mainstream theorists of the origin of the Solar System continue to use some theories that assume a conservation of angular momentum is one of the proofs of their rebellion, stubbornness, and persistence in the wrong direction. Their wandering in wrong directions has caused the scientific community to ignore reality, but to keep going deeper in useless pursuit despite evidences everywhere which are more than enough to cause them to stop and better think about their methods and much more their life! For the lack of conservation of the rotational angular momentum alone should have been enough for scientists to reject the hypothesis according to which the Solar System started with a gas which was on a "flat" surface of a disk! That is NOT how it started and we can never prove it that way unless more useless and wrong theories about precursor migration and collision are added to the mess. Collision or migration can never prove the shift from a conserved rotational angular momentum to a non-conserved angular momentum! Seeing how the rotational angular momentum of the satellites varies according to their primary planet and their turbulence zone, it is probable that the rotational angular momentum of the electron is not a constant, but a variable which could depend on the type of chemical elements. Consequently, many errors are made in nuclear physics and all the scientific domains relying on measurement of the proprieties of particles. I will revisit these topics later.

12.11. Take home message

During the split-gathering of the precursor of a system of bodies, more than 99.9% of the 9 system-additive variables (mass, orbital angular momentum, orbital moment of inertia, rotational angular momentum, rotational kinetic energy, rotational moment of inertia, total kinetic energy, translational kinetic energy, and volume) were pushed into the precursor of the primary body and less than 0.1% was pushed into the precursors of the secondary bodies. As the fluids in the aforementioned 0.1% started flowing, different precursors of secondary bodies were born. The precursor of some daughter secondary bodies went through their own breakup and gathering together by still generally following the 99.9% and less than 0.1% rule of the amount of the 9 system-additive variables that they pushed into their primary bodies and secondary bodies. Below, I used the precursor of the Solar System to illustrate what I said above:

- During the split-gathering of the precursor of the Solar System, 99.9% of the value of the 9 system-additive variables went into the precursor of the Sun and less than 0.1% went into the precursors of the bodies orbiting the Sun.
- During the split-gathering of the precursor of the bodies orbiting the Sun, more than 99.9% of the 9 system-additive variables went into the precursors of the planetary systems and less than 0.1% went into the precursors of the asteroids.
- More than 99% of the 9 system-additive variables of the planetary systems were accumulated in the 4 giant planetary systems; and the Jovian and Saturnian planetary systems together account for more than 95%.
- During the split-gathering of the precursors of the planetary systems, more than 99% of each of the 9 system-additive variables went into the precursors of the primary planets and less than 1% went into the precursors of their satellites.
- During the split-gathering of the precursors of the satellites, more than 99% of the 9 system-additive variables were accumulated in their most turbulent zone, usually Zone 3.
- Most of the 9 system-additive variables of the precursors of the satellites were concentrated in a few satellites.

'Science180 Academy' Success Strategy
SCIENCE180 PUBLISHING: AUTHORS WANTED

Science180 Publishing, the American publishing company that published the groundbreaking discovery about the origin of the universe, of life, and of chemicals spearheaded by Dr. Nathanael-Israel Israel, really wants to publish your book(s) regardless of your field of expertise. This is a unique opportunity for:

- established authors
- people aspiring to become authors
- people who have written a book or are wanting to write one and need help with anything regarding publishing
- people who are not well known, inexperienced
- people whose books are viewed as nonconformist, controversial, or unconventional
- people who do not have enough resources or knowledge to navigate the publishing process
- people who are struggling to find an affordable, experienced, and high-quality publisher

Although Science180 Publishing is based in the USA, it can publish your books within your budget regardless of your geographical location. Science180 Publishing is highly interested in your document and possibly helping you publish it. Please visit Science180Publishing.com to explore how we may assist you. No matter the content of your book, as far as it is original, not promoting anything illegal, not duplicating anyone else's idea, Science180 Publishing can help you publish it in the USA. Please contact us asap and see how we can help.

To start your journey of publishing your book with Science180 Publishing, please visit Science180Publishing.com today.

CHAPTER 13

THE MAIN REASONS I DID NOT AND YOU SHOULD NOT TREAT THE SATELLITES IN THE SOLAR SYSTEM CARELESSLY … IF YOU REALLY WANT TO ACCURATELY CRACK THE CODE OF THE UNIVERSE

13.1. Definition

Now that I am about to start specifically talking more about satellites in the Solar System, I felt like I needed to present their names and turbulence zones one by one. Without a careful study of the relationships existing between the satellites according to their types, it could have been impossible for me to scientifically unearth and explain some secrets about the formation of the universe. Although satellites are generally known as bodies orbiting planets, it is important to note that some asteroids also have their own satellites.

The satellites mentioned in this chapter are mainly those officially confirmed by the International Astronomical Union (IAU) as of 2020 to orbit the planets in the Solar System. Although since 2005, some people do not acknowledge Pluto as a planet anymore, I considered it as a planet in my writings, mainly because I was interested in doing so instead of throwing its data out. I also believe that very soon, Pluto will regain its status of a planet. Although satellites are also called moons, I prefer using the word "satellite" to avoid confusing it with the Moon, the only satellite of the Earth.

Before I continue, I need to also define planet. Indeed, the definition of a planet has been one of the most controversial topics in astronomy. For instance, since its discovery until 2005, Pluto was considered a planet. However, after the discovery of other celestial bodies which sizes are closer to that of Pluto, a lot of debates have taken place and, as of today, Pluto is not considered a planet by most astronomers. Nevertheless, in my writings, I considered Pluto as a planet. According to the IAU (International Astronomy Union), a planet is a *"a celestial body that (a) is in orbit around the Sun, (b) has sufficient mass for its self-gravity to overcome rigid body forces so that it assumes a hydrostatic equilibrium (nearly round) shape, and (c) has cleared the neighborhood around its orbit"*. As of 2016, a minor planet is an *"astronomical object in direct orbit around the Sun that is neither a planet nor exclusively classified as a comet"*. They can be dwarf planets, asteroids, trojans, centaurs, Kuiper belt objects, and other trans-Neptunian objects. As of 2016, the orbits of 709,706 minor planets were archived at the Minor Planet Center. As soon as a celestial body is discovered, it is automatically assigned a provisional designation. When the orbit of a body is secured, that body is given a formal minor planet designation which has 2 parts:

- a number, historically given in approximate order of discovery, now assigned only after the orbit has been secured by 4 well-observed oppositions and

- a name such as that assigned by the discoverer or the provisional designation.

13.2. Satellites of the asteroids and planets in the Solar System

The literature mentioned the existence of hundreds of satellites orbiting asteroids in the Solar System. Satellites are found around many asteroids in the main belt, several Trans-Neptunian objects, Kuiper belt objects, Apollo asteroids, Amor asteroids, Centaurs, and Jupiter trojans. However, because they were not studied much, I did not spend a lot of time exploring them. The first satellite discovered around an asteroid was Dactyl and it orbits 243 Ida, a main asteroid belt.

From this point onward, I discussed only the satellites of the planets. Since antiquity, human beings have been intrigued at knowing more about the celestial bodies in the sky. The Moon is the most obvious body in the night sky. With time, many other satellites were discovered, beginning by the work of Galileo, who opened the door of the exploration of satellites around planets. Indeed, using his newly made telescope, Galileo discovered the 4 biggest Jovian satellites in 1610. Afterward, it took 45 more years before Titan, the biggest Saturnian satellite, was discovered. By the end of the 19th century, only 21 satellites were discovered in the Solar System. In the entire 20th century, 47 satellites were discovered. However, in the first 20 years of the 21st century only, 140 new satellites were discovered in the Solar System making the total number of satellites discovered (as of 2020) since the time of Galileo to 210. New satellites continue to be discovered, but most of the new ones are smaller usually less than 2 km of radius. As I was about to publish this book, I learned in February 2025 that new satellites were discovered around Jupiter and Saturn, and the number will keep growing until a "complete" list will be made. Indeed, at the time I began working on the bodies in the Solar System in 2013, only 178 natural satellites were known to orbit planets in the Solar System. As of 2016, as I finished the first draft of my preliminary statistical analysis, that number did not change. However, 32 new satellites in the Solar System were discovered from 2016 to 2019, causing me to rewrite certain chapters related to satellites. Therefore, as of 2020 when I wrote this chapter, 210 satellites were discovered around planets in the Solar System and they are distributed as:

- Moon (Earth's satellite)
- 2 Martian satellites (orbiting Mars)
- 79 Jovian satellites (orbiting Jupiter)
- 82 Saturnian satellites orbiting Saturn)
- 27 Uranian satellites (orbiting Uranus)
- 14 Neptunian satellites (orbiting Neptune)
- 5 Plutonian satellites (orbiting Pluto)

No satellite is found around Mercury or Venus. About the time I published this book in 2025, it was reported in the news that more than 100 additional satellites were found in the Solar System, mainly around Saturn. This discovery does not change the core analysis in this chapter or in this book.

Although the Earth is known to have only one natural satellite (i.e. the Moon), it is possible that other natural bodies orbiting the Earth are still undetected because of their small size and remote distance from the Earth. For instance, certain bodies have been claimed to orbit the Earth, but these claims were rejected because some astronomers believe that bodies orbiting the Earth are some asteroids that the Earth may have captured and which got lost quickly. I don't personally believe in celestial bodies capturing others, for the process that formed the universe did not act that way. The satellites in the Solar System orbit their primary planets at various distances and the trajectory of these orbits are differently shaped.

13.3. Number of satellites according to their types and turbulence zones

As of 2020, about 51.9% of the 210 known satellites are found in turbulence Zone 5 (Fig. 45). In contrast, Zone 1 recorded the minimum number of satellites. The number of satellites in the turbulence zones increased as the semi major of the zones increased.

The satellites in each turbulence zone account for the following percentages of the total number of satellites in the Solar System:

- Zone 1: 7.14%
- Zone 2: 9.05%
- Zone 3: 13.33%

- Zone 4: 18.57%
- Zone 5: 51.90%

In other words, the satellites in Zones 1 and 2 account for 16.2%, while those in Zones 4 and 5 contain 70.48% of the total. The satellites in Zones 1, 2, and 3 account for 29.52%. Below, I present the details by zone (Fig. 46). The number of satellites in each zone depends on the types of satellites. Regardless of the types of satellites, the outermost zone has the highest number of satellites. With 62 satellites as of 2020, Zone 5 of the Jovian satellites contains more satellites than any other zone of any satellite system. As of 2020, after the Jovian satellites in Zone 5 come the Saturnian satellites in Zone 5 with 42 satellites. In other words, as of 2020, although the Saturnian satellites are more than that of any other planetary system, their number in Zone 5 as of 2020 is 20 less than that of the Jovian planetary system.

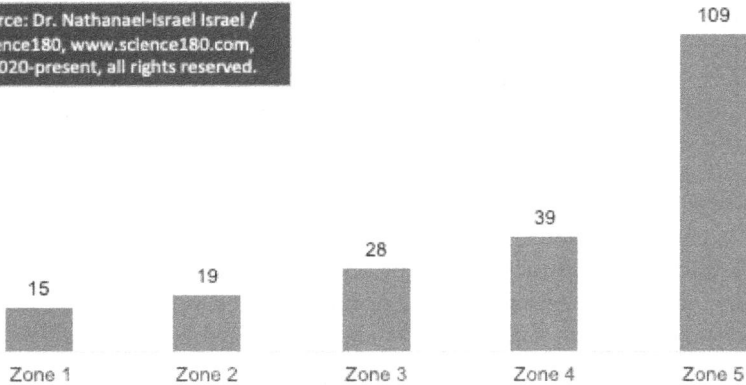

Fig. 45: Number of satellites according to the turbulence zones

Nathanael-Israel Israel: Who happens to be the World's #1 Authority on the Turbulent Origin of the Universe and Life

Fig. 46: Number of satellites according to their types and turbulence zones

■ Zone 1 ■ Zone 2 ■ Zone 3 ▪ Zone 4 ■ Zone 5

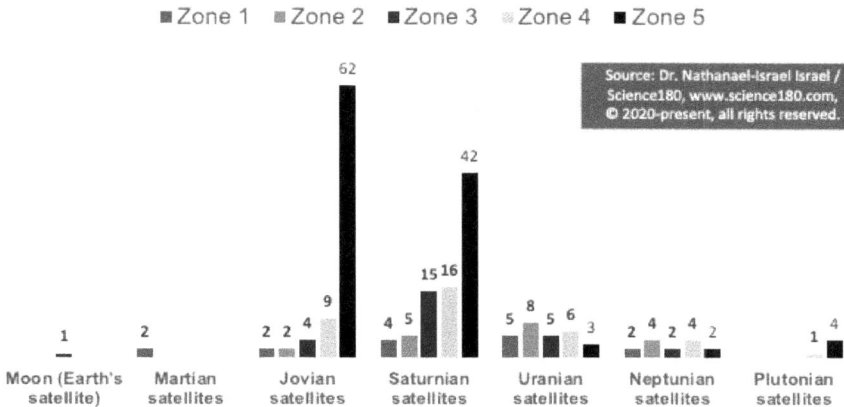

13.4. Historical facts about how on December 25, 2013, I was inspired to study the planetary systems and the satellite-to-planet ratio

At this point, I would like to recall a historical fact about the impact that the study of the satellites had on the methodology I used to unearth the secrets of the formation of the universe. Indeed, on December 25, 2013, while the entire world was celebrating Christmas, I was inspired that, to properly analyze the bodies orbiting the Sun, I needed to consider the planetary systems like "satellites" orbiting the Sun just as I analyzed the satellites of the planets with respect to their primary planets. It also appeared to me that day that, just as I would be thinking about the precursor of the Solar System to try to understand the relationship between the planetary systems and the Sun, so also, I would need to think about the precursors of the planetary systems in order to understand the relationship between the primary planets and their satellites. Toward that end, I was later led to calculate the ratio between the satellites and their primary planets and compare that with the ratio between the planets and the Sun. In those days, I had not even collected any data on the bodies in the Solar System yet. I was just relying on my intuition and the flow of ideas in my mind. But as I kept investigating the variation of the variables I was studying across the planetary systems, I realized some trends, which at that time, made me to start unraveling some underlying laws affecting the distribution of the variables. Little by little, the comparative study of the planetary systems helped me to better see the potential root cause of the variation of some variables from one planetary system to the other. To put it another way, if I did not have the early intuition to analyze the variables of the satellites all together and also separately according to the types of satellites, and then compare their trends across the planetary systems, I would not have been able to notice significant similarities and differences among the satellite systems and then expound my thinking to the scale of the entire Solar System and beyond. As you will notice in the rest of my writing, many relationships were not significant when all satellites were analyzed all together, but when they were separated by types, strong relationships were found. The strategic consideration of the turbulence zones while analyzing the satellites also tremendously increased my discernment about how the bodies could have been formed. This early discovery also caused me to divide the asteroids into different types and to look for trends among them. By doing so, I was also able to pinpoint interesting facts (about all types of bodies) which for instance explain why the main belt asteroids were not gathered together into a planet or one single bigger asteroid, but were scattered. By the time I finished analyzing the entire matrix of data according to the types of bodies, it was easier for me to see the trends and underlying laws which, added together, led to what I termed "Science180 Cosmology".

13.5. Take home message

Although many asteroids have their own satellites, in this book, when I talk about satellites, I am referring to the satellites of planets. When I began working on the celestial bodies in the Solar System in 2013, only 178 natural satellites were known to orbit planets in the Solar System. However, 32 new satellites in the Solar System were discovered from 2016 to 2019. Therefore, as of 2020 when I wrote this chapter, 210 satellites were discovered around planets in the Solar System and they are distributed as:

- Moon (Earth's satellite)
- 2 Martian satellites (orbiting Mars)
- 79 Jovian satellites (orbiting Jupiter)
- 82 Saturnian satellites (orbiting Saturn)
- 27 Uranian satellites (orbiting Uranus)
- 14 Neptunian satellites (orbiting Neptune)
- 5 Plutonian satellites (orbiting Pluto)

As of 2025, new satellites were discovered in the Solar System, but my math in this book will be based on the aforementioned 210 satellites. No satellite is found around Mercury or Venus. Using characteristics of the satellites, I was able to define the turbulence zones, which was crucial for my decoding of the turbulence secrets behind the formation of the universe. For instance, the comparison of the relationship between the variables I studied from one planetary system to the others significantly helped me to understand key rules surrounding turbulence. The radius, the rotational speed, and the semi major axis of the planets affected the number of their satellites. As of 2020, about 51.9% of the 210 known satellites are found in turbulence Zone 5. In contrast, Zone 1 recorded the minimum number of satellites. The number of satellites in the turbulence zones increased as the semi major of the zones increased. The satellites in each turbulence zone account for the following percentages of the total number of satellites in the Solar System:

- Zone 1: 7.14%
- Zone 2: 9.05%
- Zone 3: 13.33%
- Zone 4: 18.57%
- Zone 5: 51.90%

'Science180 Academy' Success Strategy
SCIENCE180 MASTER CLASS

Hear the greatest scientific and philosophic lessons from top scientists, philosophers, thinkers, and public figures who have realized historic mistakes they made in life (concerning the origin of the universe, life, and chemicals), and that they corrected thanks to the historic discovery of Nathanael-Israel Israel, the world's first 180Scientist who founded Science180 and who is known as the one who truly decrypted the universe-origin for the first time. In their own words, these renowned personalities share with the world key lessons they have learned in life and how people can learn from their experiences to improve lives instead of repeating their mistakes that many people still ignore at their own perils. To learn more, contact us at Science180.com/contact.

CHAPTER 14

THIS SCIENTIFIC ARGUMENT ABOUT THE "TURBULENT MULTIPLIERS OF THE 9 SYSTEM-ADDITIVE VARIABLES" BECOMES THE LATEST THAT IS TROUBLING THE ONCE-POPULAR UNIVERSE-ORIGIN THEORY

In the literature, a random multiplier is also called a fractional with respect to a variable of interest. For instance, a "random multiplier with respect to volume" can also be called "fractional with respect to volume". A random multiplier is one of the key variables used to explain fragmentation or breakup of fluids. Therefore, as I studied turbulence in the universe, I deeply felt that I must explore how the precursors of the celestial bodies may have been fragmentated to yield the various bodies present in the universe.

After studying the random multipliers for 2 years, I noticed that although they are extensively used and accepted in physics and even in other disciplines to describe fractional of quantities, I became very uncomfortable with them, particularly the adjective "random". Indeed, as my understanding of the origin of the universe was increasing, I realized that the multipliers or fractional of the variables were NOT random, but very precise and governed by meticulous laws beyond chance or randomness. I therefore was extremely uncomfortable to keep qualifying these multipliers as random, for saying so seemed to me as a proof of a major ignorance or a major unconsciousness. It is the inability of human beings to fully unravel the systematic approach used to design and implement the distribution of these multipliers that has caused scientists to label them as "random". Because I am the first human being to unravel the mysteries behind these fractional of variables from the perspective of the mother of all turbulences that prevailed at the origin of the universe, I could not continue to call them random while I know that they follow a very organized pattern. Hence, I chose

Nathanael-Israel Israel: Original, Universal, Accurate Universe-Origin Decoder

to call them "turbulent multipliers".

In each system, the turbulent multiplier with respect to a variable was calculated by dividing the value of that variable for the body by the sum (total) of the same variable for all the bodies in its system. By tracing the sequential cascade of breakup of the precursors, I was able to decode how mother precursors were fragmentated into daughter bodies and so on so forth. I realized that the precursors of the celestial

> Because I am the first human being to unravel the mysteries behind these fractional of variables from the perspective of the mother of all turbulences that prevailed at the origin of the universe, I could not continue to call them random while I know that they follow a very organized pattern. Hence, I chose to call them "turbulent multipliers"

bodies were not fragmentated by chance, but according to laws, some of which I found by investigating the 9 system-additive variables. Once I discovered some of those laws, I became uncomfortable using the term "random" multiplier. By the time I finished studying most of the variables, I coined the term "turbulent multiplier"; for I realized that the process of the formation of the universe was not random or by chance. In other words, I used the adjective "turbulent" because I discovered them by studying turbulence. Subsequently, I invented 9 turbulent multipliers of the 9 system-additive variables:

1. Turbulent multiplier of mass
2. Turbulent multiplier of orbital angular momentum
3. Turbulent multiplier of orbital moment of inertia
4. Turbulent multiplier of rotational angular momentum
5. Turbulent multiplier of rotational kinetic energy
6. Turbulent multiplier of rotational moment of inertia
7. Turbulent multiplier of total kinetic energy
8. Turbulent multiplier of translational kinetic energy
9. Turbulent multiplier of volume

Here, I will present a few aspects of them. Indeed, the turbulent multiplier with respect to all bodies in the Solar System aims at seeing how the precursor of the Solar System split into all the bodies in the Solar System. In contrast, the turbulent multiplier with respect to all the bodies orbiting the Sun aimed at exploring how the precursors of the bodies orbiting the Sun were split into planets, asteroids, and satellites after separating from the precursor of the Sun. The turbulent multiplier with respect to all bodies in a planetary system helped me to get more insight into how the precursor of a planetary system was split into a primary planet orbited by its satellites, if any. Finally, the turbulent multiplier of the satellites with respect to all satellites in a planetary system was used to explore how the precursor of the satellites (in a planetary system) was fragmented into its satellites. Studying these turbulent multipliers from the perspective of the aforementioned 9 variables allowed me to have a more intimate understanding of the split-gathering of the precursors of the bodies and how the scale, reference or perspective of the study can affect the outcome. I initially devoted more than 100 pages to these multipliers, but for the sake of space and relevance, I removed them.

To read the rest of this chapter and discover how the turbulent multipliers can sharpen your understanding of the universe-origin and future research, visit www.Science180.com/TurbulentMultipliers.

CHAPTER 15

DOES THE ORBITAL SPEED OF CELESTIAL BODIES HAVE ANY SCIENTIFIC POWER TO IRRESISTIBLY BRING DOWN WRONG COSMOLOGICAL THEORIES... EVEN WHEN ALBERT EINSTEIN OR ISAAC NEWTON MAY DISAGREE?

15.1. Introduction

Although a mere human eye or the most sophisticated scientific equipment in the world cannot perceive all movements in the universe, everything in nature is moving. From the electrons to the planets, satellites, asteroids, stars, planetary systems, and galaxies, everything in the universe is changing position. Some movements are easily detectable, whereas others can be noticed only by using sophisticated equipment. Although some objects in nature are very remote from Earth, scientists have tried to measure the speed of some of them.

The orbital movement can be defined as the movement of an object around a primary object. In the Solar System for instance, the movement of the planets and asteroids around the Sun is an orbital movement. For example, the Earth orbits the Sun once every year. In contrast, satellites orbit their primary planets. During their orbital movement, also called revolution, planets and asteroids travel at different speeds. The orbital speed is the speed of a celestial body when considering its orbital movement. The orbital speed of celestial bodies is not constant, but varies with their position in orbit. There is a point in the orbit where the orbital speed is at its minimum and another point where it is at its maximum. The mean orbital speed is obtained by averaging all the speeds recorded during the orbital movement. It is also the ratio of the distance traveled by the object divided by the time it takes to finish one orbit.

Indeed, the perihelion is the closest point of approach of a celestial body orbiting a primary body. For the planets and asteroids in the Solar System, the perihelion is the point in space where they are the closest to the Sun. For example, the perihelion of the Earth is the point in space where the distance between the Earth and the Sun is minimum. Similarly, the perihelion of a satellite is its closest point to its primary planet. For instance, the perihelion of the Moon is the point in space where the distance between the Moon and the Earth is at its minimum. In contrast, the aphelion is the most remote (farthest) point of approach of a celestial body orbiting a primary body. Therefore, the aphelion of the planets and asteroids in the Solar System is their farthest point in space from the Sun. Similarly, the aphelion of the satellites is the farthest point to their primary planet. In general, celestial bodies reach their maximum orbital speed at their perihelion, whereas they reach their minimum orbital speed at their aphelion. Between the aphelion and the perihelion, the orbital speed changes. It increases from the aphelion to the perihelion, whereas it decreases as a celestial body gets farther and farther away from the primary body it is orbiting. Although data on maximum and minimum orbital speed exist for the planets, it is not available for asteroids and satellites. Therefore, unless otherwise specified, the orbital speed I dealt with in this chapter is about the mean orbital speed. Below, you will discover why the orbital speed of celestial bodies is a crucial key to properly decrypt the universe-origin.

15.2. How the fluid layers in the precursors of bodies acquired their speed and transferred it to their daughter bodies

Rheological reports (Malvern, 2016) showed that, "in a shear flow, fluid layers slide over one another with each layer moving faster than the one beneath it. Due to an external force termed a shear stress which acts over the surface of the fluids, the fast movement of the upper layer generate a displacement gradient termed shear strain across the fluid, which can increase during the duration of the applied stress. The shear stress contributes to the transfer of momentum across the fluid layers through their interactions".

Like I have been explaining, during the split-gathering of the precursors of bodies, fluid layers were stacked one on top of the other and slid one over the other, with each layer moving faster than the one beneath it. Consequently, in the precursor of each system of bodies, the uppermost layers had the maximum speed, while the bottom layers had the smallest speed. The energy with which the mother precursors "pushed" away their daughters could have been one of the main forces which communicated a shear force to act on the fluids of the daughters and consequently, a shear flow could have started to take place. Because the daughter bodies occupied a certain portion of space (which was changing as the bodies were being molded), the forces (e.g. external forces) which acted on them over a certain unit area could have changed as the formless daughter precursors were being shaped. Consequently, the external forces which "propelled" or launched the daughter precursors into motion could have also taken the form of a shear stress as the precursors were being pushed or wrapped together by the processes that molded them. At the same time, the top layers were also acted on the layers that were beneath them. Indeed, a stress generally expressed the amount of force applied per area of an object. In the case of the celestial bodies, in addition to the force that propelled or pushed them away from their mothers, some fluid layers were subdued to the weight or amount of fluids laid on top of them. In other words, from the top layer to the bottom layer, the stress due to the weight increased and caused the bottom layers to be under more pressure. Like I will explain later, this stress affected certain characteristics of the bodies born from the bottom layers.

Strain is the change in form or shape of a body after a stress is applied to it. If the stress is not strong enough, no strain can occur. But if the stress is strong, the affected body can be deformed, crushed, broken, squeezed, etc. In the case of the precursors of celestial bodies, factors such as the speed of the fluid layers could have affected the stress applied to the vortical structures formed in them. For instance, in response to the force acting on them, the top layers could have moved at a given distance "x", while the bottom layers could have moved at a distance smaller than "x". Therefore, a displacement gradient could have existed across the layers according to the depth or position with respect to the entire clusters of layers of their mother precursor being split-gathered. This displacement gradient birthed a shear strain. Because the precursors of the bodies reached a point where they were like a fluid which constituents or components (e.g. precursors of subatomic particles, atoms, minerals, etc.) were moving relative to one another, the shear strain could have continued to increase in some conditions and decrease in others for the period of the applied stress until the layers split from one another, therefore freeing the daughters bodies from their mothers and later completing their formation under the influence of their environment and the characteristics they inherited from their mothers. By then, a velocity gradient could have been created and which can be explained by a shear rate or strain rate that I addressed in the next chapter. In fact, I discovered the existence of a shearing of the precursors when I was studying the derivatives of the orbital speed of the bodies with respect to their semi major. These derivatives gave me a clue into the shear rate of the layers in the precursors, which is the change of velocity of one fluid layer over its adjacent layer. Since the formation of the bodies, their systems have expanded of course, meaning that the shear rate deducted from the position and speed of the bodies as today is also impacted by the expansion of the universe and the systems of bodies it contains. I suspected that the shear rate or strain rate could be explained by the rate of change of the strain with time. As I carefully studied the speed of the celestial bodies and the distance separating them, I had a clue into these rates of change which occurred when the precursors of the celestial bodies were being molded into its daughter bodies.

From the top fluid layers to the bottom ones, the stress varies in most cases, and could have increased as the depth of the layers increased as well. As shear stress was applied to a fluid of the precursors of the bodies in the universe, momentum was transferred regardless of the size of the systems going through this process. According to their speed and mass, the moving fluid layers had different momentum, therefore giving rise to a momentum flux that can be related to the rate of change of the momentum in "horizontal" layers with respect to a direction perpendicular to the direction of the flow of the fluids in the precursors. The shear stress of the fluids in the precursors of the bodies could have been like a momentum flux or a rate of momentum transfer to fluid layers through the interactions with other fluid components. These interactions between fluid layers and the structures being formed in them could have reduced the speed of the fluid of the precursors and also affected their kinetic energy. Part of the energy communicated to the fluid layers was also used to form different bodies inside the evolving layers and the momentum was being transferred as well. During these processes, part of the available energy was also used to "compress" or increase the density of some bodies or systems of bodies being formed. The relationship, including the proportionality, between the shear and shear rate could have defined the shear viscosity or the dynamic viscosity, which can be viewed as a measure of the friction of the fluids inside each of the precursors of the bodies being shaped while many variables (including the kinetic energy, density, size, pressure, temperature, etc.) were dynamically changing in the bodies or systems of bodies being split-gathered. Later in this book, I delved into the impact of viscosity of many other variables related to the bodies.

To summarize, as the precursors of the systems of bodies were being split-gathered, they and their daughter bodies were organized into layers moving at different speeds. The bottom layers moved slowly, while those on top moved more rapidly. The speed of these movements depended on the position of the bodies. The orbital speed of the bodies as of today is a footprint of the speed with which their precursors were moving in the layers of fluids from which they were born. The turbulence that the precursors of the bodies went through shaped or communicated to them an orbital and rotational movement under the influence of the push they received from their mothers. In other words, as the precursors of the primary bodies started flowing, the turbulence they went through and the reorganization or rearrangement that their internal constitutive bodies went through affected their speed. For instance, because of the size of the precursors of the primary bodies, their layers did not flow at the same speed as they were being gathered together to form the primary bodies. Consequently, the speed of the primary bodies could have been slowed down a little bit as they were being gathered together as a whole. Put another way, although the precursors of the primary bodies did not split into other bodies, due to their size, their fluids were also organized into layers with the top layers probably being positioned at the north pole and the bottom layers of the bodies becoming the portion of the primary bodies located at the south pole. But because the primary bodies formed unified bodies, their speed was a balance between the orbital speed of the layers of fluids in their precursors. Hence, in the end, the orbital speed of the primary bodies is not always the highest of all the bodies in their system. At the same time, the fluid layers of the precursors of the secondary bodies moved at different speeds as they were being split from one another. The layers of the precursors of the secondary bodies which were beneath moved more slowly than those of the secondary bodies located in the top layers. As I explained later, the intensity of the turbulence in the turbulence zone affected also how the speed of the biggest secondary bodies were defined and why the value of their speed is close to that of the primary bodies.

Because after their split from their mother precursors, some precursors had to go through additional split(s), the orbital speed of their daughters was also affected by their position with respect to the other layers of fluids they were born from. For instance, after the precursors of the planetary systems split from the bulk of the precursors of the bodies orbiting the Sun, some of these precursors had to go through another split to yield the precursors of their primary planets and the precursors of their satellites. The speed of the precursors of the primary planets and that of the satellites is affected by the characteristics of the precursors of the planetary systems. Consequently, the speed of the primary planets is defined by their position with respect to the primary star in their stellar system, while the speed of the satellites is defined by their position with respect to their primary planet. Although what I explained above applied for the orbital speed and for the rotational angular speed, other factors were evolved which made the bodies in Zones 4 and 5 to have a rotational angular speed which followed another trend that I will explain in a subsequent chapter.

The fluids in the precursors of the bodies being formed were also compressed, meaning that the physics of incompressible fluids cannot properly explain what happened during the formation of the bodies in the universe. Unfortunately, the physics of compressible fluids which is better suited to explain the formation of the bodies in the universe is still embryonic as many things in that field are still misunderstood since the days of Euler, who is one of the pioneers that laid the old-classical foundations of such a domain. When I tried to turn to modern mathematics of complex systems, I was bitterly disappointed by its heavy equations and models, which failed to address reality, while, in the name of conventional modeling empowered by "sophisticated computational power", which amplifies the confusion of the misunderstanding of the initial complexity, which existed and has been maintaining all things together since the beginning of the universe. To make a long and multifaceted story short, I had to apply a special form of critical thinking to strategically invent methods to figure out and then explain the formation of the complex systems in the world from a realistic perspective, which many experts have failed to perceive because of their way of thinking and interpreting scientific and philosophical data. The physics behind "Science180 Cosmology" that I developed goes beyond the conventional physics

207

Science180: Understand the Origin of The Universe. Increase Your Glory and Peace of Mind of Newtonian and Non-Newtonian fluids as well as the traditional physics of compressible and incompressible fluids, which mainly turn around viscosity and density.

The understanding of the complexity of all the phenomena that occurred in the fluids of the precursors as they were being molded into their daughter bodies requires a broad and deep knowledge of many scientific disciplines. Hence, before the days of Science180 (www.Science180.com) that I forged, it was difficult to comprehend the origin of the universe and "no conventional scientist" has ever grasped its turbulent complexity. In the description I provided above, I did not even address how some daughter bodies were "flipped over" as they were "pushed" or "ejected" from their mother precursors and relative precursors; nor did I even mention how the bodies were positioned, and many more details I reserved for other chapters later so that I do not overload and complicate the account of a complex story, which requires to have an eye on many things at once. Based on the trends I recorded on various bodies in the universe, what I am describing here can also apply to all scales of systems of bodies, from the subatomic and atomic systems all the way to the galactic systems passing by the planetary systems, the satellite systems, asteroid systems, stellar systems, and any other clusters or systems of bodies in the universe. In the next sections, I elaborated more specifically on how the celestial bodies in the Solar System (e.g. Sun, asteroids, planets, and satellites) got their orbital speed.

15.3. Orbital speed of the Sun

I will address the orbital speed of the bodies in the Solar System by starting with that of the Sun. Indeed, despite being the largest body in the Solar System, the Sun is not the fastest body. Indeed, the Sun's speed is usually measured with respect to other stars. When measured with respect to nearby stars, the Sun's relative speed to nearby stars is estimated at 19.4 km/s. Some literatures claim that the Sun orbits around the center of the Milky Way at 220 km/s. Others also claim that the velocity of the Sun relative to the so-called cosmic microwave background is 370 km/s. However, I did not focus on the last two orbital speeds, but only on the 19.4 km/s. In other words, in my writing, unless I specified otherwise, when I talked about the orbital speed of the Sun, I was referring to the Sun's relative speed to nearby stars (19.4 km/s).

The accuracy of the speed of the Sun can be affected by the level of accuracy of distance measurements between stars. This means that any error in current estimation of the distance between stars could affect the perception of the speed of the Sun. Because the estimation of the distance between stars is currently based on and/or related to the speed of light, any errors in the assessment of the speed of light could also affect the estimation of the speed of the Sun and that of other stars. Therefore, because it is impossible to know for sure that the speed of light is well known and that the distance between stars is accurately calculated, I felt like it would be better to consider the calculated speed of the stars cautiously.

Consequently, I did not find it fitting to use any of the other speed calculated for the Sun. For scientific methodologies used to measure the size and speed of some bodies located very far from the Earth may be inaccurate. I did not feel comfortable using a Sun's speed of 220 km/s or 370 km/s while the distance to the referencing bodies of those measurements can be unreliable. I particularly believe that the speed of light used to estimate some of those distances is not even a constant and the more remote some objects are, the more the likelihood that their size and speed can be overestimated.

15.4. Orbital speed of the asteroids and planets in the Solar System

Before I detail the orbital speed of the asteroids and planets in the Solar System, I would first like to specify how they could have gotten their speed. Indeed, as the precursor of the Solar System was being organized into layers of the precursors of the planetary systems and of the asteroids, the bottom fluids layers were slower than those located on top. Because the fluids in the top layers of the precursor of the bodies orbiting the Sun were what became the innermost bodies in the Solar System, the orbital speed of the innermost bodies in the Solar System are higher than that of the outermost most bodies (which were born from the bottom fluid layers of the precursor of the bodies orbiting the Sun). Based on the bodies I studied in the Solar System, Mercury is the innermost body orbiting the Sun consequently, it came from the top fluid layer of the precursor of the bodies orbiting the Sun. Consequently, its speed is the highest among all the bodies orbiting the Sun. However, due to the difficulty to detect smaller bodies located between the Sun and Mercury because of the strong solar radiation, I cannot rule out that other bodies in the Solar System are closer to the Sun than Mercury. In other words, as the exploration of the Solar System continues, other innermost asteroids and even planets may be found with an orbital speed higher than that of Mercury.

In the end, because the asteroids and the planets in the Solar System directly orbit the Sun, their orbital speed is directly related to their distance from the Sun, which is generally expressed by the semi major axis. Although the position of celestial bodies as of today is not the position of their precursors, semi major axis should be positively correlated to the distance separating the position of the mother precursors and the layers of its daughter bodies after their split. In other words, the precursors of the innermost bodies were closer to the top layers, while those in the outermost bodies were also closer to the bottom layers. Because of the expansion of the universe, I did not bother to estimate the distance at which their precursors could have been located. However, based on the data that I studied, I think that no system of bodies in the universe started from a single point as some singularity theories try to paint it.

In the Solar System, I studied the orbital speed of 461 bodies and it ranges between 0.123 km/s and 47.36 km/s. The highest orbital speed was recorded with Mercury, while the smallest was with Hydra, the outermost Plutonian satellite. The mean orbital speed of the planets of the Solar System ranges from 4.67 to 47.36 km/s, the lowest value was recorded with Pluto (the outermost planet in the Solar System) and the highest value with Mercury (the innermost planet in the Solar System). The mean orbital speed of the Earth is 29.78 km/s, which is 1.54 times the orbital speed of the Sun (19.4 km/s). The orbital speed of the Sun is smaller than that of the 4 innermost planets (Mercury, Venus, Earth, and Mars) but higher than that of the 4 giant planets.

Using the maximum and minimum orbital speed of the planets, I calculated the range of orbital speed for the planets and noticed that, the highest range of the orbital speed of the planets was observed with Mercury (20.12 km/s), whereas the lowest range was recorded with Neptune (0.13 km/s). This suggested to me that in the space that Mercury (the innermost planet in the Solar System) travels, something has been causing its orbital speed to vary a lot. Mars is the planet that has the second highest orbital speed range (4.53 km/s). In contrast, the orbital speed range of the other planets in the Solar System is less than 2.29 km/s, meaning that in general, the orbital speed of these planets does not vary a lot during their course on their orbit. For instance, from Jupiter (the 5th innermost planet in the Solar System) to Pluto (the outermost planet in the Solar System), the range of the orbital speed is less than 1.3 km/s. To put it in other words, the orbital speed of the outer planets does not change much during the course of their revolution around the Sun.

The mean orbital speed of the asteroids and planets combined together ranges between 0.17 km/s and 47.36 km/s. Comets recorded the smallest speed and they are followed by the Trans-Neptunian objects. Particularly, the orbital speed of the main belt asteroids varies between 15.95 km/s and 20.26 km/s, a range which encompasses the orbital speed of the Sun. The orbital speed of Ceres (the biggest main belt asteroid) is 17.93 km/s. Of the main belt asteroids that I studied, all of the 16 main belt asteroids which are upstream of 11 Parthenope have an orbital speed higher than that of the Sun (19.4 km/s). Fig. 47 illustrates the orbital speed of the asteroids and planets.

Fig. 47: Orbital Speed of the Asteroids and Planets in the Solar System

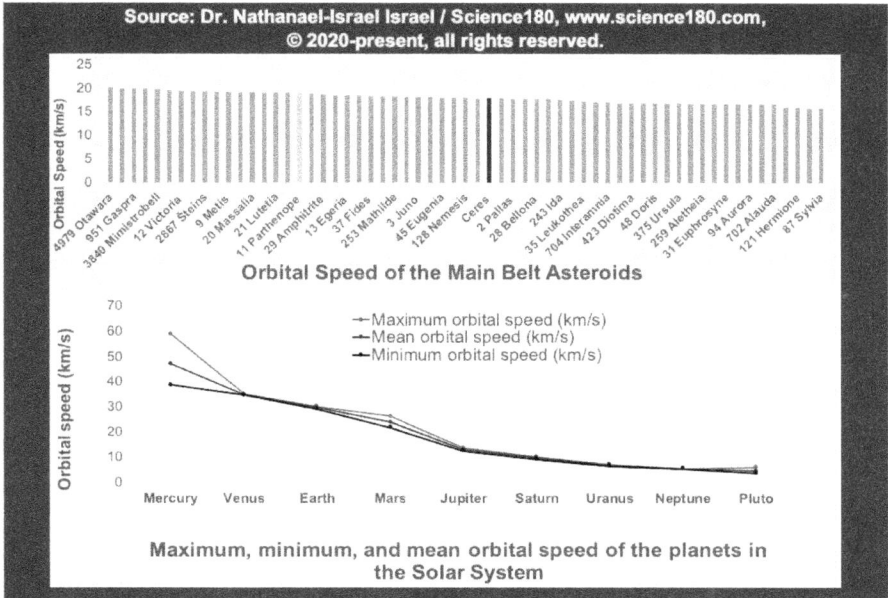

Orbital Speed of the Main Belt Asteroids

Maximum, minimum, and mean orbital speed of the planets in the Solar System

Some of the conclusions I stated above came to me as I was deeply investigating the relationship between orbital speed and semi major axis of various combinations of the bodies in the systems of bodies in the Solar System. For instance, when I combined the orbital speed of all the bodies orbiting the Sun (e.g. planets, asteroids, and satellites combined) before doing the regression, no significant regression was found between orbital speed and semi major axis ($r2<0.03$). Similarly, when I analyzed the planets and satellites together, the regression between the orbital speed and the semi major axis was not strong ($r2<0.13$). Similarly, when I analyzed all the satellites together, I did not find a strong regression either. In the sections on the orbital speed of the satellites, I presented the regressions involving satellites alone and satellites combined with planets.

I tested different models for the regressions between orbital speed and semi major axis of the bodies in the Solar System because the search for the best model to fit the speed of the bodies in turbulence has been one of the major questions regarding the modeling of turbulence. Indeed, for a long time, experts in the field of turbulence have been wondering which model (e.g. power, logarithmic, exponential, linear, etc.) better fits speed when turbulence is concerned. The next paragraphs focus on the trends I found for the speed of asteroids and planets. When I tested the asteroids and planets together, I found significant regressions between their orbital speed and semi major axis ($r2\sim1$; $p <0.02$) (Fig. 48). Among all the models I tested, the strongest coefficients of determination were found with the following functions:

- Logarithmic (r2=0.82)
- Inverse (r2=0.895)
- Power (r2=0.998)

Fig. 48: Regression between Orbital Speed and Semi Major Axis of the Asteroids and Planets in the Solar System
Source: Dr. Nathanael-Israel Israel / Science180, www.science180.com,
© 2020-present, all rights reserved.

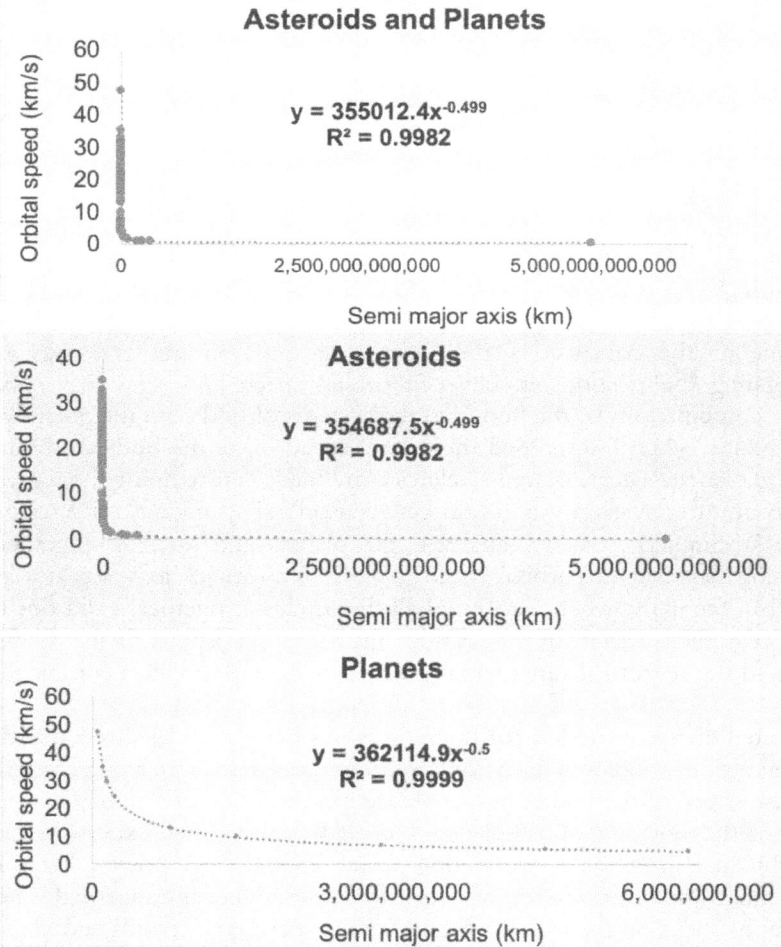

Asteroids and Planets

$y = 355012.4x^{-0.499}$
$R^2 = 0.9982$

Semi major axis (km)

Asteroids

$y = 354687.5x^{-0.499}$
$R^2 = 0.9982$

Semi major axis (km)

Planets

$y = 362114.9x^{-0.5}$
$R^2 = 0.9999$

Semi major axis (km)

Then, when I considered the asteroids altogether without adding the data of the planets, the trend is similar to the one obtained above, meaning a significant regression exists between their orbital speed and semi major axis. The power function gave the highest coefficient of determination (r2=0.998) (Fig. 48), followed by the inverse model (r2=0.918), and the logarithmic model (r2=0.83). Although significant, the other models are not very strong. Even when I did the regression after splitting the asteroids by their types (Amor asteroids, Apollo asteroids, Aten asteroids, centaurs, comets, Jupiter trojans, Mars trojans, Neptune trojans, Kuiper belt objects, main belt asteroids, and Trans Neptunian objects), significant models were also found between the orbital speed and the semi major axis. Except the linear model for the comets (r2=0.18), all these regressions were strong regardless of the model functions.

After consistently obtaining a strong regression for the regression for the asteroids no matter if I analyzed them together or by their types or by mixing all of them with planets, I then did another regression between orbital speed and semi major of the planets only. I also found a very strong correlation (r2>0.99) (Fig. 48).

15.5. General trends of the orbital speed of satellites

When I started investigating the orbital speed of the bodies in the Solar System, I heavily relied on the raw data available from the website of NASA (NASA, 2018). Unfortunately, the speed of most of the bodies I studied was not listed on that site. By July 14, 2016, when I drafted the first chapter on orbital speed, data on the orbital speed of the satellites was available in the literature for only 43 out of the 178 satellites known in the Solar System by then. Despite this limited data, I was able to understand that the mean orbital speed of these 43 satellites depended on their planetary system. I also quickly understood that I could not understand the mysteries behind the orbital speed of the satellites if I just mixed all of them together. In other words, I perceived that I must study the satellites according to their planetary system, and by knowing how their trends vary accordingly, I could deduct substantial secrets.

However, with the limited data available, I could not confidently state which satellite has the smallest or the highest orbital speed, for it can be one of those which data was missing. I hungrily searched for the missing orbital speed on other astronomical websites and reference websites including Wikipedia and many observatories (e.g. in Spain, France, Russia, Chili, etc.) but I could not find them. I read many articles but could still not find the answer. As I was analyzing other variables, I felt the need to estimate as many missing orbital speeds as I could. Therefore, using orbital period and the semi major axis, I estimated the missing orbital speed as:

Orbital speed = 2 * π * (semi major axis) / orbital period

I based this formula on the convention expression relating linear speed (V, in m/s) and orbital angular speed (ω, in rad/sec or s^{-1}) or orbital period (T, in s):

$$V = r * \omega = 2\pi * r / T$$

Here, although the orbit of the celestial bodies is not always a perfect circle, this formula gave an acceptable estimation and here, I considered radius as the semi major axis. I also used the same formula to estimate the missing orbital speed of asteroids. Without estimating the missing orbital speed of some bodies in the Solar System, it would have been very difficult to study some key variables indispensable for unearthing the laws of the formation of the universe. After estimating the orbital speed, I compared them with those already revealed in the literature including that of the 43 previously mentioned satellites. The estimation was found to be solid. Then, I started looking for the relationship between orbital speed and semi major axis of the satellites. When all of the satellites were combined together, the regression between the orbital speed and the semi major axis was significant but not very strong ($r2<0.52$).

However, when the satellites were analyzed according to their types (i.e. Martian, Jovian, Saturnian, Uranian, Neptunian, and Plutonian), very strong correlations ($r2\sim1$) were found regardless of the types of satellites, therefore pointing to the importance of separating the satellites according to their primary planets while dealing with their speed. Regardless of the types of satellites, the power function is the strongest best fit ($r2=1$) for the regression between orbital speed and semi major axis. The strength of the other equations depends on the types of satellites. For instance, for the Jovian satellites, although most of the other models are also very strong ($0.66 < r2 < 0.96$), the linear model has the smallest coefficient of determination. Similarly, the other models that illustrate the relationship between the orbital speed and semi major axis are fairly strong for the Saturnian satellites ($r2>0.8$), the Uranian satellites ($0.64 < r2 < 0.96$) and the Plutonian satellites ($r2>0.94$). For the Neptunian satellites, besides the linear model ($r2=0.56$), all of the other models are fairly strong ($0.75 < r2 < 0.95$). Fig. 49 illustrates the best fit models.

Fig. 49: Regression between orbital speed and semi major axis of the satellites according to their types

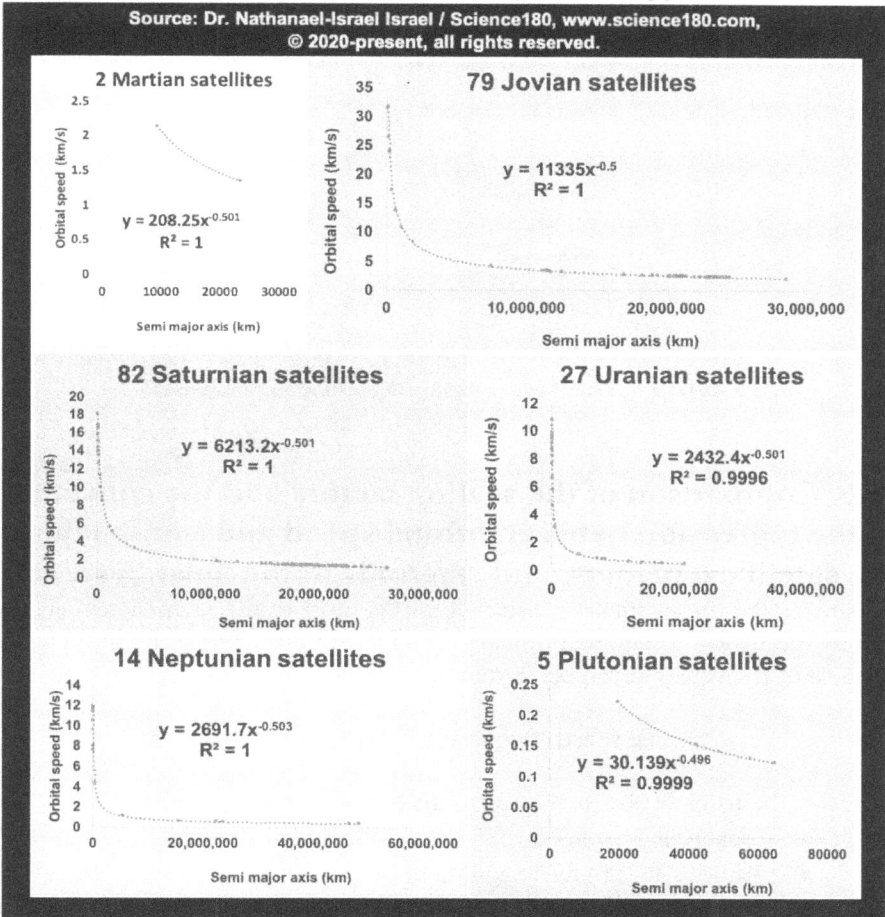

In Fig. 50, I combined the above graphs into a single one.

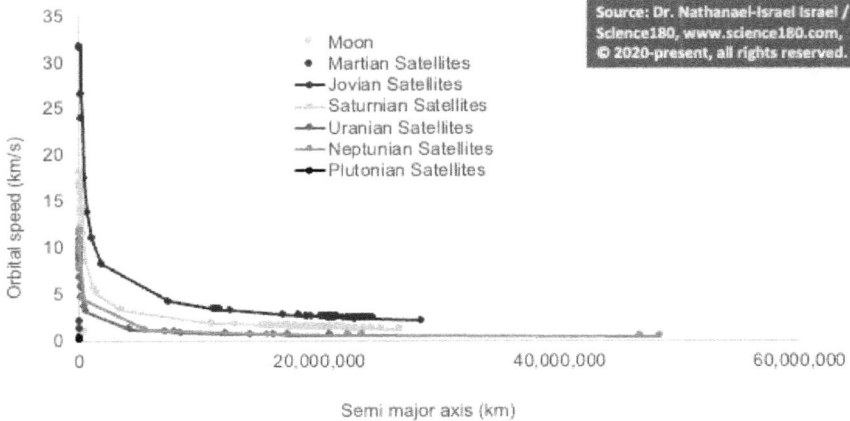

Fig. 50: Orbital speed of the satellites according to their types
and semi major axis

15.6. Comparison of the scaling factors and the constants of the regression between orbital speed and semi major of the satellites, planets, and asteroids in the Solar System

In general, the power function usually best fits most of the regressions between orbital speed (km/s) and semi major axis (km) of most of the bodies and the formula of the equation can be written as:

$$\text{Orbital speed} = \text{constant} * (\text{SemiMajor})^{\wedge}\text{exponent}$$

Using the formula above, the equations of the best regression are:
- Planets and asteroids: 355012.413*SemiMajor^-0.499
- Planets: 362114.897*SemiMajor^-0.5
- Asteroids: 354687.451*SemiMajor^-0.499
- Martian satellites: 208.147*SemiMajor^-0.501
- Jovian satellites: 11334.716*SemiMajor^-0.500
- Saturnian satellites: 6213.156*SemiMajor^-0.501
- Uranian satellites: 2431.135*SemiMajor^-0.501
- Neptunian satellites: 2690.372*SemiMajor^-0.503
- Plutonian satellites: 30.124*SemiMajor^-0.496

Where SemiMajor stands for semi major axis. Table 1 summarizes the parameters of the best regressions that I used in the next chapter to calculate the derivatives of the regression between the orbital speed and semi major axis of the bodies in the Solar System.

Table 1: Model Summary and Parameter Estimates for the Regression between Orbital Speed and Semi Major Axis of Celestial Bodies in the Solar System according to their Types

Dependent Variable: Orbital Speed (km/s) of bodies in the Solar System

Types of bodies	Equation	Model Summary					Parameter Estimates	
		R Square	F	df1	df2	Sig.	Constant	b1
Planets and asteroids	Power	0.998	138,797.72	1	248	0.000	355012.41	-0.499
Planets	Power	1	1,047,754.47	1	7	0.000	362114.9	-0.5
Asteroids	Power	0.998	129,065.67	1	239	0.000	354687.45	-0.499
Martian satellites	Power	1	.	1	0	.	208.147	-0.501
Jovian satellites	Power	9.9957E-01	176,977.73	1	77	3.43E-131	11334.716	-0.500
Saturnian satellites	Power	9.9989E-01	737,132.28	1	80	2.34E-160	6213.156	-0.501
Uranian satellites	Power	9.9981E-01	134,763.87	1	25	3.57E-48	2431.135	-0.501
Neptunian satellites	Power	9.9979E-01	56,414.60	1	12	2.09E-23	2690.372	-0.503
Plutonian satellites	Power	9.9987E-01	22,296.43	1	3	6.62E-07	30.124	-0.496

Based on the parameter estimates of the best regressions mentioned above, the exponent (b1) varies between -0.503 and -0.496, whereas the constant varies between 30.12 km and 362114.9 km. The highest exponent was recorded with the Plutonian satellites, whereas the smallest was found with the Neptunian satellite. As far as the constant was concerned, Plutonian satellites have the smallest constant, whereas the planets recorded the highest constant. The mean of the exponents of the regression is -0.5. The standard deviation is 0.0019, implying that the coefficient of variation is 0.38%, which is very small. This suggests that the rate of change of the orbital speed according to the semi major axis is almost the same for all of the bodies.

The near-constant value of the exponent appears as an expression of the continuity of the fluid flow from the mother precursors to their daughter bodies. It also suggests that in some conditions, the rate at which energy was supplied (e.g. by the fluid flow) by some mother precursors could have been similar to the rate at which it was transferred and dissipated at the scale of their daughter bodies regardless of the scale. But because the intensity of the turbulence varied according to the scales or the systems, bodies of various sizes were formed and the size of the systems varied.

However, the constants or scaling factor of these equations are highly different. The constant for the regression of the planets and asteroids altogether (355,012.4 km) is not that different from that of the regression for the asteroids only (354,687.5 km). However, the constant for the regression of the planets only (362,114.9 km) is higher than that of the asteroids or that of the planets and asteroids altogether. For satellites, the constant seems to be correlated with the size of the primary planets. For instance, the constant or the scaling factor for the regression of the Jovian satellites is 11334.7 km, whereas it is 30.1 km for the Plutonian satellites. Knowing that the constant of the equation is a scaling factor and that the exponent or power determines the rate of change (growth or decay) of the function, I deduced that the rate of change of the regressions are not scaled the same way. To put it another way, although the rate of change of the orbital speed is the same for the bodies, the scaling of the power of the semi major is very different. The difference of the shape of the function of the speed can therefore be attributed to the way the semi major is scaled. Fig. 51 illustrates the trends of the scaling factors and the exponents. Like I better illustrated later, the similarity of the value of the exponent of the model testifies of the way energy was transferred from the flowing fluid of the precursor of the Solar System to the precursors of all bodies orbiting the Sun and from the fluids in the precursors of the planetary systems to the precursors of the satellites.

Fig. 51: Constant or Scaling Factor and Exponent of the Regression (Power Function) between Orbital Speed and Semi Major Axis of the Bodies in the Solar System

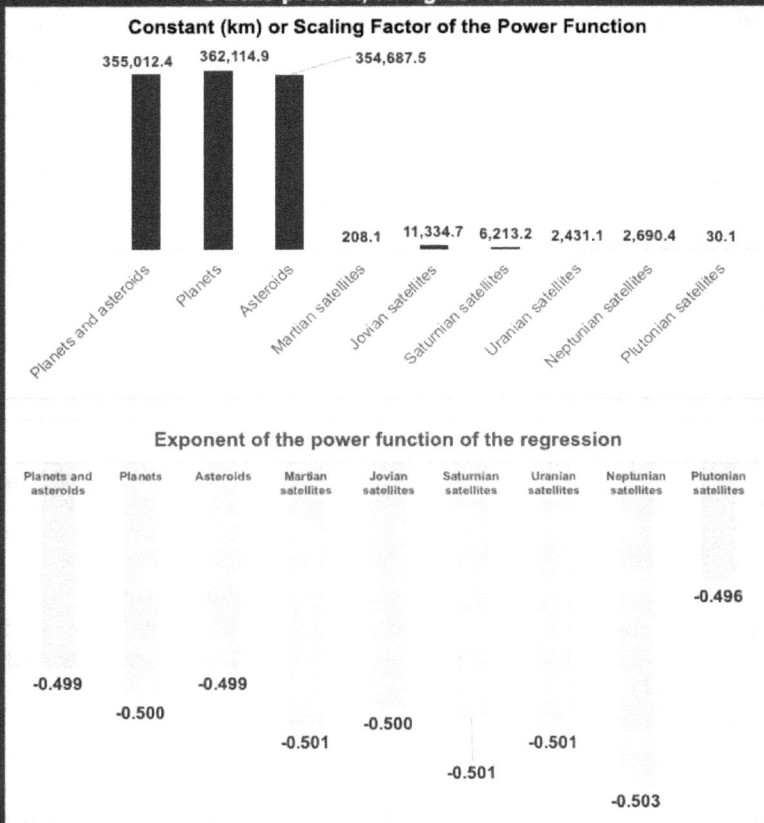

Constant (km) or Scaling Factor of the Power Function

Planets and asteroids	Planets	Asteroids	Martian satellites	Jovian satellites	Saturnian satellites	Uranian satellites	Neptunian satellites	Plutonian satellites
355,012.4	362,114.9	354,687.5	208.1	11,334.7	6,213.2	2,431.1	2,690.4	30.1

Exponent of the power function of the regression

Planets and asteroids	Planets	Asteroids	Martian satellites	Jovian satellites	Saturnian satellites	Uranian satellites	Neptunian satellites	Plutonian satellites
-0.499	-0.500	-0.499	-0.501	-0.500	-0.501	-0.501	-0.503	-0.496

15.7. Factors affecting the scaling factor of the regressions

As I kept looking at the scaling factors of the above regressions for the satellites, I noticed that their trends seemed to correlate with the energy in their planetary system. Therefore, to confirm the trends I was visually seeing, I mathematically explored what affected the scaling factors by testing the impact of the translational kinetic energy, rotational kinetic energy, and total kinetic energy on the scaling factors (Fig. 52). For each of the 3 types of kinetic energy, I considered the energy of the planets, that in all the satellites of each planetary system, and that in all the bodies (planets and satellites) in each planetary system. For instance, when I considered the translational kinetic energy, I did 3 kinds of regressions, meaning the regression between:

- the translational kinetic energy in the planets and the scaling factor,
- the translational kinetic energy in all the satellites and the scaling factor, and
- the translational kinetic energy of all the bodies (planet and satellites) in each planetary system and the scaling factor.

I did the same for the rotational kinetic energy and the total kinetic energy. The results showed that, although each of the 3 types of kinetic energy mentioned above has a significant impact on the scaling factors ($r2>0.93$), the strongest relationship was found between the scaling factor and the rotational kinetic energy of:

- all of the bodies in the planetary systems ($r2>0.99$),
- the planets ($r2>0.99$), and
- all of the satellites in the planetary systems ($r2=0.97$).

The impact of the rotational kinetic energy of the planets and that of the planetary system is similar because the rotational kinetic energy of the planets is about 99% that of the planetary systems. The distribution of the orbital speed of the satellites is connected to the rotational kinetic energy of something related to their precursors. It sounds like the precursor of the satellites in each planetary system would have escaped the precursors of their primary planets while the precursor of the planetary system was splitting and the precursor of the planet was about to rotate for the first time or early in its formation. Hence the rotational kinetic energy of those precursors affected how much energy is pushed into their daughter bodies. In the end, the kinetic energy of the precursors generally positively affected the scaling of the orbital speed of the bodies.

Considering how these regressions fit into the story of the other variables, it appeared to me that the energy that a mother precursor pushed into its daughter bodies affected the way the speed of these daughters was distributed. As I better explained later, these regressions point at the way kinetic energy was transferred and "dissipated" through the fluid layers of the precursors and the bodies they birthed after using part of that energy to mold or shape themselves.

I also found that the scaling factor of the regression between orbital speed and semi major axis of the satellites depends on the rotational angular momentum of:

- all the bodies in the planetary system (r2>0.99),
- the planets (r2=0.99), and the rotational angular momentum of
- all the satellites in the planetary system (r2=0.95).

Fig. 52: Impact of the Types of Kinetic Energy (Rotational, Translational, and Total) and of the Rotational Angular Momentum on the Constant or Scaling Factor (of the Regression between the Orbital Speed (km/s) and Semi Major Axis (km) of the Satellites)

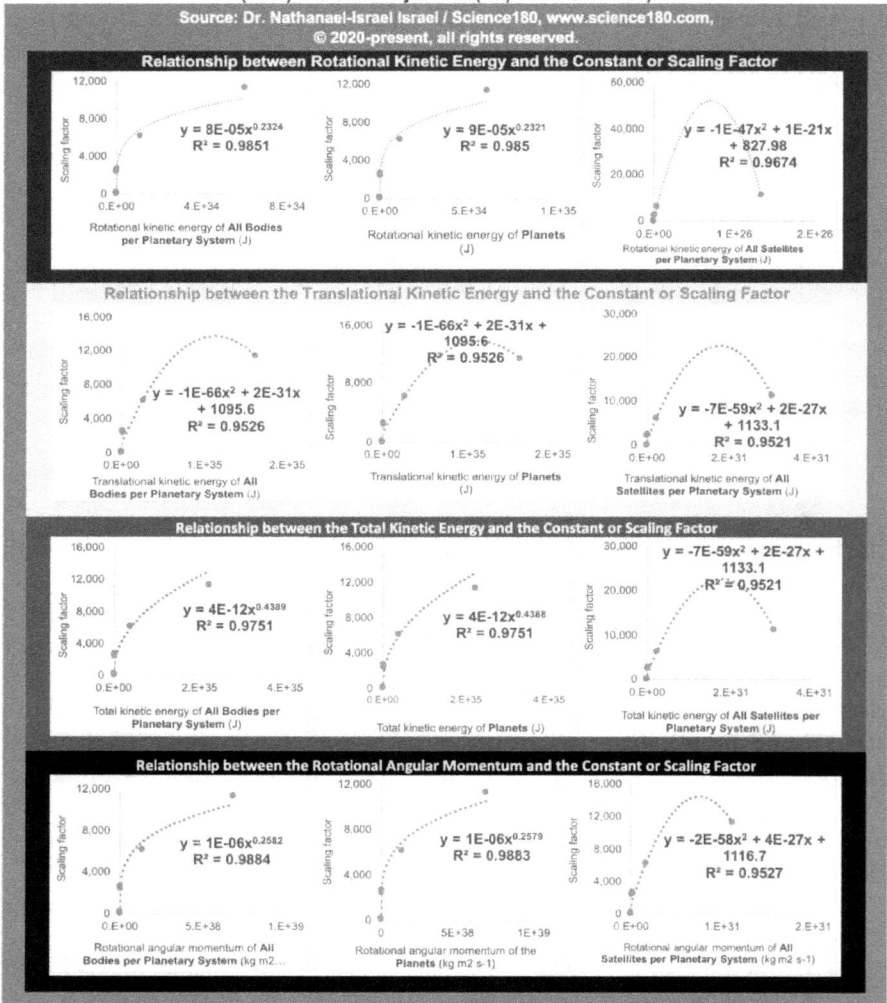

Although finding that the orbital speed of the bodies (satellites, planets, and asteroids) depends on their semi major axis was easy, discovering the impact of the kinetic energy on the scaling factor was the fruit of a lot of work and

Nathanael-Israel Israel: Original, Universal, Accurate Universe-Origin Decoder

investigation. The insight I got from that and other variables I studied caused me to start reflecting on what I now called the law of "turbulent heredity" of living and nonliving things, which I revisited in another book. That law explains how things in nature are so connected in such a way that parents of both nonliving and living things influence the characteristics of their children. The finding in this section was one of the keys to understanding movement in the universe.

15.8. Orbital speed of the satellites according to their types and turbulence zones

Varying between 0.123 km/s and 31.58 km/s, the orbital speed of the 210 satellites known in the Solar System as of 2020 (Fig. 53) depends on their type (p<0.000), the turbulence zone (p<0.000), and the interaction between these 2 variables (p<0.000). The Plutonian satellites are the slowest while the fastest satellite is Metis, the innermost Jovian satellite. The 4 innermost Jovian satellites have an orbital speed higher than that of the Sun:

- Metis: 31.58 km/s
- Adrastea (the outermost satellite in Zone 1): 31.45 km/s
- Amalthea (the innermost satellite in Zone 2): 26.48 km/s
- Thebe (the outermost satellite in Zone 2): 23.92 km/s.

In contrast, the slowest satellite is Hydra (a Plutonian satellite) whereas, the fastest is a Jovian satellite. Indeed, the 5 slowest satellites are all Plutonian satellites, suggesting that the ability of satellites to move around their primary planet decreases as the semi major axis of that primary planet decreases.

At this time, I wondered why is the orbital speed of the Plutonian satellites smaller than that of some very remote asteroids located at thousands of AU from the Sun? It appeared to me that, the smaller orbital speed of the Plutonian satellites is due to the fact that the kinetic energy of the precursor of the Plutonian planetary system was too small to allow them to be propelled with a higher speed. In other words, the orbital speed of the bodies in the Solar System depends on the energy that their precursors had. The speed of bodies in the Solar System does not depend on these bodies themselves, but on the conditions that prevailed during their formation. Similarly, the speed of daughter bodies depends on that of their mother or father. For instance, the vigor of seeds, no matter if it is plant seeds or animal seeds, depends on the genetic make-up of their parents. This is a kind of heredity law that conditions the characteristics of daughters to those of their parents. For the strength or genetics of the parents is hidden in the seeds so that their daughters or children can profit from it, for good or bad. To explain this law in other context, the heritage that children receive from their parents plays a huge role in what they will become. Those who come from a wealthy family end up making it in life quickly, whereas those who are born into a poor family usually have to face more difficulties before achieving great things in life.

In other words, how fast things can move in life for someone does not usually depend only on the person trying to cause the movement, but on how that person has been launched into his life journey. For instance, no matter how much grass a mouse can eat, it will never become as big as an elephant. Similarly, no matter how much food a tortoise will eat, it can never run faster than most animals in the wild. Although human beings may use their free will to try to move themselves or others things faster, they cannot do it beyond certain limits (e.g. imposed by their parents, education, environment) without damaging themselves or the things they try to move. For, after all, everything, including human beings and their will, desire, and technologies are limited by the constraints imposed on their environment and their elements that are in them including the particles which they consist of. And again, all particles or bodies are limited by the conditions that prevailed during their formation. Therefore, without knowing how things were formed and the laws that govern them, there is not much human beings can do to improve lives without destroying or damaging other things in the universe!

In the same manner, the movement of nonliving organisms is highly defined by how they were born or formed. Speed does not just depend on size. Particles or bodies that were born from parents full of energy end up moving faster than those whose parents were less energetic. Unless, scientists can realize this universal truth, they will continue to force some particles or bodies to move faster than they should, and in the end, these scientists can hurt themselves or the environment because of the response of these particles to the forces acting on them, against their nature. Yet, we still wonder why some diseases are appearing and devasting humankind while anthropic actions keep destroying the environment! As the French would say it, «*Science sans conscience n'est que ruine de l'âme!*» (Rabelais), which means "*Science without conscience is but the ruin of the soul*".

The orbital speed of satellites depends on their turbulence zones. For instance, the orbital speed of the satellites in turbulence Zone 1 varies between 1.35 km/s and 31.58 km/s. The slowest satellites in Zone 1 are the Martian satellites (Deimos and Phobos) whereas the fastest satellites are the Jovian satellites (Metis and Adrastea). In Zone 1, Uranian satellites are slower than Neptunian satellites. The energy used to tilt the precursor of the Uranian planetary system may have contributed to reducing the orbital speed of the Uranian satellites, particularly those in Zones 1, 2, and 3, as compared to the Neptunian satellites in the same range.

The orbital speed in Zone 2 ranges from 7.7 km/s to 26.48 km/s. The smallest speed in that zone was recorded with Mab, a Uranian satellite. Just as in Zone 1, the highest speed in Zone 2 was recorded with Jovian satellites: Amalthea (26.48 km/s) and Thebe (23.92 km/s). For the same semi major axis, the Neptunian satellites in Zone 2 orbit faster than the Uranian satellites.

The orbital speed of the satellites in turbulence Zone 3 varies between 0.22 km/s and 17.34 km/s. The slowest satellite in Zone 3 is Charon, the innermost and biggest Plutonian satellite. In contrast, the fastest satellite in Zone 3 is Io, a Jovian satellite. For the same semi major axis, the orbital speed of the Jovian satellites is higher than that of the Saturnian satellites. In contrast, the orbital speed of the Neptunian satellites for the same semi major axis is higher than that of the Uranian satellites.

Fig. 53: Orbital speed of the satellites according to their types and semi major axis

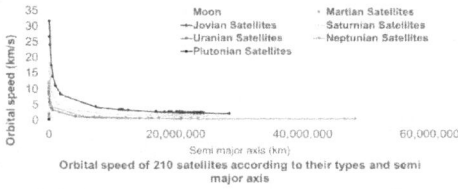

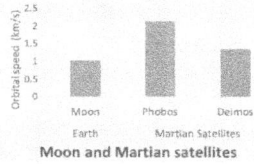

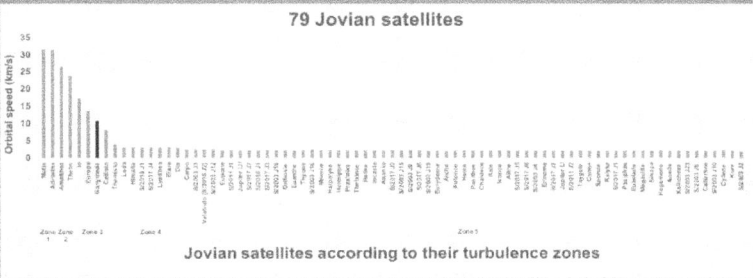

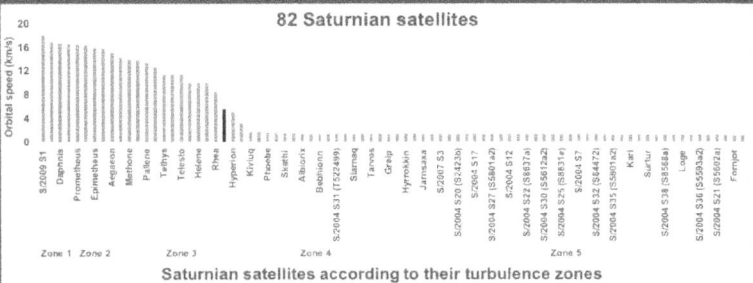

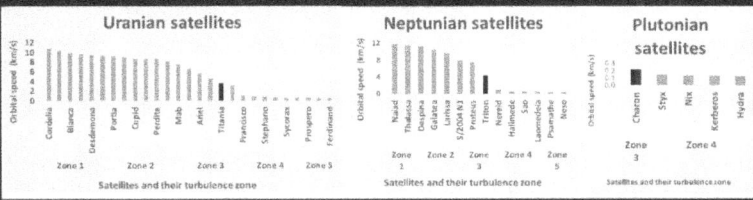

In Zone 4, the orbital speed of the satellites varies between 0.123 km/s and 4.2 km/s. The 4 smallest orbital speeds in Zone 4 were recorded with the 4 outer

Plutonian satellites while the highest speed was with Themisto, the innermost Jovian satellite in Zone 4.

Finally, in Zone 5, the orbital speed ranges between 0.37 km/s and 2.65 km/s. The slowest satellite in Zone 5 is Psamathe (a Neptunian satellite), followed by Neso (the outermost Neptunian satellite). The orbital speed of the 62 Jovian satellites in Zone 5 is higher than that of the satellites in the same zone of any other planets. Similarly, the orbital speed of the 42 Saturnian satellites in Zone 5 is higher than that of any Uranian or Neptunian satellite in Zone 5. Fig. 54 illustrates the trends I mentioned above.

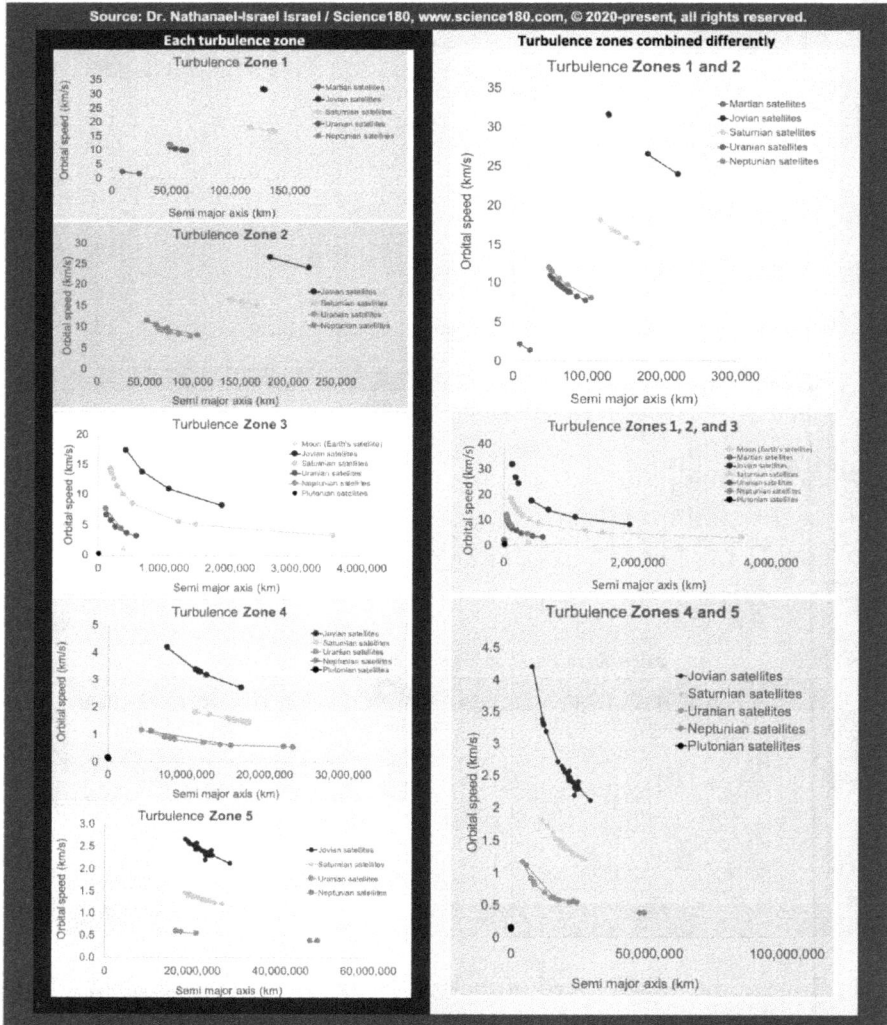

Fig. 54: Orbital speed of the satellites in each turbulence zone

Nathanael-Israel Israel: Original, Universal, Accurate Universe-Origin Decoder

15.9. Mean orbital speed of the satellites according to their types and turbulence zones

The speed of the fluid layers of the precursors of the satellites is not exactly the same as the speed of the satellites today. However, the relationship between the statistical parameters of the satellites in the turbulence zones as illustrated below gave an insight into the relationship between the fluid layers of the precursors of the bodies. In other words, the relationship found between the satellites in the turbulence zones as presented below is a witness of the probable interactions between the speed of the fluid layers of the precursors of the satellites as they could have been during the process which split-gathered them into the satellites. As of 2020, the semi major axis of the primary planets negatively affected the mean of the orbital speed of their satellites in Zone 2 ($r2=0.99$) and those in Zones 4 and 5 ($r2>0.9$) (Fig. 55).

15.10. Coefficient of variation (CV) of the orbital speed according to the types of satellites and turbulence zones

When calculated according to the planetary systems, the CV varies between 25.91% and 141.5%, the smallest CV was with the Plutonian satellites whereas the highest CV was with the Jovian satellites. According to the turbulence zones and the types of satellites, the CV of the orbital speed of the satellites varies between 0.28% and 38.7%. The smallest CV were recorded in Zone 1 of the Jovian satellites whereas the highest were in Zones 3 and 4 of the Neptunian satellites. The highest CV were found in the most turbulent zone, which is usually Zone 3 usually followed by Zone 2 (the transition zone to turbulence). In general, the smallest CV are found in Zone 5. The coefficient of variation of the orbital speed of the satellites in Zone 3 varies between 30.6% (for the Uranian satellites) and 38.1% (for the Neptunian satellites). While the CV for the Jovian satellites in Zone 3 is 31.2%, for the Saturnian satellites, it is 33.2%. The semi major axis of the primary planets significantly affected the coefficient of variation of the orbital speed of satellites mostly in Zone 2 ($r2=0.95$), Zone 3 ($r2=0.8$), and Zone 5 ($r2=0.99$) (Fig. 56).

Fig. 55: The semi major axis of the primary planets negatively affected the mean orbital speed of their satellites in Zones 2, 4, and 5

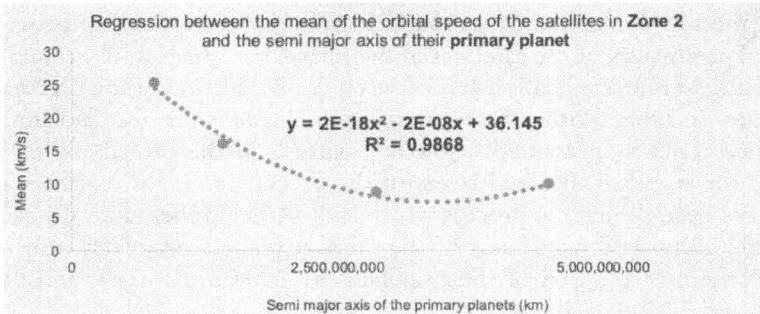

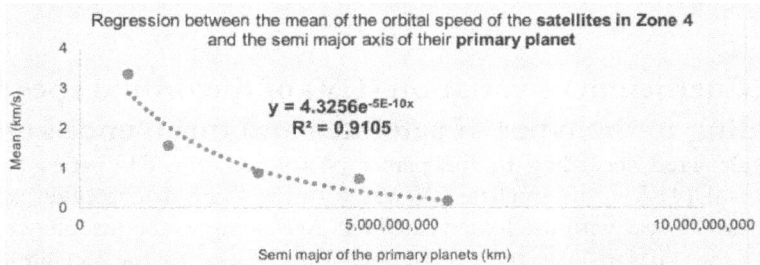

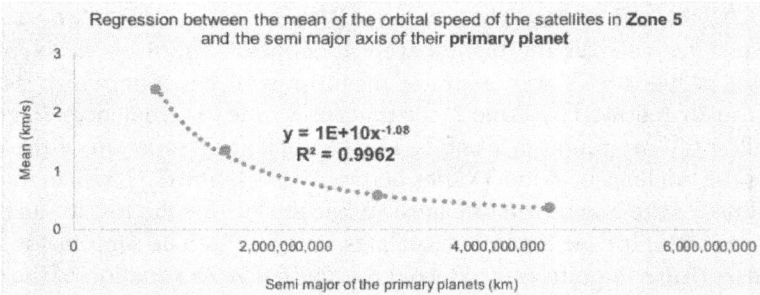

Regression between the mean of the orbital speed of the satellites in Zone 2 and the semi major axis of their **primary planet**

$$y = 2E\text{-}18x^2 - 2E\text{-}08x + 36.145$$
$$R^2 = 0.9868$$

Regression between the mean of the orbital speed of the **satellites in Zone 4** and the semi major axis of their **primary planet**

$$y = 4.3256e^{-5E\text{-}10x}$$
$$R^2 = 0.9105$$

Regression between the mean of the orbital speed of the satellites in **Zone 5** and the semi major axis of their **primary planet**

$$y = 1E\text{+}10x^{-1.08}$$
$$R^2 = 0.9962$$

Nathanael-Israel Israel: Original, Universal, Accurate Universe-Origin Decoder

Fig. 56: Regression between the semi major axis of the primary planets and the coefficient of variation of the orbital speed of their satellites in the zones

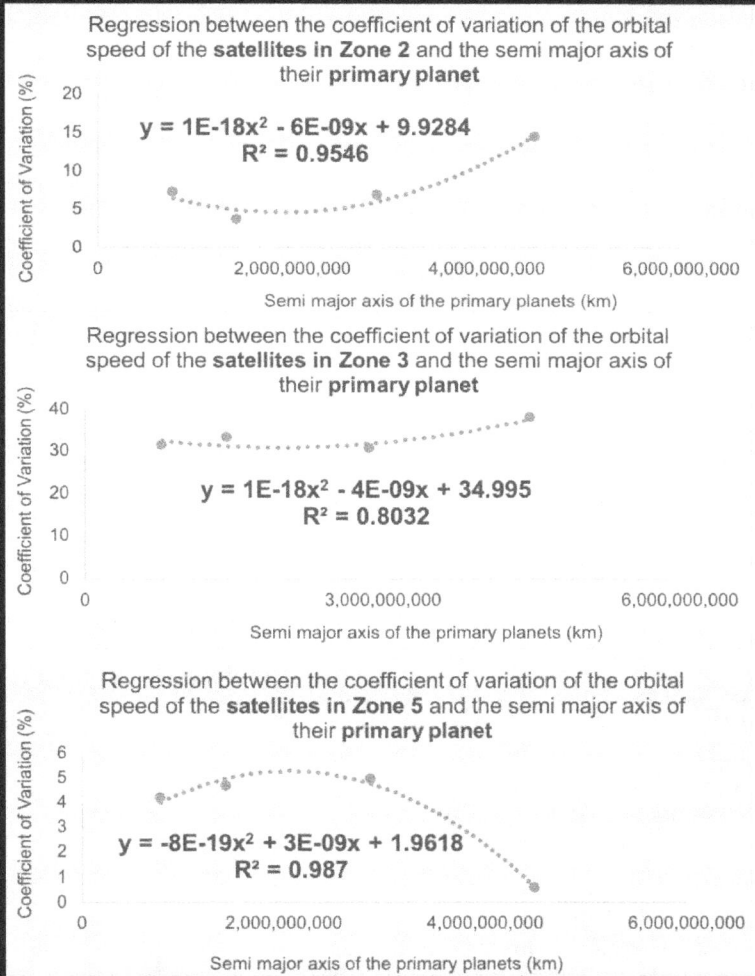

Regression between the coefficient of variation of the orbital speed of the **satellites in Zone 2** and the semi major axis of their **primary planet**

$y = 1E-18x^2 - 6E-09x + 9.9284$
$R^2 = 0.9546$

Semi major axis of the primary planets (km)

Regression between the coefficient of variation of the orbital speed of the **satellites in Zone 3** and the semi major axis of their **primary planet**

$y = 1E-18x^2 - 4E-09x + 34.995$
$R^2 = 0.8032$

Semi major axis of the primary planets (km)

Regression between the coefficient of variation of the orbital speed of the **satellites in Zone 5** and the semi major axis of their **primary planet**

$y = -8E-19x^2 + 3E-09x + 1.9618$
$R^2 = 0.987$

Semi major axis of the primary planets (km)

15.11. Take home message

As the precursors of the systems of bodies were split-gathering, they were organized into layers moving at different speeds. The bottom layers moved slowly while those on top were moved more rapidly. The orbital speed of the bodies as of today is a footprint of the speed with which their precursors moved in the layers of fluids from which they were born. In each system of bodies, the orbital speed is negatively correlated with semi major axis. Hence, the innermost satellites orbited their primary planets faster than the outermost satellites, just as the innermost planets move faster than the outermost ones. The power function best fits the aforementioned regression between orbital speed and semi major axis of the bodies in the Solar System. Although the rate of change of the orbital speed is the same for the bodies, the scaling of the power of the semi major is very different. The mean of the exponents of the regression according to the type of bodies is -0.5 and its variation is very small, suggesting that the rate of change of the orbital speed according to the semi major axis is almost the same for all of the bodies. The near-constant value of the exponent appears as an expression of the continuity of the fluid flow from the mother precursors to their daughter bodies. It also suggests that, in some conditions, the rate at which energy may have been supplied by some mother precursors could have been similar to the rate at which it could have been transferred and dissipated at the scale of their daughter bodies regardless of the scale. The similarity of the value of the exponent of the model testifies of the way energy was transferred from the flowing fluid of the precursor of the Solar System to the precursors of all bodies orbiting the Sun and from the fluids in the precursors of the planetary systems to the precursors of the satellites. The translational kinetic energy, rotational kinetic energy, total kinetic energy, and the rotational angular momentum of the bodies in the planetary systems affected the scaling factors of the regression between orbital speed and semi major axis.

- Accurately answer the most critical universe-origin and life-origin questions so you can stop standing in tension with consequential question marks including those related to religion and reason or the so-called war between science and the Bible
- Discover the errors in the scientific and religious theories about the universe-origin and life-origin that are putting you at a high risk you will never recover from if you don't quickly and confidently learn how to rationally take control over threats lurking at the edge of your efforts to understand the universe and life today
- Challenge the cosmological status quo and embrace the real change that will disrupt the hidden cages that may be holding you and that you ignore
- Definitively answer all your doubts about the source or author of the universe and life … (learn more at Science180.com/godproof)
- Understand that religion or faith, reason or science can coexist and can be properly reconciled to accurately lead you to the correct source of everything in the universe
- Satisfy your burning desire for freedom from beliefs and scientific theories about the universe-origin and life-origin that suffocate you and bind your mind, faith, unbelief, heart, and education
- Scientifically set on fire all false theories or dogmas about the existence of God, the Creator, that are enslaving humankind

Whether you are a believer, unbeliever, freethinker, administrator, politician, curriculum designer, curriculum specialist, education policymaker, teacher, librarian, school board member, researcher, parent, student, clergy, or a layperson, as long as you are really seeking to scientifically understand the rational proof of the existence of God, *"Science180 Accurate Scientific Proof of God"* is the much-admired book written for great people just like you! Grab your copy today and start reading it! Don't wait any longer!

Dr. Nathanael-Israel Israel is a Beninese-American scientist, entrepreneur, and international consultant, who shows people of all ages and educational backgrounds how to scientifically decode the formation of the universe and of life, and who is acknowledged as the creator of the Chemicals Turbulent Origin Formula™, the inventor of the Life Turbulent Origin Formula™, and the discoverer of the Universe Creation Formula™. He is the Founder of Science180 Academy, which is trailblazing the reconciliation between science and the creation.

CHAPTER 16

HOW THE DERIVATIVES OF THE REGRESSION BETWEEN ORBITAL SPEED AND SEMI MAJOR AXIS OF CELESTIAL BODIES POINTS AT THE SHEARING RATE OF THE FLUIDS DURING THE UNIVERSE FORMATION

The first derivative is like a shear rate or rate of shear of the flow of the fluids of the precursors of bodies. Beside the 1st derivative, I also calculated 4 more: 2nd, 3rd, 4th, and 5th order of derivatives. In other words, to study the rate of change of the orbital speed of the bodies in the Solar System, I calculated and studied 5 types of derivatives:

- 1st derivative of the regression between orbital speed and semi major axis
- 2nd derivative of the regression between orbital speed and semi major axis
- 3rd derivative of the regression between orbital speed and semi major axis
- 4th derivative of the regression between orbital speed and semi major axis
- 5th derivative of the regression between orbital speed and semi major axis

Wherever I talked about the "derivative of the orbital speed regression", I meant the "derivative of the regression between orbital speed and semi major axis". Finally, to ease the reading of this chapter, I sometimes did not add the units (s^{-1}) of the 1st derivatives to their values. My goal was to explore factors which affected the deceleration of the orbital speed of the bodies according to their systems and to investigate how they can explain the formation of the bodies in the universe. Because the first derivatives are negative, a smaller value means a higher deceleration of the orbital speed while a higher derivative means a smaller deceleration of the orbital speed. Similarly, a deceleration of the orbital speed of the bodies with an increasing semi major axis could also mean or imply a deceleration of the flow in the fluids of the precursors of these bodies according to their depth with respect to the top layers of these fluids.

The 1st derivative could have been the rate of change of the speed at which the fluid layers of a precursor could have passed over adjacent layers. As these fluid layers were wrapped into specific bodies, they could have conferred some of their orbital speed to their daughter bodies, meaning that the orbital speed and their derivatives are a reminder of what could have happened to the fluids of the precursors of the bodies. The trends of the 1st derivatives also inspired me greatly concerning the law of heredity of living and nonliving things that I apostrophized in the previous chapter. Before my discovery on the origin of the universe, there was no theory which could explain the "heredity" of nonliving things. I am glad my work is able to shed some light onto this controversial phenomenon that I explained in other books.

Because the trends of the orbital speed according to the semi major relate to a turbulent shear flow, I also suspected that the interpretation of the first derivative as a shear rate can help determine the viscosity of the fluids in the precursors of the celestial bodies. However, I refrained from using terms like shear, shear rate, and many other terms associated with a shear flow as I dealt with the derivatives because 1) I suspected that some people may not understand or believe quickly or ever believe that the fluid flow that led to the formation of the bodies in the universe is related to a turbulence shearing and 2) because I want to reach out to as many people as possible. Indeed, some studies were done to link shear rate to viscosity and the width or size of a flow. For instance, some works suggested that shear rate, viscosity, and shear stress are highly connected. Shear stress is calculated as the ratio between an applied force and the area that the force is applied to. Knowing that the form, shape, and size of the precursors of the bodies changed during the process that led to the formation of their daughter bodies, it may be difficult to properly calculate the force or the area which should be considered to access shear stress without too many assumptions. Although such itineration may help to estimate the viscosity of the flow using the derivatives, I chose not to engage into this kind of investigation at this point because I do not like to assume things. Therefore, for now, I will not assume anything while dealing with the complex compressive fluids involved in the turbulence of the formation of the universe.

For the sake of space and time, I placed this chapter in the electronic chapter of this book and you can get it at www.Science180.com/OrbitalSpeedDerivatives

231

Science180: Understand the Origin of The Universe. Increase Your Glory and Peace of Mind

CHAPTER 17

HOW CAN YOU RESET YOUR MIND TO THINK BETTER AND UNDERSTAND THAT THE ROTATIONAL ANGULAR SPEED OF CELESTIAL BODIES HOLDS A BIG SECRET—THAT YOU MOST LIKELY IGNORE?

17.1. How rotational angular speed can help explain the formation of the universe

Unlike most of the other variables I studied, data about rotational angular speed was available nowhere. I had to use the data available on rotational period to calculate it. The rotational angular speed of a body can be calculated using its rotational period (T_{rot}):

$$\text{Rotational angular speed} = 2\pi \ / \ T_{rot}$$

When I looked at the data on rotational angular speed and its variation across the turbulence zones, and from one planetary system to the other, I felt it can help explain the way the rotation of the bodies was acquired and how the precursors of the bodies could have been compressed into the current chapter of their daughter bodies. I felt like, as the layers of fluids of the precursors of the bodies were moving, the fluids in them started collecting themselves and progressively acquiring a rotational movement. By the time the formation of the bodies was complete, their orbital and rotational movement were established. In the following segments, as I explain the distribution and characteristics of the rotational angular speed of the bodies in the Solar System today, I also present how they fit into the general picture of the events that happened during the formation of the universe.

Rotation was imparted onto the bodies as a consequence of the spiraling, winding up, and other aspects of the split-gathering of the fluids in the precursors of the bodies. As the particles and the bodies were gathering together, rotational movement and orbital movement was birthed as the balancing of the motion of all their components. This is not just a matter of particles or components of the bodies averaging out their momentum, but a consequence of the gathering together under the influence of a turbulence traceable back to a precursor, which at its turn was also birthed by a mother and so on and so forth until the turbulent prima materia, the starting point of all materials in nature. Below, I will present more supporting data.

17.2. General trends of rotational angular speed in the Solar System

Studied for 224 bodies in the Solar System, the rotational angular speed depends on the types of bodies (p=0.016) and varies from 2.99E-07 s^{-1} to 0.0346 s^{-1}. The smallest value was recorded with Venus, whereas the highest was with 2004 FH, an Aten asteroid. With a rotational angular speed of 2.86E-06 s^{-1}, the fourth smallest in the Solar System, the Sun is one of the bodies which has a small rotational angular speed; but, because of its radius, its rotational speed is much higher as compared to that of other bodies in the Solar System. The 4 bodies which have a rotational angular speed smaller than that of the Sun are presented in Fig. 62.

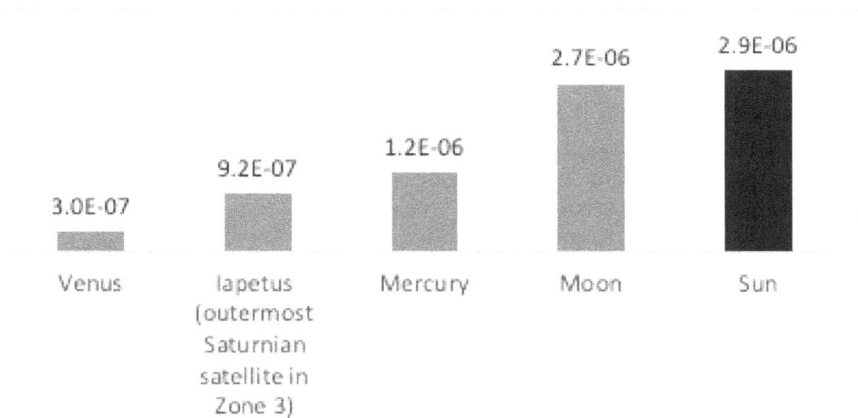

Fig. 62: Bodies in the Solar System which rotational angular speed (s^{-1}) is smaller than that of the Sun

For many years, I have wondered why the rotational speed of the Sun, Venus, Mercury, Moon, and Iapetus are the 5th smallest in the Solar System. On my journey to answer that question, I noticed that, with a rotational angular speed of

2.86E-06 s^{-1} (the fourth smallest in the Solar System), the Sun is one of the bodies that have a small rotational angular speed; but, because of its radius, its rotational speed is much higher as compared to that of other bodies in the Solar System. It seemed to me that, in the precursor of a star like the Sun, the turbulence may have not been too strong to cause a strong rotation. It could have been possible that the precursors of Mercury, Venus, and the Moon were very viscous, while the turbulence in their fluids could have been weak; hence their small rotational speed. In other words, the high viscosity of the precursors of Venus, Mercury, and the Moon could have prevented them from inculcating a strong rotation to them, but just translated their speed into an orbital speed. In the case of Iapetus (the outermost Saturnian satellite in Zone 3), its position at the bottom of the layers of fluids in Zones 1, 2 and 3 of the Saturnian satellites could have caused it not to rotate much. When I omit the unit of the rotational angular speed, please be advised that it is s^{-1}. I will provide more details in the chapter on vortex and ligament.

17.3. Rotational angular speed of planets

The rotational angular speed of the planets ranges from 2.99E-07 s^{-1} to 1.76E-04 s^{-1}, the smallest value was observed with Venus while the highest was with Jupiter (Fig. 63). The rotational angular speed of Earth (7.29E-05 s^{-1}) is in the same order of magnitude as that of Mars (7.08E-05 s^{-1}). The rotational angular speed of the giant planets is 1.4 to 2.4 times that of the Earth. Among the giant planets, Uranus has the lowest rotational angular speed. The tilting of the axial tilt could have reduced the ability of Uranus to rotate faster. A similar logic can be applied to explain why the rotational angular speed of Venus is the smallest among the planets. In fact, the axial tilt of Venus is 177.36° whereas that of Pluto is 122.53°. The smaller the axial tilt, the greater the rotational angular speed. For the more the rotational axis is tilted, the more the energy required to tilt it could have been and consequently, leaving less energy to do other things such as rotating the planet itself. However, because of the trends of the rotational angular speed and axial tilt, other variables besides axial tilt could have affected the rotational angular speed. Some of these variables could be the turbulence intensity and the size of the bodies.

Nathanael-Israel Israel: Universe-origin Doctor / Expert, Consultant, Speaker, Scientist, Author

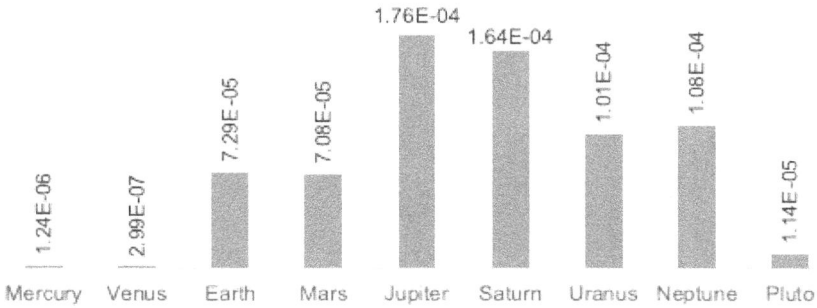

Fig. 63: Rotational angular speed of Planets in the Solar System

17.4. Rotational angular speed of satellites

17.4.1. Rotational angular speed of the satellites according to their types

The rotational angular speed of the satellites in the Solar System significantly depends on their types ($p < 0.05$) and their turbulence zone ($p < 0.05$) whether all of the satellites only or those in Zones 1, 2, and 3 only were analyzed. The 81 rotational angular speeds available for the satellites varies between 9.16E-07 s^{-1} and 6.46E-04 s^{-1} (Fig. 64).

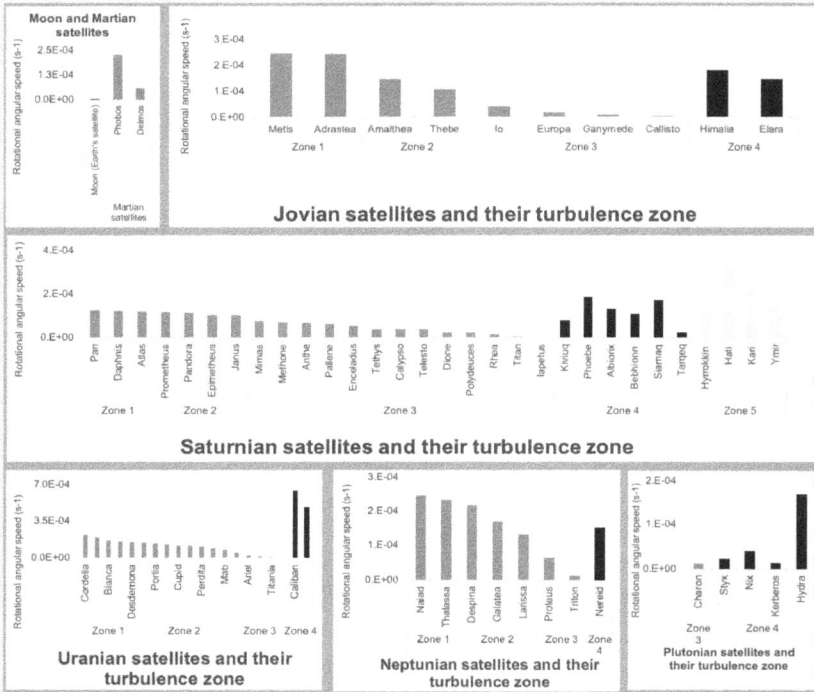

Fig. 64: Rotational Angular Speed of Satellites according to their Types and Turbulence Zones

The lowest value was recorded with Iapetus, the outermost Saturnian satellite in Zone 3. The second lowest rotational angular speed (2.65274E-06 s⁻¹) was recorded on the Moon, while the highest rotational angular speed was recorded with Caliban, a Uranian satellite in Zone 4. The Saturnian satellites are the only satellites which rotational angular speed is reported for Zone 5. In general, rotation angular speed of the satellites located downstream of the biggest satellites are usually higher than those of the satellites located upstream (i.e. Zones 1, 2, and 3). The top 3 rotational angular speeds were reported in Zones 4 and 5:

- $6.46E-04$ s⁻¹: Caliban (a Uranian satellite in Zone 4)
- $4.85E-04$ s⁻¹: Sycorax (a Uranian satellite in Zone 4)
- $3.17E-04$ s⁻¹: Hati (a Saturnian satellite in Zone 5).

Then, the next 4 highest values were in Zone 1.

Nathanael-Israel Israel: Universe-origin Doctor / Expert, Consultant, Speaker, Scientist, Author

17.4.2. Rotational angular speed of the satellites according to their turbulence zone

As I was doing the regression between rotational angular speed and semi major axis, I found a very strong relationship ($r2=1$) between these variables in Zones 1, 2, and 3. However, in Zones 4 and 5, no significant relationship exists between the variables. Therefore, so that the description of the moment in turbulence Zones 1, 2, and 3 could be exhaustive, I used the regression between rotational angular speed and semi major, as explained in the section on regression below, to estimate the missing rotational angular speeds in Zones 1, 2, and 3. These missing values in Zones 1, 2, and 3 were for 4 Saturnian satellites and one Neptunian satellites.

The rotational angular speed of the satellites in Zone 1 varies between 5.76E-05 s^{-1} and 2.47E-04 s^{-1}. The smallest value was recorded with Deimos (the outermost Martian satellite) while the highest was with Naiad (the innermost Neptunian satellite). The 4 highest rotational angular speeds in Zone 1 were observed with the 2 innermost Jovian and Neptunian satellites. The rotational angular speed of the innermost Martian satellite (Phobos) is higher than any of the Uranian and Saturnian satellites in Zone 1. The ratio between the highest and the smallest rotational angular speed in Zone 1 is 4.29. The mean of the rotational angular speed in Zone 1 is 1.78E-04 s^{-1}, and the standard deviation is 5.75E-05 s^{-1} and the CV is 32.4%.

In Zone 2, the rotational angular speed varies between 7.65E-05 s^{-1} and 2.17E-04 s^{-1}. The smallest was with S/2004 N1 (the outermost Neptunian satellite in Zone 2) whereas the highest was with Despina (the innermost Neptunian satellite in Zone 2). The maximum value in Zone 2 is 2.84 times the smallest one.

The rotational angular speed of the satellites in Zone 3 ranges between 9.16E-07 s^{-1} and 7.72E-05 s^{-1}. The smallest was found with Iapetus (the outermost Saturnian satellite in Zone 3), while the highest was with Mimas (the innermost Saturnian satellite in Zone 3). In Zone 3, the highest rotational angular speed in Zone 3 is 84.2 times that of the smaller value. Furthermore, the second smallest value was recorded with the Moon (2.65E-06 s^{-1}) which is 2.9 times that of the smallest value recorded in Zone 3.

The rotational angular speed of the biggest satellite in each planetary system varies between 2.65E-06 and 1.24E-05, the smallest value being obtained with the Moon whereas the highest was with Triton, the biggest Neptunian satellite:

- 2.65E-06 (Moon)
- 1.02E-05 Ganymede (the biggest Jovian satellite)
- 4.56E-06 Titan (the biggest Saturnian satellite)
- 8.35E-06 Titania (the biggest Uranian satellite)
- 1.24E-05 Triton (the biggest Neptunian satellite)
- 1.14E-05 Charon (the biggest Plutonian satellite)

The ratio between the highest and the smallest rotational angular speed of the biggest satellites mentioned above is 4.66.

Varying between 9.16E-07 and 1.24E-05, the rotational angular speed of the outermost satellite in Zone 3 is not a constant but depends on the types of planetary system (Fig. 65); the highest value being with Triton (the outermost Neptunian satellite in Zone 3) whereas the smallest value was with Iapetus, the outermost Saturnian satellite in Zone 3:

- 2.65E-06 (Moon)
- 4.36E-06 Callisto (the outermost Jovian satellite in Zone 3)
- 9.16E-07 Iapetus (the outermost Saturnian satellite in Zone 3)
- 5.40E-06 Oberon (the outermost Uranian satellite in Zone 3)
- 1.24E-05 Triton (the outermost Neptunian satellite in Zone 3)
- 1.14E-05 Charon (the Plutonian satellite in Zone 3).

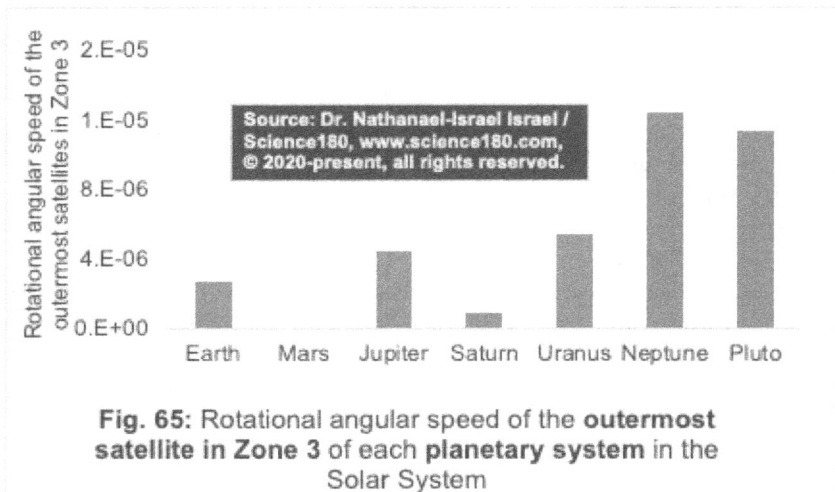

Fig. 65: Rotational angular speed of the **outermost satellite in Zone 3** of each **planetary system** in the Solar System

The rotational angular speed of the satellites in Zone 4 varies between 1.37E-05 and 6.46E-04. The smallest value was recorded with Kerberos, a Plutonian satellite, whereas the highest value was with Caliban, a Uranian satellite. The ratio between the highest and the smallest rotational angular speed reported in Zone 4 is 47.2. The 4 rotational angular speeds reported in Zone 5 vary between 1.37E-04 and 3.17E-04. Next, I will present the graph of the rotational angular speed of the satellites according to the zones (Fig. 66).

Nathanael-Israel Israel: Universe-origin Doctor / Expert, Consultant, Speaker, Scientist, Author

Fig. 66: Rotational Angular Speed of Satellites according to their Types and Turbulence Zones

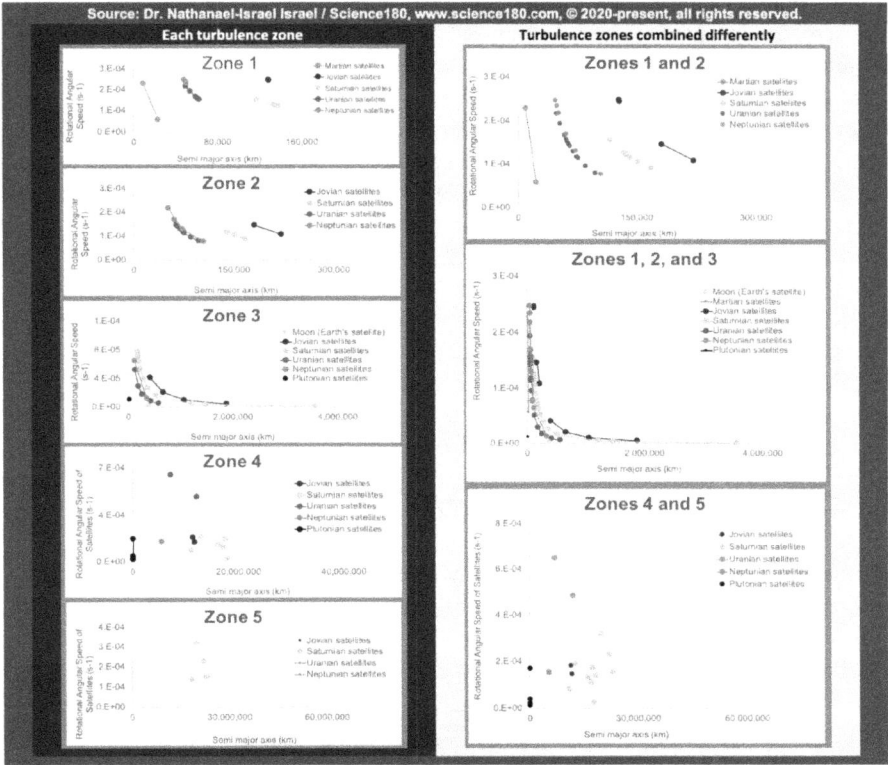

17.5. Regressions between rotational angular speed and semi major and factors affecting its scaling

The energy which the precursors of the planetary systems pushed into the precursors of the satellites defined how the rotational angular speed was scaled among the satellites from one planetary system to another one. Within a planetary system, the rotational angular speed of the satellites in Zones 1, 2, and 3 was negatively affected by their semi major axis. Then, in Zones 4 and 5, the rotational angular speed jumped, suggesting that something could have happened to the precursors of the satellites in Zones 4 and 5 so that their rotational angular speed stopped depending on the semi major axis as in the case for the satellites in Zones 1, 2, and 3. Considering the trends of the other variables, I think that, in addition to how the bodies were split-gathered, the way the precursors of the satellites in Zones 4 and 5 were ejected from that of the satellites in Zones 1, 2, and 3 could explain why they gained rotational angular speed. After breaking from the precursors of the bodies in Zones 1, 2, and 3, the precursors of the bodies in Zones 4 and 5 gained momentum and consequently transferred that to their daughter bodies as an increase in rotational angular speed.

Now, I will provide details about regressions between rotational angular speed and semi major axis. When I analyzed the rotational angular speed of all the satellites without distinguishing between the types of bodies, the regression between rotational angular speed and semi major axis was not very strong. For, either the probability of the significance level was high or the coefficient of determination was too small. When the satellites were analyzed according to their turbulence zone, strong regressions were found according to their types only when the regression focused on the satellites in Zones 1, 2, and 3 only. In other words, the regression between rotational angular speed and semi major axis is not strong for all the satellites when the data of the zones were combined. Although all the models used to test the regression between rotational angular speed and semi major axis of the satellites in Zones 1, 2, and 3 were significant, the power function yielded the most significant and strongest regression all the time. Because the Moon and Charon are the only satellites in Zones 1, 2, and 3 of their planetary system, no regression could be done for them. Fig. 67 and Table 3 summarize the best models for the regression between rotational angular speed and semi major axis of the satellites in Zones 1, 2, and 3.

Table 3: Model Summary and Parameter Estimates of the best models for the regression between rotational angular speed (s-1) and semi major axis (km) of the satellites in Zones 1, 2, and 3

Dependent Variable: Rotational angular speed of satellites in Zones 1, 2, and 3 (s-1)

Satellite Type	Equation	Model Summary					Parameter Estimates			
		R Square	F	df1	df2	Sig.	Constant	b1	b2	b3
Martian satellites	Power	1.000		1	0		208.15	-1.501		
Jovian satellites	Power	1.000	2.50E+07	1	6	4.32E-21	11460.06	-1.501		
Saturnian satellites	Power	1.000	5.58E+06	1	18	7.03E-51	6203.05	-1.501		
Uranian satellites	Power	1.000	7.54E+04	1	16	8.08E-31	2283.01	-1.495		
Neptunian satellites	Power	1.000	3.23E+08	1	5	1.01E-20	2622.62	-1.500		

The independent variable is Semi Major (km) of Satellites.

Based on these parameters, the formula of the best equations of the rotational angular speed according to the semi major of the satellites are:
- Martian satellites: $208.147*SemiMajor^{-1.5006112}$
- Jovian satellites: $11460.06*SemiMajor^{-1.5013310}$
- Saturnian satellites: $6203.05*SemiMajor^{-1.5005618}$
- Uranian satellites: $2283.01*SemiMajor^{-1.4950973}$
- Neptunian satellites: $2622.62*SemiMajor^{-1.5002857}$

The constant of the regression between rotational angular speed and semi major axis ranges from 208.147 to 11460.1. The lowest constant was found with the Martian satellites. In contrast, the highest constant was recorded with the Jovian satellites which is almost twice that of the Saturnian satellites. Because the rotational angular speed decreases with the semi major axis, a higher constant for Jovian satellites implies that the rotational angular speed of these satellites decreased more slowly as the semi major axis increased. In other words, the deceleration of the rotational angular speed of the Jovian satellites might have been very small.

The average power or exponent of the regression between rotational angular speed and semi major axis is -1.5 and the coefficient of variation is very small (-0.169%), meaning that the power of the regression is almost a reliable constant. This suggests that the fluid flow that birthed the rotation of the precursors of the satellites, regardless of their types, have something in common which can be traced back to how the precursor of the entire Solar System was fragmented and launched its daughter bodies into motion.

To calculate the turbulence intensity and the increment of the rotational angular speed, I needed the least missing data as much as possible. For, the more data are missing, the more biased would be the increment of the rotational angular speed and the turbulence intensity based on the rotational angular speed. To avoid this bias, I estimated the missing rotational angular speed in Zones 1, 2, and 3 using the regressions mentioned above. I also needed the rotational angular speed in Zones 4 and 5 to later estimate the rotational kinetic energy and the rotational angular momentum. Although more Saturnian satellites and Uranian satellites in Zones 4 and 5 have a rotational angular speed, I could not estimate the missing rotational angular speed in Zones 4 and 5 using a regression because no strong regression was found with the limited data existing in those zones. In other words, considering the trends of the rotational angular speed for the Saturnian satellites and the Uranian satellites which are the satellites which have more data in Zone 4, it appeared that no strong regression can be used to estimate the missing rotational angular speed in that area. Therefore, the missing rotational angular speed of any satellite in Zone 4 of any planet is estimated as the mean calculated using the available rotational angular speed in Zone 4. Similarly, using the available rotational angular speed for the Saturnian satellites in Zone 5, the only satellites which have some rotational angular speed in Zone 5, no clear regression could allow to estimate the rotational angular speed in those zones. Therefore, except for the Saturnian satellites, I estimated the missing rotational angular speed in Zone 5 to be the same value for the missing rotational angular speed in Zone 4. I personally believe that the rotational angular speed of the satellites in Zones 4 and 5 is reported for just a few satellites because the people who measured them focused on the bigger satellites in those zones and/or they thought that it could not vary much and should be a constant. For the reported values in Zones 4 and 5 are for the biggest satellites in those zones. Let us not forget that the radius and speed of the satellites in Zones 4 and 5 did not vary much. Therefore, it can be understandable that the rotational angular speed is almost constant in those zones.

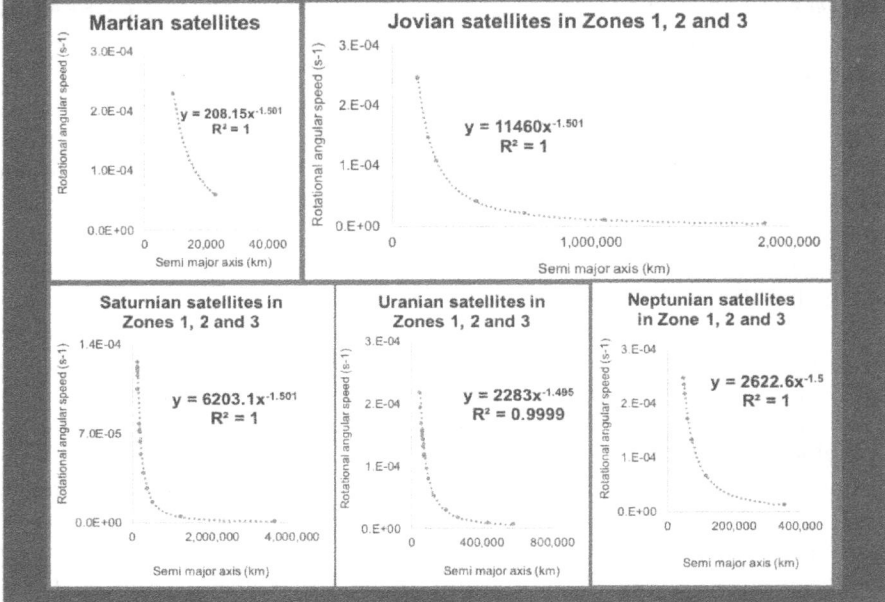

Fig. 67: Regression between rotational angular speed and semi major axis of the satellites

17.6. Factors affecting the scaling factor of the regression between rotational angular speed and semi major axis

I explored how the scaling of the regression between rotational angular speed (s^{-1}) and semi major axis (km) of the satellites is impacted by the:

- Rotational kinetic energy
- Translational kinetic energy
- Total kinetic energy, and
- Rotational angular momentum (Fig. 68)

As I was about to do the regression between the kinetic energies and the scaling factor (of the regression between rotational angular speed (s^{-1}) and semi major (km) of the satellites), I noticed that the scaling factors are almost the same as those I found when I did the regression between orbital speed and semi major axis of the satellites. Therefore, I tested the relationship between the scaling factor of the regression of the orbital speed and that of the rotational angular speed, each of them with respect to the semi major axis.

The results showed a positive correlation ($r2>0.99$) between the scaling factor, suggesting that the event that defined the scaling of the orbital speed and the rotational angular speed could have been linked, and positively affected each other. The strong connection between the scaling of the rotational angular speed and the orbital speed suggests that the processes that caused the bodies to rotate or to orbit others were connected. In the chapters on vortex, fluid ligaments, and fundamental forces in nature, I expounded on the acquisition of the rotation and revolution of the celestial bodies. I also introduced this issue in a previous chapter where I talked about the precursors of speed.

The scaling of the regression between rotational angular speed (s^{-1}) and semi major axis (km) of the satellites is highly affected by the rotational kinetic energy of:

- All bodies (planets and satellites) in their planetary system ($r2>0.99$),
- Planets ($r2>0.99$), and of all
- Satellites in the planetary system($r2=0.96$).

The scaling of the regression between rotational angular speed (s^{-1}) and semi major axis (km) of the satellites is highly affected by the translational kinetic energy of:

- All bodies (planets and satellites) in their planetary system ($r2>0.97$),
- the planets ($r2>0.97$), and of all
- the satellites in the planetary system ($r2=0.97$).

The scaling of the regression between rotational angular speed (s^{-1}) and semi major axis (km) of the satellites is highly affected by the total kinetic energy of:

- All bodies (planets and satellites) in their planetary system ($r2=0.96$),
- the planets ($r2=0.96$), and of all
- the satellites in the planetary system ($r2=0.97$).

Finally, the scaling of the regression between the rotational angular speed (s^{-1}) and semi major axis (km) of the satellites is highly affected by the rotational angular momentum of:

- All bodies (planets and satellites) in their planetary system ($r2>0.99$),
- the planets ($r2>0.99$), and of all
- the satellites in the planetary system ($r2=0.96$).

The most striking fact in this chapter is that the rotational angular speed decreased from the innermost satellite to the outermost satellite in Zone 3. Then it increased in Zones 4 and 5. In the chapters on vortex and fluid ligament, I explained why the rotational angular speed increased in Zones 4 and 5. Moreover, the rotational angular speed of most of the satellites in Zones 4 and 5 are missing. The outermost satellites in Zone 3 are like an inflection point of the graph relating rotational angular speed to semi major axis. It was also interesting to find out that the variables which control the scaling of the regression between rotational angular speed and semi major axis of the satellites is mostly the kinetic energy (total, translational, and rotational) and the rotational angular moment. More specifically, the rotational kinetic energy shows the strongest impact($r2>0.99$).

Fig. 68: Impact of the Types of Kinetic Energy (Rotational, Translational, and Total) and of the Rotational Angular Momentum on the Constant or Scaling Factor (of the Regression between the Rotational Angular Speed (s-1) and Semi Major Axis (km) of the Satellites)

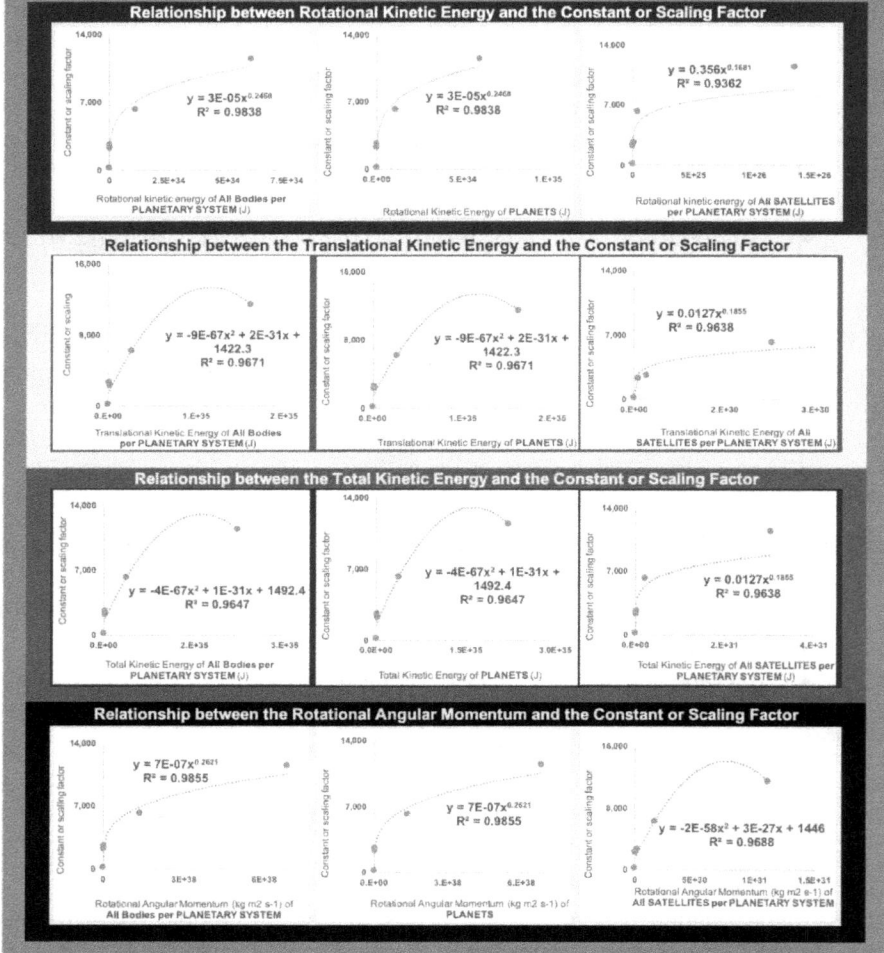

17.7. Take home message

Rotation was imparted onto the bodies as a consequence of the spiraling, winding up, and other aspects of the split-gathering of the fluids in the precursors of the bodies. As the precursors of bodies were being gathered together, rotational movement and orbital movement was birthed as the balancing of the motion of all their components. This is not just a matter of particles or components of the bodies averaging out their momentum, but a consequence of the gathering together under the influence of a turbulence traceable back to a precursor, which was also birthed by a mother and so on and so forth until the turbulent prima materia (the starting point of all materials in nature). In the Solar System, the 4 bodies which have a rotational angular speed smaller than that of the Sun are Venus, Iapetus (the outermost Saturnian satellite in Zone 3), Mercury, and the Moon. In the precursor of the Sun, the turbulence may not have been strong enough to cause a strong rotation. The high viscosity of the precursors of Venus, Mercury, and the Moon could have prevented them from inculcating a strong rotation to those planets, but just translated their speed into an orbital speed. In the case of Iapetus (the outermost Saturnian satellite in Zone 3), its position at the bottom of the layers of fluids in Zones 1, 2, and 3 of the Saturnian satellites could have caused it not to rotate much. The more the rotational axis is tilted, the more the energy required to tilt it could have been, and consequently, leaving less energy to do other things such as rotating the planet itself. However, because of the trends of the rotational angular speed and axial tilt, other variables besides axial tilt could have affected the rotational angular speed. Thanks to my delimitation of the turbulence zones, I found that the rotational angular speed decreased from the innermost satellites to the outermost body in Zone 3, and then, it picked up with the innermost satellite in Zone 4. The data suggested that the fluid flow that birthed the rotation of the precursors of the satellites regardless of their primary satellites, has something in common that can be traced back to how the precursor of the entire Solar System "propelled" or "ejected" and launched its daughter bodies into motion. The scaling of the regression between rotational angular speed and semi major axis of the satellites is mostly affected by the kinetic energy (total, translational, and rotational) and the rotational angular moment.

Another Book by Nathanael-Israel Israel:
TURBULENT ORIGIN OF LIFE
THE ONLY ACCURATE FORMULA TO SCIENTIFICALLY EXPLAIN THE FORMATION OF ALL FORMS OF LIFE QUICKLY

Every human being will benefit from understanding the real origin of life. But the problem is that most efforts to explain the origin of life are complex, inaccurate, confusing, partisan, complicated, therefore, creating serious challenges to those who are eager to scientifically decrypt where all forms of life came from. Most people want an accurate, simple, straightforward, nonpartisan life-origin book that is free from jargons and difficult concepts only known by the experts. This elegant scientific book breaks down the technicality of the origin of life in a language that even the nonscientists can easily comprehend.

It is a trustworthy book that will help you to quickly, cheaply, easily, and efficiently navigate everything you need to know to finally decode and solve the puzzling problems about the origin of life, while also giving you a crash course on the universe-origin.

Unlike any book you have ever read on the origin of life, this historic masterpiece (that distills complex scientific data down to simple explanations that make sense) is the starting point of any smart person wanting to rationally understand the formation of all living things. By the time you finish reading *"Turbulent Origin of Life"*, you will discover:

- Why in spite of the massive amount of scientific data collected on living things, scientists have misunderstood the formation of life until now, and then uncover in a simple language the one thing that was needed to accurately crack the code of life but that scientists have missed and that has been causing them headaches, overwhelm, and burnout

- Step-by-step pathway to decode the origin of life and get the power, freedom, and boldness to take advantage of the opportunities that accurate understanding of the origin of life creates (*Science180.com*/life)

- The high connection between the code of the universe formation and the process by which life on Earth was formed so you can become a fulfilled thought leader in your field of expertise

- Tools to stand as a lighting bolt that electrifies those who are still struggling to understand the formation of all forms of life in the universe

- Strategies to push the boundaries of human abilities to properly understand what is perceived as un-understandable, mysterious, supernatural, unimaginable, impossible, and unthinkable that hold people back

- Scientific approach to holistically detect, correct, and remove all misinformation, ambiguity, and misleading claims and theories surrounding the origin of life

Whether you are a scientist or a layperson, a believer, or a skeptic, you cannot afford to ignore the greater, better, faster, simpler, cheaper, easier, and accurate formula unlocked in this important book that successfully decoded the origin of life. Get *"Turbulence Origin of Life"* today and change lives! Don't wait!

Dr. Nathanael-Israel Israel is the Father of Science180 Cosmology and the Founder of Science180 Academy. He is fortunate to be known as the source of unconventional wisdom and knowledge that help people accurately crack the code of the formation of the universe, of life, and of chemicals. Get some resources by visiting his personal website at Israel120.com

CHAPTER 18

HOW THE ROTATION PERIOD OF CELESTIAL BODIES FITS INTO THE BIG PICTURE OF THE UNIVERSE FORMATION

I studied the rotation period of 224 bodies in the Solar System. The content of this chapter is exclusively available online at www.Science180.com/OrbitalPeriod

CHAPTER 19

WHY DON'T MOST SCIENTISTS TAKE THE ROTATIONAL SPEED OF CELESTIAL BODIES SERIOUSLY?

19.1. Rotational speed of the Sun, planets, and asteroids

Because in the literature (e.g. NASA, 2018), data was not available on the rotation speed of all the bodies I studied, but for only 9 planets, 3 satellites, and 7 asteroids, I was not initially able to do much statistical analysis of these data. Therefore, because data on rotation period (T_{rot}) is available for hundreds of bodies, I used the following formula to estimate the rotation speed of most of the bodies:

$$\text{Rotational speed} = \text{Radius} * 2 * \pi \, / \, (\text{Rotational period})$$

The rotation speed of 322 bodies in the Solar System varies between 8.24E-06 km/s and 12.57 km/s. The lowest value was measured on 2005 YU55, an Apollo asteroid, whereas the highest value was with Jupiter. The second highest rotation speed was measured on Saturn, the second biggest planet in the Solar System. The rotation speed of Uranus (2.59 km/s) and Neptune (2.68 km/s) are even higher than that of the Sun (1.99 km/s). In other words, considering the rotation speed data that was available as of 2020, the rotation speed of the Sun is the 5th highest in the Solar System, coming after that of the 4 giant planets: Jupiter, Saturn, Neptune, and Uranus. The rotation speed of the Sun is 4.29 times that of the Earth and 8.28 times that of Mars. I will very soon explain why the rotation speed of the Sun (the largest body in the Solar System) is smaller than that of many planets.

The rotation speed of the planets ranges from 0.0018 km/s to 12.66 km/s (Fig. 73). Jupiter is the planet that has the fastest rotation speed (12.6 km/s) in the Solar System, whereas Venus has the lowest value. All of the planets that are smaller than Earth (i.e. Mercury, Venus, Mars, and Pluto) have a rotation speed smaller than that of the Earth. The rotation speed of the 4 giant planets in the Solar System (i.e. Neptune, Uranus, Saturn, and Jupiter) is 5.8 to 27.23 times that of Earth. In general, except for Uranus, the rotation speed reported by NASA agree very well with that I calculated using the radius and rotational period. In fact, according to NASA, the rotation speed of Uranus is 4.1 km/s whereas according to the estimation is 2.6 km/s.

Fig. 73: Rotational speed of planets in the Solar System

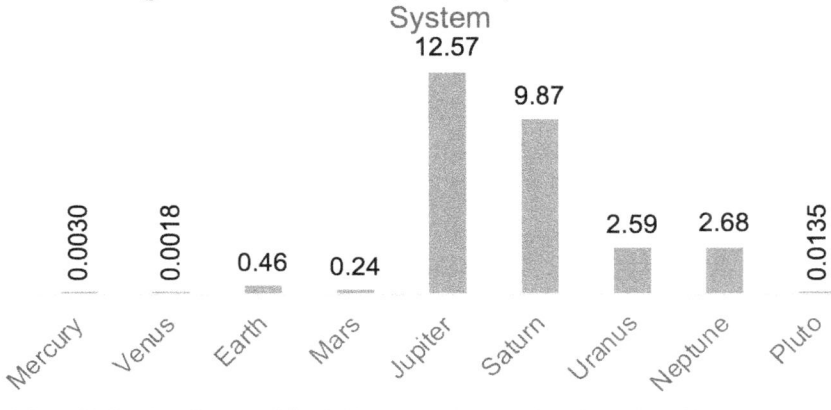

In general, the rotation speed of the asteroids is smaller than that of the giant planets (Fig. 74). The innermost asteroids have a smaller rotation speed than the outermost ones. In general, the bigger asteroids seem to have a higher rotation speed. The rotational speed of the asteroids is generally higher than that of the satellites.

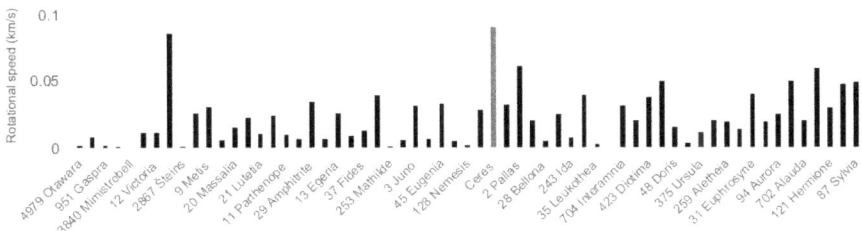

Fig. 74: Rotational Speed of Asteroids of the Main Belt

Science180: Unconventional Approach. Different Perspective. Easy Understanding

19.2. Rotational speed of the satellites

The rotational speed of the satellites ranges between 3.28E-05 and 0.075 km/s (Fig. 75). The lowest value was with Polydeuces, a Saturnian satellite, whereas the highest value was with Io, a Jovian satellite.

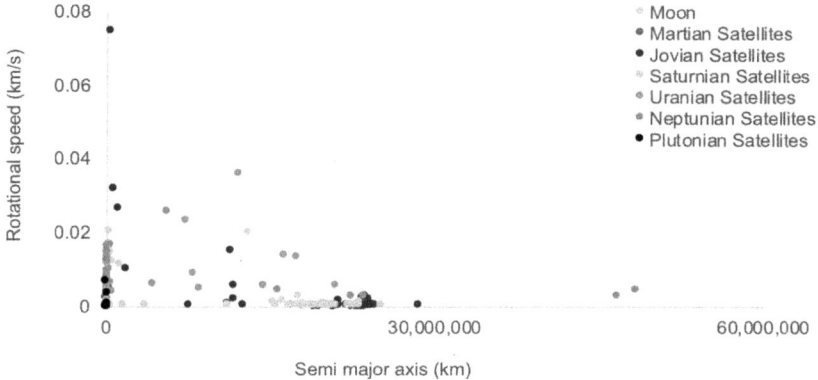

Fig. 75: Rotational Speed of the Satellites according to their Primary Planet and Semi Major Axis

The Jovian satellites with the highest rotational speed are found in Zone 3 (Fig. 76). In Zone 5, most of the rotational speed are very small. The highest rotational speed was not found with the biggest satellite nor with the satellite with the highest rotational angular speed or highest orbital speed, but with the innermost satellite in Zone 3. The highest rotational speed is 2.8 times that of Ganymede, the biggest Jovian satellite. The interaction between rotational angular speed and radius could have contributed to explaining why Io instead of Ganymede (the biggest Jovian satellite) has the highest rotation speed. The 3 highest rotation speeds of the Jovian satellites are in Zone 3. Some satellites in Zone 4 (e.g. Himalia) rotate faster than some satellites in Zones 1, 2, and 3. Nevertheless, no Jovian satellite in Zone 5 rotates as fast as any in Zone 3.

The 6 highest rotational speeds of Saturnian satellites are found in Zone 3. The highest rotating satellite is with Tethys. Phoebe, a Saturnian satellite in Zone 4 rotates almost as fast as the fastest rotating Saturnian satellite. In fact, the rotation speed of Phoebe is 98.2% that of Tethys. Beside Tethys, no other Saturnian satellite rotates as fast as Phoebe, the innermost retrograde Saturnian satellite. Titan, the biggest Saturnian satellite comes in position 6 as far as the fastest rotational speed of Saturnian satellites is concerned. Yet, Tethys is 20.6% the radius of Titan whereas Phoebe is 4.1% the radius of Titan. Unlike the Jovian satellite system where the innermost satellite in Zone 3 was the one with the highest rotational speed, the innermost Saturnian satellite in Zone 3 is the 4th fastest. The rotational speed of Titan is 57.5% that of the satellite with the highest value. These data suggest that the size of the satellites may not be the main driver of their rotation speed. For both the Jovian satellites and the Saturnian satellites, the satellite with the fastest rotational speed in Zone 4 is not the innermost satellite in that zone, but the 3rd.

In contrast to the Jovian and Saturnian satellites where the satellites with the highest rotational speed were found in Zone 3, for the Uranian satellites, the first 2 highest rotational speeds are in Zone 4:

- Sycorax, the Uranian satellite with the highest rotational speed is the 5th satellite in Zone 4, whereas
- Caliban, the Saturnian satellite with the second highest rotational speed, is the second Saturnian satellite in Zone 4.

Eleven Uranian satellites have a rotational speed higher than that of Titania, the biggest Uranian satellite. The rotational speed of Titania is 18.1% that of the satellite with the highest value. Three Uranian satellites in Zone 2, three in Zone 3, three in Zone 4 and even two in Zone 5 have a rotational speed higher than that of the biggest Uranian satellite. The second smallest coefficient of variation of the rotational speed was found with the Uranian satellites, just after the Neptunian satellites, suggesting that the intermittency of the rotational speed of the Neptunian satellites and the Uranian satellites is smaller than that of the Jovian satellites and the Saturnian satellites.

- One may wonder why the Uranian satellites in Zones 4 and 5 rotate so much faster than their peers in the Jovian and Saturnian systems!
- Can this be affected by the small axial tilt of the Uranian satellites?
- Is it related to the high orbital inclination of Uranus?
- Is that related to the precursor of the Uranian planetary system?

I later addressed these questions in other chapters of this book including those on vortex and fluid ligaments. Unlike the Jovian and Saturnian satellites for which the satellite with the highest rotational speed was found in Zone 3, the Neptunian satellite with the highest rotational speed, Nereid, was the innermost Neptunian satellite in Zone 4. Just as seen with the satellites of the other planets, the biggest satellite of Neptune, Triton, is not the satellite with the highest rotational speed. For the rotational speed of Triton is 65% that of the satellites with the highest rotational speed. The coefficient of variation (CV) of the rotational speed of the Neptunian satellites (72.1%) is smaller than that of the satellites of any other planets.

Can anything related to the rotational speed of the Neptunian satellites partially explain why the range of the semi major of the Neptunian satellites is the biggest?

Unlike the other planets, Pluto is the only planet where the highest rotational speed of a satellite was found in Zone 3. Charon is that satellite. Hydra, is the satellite with the second highest rotational speed and it is the outermost satellite of Pluto. It is possible that the binary system formed by Pluto and Charon could have increased the rotational speed of Charon.

Fig. 76: Rotational Speed of the Satellites according to their types and turbulence zones

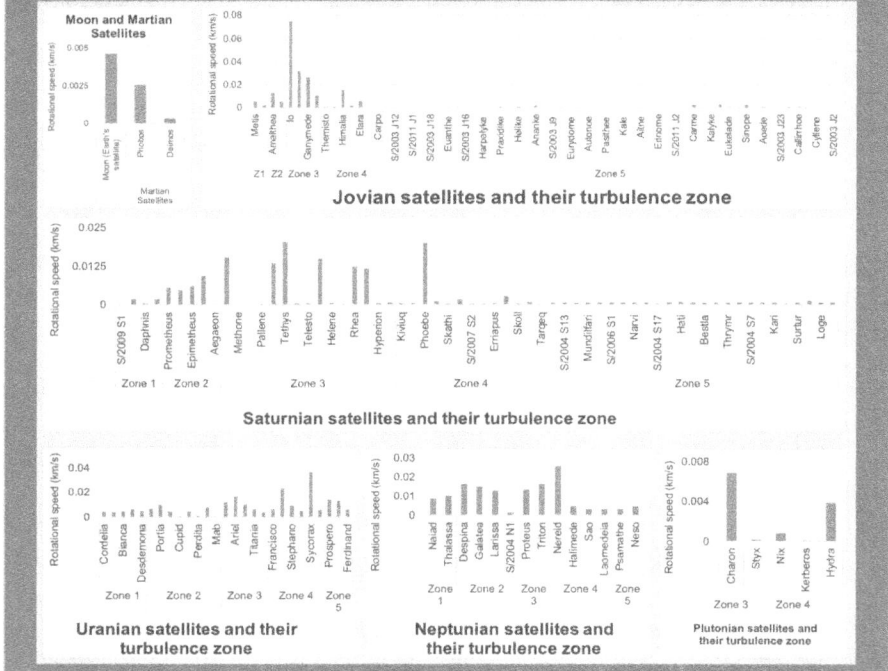

Although some of them are significant, most of the regressions between the rotational speed and the semi major axis in Zones 1, 2, and 3 of the satellites are not strong (Table 6). Therefore, I did not calculate any derivative for these regressions.

Table 6: Model summary and parameter estimates of the regression between rotational speed and semi major axis of satellites

Dependent Variable: Rotational speed (km/s) of satellites in Zones 1, 2 and 3

Satellite Types	Equation	Model Summary					Parameter Estimates			
		R Square	F	df1	df2	Sig.	Constant	b1	b2	b3
Martian satellites	Power	1.000		1	0		9.07E+05	-2.152		
Jovian satellites	S	0.523	6.586	1	6	0.043	-3.308	-2.83E+05		
Saturnian satellites	Quadratic	0.146	1.456	2	17	0.261	0.003	1.39E-08	-4.10E-15	
Uranian satellites	Quadratic	0.367	4.342	2	15	0.033	0.000	7.56E-08	-1.23E-13	
Neptunian satellites	Logarithmic	0.303	2.174	1	5	0.200	-0.014	0.002		

The independent variable is Semi Major (km) of satellites.

19.3. Take home message

The rotation speed of the Sun is the 5th highest in the Solar System, coming after that of the 4 giant planets: Jupiter, Saturn, Neptune, and Uranus. The rotation speed of Uranus (2.59 km/s) and Neptune (2.68 km/s) are even higher than that of the Sun (1.99 km/s). The rotation speed of the Sun is 4.29 times that of the Earth and 8.28 times that of Mars. Jupiter is the planet that has the fastest rotation speed (12.6 km/s) in the Solar System, whereas Venus has the slowest value. In general, the rotation speed of the asteroids is smaller than that of the giant planets. The innermost asteroids have a smaller rotation speed than the outermost ones. The rotational speed of the asteroids is generally higher than that of the satellites. No satellite rotates more than 20.2% as its primary planet.

CHAPTER 20

DISCOVER THE RATIOS THAT ACCURATELY PREDICT THE LOCALIZATION AND OTHER KEY DETAILS OF THE BIGGEST BODIES IN A SYSTEM OF SECONDARY BODIES, A MAJOR BREAKTHROUGH TO DECODE THE COSMOS

As I was working on the orbital speed and rotational angular speed chapters, I realized that the localization of the biggest bodies seemed to depend on some ratios. Therefore, in an attempt to explore the relationship between revolution (orbital movement) and rotation and to unravel some secrets controlling the movement of celestial bodies, I studied the many ratios including:

Ratio "Orbital speed / Rotational speed"
Ratio "Orbital Angular speed / Rotational angular speed"
Ratio "Orbital period / Rotational period"

In this chapter, I explored these ratios.

20.1. Ratio "orbital speed / rotational speed"

The ratio "orbital speed / rotational speed" depends on the types of bodies. It varies between 0.98 and 3.38E+06 (Fig. 77). The lowest value was recorded by Saturn, whereas the highest value was with 2005 YU55, an Apollo asteroid. The 4 lowest ratios were recorded with the giant planets followed by that of the Sun. For instance, for Saturn and Jupiter, that ratio is almost 1, whereas for Neptune and Uranus, it is almost 2. For the Earth, that ratio is 64, whereas it is 100 for Mars.

Fig. 77: Planets which **ratio** "Orbital speed / Rotational speed" is **smaller than that of the Sun**

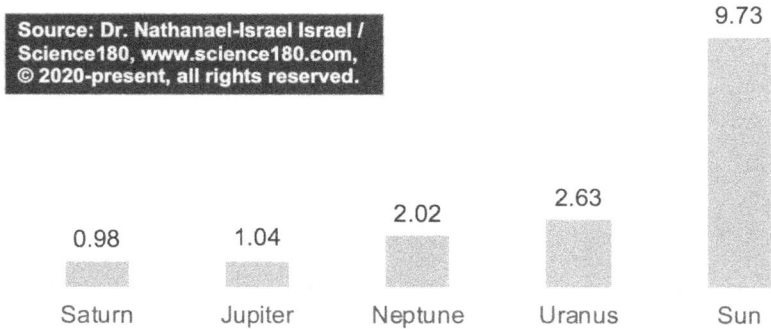

0.98	1.04	2.02	2.63	9.73
Saturn	Jupiter	Neptune	Uranus	Sun

Some bodies that are smaller than the Earth have a smaller ratio:

- Haumea, a Trans-Neptunian object: 13.7
- Sedna, a Trans-Neptunian object: 15.7
- Sycorax, a Uranian satellite: 19
- Makemake, a Trans-Neptunian object: 27.2
- Charon, the biggest Plutonian satellite: 32.4
- Hydra, the outermost Plutonian satellite: 32.5

Although the biggest bodies in the Solar System have the smallest ratio "orbital speed / rotational speed", the examples mentioned above suggest that the ratio "orbital speed / rotational speed" alone may not be sufficient to explain the sizing of all the bodies across different types of bodies.

The ratio "orbital speed / rotational speed" of the planets varies between 0.95 (for Saturn) and 19336.2 (for Venus) (Fig. 78). In general, the smallest values were found with the biggest satellites:

- Jupiter: 1.03
- Saturn: 0.95
- Uranus: 1.66
- Neptune: 2.01

In contrast, the highest ratio "orbital speed / rotational speed" of the planets were found with the smallest planets, particularly the innermost planets:

- Mercury: 15742.94
- Venus: 19336.2

Fig. 78: Ratio "**Orbital speed / Rotation speed**" of planets in the Solar System

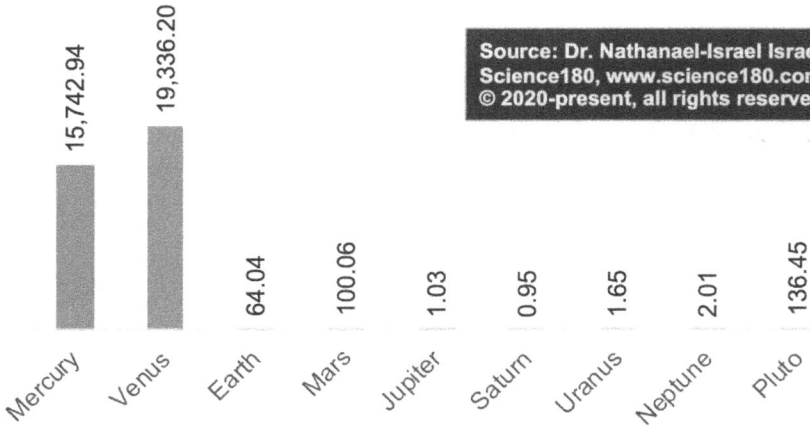

As I carefully pondered on why the ratio "orbital speed / rotational speed" of the biggest planets is smaller than that of the smaller planets, it appeared to me that the highest orbital speed of the innermost planets and the way their fluids were gathered once could have contributed to preventing them from rotating faster. I also felt like the way the energy in the precursors was split-gathered into translational movement and rotational movement could have significantly contributed to defining the shape and form of their daughter bodies. During the formation of the bodies, the force which "pushed" the biggest bodies into orbit could have also affected their rotation. The fluids that "swirled" or wound up to birth the biggest bodies could have been moved at a speed similar to their orbital speed as they were being collected to form and shape (e.g. spherical shape) the bodies. To put it another way, an orbiting fluid would have swirled to confer to its daughters a rotational movement. Some people may ask why and how an orbiting fluid could have started rotating. The turbulence that subdued the precursors birthed and affected the rotation of their daughter bodies. Hence, for the largest planets, the orbital speed and the rotational speed are nearly similar, for the rotational speed was developed out of the turbulence which was going on in the moving fluids.

The ability of Saturn to rotate faster than it revolves around the Sun could have also partially affected its oblate shape. By the way, Saturn is the planet that has the highest ellipticity. The combination of the rotational speed of Saturn and the probable low viscosity and density of its precursor as compared to that of Jupiter, could explain why Saturn ended up being so much oblate. For because the viscosity of Saturn's precursor could have been smaller than that of Jupiter's precursor, and that the rotational speed of Saturn and Jupiter are not that different, Saturn could have been more squeezed at the pole than Jupiter. Like I

256

explained in other chapters, as bodies were rotating, air could have flown to their poles. The effect of the air coming at the North and South poles could have also compressed the sphere of the planets and caused it to be oblate.

As for the main belt asteroids, the ratio "orbital speed / rotational speed" ranges from 198.7 to 166,115. Furthermore, the range of the semi major axis of the main belt asteroids is about 1.32 AU, whereas the highest range of the semi major axis of the satellites (0.32 AU) is encountered with the Neptunian satellites. Yet, the range of the ratio in the main belt asteroid is not as high as that of the satellites of some planets (e.g. Saturn).

To some extent, the ratio "orbital speed / rotational speed" is smaller for the biggest asteroids in the main belt. For instance, the smallest ratio "orbital speed / rotational speed" was recorded with Ceres, the biggest main belt asteroid (radius = 469.5 km), whereas the highest ratios were observed with some of the smallest main belt asteroids. The high value of the ratio "orbital speed / rotational speed" of most asteroids could not have allowed the "eddies" or vortical structures in the precursors to be gathered into one single body to form a unified planet. Therefore, they were spread over the portion of space occupied by their precursors.

Finally, as for the satellites, the ratio "orbital speed / rotational speed" depends on the types of satellites (p=0.001), but not on the turbulence zones (p=0.39). The ratio for the satellites varies between 19 and 390,633.5 (Fig. 79). The highest value was observed with the Saturnian satellites, whereas the lowest value was with the Uranian satellites and the Plutonian satellites. While some Uranian satellites have a very small ratio, others have a high ratio. While some of the smallest ratios are present in Zones 4 and 5 of the Jovian satellites, the first smallest ratios are found in Zone 3. Four out of the 6 smallest ratios are found in Zone 3, whereas the other 2 smallest numbers out of the 6 are found in Zone 4. For the Uranian satellites, the highest ratio was in Zone 2.

The ratios for the biggest bodies are neither the highest ones nor the smallest ones. These ratios are somewhere toward the lowest edges. This suggests that the extremes of these ratios are not what controls the size of the bodies, but a certain moderate range. Except for Charon (the largest Plutonian satellite) which ratio is 32.4, most of the largest satellites in Zone 3 have a ratio ranging between 221.9 and 4851. The lowest value is for the Moon, whereas the highest is for a Saturnian satellite. The ratio of Nereid is very small (43.2), that of Hyperion is very high (10693.4), and that of Phoebe is very small (85.6). Nereid and Hyperion are some of the satellites that have some of the most chaotic rotation in the Solar System.

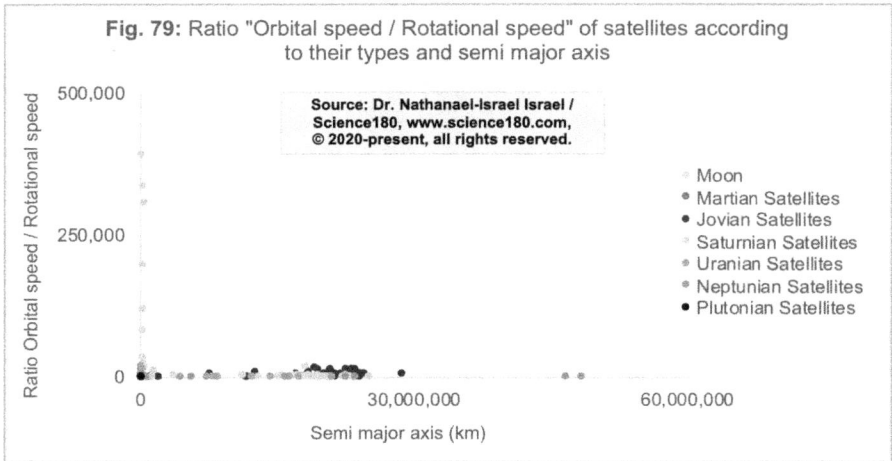

Fig. 79: Ratio "Orbital speed / Rotational speed" of satellites according to their types and semi major axis

20.2. Ratio "orbital angular speed / rotational angular speed" of the bodies in the Solar System

Dependent on the types of bodies (p=0.000), the ratio "orbital angular speed / rotational angular speed" of the bodies in the Solar System varies between 1.7E-08 and 1.08 (Fig. 80). The smallest value was found with Hyakutake, a Comet, whereas the highest ratio was with Venus. Besides Venus, all of the highest values were found with satellites. Most of the smallest values are found with the asteroids. As far as planets were concerned, the first 2 innermost planets in the Solar System have the highest ratio.

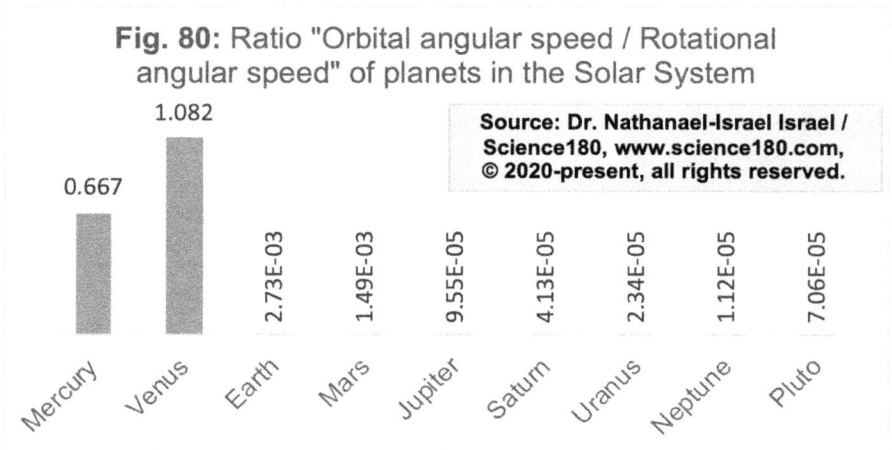

Fig. 80: Ratio "Orbital angular speed / Rotational angular speed" of planets in the Solar System

Nathanael-Israel Israel: Discoverer of the Universe Creation Formula™

Fig. 81 is made from the inverse of the ratio calculated above, meaning that it shows the ratio "rotational angular speed / orbital angular speed" instead of the ratio "orbital angular speed / rotational angular speed" as done above. It clarifies the trends for the outer planets.

Fig. 81: Ratio "Rotational angular speed / Orbital Angular speed" of planets in the Solar System

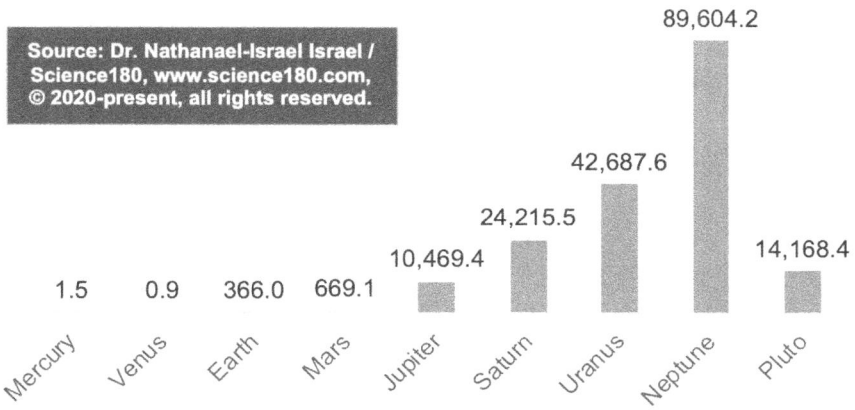

The ratio for the main belt asteroids is relatively smaller as compared with that of the satellites (Fig. 82). For instance, the highest ratio found in the main belt asteroids is 0.01 which is 100 times smaller than the highest ratio found for the satellites. In other words, the rotational angular speed of the asteroids in the main belt was too much higher than the orbital angular speed.

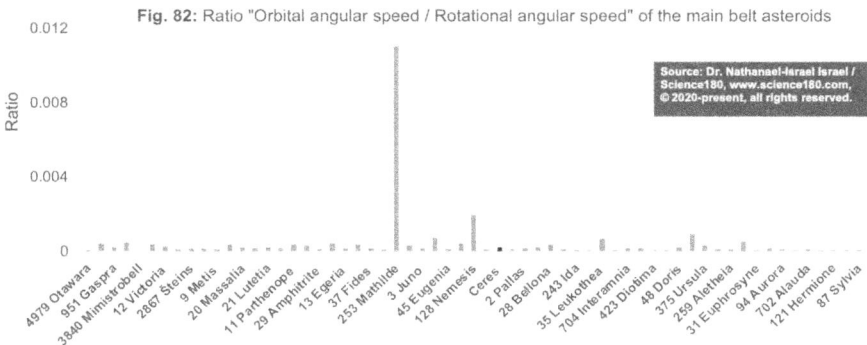

Fig. 82: Ratio "Orbital angular speed / Rotational angular speed" of the main belt asteroids

When I narrowed the types of bodies to satellites only, I found that the ratio "orbital angular speed / rotational angular speed" depends on the types of satellites (p<0.000), the turbulence zones (p=0.000), and the interaction between the types of satellites and turbulence zones (p<0.000). From the innermost satellite in Zone 1 until the outermost satellite in Zone 3, meaning for the satellites in Zones 1, 2, and 3, the ratio "orbital angular speed / rotational angular speed" is almost equal to 1 regardless of the types of satellites or the primary planet. Downstream of the outermost satellite of Zone 3, that ratio significantly dropped (Fig. 83-84).

Since I noticed this trend, I knew it would not be by chance, but must be related to how the bodies acquired their movement. As I was trying to understand why this ratio is 1 for the satellites in Zones 1, 2, and 3, I felt like as the fluids in the precursors of the bodies in Zones 1, 2, and 3 were being split-gathered, the same movement of the fluids was translated in a rotation and a revolution. In other words, as the fluids of the precursors of the bodies in Zones 1, 2, and 3 were moving to be positioned into orbit, they were also being "wrapped" or "wound up" into the daughter bodies in such a way that the orbital angular speed of these fluids was the same as their rotational angular speed. The main difference was the reference of the revolution and the rotation. For the orbital angular speed was defined with respect to the position of the primary bodies, whereas the rotational angular speed was with respect to the rotation axis of the satellites. After the disappearance of the precursors of the bodies, the rotational angular speed of the daughter bodies in Zones 1, 2, and 3 is the same as the rotational angular speed. To put it another way, a rotation and a revolution were imparted onto the daughter bodies by similar processes which acted on the fluids of their mother. However, as explained in other chapters (e.g. orbital inclination, semi major axis, axial tilt), as the precursors of the satellites in Zones 4 and 5 split from those of the satellites in Zones 1, 2, and 3, they were "ejected" due to a change of momentum, vortex stretching and change of some components of the vorticities of their fluid layers, therefore breaking the perfect relationship found in Zones 1, 2, and 3. I revisited these concepts of change in vorticity in the chapter consecrated to vortex and fluid ligaments. Hence for the satellites in Zones 4 and 5, the ratio is not 1. For instance, while considering the ratios for all the satellites in Zones 4 and 5, the 4 Plutonian satellites in Zone 4 have the highest values, ranging from 0.01 to 0.17. The next highest ratio (0.0034) for the satellites in Zones 4 or 5 is found with Themisto, a Jovian satellite, and with Tarqeq, a Saturnian satellite.

After I finished my writing, I later learned that as of 2020, Hyperion is the only natural satellite in Zones 1, 2, and 3 of the planetary systems in the Solar System known to have a rotational period different from its orbital period, which means that its ratio as calculated above should be different than 1. Because the motion of Hyperion is chaotic, I deducted that bodies having a ratio different than 1 may likely have a chaotic motion.

Fig. 83: Ratio "Orbital angular speed / Rotational angular speed" of the satellites according to their types and semi major axis

Fig. 84: Ratio "Orbital angular speed / Rotational angular speed" of the satellites according to their types and turbulence zones

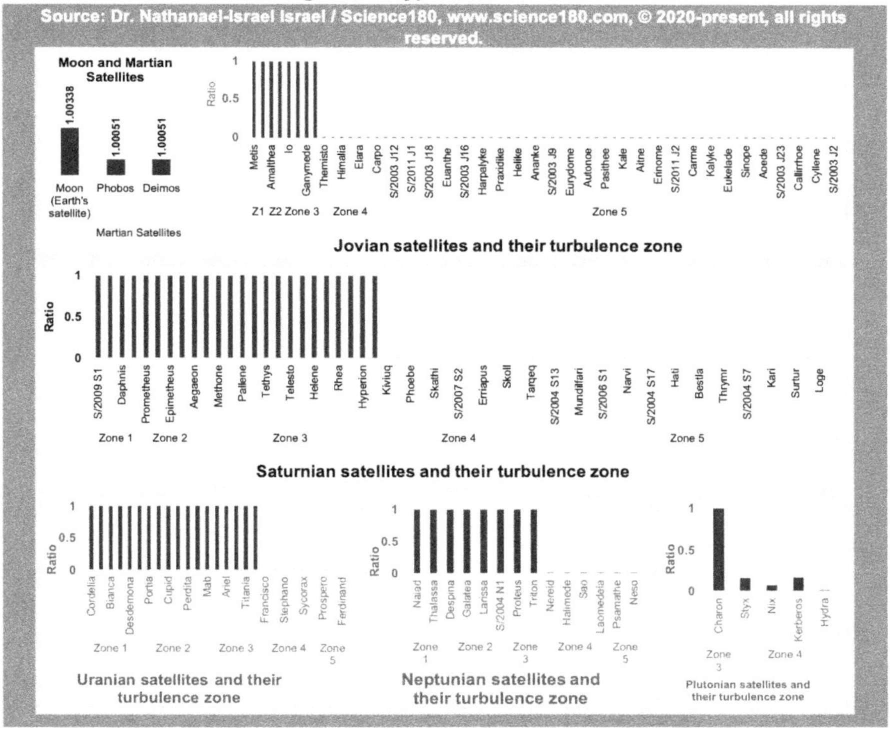

Science180: The Place Where the Accurate Interpretation of Universe-Origin Data Matters

20.3. Ratio "orbital period / rotational period" of the bodies in the Solar System

Orbital period and rotation period are key variables used to estimate the orbital speed and rotational angular speed respectively of some bodies. They played a crucial role in my study of turbulence and the origin of the universe. As a preamble to this segment, I would like to mention that in the literature, the word spin-orbit resonance is used to address the ratio "orbital period / rotational period". When the "orbital period / rotational period" = 1, some literatures say that the spin-orbit resonance is 1. For instance, because the ratio for Mercury is about 1.5, which is equivalent to 3/2, meaning that Mercury rotates 3 times for every 2 orbits, some literatures say that Mercury has a 3:2 spin-orbit resonance. Likewise, "tidal locking" I used for bodies which ratio "orbital period / rotational period" is 1; and the satellites which ratio "orbital period / rotational period" is 1 are said to be "tidally locked". However, because of the errors of many theories related to resonance and their implications, and because I do not want to confuse some people and make them think about disagreeable things, I chose not to use that term in my writings. Therefore, instead of saying satellites tidally locked, I will keep the simple term "satellites whose ratio "orbital period / rotational period" is 1".

The ratio "orbital period / rotational period" significantly depends on the types of bodies in the Solar System ($p<0.000$). Based on the data I investigated, the highest ratio "orbital period / rotational period" (5.84E+07) in the Solar System was held by Hyakutake (a comet), whereas the smallest ratio (0.92) was observed with Venus. Most of the highest values were recorded with the asteroids (Table 7).

Table 7: Ratio "Orbital period / Rotational period" of celestial bodies in the Solar System

Types of bodies	N	Mean	Minimum	Maximum	Range	Std. Deviation	Variance	Skewness	Kurtosis	CV (%)
PlanetsAsteroidsSatellites	325	232,730.9	0.92	5.84E+07	5.84E+07	3.28E+06	1.08E+13	17.363	307.400	1410.22
Planets	9	20,242.5	0.92	8.96E+04	8.96E+04	2.97E+04	8.83E+08	1.912	3.698	146.76
PlanetsAndAsteroids	147	512,293.8	0.92	5.84E+07	5.84E+07	4.87E+06	2.38E+13	11.676	138.978	951.50
PlanetsAndSatellites	187	2,740.8	0.92	8.96E+04	8.96E+04	8.01E+03	6.42E+07	7.941	77.834	292.40
Asteroids	138	544,384.1	90.39	5.84E+07	5.84E+07	5.03E+06	2.53E+13	11.312	130.461	924.04
Amor asteroids	8	4,153.7	461.86	1.06E+04	1.01E+04	3.25E+03	1.06E+07	1.078	1.165	78.30
Apollo asteroids	21	3,759.6	268.19	1.05E+04	1.02E+04	3.31E+03	1.09E+07	0.857	-0.648	87.96
Aten asteroids	8	18,666.5	204.12	1.29E+05	1.28E+05	4.47E+04	2.00E+09	2.755	7.661	239.67
Centaurs	6	80,992.1	64,118.6	1.13E+05	4.87E+04	1.66E+04	2.76E+08	1.771	3.988	20.52
Comets	4	14,604,612.3	1,183.8	5.84E+07	5.84E+07	2.92E+07	8.52E+14	2.000	4.000	199.92
Jupiter trojans	12	9,535.4	1,015.8	1.61E+04	1.51E+04	5.30E+03	2.81E+07	-0.469	-1.428	55.57
Kuiper belt objects	3	231,945.5	141,542.1	3.90E+05	2.48E+05	1.37E+05	1.88E+10	1.676		59.07
Main asteroid belt	60	5,326.6	90.4	1.38E+04	1.37E+04	3.06E+03	9.38E+06	0.603	-0.101	57.51
Mars trojans	1	6,122.2								
Trans-Neptunian objects	15	988,207.8	107,033.9	9.72E+06	9.61E+06	2.42E+06	5.86E+12	3.840	14.820	245.07
Satellites	178	1855.873	0.988	21959.8	21958.8	3352.1	1.12E+07	4.021	18.104	180.62
Moon (Earth's satellite)	1	0.997								
Martian satellites	2	1	1	1	0	0	0			0.00
Jovian satellites	67	1276.549	1	2210.6	2209.6	582.8	3.40E+05	-1.303	0.430	45.65
Saturnian satellites	62	1517.831	9.88E-01	4534.0	4533.0	1476.5	2.18E+06	0.320	-1.426	97.28
Uranian satellites	27	3508.388	1	21959.8	21958.8	6275.0	3.94E+07	1.817	2.346	178.86
Neptunian satellites	14	3990.413	1	19529.2	19528.2	6880.6	4.73E+07	1.822	2.228	172.43
Plutonian satellites	5	23.163	1	88.9	87.9	37.0	1.37E+03	2.156	4.704	159.94

As far as the planets in the Solar System are concerned, Venus has the lowest ratio "orbital period / rotational period" (0.92), whereas Neptune has the highest value (89604.2 (Fig. 85). For the Earth, that ratio is equal to 365 which is the number of days in a year.

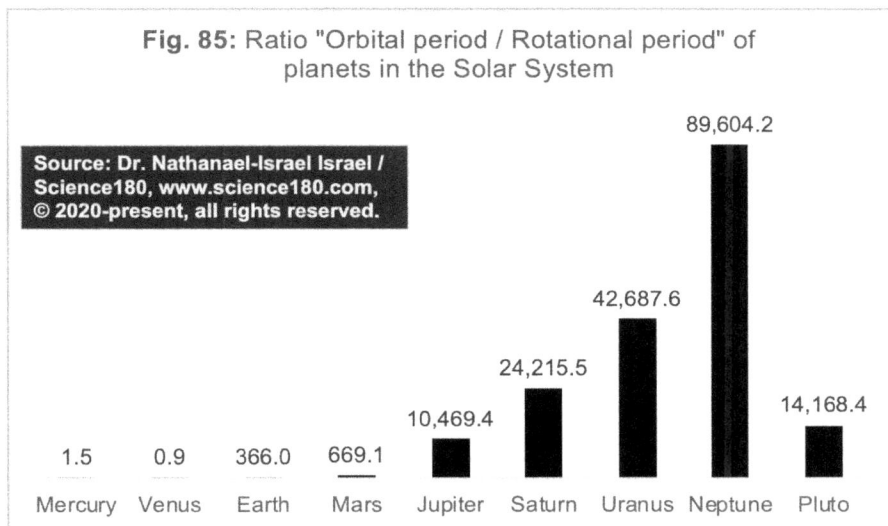

Fig. 85: Ratio "Orbital period / Rotational period" of planets in the Solar System

Source: Dr. Nathanael-Israel Israel / Science180, www.science180.com, © 2020-present, all rights reserved.

For the asteroids, the ratio between "orbital period / rotational period" varies between 90.4 and 58,400,000. The lowest value was recorded with Mathilde, an asteroid in the main asteroid belt which semi major is 2.6 AU. In contrast, the highest ratio was recorded with Hyakutake, a comet, whose semi major axis is 1165 AU. Considering the value of the semi major axis of the comet Hyakutake, it is possible that a mistake could have been made in estimating not only the semi major axis but also the rotation period.

For the main belt asteroids (Fig. 86), the intermittence of the value is huge. In other words, smaller ratios are found between higher values, suggesting that the ratio is not related to semi major axis.

The ratio "orbital period / rotational period" of the satellites in the Solar System varies between 1 and 21959.8 (Fig. 87). That ratio is 1 for more than 99% of the satellites in Zones 1, 2, and 3. Indeed, as of 2020, Hyperion is the only natural satellite in Zones 1, 2, and 3 known to have a rotational period different from that of its orbital period. In contrast, in Zones 4 and 5, that ratio jumped, reaching several thousands in some cases according to the satellites. Toward the end of this section, I gave more details about the ratio in Zones 4 and 5. That ratio for the Moon (Earth's satellite), the Martian satellites, and Charon (the biggest satellite of Pluto) is 1, suggesting that those satellites deserve not to be put in Zones 4 or 5. As I explained in the previous ratio, the process which formed the bodies explained why the ratio is 1 for all of the satellites in Zones 1, 2, and 3.

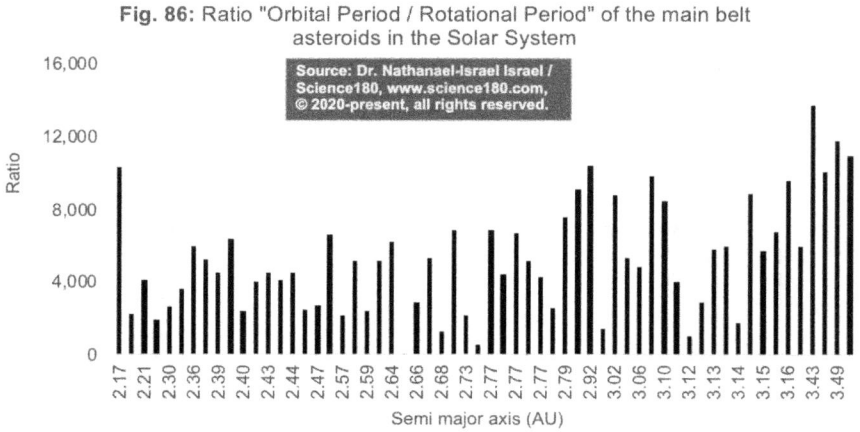

Fig. 86: Ratio "Orbital Period / Rotational Period" of the main belt asteroids in the Solar System

Considering the formula of orbital period and rotational period, I then tried to see if I can find any relationship between certain variables in Zones 1, 2, and 3.

Orbital period = Time to complete one revolution or orbit = (Circumference of orbit / mean orbital speed)

Circumference of orbit = 2 * π * (Semi major axis)

Therefore,
Orbital period = (2 * π * Semi major axis / Orbital speed)

On another hand,

Rotation period = Time to complete one rotation = (Circumference of the body / Rotational speed)

Circumference = 2 * π * Radius

Therefore,

Rotational period = (2 * π * Radius / Rotational speed)

Considering the aforementioned equations of the orbital period and the rotational period, the fact that the ratio "orbital period / rotational period" = 1 in Zones 1, 2, and 3 implies that in those zones:

264

Nathanael-Israel Israel: Discoverer of the Universe Creation Formula™

$$(2 * \pi * \text{Semi major axis} / \text{Orbital speed}) = (2 * \pi * \text{Radius} / \text{Rotational speed})$$

A little simplification of this formula leads to:

$$(\text{Semi major axis} / \text{Orbital speed}) = (\text{Radius} / \text{Rotational speed})$$

Therefore,

Radius of the bodies in Zones 1, 2, and 3 = (Semi major axis) * (Rotational speed) / (Orbital speed)

This formula can have a major implication, for it can help to generalize a lot of things, at least in Zones 1, 2, and 3. For instance, because orbital speed is highly related to semi major, it is possible to replace the formula of the orbital speed and better simplify the relationship. By doing so, in Zones 1, 2, and 3, the radius of the satellites is (Table 8):

Table 8: Formula of the radius of the satellites in Zones 1, 2, and 3

Radius	Type of body	Equation of best Model of the regression between orbital speed and semi major axis	Radius of satellites in Zones 1, 2 and 3
Radius of satellites in Zones 1, 2 and 3 = (Semimajor * Rotational Speed)/Orbital speed	Martian satellites	208.147*SemiMajor^-0.501	(Rotational speed/208.147)*SemiMajor^1.501
	Jovian satellites	11358.893*SemiMajor^-0.501	(Rotational speed/11358.893)*SemiMajor^1.501
	Saturnian satellites	6278.97*SemiMajor^-0.502	(Rotational speed/6278.97)*SemiMajor^1.502
	Uranian satellites	2431.135*SemiMajor^-0.501	(Rotational speed/2431.135)*SemiMajor^1.501
	Neptunian satellites	2690.372*SemiMajor^-0.503	(Rotational speed/2690.372)*SemiMajor^1.503
	Plutonian satellites	30.124*SemiMajor^-0.496	(Rotational speed/30.124)*SemiMajor^1.496

Unlike Zones 1, 2, and 3 where that ratio "orbital period / rotational period" is 1, in Zones 4 and 5, it varies between 6.1 and 21959.8 according to the primary planets and the satellites. The lowest ratios found in Zones 4 and 5 were recorded with the 4 Plutonian satellites, ranging between 6.1 and 88.9 with the highest value obtained with the outermost Plutonian satellite and the lowest value with the innermost Plutonian satellite. In contrast, the highest value was with Ferdinand, a Uranian satellite. In general, the 12 highest ratios were recorded with 8 Uranian satellites and 4 Neptunian satellites.

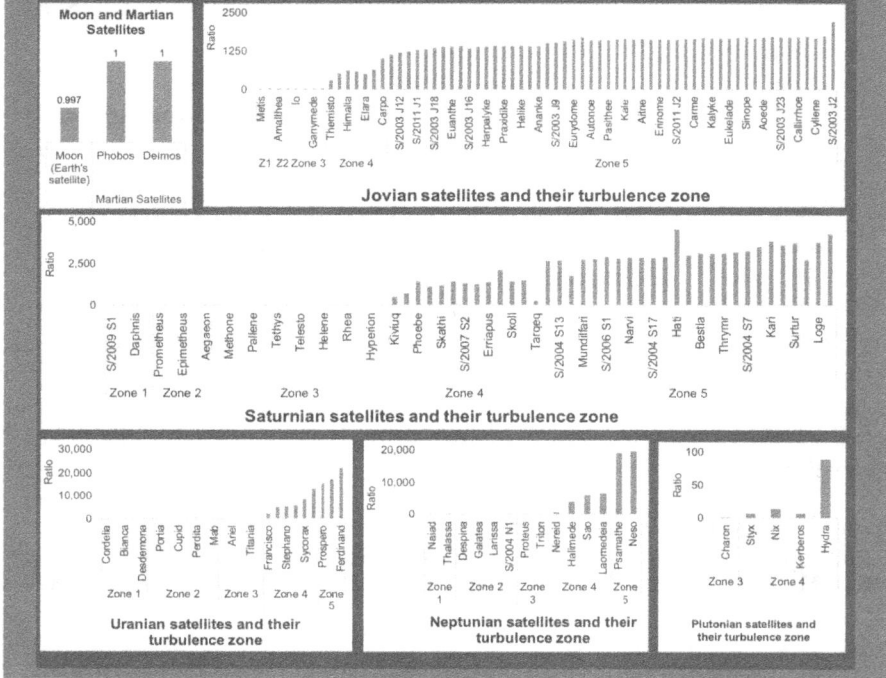

Fig. 87: Ratio "Orbital period / Rotational period" of the satellites according to their types and turbulence zones

20.4. Take home message

The highest orbital speed of the innermost planets and the way their precursor was shaped by turbulence and "ejected" from their mother precursor could have contributed to preventing them from rotating faster. The way the energy in the precursors was split-gathered into translational movement and rotational movement could have significantly contributed to defining the shape and form of their daughter bodies. During the formation of the bodies, the force which "pushed" the biggest bodies into orbit could have also affected their rotation. The fluid which was "swirled" or wound up to birth the biggest bodies could have been moving at a speed similar to that of its orbital speed as it was being collected to form and shape (e.g. spherical shape) the bodies.

From the innermost satellite in Zone 1 to the outermost satellite in Zone 3, meaning for the satellites in Zones 1, 2, and 3, the ratio "orbital angular speed / rotational angular speed" is almost equal to 1 regardless of the types of satellites or the primary planet. Downstream of the outermost satellite of Zone 3, that ratio significantly dropped. As the fluids in the precursors of the bodies in Zones 1, 2, and 3 were being split-gathered, a same movement of the fluids was translated into a rotation and a revolution. In other words, as the fluids of the precursors of

266

Nathanael-Israel Israel: Discoverer of the Universe Creation Formula™

the bodies in Zones 1, 2, and 3 were moving and then positioned into orbit they were also being "wrapped" or "wound up" into the daughter bodies in such a way that the orbital angular speed of these fluids was the same as their rotational angular speed. As of 2020, Hyperion was the only natural satellite in Zones 1, 2, and 3 of the planetary systems in the Solar System known to have a rotational period different than its orbital period. Because the motion of Hyperion is chaotic, I deduced that the unbalance between the rotational speed and orbital speed of the bodies can explain their chaotic motion. Finally, in Zones 1, 2, and 3, the ratio "orbital period / rotational period" is 1.

CHAPTER 21

DISCOVER HOW SMART PEOPLE THINK ABOUT THE ECCENTRICITY OF CELESTIAL BODIES AS THEY PREPARE TO FIX WRONG SCIENTIFIC THEORIES

21.1. Generalities

Eccentricity is a dimensionless variable (meaning it has no unit) used to measure the shape or the circularity of an orbit. It tells how round or flat an orbit is. Eccentricity and semi major axis are the two main elements used to define the shape and size of the orbit of celestial bodies. When an orbit is a perfect circle, its eccentricity is 0. The orbit of the bodies which have a larger eccentricity are more elongated. The eccentricity of an ellipse is greater than 0 and less than 1. The eccentricity of a parabola is 1. When the eccentricity is greater than 1, the orbit is a hyperbola. Finally, the eccentricity of a pair of lines is infinite. When eccentricity is closer to 1, the orbit becomes flatter. That may be the case for some outermost secondary bodies that lost their ability to orbit their primary body.

In this chapter, I explained how the way some bodies were separated from others can explain the elongation of their orbit. Indeed, I studied the eccentricity of 462 celestial bodies in the Solar System. Eccentricity depends on the types of bodies ($p < 0.000$) and ranges from 0 to 1.057. The lowest value was measured on a satellite, whereas the highest value was measured on a comet. Besides planets and trojans also have a very small eccentricity. To answer the question "Why is the eccentricity of the bodies different?", the shortest answer I could give as of 2020 was that, as the bodies were being formed, the precursors of some were ejected or propelled from their peers more than others and in the end, their orbit was stretched. In this chapter, I provided a glimpse at that answer, but more details will come in other chapters including that on vortex and fluid ligaments, as I delve into other variables.

21.2. Eccentricity of the asteroids and planets in the Solar System

The eccentricity of the asteroids and planets ranges between 0.007 and 1.057 and seems not to depend on their types or semi major axis, suggesting that, on a global scale, something else besides semi major axis may explain the variation among the eccentricity (Fig. 88).

Fig. 88: Eccentricity of the Planets and Asteroids in the Solar System

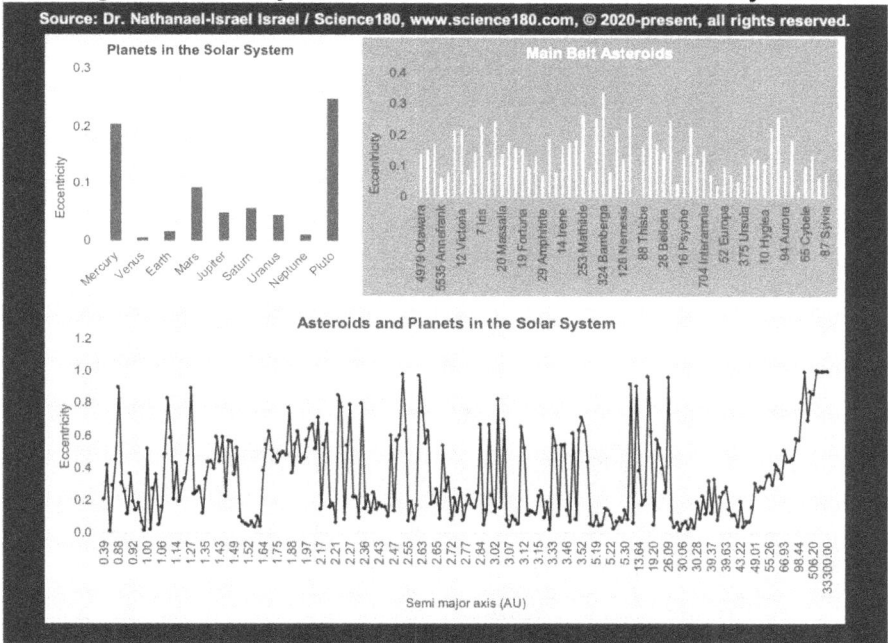

The eccentricity of the asteroids in the main belt ranges between 0.019 and 0.340 (Fig. 88). The eccentricity of the planets in the Solar System ranges between 0.007 and 0.249 with the highest value recorded with Pluto followed by Mercury (0.2056) (Fig. 88). In contrast, the orbit of Venus is the least eccentric among the planets in the Solar System. Why? Among other things, the high axial tilt of Venus might have contributed to preventing that planet from having a very elongated orbit, hence a lower eccentricity. Probably, the more the precursor of a planet spent its energy to tilt its axis of rotation, the lesser the tendency of that orbit to be elongated. Although imperfect, a negative correlation was found between the axial tilt of the planets and their eccentricity (r2=0.566). In the case of the satellites, this relationship is stronger (r2=0.923). Probably, the forces that tilted the celestial bodies and their orbital planes into different directions and angles also contributed to define the shape and circularity of their orbits differently.

Hence, a significant correlation between these variables. Although the strength of these correlations is not always very high, they partially add up to explain the variation in the shape of the orbit and the inclination of the celestial bodies.

21.3. Eccentricity of satellites

The eccentricity of the satellites depended on their types ($p=0.02$), the turbulence zone ($p<0.000$), and the interaction between these 2 variables ($p=0.003$). It ranges between 0 and 0.751 (Fig. 89-90).

Fig. 89: Eccentricity of 210 Satellites according to their types and rank

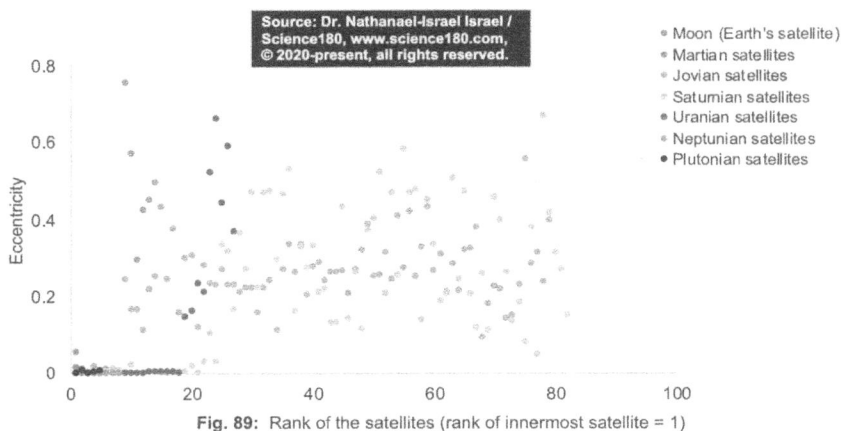

Legend:
- Moon (Earth's satellite)
- Martian satellites
- Jovian satellites
- Saturnian satellites
- Uranian satellites
- Neptunian satellites
- Plutonian satellites

Fig. 89: Rank of the satellites (rank of innermost satellite = 1)

Fig. 90 allows a quick comparison of the trends of the eccentricity of all the satellites according to their types.

Among the satellites in the Solar System. the highest eccentricity was observed with Nereid, the innermost Neptunian satellite in Zone 4. The high energy with which the precursors of the Neptunian satellites in Zones 4 and 5 were pushed away from their peers in Zones 3, 2, and 1 could have contributed to explaining why the eccentricity of Nereid was so high. Furthermore, in the chapters of vortex and fluid ligament, I explained how the stretching of the vortices of the precursors of satellites in Zones 4 and 5 under the influence of the precursors of bodies in Zone 3 (which rotational angular speed was highly reduced) caused the precursors of satellites in Zones 4 and 5 to be tilted and to have elongated orbits. Hence, their higher eccentricity and orbital inclination.

Fig. 91 presents the eccentricities according to the turbulence zones.

Fig. 90: Eccentricity of the satellites in the Solar System according to their types and turbulence zones

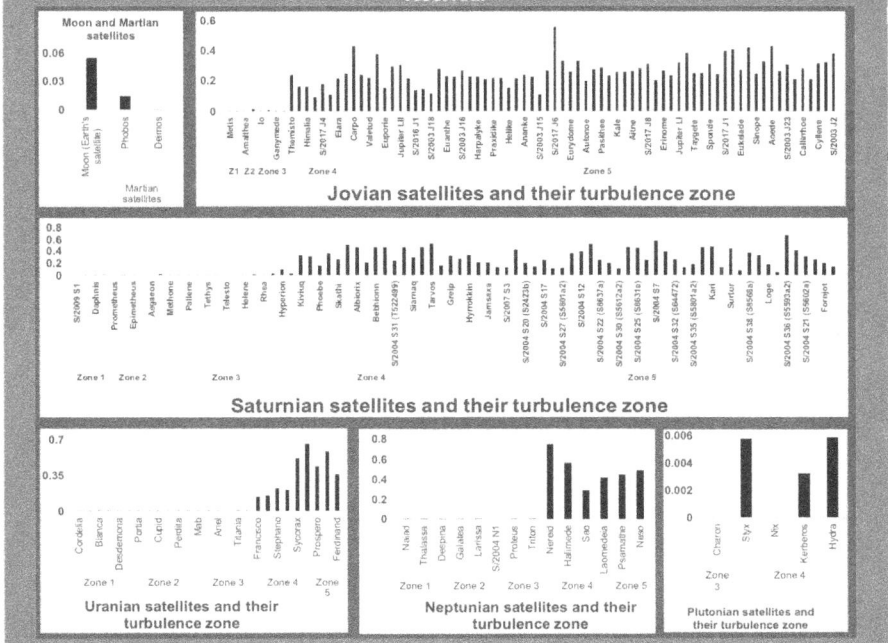

Jovian satellites and their turbulence zone

Saturnian satellites and their turbulence zone

Uranian satellites and their turbulence zone

Neptunian satellites and their turbulence zone

Plutonian satellites and their turbulence zone

In all planetary systems, no satellite in Zones 1, 2, or 3 has an eccentricity higher than that of the innermost satellite in Zone 4 (Fig. 91). In other words, from the outermost satellite in Zone 3 to the innermost satellite in Zone 4, there is a huge leap in the eccentricity, suggesting that something must have happened between the outermost satellite in Zone 3 and the innermost satellite in Zone 4 that significantly increased the eccentricity of the orbit of the innermost satellite in Zone 4 and of that of most satellites afterwards or downstream of it. Like I explained in the chapter on semi major increment, the distance separating the innermost satellite in Zone 4 and the satellite upstream of it is usually among the highest in each planetary system. As I explained in other chapters as well, the way the precursors of the outermost satellite in Zone 3 and that of the innermost satellite in Zone 4 were separated from one other was responsible for the shift in the eccentricity. For the precursors of the satellites downstream of the outermost satellite in Zone 3 was kind of expelled, propelled, or stretched as it was splitting, and/or after it split from the precursor of the bodies that were upstream of them.

The highest eccentricities of the satellites in most of the planetary systems were usually observed in Zones 4 or 5. However, the position of the satellite with the highest eccentricity varied according to the planetary system. For instance, Carpo, the Jovian satellite with the highest eccentricity, is the outermost satellite in Zone 4. Similarly, Margaret (the Uranian satellite with the highest eccentricity) is the outermost Uranian satellite in Zone 4. In contrast, S/2004 S7 (the Saturnian satellite with the most eccentric orbit) is located in Zone 5. Hydra (the Plutonian satellite with the highest eccentricity) is the outermost satellite in Zone 4.

Fig. 91: Eccentricity of the satellites in the Solar System according to their types, semi major axis and turbulence zones

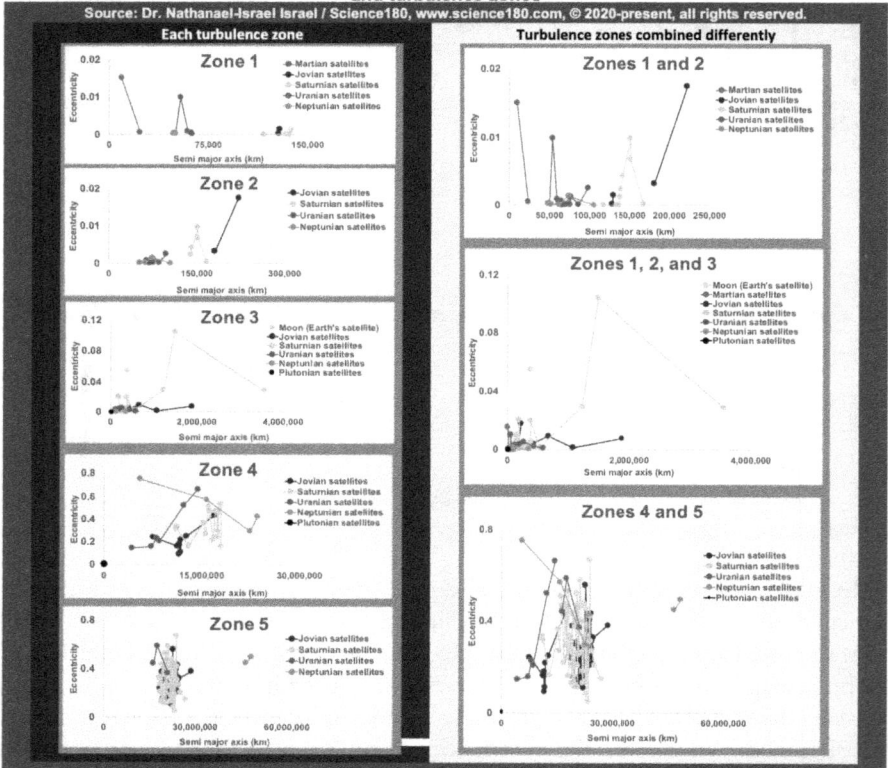

The localization of the satellites with the smallest eccentricity depends on the planetary system. For the Saturnian and Jovian planetary systems, the satellite with the smallest eccentricity was observed in Zone 1. In contrast, for the Neptunian and Uranian satellites, the smallest eccentricity was observed with a satellite in Zone 2. Because the Plutonian planetary system has no satellite in Zones 1 or 2, its smallest eccentricity was observed in Zone 3. Although some of the smallest eccentricities were recorded with satellites in Zone 3, that zone is not where the smallest eccentricities are always found.

Furthermore, the variation of the eccentricity is also smaller in Zones 4 and 5 than in the other zones. All of the satellites which semi major axis is equal to or less than 0.02380 AU (3,561,300 km) have an eccentricity smaller than 0.03. Any satellite which semi major is higher than 0.02380 AU has an eccentricity higher than 0.1. In general, the inner satellites usually have the least eccentricity, whereas the eccentricity of the outer satellites fluctuates and the shape of their graphical representation is like a zigzag. This suggests that, in addition to the semi major axis, something else is controlling the eccentricity of the satellites. According to the rank of the satellites, the eccentricities of the satellites look like the following.

21.4. Take home message

The highest eccentricities in the Solar System were observed with asteroids. The orbit of most satellites in the Solar System is more eccentric than that of their primary planet. The eccentricity of most satellites in Zones 1, 2, and 3 is less than that of their primary planet. However, the highest eccentricity in each planetary system was observed in Zones 4 and 5. From the outermost satellite in Zone 3 to the innermost satellite in Zone 4, there is a huge jump or leap in the value of the eccentricity. A main question is to know why the orbits of the satellites in Zones 4 and 5 are so eccentric and what can explain the sharp increase of eccentricity from the outermost satellite in Zone 3 to the innermost satellite in Zone 4. The increase of the eccentricity from the outermost satellite in Zone 3 to the innermost satellite in Zone 4 could be due to how the precursors of the bodies in Zone 3 affected those in Zone 4. As the precursor of the bodies in Zones 4 and 5 were detached from the precursors of the bodies in Zones 1, 2, and 3, it was ejected with so much energy that it propelled its daughter bodies very far. In their propulsion, the shape and plane of their orbit were affected. Hence the eccentricity of the satellites in Zones 4 and 5 and as explained in the chapter on orbital inclination, bodies in Zones 4 and 5 also have the highest orbital inclination. The way the layers of fluids in some precursors of the bodies (particularly in Zones 4 and 5) were separated from the remainder of the fluid layers upstream and downstream of them contributed to elongating the orbit of their daughter bodies, hence their higher eccentricity. The data of eccentricity suggested to me that a connection exists between eccentricity, orbital inclination, and axial tilt of the celestial bodies. In other words, the eccentricity of the orbit of the bodies in the Solar System has something to do with the way their movement was launched during their formation. For instance, it is interesting that the axial tilt of Mercury is among the least, yet its eccentricity and orbital inclination are among the highest. The energy of the precursor of each body may have been split into the energy used to form and pull/push the daughter bodies into their orbit and the energy that pushed them into different positions, leading to a different inclination, tilt, orbital inclination, and eccentricity. In the chapters on vortex and fluid ligament, I elucidated how the stretching, tilting, and changes in the vorticity components of the fluid layers explain the trends of the eccentricity as well as the orbital inclination in Zones 4 and 5.

CHAPTER 22

AXIAL TILT OF CELESTIAL BODIES CAN ASSIST YOU TO UNDERSTAND THE UNIVERSE-ORIGIN PROPERLY

22.1. Definition

Also called "obliquity" or "obliquity to orbit", the axial tilt is a physical characteristic that defines the angle between a celestial body's rotational axis and its orbital axis. It is equivalent to the angle between the equatorial plane and the orbital plane. When the axial tilt of a celestial body is less than 90°, the motion of that body is said to be prograde. In contrast, when the axial tilt is over 90°, the motion of that body is retrograde.

When ideas about the formation of the universe started flowing in my mind without me planning it, images of the axial tilt of the planets were one of the few trends that shaped some of my early perspective on the origin of celestial bodies. Indeed, when I was reading about the rotation of planets on 18 November 2013, I came across some axial tilt pictures (Fig. 92) that gave me an instant insight into what may have tilted the planets.

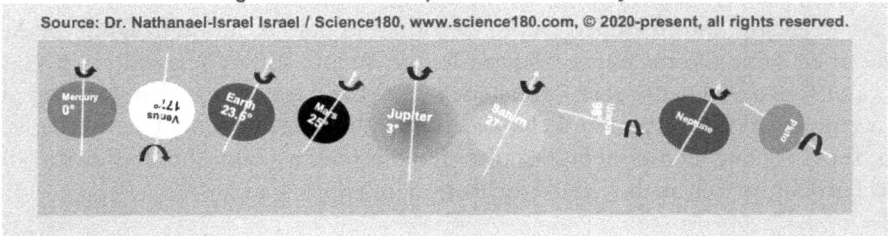

Fig. 92: Axial tilt of the planets in the Solar System

When I was looking at these pictures (that I sketched above), I felt like something must have pushed and/or pulled the planets, hence the variation of their rotational axis, making the motion of some prograde and that of others retrograde as if these bodies were rolled over. I immediately knew that the axial tilt of the planets must have been related to how they were formed. As I was carefully looking at these pictures in 2013, I realized that the precursor of the Solar System at one point was located somewhere about the Sun and that something happened to move it and split into other precursors which, after going through some changes, led to the formation of the current bodies in the Solar System.

22.2. General trend of the axial tilt of the bodies in the Solar System

Axial tilt data was available only for 71 celestial bodies in the Solar System: the Sun, 9 planets, 52 satellites, and 9 asteroids. The analysis of variance of the data showed that the axial tilt depends on the type of body (p=0.006). The axial tilt of these celestial bodies ranged from 0° to 177.36°. The lowest value was measured on satellites, whereas the highest value was measured on a planet.

22.3. Axial tilt of the asteroids and planets

The axial tilt of the asteroids in the main belt ranges between 4° and 117°. The highest axial tilt was recorded with 45 Eugenia whereas the smallest was recorded with Ceres, the biggest main belt asteroid. When all the asteroids were considered, 1993 KA2, an Apollo asteroid, recorded an axial tilt (3.19°) smaller than that of Ceres. These data suggested that both prograde and retrograde asteroids exist in the main asteroid belt.

The axial tilt of the planets in the Solar System ranges from 0.034° to 177.36°. The least tilted planet is Mercury, whereas the most tilted planet is Venus. The tilt of the Earth is 23.44°. Although most of the planets in the Solar System orbit or revolve around the Sun in the same eastward direction, which is the direction of the rotation of the Sun itself, they do not rotate their rotational axis in the same direction. The direction in which bodies rotate around their rotational axis is used to differentiate the prograde bodies from the retrograde ones. Six out of the 9 planets in the Solar System have a direct rotation, meaning they rotate west to east. In contrast, 3 planets (Venus, Uranus, and Pluto) are retrograde because they rotate westward (from east to west).

22.4. Axial tilt of the satellites

The axial tilt of the satellites does not depend on their types (p=0.079), but on their turbulence zone (p=0.000), and the interaction between these 2 variables (p=0.015). The axial tilt that was available for 52 satellites ranged from 0 to 157°. The highest axial tilt was recorded with Triton, the biggest Neptunian satellite, whereas the axial tilt of 86.6% of these 52 satellites was zero. Table 9 presents the axial tilt of the satellites in the Solar System according to their types and turbulence zone.

Table 9: Axial tilt (°) and satellite-to-planet axial tilt ratio of the satellites in the Solar System according to their types and turbulence zones

Type of Satellites	Turbulence Zone	Name	Key characteristics of the satellite	Axial Tilt of satellite (°)	Satellite-to-planet axial tilt ratio
Moon	Zone 3	Moon	Biggest satellite	6.687	0.28528
Martian satellites	Zone 1	Phobos	Biggest satellite	0	0
Jovian satellites	Zone 1	Metis	Innermost satellite	0	0
		Adrastea	Zone 1 Last satellite	0	0
	Zone 2	Amalthea	Zone 2 First satellite	0	0
		Thebe	Zone 2 Last satellite	0	0
	Zone 3	Europa		0.1	0.03195
		Ganymede	Biggest satellite	0.16	0.05112
		Callisto	Zone 3 Last satellite	0	0
Saturnian satellites	Zone 1	Pan		0	0
		Atlas	Zone 1 Last satellite	0	0
	Zone 2	Prometheus	Zone 2 First satellite	0	0
		Pandora		0	0
		Epimetheus		0	0
		Janus		0	0
	Zone 3	Mimas	Zone 3 First satellite	0	0
		Methone		0	0
		Pallene		0	0
		Enceladus		0	0
		Tethys		0	0
		Calypso		0	0
		Telesto		0	0
		Dione		0	0
		Rhea		0	0
		Titan	Biggest satellite	0	0
		Iapetus	Zone 3 Last satellite	0	0
	Zone 4	Phoebe		152.14	5.69173
Uranian satellites	Zone 1	Cordelia	Innermost satellite	0	0
		Ophelia		0	0
		Bianca		0	0
		Cressida		0	0
		Desdemona	Zone 1 Last satellite	0	0
	Zone 2	Juliet	Zone 2 First satellite	0	0
		Portia		0	0
		Rosalind		0	0
		Cupid		0	0
		Belinda		0	0
		Perdita		0	0
		Puck		0	0
		Mab	Zone 2 Last satellite	0	0
	Zone 3	Miranda	Zone 3 First satellite	0	0
		Umbriel		0	0
Neptunian satellites	Zone 1	Naiad	Innermost satellite	0	0
		Thalassa	Zone 1 Last satellite	0	0
	Zone 2	Despina	Zone 2 First satellite	0	0
		Galatea		0	0
		Larissa		0	0
	Zone 3	Proteus	Zone 3 First satellite	0	0
		Triton	Biggest satellite	157	5.54379
Plutonian satellites	Zone 3	Charon	Biggest satellite	0	0
	Zone 4	Styx	Zone 4 First satellite	82	0.66922
		Nix		132	1.07729

While the axial tilt of most satellites is zero, that of other satellites is more than 5 times that of their primary planet. For instance, the axial tilt of the Earth

Nathanael-Israel Israel: Historic Discoverer of the Life Turbulent Origin Formula™

(23.44°) is 3.5 times that of the Moon. Nix (a satellite of Pluto) is more tilted than Pluto. Similarly, the axial tilt (152.14°) of Phoebe (a Saturn satellite) is 5.69 times that of Saturn (26.73°). The axial tilt of Triton is 5.54 times that of its primary planet, Saturn. Although a few satellites have a retrograde motion, Triton is unusual as it is the largest natural satellite of Neptune. As of 2025, Triton is the only large satellite in the Solar System with a retrograde orbit. It is the seventh-largest satellite in the Solar System and, because of its size, people do not expect it to have such a big axial tilt, which put it into the category of the retrograde satellites. To put it in other words, because the retrograde motion of Triton has been puzzling scientists for a long time and they did not find any explanation for it, most theories consider the axial tilt of Triton as abnormal.

However, as I was exploring the data, I realized that nothing was abnormal about Triton if it is not the theories which are labeling this satellite as abnormal. I therefore knew that something is very wrong about the theories which tried to explain the axial tilt of Triton. Early in my investigation, although I did not know how to scientifically explain the tilt of that satellite, I sensed that the retrograde motion of some celestial bodies may not be due to their radius or size alone, for some small bodies are not tilted while other big ones like Triton are tilted. In 2016, it appeared to me that it would be erroneous to use the regression data of the 6 non-null axial tilts to conclude what tilted the satellites. To reduce the bias, I referred to the data available on the motion of the celestial bodies as I explained below. But that did not yield much until my discovery in November 2019. Fig. 93 is a graph of the axial tilts of satellites which are not null.

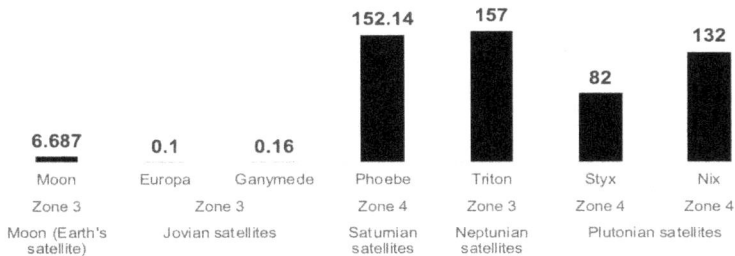

Fig. 93: Axial Tilt (°) which are not zero (0) of satellites according to their turbulence zones and types

The axial tilt of the Moon is more than that of the Martian satellites. Axial tilt data exists for the Jovian satellites in Zones 1, 2, and 3. Although very small (0.16°), the highest axial tilt recorded on the Jovian satellites was with Ganymede, the biggest Jovian satellite. The other Jovian satellite which axial tilt is not null is Europa (tilt = 0.1°). Besides the Jovian planetary system, none of the other 3 giant planetary systems has more than one satellite with a tilt that is not null. In other words, among the 4-giant planetary systems, only the Jovian one has 2 satellites which axial tilt is not zero. Nevertheless, these tilts are near zero. The Jovian satellites with the highest tilt are in turbulence Zone 3.

The axial tilt of the Saturnian satellites is zero except that (152.14°) of Phoebe, the biggest Jovian satellite in Zone 4. Having a radius of 106.7 km, Phoebe is located at about 11,722,170 km from Titan, the biggest Saturnian satellite. Discovered in August 1898 by the American astronomer William Henry Pickering, Phoebe is not the innermost satellite in Zone 4, but the third, according to the data I collected on the 62 Saturnian satellites known as of 2016. As of 2016, the 2 innermost satellites in Zone 4 located upstream of Phoebe are very small: Kiviuq (7 km of radius) and Ijiraq (5 km of radius). One of Saturn's most intriguing moons, Phoebe is located at 1,834,000 km from Kiviuq, the innermost Saturnian satellite in Zone 4. As explained in another section below and in the chapters on vortex and fluid ligament, other information related to Phoebe helped unravel the secret behind axial tilt.

The axial tilt of the Uranian satellites is 0. By the way, it is important to keep in mind that the axial tilt of Uranus itself is the third highest among all 9 planets in the Solar System, coming after that of Venus and Pluto. In other words, the axial tilt of Uranus is the highest among all of the 4-giant planets. This may explain why its satellites are not tilted. However, as explained below, considering the fact that Pluto itself is very tilted, yet some of its satellites are also very tilted, I deduced that the axial tilt of a primary planet is not enough to predict that of its satellites. The orbits of the Uranian satellites are almost perpendicular to the 'ecliptic' (the orbital plane of the Earth). The rotational axis of Uranus itself is inclined and Uranus' equator is at 98° to the ecliptic. Instead of rotating like the other planets in the Solar System, Uranus rolls along on its side. Consequently, during the revolution of Saturn, its pole (north or south) faces the Sun.

The rotational axis of Neptunian satellites is not tilted except that of Triton (157°), the biggest Neptunian satellite which is also the outermost Neptunian satellite in Zone 3. Axial tilt data is not reported for any of the Neptunian satellites in Zones 4 and 5. As explained above, the fact that a big satellite such as Triton is retrograde has been puzzling scientists for a long time and they think that axial tilt is abnormal.

The Plutonian satellite with the biggest axial tilt (132°) is not the biggest Plutonian satellite, but Nix, the second Plutonian satellite in Zone 4. Its radius, 19.7 km, is not the highest in Zone 4, but the second highest for the satellite with the highest radius (22.5 km) in Zone 4 is Hydra, the outermost Plutonian satellite. Between Nix (the most tilted Plutonian satellite) and Charon (the biggest Plutonian satellite whose radius is 606 km), there is a small satellite called Styx (the innermost Plutonian satellite in Zone 4) and its radius is 5.5 km. These data also suggested that the size of the satellites alone is not the crucial factor which defines their axial tilt.

22.5. Regression between axial tilt and other variables

Because I sensed early that axial tilt is one of the variables which holds a secret of the formation of the bodies in the Solar System, I spent a lot of time trying to understand what it means. By exploring the variables that I collected on the bodies, I thought I would find some of the root causes of the tilt of the rotational axis of the bodies. In other words, I deeply believed that an intimate knowledge of most of the variables that affect axial tilt could give me a clue into the formation of the celestial bodies. Besides doing some descriptive analysis, regression between the variables was one of the analyses I did. Unfortunately, as I explained below, these regressions did not unravel the secrets behind what tilted the rotational axis of the bodies in the Solar System.

To do these regressions, I was a little bit limited by the sample number per type of body. For instance, because it seems inappropriate to do a statistical analysis with 1 Apollo asteroids and 1 comet, I focused the regression analysis of axial tilt on the planets, satellites, and the main belt asteroids. Then, I used the knowledge gained on them to judge how the axial tilt fit into the big picture.

After a preliminary regression in 2016, I realized that the axial tilt of the asteroids in the main belt is correlated with their equatorial radius. For the equatorial radius is one of the variables that have the strongest negative effect on the axial tilt of the main asteroid belt ($r2=0.77$, $p=0.009$), meaning that the smallest asteroids in the main belt generally have the highest axial tilt. In contrast, no significant relationship was found between radius and axial tilt of the planets ($r2=0.14$, $p=0.33$) and satellites ($r2=0.002$, $p=0.74$). I tried to do some regressions with the axial tilt of the satellites, but I could not get significant trends as most of them are 0 (Table 10).

Table 10: Model summary and parameter estimates of the regression between axial tilt and equatorial radius of some bodies in the Solar System

Source: Dr. Nathanael-Israel Israel / Science180, www.science180.com, © 2020-present, all rights reserved.										
Dependent Variable: Axial Tilt (°)										
Types of bodies	Equation	Model Summary					Parameter Estimates			
		R Square	F	df1	df2	Sig.	Constant	b1	b2	b3
Planets	Linear	.136	1.099	1	7	.329	75.307	-.001		
Main Asteroid Belt	Exponential	.770	16.752	1	5	.009	210.903	-.007		
Satellites	Linear	.002	.115	1	49	.735	9.485	.003		

The independent variable is Equatorial Radius (km).

Because the coefficient of determination (r^2) is not 1, these regressions suggested that radius is not the only variable that explains the variation among the axial tilt of the bodies. I suspected that other variables should also explain the variation of the axial tilt.

At the early stage of this investigation, I wished that the axial tilt of most satellites was not null, for they were making some of the regressions I was trying to yield nothing. In those days, I did not know why most of the axial tilts were 0 for I was dealing with so many data, and I was not fully grasping their meaning nor remembering what they meant. At that time, because I was not getting much out of the analysis of axial tilt, I put them aside and continued to analyze other aspects of the massive data I had. It took me almost 5 more years before I realized why only a handful of satellites have an axial tilt which is not zero, while most of them are not tilted.

When I explored the data available on celestial bodies (NASA, 2018), I noticed most of the 178 satellites known as of 2016 have a data on motion. Because I was not getting much info from the underlying cause of the axial tilt of the bodies, I turned my attention to their motion, thinking that using the terminology "prograde" and "retrograde" that is adopted to label the motion of the bodies, I could make a better use of the data and have better regressions of the trends instead of the 0 tilt which seemed to be causing problems. Therefore, I investigated the motion of the celestial bodies and wrote a chapter of about 40 pages. Although that chapter on motion did not allow me to clearly see the cause of the axial tilt, it was the one which for the first time allowed me to know that if I better analyze the satellites not as a dependent of the planets, but to analyze both the satellites and planets as dependent of their common precursor, which was the precursor of the planetary system, I could unearth great information. Although by July 2016 I was already able to start clearly seeing the planets and their satellites as daughters of the precursors of their planetary system, I still did not know how those precursors gave birth to their daughters.

In those days, at one point, as I was trying to study the motion of the bodies, I ended up mixing the motion's nomenclature based on axial tilt with that based on orbital inclination. For the satellites which axial tilt or orbital inclination is above 90° are categorized as retrograde and those which value is inferior to 90° are called prograde. But some satellites can be prograde based on their axial tilt, but not based on their orbital inclination. As I was carefully studying the variables, I learned that it is important to differentiate whether the motion is defined using the axial tilt or using the orbital inclination, for both do not mean the same thing. For instance, a retrograde orbit does not mean that axial tilt is necessary more than 90°, but that orbital inclination is less or more than 90°. To put it into other words, orbital inclination refers to the inclination of the orbit whereas axial tilt refers to the inclination of the rotational axis. By trying to implement the nuance of this detail, I ended up with a small sample, therefore, I understood that I needed to wait for another time to find the best way to explain the cause of the axial tilt of the bodies. It would take a few more years before I figured out the answer as I described in the chapters on vortex and fluid ligament.

22.6. Insights from my review of the axial tilt of Phoebe and Triton in conjunction with their eccentricity, size, rotational speed, and orbital inclination

A careful review of the high axial tilt of Phoebe and Triton in conjunction with their eccentricity, size, rotational speed, and orbital inclination contributed to unearthing the cause of their axial tilt. Indeed, from 2013 to 2019, and particularly in 2016 (when I did the preliminary regressions), I did not have a clear understanding of what may have caused the rotational axis of the bodies to be tilted. It was on November 7, 2019, that the meaning of the axial tilt became more real to me. By that time, I already realized that the data I collected on the bodies in the Solar System was about turbulence.

Indeed, as I was carefully exploring the axial tilt of the satellites, that of two satellites appeared most puzzling: that of Phoebe (a Saturnian satellite) and that of Triton (a Neptunian satellite). I will first review some key trends surrounding Phoebe, known as one of Saturn's most mysterious moons. The highest rotational speed (0.02 km/s) of the Saturnian satellites was observed with Phoebe (located in Zone 4) and Tethys (located in Zone 3). The Saturnian satellite with the highest axial tilt, Phoebe, is the biggest Saturnian satellite in Zones 4 and 5, having a radius of 106.7 km, while the radius of the other 37 Saturnian satellites known in Zones 4 and 5 as of 2016 have a radius ranging between 2 km and 16 km, with 81% of these 37 other satellites having a radius inferior to 5 km. In other words, the major satellite in Zones 4 and 5 of the Saturnian satellites is Phoebe. I used to think that the way the precursor of Phoebe was pushed away, propelled, expelled, or pulled from the ligament connecting it to Zone 3 threw it away in such a way that its rotational axis was much tilted, hence Phoebe has the highest axial tilt among all of the Saturnian satellites. All the axial tilts reported for the Saturnian satellites is 0 except for Phoebe which is 152.14°. Much more, Phoebe is the innermost retrograde Neptunian satellite, suggesting that it may have been the innermost satellite formed from the ligament of Zones 4 and 5 which was detached from that of Zone 3. This also suggests that the satellites of Zone 4 which semi major axis is less than that of Phoebe may have been formed from the recoiling of the neck of the fluid ligament that formed Phoebe. It is also possible that the ability of Phoebe to rotate very fast could have also contributed to throwing its rotational axis. The shear rate of the precursor of the bodies in Zones 4 and 5 around Phoebe could have affected its axial tilt. Therefore, I felt like the understanding of the density, axial tilt, radius, and positioning of Phoebe will help unlock some of the secrets surrounding the formation of the bodies in the Solar System.

Triton, a Neptunian satellite, presents also interesting features similar to those of Phoebe. By embodying mysteries that no theory has solved except Science180 Cosmology that I pioneered, Triton has given so much problems to theorists; therefore, people labeled it by names such as: "Neptune's oddball moon", "weird moon", "weird object", "unusual satellite", "bizarre moon", yet that moon detains one of the keys to unlocking the secrets of the formation of the Solar System.

As I was trying to understand axial tilt, it appeared to me that when the precursors of the bodies were ejected or separated from their mother precursors, they were connected as fluids in a flow in which many distinct interconnected eddies or structures were present. As these rotating structures were being dragged downstream away from the position of their mother precursor, it came to a point that, for reasons I detailed in the chapters on vortex and fluid ligaments, the thread or fluid ligament connecting the block downstream of Zone 3 was stretched and tilted in the direction of the flow. Because the fluids of the distinct bodies were not strong enough to prevent them from being moved, they yielded to the shear and to the other factors shaping their "destiny" by letting their rotational axis to be tilted. The direction of the force that was moving the blocks of the precursors of the satellites downstream of the biggest satellites like a kind of swirl or spiral "bent" the rotational axis of the biggest body around the zone where the "bending" occurred. This "bending" of the flow current in the case of the precursor of the satellites occurred around the end of Zone 3 and near the beginning of Zone 4. According to the energy in the precursors, their viscosity, the shear rate of the fluid in motion, and other factors, the position where the "bending" force was able to tilt the bodies varies. In the case of the satellites, the tilting occurred around the innermost biggest satellites and some of the outermost satellites in Zone 4. For the Jovian and Neptunian satellites, the inclination of the rotational axis of the satellites occurred in Zone 3, around the biggest satellites. Another way to express this is that when the fluids were being slowed down, stacked, or stuck in the biggest satellites in Zone 3, bodies downstream were subjected to a shearing force dragging them toward the outer edge of their system. Downstream of the satellites with the biggest axial tilt, are bodies which orbital inclination are the highest because, as the axial tilt of a satellite was inclined upstream of them, their orbit was thrown into space as a result of the breakup of their ligament from that upstream. In the chapter on orbital speed and fluid layers, I already explained how the squashing of fluid affected the bottom fluids.

The satellites with the highest axial tilt are for the satellites in their planetary system what a knee is for the leg and what an elbow is for an arm. They are like the hinge around which the ligament of the precursor of the satellites in Zones 4 and 5 were ejected or pushed toward the edge of their planetary system. This is also why I called the satellites with the highest axial tilt, "hinge satellites", "knee satellites", "elbow satellites". For other systems of secondary bodies, the bodies with the highest axial tilt can be called "hinge bodies", "knee bodies", "elbow bodies". That is why around the satellites with the highest axial tilt, the orbital inclination of the satellites downstream of them is the highest. Around the end of Zone 3 and near the start of Zone 4, the trends of the orbital inclination, eccentricity, and size of the bodies suggests these variables are related to an event that separated the bodies.

The same logic used for the satellites can be applied for other bodies such as planets. This suggests that the axial tilt of the planets may be due to the way the structures in their precursors rotated and were separated from one another. For instance, although the axial tilt of Mercury is null, its eccentricity is the highest among the 9 planets, suggesting that instead of tilting its rotational axis, the force that pulled the precursors of the other planetary systems away from it pulled it into orbit, hence a higher eccentricity. Venus has a retrograde motion because its precursors were pulled too much and not being able to resist the force which sheared the other precursor away from it, it was thrown around its rotational axis, hence a higher axial tilt. The massive energy of Jupiter could have prevented its rotational axis from being pulled or tilted too much, hence its smaller axial tilt.

It is possible that the axial tilt of the planets was caused by the impact of the energy or force with which the precursors of the satellites pulled away from them and vice versa. Because at one point, the precursor of each planetary system was a unified fluid body before being split into a primary planet and its satellites, it is possible that during their split from one another, the ligament which attached them served as an intermediate through which they were pulled away.

Considering all the data I analyzed, the axial tilt of the bodies may be related to:

- the rotating vortex or eddies formed by the precursors of these bodies,
- the shear of the fluids which were spread from the position of the main precursor toward the downstream of its position by the edge,
- the energy in the current or fluid flow,
- the energy in the current downstream and upstream of each vortex,
- the viscosity of the fluids in these precursors,
- the way the bodies in Zones 4 and 5 were separated or ejected from those in Zones 1, 2, and 3, etc.

22.7. Take home message

While the axial tilt of most satellites is zero, that of other satellites is more than 5 times that of their primary planet. Triton is the only large satellite in the Solar System with a retrograde orbit. I felt like the data on Triton and Phoebe must hold a secret of the cause of the inclination of the rotational axis of the satellites. Similarly, as far as planets are concerned, the axial tilt of Venus, Uranus, and Pluto holds a secret of how these bodies were formed. An intimate understanding of axial tilt data would help unearth how the eddies or vortex of the precursors of the bodies in the universe were split and fragmented. The bodies with the highest axial tilt are for their system what a knee is for the leg and what an elbow is for an arm. They are like the hinge around which the ligament of the precursor of the bodies downstream of them were ejected, pushed, and titled toward the edge of their system. That is why around the bodies with the highest axial tilt, the orbital inclination of the bodies downstream of them is usually high. I provided ample details about the origin and meaning of axial tilt in the chapters I devoted to vortex and fluid ligament.

CHAPTER 23

UNEARTH WHY A LITTLE-KNOWN VARIABLE LIKE THE ORBITAL INCLINATION HIDES A KEY OF SOMETHING CRITICAL ABOUT THE MOVEMENT OF THE PRECURSORS OF CELESTIAL BODIES

23.1. Definition and illustration of orbital inclination

Celestial bodies in the Solar System moved on different orbits. These orbits are generally defined by a plane. Orbital inclination is a parameter or variable used to define the tilt of the orbit. Besides orbital inclination, longitude of the ascending node (which I covered elsewhere) is one of the main elements used to define the orientation of the orbit of celestial bodies. Orbital inclination is characterized using a reference plane. Called the elliptic plane or the elliptic, the plane of the Earth's orbit is the reference plane used to define the orbital inclination of the planets and asteroids in the Solar System. In other words, the orbital inclination of the planets and asteroids in the Solar System are defined by the angle between the elliptic and the plane of their orbit. Hence the orbital inclination of Earth is generally considered to be zero.

Sometimes, instead of using the elliptic as the reference plane to define orbital inclination of the bodies (e.g. planets) in the Solar System, the Sun's equator is used. In such a case, the inclination is called "inclination to the Sun's equator". The inclination of the celestial bodies can also be measured with respect to their equator and in that case, it is called the "inclination of the equator". Moreover, the inclination is sometimes defined with respect to an invariable plane and in that case, it is called the "inclination to the invariable plane". The invariable plane is also called Laplace's invariable plane, invariable plane of Laplace, or Laplacian plane.

Although some people consider the elliptic plane and the equatorial plane of the Sun to be the same, they are different, for the plane of the orbit of Earth is not in the same plane of the equator of the Sun. This implies that those who say that the orbit of the planets and asteroids lay in the same plane as the equator of the Sun are not telling the truth, meaning that their statement is misleading for it does not capture the whole truth.

In contrast to planets and asteroids that directly orbit the Sun, the orbital plane of the satellites is usually measured with respect to the equator of their primary planet. Therefore, the orbital inclination of a satellite is the angle between the orbital plane of that satellite and the equator of its primary planet. To put it another way, the orbital inclination of the satellites is not usually measured with respect to the elliptic nor with respect to the Sun's equator, but with respect to the equator of their primary planet. Sometimes, when the orbital inclination of the satellites is defined in relationship with the elliptic plane, it is the inclination to the elliptic.

Orbital inclination is also used to define the motion of celestial bodies. Indeed, when the orbital inclination is between 0° and 90°, the motion of that body is prograde, which means that the body and its primary body are rotating in the same direction. In contrast, when the orbital inclination of a body is between 90° and 180°, the motion of that body is retrograde, which gives a perception that the body is rotating in a retrograde or reverse direction. In other words, a retrograde planet orbits its primary planet in a direction opposite to that of the planet's rotation. When the inclination is 90°, the orbit is polar.

To summarize, the inclination of the bodies in the Solar System can be defined using:

- Orbital inclination (with respect to the equator of their planet)
- Orbital inclination with respect to the elliptic, also called "inclination to the elliptic"
- Inclination of the equator
- Inclination to the Sun's equator

Because some of my preliminary work showed that the inclination to the elliptic is highly correlated with orbital inclination ($r2\sim1$), I did not devote too much time to the inclination to the elliptic in this book. Because the orbital inclination of some planets is almost null, the orbital inclination and the inclination of the elliptic of some satellites are the same.

One of the unanswered questions in astronomy is the cause of the variation of the inclination of celestial bodies. No profound work was done to explain why some bodies are more inclined than others or why they are inclined at all. So far, all of the theories elaborated before mine just speculated that the orbital inclination and the axial tilt (inclination of the rotational axis) of celestial bodies were caused by random collisions during the formation of the universe. Because these theories do not explain the true origin of the universe, some bodies are labeled as abnormal when their orbital inclination is too high in the environments where those theories think motion must be prograde. However, I personally never thought that the variation of the inclination was caused by chance, but by factors that I carefully studied and decrypted over the years. Before I quantitatively analyzed the orbital inclination, I first focused on the inclination of the equator and the inclination to the Sun's equator of the planets.

23.2. Inclination to the Sun's equator of the planets

Although it is commonly said that the planets in the Solar System lay on the same plane as the Sun's equator, it is important to note that significant differences exist. The orbital planes of the planets are inclined differently with respect to the Sun's equator. These inclinations vary from 3.38° to 11.88°, the lowest value being recorded with Mercury while the highest value was found with Pluto (Fig. 94). The Earth's orbital plane is tilted 7.25° with respect to the Sun's equator. However, because the Sun wobbles during its motion, the inclination to the Sun's equator slightly vary with time.

As a reminder, Venus rotational axial tilt is 177.36°, whereas that of Pluto is 122.53°. With Pluto being the second most tilted planet in the Solar System after Venus, it should not be surprising that Pluto's orbital plane is the most tilted with regard to the Sun's equator. However, one may wonder why Venus' inclination to the Sun's equator is not as high as that of Pluto? I will come back to that problem later. Because the orbits of the planets in the Solar System are not in the same plane as the Sun's equator, all of the theories which claim that the planets and the Sun are in the same plane are simply false.

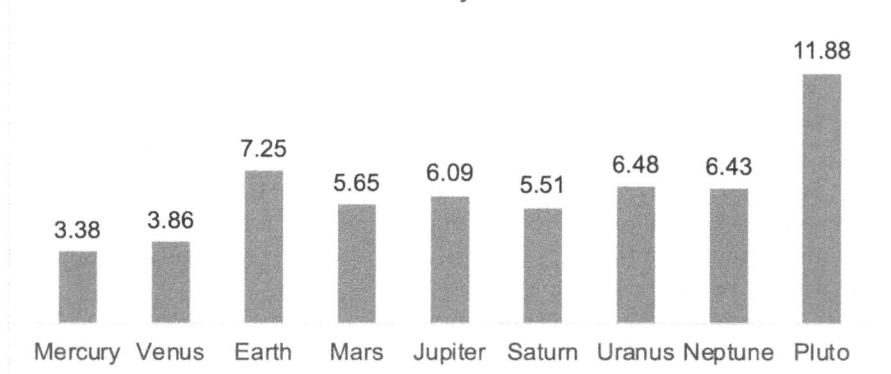

Fig. 94: Inclination to Sun's Equator of Planets in the Solar System

23.3. Inclination of the equator of the planets in the Solar System

According to NASA (2018), the inclination of equator is the "angle between a body's equator and the body's orbital plane with north defined as the pole axis above (north of) the plane of the Solar System, and it is also denoted as axial tilt".

This definition sounds like the inclination to equator is the same as axial tilt. However, looking at their values, axial tilt and inclination of the equator are different.

Data on the inclination of equator was available only for the planets. The inclination of equator of the 9 planets in the Solar System ranges from 0.034° to 82.23° (Fig. 95). The highest value was recorded with Uranus, whereas the smallest value was recorded with Mercury.

Fig. 95: Inclination of Equator of Planets in the Solar System

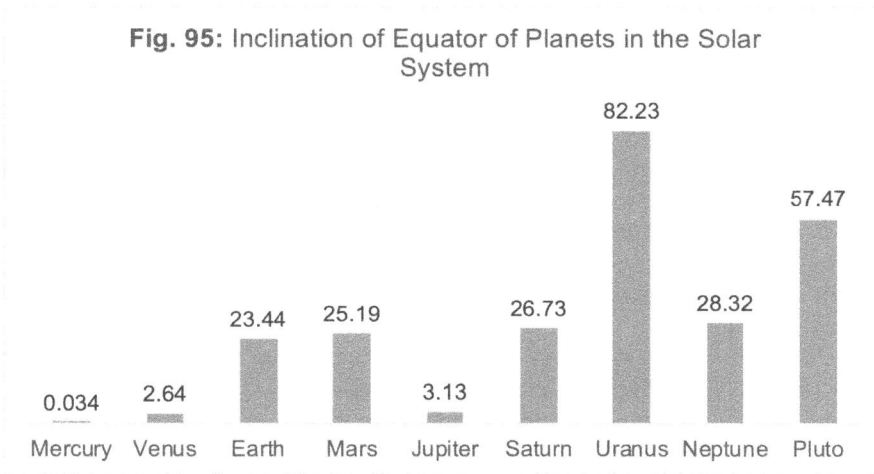

23.4. Motion of celestial bodies: Prograde, retrograde, and chaotic

23.4.1. General trends on the motion of the bodies in the Solar System

The motion of the celestial bodies can be defined by referring to the orientation of their orbit or of their rotational axis. Motion can be generalized into two categories:

- prograde or direct motion when the orbital inclination of that body is between 0° and 90°
- retrograde or indirect motion when the orbital inclination is between 90° and 180°.

Most of the planets in the Solar System orbit or revolve around the Sun in the same direction that the Sun is rotating, which is counterclockwise when viewed from above the Sun's North Pole. The motion of the celestial bodies was one of the first things that opened my mind to the world of modeling the planets and their satellites as daughters of the same precursor. Indeed, I studied the motion for 180 bodies in the Solar System, the majority of which were satellites and planets. Considered with respect the orbital inclination of the bodies, the motion of the bodies in the Solar System depends on their types ($p<0.000$). Because the orbital inclination of the planets in the Solar System is less than 90°, their motion with respect to orbital inclination is prograde. However, when motion is viewed with perspective of the axial tilt, the rotational motion of some planets is prograde while for others, it is retrograde (e.g. Venus, Uranus, and Pluto). The remaining 6 planets in the Solar System have a prograde rotation.

The motion of satellites depends on their types ($p=0.013$), the turbulence zones ($p=0.000$), and the interaction between these 2 variables ($p=0.000$). Here, I focused on the 178 satellites known in the Solar System before 2016. The orbit of 47.8% of these satellites is prograde compared to 52.2% which are retrograde. For planets which have both prograde and retrograde satellites, the percentage of retrograde satellites ranges between 0 and 78%, with the highest percentage obtained with the Jovian satellites. In contrast, as of 2016, Saturn has an almost even number of prograde (47%) and retrograde satellites (53%). All of the satellites of Pluto, Mars, and Earth have a prograde orbit. In general, most of the outer satellites which semi major axis is superior to 0.12173 AU have a retrograde orbit (only 2 Neptunian satellites beyond that semi major axis are prograde). The orbit of the outer satellites has a higher probability to be retrograde than that of the inner satellites.

A chaotic rotation is the irregular and unpredictable rotation of a celestial body which does not usually have a fixed rotational axis or period. Some satellites having a chaotic motion are so close to one another that they sometimes share the same orbit. That is the case for Prometheus and Pandora on one hand, and on the other hand the case for Janus and Epimetheus, all of which are Saturnian satellites. The distance between the satellites can affect their inclination or tilt, hence the impact on motion. A chaotic rotation has been observed with some satellites including: Hyperion (a Saturnian satellite), Nereid (a Neptunian satellite), and some Plutonian satellites (e.g. Nix and Hydra and possibly Styx and Kerberos).

Hyperion is located downstream of Titan, the biggest satellite of Saturn. It is next to Titan. Hyperion is also near Iapetus, the Saturnian satellite which has the highest rotational period as far as the Saturnian satellites are concerned. Since the distance separating Hyperion from Titan is not as high as that separating other satellites (see chapters on semi major axis and semi major axis increment), it is also possible that the motion of Titan and the waves which accompanied it could be affecting the rotation of Hyperion, hence its chaotic movement. Iapetus may also be affecting the motion of Hyperion. Furthermore, the small rotational angular speed of Hyperion which could have also predisposed that satellite to move chaotically. Likewise, Hyperion is also one of the only 2 satellites in Zones 1, 2, and 3 that have the highest eccentricity as compared to that of their primary planet, meaning that its orbit was elongated. Nereid is the next satellite downstream of Triton, the biggest satellite of Neptune. It is small in size (170 km in radius) as compared to Triton (1353.4 km). Later in this book (see chapter on vortex and fluid ligament), I expounded on why the interaction between the precursor of Nereid and that of Triton could explain the chaotic motion of Nereid.

Talking about chaotic motions, Showalter and Hamilton (2015) reported that "*Nix* [a Plutonian satellite] *can flip its entire pole. It could actually be possible to spend a day on Nix in which the Sun rises in the east and sets in the north. It is almost random-looking in the way it rotates.*" (Showalter and Hamilton, 2015). In the chapter of vortex, I provided more information on the chaotic motion of celestial bodies.

From Jupiter to Neptune, the percentage of prograde satellites increased, whereas the percentage of retrograde satellites decreased (Fig. 96).

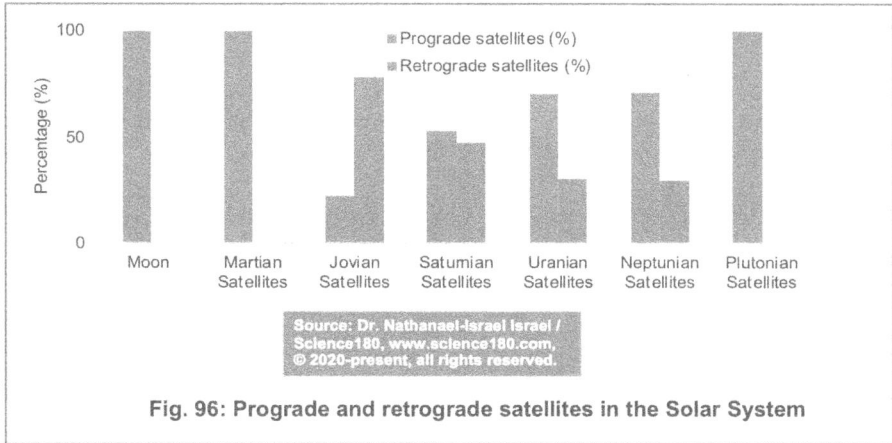

Fig. 96: Prograde and retrograde satellites in the Solar System

Fig. 97 and Fig. 98 illustrate the motion of the satellites according to their types. For the prograde or direct motion, motion = 1 whereas for the retrograde or indirect motion, motion =0.

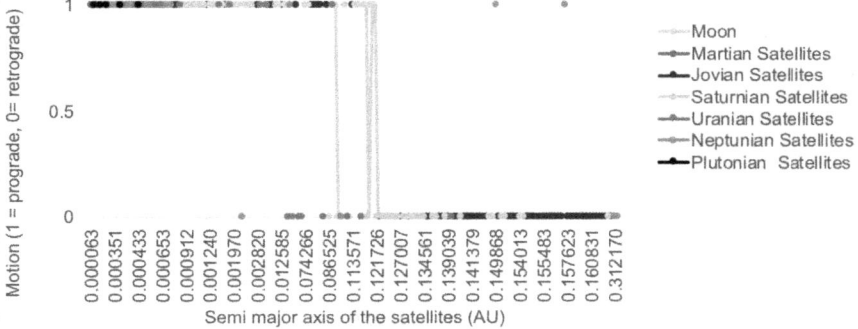

Fig. 97: Motion (Prograde = 1 and Retrograde =0) of the 178
satellites in the Solar System

Fig. 98: Motion of 178 satellites in the Solar System according to their types and semi major axis

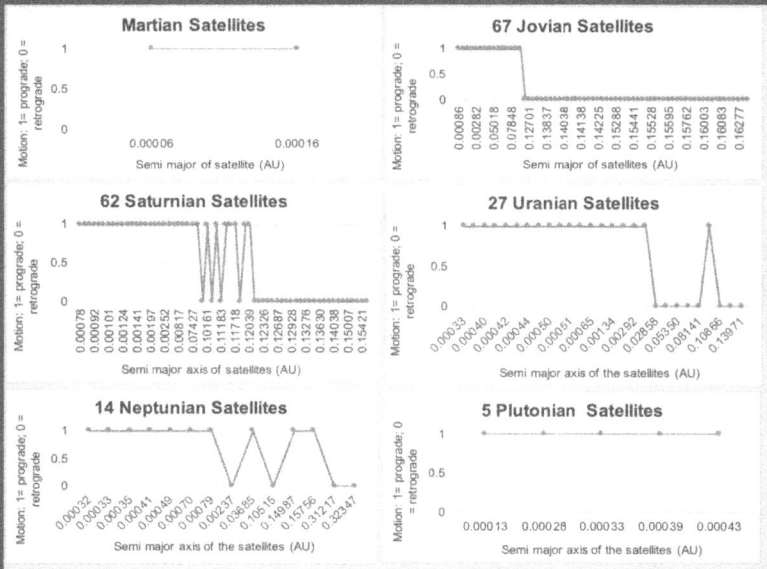

23.4.2. How I first had a glimpse at studying the celestial bodies by considering the precursor of the planetary systems instead of just the planets and satellites isolated

As early as 2016, when I was working on the motion of the satellites, the zigzag trends of their graphs caught my attention. As I deeply thought about it, I was led to do the regression between the radius of the planets, the number and percentage of retrograde and prograde satellites. As I did so, I found that the number of satellites per planet is highly correlated with the radius of the planet ($r2= 0.97$) (Fig. 99). In other words, the data suggested that the percentage of retrograde satellites a planet is inversely proportional to the size of that planet, meaning that the smaller planets have a higher proportion of prograde satellites than the bigger planets.

Fig. 99: Regression between radius of planets and the number and percentage of their prograde and retrograde satellites

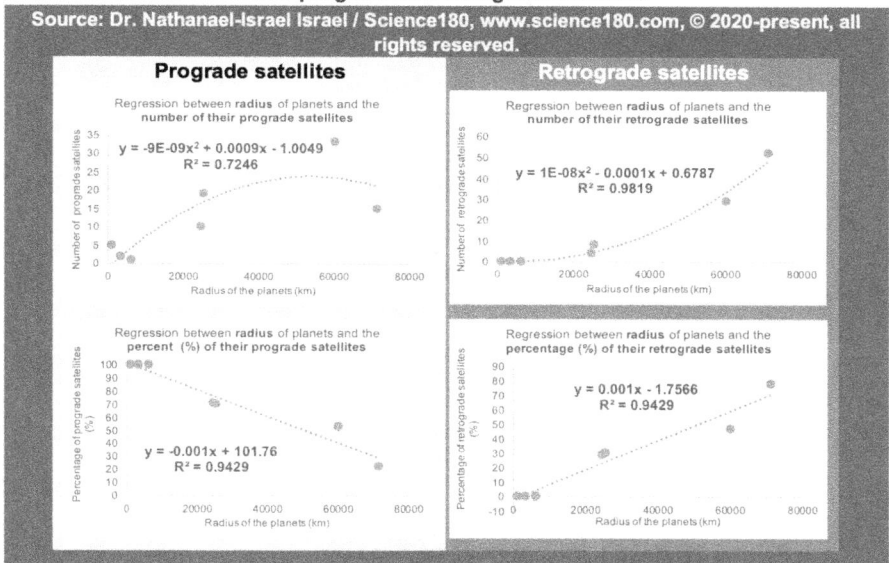

The radius of the planets positively impacts their number of prograde satellites and number of retrograde satellites. However, the strength of these regressions is significantly different. For instance, the coefficient of determination (r^2) for the regression between radius of the planets and number of their prograde satellites is 0.72. In contrast, the coefficient of determination (r^2) for the regression between radius of the planets and number of retrograde satellites is 0.98. These regressions suggest that an increase of the radius of the planets is more likely increasing the number of the retrograde satellites than the number of prograde satellites. To confirm these trends, a regression between the radius of the planets and the proportion of prograde satellites and the proportion of retrograde satellites was done ($r^2=0.94$). The above graph proved that while the regression between the radius of the planets and the percentage of prograde satellites is negative, the regression between the radius of the planets and the percentage of retrograde satellites is positive. These regressions imply that as the radius of the planets increases, the percentage of their retrograde satellites also increases, whereas that of their prograde satellites decreases. Although retrograde satellites are generally smaller than the prograde ones, the retrograde and prograde motions seem to be caused by other variables or events. Although the above regressions seem very simple, the secret behind them as to how the precursors of the planetary systems split into planet and satellites have shaped my initial thinking about how the precursors birthed their daughter bodies. In other words, these regressions introduced me into the laws that governed the split-gathering of the precursors of planetary systems into planets and satellites. Therefore, based on these regressions, since 2016, I started analyzing the data by not just considering the planets and satellites separately, but by visualizing what could have happened if they both came from a common precursor. When I looked at the motion of the satellites, I noticed that their types (retrograde or prograde) may be related to the environment of the bodies and the circumstances of their formation rather than just a random factor.

23.5. Orbital inclination of the celestial bodies in the Solar System

23.5.1. General trends of the orbital inclination

Orbital inclination was studied for 463 bodies (including the satellites discovered between 2016-2019) in the Solar System. It depended on their types ($p<0.000$) and varied from 0° to 179.8°. Because the orbital plane of the Earth is the reference for measuring orbital inclination, the orbital inclination of the Earth is obviously 0. The minimum inclination was recorded on a Saturnian satellite. Most of the highest orbital inclinations were recorded on Saturnian satellites, meaning that Saturnian satellites have both the extreme values of orbital inclination encountered on satellites in the Solar System. The orbital inclination of the planets is generally small, whereas that of the comets and the other outermost celestial bodies are high.

23.5.2. Orbital inclination of asteroids and planets

The orbital inclination of the asteroids in the Solar System ranges between 0.02° and 162.5° (Fig. 100). The highest inclinations were recorded with some outermost asteroids. The smallest values were found with the Aten asteroids. Most of the highest orbital inclinations were observed with the biggest asteroids.

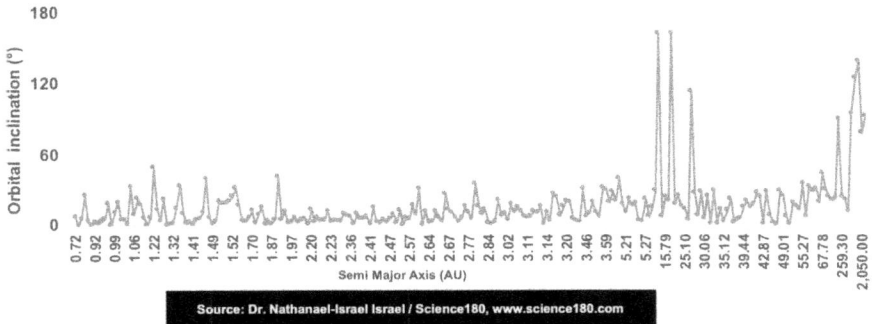

Fig. 100: Orbital Inclination of the Asteroids in the Solar System according to the semi major axis

The orbital inclination of the asteroids in the main belt average 9.11° and ranges from 0.71° to 34.84° (Fig. 101). The highest value was recorded with 2 Pallas (the second biggest asteroid in the main belt, with a radius of 272.5 km).

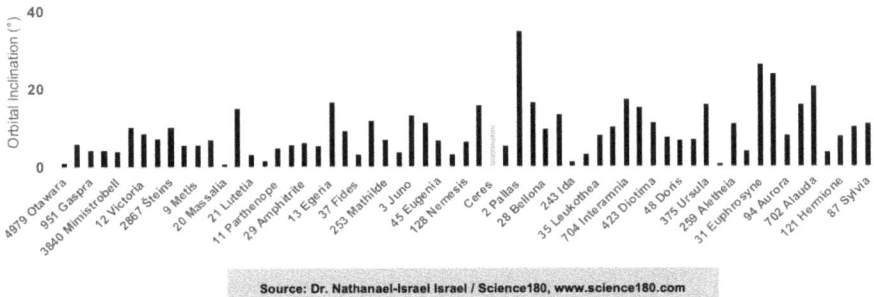

Fig. 101: Orbital Inclination of the Main Belt Asteroids

After the Earth, the next least inclined orbit of a planet was observed with Uranus (Fig. 102). Pluto is the planet that has the highest orbital inclination (17.16°). Mercury, the planet with the second most titled orbit (7°) in the Solar System has an orbit that is more inclined than that of any other planet besides Pluto.

23.5.3. Orbital inclination of the satellites

The orbital inclination of the satellites in the Solar System depends on their types (p=0.006), the turbulence zones (p<0.000), and the interaction between these 2 variables (p<0.000). The orbit of the satellites in the Solar System is not in the same plane as that of their primary planets. Satellites have the highest orbital inclination in the Solar System, ranging from 0° to 179.8° (Fig. 103-104). Most of the lowest and highest values were measured on Saturnian satellites, making

Saturn a planet surrounded by satellites with inclination covering the entire spectrum. Indeed, the highest value was measured on Greip, a Saturnian satellite. Many other Saturnian satellites have a very high orbital inclination. At the same time, most of the smallest orbital inclinations were observed with the Saturnian satellites.

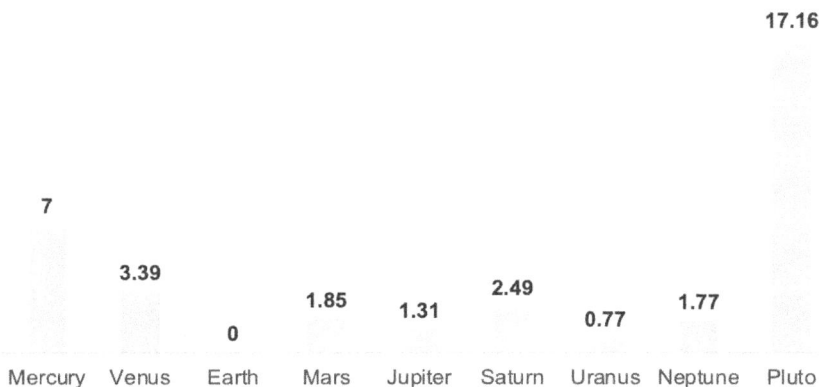

Fig. 102: Orbital Inclination of Planets in the Solar System

In general, the biggest planets seem surrounded by satellites that have the highest orbital inclination. For instance, the maximum orbital inclination of the satellites of the 4 giant planets (Jovian, Saturnian, Uranus, and Neptunian) ranged from 157.35° to 179.8°. In general, the smallest orbital inclinations are in Zones 1, 2, and 3; whereas the highest were in Zone 5 (Fig. 105). Zone 4 is a transition zone were both prograde and retrograde are generally found. All satellites in Zones 1 and 2 are prograde.

The following graph shows that the outermost satellites have the highest orbital inclination, whereas the innermost satellites have the smallest orbital inclination. The high value of the orbital inclination of the outermost satellites suggests that as the precursors of the satellites got farther from the precursor of their primary planets, something acted on them at one point to tilt their orbital plane.

Fig. 104 compares the trends of the orbital inclination of the satellites in the Solar System according to their types and semi major axis. When I first looked at the orbital inclination data early in the days of my investigation of the origin of the universe, the zigzag shape of their fluctuations according to the semi major axis caught my attention.

Nathanael-Israel Israel: Member of the American Society for Microbiology

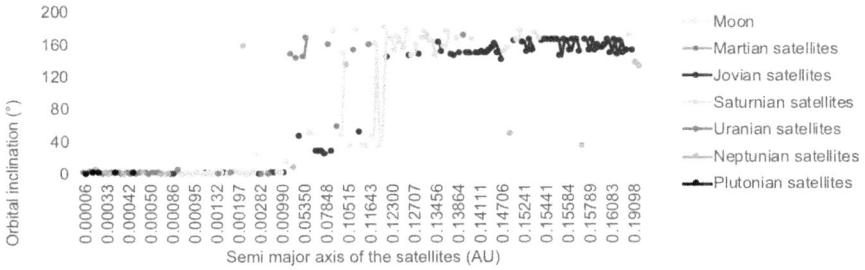

Fig. 103: Distribution of the orbital inclination of the satellites according to their types and semi major axis

Fig. 104: Orbital Inclination of satellites according to their types and semi major axis

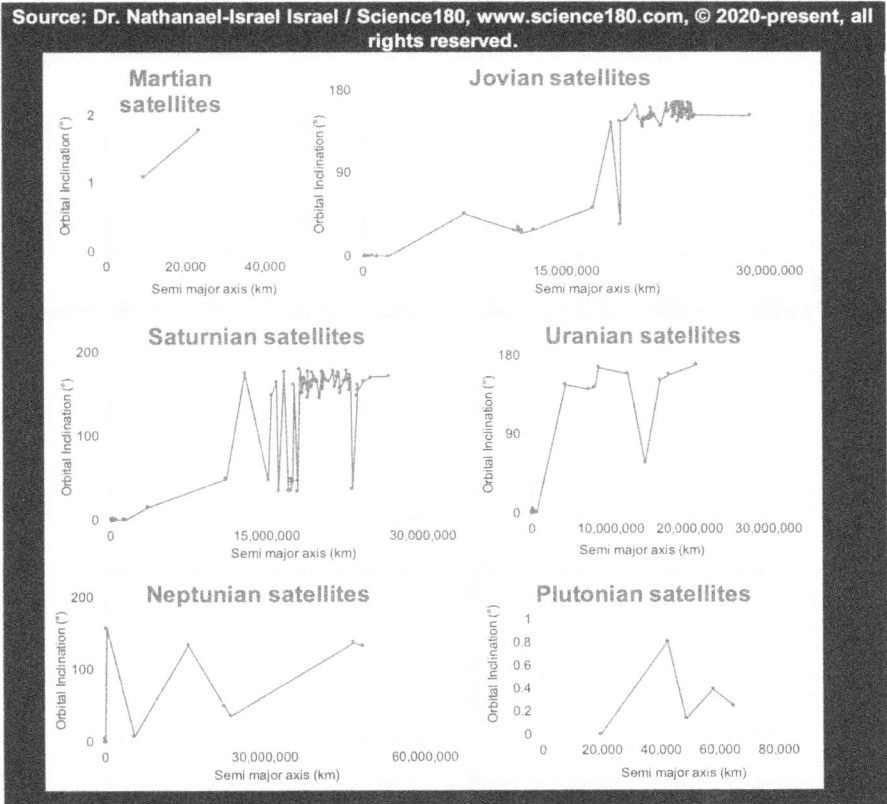

The orbital inclination of the Moon (23.43°) is higher than that of the Martian

satellites: 1.08° for Phobos (the biggest and innermost Martian satellite) and 1.79° for Deimos (the smallest and the outermost Martian satellites). Furthermore, according to NASA (2018), the inclination of the Moon to the ecliptic is 5.145°, while its inclination to the equator ranges between 18.28° and 28.58°.

The orbital inclination of the Jovian satellites varies between 0.03° and 166.4°. The smallest inclinations were recorded on satellites in Zones 1, 2, and 3 (Fig. 105). In contrast, the highest orbital inclinations were with the satellites in Zone 5. Satellites in Zone 4 have a higher orbital inclination than those in Zones 1, 2, and 3, but less than those in Zone 5. In other words, the orbital inclination of the satellites in Zone 4 is intermediate between that of the satellites downstream of them (in Zone 5) and those upstream of them (in Zones 1, 2, and 3). To put it in different words, Zone 4 encompassed satellites which motion is transitioning from prograde to retrograde in such a way that the conditions were not met for some of them to have a retrograde motion yet their orbit is not as highly inclined as that of the satellites in Zone 5. Although many satellites were discovered between 2016 and 2019, at the moment I was finalizing this chapter in 2020, the orbital inclination of these newly discovered Jovian satellites in Zones 4 and 5 did not change the global trend explained here.

The least titled orbits of the Saturnian satellites are in turbulence Zones 1, 2, and 3. In Zone 4, prograde and retrograde satellites are found. In other words, Zone 4 is a transition zone where the precursors of the satellites progressed from a predominantly prograde motion to a predominantly retrograde one. The satellite with the highest orbital inclination is Greip, a first Saturnian satellite in Zone 5. As of 2020, all Saturnian satellites in Zone 5 are retrograde, meaning that their orbital inclination is above 90°. Based on the orbital inclination available as of 2016, the innermost retrograde Saturnian satellite was Phoebe, the satellite which has the highest axial tilt among all of the Saturnian satellites. Some studies have shown that the orbit of Pandora and of Prometheus, both inner satellites of Saturn, appear to be chaotic.

Just as the satellites in the other planetary systems, the smallest orbital inclinations of the Uranian satellites are in Zones 1, 2, and 3. In Zone 4, most of the Uranian satellites are retrograde while some few are prograde. All of the Uranian satellites in Zone 5 are retrograde just as observed in Zone 5 of the other planetary systems.

Unlike the other planetary systems, a Neptunian satellite in Zone 3 is retrograde. In fact, the biggest Neptunian satellite (Triton) has an orbital inclination of 157.35°, making Triton one of the satellites which have been puzzling scientists for centuries. Until my discoveries, no scientific theory was able to properly explain why a big satellite such as Triton can have a retrograde motion. The turbulence Zone 4 of the Neptunian planetary system has 3 prograde satellites and one retrograde satellite.

Just as the other planetary systems, Zone 5 of the Uranian planetary system is populated by retrograde satellites only and they are just 2.

While the orbital inclination of Charon (the biggest Plutonian satellite) is 0, that of none of the 4 satellites in Zone 4 is higher than 0.81°, therefore making the orbit of the Plutonian satellites some of the least inclined in the Solar System. Although the orbital inclination of these Plutonian satellites is almost null, it is important to keep in mind that the Plutonian satellites in Zone 4 like Nix and Hydra are famously known for their chaotic rotation, which suggests that, to some extent, the rotational axis and/or the orbit of these satellites is subject to something that makes the movement of these satellites very chaotic. In fact, as of 2020, besides these 2 Plutonian satellites, only one more satellite, Hyperion, in the Solar System is famously known for having a very chaotic rotation. In the chapter on semi major increment, I also explained the real reason of the chaotic motion of some bodies. Fig. 105 illustrates the orbital inclination of the satellites according to their types and turbulence zones.

Fig. 105: Orbital Inclination (°) of satellites according to their types and turbulence zones

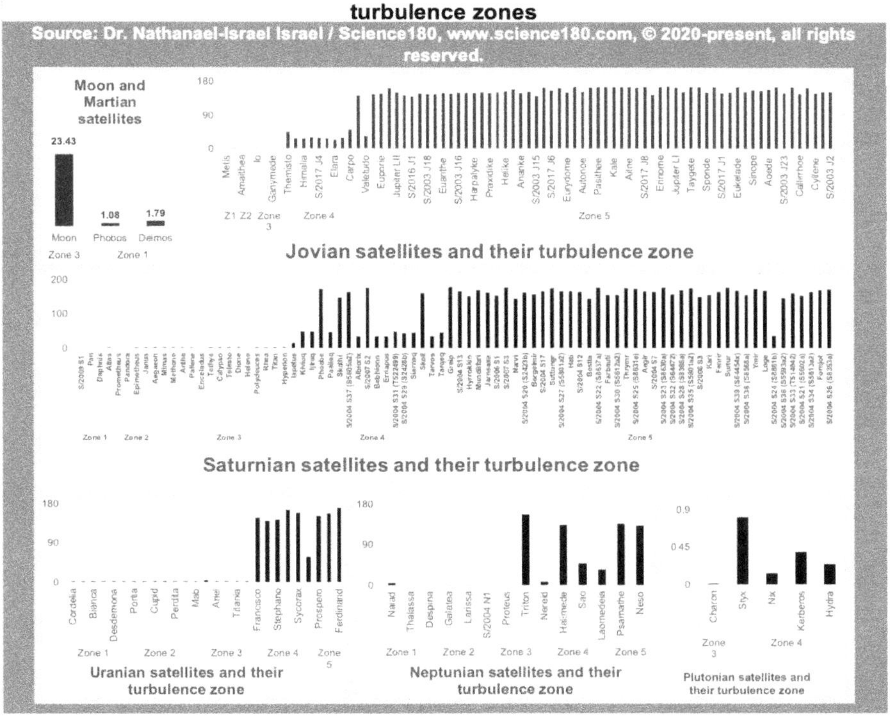

23.5.4. Orbital inclination of the satellites according to their radius

Fig. 106 showed that the orbital inclination of the smallest satellites is generally high. In contrast, while the orbital inclination of most of the biggest satellites is low, many small satellites also have a very small orbital inclination. This means that the force that inclined the orbit of the satellites is not just related to the radius of these satellites, but also to something else. That force was also able to significantly act on some big planets. In other words, although the relationship between orbital inclination and the radius is not perfect, the orbital inclination of the smaller satellites is usually among the highest. Nevertheless, the orbital inclinations of the bigger satellites have mixed values, meaning that some bigger satellites have a smaller orbital inclination, whereas others have a higher orbital inclination. This trend is like that between axial tilt and equatorial radius, therefore suggesting that, rather than being caused by a chance, the variation among tilt, radius, and inclination of the satellites was controlled by a precise factor that, as of 2016, I ignored, but that became clear to me when I understood the role of turbulence in the formation of celestial bodies. Considering the graphs of the relationship between orbital inclination, semi major, and radius, it sounds like the orbital inclination of the big satellites is affected by an interaction between their radius and semi major axis. That is why, because of their positioning, the orbit of some big satellites is sometimes very inclined.

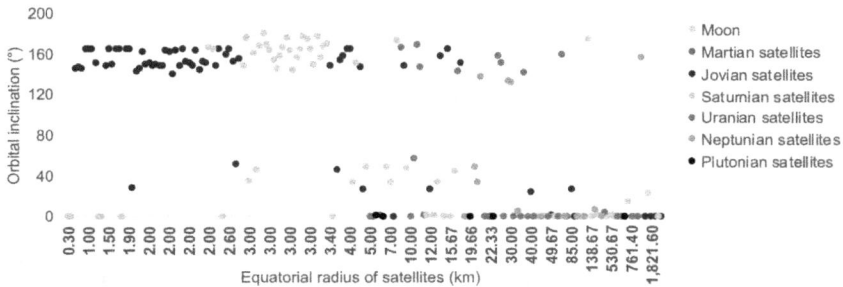

Fig. 106: Orbital inclination of the satellites in the Solar System according to their equatorial radius

23.4.5. What may have caused the tilting or inclination of the orbit?

The processes that split-gathered the fluid layers of the precursors of the bodies into specific bodies (e.g. spherical for the biggest bodies) had affected the orientation of the bodies. The orbit of most outer bodies, meaning those which precursors were located beneath the fluid layers of the biggest bodies (which are in Zone 3), are usually tilted. In other words, the most tilted orbits are usually located outward of the satellites in Zone 3. Indeed, the data on orbital inclination suggests to me that the precursors of the bodies in Zones 4 and 5 were flexed at a location between some of the outermost bodies in Zone 3 and some of the innermost retrograde bodies in Zone 4, particularly around the body with the most tilted rotational axis. The circumstances of this "flexion" could have inclined the orbit of the bodies located downstream of that flexion point, while increasing the axial tilt of the bodies located in the zones where the flexion occurred.

CHAPTER 23: ORBITAL INCLINATION AND MOTION OF CELESTIAL BODIES

When 2 people hold hands and then pull each other away in different directions, a kind of tension can be created between them. When they break the connection between them, they can fall backward depending on their mass and the force with which they were pulling one another. The one who is heavier will experience a smaller fall than the one who is lighter. The one who is lighter may even be pulled away and roll over if the force acting on him is very high. When two persons are moving in a same direction and aligned with one person in front and the other in the back, the movement of the one in front can change if the person in the back pushes him. For instance, if the person in the back pushes the one in front too hard, the one in the back can communicate an additional momentum to the one in front. If that push is strong and applied in a certain direction, it can change the direction of the movement of the body in front. Similarly, when the precursors of the bodies were moving, consecutive precursors affected one another and, in the end, they all affected one another by the way energy was transferred among them, and how they were split or ejected from their mother precursors and from one another. The precursors of some upstream bodies affected the movement of those downstream and communicated a certain momentum that affected the inclination of the orbit and rotational axis of the daughter bodies. Particularly, as the precursors of the bodies were moving and were being shaped, each time they broke up or split from the peers in a flow, the movement of some bodies near the breaking or splitting point was affected. As you will see in the chapter on semi major axis increment, a major break up occurred between the outermost satellite in Zone 3 and the innermost satellite in Zone 4 in most systems.

Looking carefully at the value of the eccentricity, tilt, and orbital inclination of the bodies suggests that some of the inner satellites in Zone 4 could have been formed by the recoiling of the fluid ligament that formed the satellites in Zone 4. Before I discovered the impact of the squashing of top fluid layers on bottom fluid layers, I thought that the innermost Saturnian satellites in Zone 4 upstream of Phoebe were formed from the recoil of the ligament which birthed Phoebe. As the ligament connecting Zone 3 and Zone 4 broke, an additional momentum could have been communicated to the ligament that was the precursor of the bodies in Zones 4 and 5. I once thought that the pulling away of that ligament could have ejected and inclined the orbit of the satellites which came out of it and also could have tilted the rotational axis of the first major satellites in Zone 4. Hence the higher axial tilt and higher semi major axis increment are around the end of Zone 3 and the beginning of Zone 4. But as you will see in the chapter on vortex and fluid ligament, I finally understood that the trend of the orbital inclination is connected to how the characteristics of the bottom fluid layers were affected by the squashing of the fluid layers above them. I will revisit this topic later.

Before 2018, I used to think that, because the precursor of the Jovian planetary system could have been very viscous, the recoiling of the precursor of the bodies

Science180: The New Physics That Will Revolutionize Science Forever in Zone 3 after it split from the precursor of bodies in Zone 4 did not yield any satellites. However, because the precursor of the Saturnian satellites could have been less viscous, the recoiling of the precursor of bodies in Zone 3 led to the formation of smaller satellites between the biggest satellites and the end of Zone 3. That is why I used to think before 2018 that some Saturnian satellites in Zone 3 which are downstream of Titan may have been formed after a recoil of some of the fluids in its turbulence zone, while others could have also been born from some fluid layers. Some of these satellites are Hyperion and Iapetus, and I thought that this may help explain why the rotation of Hyperion is so chaotic. As you will read later, the reality was more than that.

For many years, I also wondered why the orbital inclination of Mercury is higher than that of all the other planets except Pluto. For I found it interesting that Mercury (which rotational axis is not tilted) has one of the highest orbital inclinations among the planets. Initially, I thought that the high eccentricity and orbital inclination of Mercury counter powered its axial tilt. In other words, because its rotational axis was not tilted, the energy used to pull and lift its orbit was higher increased its high eccentricity and orbital inclination. To put in in other words, the energy that pushed the planets into motion could have affected each planet in a different way according to its position, radius, speed, etc. Among the planets in the Solar System, Mercury has the smallest axial tilt; and because all the kinetic energy applied to Mercury was not used to significantly tilt its axis of rotation, a bigger portion of that energy was used to tilt its orbital plane.

Before 2020, although I tried to describe the orbital inclination, I was still not satisfied with why some bodies are inclined at all. On December 6, 2018 for instance, I felt like, after the fluid ligament of some precursors broke up, some daughter bodies were thrown during a "relaxation" stage. Depending on their position, energy, or mass, some bodies could have been heavily shaken and pushed away. The force involved in the stretching and the breakup overcame some of the constraints that weight or energy may have had. Then, I realized that energy and weight alone were not the only major factors involved in the tilting of the orbital plane, else Triton (the innermost retrograde satellite of Neptune) could not be retrograde despite its massive size. I felt like the direction in which the orbital inclinations are inclined is aligned with the direction in which the precursor of the bodies landed after breaking from neighboring precursors. I also felt like a careful study of the orbital inclination could help pinpoint some key aspects of how the precursors were propelled after split-gathering. I felt like, as the precursors of the bodies were breaking from one another, the way their precursors were connected, the force that pushed and pulled them, and their viscosity, could have contributed to defining the direction into which they were moved or resisted the tendency to be moved in a certain direction. Indeed, on December 6, 2018, I thought that, to understand the secret behind orbital inclination, I needed to take a look at many variables (e.g. turbulence, orbital speed, rotation speed, fluid shear, stretching, and splitting). In those days, I sensed that viscosity could have affected the shear of the precursors. I also perceived that the high viscosity of the precursors of Mercury, Venus, and Earth could have affected how they split-gathered, flowed, and stretched. In those days, I felt like, during its break up from neighboring bodies, the precursor of Venus could have been pulled and/or pushed and, in the process, its rotational axis was affected. Because the highest orbital inclinations of satellites are in Zones 4 and 5, and because it is also downstream of the largest satellites that the orbital inclination of the satellites jumped, I felt in those days that, downstream of Zone 3, part of the energy communicated to the satellites in Zones 4 and 5 could have been used to increase their inclination and eccentricity. Part of the energy which could have been used to increase the orbital inclination of the Neptunian satellites was used to increase their semi major axis. The position of the precursors of the bodies affected the tilt, inclination, speed, and spatial distribution of their daughter bodies. As I tried to explain the inclination, tilt, eccentricity, and other variables of the celestial bodies, I realized that, the laws that elucidate how forces acted on the celestial bodies may not be explained by the current mathematical models mostly hidden in the common equations: linear, exponential, logarithmic, power, polynomial, etc. I felt like a new type of logic or reasoning was needed to explain what happened to the precursors of the celestial bodies. After presenting other key variables crucial to understanding the big picture of the orbital inclination, I will revisit it later in this book. For I need to walk you through other variables before I can provide the systemic interpretation of the origin of the orbital inclination in the chapters on vortex and fluid ligaments.

23.6. Take home message

Because the orbital plane of the planets is inclined differently with respect to the Sun's equator, the theories which claim that the planets and the Sun are in the same plane are simply false. The orbital inclination of the planets is generally small, whereas that of the comets and the other outermost celestial bodies is high. In most planetary systems, the outermost satellites have a higher orbital inclination, whereas the innermost satellites have a smaller orbital inclination. The highest orbital inclinations of the satellites are found in Zones 4 and 5. In contrast, the orbital inclination of the satellites in Zones 1, 2, and 3 are usually the smallest, except for Triton, the largest Neptunian satellite. The processes which split-gathered the fluid layers of the precursors of the bodies affected the orientation of the orbital plane of their bodies. The data on orbital inclination suggests that the precursors of the bodies in Zones 4 and 5 were flexed at a location between some of the outermost bodies in Zone 3 and some of the innermost retrograde bodies in Zone 4, particularly around the body with the most tilted rotational axis. The cause of this "flexion" could have also inclined the orbital plane of the bodies located downstream of that flexion point while increasing the axial tilt of one of the bodies located in the zones where the flexion occurred. Later in this book, I will provide the systemic interpretation of the origin of the orbital inclination.

CHAPTER 24

CAN THE DENSITY OF THE CELESTIAL BODIES HELP END ANY ASPECT OF THE UNIVERSE FORMATION DECODING TROUBLES?

24.1. General trend of the density of the bodies in the Solar System

I studied the density of 163 celestial bodies in the Solar System. It depends on the types of bodies ($p<0.000$) and ranges from 250 kg/m^3 to 6730 kg/m^3. The smallest density was observed with Pallene, a Saturnian satellite, whereas the highest density was recorded with Psyche, a main belt asteroid. Many of the smallest densities were recorded either on Saturnian satellites or on comets. Saturn itself is among the least dense bodies as, among the bodies I studied in the Solar System, it is the 16th least dense. In general, the densest bodies were recorded with the main belt asteroids and planets. For instance, out of the 30 densest bodies I studied in the Solar System, 70% were asteroids in the main belt and 13.3% were planets.

About 58% of the density of celestial bodies I studied is greater than that of the Sun, which is estimated as 1408 kg/m^3. Some theories also predicted that the central density of the Sun could be 162,200 kg/m^3. That density of the Sun is less than the mean density of the 163 bodies studied in the Solar System, suggesting that the precursor of the Sun was not compressed or squeezed much. The coefficient of variation of the density of the 163 bodies is 58.41% which is smaller than that of most of the other variables I studied, some of which CV is several thousand percent. This trend suggests that density is one of the least diverse variables I studied in the Solar System. In other words, the precursors of the bodies could not have been compressed beyond certain limits.

Nathanael-Israel Israel: Member of the American Society for Microbiology

24.2. Density of the asteroids and planets in the Solar System

The density of the asteroids and planets combined together varies between 300 kg/m³ and 6730 kg/m³ (Fig. 107). The smallest densities were recorded with comets and asteroids in the main belt. The planets scored the highest density. The density of the outermost bodies in the Solar System is not the highest, but among the smallest. The data shows an intermittence of the density, meaning high densities are separated by small densities. The density of the asteroids in the main belt varies between 990 kg/m³ and 6730 kg/m³. An interesting question would be why the asteroids in the main belt are so dense, even denser than most of the other types of bodies which density is available. Their rotational angular speed is among the highest and could have helped to tightly squeeze their precursors, which compressed their constituent more to yield a higher density. The precursor of the main belt asteroids could have rotated very fast, therefore preventing its daughter bodies from being able to collect into a single body in the main belt, but instead spreading them into small asteroids which ended up being very compressed hence their high density.

Fig. 107: Density (kg/m³) of asteroids and planets in the Solar System

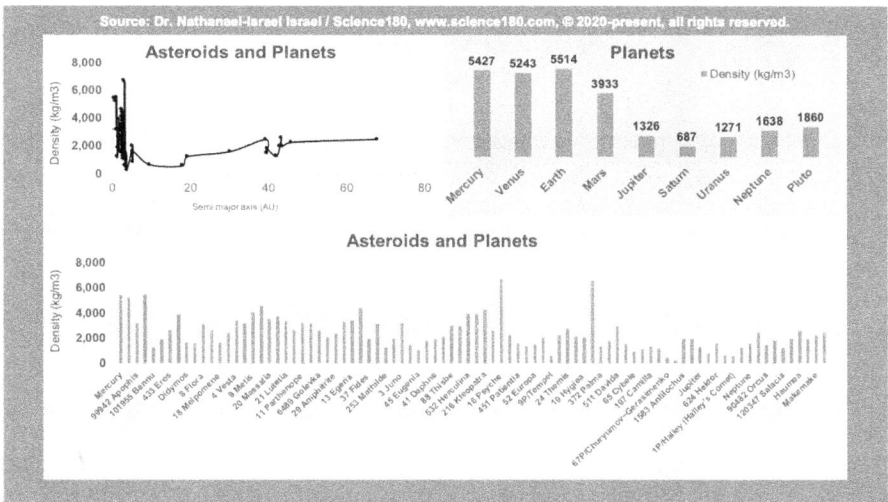

The density of the planets in the Solar System ranges from 687 kg/m3 to 5514 kg/m3. The densest planet is the Earth, whereas the least dense planet is Saturn. All of the terrestrial planets in the Solar System are denser than the Sun. Among the planets, the smallest densities were found with the giant planets. The small density of Saturn can partially explain the huge size of the Saturnian rings. In the chapter devoted to rings, I detailed their formation.

24.3. Density of the satellites

The density of the satellites in the Solar System depends on their types (p=0.000), their turbulence zones (p=0.000), and the interaction between these 2 variables (p=0.004). Density data was available for only 76 satellites in the Solar System and it ranges between 250 kg/m³ and 3530 kg/m³ (Fig. 108). The smallest densities were found with the Saturnian satellites, whereas the highest density was with Io (the innermost Jovian satellite in Zone 3). The second highest density (3344 kg/m³) was recorded with the Moon. In general, the density of many Jovian satellites is among the highest. The densest satellite in each planetary system is found in the most turbulent zone, usually Zone 3. In general, the density of the outer satellites of each planetary system is generally almost the same.

Fig. 108: Density of the satellites according to their types and semi major axis

The Moon is denser than any of the 2 Martian satellites. The density of the Jovian satellites in Zones 4 and 5 is the same (2600 kg/m3). The biggest variation occurred among the density in Zones 1, 2, and 3 (Fig. 109). The density of the satellites in Zone 1 is less than that in Zone 2. The densest satellite in each planetary system is usually but not always the biggest satellite, suggesting that the conditions that defined the size of the bodies also affected their density. For instance, the densest Jovian satellite is not Ganymede, the biggest Jovian satellite, but Io, the innermost Jovian satellite in Zone 3. Titan (the biggest Saturnian satellite) is the densest Saturnian satellite. Its density is followed by that of Phoebe, a Saturnian satellite in turbulence Zone 4.

Titania, the biggest Uranian satellite, has the highest density (1710 kg/m3) among the Uranian satellites. The density of the Uranian satellites in Zones 1, 2, 4, and 5 is reported as 1300 kg/m3. The highest density was found with Triton, the biggest Neptunian satellite. The density of the Plutonian satellites in Zone 4 is not reported in the literature. The only density available for the Plutonian satellites is that of Charon (1700 kg/m3).

Fig. 109: Density of satellites according to their types and turbulence zone

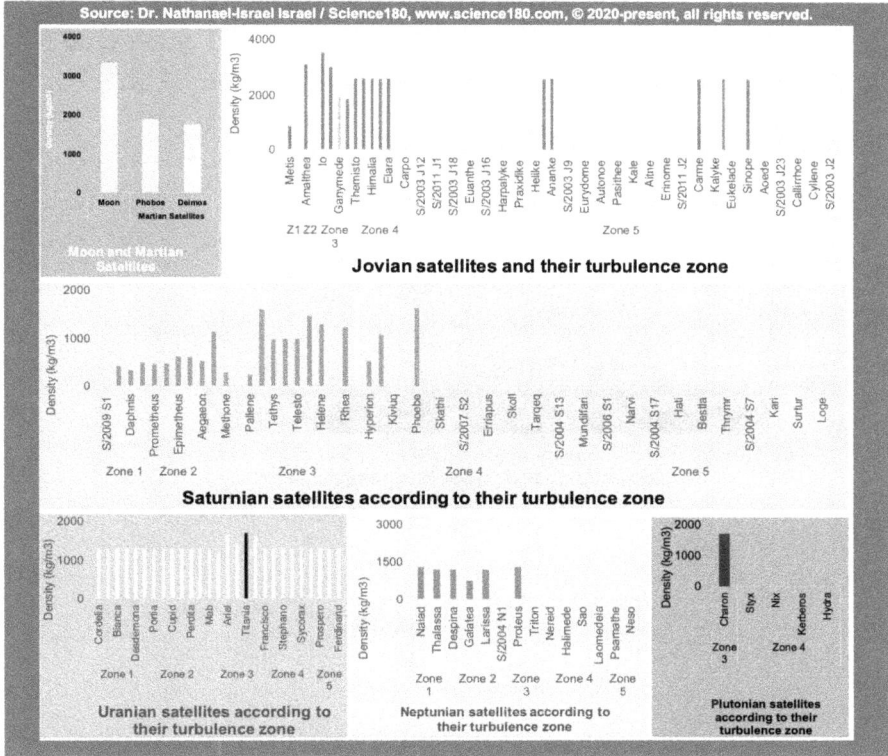

24.4. Ratio of the density of the satellites to that of their primary planets

The ratio of the density of the satellites to that of their planets in the Solar System depends on the types of satellites, the turbulence zones, and the interaction between these 2 variables (p=0.00). That ratio ranged between 0.36 and 2.74 (Fig. 110). The smallest ratio was measured with a Saturnian satellite, whereas the highest ratio was obtained with Titan, the biggest Saturnian satellite. Most of the highest ratios were recorded with the Saturnian and Jovian satellites. The ratio for 70% of the satellites is higher than 1, meaning that the density of 70% of the satellites is higher than that of their primary planet. The satellites that are denser than their primary planets were found with the 4 giant planetary systems, suggesting that the density of the satellites could be related to the level of turbulence of their precursors. The terrestrial planets are denser than their satellites. As far as the turbulence zones are concerned, the highest satellite-to-planet density ratios in the Solar System were recorded with the satellites in Zone

3. Fig. 110-111 describe the satellite-to-planet density ratio according to the types of satellites and their turbulence zones.

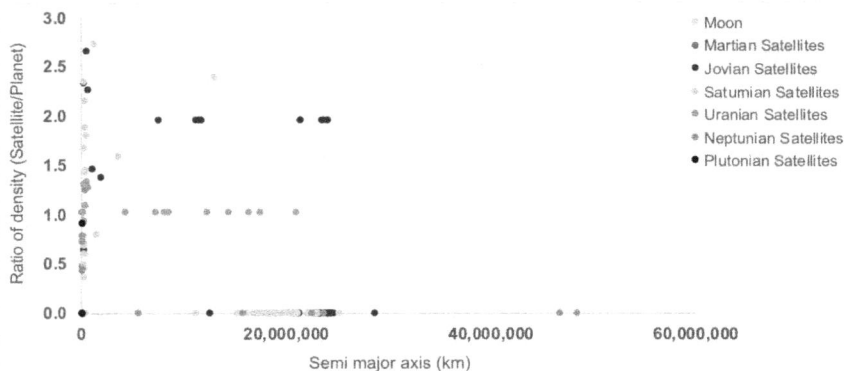

Fig. 110: Ratio of the density of the satellites to that of their primary planet according to their types and semi major axis

Fig. 111: Satellite-to-planet density ratio according to the types of satellites and their turbulence zones

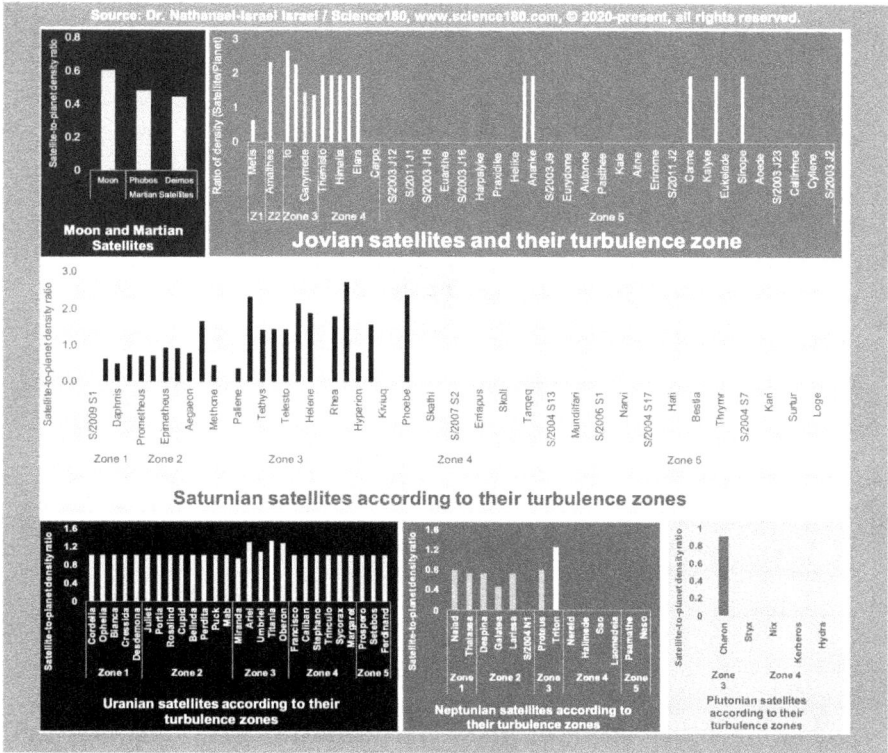

As I explained above, density data was not available in the literature for 62.3% of the bodies I studied in the Solar System. To be able to estimate the mass and other key variables such as kinetic energy, I needed a density for each body. Therefore, so I didn't end up having to deal with a very limited data for certain crucial variables, I had to estimate the missing densities based on the trends I found with the available data.

24.5. How I learned for the first time that the density of the celestial bodies could depend on their size and position

It was Thursday, July 14, 2014. I opened my computer to check my email. As I was browsing the news of Yahoo's homepage, I came across an article talking about a comet. Because, since awhile, I had been interested in astronomy, I quickly read that article. Afterward, I decided to learn more about the difference between comets and asteroids. This led me to Wikipedia. While I was reading on asteroids, I came across some studies done on the main belt asteroids. Looking at a picture of the main belt asteroids, I immediately understood that their position would be related to how the Solar System was formed. This thought crossed my mind all throughout that day. When the night came, I went to bed. During the night of July 17 leading to July 18, 2014, I lost sleep around 3 AM, when my little girl woke up from her sleep and started crying as most babies would do. For almost one hour, I tried to fall asleep again, but could not. Meanwhile, many thoughts fell into my mind concerning the main belt asteroids I learned about 3 days ago. I immediately felt like the precursor of the main belt asteroids could have formed a planet between Mars and Jupiter, but the conditions were not met for it to amass all its rocks into a single, unified body. Hence, many rocks are dispersed in that area. At the same time, I perceived that the density of the planets that are close to the Sun would be generally higher than that of those farther from the Sun.

As I was nurturing these thoughts on my bed, the curiosity to learn more about them covered me. I therefore decided to get up and read about the density of planets on the internet. As I was doing so, I found that the density of the 4 innermost planets (Mercury, Venus, Earth, and Mars) are denser than that of the giant planets in the Solar System. I later learned that some planets are giant gas, while others are giant ice, while others still are very dense like the Earth. I felt like some conditions were met for the precursors of some planets to be solid, gas, or ice. I felt like the dimension of the bodies could have affected their density. A strong conviction filled my head that, by reading and doing some research about the density of the bodies, I could be enlightened about the formation of celestial bodies. Therefore, I left my home office that night and went back to bed, where I managed to fall asleep again around 5 AM. Two years later, while working on the regression between density and other variables I was studying, I realized that density is partially connected to the radius of the celestial bodies.

The same day that I first learned about main belt asteroids, I also felt like the organization of these asteroids implies that the force that unified or gathered together the precursors of the celestial bodies and their chemical particles could have intervened while gravity was being established. I felt like the size of the precursor of the main belt asteroids could have been higher than what the forces that acted on it could have been able to squeeze into a single celestial body like a planet. Hence, beyond the main belt asteroids, the giant planets are not very dense. Pluto was formed very dense because the volume of its precursors was relatively small and the force which gathered it together was able to control it in such a way that, in the end, the density of Pluto is higher than that of the giant planets. On that same night of July 17. 2017, I also realized had the first glimpse at what I called "the force of union and cohesion", and which I detailed in another book.

24.6. Inspiration from the building of houses with mud balls made with soil or sand and water

On September 12, 2017, I was inspired to ponder on how houses are built in Africa, where I grew up. For I felt like, by referring to how houses are built, it can be possible to explain the shape and density of celestial bodies. As I was growing up, people in my town used to build houses by first digging the ground, preparing a foundation, and collecting the soil (which they had dug) and adding water to it to make mud that is used to erect buildings. The mud is usually rolled on the ground using hands and, then, collected into small spherical bodies like the size of a soccer ball. Then, those balls are carried to the location of the building and laid one on top of the other according to the design of the building (Fig. 112).

To make these balls, the builders use their hands to apply force to the quantity of mud that matches their age and strength. People who are stronger make larger amounts of mud, while those who are weak or young make small mud balls. Depending on how the hands apply a force to the mud being rolled on the ground, some balls end up being perfect spheres, while others are oblate, prolate, and some do not even have a defined shape. The density of the balls depends on the compression applied to them as the mud is rolled. The balls are turned and rolled in many directions (e.g. north to south, east to west) to ensure their shape is near spherical. As the builders are making and shaping the balls, they also simultaneously press them. In the end, the size of the balls and compression they receive is generally inversely proportional. For the bigger the mud collected, the harder it is to apply a strong force to it. But the smaller the mud used to make a ball, the easier it is to compress and roll it on the ground to make a denser ball. After the mud balls are made, the builder stacks them one on top of the other, compressing them and building the house. During the process, layers of the mud balls are compacted and left to dry so the walls of the buildings do not collapse due to its weight and the wetness of the mud. People who are rich use bricks made with concrete or cement, which not only stand longer and are stronger, but which also better resist other challenges such as rain that can bring the ongoing building down to its original material: soil and water.

Fig. 112: Building of a house with the mud and bricks made with soil and water

By recalling these constructions that I also did when I was growing up, I felt like a similar process could have been used to make the celestial bodies:

- Create the chemical particles
- Mix the particles to make complex ones
- Mix the particles and compounds with fluids (e.g. water) to lay some foundations
- Compress the bodies.

As of 2021, I perceived that the strata found in some planets like the Earth are like the bricks a builder lays while erecting a house. Driving on some roads in some rocky places can also allow those passing through to easily see how layers of materials deposited like brick fill the Earth's crust.

Because some people may still be wondering what the take home message is regarding "why celestial bodies have different densities and why some are denser than others", I would say the following. As the fluid layers of the precursors of the bodies in the Solar System started gathering together to shape the form of their bodies, their fluids were spiraled and wrapped together as vortexes of various sizes were formed inside of them. The speed of the fluid layers affected the way the spirals squeezed their daughter bodies that were being formed but which ability to rotate also affected the intensity with which they were squeezed. The higher the speed of the fluids in the layers, the higher the likelihood that their daughter bodies would be more squeezed. But when the daughter bodies could not rotate faster, their ability to squeeze their content and be denser decreased. Consequently, the densest bodies are those for which the orbital speed and the rotational speed are balanced and relatively high. For instance, although the densest bodies are usually found in Zones 1, 2, and 3, the densest ones are not usually in Zone 1. For, although having the higher orbital speed in their system, the bodies in Zone 1 do not always have the highest rotational speed. Hence, they are not always the densest. The higher orbital speed of the precursors balanced with the ability of their daughter bodies to rotate faster could have affected the density of the daughter bodies. That is why although the terrestrial planets are the densest, the Earth is denser than all of them because it has a relatively higher balance between its orbital speed and rotational speed. In contrast, although Mercury and Venus have a higher orbital speed, their density is smaller than that of Earth because their precursors were not able to rotate fast, finishing a single rotation in months, while Earth completes one full rotation every day. Similarly, although the giant planets have a higher rotational speed, their densities are not the highest. Furthermore, as I explained in other chapters later, the turbulence that the precursors of the bodies went through affected their rotational movement.

24.7. Take home message

About 58% of the density of the bodies that I studied in the Solar System is greater than that of the Sun. The densest planet is the Earth, whereas the least dense planet is Saturn. All of the terrestrial planets are denser than the Sun. The smallest densities were found with the giant planets. As for the satellites, the smallest densities were found with the Saturnian satellites whereas the highest density was with a Jovian satellite. The second highest density was recorded with the Moon. In general, the density of many Jovian satellites is among the highest. The densest satellite in each planetary system is found in the most turbulent zone, usually Zone 3. The density of 70% of the satellites in the Solar System is higher than that of their primary planet. The terrestrial planets are denser than their satellites. The density of the celestial bodies can be explained by how the fluid layers of the precursors were squeezed during their split-gathering. The speed of the fluid layers and their ability to rotate affected the way their daughter bodies were squeezed. I will provide more details later.

CHAPTER 25

DOES THE UNIVERSE HAVE A BIRTHDAY? LEARN HOW THE SEMI MAJOR AXIS (I.E. AVERAGE DISTANCE SEPARATING CELESTIAL BODIES FROM THEIR PRIMARY BODY) GIVES A PRICELESS CLUE TO THE BIRTHDATE OF ALL CELESTIAL BODIES IN THE UNIVERSE

25.1. General trends for all celestial bodies in the Solar System

Although celestial bodies can be subject to many kinds of movement, rotation and revolution are usually the most common. The position of celestial bodies in their obit changes. Their position with respect to their primary bodies is usually characterized by 3 variables: aphelion, perihelion, and semi major axis. Each of these variables reflects the journey that the precursors of these bodies had during their formation. Of these 3 variables, semi major axis is the most reported and the most used. Although it would have been great if data on aphelion and perihelion were reported for all bodies I studied, they are reported mostly for the planets and a few asteroids. Therefore, to avoid leaving out many celestial bodies, I focused on semi major axis in my writing as far as distance is concerned. The next graph illustrates the definition (Fig. 113).

NASA (2018) has defined "semi major axis" as the mean distance from the Sun (or from the primary planet in the case of the satellites) from center to center. It is a measure of the size of an orbit and can also be defined as the longest radius of the ellipse of the orbit. Besides eccentricity, semi major axis is the other main element used to define the shape and size of the orbit of celestial bodies. The semi major axis is usually measured in kilometers (km).

Fig. 113: Semi major axis, perihelion, aphelion, major axis, minor axis, and orbit of a celestial body

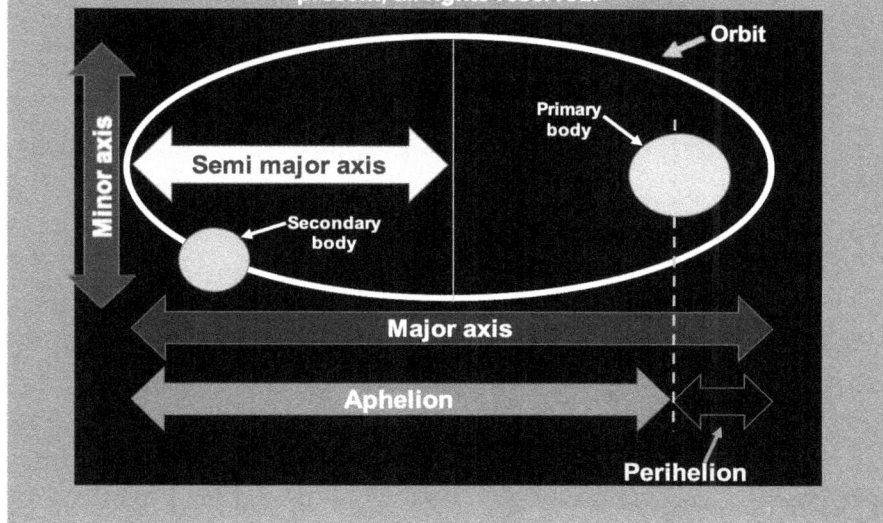

However, because some celestial bodies are located thousands of kilometers from their primary body, the Astronomical Unit (AU), which is the mean distance from the Sun to the Earth (149,597,900 km), is used to simplify the number. Throughout my writing, I also sometimes used the term "semi major" to refer to "semi major axis". Can the semi major axis (i.e. average distance separating celestial bodies from their primary body) give a clue to the birthdate of all celestial bodies in the universe?

I studied the semi major axis of 461 celestial bodies in the Solar System. It did not depend on their types (p=0.16) and ranged from 9,378 km to 4.98E+12 km. The smallest semi major axis was recorded with Phobos, the innermost Martian satellite, followed by Charon (19,596 km), the biggest Plutonian satellite. In general, the smallest semi majors were recorded with satellites of either the smallest planets and the most remote planets. In contrast, the highest semi major axis was recorded with comets with the record (33300 AU) held by C/1999 F1. Asteroids with the smaller semi major axis are Aten asteroids followed by Apollo asteroids. The semi major axis of the main belt asteroids range between 3.24E+08 km (2.17 AU) and 5.22E+08 km (3.49 AU), meaning that the main belt asteroids are spread over a portion of space wider than the distance between the Earth and the Sun (1 AU). It is also important to remark that the orbit of the main belt asteroids is crossed by some Apollo asteroids, Amor asteroids, and Comets, meaning that the semi major axis of these asteroids cross the range of space covered by the main belt asteroids.

The semi major axis of the planets in the Solar System ranges between 0.387 AU and 39.482 AU, the smallest value being recorded with Mercury and the highest value with Pluto (Fig. 114). The semi major axis of the planets in the Solar System is higher than that of any of the satellites.

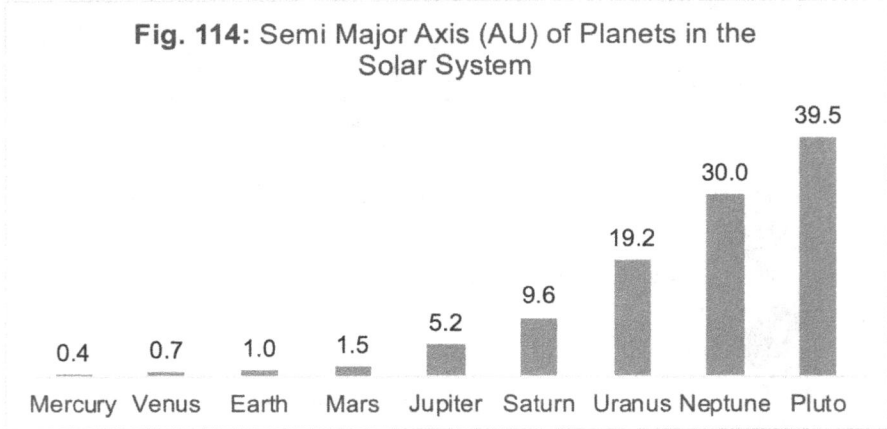

Fig. 114: Semi Major Axis (AU) of Planets in the Solar System

25.2. Ranking of the semi major axis of the satellites according to their types

To show the general picture of the variables I studied on the satellites according to their primary planets, I used to do a graph where the abscissa (X axis) is the semi major axis and the Y ordinate is the variable of interest split according to the types of satellites. Now that the semi major axis is the variable of interest in this chapter, I could not put it on the X abscissa. Therefore, I decided to rank the satellites from the innermost to the outermost with the innermost having a rank of 1 and the outermost having a rank corresponding to the number of satellites in each planetary system. I then plotted the rank on the X axis and the semi major axis on the Y axis. Fig. 115 gave a better view of the variation of the semi major axis of the satellites according to their position in each planetary system.

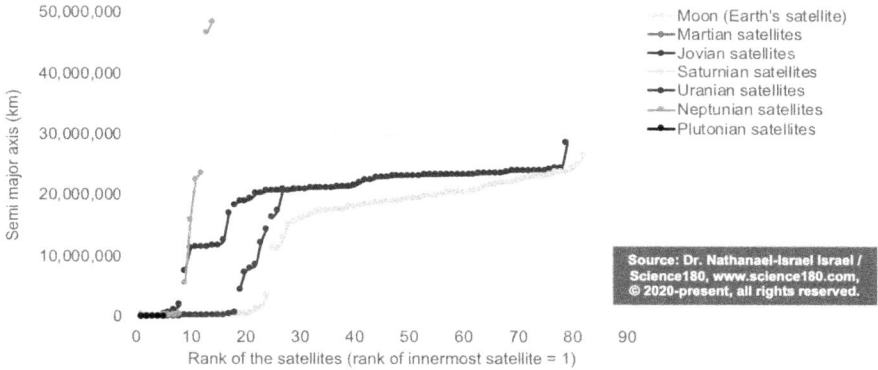

Fig. 115: Semi Major Axis of 210 satellites according to their types and rank

For a same rank number, particularly beyond 17, the Saturnian satellites generally have a smaller semi major axis than the satellites of the other planets. Because the above graph does not properly show the trend of the semi major for the innermost satellites and because I needed to clearly see it, I zoomed into it to get Fig. 116 and Fig. 117. Fig. 116 showed that the semi major axis of the Moon is higher than the semi major axis of the innermost satellite of any other planetary system.

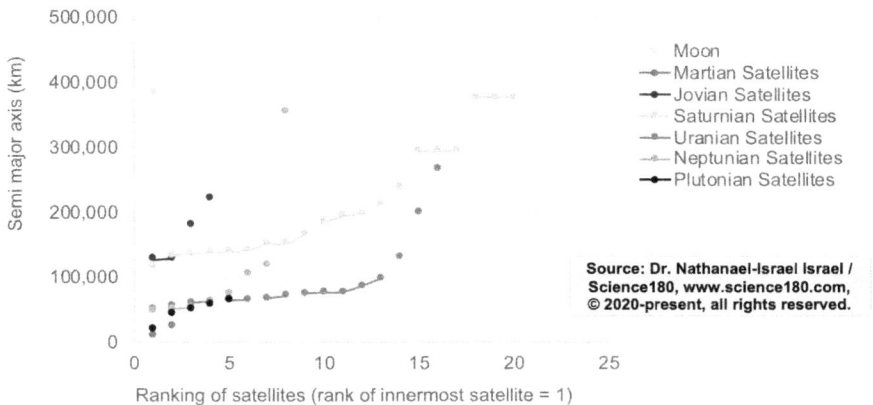

Fig. 116: Semi major axis of the satellites which semi major axis is smaller or equal to that (384, 400 km) of the Moon

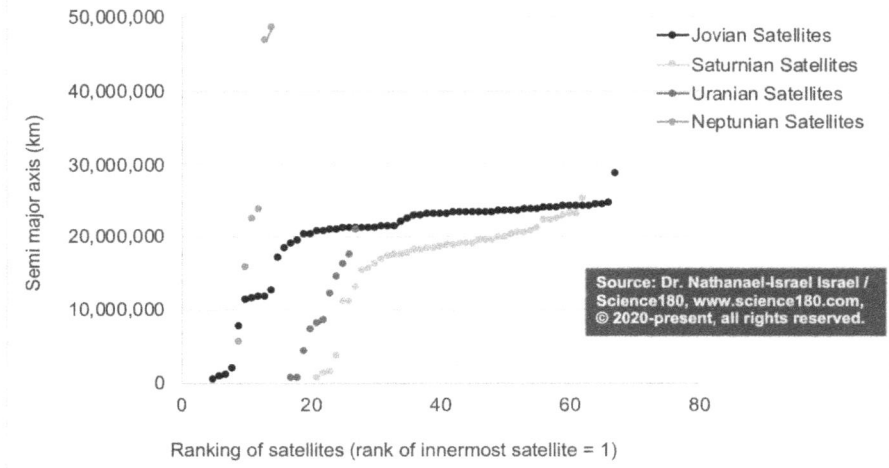

Fig. 117: Semi major axis of the satellites which semi major axis is higher than that (384, 400 km) of the Moon

When I was deeply looking at these graphs, many questions crossed my mind including:

- Why are the Neptunian satellites so dispersed or distant from each other?
- Why are the Jovian satellites so close to one another?
- Why despite the size of the Jovian and Saturnian planetary systems, are the semi major of the Jovian and Saturnian satellites not as high as or higher than that of some of the Neptunian satellites?
- Why is the semi major axis of the Moon higher than that of the innermost satellite of the other planets?

Although I did not initially have a clear answer for these questions, they kept coming into my mind for many years. One of the questions that stuck with me the most and on which I pondered a lot with the conviction that a theory would be born about of it was: Why is the Moon not as close to the Earth as the innermost satellite of the other planets? In 2016, the first time I pondered on the reason of the high semi major of the Moon, I thought that the orbital speed of the precursor of the Earth-Moon planetary system may have probably been able to strongly push the precursor of the Moon very far and even farther than the precursor of the innermost satellite of any other planet. In contrast, although the precursor of the Martian planetary system would have been about to split into its daughter bodies, it may have not been able to push its innermost satellite much farther away, and because its orbital speed was very small and instead of wrapping the Martian satellites together with Mars to form one body, it formed Mars and the satellites were closer. I suspected that the viscosity and energy of the precursor of that planetary system may partially explain the spatial and environmental

316
Nathanael-Israel Israel: Who is Told by People That He is the Universe-Origin, Life-Origin & Chemicals-Origin Accurate Decoder

conditions in that area. From Jupiter until Pluto, I was thinking that the semi major of the innermost satellites decreased because of the decrease in the orbital speed of the precursors of their planetary systems. I was thinking of the existence of a law that could define the location and size of a satellite based on some variables of its primary planet. I was having these ideas long before I realized that the data could be explained by turbulence. I also thought about investigating whether a force that connected the satellites to their primary satellites may help explain the data. Years later, I realized that although these ideas seemed to point at some of the reasons explaining the variation of the semi major of the bodies in the Solar System, they are not the real cause or explanation. As you keep reading this book, you will see how I clarified how celestial bodies got their semi major axis.

In 2016, I was strongly interested in doing a regression between the semi major of the innermost and outermost satellites as a dependent of the:

- semi major of the primary
- number of satellites of the primary planet
- range of the semi major of the satellites in each planetary system
- radius of the primary planet
- etc.

As I was reflecting on these questions, I realized that instead of thinking that the primary planet affected the satellites, I would rather think about how the precursor of a planetary system (primary planet + satellites) was split into a primary planet and its satellite(s) under the influence of factors that I was eager to search, find, determine, and define. This intuition about viewing the data in the perspective of a precursor splitting into daughter bodies was one of the first experiences that started shaping my exploration of the raw data I collected with a different perspective. My experience with analyzing the data on the motion of the satellites also inspired me a lot to start thinking about the precursors of the bodies and the systems they formed.

On November 1, 2017, as I was looking at the aforementioned graphs related to the semi major of the satellites according to their rank and trying to figure out some of the reasons the satellites were so differently distributed according to their planetary planet, it instantly came to my attention that the spatial distribution of the satellites holds a secret about the conditions and laws that drove and sustained their formation. In fact, in the last trimester of 2017, when I started getting some clues that the bodies in the universe are products of a turbulence, I started looking at the above graph differently. On one of those days, as I was learning about how the viscosity of fluids affects their movement, I suspected that if that is true, the increment of the semi major between the satellites may vary. It was one of these ideas that made me pay close attention to the distribution of the semi major according to their ranking from the innermost satellite to the outermost satellite. At that time, although I did not know the variables which can be used to explain the information coded in the above graphs, I felt like I must keep looking at them carefully until I figured out the secret they hold. I shed more light on the distribution of the semi major in the segments related to viscosity.

During my reading, I learned that diffusion is one of the phenomena used to describe the mechanics of continuous environments. Diffusion relates to the ability of a fluid to diffuse, move from one place to its outskirts, like from a center to a periphery, or from a region of high concentration to a region of low concentration. Viscosity of the precursors of the celestial bodies could have affected their diffusion and consequently their semi major axis. The precursors of some innermost celestial bodies could have been more viscous. The turbulence intensity and the chemical composition of the precursors of the celestial bodies could have also affected their viscosity. Because of its higher turbulence, the precursor of the Jovian planetary system could have been more viscous than that of the other giant planetary systems.

Viscosity of the precursors of the satellites could have probably had a negative impact on their diffusion and consequently on the semi major of the satellites. In other words, the more viscous a fluid is the less it can diffuse. In contrast, the least viscous fluids diffuse more. This can explain for instance why the precursor of the Jovian satellites diffused less than that of the Neptunian satellites. Consequently, the distance between adjacent Jovian satellites is smaller than that between Neptunian satellites. The precursor of Neptune and its satellites could have diffused more because it was less dense and also less viscous. Consequently, Neptunian satellites are very distant from one another. Because the precursor of Saturnian satellites could have been less viscous than that of the Jovian satellites they could have diffused more, which translated into the relatively higher semi major. The smaller viscosity and density of the precursor of Saturn can also explain why it is more oblate and also has larger rings. In the chapter on rings, I will elaborate more on that.

I later realized that, on top of its small kinetic energy, the high density and viscosity of the precursor of the Martian planetary system could have prevented the precursor of the Martian satellites to diffuse too much, therefore forcing the Martian satellites to be formed very close to Mars. In contrast, the precursors of the Earth-Moon system would have been more viscous than that of Mars and Jupiter, but its high orbital and rotational speed would have pushed the Moon farther away. This suggests that the distance between the satellites and their primary planets would have been affected by the orbital speed, density, and viscosity of their precursors.

25.3. Semi major axis of the satellites according to the turbulence zones

The smallest semi major axis of the satellites is in Zone 1 whereas the highest semi major axis is in Zone 5. The highest semi major axis in Zone 3 was recorded with the Saturnian satellites. The highest semi major axis of the satellites in Zone 1 were observed with the Saturnian satellites followed by the Jovian satellites. The semi major of the Martian satellites in Zone 1 are the smallest. In contrast, in Zone 2, the Jovian satellites took the lead of the highest semi major axis, whereas the Neptunian and Uranian satellites held the smallest values. The semi major axis of the Saturnian satellites in Zone 2 were intermediate between the highest and smallest values recorded on the satellites of other planets. The fact that the semi major of the outermost Jovian satellite in Zone 2 was higher than that of any other outermost satellite in the same zone, yet in Zones 1 and 2 Jupiter has only 4 satellites compared to 9 for Saturn, 13 for Uranus, and 6 for Neptune; suggested to me that the precursor of the Jovian satellites was able to stretch without quickly breaking into daughter bodies. The high viscosity of the precursor of the Jovian satellites could have contributed to its long stretch while resisting being broken.

In Zone 3, the Saturnian satellites experienced the highest semi major axis. The distance between the outermost Jovian and Saturnian satellites and the satellites preceding them in Zone 3 is higher than that between any other consecutive satellites in Zones 1 or 2. This suggests that the precursor of the outermost Jovian and Saturnian satellites in Zone 2 could have been stretched more than the precursors of any other satellite in Zones 1 and 2, and farther away from the satellites which precede it than any other consecutive satellite in Zone 3. Such stretch is not pronounced with the Uranian and Neptunian satellites in Zone 3.

Between the outermost satellite in Zone 3 and the innermost satellite in Zone 4, the stretching continued and may have even ended with the ejection of the precursor of the satellites in Zones 4 and 5. Because of the importance of the semi major increment between the outermost satellite in Zone 3 and the innermost satellite in Zone 4, I detailed their trends in the next chapter, which is about semi major increment. In Zone 4, the Neptunian satellites were spread over a longer range of semi major.

The semi major axis of the satellites in Zone 5 varies between 16,256,000 km and 48,390,000 km. The highest value was with Neso, a Neptunian satellite, whereas the smallest value was with Prospero, a Uranian satellite. No turbulence zone contains as many satellites as Zone 5. Yet, the range of the semi major in Zone 5 is not the highest among all of the 5 turbulence zones. It is only with the Jovian satellites that no turbulence zone has a semi major range higher than that of Zone 5.

25.4. General trends of the semi major of the satellites according to the types, turbulence zones, and semi major axis

Varying between 9,378 km and 4.84E+07 km (Fig. 118), the semi major axis of the satellites depended on their types ($p<0.000$), the turbulence zone ($p<0.000$), and the interaction between these 2 variables ($p<0.000$). Neso, the outermost Neptunian satellite, is the satellite that holds the highest semi major axis of the satellites in the Solar System. No satellite has a semi major axis higher than that of the innermost planet or asteroid in the Solar System. This also means that the orbits of the satellites are much closer to their primary planets than the orbit of Mercury is from the Sun. The forces that positioned the satellites around their primary planets tied the satellites more closely to their primary planets than the planets are tied around the Sun. To put it another way, the force or energy that propelled the precursors of the satellites away from the precursor of their planetary planet would not have been as strong as the force or energy that propelled the precursors of the planetary systems away from the precursors of the Sun.

Fig. 118: Semi major axis (km) of 210 satellites according to their types and turbulence zones

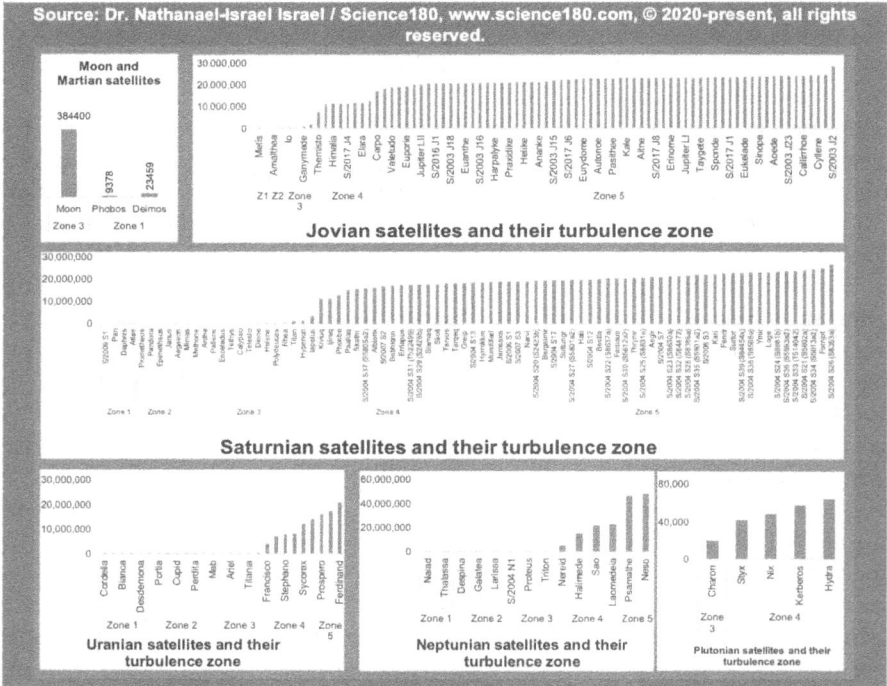

The semi major of the innermost satellite of each planet depends on their planetary system. The semi major of the Moon (the only satellite of the Earth) is higher than that of the innermost satellite of any other planet in the Solar System. Another way of putting this is that the Moon orbits the Earth at a greater distance than the innermost satellite of any of the other planets in the Solar System. Indeed, the semi major of the Moon is almost 3 times that of the innermost satellite of Jupiter (the biggest planet in the Solar System). Additionally, the semi major axis of the Moon is almost 20 times that of the innermost satellite of Pluto and almost 6 times that of the outermost satellite of Pluto. Similarly, the Moon is farther from the Earth than any of the Martian satellites are from Mars. Moreover, the Moon is even farther from the Earth than Umbriel (the 16th innermost satellite of Uranus) is from Uranus, meaning that the distance separating Uranus from its 16th innermost satellite is smaller than the distance separating the Moon from the Earth. Polydeuces (the 20th innermost satellite of Saturn) is closer to Saturn than the Moon is from the Earth. In other words, the space between the Earth and the Moon would have been enough to squeeze in the 20 innermost satellites of Saturn. Therefore, I felt like, during the formation of the satellites in the Solar System, the precursor of the Moon would have been pushed away from the precursor of the Earth much farther than the innermost satellite of any of the other planets.

The semi major of the Moon (384400 km) is 41 times that of Phobos and 16.4 times that of Deimos. The semi major of Deimos is 2.5 times that of Phobos, meaning that the outermost Martian satellite is 2.5 times further from Mars than the innermost Martian satellite. Furthermore, the semi major axis of the 2 Martian satellites are the smallest and third smallest recorded for the satellites in the Solar System. In other words, the semi major axis of the other satellites in the Solar System, except Charon, is higher than that of any of the Martian satellites.

As for the Jovian satellites, the semi major axis ranges from 128,000 km to 28,570,000 km, meaning that the ratio of the semi major of the outermost satellite to that of the innermost satellite is 223.2. The innermost Jovian satellite is Metis, whereas the outermost satellite based on the data available as of 2020 is S/2003 J2. The semi major of the outermost satellite in Zone 3 (Callisto) is 14.7 times that of the closest satellite to Jupiter. The semi major of the biggest Jovian satellite, Ganymede, is 1,070,400 km, which is 8.36 times the semi major of the innermost Jovian satellite. This means that the most remote Jovian satellite is 26.7 times further from Jupiter than the biggest Jovian satellite.

The semi major axis of the Saturnian satellites ranges between 117,000 km and 26,738,000 km. The ratio of the semi major of the outermost Saturnian satellite to that of the innermost satellite based on the data available as of 2016 was 214.6. That ratio is in the same order of magnitude than that obtained with the Jovian satellite (223.2). The semi major axis of the biggest Saturnian satellite (Titan) is 1,221,830 km, which is 1/20.55th the semi major of the outermost Saturnian satellite known as of 2016. Furthermore, the semi major of the biggest Saturnian satellite is 10.4 times the semi major of the innermost Saturnian satellite. The semi major (25,110,000 km) of Fornjot (the outermost Saturnian known as of 2016) is 7.05 times that of Iapetus (the outermost satellite in Zone 3). One of the Saturnian satellites discovered between 2016-2019 has a semi major higher than that of the outermost Saturnian satellite known before 2016. But the semi major (26,738,000 km) of a newly discovered Saturnian satellite (S/2004 S26 (S8353a)) is 1,628,000 km farther than that of Fornjot, the Saturnian satellite which as of 2016 held the record of the outermost satellite. In other words, as of 2020, the new outermost Saturnian satellite is S/2004 S26 (S8353a).

The mean distance between the Uranian satellites and their primary planet,

Uranus, ranges between 49,800 km and 20,901,000 km. This infers that the ratio of the outermost Uranian satellite known as of 2020 and the innermost satellite known as of 2016 is 419.7. That ratio is also twice that observed with the innermost and outermost satellites of Jupiter and Saturn. The semi major axis of the biggest Uranian satellite (Titania) is 436,300, which is 8.76 times that of the innermost Uranian satellite, Cordelia. This means that the semi major of the most remote Uranian satellite is 47.9 times that of the biggest Saturnian satellite. The semi major of Ferdinand (the outermost Uranian satellite) is 35.8 times that of Oberon (the outermost Uranian satellite in Zone 3).

The semi major axis of the Neptunian satellites varies between 48,227 km and 48,390,000 km, making the outermost Neptunian satellites the record keeper of the highest semi major observed on the satellites in the Solar System. To put it into other words, as of 2020, no planet in the Solar System has a satellite as distant from its primary planet as the outermost Neptunian satellites are from Neptune. This wide range of the semi major of the Neptunian satellites explains why the ratio between the semi major of the outermost Neptunian satellite (Neso) and that of the innermost Neptunian satellite (Naiad) is 1003.38. This ratio is more than 4 times that observed with the Jovian or Saturnian satellites and more than twice that recorded with the Uranian satellites.

The semi major (354,760 km) of the biggest Neptunian satellite (Triton) is 7.36 times that of the innermost Neptunian satellite. In contrast, the semi major axis of the outermost Neptunian satellite is 136.4 times that of the biggest Neptunian satellite. Because Triton is the outermost Neptunian satellite in Zone 3, the semi major of the outermost Neptunian satellite is also 136.4 times that of the outermost Neptunian satellite in Zone 3. In fact, the semi major axis of the two outermost Neptunian satellites is higher than that of the outermost Jovian and Saturnian satellites:

- Outermost Neptunian satellite, Neso: 48,390,000 km
- Second outermost Neptunian satellite, Psamathe: 46,700,000 km
- Outermost Jovian satellite (known as of 2016), S/2003 J2: 28,570,000 km
- Outermost Saturnian satellite, Fornjot: 25,110,000 km

An interesting question is: Why are the outermost Neptunian satellites so distant from their primary planet?

Finally, the Plutonian satellites have some of the smallest semi major axis in the Solar System, ranging between 19,596 km and 64,720 km. The semi major (19,596 km) of Charon, the biggest Plutonian satellite, is the second smallest semi major of the satellites in the Solar System. The semi major of the innermost Saturnian and Jovian satellites is higher than the semi major of the outermost Plutonian satellites. In other words, all of the Jovian and Saturnian satellites are more distant from their primary planets than the outermost Plutonian satellite is from Pluto. The precursor of the Plutonian planetary system would have had a "hard time" pushing the precursors of its satellites farther into space. This "difficulty" which seems to be also encountered by the precursor of the Martian planetary system may be related to the energy they could have had. The sections on the kinetic energy of the bodies in the Solar System, that I handled toward the end of this chapter, shine more light onto this issue. The semi major axis (64720 km) of the outermost Plutonian satellite, Hydra, is 3.3 times that of Charon. Because Charon is the innermost Plutonian satellite, the ratio between the semi major axis of the outermost Plutonian satellite and the innermost Plutonian satellite is also 3.3, therefore making the Plutonian planetary system the planetary system which has the second smallest ratio "semi major of the outermost satellite / semi major of the innermost satellite". As a reminder, a small ratio was observed with the Martian satellites.

In Table 11 and in the following graphs, I summarized the semi major of the innermost and outermost satellites according to their primary planet.

Table 11: Semi major axis of the innermost and outermost satellites of each planetary system

Source: Dr. Nathanael-Israel Israel / Science180, www.science180.com, © 2020-present, all rights reserved.						
Primary Planet	Number of satellites	Types of satellites	Innermost satellite		Outermost satellite	
			Name	Semi major (km)	Name	Semi major (km)
Earth	2	Earth's satellite	Moon	384,400	Moon	384,400
Mars	2	Martian satellites	Phobos	9,378	Deimos	23,459
Jupiter	79	Jovian satellites	Metis	128,000	S/2003 J2	28,570,000
Saturn	82	Saturnian satellites	S/2009 S1	117,000	S/2004 S26 (S8353a)	26738000
Uranus	27	Uranian satellites	Cordelia	49,800	Ferdinand	20,901,000
Neptune	14	Neptunian satellites	Naiad	48,227	Neso	48,390,000
Pluto	5	Plutonian satellites	Charon	19,596	Hydra	64,720

25.5. Range of the semi major according to the types of satellites and turbulence zones

The range of the semi major axis of the satellites varies according to their types (Fig. 119). The highest range of the semi major was observed with the Neptunian planetary system followed by the Jovian planetary system. For instance, the range of the semi major of the Neptunian satellites is:

- 1.70 times that of the Jovian satellites,
- 1.93 times that of the Saturnian satellites,
- 2.32 times that of the Uranian satellites,
- 1071.31 times that of the Plutonian satellites, and
- 3433.12 times that of the Martian satellites

Fig. 119: **Range** of the semi major axis (km) of 210 satellites according to their types

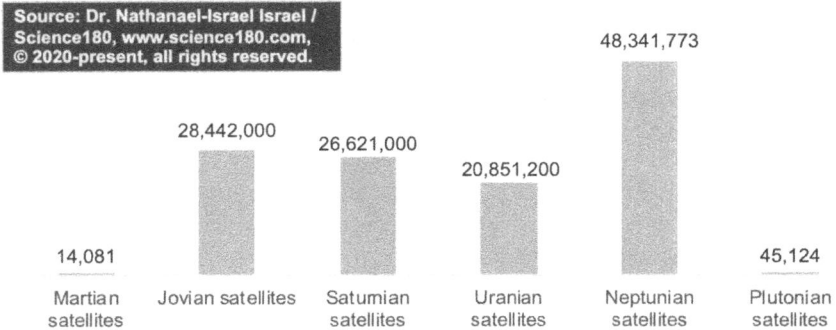

48,341,773

28,442,000

26,621,000

20,851,200

14,081   45,124

| Martian satellites | Jovian satellites | Saturnian satellites | Uranian satellites | Neptunian satellites | Plutonian satellites |

The range of the semi major of the turbulence zone depends on the types of satellites (Fig. 120 and Fig. 121). For instance, for the Jovian and Saturnian satellites, the higher ranges were observed in Zones 4 and 5. For the Jovian and Saturnian satellites, the semi major range for turbulence Zones 4 and 5 are not that different. In contrast, for the Uranian and Neptunian satellites, the highest ranges were in Zone 4 followed by Zone 5. Therefore, in general, the highest range per turbulence zone is in Zone 4, the transition zone out of turbulence.

Fig. 120: **Range** of the semi major axis (km) of the satellites according to their types and turbulence zones

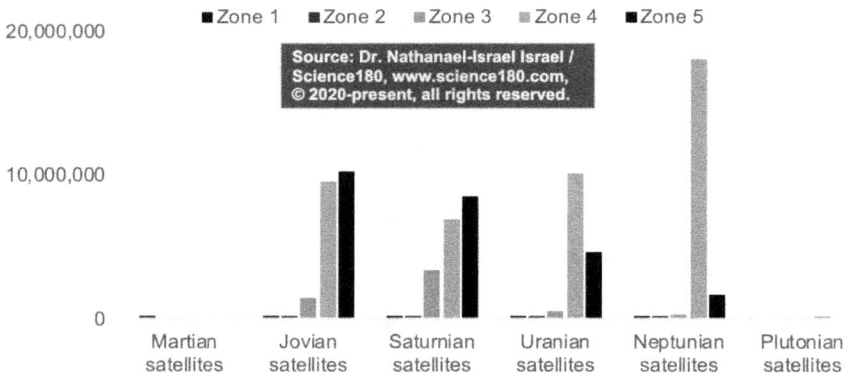

■ Zone 1   ■ Zone 2   ▓ Zone 3   ▒ Zone 4   ■ Zone 5

20,000,000

10,000,000

0

| Martian satellites | Jovian satellites | Saturnian satellites | Uranian satellites | Neptunian satellites | Plutonian satellites |

The shortest range in Zone 1 is that of the Jovian satellites, followed by that of the Neptunian satellites. The Saturnian satellites have the highest semi major range. Unlike in Zone 1 where they have the smallest semi major range, the Neptunian and Jovian satellites recorded the highest semi major range in Zone 2 as if the expansion of their semi major that they failed to achieve in Zone 1 was

Nathanael-Israel Israel: Who is Told by People That He is the Universe-Origin, Life-Origin & Chemicals-Origin Accurate Decoder

recovered in Zone 2. The Saturnian satellites in Zone 3 were spread over a wider range of semi major than the satellites in Zone 3 of any other planetary system. The second highest range of the semi major was observed with the Jovian satellites. The range (3,375,780 km) of the semi major of the Saturnian satellites in Zone 3 is:

- 2.3 times that of the Jovian satellites in Zone 3
- 14.23 times that of the Neptunian satellites in Zone 3

The turbulence in the Zone 3 of the Saturnian satellites was developed over a wider range of semi major than that of any other planetary system in the Solar System (Fig. 121). The combination of the viscosity and the speed of the precursors of the satellites in the Saturnian planetary system could have favored such an expansion of the semi major in the Saturnian satellites in Zone 3. In contrast, a higher viscosity of the precursor of the Jovian satellites slowed down its expansion or shear of turbulence Zone 3 as compared to that in the Saturnian satellites in Zone 3. For the high viscosity of the precursor of the Jovian System would have resisted its movement, therefore its flow was slowed.

Despite the fact that the Neptunian satellites hold the record for the highest range of satellites in the Solar System, the range of the semi major in Zone 3 is among the smallest. Because Charon and the Moon are the only satellites in Zone 3 of their planetary system, no semi major range exists for the Plutonian planetary system and the planetary system formed by the Moon and the Earth. In Zone 4, the Neptunian satellites have the highest range of semi major axis.

The Jovian satellites in Zone 5 were spread over a higher distance than the satellites in Zone 5 of any other planetary system. In fact, the range of the semi major of Zone 5 of the Jovian satellites is:

- 1.48 times that of the Saturnian satellites in Zone 5,
- 2.2 times that of the Uranian satellites in Zone 5, and
- 6.05 times that of the Neptunian satellites in Zone 5.

25.6. Factors affecting the semi major axis of the innermost satellite, biggest satellite, and outermost satellite in each planetary system

As I "finished" investigating the semi major of the bodies in the Solar System, I was wondering what is affecting their values for instance for the biggest satellite in each planetary system. As I was working on the derivatives of the regression between speed and semi major axis on February 23, 2020, it appeared to me that the kinetic energy of the precursors of the planetary systems could have affected the distance that the precursor of their satellites could have travelled before a moderate decrease of their orbital speed could have been significant enough to generate a major turbulence able to fit Zone 3. Therefore, during the search of the factors which could have defined the semi major axis of the satellites, I tested the impact of the translational kinetic energy, rotational kinetic energy, total kinetic energy, and the rotational angular momentum on the semi major axis of the innermost, biggest, and outermost satellite in each planetary system. For each of the 3 types of kinetic energy, I considered the energy of the primary planet, that of all the satellites in each planetary system, and that in the planet and satellite(s) in each planetary system. As I explained in the next segments, the impact is not always significant. Based on the following results, I suspected that in the stellar systems that are very big, the biggest planet could be farther away than Jupiter is from the Sun. Similarly, in other stellar systems, the biggest planet can be closer to or farther away from the primary star than Jupiter is from the Sun.

Fig. 121: Range of the semi major axis (km) of the satellites according to their types and turbulence zone

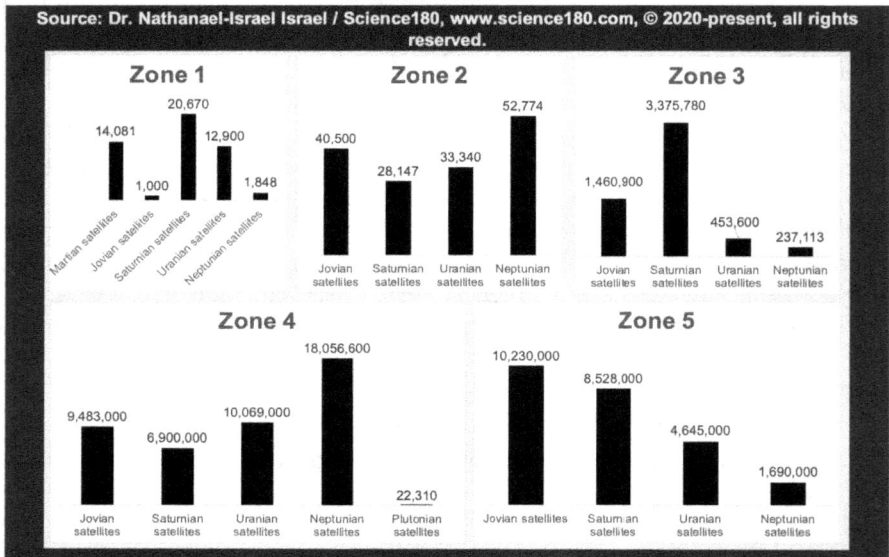

25.6.1. Impact of the kinetic energy (rotational, translational, and total) and of the rotational angular momentum on the semi major axis of the innermost satellite in the planetary systems

The semi major axis of the innermost satellite in the planetary systems is not affected ($r2 < 0.4$) by the rotational kinetic energy, translational kinetic energy, total kinetic energy, or rotational angular momentum of the planets or of all the bodies in the planetary systems. Only the rotational kinetic energy, translational kinetic energy, total kinetic energy, or rotational angular momentum of all the satellites in the planetary system have a slight impact ($r2 = 0.6$) on the semi major of the innermost satellite.

25.6.2. Impact of the kinetic energy (rotational, translational, and total) and of the rotational angular momentum on the semi major axis of the largest or biggest satellite in the planetary systems

The semi major axis of the largest satellite in the planetary systems is significantly affected (r2=0.9) by the rotational kinetic energy, translational kinetic energy, total kinetic energy, or rotational angular momentum of the planets or of all the bodies in the planetary systems (Fig. 122). The highest impact was caused by the translational kinetic energy of all satellites (r2=0.94).

Fig. 122: Impact of the Types of Kinetic Energy (Rotational, Translational, and Total) and of the Rotational Angular Momentum on the Semi Major Axis (km) of the Largest Satellite in the Planetary Systems

Relationship between Rotational Kinetic Energy and the Semi Major Axis (km) of the Largest Satellite

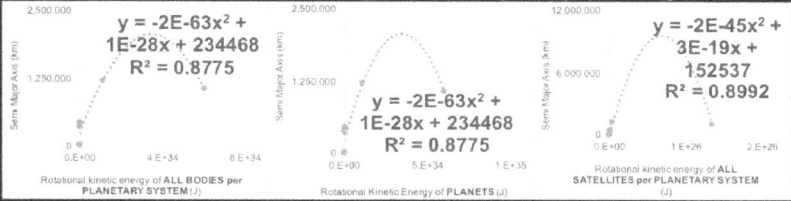

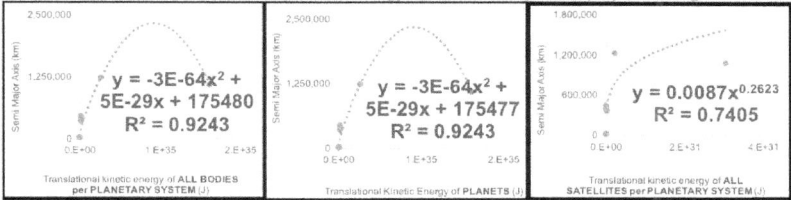

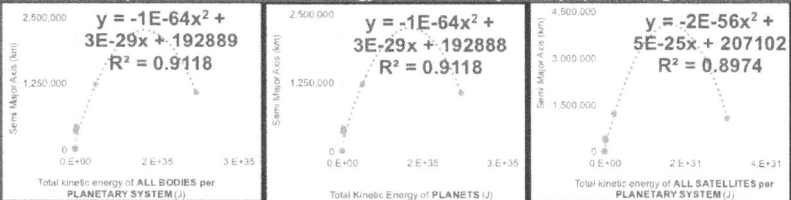

$y = -2E-63x^2 + 1E-28x + 234468$
$R^2 = 0.8775$

$y = -2E-63x^2 + 1E-28x + 234468$
$R^2 = 0.8775$

$y = -2E-45x^2 + 3E-19x + 152537$
$R^2 = 0.8992$

Relationship between the Translational Kinetic Energy and the Semi Major Axis (km) of the Largest Satellite

$y = -3E-64x^2 + 5E-29x + 175480$
$R^2 = 0.9243$

$y = -3E-64x^2 + 5E-29x + 175477$
$R^2 = 0.9243$

$y = 0.0087x^{0.2623}$
$R^2 = 0.7405$

Relationship between the Total Kinetic Energy and the Semi Major Axis (km) of the Largest Satellite

$y = -1E-64x^2 + 3E-29x + 192889$
$R^2 = 0.9118$

$y = -1E-64x^2 + 3E-29x + 192888$
$R^2 = 0.9118$

$y = -2E-56x^2 + 5E-25x + 207102$
$R^2 = 0.8974$

Relationship between the Rotational Angular Momentum and the Semi Major Axis (km) of the Largest Satellite

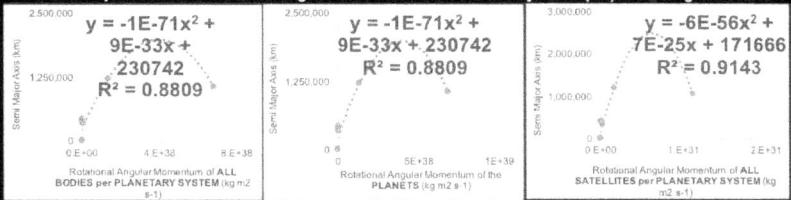

$y = -1E-71x^2 + 9E-33x + 230742$
$R^2 = 0.8809$

$y = -1E-71x^2 + 9E-33x + 230742$
$R^2 = 0.8809$

$y = -6E-56x^2 + 7E-25x + 171666$
$R^2 = 0.9143$

Science180: Irrefutable Scientific Demonstration of the Universe-Origin

25.6.3. Impact of the kinetic energy (rotational, translational, and total) and of the rotational angular momentum on the semi major axis of the outermost satellite in the planetary systems

The semi major axis of the outermost satellite in the planetary systems is significantly affected (r2=0.8) by the translational kinetic energy, total kinetic energy, and rotational angular momentum of the satellites in the planetary systems. The impact of rotational kinetic energy on the semi major of the outermost satellite is weak (r2=0.6). In contrast, the impact of the rotational kinetic energy of the planets and all of the bodies in the planetary systems is stronger (r2=0.8). The translational kinetic energy and the total kinetic energy of the planets and of all the bodies in the planetary system is not strong (r2~0.5).

25.7. Take home message

Calculated using the aphelion and perihelion, the semi major axis is the average distance of the celestial bodies from their primary body. The smallest semi major axis in the Solar System was recorded with Phobos (the innermost Martian satellite), followed by Charon (the biggest Plutonian satellite). Satellites recorded the smallest semi majors, whereas the highest semi major axis was recorded with comets. The semi major axis of the planets in the Solar System is higher than that of any of the satellites. The smallest semi major axis of the satellites is in Zone 1 whereas the highest semi major axis is in Zone 5. The viscosity of the precursor of the celestial bodies could have affected the distribution of the semi major of the bodies they birthed. The semi major axis of the largest satellite and of the outermost satellite in the planetary systems is significantly affected by the rotational kinetic energy, translational kinetic energy, total kinetic energy, or rotational angular momentum of the bodies in the planetary systems.

CHAPTER 26

WHAT CAN THE SEMI MAJOR AXIS INCREMENT (I.E. THE DISTANCE BETWEEN CONSECUTIVE CELESTIAL BODIES) PROVIDE THAT OTHER VARIABLES CANNOT?

26.1. Generalities

The semi major increment between two points A and B is the difference of the semi major of B and the semi major of A. In other words, I calculated the semi major increment by subtracting the semi major of the body which is upstream from that which is immediately downstream. The semi major increment can help to determine the distance separating the bodies from one another. Although it is not the distance separating one fluid layer from the other in the precursors of the bodies, it allowed to estimate the distance at which the downstream layers could have traveled after splitting from the one upstream of them.

The semi major axis increments of the bodies studied in the Solar System ranged from 0 to millions of kilometers, meaning that some bodies shared almost the same orbit during their revolution while some are millions of kilometers farther from others. Below, the semi major axis increment of the planets was calculated by mixing them with asteroids, meaning that their value is not the distance between the planets but between them and their closest bodies orbiting the Sun including the asteroids.

The semi major increment of the bodies in the Solar System did not depend on their types (p=0.34). Most of the smallest semi major increments in the Solar System were recorded with the satellites. Asteroids are more distant from one another than any other type of body.

26.2. Semi major increment of the main belt asteroids

In the main belt, the increment of the semi major ranges from 0 to 19,298,129 km. The main belt asteroid with the highest semi major axis increment is 65 Cybele. That semi major increment is 2.8 to 476 times that of other asteroids in the main belt, suggesting that something "strange" may have happened between that asteroid and the ones surrounding it. Something may have happened to the precursor of that asteroid for it to be so distant from the asteroids upstream of it.

In contrast, the smallest semi major axis increment in the main belt was recorded with 88 Thisbe, the asteroid which is immediately downstream of Ceres (the biggest asteroid in the main belt). This suggests that a certain physical law explained that the asteroid just after Ceres was not that pushed away. In fact, 65 Cybele is one of the outermost asteroids in the main belt. In fact, based on the asteroids discovered in the main belt by 2016, only 3 main belt asteroids were found downstream of 65 Cybele:

- 121 Hermione located 3590349.6 km downstream of 65 Cybele
- 107 Camilla located 4824532.275 km downstream of Hermione
- 87 Sylvia (the outermost asteroid in the main belt), located 112198.425 km downstream of Camilla.

As explained in the section on satellites, the highest semi major increments were observed downstream of the biggest bodies in each planetary system and toward the outskirts of their system. Even with the planets, the highest increment was found downstream of the biggest bodies. The trend observed with 65 Cybele corroborates the general trend of the positioning of the highest semi major increment with respect to the biggest bodies and the semi major distribution of the bodies in each system. Maybe, if the precursor of the bodies in the main belt was collected into one primary body orbited by its satellites, 65 Cybele may be around the innermost satellite in Zone 4. For, as I explained below, the highest semi major increment is usually obtained with the innermost secondary body in Zone 4.

26.3. Semi major increment of the planets in the Solar System

If the semi major of the planets is calculated with respect to whatever celestial bodies is near them, meaning by considering the distance between the asteroids or planets whichever are nearer, the increment of their semi major axis ranges between 344,679 km and 181,792,171 (Fig. 123). However, when analyzed by considering the distance separating the planets from their nearing planets, the increment of the semi major of the planets ranges from 41,390,000 km and 1,622,600,000 km. The smallest semi major increment was recorded between

Earth and Venus, whereas the highest semi major increment was between Uranus and Neptune. In other words, the distance between Uranus and Neptune is higher than that between any other consecutive planets in the Solar System, whereas the distance between Venus and Earth is the smallest among the semi major increment of the planets. Estimated using the astronomical unit (the distance separating the Earth and the Sun), the distance between Neptune and Uranus is 10.85 AU. That distance is not that different from the 9.43 AU separating Neptune and Pluto or from the 9.62 AU between Saturn and Uranus. An interesting question would be why the 3.68 AU separating Mars and Jupiter or the 4.38 AU separating Jupiter and Saturn is not as high as that separating the planets downstream of Saturn.

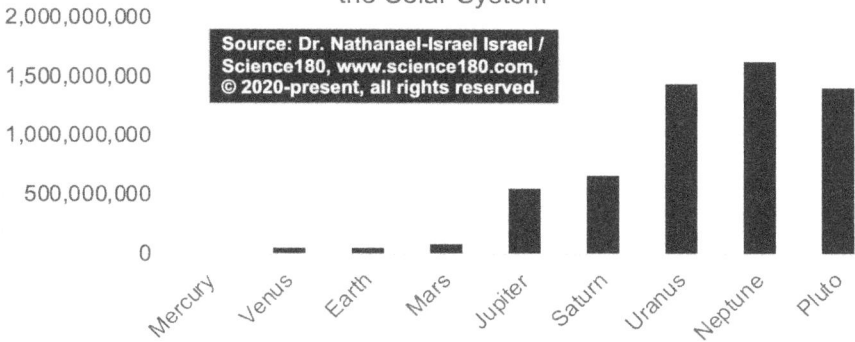

Fig. 123: Semi Major Axis Increment (km) of Planets in the Solar System

As I looked at Fig. 123, I wondered why Uranus can be classified as belonging to in Zone 4. Considering the fact that most of the highest semi major increments were observed with the innermost satellites in Zone 4 (see the sections below on satellites), it may not be wrong to consider Neptune as the innermost planet in turbulence Zone 4 of the bodies orbiting the Sun. In that case, Jupiter, Saturn, and Uranus can be classified in Zone 3 and their size and other characteristics well meet those of bodies in Zone 3 of the bodies orbiting the Sun. However, the high axial tilt of Uranus may work against placing it in Zone 3 for most of the satellites in Zone 3 do not usually have an axial tilt higher than 90 degrees. But considering that Triton, a satellite in Zone 3 has an axial tilt higher than 90 degrees, it may not be misleading to place Uranus in Zone 3 either. Anyhow, the problem that Triton had with its retrograde motion although it is placed in Zone 3 may also explain why Uranus, a retrograde planet, may also be placed in Zone 3. The case can be made that Uranus has all the qualifications to be placed in either Zone 3 or 4, but if I can choose, I would place it in Zone 4. That way, Uranus and Neptune can be in Zone 4, while I can keep Pluto in Zone 5.

Looking at the semi major increment between Uranus and Saturn made me believe that the precursor of the Uranian planetary system could have travelled over a longer distance after breaking from the precursor of the Saturnian planetary system. The precursor of the Uranian planetary system could have been ejected from the precursor of the Saturnian planetary system. Also, the way the precursor of the Uranian planetary system separated from the precursor of its neighboring planetary systems (e.g. precursor of the planetary system downstream and upstream of the Uranian planetary system) and the gain of momentum associated with it could have tilted the precursor of the Uranian planetary system and its daughter bodies (Uranus and its satellites). Later, I will give more details.

26.4. General trends for the semi major increment of the satellites

Ranging between 0 and 23,130,000 km, the semi major increment of the satellites (Fig. 124) depends on their types ($p < 0.000$), the turbulence zone ($p < 0.000$), and the interaction between these 2 variables ($p < 0.000$). The highest semi major increments were observed with the Neptunian satellites, whereas the smallest were with the Jovian and Saturnian satellites. The standard deviation of the semi major increment of the satellites increased as the semi major of their primary planet increased. For instance, the standard deviation of the semi major increment of the Jovian and Saturnian satellites is smaller than that of the Uranian and Neptunian satellites. Satellites which semi major increment is 0 could have been born from a special breakup whereby a fluid layer split to yield their bodies instead of them coming from different layers of fluids. For if these bodies came from different layers, a distance could have separated them. It can also be possible that these bodies came from different layers which did not distance from one another before being gathered into their daughter bodies.

Nathanael-Israel Israel: Historic Discoverer of the Formula that Accurately Decode the Formation of the Universe, of Life, and of Chemicals in a Few Days

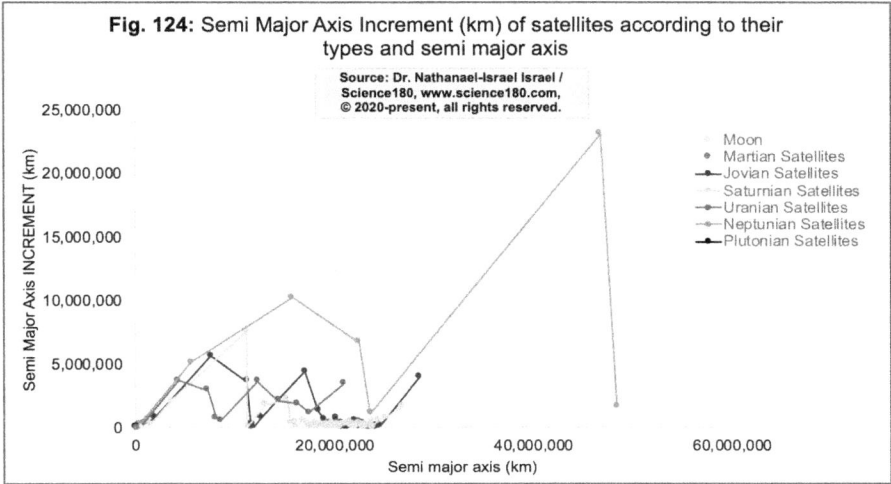

Fig. 124: Semi Major Axis Increment (km) of satellites according to their types and semi major axis

Source: Dr. Nathanael-Israel Israel / Science180, www.science180.com, © 2020-present, all rights reserved.

26.5. Ranking of the satellites according to their semi major axis increment

For each of the planetary systems, the semi major increment of the inner and outer planets is generally smaller than that of the bodies located between them. In Fig. 125, around the areas where the number of satellites is 10 to 30, the shape of the graph is like a zig zag or a "periodic" shape as if the increment increased until it reached a certain maximum value, then it went down until it became almost zero, then it picked up again. For all of the planetary systems, the increment of the outermost satellite is usually but not always higher than that of the satellites immediately upstream of it. Indeed, considering the 67 Jovian satellites known as of 2016, the semi major increment (4,030,000 km) of the outermost S/2003 J2 is 21.1 times that (190,000 km) of Kore, the satellite just upstream of it. This means that, before being gathered into a satellite, the precursor of the Jovian outermost satellite travelled over a very long distance after splitting from the precursor of the satellite upstream of it. It is also possible that many satellites located between those I just mentioned are not discovered yet, and as they would be, the large increments previously discussed will be reduced.

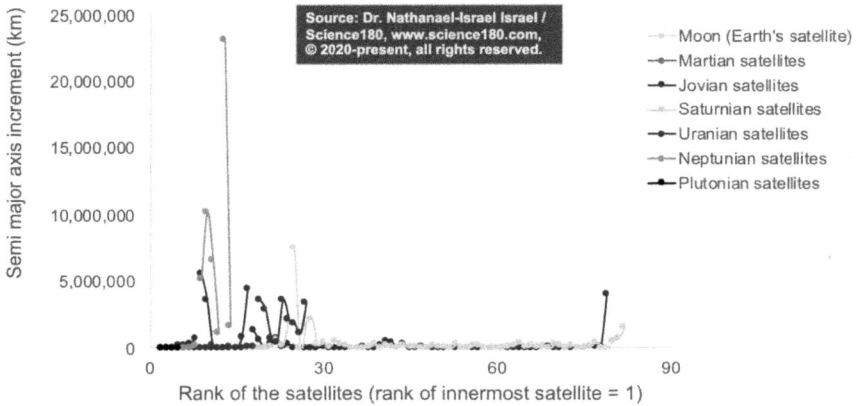

Fig. 125: Semi Major Axis Increment (km) of satellites according to their types and rank

For the Saturnian satellites known as of 2016, the semi major increment (2,040,000 km) of the outermost satellite (Fornjot) is 68 times that of the satellite located just upstream of it (Loge). However, between 2016 and 2019, out of the 20 more satellites discovered, not only were 5 discovered between Fornjot and Loge but also one, S/2004 S26 (S8353a), was discovered 1,628,000 km downstream of Fornjot. Therefore, the outermost Saturnian satellite as of 2020 was no longer Fornjot but S/2004 S26 (S8353a). The semi major increment of the new outermost Saturnian satellite, S/2004 S26 (S8353a), is 2.17 times that of Fornjot, the satellite which is just upstream of it. For the Uranian satellites, the ratio of the semi major increment (3,483,000 km) of the outermost satellite (Ferdinand) to that of the satellite before (Setebos) it is 3. Unlike the other planetary systems, for the Neptunian system, the semi major increment of the outermost satellite (Neso) is not higher than that of the satellite upstream of it (Psamathe). For the semi major increment of Neso (1,690,000 km) is 7.3% that of Psamathe. In contrast the semi major increment of Psamathe is 20.1 times that of the satellite upstream of it (Laomedeia). Similarly, the semi major increment (6,970 km) of the outermost Plutonian satellite (Hydra) is 76.9% that of the satellites located just before it (Kerberos). This suggests that as the semi major of the planet increases, the ratio between the semi major increment of the outermost satellite and the satellite immediately downstream of it changed, and in the current case, it decreased.

26.6. Semi major increment of satellites in Zone 3

Because the Moon and Charon are the only satellites in the Zone 3 of their planetary system, their semi major increment were not calculated. Therefore, in

Zone 3, the focus was on the satellites of the 4 giant planets. The semi major increment in Zone 3 ranges between 0 and 2,080,200 km, the highest value was observed with Iapetus, the 24th innermost Saturnian satellite, which is also the outermost Saturnian satellite in Zone 3. The satellites in Zone 3 which semi major increment is zero are all Saturnian:

- Calypso (the 16th innermost Saturnian satellite),
- Telesto (the 17th innermost Saturnian satellite),
- Helene (the 19th innermost Saturnian satellite), and
- Polydeuces (the 20th innermost Saturnian satellite).

Between the 17th and 19th Saturnian satellites is located Dione (the 18th innermost Saturnian satellite) just 82,740 km downstream of Telesto (the 17th innermost Saturnian satellite). Why is the semi major increment of Dione so high, while that of the other satellites around it is zero? These data suggested that between the 16th and 20th innermost Saturnian satellites something may have happened so that the semi major increment of most of the satellites in the region is zero. I will revisit these observations later.

The top 9 semi major increments recorded in Zone 3 include that of the biggest satellites, suggesting that the highest semi major axis increments in Zone 3 were generally recorded not very far from the region of the biggest satellite. However, the highest semi major increment in Zone 3 of each planetary system was not always recorded with the biggest satellite. For instance, the radius of Iapetus (the Saturnian satellite which recorded the highest semi major axis increment) is just 734.5 km as compared to Titan which is 3.5 times bigger (radius = 2575.5 km). Yet, the semi major increment of Iapetus is 3 times that of Titan (694,790 km). Similarly, Callisto which is the Jovian satellite with the highest semi major increment (812,300 km) in Zone 3 has a radius of 2410.3 km which is 91.5% that of Ganymede, the biggest Jovian satellite. Yet, the semi major increment of Callisto is more than twice that of Ganymede.

In contrast, for the Neptunian satellites, the largest semi major increment (237,113 km) in Zone 3 was recorded with Triton, the biggest satellite in that system. Similarly, the largest semi major axis increment (170,300 km) of the Uranian satellites in Zone 3 was recorded with Titania, the biggest Uranian satellite.

The semi major increment of the outermost satellite in Zone 3 is usually higher than that of the satellite upstream of it. The only exception was with Oberon, the outermost Uranian satellite which semi major increment (147,200 km) is 86.4% that of Titania (the biggest Uranian satellite) which is located upstream or just before it. Looking at the trends of the semi major axis increment, I felt like it is possible to even estimate where the precursors of the satellites could have been broken and recoiled and even which satellites were born from the recoiling of some major precursors following their breakup. I revisited that idea later in this book.

26.7. Increment of the semi major axis between the biggest satellite and the satellite immediately upstream of it in each planetary system

As I was carefully looking at the trends of the semi major increments of some key bodies including the biggest satellites in each planetary system, I found the distance separating the biggest satellite in each planetary system from the satellite upstream of them is enough to squeeze many satellites located upstream of them. Indeed, although the ratio between the semi major of Ganymede and Europa (the satellite upstream of Ganymede) is 1.6, the ratio between the biggest satellite of the other satellites and the satellite which is upstream of them is much higher. For instance, the ratio between the semi major increment of Titan (the biggest Saturnian satellite) and Rhea (the satellite upstream of it) is 2.3, meaning that the distance (694,790 km) between Titan and the satellite upstream of it is more than the distance (527,040 km) between Saturn and Rhea. Yet between Saturn and Rhea, there are more than 20 satellites. In other words, the distance between Titan and the satellite preceding it is more than the distance that covers more than 20 Saturnian satellites before Titan. This suggests that the precursor of Titan had to travel over a very long distance before its fluid layers could be collected into Titan, the largest Saturnian satellite. This explanation is also true for the other satellites which have a very high semi major increment.

Similarly, the ratio of the distance (170,300 km) between Titania (the largest Uranian satellite) and Umbriel (the satellite upstream of it) is more than the distance separating Uranus and its 14th innermost satellite (Miranda) which is the innermost satellite in Zone 3. The ratio between the semi major of Titania and Umbriel is 1.64.

The semi major increment of Triton (the biggest Neptunian satellite) is 3.02 times that of Proteus (the satellite immediately upstream of it). In other words, the distance (237,113 km) between Triton and Proteus is more than twice the distance (117,647 km) separating Neptune from Proteus and this distance encompasses all of the 7 Neptunian satellites upstream of Triton. Finally, because the Moon, Phobos, and Charon do not have any satellite upstream of them, these increments of semi major of the biggest satellites and the satellites upstream of them as discussed here do not apply to them.

An important question is why the distance between the largest satellite and the satellite preceding it in each planetary system is generally higher than the distance between any 2-consecutive satellites before them. Something may have caused the precursor of the biggest satellite to travel over a longer distance more than the precursor of the satellites before it. As compared to the satellites upstream of them, the precursor of the biggest satellites could have travelled over a longer distance before gathering together all fluids into a single satellite. In other words, the semi major increment of the largest satellites in Zones 1, 2, and 3 is generally higher than that of any other satellites in that zone because the precursor of the largest satellites could have taken more time to gather themselves together than the precursors of the other bodies did, at least in Zones 1, 2, and 3. Consequently, the distance they travelled ended up being higher. In the chapter on timescale increment, I shed light onto the time it took to travel the increment distance.

26.8. Semi major increment of the satellites in Zone 4

Most of the highest semi major increments are found in Zone 4. When I considered all the semi major increment data, it appeared to me that this trend may be due to the position of the innermost satellite in Zone 4 with respect to the location of the major breaking point and force that could have detached or ejected the precursor of satellites in Zone 4 from those in Zone 3 during the breakup of the ligament of the satellites. Indeed, except for the Neptunian satellites where the highest increment was in Zone 5, for the other planetary systems, the highest increment was in Zone 4, and more specifically with the innermost satellite in Zone 4. The 3 smallest semi major increments recorded in Zone 4 were with the Plutonian satellites:

- Kerberos (semi major increment is 9060 km),
- Hydra (the outermost satellite, with a semi major increment of 6,970 km)
- Nix (semi major increment is 6,280 km).

Because of the importance of the semi major increment, I gave more details about the semi major increment of the innermost satellite in Zone 4 in another section below. The semi major increment of the satellites in Zone 4 varies between 6,280 km and 10,216,600 km. The highest value was with Halimede, the 10th innermost Neptunian satellite. In general, the semi major increment of the satellites in Zone 4 followed a kind of zig zag shape with more than one peak, meaning the distribution of the semi major axis in Zone 4 is multimodal. For instance, for the Uranian satellites in Zone 4, two modes exist:

- Francisco, the 19th Uranian satellite (increment is 3692500 km) and
- Sycorax, the 23rd Uranian satellite (increment is 3,675,000)

Similarly, the distribution of the semi major axis increments of the Jovian and Saturnian satellites is multimodal. For the Saturnian satellites, the value of the peaks of the semi major increment are not the same, but decreased as if something was attenuated or slowed down. Indeed, the 2 peaks observed in the trends of the semi major increment of the Saturnian satellites in Zone 4 were:

- the first one with Kiviuq (the 25th innermost Saturnian satellite corresponding to a semi major increment of 7,548,700 km) and
- the second one, downstream of the first one, recorded with Paaliaq (the 28th innermost Saturnian satellite, which semi major increment

corresponds to 2,256,000 km).

Yet the semi major increment of the second peak is 3.3 times that of the first peak, suggesting that the semi major increment slowed down as the precursors of the bodies were being moved downstream of the bodies that birthed them.

For the Jovian satellites in Zone 4,

- the upstream peak of the semi major increment was recorded with Themisto (the 9th innermost Jovian satellite which semi major axis increment is 5,624,300 km). In contrast,
- the second semi major axis increment peak of the Jovian satellites in Zone 4 was found with Carpo (the 15th innermost Jovian satellite which semi major increment is 4,430,000 km).

In other words, the upstream semi major increment peak in Zone 4 of the Jovian satellites is 1.27 times that of the second or the downstream semi major peak. Recalling that the ratio between the first peak and the second peak for the Saturnian satellite is 3.3, the smaller value (1.27) observed for the same ratio for the Saturnian satellites in Zone 4 implies that the deceleration of the semi major axis increment for the Jovian satellites is slower than that of the Saturnian satellites. The distance between the first peak and the second peak of the semi major increment of the Jovian satellites is 9,483,000 km, which is 2.32 times the distance between the first and second semi major increment peaks of the Saturnian satellites. In other words, the precursor of the Jovian satellites between the first and second peaks could have been stretched or moved over a longer distance (2.32 times) that of the distance over which the precursor of the Saturnian satellites between the upstream and downstream semi major increment peak. Yet, the Jovian satellites did not lose as much increment as did the Saturnian satellites. Between the first and second peaks of the increment of the Jovian satellites in Zone 4, there are 5 satellites as compared to 2 for the Saturnian satellites. These data suggested to me that the precursor of the Jovian satellites could have resisted stretching or flowing less than that of the Saturnian satellites and could have been broken into many pieces (satellites) after its stretch. Furthermore, it is also important not to view the semi major increment as a result of a simple stretch, flow, or movement of the precursor of the bodies, but also as a product of the energy or force which moved the precursors of the bodies away from their mother precursors. As I explained in other chapters, the high viscosity of the precursor of the Jovian satellites could have played a significant role in slowing down the movement of the precursor of some Jovian satellites. Even when I considered the satellites discovered between 2016-2019, the trend did not change much. The main difference was that most of the satellites discovered between 2016-2019 in Zone 4 are around the space where the semi major increment is minimum.

To sum it up, the distribution of the semi major axis in Zone 4 is usually multimodal, but the value of the modes decreased as the semi major of the satellites increased, suggesting that something (e.g. the energy in the precursors which led to the formation of the satellites in Zones 4 and 5) could have been attenuating as their precursors were traveling far away from their source. The decrease of the semi major increment of the modes is smaller for the Jovian satellites than for the Saturnian satellites. The upstream mode is not always higher than the downstream ones. The modes of the semi major increments in Zones 4 and 5 are usually located between a crescendo of increments upstream and a decrescendo of increments downstream of them. Considering the radius of celestial bodies (see the next chapter), I thought that the breakup and coalescence of fluids in the precursors in certain turbulence zones (e.g. Zones 4 and 5) could explain the formation of some relatively big satellites between smaller ones and the positioning of the highest semi major increment around the smallest satellites. I will revisit this concept later.

26.9. Semi major increment of satellites in Zone 5

The semi major axis increments of satellites in Zone 5 ranges from 0 km to 23,130,000 km. As of 2020, the smallest increment was observed with Isonoe (the 44th innermost Jovian satellite) and S/2007 S3 (the 44th innermost Saturnian satellite). While the increment of the Neptunian satellites is generally higher than those of the Uranian satellites, the trends between the Jovian and Saturnian satellites is not very clear at first sight (Fig. 126). Particularly, around the semi major of 20,000,000 km, many small Jovian and Saturnian satellites are concentrated. While the semi major increment of the Jovian satellites in Zone 5 ranges between 0 km and 4,030,000 km, that of the Saturnian satellites ranges from 0 km to 2,040,000 km, with the highest value observed with Fornjot, the outermost Saturnian satellite known by 2016.

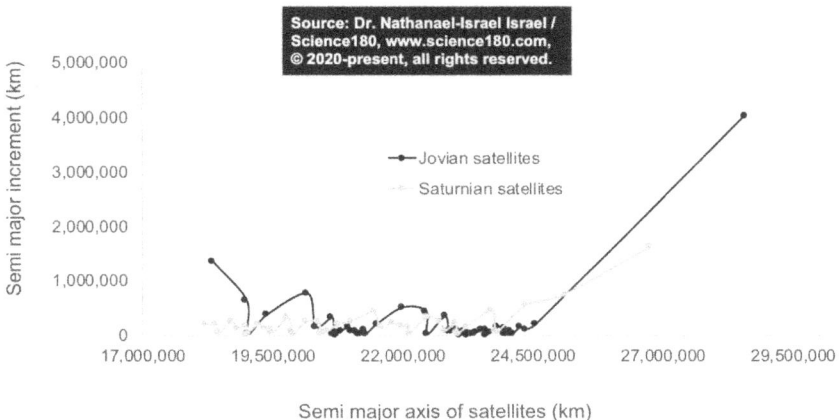

Fig. 126: Semi major axis increment (km) of **Jovian** and **Saturnian** satellites in **Zone 5**

Science180: The Most Accurate, Reliable, Safest, Best Explanation of the Universe-Origin Ever

26.10. Sum of the semi major increment of the satellites according to their types

To explore the proportion of the semi major increment of each turbulence zone and of each satellite with respect to the distance separating the innermost satellite from the outermost satellite in each planetary system, I studied the following ratio:

- Ratio of the sum of the semi major increment of the satellites in each turbulence zone to the sum of the semi major increment of all satellites in their planetary system.
- Ratio of the semi major increment of each satellite to the sum of the semi major increment of all satellites in each planetary system.

These ratios are like a random multiplier of the semi major increment of the bodies in each planetary system. The distance between the innermost and outermost satellite in each planetary system depends on the type of planetary system and varies between 14,081 km and 48,341,773 km. The longest distance was recorded with the Neptunian satellites whereas the shortest distance was with the Martian satellites. The distance over which the Neptunian satellites were spread is 3433.12 times that over which the Martian satellites were spread. In contrast, the distance (26,621,000 km) over which the 82 Saturnian satellites known as of 2019 were spread is 93.6% that (28,442,000 km) over which the 79 Jovian satellites known as of 2019 were spread. Furthermore, the range of the semi major increment of the Jovian satellites is 58.83% that of the Neptunian satellites. The distance between the outermost and the innermost Uranian satellites (20,851,200 km) is 73.31% that separating the innermost and outermost Jovian satellites.

The semi major increment for the Moon was not calculated because it is the only known satellite of the Earth as of 2019. However, it may be useful to mention that the semi major of the Moon (384,400 km) is:

- 27.3 times the distance between the innermost and outermost Martian satellite,
- 1.35% of the distance between the innermost and outermost Jovian satellite,
- 1.44% of the distance between the innermost and outermost Saturnian satellite,
- 1.84% of the distance between the innermost and outermost Uranian satellite,
- 0.8% of the distance between the innermost and outermost Neptunian satellite, and
- 8.52 times the distance between the innermost and outermost Plutonian satellite.

26.11. Semi major increment of the turbulence zones according to the types of satellites

The length of the turbulence zones ranges from 1,000 km to 24,820,000 km according to the types of satellites (Fig. 127). The shortest turbulence zone was recorded with the Jovian satellites, whereas the longest turbulence zone was found with the Neptunian satellites. The shortest turbulence zone is Zone 1. Although the longest turbulence zone was found in Zone 5 of the Neptunian satellites, for the other types of satellites, the longest turbulence zone is Zone 4. The length of Zone 1 of the Martian satellites (14,081 km) is higher than that of Zone 1 of the following types of satellites:

- Jovian satellites: 1,000 km,
- Neptunian satellites: 1,848 km, and
- Uranian satellites: 12,900 km.

The distance between the innermost and outermost Plutonian satellite is 3.2 times that between the two Martian satellites.

The relative semi major increment of each satellite was calculated by dividing their semi major increment by the total semi major increment of the satellites in their planetary system. This ratio is like a turbulent multiplier of the semi major increments as it defined the proportion that the semi major increment of a satellite accounts for the total semi major increment of all satellites in its planetary system. Similarly, I calculated the relative semi major increment of each of the turbulence zones by referring to the sum of the semi major increment of their satellites as compared to the total semi major increment of all satellites in their corresponding planetary system.

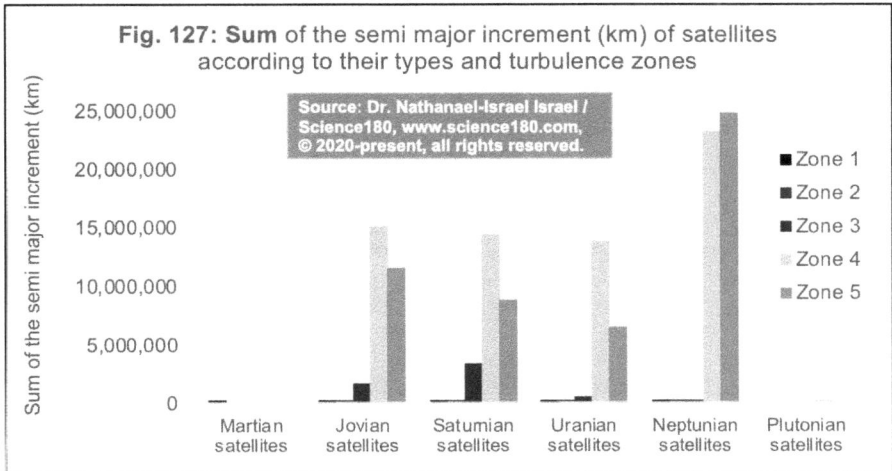

Fig. 127: **Sum** of the semi major increment (km) of satellites according to their types and turbulence zones

The relative semi major increment of the turbulence zones depends on the types of satellites. The highest semi major increments of turbulence zones were observed in Zones 4 and 5, whereas the lowest were in Zones 1, 2, and 3. Because the semi major increment in Zones 4 and 5 are the highest, I was interested in combining the data for Zones 4 and 5 on one hand and that in Zones 1, 2, and 3 on the other hand. Below is what I found.

26.12. Relative semi major increment in Zones 1, 2, and 3 combined vs. that in Zones 4 and 5

Because the Martian satellites are only in Zone 1, their ratio is 100%. Similarly, because the Plutonian satellites have only 1 satellite in Zone 3, meaning all the semi major increments were calculated for Zone 4, their ratio is also 100%. However, for the satellites of the giant planets where all 5 zones are present, 87.1% to 99.37% of the distance between the innermost and the outermost satellites is found in Zones 4 and 5 (Fig. 128). In other words, the distance covered by Zones 1, 2, and 3 altogether is 0.63% to 12.94% of the range of the semi major of the satellites of the giant planets. This suggests that in a fully developed turbulence, where all 5 turbulence zones are distinguishable, such as for the 4 giant planetary systems, more than 80% of the semi major increment is found in Zones 4 and 5.

Fig. 128: Relative semi major increment of satellites according to the turbulence zones and types of satellites

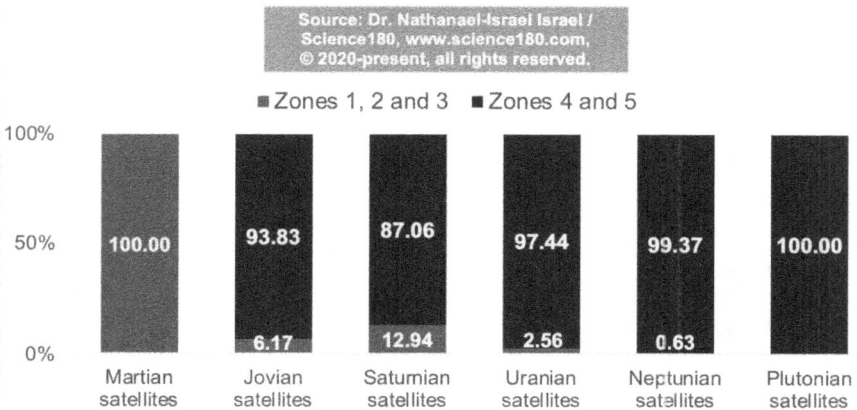

To ease their visualization, details about Fig. 128 are presented in Fig. 129 according to the turbulence zone. In Zone 1, the Saturnian satellites occupied more domain in space, whereas the shortest distance was travelled by the Jovian satellites. This suggested that the precursor of the Jovian satellites did not travel too far away from its mother precursor before split-gathering into the satellites in Zone 1. The high viscosity of the precursors of the Jovian satellites could have explained their resistance to movement (e.g. shearing), hence their shortest semi major increment.

In Zone 2 (the transition zone into the most developed turbulence), the Jovian satellites were spread over a longer semi major axis, suggesting that their precursor could have been elongated, or flowed over a long distance before breaking. In contrast, the Saturnian satellites recorded the shortest semi major increment in Zone 2. The shortness of Zone 2 (the transition zone to turbulence) of the Saturnian satellites could explain why its precursor was broken up quickly, leading to some satellites which share the same orbit and consequently have a chaotic movement. In fact, it is said that the Saturnian satellites Prometheus and Pandora which semi major are separated by about 2347 km sometimes share the same orbit. Similarly, the Saturnian satellites Janus and Epimetheus (separated by about 50 km) also are said to share the same orbit sometimes. The radius of any of these 4 Saturnian satellites in turbulence Zone 2 is higher than 40 km. Knowing that the two Jovian satellites in Zone 2 are even bigger than any of these 4 Saturnian satellites, but that the Jovian satellites are located more than 40,000 km from one another, it sounds like the chaotic movement of the Saturnian satellites in Zone 2 is caused by the short distance between them despite their relatively big size. This suggests that the conjugation of the size of the satellites and the distance separating them from their neighbors can explain some chaotic movement. The flow of air or the waves in the field around them could have played a role.

Fig. 129: Sum of the semi major increment (km) of satellites according to their types and turbulence zones

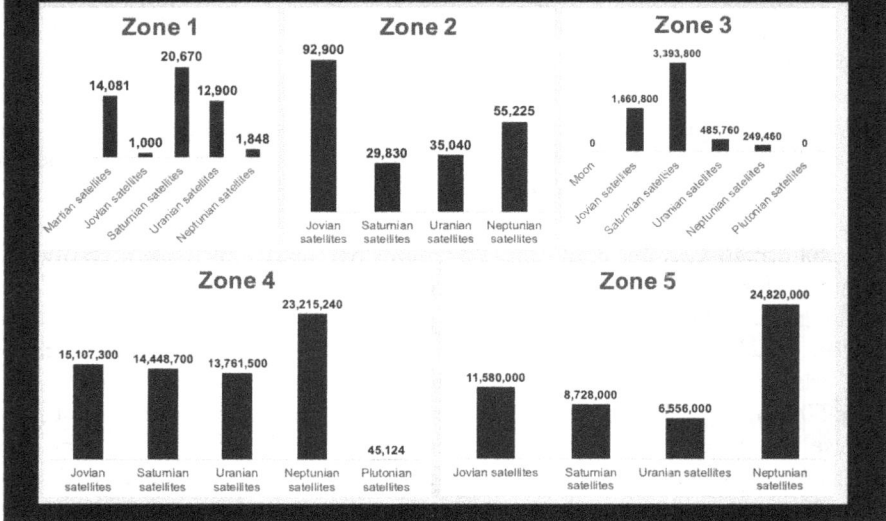

The Saturnian satellites in Zone 3 were spread over a larger semi major range (more than twice) than the satellites of any other planetary system in a similar zone. The difference between the semi major increments in Zone 4 for the 3-innermost giant planetary systems is very small. The highest semi major axis increment in Zone 5 was recorded with the Neptunian satellites.

The high viscosity of the precursors of the Jovian and Saturnian satellites in Zone 5 could have affected their stretching or spacing particularly in Zone 5, hence their smaller semi major increment. In contrast, because the precursors of the Uranian and Neptunian satellites could have been less viscous, they were better stretched or they resisted the stretching less, hence a higher spacing finally reflected in the higher semi major increment of their satellites in Zone 5.

26.13. Semi major increments of the satellites according to their types

Because no satellite is upstream of the Moon and Phobos, their semi major increment was not calculated. The increment of semi major axis of the Jovian satellites ranges between 0 km and 5,624,300 km, the highest was recorded with Themisto, the innermost Jovian satellite in Zone 4. Most of the smaller semi major increments were in Zones 1 and 2. The semi major increment of the Jovian satellites has at least 3 modes located at:

- The innermost satellite in Zone 4

- The outermost satellite in Zone 4
- The outermost satellite in Zone 5

Themisto, the innermost Jovian satellite in Zone 4 accounts for 19.8% of the total semi major increment of the 79 Jovian satellites. The second highest semi major increment for the Jovian satellites was found with Carpo (15.6%) whereas the outermost Jovian satellite has the third highest semi major increment (14.2%). About 53.1% of the semi major increments of the Jovian satellites were in Zone 4 followed by 40.7% in Zone 5. About 6.2% of the semi major increment of the Jovian satellites was accumulated by Zones 1, 2, and 3 altogether (Fig. 130-133), meaning that the biggest Jovian satellites were formed by the time their precursors travelled about 6% of the total distance travelled by the precursor of the outermost Jovian satellite.

The semi major increment for the Saturnian satellites varies between 0 km and 7,548,700 km (Fig. 131). The highest value was observed with Kiviuq, the innermost Saturnian satellite in Zone 4. The semi major increment of Kiviuq is 28.4% of the total semi major increments covered by the 82 Saturnian satellites known as of 2019. Four of the 5 Saturnian satellites which semi major increment is 0 are found in Zone 3. The precursors of these satellites could not have been pushed away from one another much before their daughter satellites were born, hence they are very close to one another. For the Saturnian satellites, at least 3 main peaks (modes) of the semi major increment were observed, the highest of which was recorded with Kiviuq, the innermost satellite in Zone 4. The other Saturnian satellites with a higher semi major increment are:

- Iapetus: 7.8%
- Phoebe: 6.9%
- Paaliaq: 8.5%
- S/2004 S26 (S8353a), the outermost Saturnian satellite: 6.1%

Although each of these 4 Saturnian satellites could be considered as a mode, their positions as compared to the semi major increment of Kiviuq (the innermost Saturnian satellite in Zone 4) make it very difficult to define. I also addressed the complexity of this modal analysis in the chapter related to radius. More than 87% of the semi major increment of the Saturnian satellites is concentrated with satellites in Zones 4 and 5. Just like the Jovian satellites, the Saturnian satellites in Zone 4 account for the highest proportion, 54.27%. However, the Saturnian satellites in Zone 3 account for a higher proportion of the semi major increment than the Jovian satellites in a similar zone. In fact, the semi major increment for the Saturnian satellites in Zone 3 is 12.75% of the total semi major increment while for the Jovian satellites, that value is 5.83%.

Fig. 130: Relative semi major increment of the turbulence zones of the satellites of the giant planets

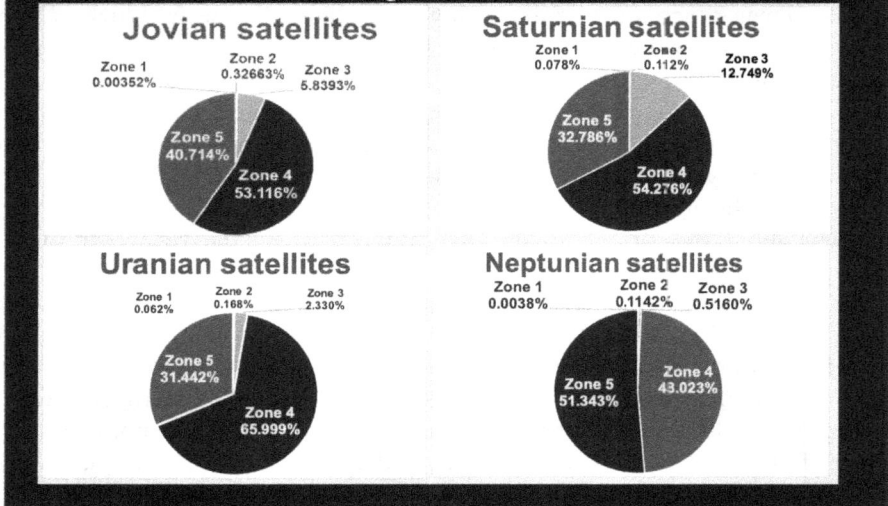

For the Uranian satellites, the semi major ranges from 900 km to 3,692,500 km. Francisco (the innermost Uranian satellite in Zone 4) recorded the highest semi major axis increment. The semi major axis increments of 2 other Uranian satellites is very near that of Francisco:

- Ferdinand, the outermost Uranian satellite having a semi major increment of 3,483,000 km, which is 94.3% that of Francisco,
- Sycorax, a Uranian satellite located in Zone 4 which semi major increment is 3,675,000 km, which is 99.5% that of Francisco.

The highest relative semi major increment of the Uranian satellites was observed with Francisco (17.7%) followed by Sycorax (17.6%) and Ferdinand (16.7%), therefore making the distribution of the semi major increment of the Uranian satellites tri-modal. The semi major increment ratio of Zone 4 of the Uranian satellites (65.99%) is much higher than that in Zone 4 of any of the other 3 giant planetary systems. Although the ratio of the semi major increment of Zone 5 is relatively similar to that of the Jovian and Saturnian satellites, the Zones 1, 2, and 3 altogether are relatively much shorter, occupying about 2.6% of the total semi major increment.

Nathanael-Israel Israel: Historic Discoverer of the Formula that Accurately Decode the Formation of the Universe, of Life, and of Chemicals in a Few Days

Fig. 131: Semi Major Axis Increment (km) of satellites according to their types and turbulence zones

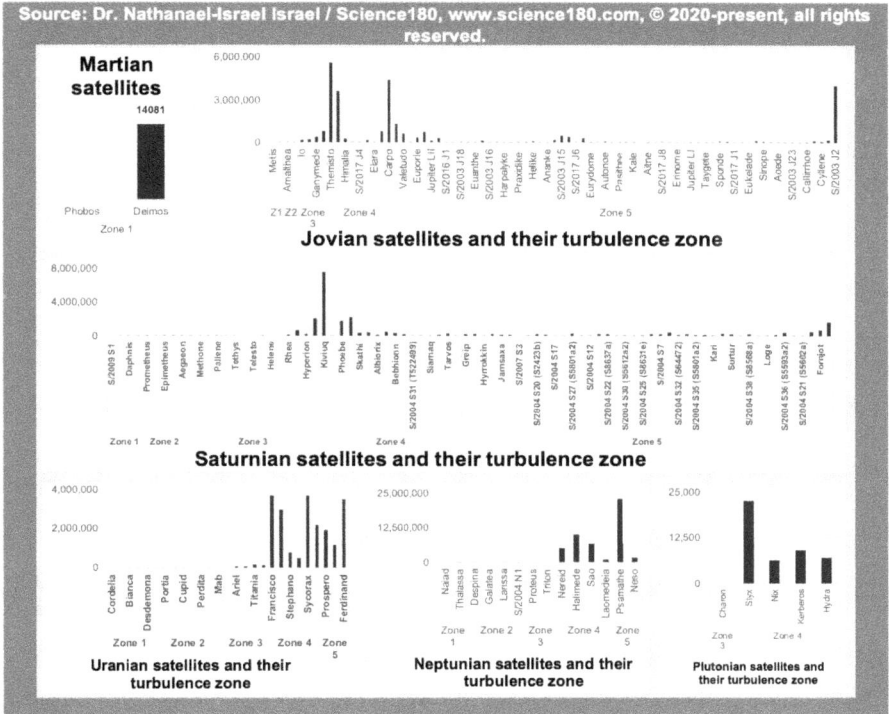

Fig. 132 illustrates the semi major increments of the satellites according to their semi major.

The semi major axis increment of the Neptunian satellites varies between 1,848 km and 23,130,000 km. The smallest value was obtained in Zone 1 whereas the highest value was with Psamathe, the innermost satellite in Zone 5. Psamathe alone contributes 47.8% of the total semi major increment of the Neptunian satellites (Fig. 133). It can also be said that the distribution of the increment for the Neptune satellites is bimodal with the highest mode recorded with Psamathe, and the second highest mode with Halimede, which semi major increment is 44.2% that of Psamathe. Unlike the satellites of the other 3 giant planetary systems, the Neptunian satellites in Zone 5 recorded the highest semi major increment. They account for 51.3% of the total semi major increment. Together, Zones 4 and 5 of the Neptunian satellites accounted for 97.43% of the total semi major increment of all the Neptunian satellites. In contrast, Zones 1, 2, and 3 of the Neptunian satellites account for less than 1% of the semi major increment.

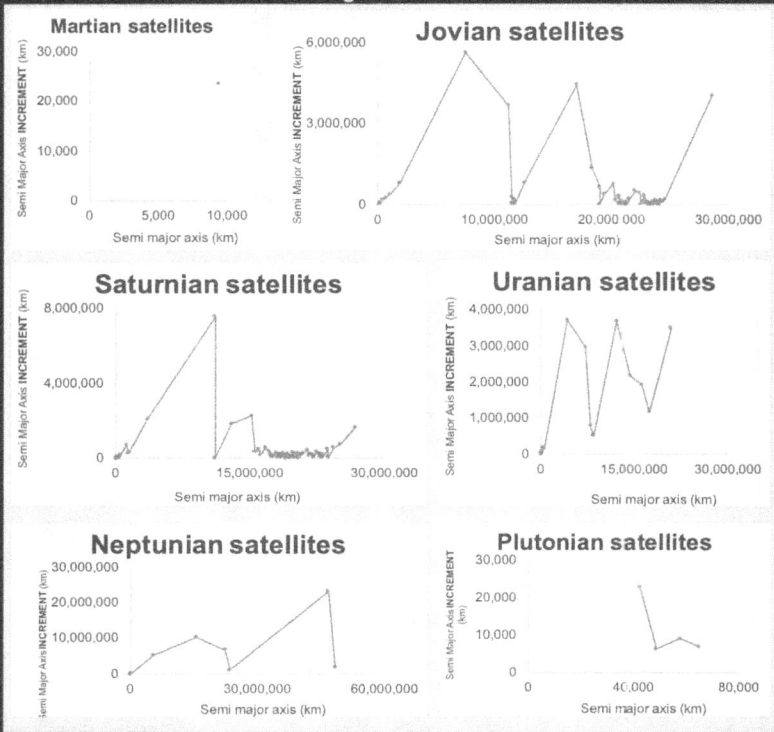

Fig. 132: Semi Major Axis Increment (km) of satellites according to their types and semi major axis

Finally, the Plutonian satellites recorded some of the smallest semi major increments. Just like most of the other planetary systems, the highest semi major increment of the Plutonian satellites was observed with Styx, the innermost satellite in Zone 4. Styx contributed 50.6% of the semi major increment of the Plutonian satellites. In other words, the distance between Charon and Styx is almost the same as that separating Styx from Hydra, the outermost Plutonian satellite. Just like most of the other planetary systems, the highest semi major increment of the Plutonian satellites was observed with Styx, the innermost satellite in Zone 4. Indeed, Styx contributed 50.6% of the semi major increment of the Plutonian satellites. In other words, the distance between Charon and Styx is almost the same as that separating Styx from Hydra, the outermost Plutonian satellite.

Nathanael-Israel Israel: Historic Discoverer of the Formula that Accurately Decode the Formation of the Universe, of Life, and of Chemicals in a Few Days

Fig. 133: Relative Semi Major Axis Increment of satellites according to their types

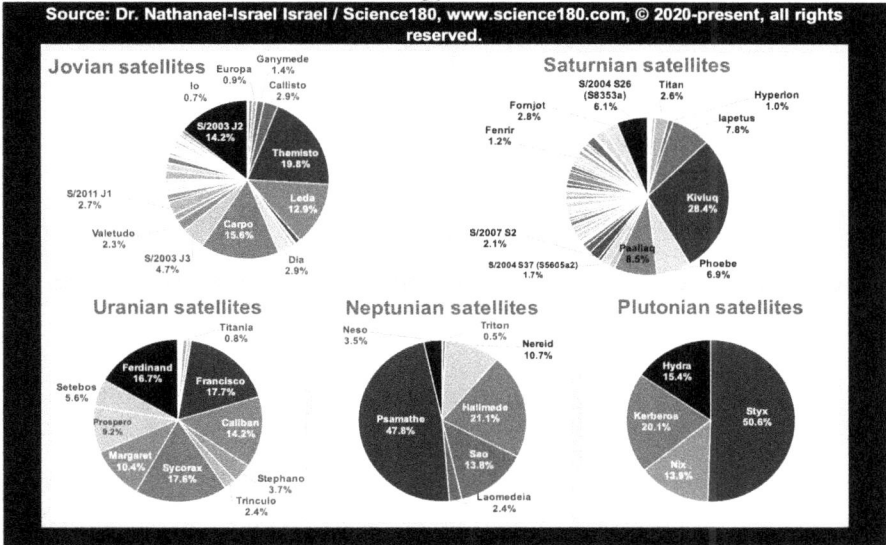

26.14. Regression between semi major increment and semi major axis of the planets and satellites

Because I was interested in knowing how the semi major increment varies from the innermost satellites to the outer ones, I did the regression between semi major increment and semi major axis of the bodies. I found significant relationships for the planets and satellites but not for the main belt asteroids. Indeed, a positive relationship (Fig. 134) exists between semi major increment and semi major axis of the planets ($r2=0.98$). The semi major increment increased from Mercury to Neptune but decreased from Neptune to Pluto. As you will see in the next paragraphs, this trend also matched what I got for the satellites. For the semi major increment of the satellites did not increase indefinitely but reached a point after which it dropped.

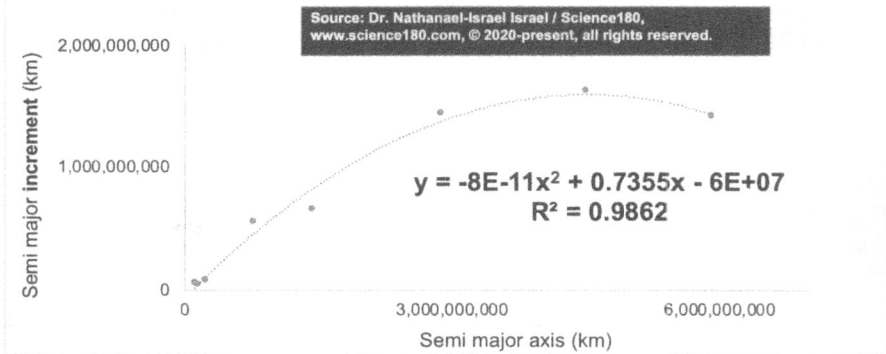

Fig. 134: Regression between semi major **increment** and semi major of **planets** in the Solar System

$$y = -8E-11x^2 + 0.7355x - 6E+07$$
$$R^2 = 0.9862$$

When I combined all of the 210 satellites discovered in the Solar System by 2019, a significant but weak relationship ($r2<0.4$) was found between semi major increment and semi major. When the regression between semi major axis increment and semi major were done after splitting the satellites according to their types, a significant relationship was found except for the Plutonian satellites. Although for the Jovian and Saturnian satellites the regressions were significant ($p<0.05$), they were not strong ($r2<0.4$). However, for the Uranian and Neptunian satellites, stronger regressions were found ($r2= 0.89$ and $p<0.05$) with the power model found for each of these types of satellites. To explore the data on another angle, I split the satellites according to their turbulence zones, and I presented the results below.

Because the coefficients of determinations, the above regressions were not perfect and some equations were not applicable to some types of planetary systems because the existence of non-positive values, I split the satellites according to their planetary systems before doing the regression for each turbulence zone. As illustrated below, a strong and positive linear model ($r2=0.92$) (Fig. 135) and a quadratic (polynomial) model ($r2=0.95$) (Fig. 136) were found between semi major increment and semi major of the satellites in Zone 3.

Nathanael-Israel Israel: Historic Discoverer of the Formula that Accurately Decode the Formation of the Universe, of Life, and of Chemicals in a Few Days

CHAPTER 26: INCREMENT OF THE SEMI MAJOR AXIS OF CELESTIAL BODIES

Fig. 135: **Regression** between semi major **increment** and semi major of satellites in turbulence **Zone 3**

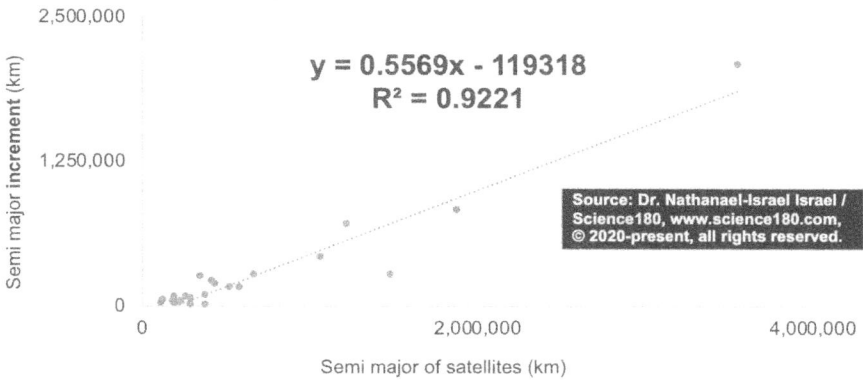

$$y = 0.5569x - 119318$$
$$R^2 = 0.9221$$

Fig. 136: **Regression** between semi major **increment** and semi major axis of satellites in Turbulence **Zone 3**

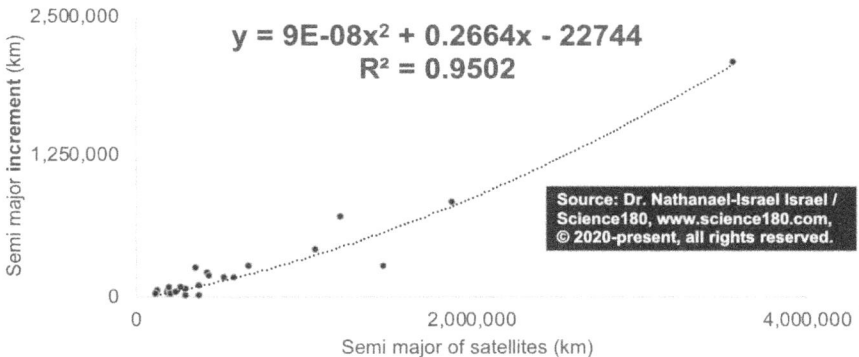

$$y = 9E-08x^2 + 0.2664x - 22744$$
$$R^2 = 0.9502$$

When I did the regression between semi major increment and semi major of the satellites in each of the turbulence zones, I found no significant regression in Zone 1 ($p > 0.26$). Although I found significant regressions ($p < 0.005$) in the other 4 turbulence zones (Zones 2, 3, 4, and 5), the strongest regression was only in Zone 3 ($r2 > 0.92$, $p < 0.000$). As I was reflecting on the positive relationship between semi major increment and semi major in Zone 3, I noticed that the positive relationship does not stop just at the end of Zone 3, but continues beyond Zone 3 until a peak of the semi major increment is reached in Zone 4. I also noticed that once the semi major increment peak is reached, the semi major increment started decreasing until it reached a minimal value before picking up again, describing a kind of cone upstream and downstream of the peaks. In other words, these trends give a kind of conical shape to the increment upstream and

Science180: The Most Accurate, Reliable, Safest, Best Explanation of the Universe-Origin Ever downstream of the peaks. I elaborated on that conical shape of the semi major increment below and also in the chapter on radius. Therefore, I decided to create a new group of satellites that will include not only those in Zone 3 but also those of the satellites in Zone 4 located before the first peak of the semi major increment is reached. For the Jovian, Saturnian, and Uranian satellites, the first peak of the semi major increment was reached with the innermost satellite in Zone 4. However, for the Neptunian satellites, the peak is reached with the second innermost satellite in Zone 4. Therefore, to explore the big picture, instead of doing the regressions between the semi major increment and the semi major in just 5 turbulence zones, I did it for the following zones or groups of zones:

- Zone 1
- Zone 2
- Zone 3
- Zone 4
- Zone 5
- Zones 1 and 2
- Zones 1, 2, and 3
- From Zone 1 to the first peak of the semi major increment
- From Zone 3 to the first peak of the semi major increment
- Zones 4 and 5

The data (Table 12) showed that, out of the significant regressions, the strongest relationships ($p < 0.05$, $r2 > 0.9$) were found for the following groups of zones:

- Zones 1, 2, and 3 studied altogether ($r2 = 0.955$) which is similar to what I found above with Zone 3 alone ($r2 = 0.95$)
- From Zone 1 to the first peak of the semi major increment ($r2 = 0.985$)
- From Zone 3 to the first peak of the semi major increment ($r2 = 0.986$)

Table 12: Model summary and parameter estimates for the regression between semi major increment and semi major of the satellites (including those discovered between 2016-2019) in the Solar System according to their turbulence zones.

Dependent Variable: Semi Major Increment (km) of satellites

Turbulence Zones	Equation	Model Summary					Parameter Estimates		
		R Square	F	df1	df2	Sig.	Constant	b1	b2
Zone 1	Inverse	0.155	1.47	1	8	0.26054	1,726.462	2.077E+08	
	Power	0.108	0.97	1	8	0.35421	1,949,736.250	-0.57586	
Zone 2	Linear	0.385	10.63	1	17	0.00461	-8,921.629	0.18583	
	Quadratic	0.464	6.91	2	16	0.00686	16,815.258	-0.29663	1.895E-06
Zone 3	Linear	0.922	284.17	1	24	8.33E-15	-119,317.929	0.55688	
	Quadratic	0.950	219.33	2	23	1.05E-15	-22,743.844	0.26637	8.992E-08
Zone 4	Power	0.245	12.03	1	37	0.00135	15.381	0.63407	
	S	0.285	14.77	1	37	0.00046	13.166	-198,301.73	
Zone 5	Linear	0.377	64.68	1	107	1.28E-12	-7,243,114.524	0.34601	
	Quadratic	0.476	48.20	2	106	1.30E-15	6,625,567.220	-0.63679	1.568E-08
Zones 1 and 2	Linear	0.271	10.06	1	27	0.00375	-4,594.590	0.13652	
	Quadratic	0.457	10.94	2	26	0.00036	19,472.097	-0.37216	2.196E-06
Zones 1, 2, and 3	Linear	0.918	594.22	1	53	1.80E-30	-70,505.760	0.52619	
	Quadratic	0.955	554.94	2	52	8.35E-36	-18,413.266	0.25665	9.251E-08
From Zone 1 to the First Peak Semi Major Increment	Linear	0.980	2774.81	1	57	4.84E-50	-95,013.642	0.69675	
	Quadratic	0.985	1849.77	2	56	7.22E-52	-154,086.917	0.84597	-1.183E-08
From Zone 3 to the First Peak of Semi Major Increment	Linear	0.978	1290.26	1	29	1.35E-25	-132,622.596	0.70115	
	Quadratic	0.986	957.51	2	28	1.67E-26	-265,528.583	0.88835	-1.427E-08
Zones 4 and 5	Linear	0.016	2.40	1	146	0.12386	-140,887.589	0.04773	
	Quadratic	0.258	25.27	2	145	3.84E-10	3,447,512.107	-0.36411	1.057E-08

The independent variable is Semi Major (km) of satellites.

In Turbulence Zone 3 and 5, the dependent variable (IncrementSemiMajor) contains non-positive values. The minimum value is 0. Log transform cannot be applied. The Compound, Power, S, Growth, Exponential, and Logistic models cannot be calculated for this variable.

To put it another way, the regression between semi major axis increment and semi major is very strong for the satellites not only in Zone 3, but also for the satellites "From Zone 1 to the first peak of the semi major increment" (Table 12 and Fig. 137). By considering all the satellites from Zone 1 until the first peak of the semi major increment, the weakness of the regression found in Zones 1 and 2 only were overpowered by the strength brought by the data from Zone 3 until the first peak of the semi major increment.

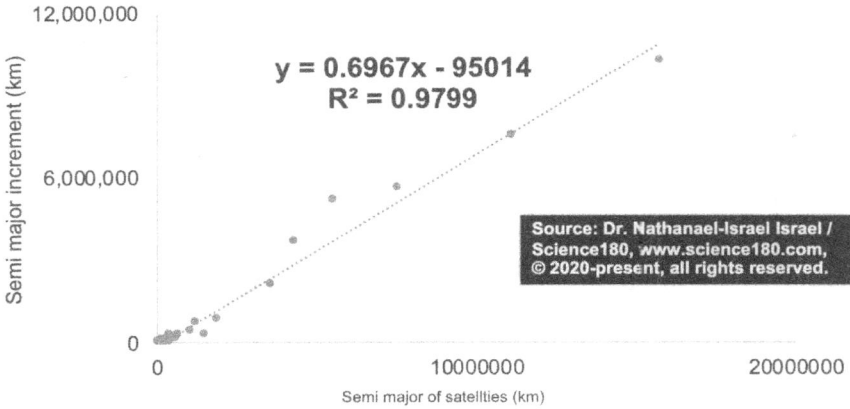

Fig. 137: Regression between semi major **increment** and semi major of satellites from **Zone 1 untill the First Peak** of the semi major increment

I deepened the analysis to explore the above trends for each type of satellites. When the regression in the turbulence zones and the way I grouped some of them were investigated according to the types of satellites, I found regressions even stronger than those I mentioned above, which were already very satisfactory. Most of the regressions yielded a perfect relationship (r2>0.99). In general, unlike the other zones where positive and negative relationships were found according to the types of satellites, in Zone 3, the relationships between semi major increment and semi major of the satellites are positive. The coefficient of determination of the best models in Zone 3 varies between 0.9 and 1, attesting that these regressions are very strong. In other words, in Zone 3, as the semi major of the satellites increased, the distance separating them increased as well. This may be due to the fact that, after the fluids of the precursor of the satellite immediately upstream of them split, the precursors of the satellites in Zone 3 were more stretched and travelled over a longer period of time before gathering together into a body during the formation of the satellites. When all the turbulence zones are properly analyzed according to the types of satellites, the best regression was still found when considering satellites "From Zone 1 to the First Peak of the Semi Major Increment" (Table 13).

Nathanael-Israel Israel: Historic Discoverer of the Formula that Accurately Decode the Formation of the Universe, of Life, and of Chemicals in a Few Days

Table 13: Model summary and parameter estimates for the regression between Semi Major Increment (km) of the satellites "From Zone 1 to the First Peak of the Semi Major Increment" according to the types of satellites (including those discovered between 2016-2019) in the Solar System

Dependent Variable: Semi Major Increment (km) of satellites

Turbulence Zone	Types of	Equation	Summary					Estimates		
			R Square	F	df1	df2	Sig.	Constant	b1	b2
From Zone 1 to the First Peak of Semi Major Increment	Jovian satellites	Linear	0.990	571.441	1	6	3.52E-07	-233,696.85	0.76527	
		Quadratic	1.000	11,716.60	2	5	6.65E-10	-13,549.12	0.33918	5.485E-08
	Satumian satellites	Linear	0.992	2,649.13	1	22	2.00E-24	-171,207.24	0.68326	
		Quadratic	0.995	2,237.65	2	21	3.38E-25	-106,089.45	0.50780	1.640E-08
	Uranian satellites	Linear	0.991	1,824.16	1	16	6.44E-18	-92,493.25	0.87413	
		Quadratic	1.000	22,284.22	2	15	8.95E-27	-11,834.77	0.25677	1.425E-07
	Neptunian satellites	Linear	0.979	328.12	1	7	3.87E-07	91,793.63	0.67344	
		Quadratic	1.000	36,906.97	2	6	5.37E-13	-83,978.61	1.10889	-2.886E-08

The independent variable is Semi Major (km) of satellites.

26.15. Regression between semi major increment and kinetic energy (total, rotational, and translational) of the satellites

Although the semi major axis of the satellites has a significant impact on the distance separating the satellites, I wondered if the energy of these satellites was responsible for their spacing. To test that idea, I did the regression between the semi major increment of the satellites and the 3 types of kinetic energy (total, rotational, and translational) that I studied. The result did not yield any strong regression. Most of these regressions were not significant and the few that were significant gave a very small coefficient of determination. I did not report those data in this book, for they do not add much new information and will just take up more space and time. This suggested that the energy in the precursors of the individual celestial bodies was not the driving force behind the spacing between them. In contrast, the energy of their mother precursors and the way their precursors were launched into motion affected their flow and how their daughter bodies were formed. To put it in other words, the strong relationship between semi major increment and semi major regardless of the other variables suggested that the way the bodies were spaced did not depend on the bodies themselves, but on something related to their mother precursors and the fluid flow. I came to realize that the way the mother precursors were put into motion and the energy in them and with which they were launched as a whole could have affected how the precursors of their daughter bodies flowed from the position of the innermost daughter to the position of the outermost daughter passing by the position of the precursors of the bodies between. As the fluids of the precursors were flowing, they were clustered into different bodies that were spaced according to the events that surrounded their birth. In the chapters on vortex and fluid ligaments, I expounded on other factors (e.g. diffusion, stretching of the fluids in the precursors) that could have also affected the distance separating some bodies.

26.16. Thoughts about the expansion of the universe as the consequence of how the universe was formed

As I was finishing this chapter, I felt like I should say something about the expansion of the universe as it is related to the increase of the semi major as well. It came to my understanding that the range and increment of the semi major axis of the bodies can give a clue not only about the way their precursors would have been stretched, but also why the universe is expanding as scientists have claimed. The range of the semi major axis of the satellites can be a good indicator of how hard or easy it was for their precursors to flow or stretch before breaking up. The semi major axis as of today is the combination of the:

- semi major axis at the "end" of the formation of the universe and the
- expansion of the semi major axis after the "end" of the formation of the bodies.

In other words, because of the expansion of the universe (which also implies an expansion of some of its constituents), the distance between some constituents in the universe has been increasing. Similarly, the semi major has also been increasing, although the increase may not be always properly seen or estimated. Therefore, any mathematical itineration which may try to estimate distance, timing or duration, or speed of things or events during or after the formation of the universe must also consider the expansion of the universe. Although the major stretching of the precursors of the bodies occurred early during the formation of the universe, the force applied to them and the mapping of space with respect to the position of hot versus cold places on one hand, and on the other hand, the interaction between darkness and light could have also contributed to the expansion to be expanding. If accurate measurements can be made, it will be proven that even the distance between matter at the subatomic and atomic scale may also be expanding but to an extent constrained by the systems that contain them. For the universe cannot be expanding without its constituents at different levels doing the same according to their scales and the factors affecting those systems.

It also appeared to me that the expansion of the universe is a consequence of the "original mysterious scattering" and the fluid shear flow that took place during the turbulence at the formation of the universe. Since then, the natural tendency of the things formed in the universe is to expand although the expansion rate may be changing, probably decreasing more than increasing. As a body loses more energy, the likelihood that it gets farther from the position of its precursor will probably increase. In the end, the expansion of the universe is a proof that most bigger bodies in the universe are losing their energy which in return is warming the outermost part of the corresponding systems. Although most stars may be losing their energy, other bodies or portions in the universe may be gaining more energy. But because the universe is huge, these transfers of energy may not be easily detectable.

26.17. Take home message

The semi major axis increments of the bodies that I studied in the Solar System ranged from 0 to millions of kilometers, meaning that some bodies shared almost the same orbit during their revolution, while some are millions of kilometers away from others. The data of the semi major axis increment suggested to me that it is possible to even estimate where the precursors of the satellites could have been broken and recoiled and even the satellites which were born from the recoiling of some major precursors following their break-up. I found that, the semi major increment of the largest satellites in Zones 1, 2, and 3 is generally higher than that of any other satellite in those zones because the precursors of the largest satellites could have taken more time to gather themselves together than the precursors of the other bodies did. Consequently, the distance they travelled ended up being higher. Except for the Neptunian satellites where the highest increment was in Zone 5, for the other planetary systems, the highest increment was in Zone 4, and more specifically with the innermost satellite in Zone 4. The breakup and coalescence of fluids in the precursors in certain turbulence zones (particularly Zones 4 and 5) could explain the formation of some relatively big satellites between smaller ones and the positioning of the highest semi major increment around the smallest satellites. The way the precursors of the satellites in Zone 5 flowed, split, and recoiled can explain why some are so close to one another, while others are so distant from their peers. The precursor of the Jovian satellites did not travel too far away from the precursor of Jupiter before split-gathering into the satellites in Zone 1. The high viscosity of the precursors of the Jovian satellites could have explained their resistance to movement (e.g. shearing), hence their shortest semi major increment. The conjugation or combination of the size of the satellites and the distance separating them from their neighbors can explain their chaotic movement. The high viscosity of the precursors of the Jovian and Saturnian satellites in Zone 5 could have affected their stretching or spacing particularly in Zone 5, hence their smaller semi major increment. In contrast as the precursors of the Uranian and Neptunian satellites could have been less viscous, they were better stretched or they resisted the stretching less, hence a higher spacing finally reflected in the higher semi major increment of their satellites in Zone 5. The regression between semi major axis increment and semi major is very strong for the satellites in Zones 1, 2, and 3 all the way to the first peak of the semi major increment found in Zone 4. I also found that the expansion of the universe is a consequence of the "original mysterious scattering" and the fluid shear flow that took place during the turbulence at the origin or during the formation of the universe. Since then, the natural tendency of the things formed in the universe is to expand although the expansion rate may be changing, I would say that it is probably decreasing more than increasing.

CHAPTER 27

A LITTLE-KNOWN SECRET ABOUT THE RADIUS OF CELESTIAL BODIES THAT "THEY" WON'T TEACH YOU AT HARVARD UNIVERSITY OR AT ANY OTHER TOP UNIVERSITY IN THE WORLD

Although I studied many types of bodies in the Solar System (e.g. chemical elements, minerals, rocks, celestial bodies), in this chapter on radius, I focused on celestial bodies only. In my book *"Turbulent Origin of Chemical Particles"*, I addressed the radius of microscopic particles.

The radius is the main variable used to address the dimension of most celestial bodies. Two types of radii are usually taken:

- The equatorial radius which is the radius of the body at the equator and
- The polar radius which is the radius of the body at the poles.

Because the shape of some bodies is irregular, their radius was estimated by averaging all their sizes. Throughout this book, when I used the word radius without specifying whether it is equatorial or polar, I am referring to the equatorial radius or the average radius. Because no one is able to take a ruler and manually measure the radius of the celestial bodies, their values are theoretical and sometimes change based on the "advancement" of scientific measurements. The radius of celestial bodies is sometimes measured using theoretical models that consider the brightness of such bodies. Therefore, the more remote a celestial body is from an observer, the more the bias in measurement its radius could be. Also, the smaller a celestial body, the lesser its brightness, and the more biased the estimation of its radius could be. Many celestial bodies cannot be discovered because they are just too small to be seen even by a telescope. To improve the precision of some measurements and to take more accurate data, probes have been sent to certain celestial bodies (e.g. planets).

27.1. General trends of the radius of the bodies in the Solar System

I studied the radius of 440 bodies (including the satellites discovered between 2016-2019) in the Solar System, and it depends on their types (p<0.001). With a radius of 696,000 km, the Sun is the largest object in the Solar System. The radius of the smallest celestial bodies I studied in the context of this work was 2 meters. Planets are the biggest objects orbiting the Sun. The radius of Jupiter, the biggest planet in the Solar System, is 71,492 km which is 10.27% that of the Sun. Yet, that radius of Jupiter is about 11.2 times the radius of the Earth.

27.2. Radius of the asteroids and planets in the Solar System

The radius of the asteroids that I studied in the Solar System varies between 2 meters and 1,163 km. The radius of the asteroids in the main belt varies between 2 km and 469.5 km (Fig. 138). The biggest body in the main belt is Ceres.

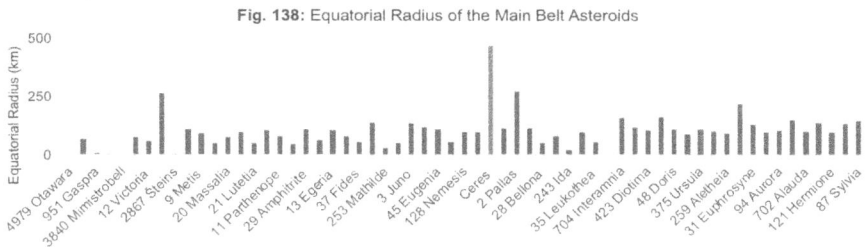

Fig. 138: Equatorial Radius of the Main Belt Asteroids

In each planetary system, planets are the bigger bodies. Although planets are usually characterized by 3 types of radii (equatorial radius, polar radius, and volumetric radius), here I focused on only their equatorial radius. The radius of the planets in the Solar System ranges between 1,187 km and 71,492 km:

- Mercury: 2439.7 km
- Venus: 6051.8 km
- Earth: 6378.137 km
- Mars: 3396.2 km
- Jupiter: 71492 km
- Saturn: 60268 km
- Uranus: 25559 km
- Neptune: 24764 km
- Pluto: 1187 km

The radius of Jupiter (the biggest planet in the Solar System) is:
- 29.30 times that of Mercury

- 11.81 times that of Venus
- 11.21 times that of Earth
- 21.1 times that of Mars
- 1.19 times that of Saturn
- 2.8 times that of Uranus
- 2.89 times that of Neptune
- 60.23 times that of Pluto

The radius of the Earth is 6378.14 km, which is $1/109.13^{th}$ the radius of the Sun.

27.3. Radius of the satellites

27.3.1. General trends

The radius of the satellites in the Solar System depends on their types (p<0.000), the turbulence zone (p<0.000), and the interaction between these 2 variables (p<0.000). As of 2020, the radius of the 210 satellites known in the Solar System varied between 300 meters and 2634.1 km (Fig. 139). The smallest satellite (S/2009 S1) is the innermost Saturnian satellite, whereas the biggest satellite is Ganymede, a Jovian satellite. In general, bigger satellites are generally separated by smaller satellites or clusters of smaller satellites. The biggest satellite in each planetary system is usually toward the edge of turbulence Zone 3 while most of the smaller satellites are found at the edge of their planetary system and sometimes between larger satellites. The biggest satellite of any planet in the Solar System is located at a semi major axis no higher than 1,222,214.8 km. The Moon is bigger than the smallest satellite of any of the other planets in the Solar System. Yet, the semi major of the Moon is higher than that of many satellites in the Solar System. As I pondered on why this semi major is so high despite its size, it appeared to me that the precursor of the Earth-Moon system was able to push the precursor of the Moon farther away from the precursor of the Earth compared to the precursors of the satellites of the other planets because it could have had a lot of energy due to its higher orbital speed as compared to the speed of other satellite's precursors.

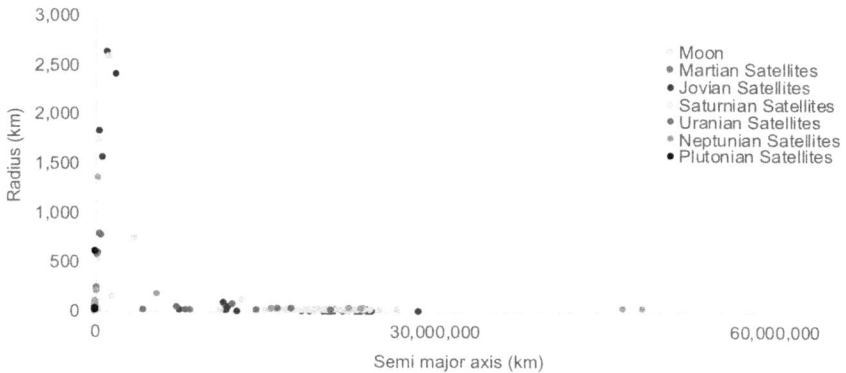

Fig. 139: Radius of satellites according to their types and semi major axis

In Fig. 140, I presented the same data but differently to show the trends of the bodies clustered at the origin of the axis.

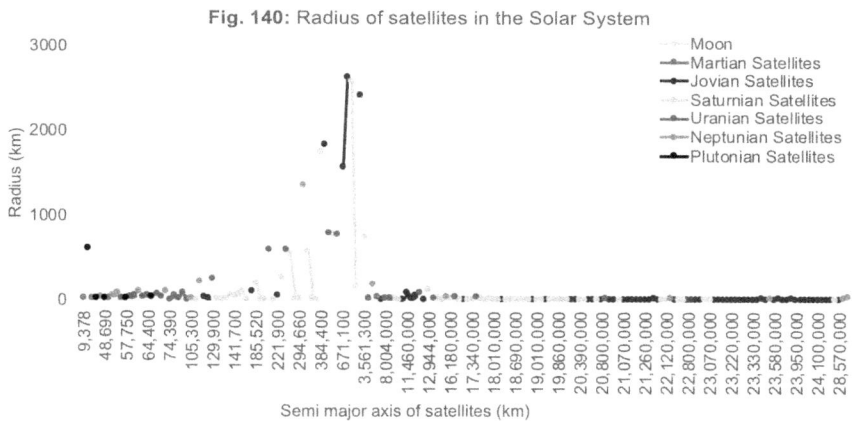

Fig. 140: Radius of satellites in the Solar System

One of the first things that appealed to me the first time I made these graphs in 2015 was that the biggest satellites (Fig. 141) are not the innermost nor the outermost ones but are located some distance in between. The biggest satellites of Jupiter and Saturn are located at a semi major axis higher than that of the biggest satellite of any other planet. Some of the questions that crossed my mind were:

- Why are the largest satellites of each planetary system neither the innermost nor the most outermost satellite?
- Why are the biggest Martian and Plutonian satellites the closest to their primary planet?
- Why unlike the satellites of the others planetary systems, is the innermost

satellite of Pluto and Mars the biggest satellites of those planets?

- Was the split-gathering of the bodies arbitrary or did it follow a certain specific law? If so, what is the law that gave the bodies a specific size?

I suspected that the semi major of Charon is so small because when the fluid in the precursor of the Plutonian satellites started moving away from the precursor of Pluto, its speed could have decreased quick and the quick variation of its speed led to a quicker turbulence that birthed Charon. I felt like the higher the speed of the fluids in a precursor, the longer they could have flown away from their mother precursor before encountering the type of variation needed for them to yield a turbulence worthy of that of satellites in Zone 3. That is why the semi major of the Moon for instance is very high compared to the semi major of the innermost satellites in Zone 3 of other planetary systems except the Jovian planetary system. The Jovian planetary system is the only system where the innermost satellite in Zone 3 has a semi major higher than that of the Moon. This could have been so because the high viscosity and energy in the precursor of the Jovian satellites could have decreased its speed more slowly, causing it to travel over a longer distance before the speed variation could be in the range that allowed the eddies to stumble upon one another to form bigger ones and cause the major turbulence known in Zone 3. The precursor of the biggest Jovian and Saturnian satellites could have had to travel more than twice the semi major of the Moon before being formed.

Fig. 141: Radius (km) of the biggest satelllite of each planet in the Solar System

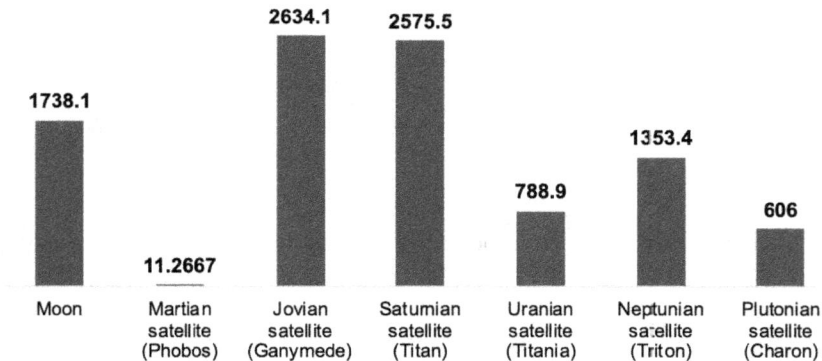

The equatorial radius of the Moon is 1738.1 km, whereas its polar radius is 1736 km. The radius of the Moon is in the range of the radius of the satellites of other planets which have a semi major near that of the Moon. The radius (6.2 km) of Deimos, the outermost Martian satellite, is 55% that of Phobos, the biggest Martian satellite. This means that the Moon is 154.27 times bigger than the biggest Martian satellite.

Usually referred to as the Galilean moons, the 4 biggest Jovian satellites are nearly spherical and were the first objects found to orbit a body that was neither Earth nor the Sun. Their radii are:

- Ganymede: 2634.1 km (the 9th largest object in the Solar System and bigger than Mercury)
- Callisto: 2410.3 km
- Io: 1821.6 km
- Europa: 1560.8 km

In general, the Jovian satellites in Zone 1 are smaller than those in Zone 2 (Fig. 142). In contrast, as the precursors transitioned out of turbulence (Zone 4), the size of their daughter bodies decreased. The biggest Jovian satellites are in turbulence Zone 3. Unlike satellites in Zone 3 of the other planetary systems, the Galilean satellites are not separated by any small satellites. The Jovian satellites in Zone 3 are all big and not separated by smaller ones probably because, when their precursor was splitting to yield its daughters bodies, its high viscosity, and its way of collecting together into bodies did not allow its resulting daughter bodies to leak any of their constituents which could have formed smaller satellites between them. In the chapters on vortex and fluid ligaments, I expounded on this topic.

However, in Zone 4 (Fig. 142), the intermittence of the radius is more pronounced. In that zone, bigger satellites are separated by smaller ones. Because of its importance in understanding the formation of celestial bodies, particularly how the precursors were broken up into daughter bodies, I detailed below the variation of the radius of the satellites in Zones 4 and 5.

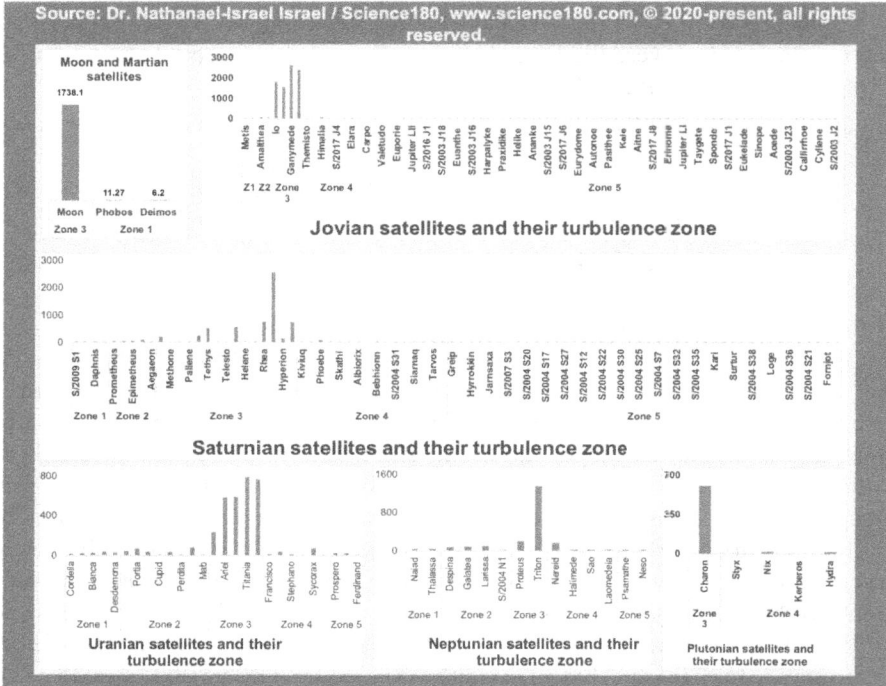

In Zone 5, although there are relatively big satellites, they are separated by smaller ones. Indeed, besides the Galilean satellites, the other Jovian satellites have a radius inferior to 90 km. In fact, 64.6% of the 79 Jovian satellites known by 2020 have a radius less than 3 km. Immediately downstream of the outermost biggest Jovian satellites in Zone 3 are 2 small satellites (Themisto and Leda which radius is 4 km and 5 km respectively) (Fig. 143). Then, comes a bigger satellite, Himalia, which radius is 85 km, then come 2 of the new satellites discovered between 2016-2019 which radius is 3 km each. Then, come 2 relatively bigger satellites: Lysithea (12 km) and Elara (radius = 40 km).

None of the 64 Jovian satellites downstream of Elara has a radius superior to 20 km. Indeed, the 23 Jovian satellites immediately downstream of Elara have a radius smaller than 5 km. Then, comes Ananke, which radius is 10 km. Downstream of Ananke are 22 Jovian satellites which radius is inferior to 4 km. Then comes Carme which radius is 15 km. Three Jovian satellites which diameter is inferior to 3 km are located downstream of Carme. After these 3 satellites, comes Pasiphae, a bigger satellite which radius is 18 km. After Pasiphae is 2 satellites which radius is inferior to 5 km, then comes Sinope which radius is 15 km. All of the 10 Jovian satellites downstream of Sinope have a radius inferior to 5 km. In other words, from the innermost Jovian satellite in Zone 4 to the outermost Jovian satellite (Fig. 143), there are relatively big satellites which size is inferior to 90 km separated by groups of much smaller satellites which radius is less than 5 km. The biggest satellites in Zones 4 and 5 are separated by smaller satellites in such a way that downstream and upstream of the relatively bigger satellites in Zones 4 and 5 there are smaller satellites.

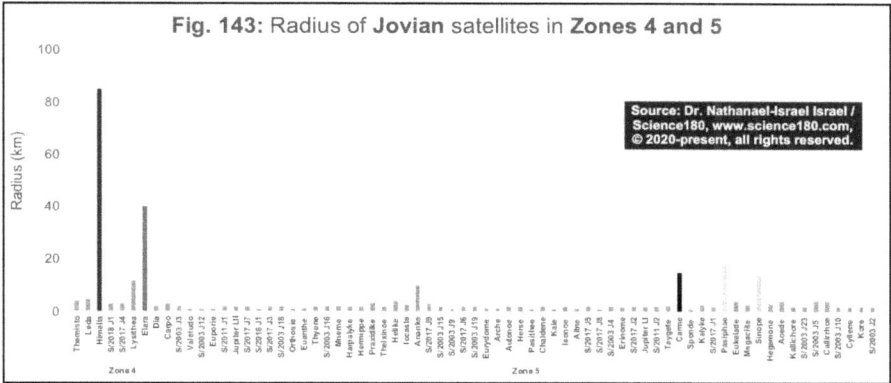

Fig. 143: Radius of **Jovian** satellites in **Zones 4 and 5**

One of the main questions about the distribution of the radius in Zones 4 and 5 is why the bigger satellites are generally separated by smaller ones or clusters of smaller satellites? The breakup of the precursors of the satellites and the collection of some of their fluid can explain the distribution of some satellites, particularly in Zones 4 and 5. For, the coalescence of the precursors of the satellites after their mother precursor broke up from the ligament holding them can also help explain why some bigger satellites in Zones 4 and 5 are surrounded by smaller ones downstream and upstream of them. In the chapters on the precursors of celestial bodies and on the fluid layers, I already explained the root cause of the size intermittence of the bodies in the universe. The data on semi major increment and orbital inclination also helped me to better understand and explain the distribution of the radius with respect to other variables.

The radius of the Saturnian satellites ranges between 0.3 km and 2575.5 km. The biggest Saturnian satellite is Titan, the second largest satellite in the Solar System. Below, I detailed the distribution of the satellites according to their turbulence zones. Although the biggest Saturnian satellites are in Zone 3, they are separated by much smaller satellites (Fig. 144). For instance, the radius of the Saturnian satellites in Zone 3 ranges between 1 km and 2575.5 km. Most of the biggest Saturnian satellites in Zone 3 are located toward the edge of that zone.

Fig. 144: Radius (km) of **Saturnian** satellites in **Zone 3**

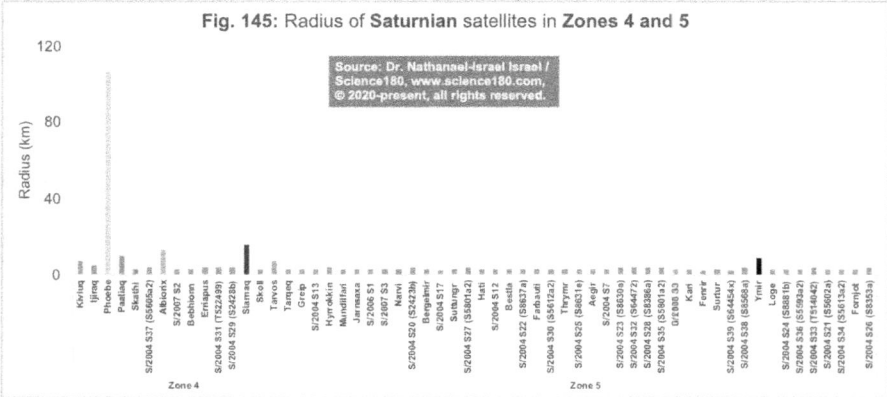

The Saturnian satellites in Zones 4 and 5 (Fig. 145) are much smaller, the biggest of them is Phoebe (radius = 106.67 km). Between Phoebe and Iapetus (the outermost Saturnian satellite in Zone 3), there are 2 small satellites:

- Kiviuq (the innermost satellite in Zone 4): radius = 7 km and
- Ijiraq, the second innermost satellite in Zone 4, radius = 5 km.

All of the 55 Saturnian satellites downstream of Phoebe have a radius inferior to 20 km. The biggest satellite beyond Phoebe is Siarnaq (radius = 16). Between Phoebe and Siarnaq are 9 satellites out of which Albiorix (radius 13 km) is the biggest. The biggest satellite beyond Siarnaq is Ymir (9 km). Between Siarnaq and Ymir, there are 36 satellites (average radius = 3.25 km and standard deviation = 0.84 km). Six satellites are located downstream of Ymir and although their mean radius (3.25 km) is the same as that of the 36 satellites located upstream of Ymir, their standard deviation (0.46 km) is 55.05% that of the 36 satellites, meaning that the intermittence is smaller for the outermost satellites.

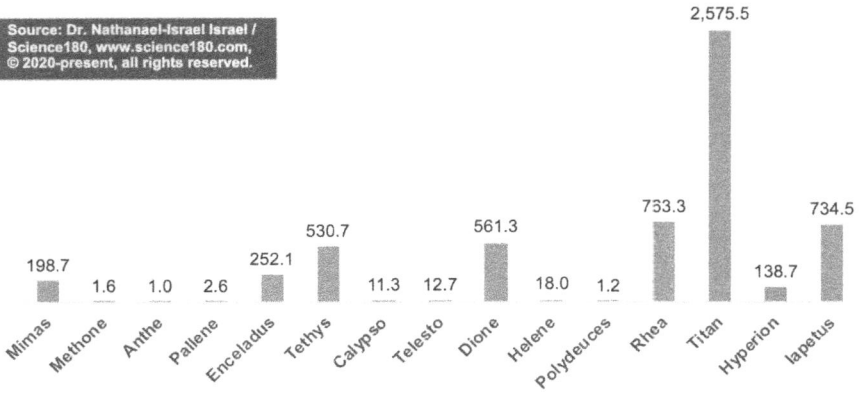

Fig. 145: Radius of **Saturnian** satellites in **Zones 4 and 5**

Out of the 27 Uranian satellites known by 2020, five are big and they all belong to Zone 3 and were separated by no small satellite:

- Miranda: 235.7 km
- Ariel: 578.9 km
- Umbriel: 584.7 km
- Titania: 788.9 km
- Oberon: 761.4 km

This trend of big satellites separated by no smaller one is also found with the Jovian satellites in Zone 3. The biggest Uranian satellite in Zones 4 and 5 is Sycorax and its radius is 75 km (Fig. 146). The 2 biggest Uranian satellites, Caliban (radius = 36 km) and Sycorax (radius = 75 km) are surrounded by smaller satellites upstream and downstream of them. The first 2 innermost satellites in Zone 5 have almost an equal radius.

Fig. 146: Radius of **Uranian** satellites in **Zones 4 and 5**

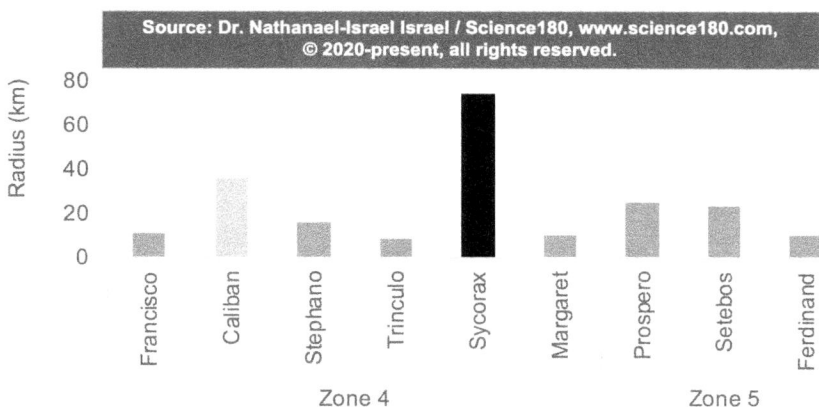

Two of the 3 biggest Neptunian satellites belong to Zone 3 while the third one, Nereid, is the innermost satellite in Zone 4:

- Proteus (the innermost satellite in Zone 3): radius is 210 km
- Triton (the biggest Neptunian satellite): radius is 1353.4 km
- Nereid (the innermost satellite in Zone 4): radius is 170 km

None of the Neptunian satellites downstream on Nereid has a radius superior to 30 km. However, in Zones 2 and 3, the radius of the satellites is 30 km. Fig. 147 is just a zoom into the Neptunian satellites in Zones 4 and 5. The Neptunian satellites downstream of Nereid have a mean radius of 24 km and a standard deviation of 5.48 km.

Fig. 147: Radius of **Neptunian** satellites in **Zones 4 and 5**

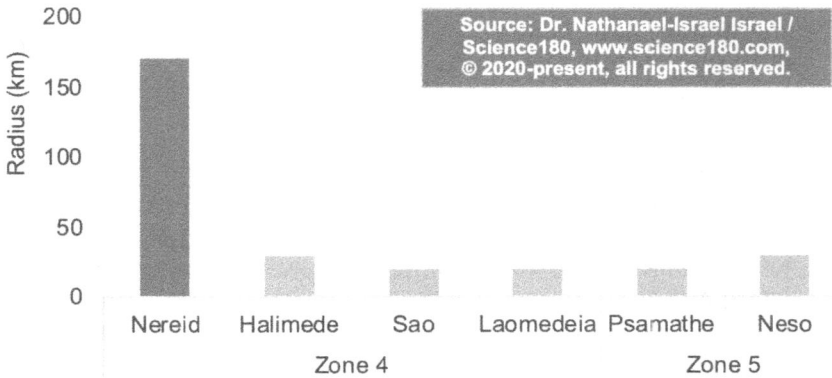

The size of Charon, the biggest Plutonian satellite, is 51.1% that of Pluto, its primary planet. The biggest Plutonian satellite in Zone 4, Nix (radius = 19.66 km), is downstream of Styx, the innermost satellite in Zone 4. Downstream of Nix are two small satellites:

- Kerberos: radius = 6.33 km
- Hydra, the outermost Plutonian satellite: radius = 22.5 km

Because the size of Charon is 26.93 to 110.18 times that of the other Plutonian satellites, it did not allow to see some details in the trends of the satellites in Zone 4. Therefore, I did Fig. 148 to better show the radius. Just as seen in the other types of satellites, the big Plutonian satellites are separated by small ones.

Fig. 148: Radius of **Plutonian** satellites in **Zone 4**

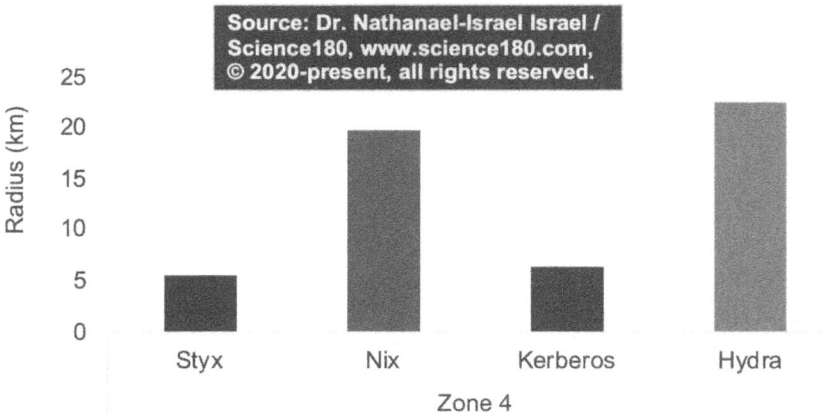

27.3.2. Radius of the satellites according to their turbulence zones

The biggest satellites in Zone 1 are Neptunian, followed by the Uranian ones (Fig. 149). The radius of the Neptunian and Uranian satellites in Zone 1 is higher than that of the 2 Martian satellites. The biggest satellites in each planetary system are usually found in Zone 3. The biggest satellite in each planetary system is usually toward the edge of Zone 3.

The radius of Titan, the biggest Saturnian satellite, is 97.78% that of Ganymede, the biggest Jovian satellite. The radius of the Moon (1738.1 km) is:

- 65.98% that of Ganymede (the biggest Jovian satellite),
- 67.49% that of Titan (the biggest Saturnian satellite),
- 2.2 times that of Titania (the biggest Uranian satellite),
- 1.28 times that of Triton (the biggest Neptunian satellite) and
- 2.87 times that of Charon (the biggest Plutonian satellite).

The innermost satellite in Zone 4 (Fig. 149) is usually among the smallest. It was only for the Neptunian satellites that the innermost satellite in Zone 4 (Nereid) has a radius higher than that of any of the other satellites in Zones 4 and 5. In fact, the radius of Nereid is 170 km whereas no other satellite in Zones 4 or 5 has a radius higher than 30 km. The 2 Neptunian satellites (Proteus and Triton) upstream of Nereid are the only Neptunian satellites in Zone 3 and their radii are 210 km and 1353.4 km respectively. This data suggested that it could have been ok to place Nereid in Zone 3, but I chose to place it in Zone 4 because of its high rotation period and eccentricity. In my book on the delimitation of the turbulence zones, I provided more details about the logic used to put Nereid in turbulence Zone 4. A key characteristic of Zone 4 is that the highest coefficient of variation (CV) of the radius is found there regardless of the types of satellites. Details about the CV are provided below. The outer satellites in Zone 5 are smaller than the innermost satellites in Zone 4.

Fig. 149: Equatorial Radius of the satellites according to the turbulence zones and semi major axis

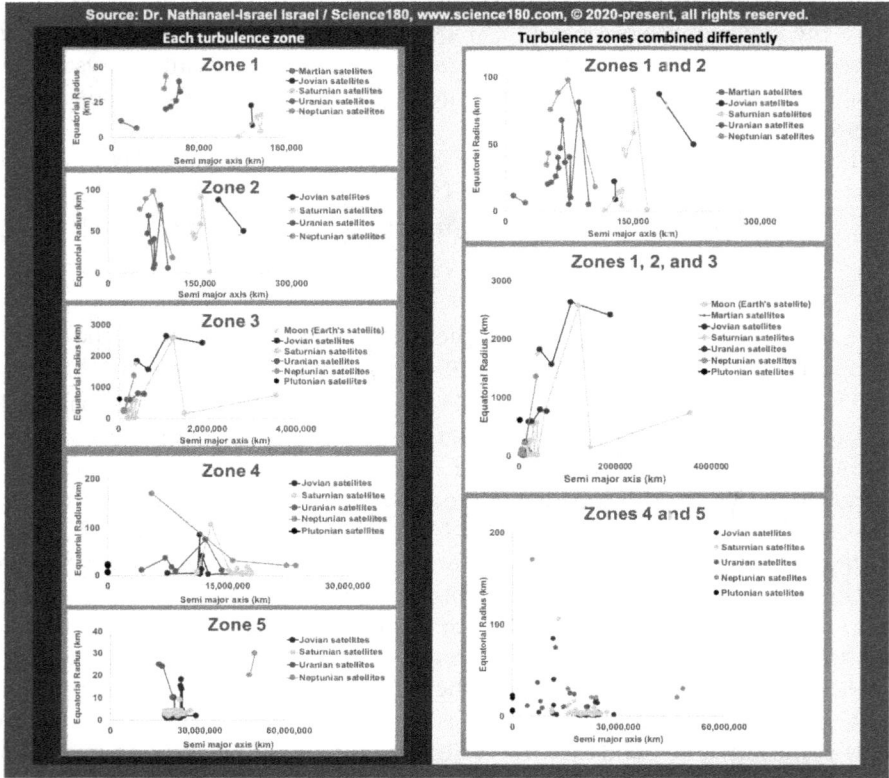

27.3.3. Mean of the radius of the satellites according to their types and turbulence zones

Although the mean of the radius in the turbulence zones depends on the types of satellites, the highest mean of the radius was found in Zone 3. The highest value was observed with Jovian satellites. The mean radius of the Neptunian satellites in Zone 3 is higher than that of the Uranian and Saturnian satellites. The highest mean radius in Zones 1, 2, 4, and 5 were recorded with the Neptunian satellites. In turbulence Zone 3, the highest mean of the radius of the satellites was found with the Jovian satellites. The mean was higher in Zone 3 of the Jovian satellites because these satellites are all big and are not separated by smaller ones as is the case for most of the other planetary systems. The mean radius of the satellites in Zone 5 (Fig. 150) is positively correlated with the semi major of their primary planets (r2=0.93) (Fig. 151). This implies that as the turbulence decreased, the radius of the satellites in Zone 5 increased.

Fig. 150: Mean Equatorial **Radius** (km) of satellites in Turbulence **Zone 5**

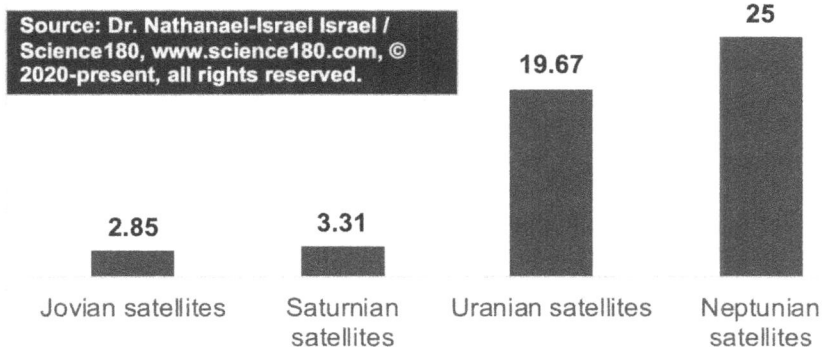

			25
		19.67	
2.85	3.31		
Jovian satellites	Saturnian satellites	Uranian satellites	Neptunian satellites

Fig. 151: Regression between **mean** radius of satellites in **Zone 5** and semi major of their primary **planet**

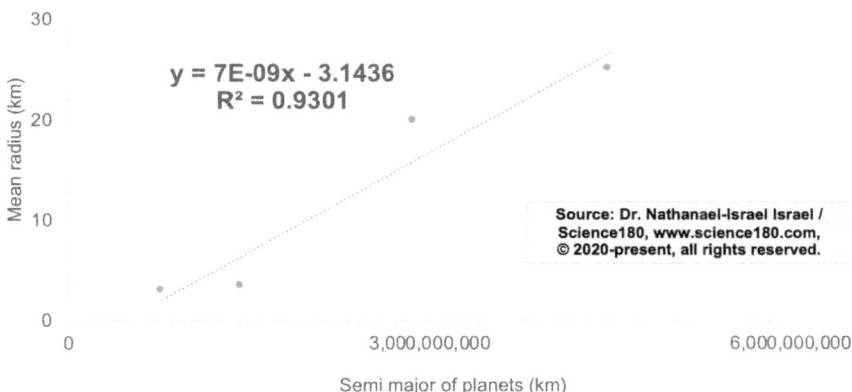

$$y = 7E-09x - 3.1436$$
$$R^2 = 0.9301$$

Mean radius (km)

Semi major of planets (km)

Because in some turbulence literatures, the ratio "radius of the biggest body / mean radius of the bodies in the dissipation zone" is studied as if it is a big deal, I also investigated the ratio "radius of the biggest satellites / mean radius of the satellites in turbulence Zone 5" for the 4 giant planetary systems which have Zone 5. Varying between 54.14 and 924.25, that ratio decreased as the semi major of the primary planet increased. For instance, the highest ratio was observed with the Jovian satellites, whereas the smallest ratio was recorded with the Neptunian satellites.

27.4. Relationship between the size of the precursors of celestial bodies and that of their daughter bodies

Because the precursors of each body or system of bodies went through a lot of changes (e.g. topology, constituent changes, density changes), it is difficult to know what their exact volume, size, and shape could have been. For instance, some bodies were more compressed than others. However, the order of magnitude of the size of a precursor of the bodies can be estimated by referring to the size of its daughter bodies. Therefore, the size of the precursor of the entire Solar System could have been in the order of magnitude to the size of all the bodies in the Solar System. Similarly, the size of the Sun, planets, satellites indicate the size of their precursors. The size of the precursor of each planetary system could be related to the size of its primary planet and satellites. Because in each system (e.g. planetary, stellar), the primary system generally accumulates more than 99% of the 9 system-additive variables (e.g. mass, momentum, moment of inertia, kinetic energy, and volume) of the system, it can be fair to even estimate the 9 system-additive variables of the precursors of the systems of bodies by focusing on the 9 system-additive variables of the primary bodies of those systems. In other words, the 9 system-additive variables of the precursor of a planetary system can be in the same order of magnitude as those of the primary planet, just as the size of the precursor of the Solar System can be in the order of magnitude of the size of the precursor of the Sun or the Sun itself.

27.5. Take home message

The careful study of the radius of the celestial bodies according to their semi major axis contributed to helping to unearth fundamental facts about the formation of the bodies in the universe with the perspective of turbulence. Because of their size and the inability to see them using telescopes, many small bodies in the Solar System were just recently discovered after spacecrafts were sent to or near their place. For instance, although the biggest Jovian satellites were known since the 17th century, most of the smaller Jovian satellites were discovered late in the 20th century and early 21st century. In each system (e.g. Solar System, planetary system, and asteroid system), the largest body is the primary body, while the largest secondary bodies are usually located in the most turbulent zone, usually Zone 3. Hence, the Sun is larger than any other body in the Solar System, while the planets are also larger than any other bodies in their planetary system. Likewise, the largest planets in the Solar System are found in the turbulence Zone 3 of the precursors of the bodies orbiting the Sun. Likewise, in each planetary system, the largest satellites are found in turbulence Zone 3 of the precursor of the satellites. In general, in each planetary system, the number of the bigger bodies is smaller than that of smaller bodies. Although satellites exist in all turbulence zones, they are more predominant in Zones 4 and 5. Because of its connection with other variables that I will explore later in this book, as you keep reading this book, you will see more data on the significance of the radius of the bodies in unearthing the code of the formation of the universe. The viscosity, energy, and the size of the fluid layers in the precursor of the celestial bodies affected not only their size but also how far they traveled before forming their daughter bodies. Some fluid layers traveled over a longer distance before the variation of their speed could be in the range that allows the eddies to stumble upon one another to form bigger ones and cause the major turbulence known in Zone 3. In other cases, because of the small energy in them, some precursors were not able to travel much before exhausting their ability to go farther, therefore forming their bodies very close to their source. For instance, the Jovian satellites in Zone 3 are all big and not separated by smaller ones probably because, when their precursor was splitting to yield its daughters bodies, its high viscosity, and its way of collecting together into bodies did not allow its resulting daughter bodies to leak any of their constituents which could have formed smaller satellites between them. The breakup of the precursors of the satellites and the coalescence of some fluids can explain the distribution of the satellites, particularly in Zones 4 and 5. For, the coalescence of the precursors of some satellites after their mother precursor broke up from the ligament holding them can explain why some bigger satellites in Zones 4 and 5 are surrounded by smaller ones downstream and upstream of them. The mean radius of the satellites in Zone 5 is positively correlated with the semi major of their primary planets implying that as the turbulence decreased, the radius of the satellites in Zone 5 increased.

CHAPTER 28

IGNORE THESE 7 THINGS AND YOU WILL NEVER CRACK THE CODE OF THE UNIVERSE FORMATION SCIENTIFICALLY – IF YOU THINK I AM JOKING, READ THIS AND THANK ME LATER

Do you know or did you ever imagine that a wonderful secret at the crossroad of what is termed "vortex stretching, tilting, inclination, elongation, rotation, density, and size of bodies" can work together to detect, denounce, and bring down all useless cosmological theories that have fouled human beings to this day? If you answered no, you are an honest person and you are not alone; for until I discovered this, nobody before me has invited or welcomed that idea into their mind! Now, just relax and enjoy the ride of how, little by little, some variables begged me to pay attention to them, and without them, I would not have done what the world is now celebrating as a major scientific breakthrough! When I leave Science180 Seminars (www.Science180Seminars.com), people always tell me that they have never heard of anything like this before, and that the secrets I am about to share with you have changed their life and understanding of the universe. Without wasting any second, let's dig into these secrets now.

In fluid dynamics, a vortex (pleural = vortices or vortexes) is a region in a fluid in which the flow rotates around an axis line, which may be straight or curved (Ting, 1991; Kida, 2001). One of the most difficult tasks I dealt with in the writing of this book was how to scientifically explain why the axial tilt, orbital inclination, and eccentricity is very high for some celestial bodies and small or null for others. For, I felt like before completing my investigation on the formation of the universe, I needed to properly answer the following questions:

- What tilted the rotational axis of some celestial bodies?
- What inclined some orbital planes?
- What elongated some orbits to give them a higher eccentricity?

- What accelerated the rotational angular speed of some bodies in Zones 4 and 5?
- Why celestial bodies have to rotate at all?

As these questions and many more raced through my mind for almost 7 years, I ended up feeling like I must properly address the relationship between axial tilt, eccentricity, and orbital inclination together in conjunction with the shearing, stretching, and break-up of the fluids in the precursors of the bodies. To quench my thirst and satisfy my hunger for an answer, I turned to my strategic thinking. By addressing the link between stretching and tilting of a vortex, orbital inclination, axial tilt, eccentricity, rotation, radius, and density of the celestial bodies, this crucial chapter that also applies to chemical particles, is an introduction to how I patiently found the answer to the aforementioned questions like one who is peeling off the layers of an onion.

28.1. My first glimpse at the connection between turbulence, rotational period, size, orbital inclination, axial tilt, and eccentricity of bodies

As of December 12, 2016, it appeared to me that, when the ligament of the precursors of bodies broke up, the precursor of a body which is downstream of the breaking point could have been propelled downstream while the precursor of a body upstream of the breaking point may tend to move upstream. As the ligaments were retracting, the vortical structures they contained could have still been rotating and the intensity of their rotation may also have affected their motion. In those days, I felt like the way the eddies of the precursors were stretched and pinched off or broke up from one another, the forces involved, the mass of the bodies stretched, the intensity or strength of the shearing, and the viscosity of their constituents can explain the variation of the axial tilt and inclination among the celestial bodies. For instance, the rotational axis of Jupiter and Saturn is not that tilted nor their orbital plane inclined much because of the size and dominance of their precursors, which "resisted" the pushing or pulling by the flow of the fluids of the bodies surrounding them in the huge ligament of the precursor of the bodies orbiting the Sun. But in 2016, meaning 9 years before I published this book, I could not properly articulate the underlying cause of the inclination or tilting of the orbital plane and rotational axis of the bodies.

Two years after the inspiration I explained in the previous section, I got another thought on the tilting of the bodies. Indeed, during the entire day of October 23, 2018, as I was trying to calculate the rotational kinetic energy of the satellites, I noticed that most of the rotational speeds were not reported in the literature. Therefore, I decided to calculate the missing rotational speed using the rotational period so I can use it to estimate other variables such as the rotational kinetic energy for all the satellites so I could be able to add them to the translational kinetic energy to have more data points when addressing rotational kinetic energy and total kinetic energy. I was tempted to find a correlation between the limited rotational period provided by NASA and other variables which could allow me to estimate the missing rotational speed of many satellites. In fact, in those days, besides the Moon, NASA's website (NASA, 2018) did not report the fact sheet of the rotational speed of any other satellites in the Solar System. I then began reviewing the regression between rotational period and other variables. I did not find any strong regression that I could use to estimate the missing rotational period or rotational speed. I therefore decided to see if I could average the available rotation period or rotation speed for all the turbulence zones in each planetary system in order to apply that average for each satellite lacking that data, according to their planetary system. As I started exploring the data, no trend quickly came to my mind. Therefore, I decided to work on the rotational kinetic energy using two scenarios:

- Calculate the rotational kinetic energy using the estimated rotational period provided by NASA (2018), which means many data would be missing; and
- Calculate the rotational kinetic energy of the satellites by finding an average value for the missing rotational period or rotational speed of the concerned satellites.

My goal was to explore how these two scenarios differ and what kind of information I could get from them. After coming up with that strategy, I calculated the rotational speed of the satellites which have a rotational period using the formula:

Rotational speed = Radius * Rotational angular speed = Radius * 2 * π / Rotational period

By the time I finished doing that calculation on October 23, 2018, it was about 8 pm, the time I usually stopped working on Science180 in those days so I could study with my daughter Josephine, who at that time was in Pre-K. I therefore closed my computer, took my 3 children to the couch so I could teach them before sending them to bed. In those days, my focus for Josephine's education was to teach her to pronounce and write the alphabets, count numbers from 1 to 100, and write numbers from 1 to 30. As I was doing some phonetics with her, I noticed that she was struggling to pronounce the letters "L" and "R". I spent about 15 minutes teaching her phonetics. Then, I asked Josephine to write for me the numbers from 1 to 10. She did and I noticed that her writing of 2, 3, and 8 were not clean. I showed her how to write these numbers and asked her to write them from the left margin of the page all the way to the end of the line. As she was writing, and I was watching her, my mind just spontaneously went to the data I was working on concerning angular momentum and rotational speed. I heard a voice in my heart asking me, "Do you know why the bodies in the Solar System do not have the same rotational period or rotational speed?". I answered "no, but probably because they have different radii". The voice replied, "How can radius be related to rotational speed or rotational period?" I did not know what to say. Immediately, I felt in my heart that "the turbulence zone where the larger bodies are located in each system could have been a center of a huge turbulence. I felt like, when the fluids of the precursors of the bodies were going through turbulence, the balance of the movement of the vortical structures formed in them as a whole, the interactions between the fluid layers, and how the fluids were stretched and split could have contributed to birthing their rotation and affecting the elongation of the orbit and orbital plane, and the tilting of the rotational axis of some daughter bodies. The position of the fluid layers affected their speed. For instance, from the top layer to the bottom layer, the speed of the precursors decreased. As the fluid layers of the precursors were being molded into celestial bodies and positioned into their orbit, their speed was progressively converted into an orbital speed and a rotational speed. Because the speed of the fluid layers decreased from the top layers to the bottom ones, the orbital speed of the daughter bodies also decreased. But in the case of the rotational angular speed, things were a little different. Indeed, as the fluid layers were splitting, they were also being wrapped around or wound up. The speed with which they were wound up decreased from the top layers until a point where their ability to rotate was very small, and consequently, the fluids which were around the position where the rotational ability was nearing zero were "stuck", leading to the formation of larger or thicker fluid layers and consequently the formation of bigger bodies around and downstream of them. The zone where the fluids were stuck were usually toward the outermost region of Zone 3. Although the ability of the fluid layers of the precursors of the bodies in Zone 3 to rotate was decreasing, meaning their rotational angular speed was decreasing, as a whole, they were still moved in the direction of the flow. In other words, the rotation of the fluids which were being gathered was happening as the layers as a whole were being pushed by what was the precursor of the orbital speed and escorted into orbit. Another way of saying this is that, although the rotational angular speed of the fluids in Zone 3 was decreasing significantly, they were still being pushed and ushered onto orbit by their translational motion.

It appeared to me that, as the fluid layers in Zone 3 separated from those beneath them (meaning the fluid layers in Zones 4 and 5), the bodies being formed in the fluid layers of Zones 4 and 5 gained a rotational momentum that was translated into an increase of their rotational angular speed. In other words, the increase in rotational angular speed of the bodies in Zones 4 and 5 can be explained by an event which displaced them quickly as if they were being ejected from the precursors of the bodies in the inward zones which were "sitting" on top of them and/or were "failing" to rotate fast. In contrast, the precursors of the satellites in Zones 4 and 5 did not gain orbital speed because the fluid flow carrying them was not affected by the decrease of the rotational angular speed in the previous zones. All of this suggests that the precursors of the bodies were being shaped inside a fluid flow carrying them and which defined their orbital speed.

To put it another way, for the satellites in Zones 1, 2, and 3, the rotational angular speed decreased while the rotational period increased as their distance from their primary planet increased because, due to the decrease of the speed of the fluids from the top to the bottom layers, the fluids of the precursors ended up being stuck, trapped, and unable to rotate much as their position with respect to the top fluid layer increased. In Zone 3, because the speed of the fluids decreased a lot as compared to Zones 1 and 2, more fluids were stuck or trapped, and in the end, the biggest or largest satellites are usually found in Zone 3. The rotational angular speed of the satellites in Zones 1, 2, and 3 did not depend on any other variable but on the distance from their primary planets because that decrease was controlled by how the fluid layers of the precursors of the satellites were positioned as they moved inside the bulk of fluid layers of the precursor of the satellites. The smaller the speed of the fluid layer, the lower its ability to quickly wrap it or "convert" part of that speed into a rotational angular speed to inculcate a strong rotation to its moving body. In contrast, the higher the speed of the fluid layer of a precursor, the higher its ability to make its daughter body to have a higher rotational angular speed.

As my mind was racing while going over all these things on that night of October 23, 2018, I felt good but I was shocked. I just started worshipping God for who He is, and for the connection I was discovering between turbulence and the movement of the precursors of celestial bodies. I then started speaking in tongues and asked my children to come close to me. My 1-year-old son Joshua-Enoch was laying by my left, my oldest daughter (4 years old) Josephine was sitting at my right; I asked my youngest daughter Joelle-Major (3 years old) to lay on my belly so I could hold them together and we prayed in tongues. Afterwards, I anointed them with oil and sent them to bed; for on that day, my wife was at work and would be off at midnight. After putting the children to bed on that October 23, 2018, I quickly rushed to my computer to look at the characteristics of the celestial bodies and see if they reflected what just passed through my mind. To my surprise, I noticed that for each planetary system, the biggest satellites are usually downstream of the one having the maximum rotational period. I also noticed that the rotation period increased from the innermost satellite until the maximum rotational period is reached with the outermost satellite in Zone 3. Then, the rotational period dropped significantly. I also observed that, besides the biggest Plutonian satellite (Charon) and the biggest Neptunian satellite (Triton), the biggest satellite in Zone 3 of the other planetary systems is followed by 1 or 2 other satellites in Zone 3. For instance, no Neptunian satellite in Zones 1, 2, and 3 has a rotational angular speed smaller than that of Triton, the outermost Neptunian satellite in Zone 3 and the largest Neptunian satellite. That day, I felt like the inability of Triton to rotate faster than any other outermost satellite in Zone 3 may have also played a role in tilting its rotational axis and inclining its orbital plane. Therefore, I felt like, by better studying the connection between orbital inclination, axial tilt, and eccentricity of the celestial bodies, I could unearth a big secret behind their formation. By this time, it appeared to me that, as the precursors of the celestial bodies were being fragmentated, the energy communicated or transferred to them by their mother precursor was not just used to form and move them (e.g. rotationally and translationally) but to also tilt their rotational axis, change their orbital inclination, and elongate their orbit, which effect can be felt on their eccentricity. Depending on some characteristics (that I explained later) of the bodies, the energy transferred to them by their mother affected differently the parameters I just mentioned. For instance, while the orbit of Mercury was elongated very much, the rotational axis of Venus, Uranus, and Pluto was highly tilted. In contrast, for the bigger planets, the rotational axis was not tilted much, but the rotation was sped up. Having the least rotational angular speed in Zones 1, 2, and 3, the outermost satellite in Zone 3 appears to be at a place where a special instability could have occurred in the fluid layer of the precursors of the satellites located downstream of it, and this instability could explain the behavior of the precursor of the bodies downstream of the outermost satellite in Zone 3. That was how, for the first time, I felt a deep connection between eccentricity, orbital inclination, axial tilt, size, and speed of celestial bodies. But as days went by, I got a clearer explanation of what happened.

To try to explain some of the concepts I addressed above, I would like to give an example. When two people are holding their hands and pulling away from each other, they can feel a tension of pull in their arm until a moment when they separate their hands and then the force that they were using to pull away can cause them to fall to the ground if that force was too high. But, when the pulling force is not too high, the people may not fall on the ground. The falling on the ground involved the tilting of their axis. To put it in another way, if instead of holding their hands, 2 people are holding the edge of a plastic rubber band which can expand, as soon as they start pulling, that plastic rubber band can stretch until it breaks if the pulling force is sufficient. Then, they can fall. This reminds me of a game called "tug of rope" that kids play. It consists of taking a group of kids and dividing them in half. Then, take the end of the rope and give each group to hold onto. Usually, a line transverse to the rope is drawn or place on the ground, indicating the halfway point between the two groups. When the teacher says "go", each group starts pulling back using as much force as it can. The stronger group pulls the other group of kids across the line, but as the weaker group starts to lose the "battle", the kids' feet start slipping or sliding and the stronger group gains momentum. These examples were not exactly what happened during the formation of the universe but they illustrate the fact that, when the fluids of the precursors of the celestial bodies were shearing and spiraling to form and shape their daughter bodies, some of them were stretched, pulled, or pushed by others, and then separated from their neighbors. Because these precursors were also rotating and moving along the path taken by their fluid layers, some of them could have been "thrown" away in a direction which balanced the direction of the shear flow, that of the orbital movement, and the rotation they were going through. Hence, a modification of many angles occurred (axial tilt, inclination of equator, and orbital inclination). In some cases, the movement can be chaotic. As these thoughts were racing through my mind, I felt like I better properly address the relationship between axial tilt, eccentricity, and orbital inclination together in conjunction with the shear, stretching, and breakup of the ligament, which is what I will do in the next segments and in the following chapter.

After the inspiration I got and explained above, I put aside the search for the connection between tilt, eccentricity, and orbital inclination and I focused on other aspects of my investigation. More than a year later, I had another flash of inspiration and ideas regarding the connection I was trying to make between axial tilt, orbital inclination, and eccentricity of the celestial bodies. Indeed, in those days, I usually went to bed around 11:30 PM, but on Friday November 22, 2019, after a very busy day during which I worked on the semi major axis of the satellites, I was more exhausted than usual that I had to close my computer at 8 PM, meaning more than 3 hours earlier than usual. The day before, I was so tired of working hard that I even went to bed at 9 PM. As I was about to close my computer that Friday, I decided to check a file I wrote in 2013 when I was first inspired to view the bodies in the Solar System as a product of an event like a hurricane. As a reminder, I did not embark into this investigation on the origin of the universe with a protocol of research in my hands after sitting down to elaborate hypotheses. Most of the ideas I got just came to my mind, even at times I was not even thinking about physics, which is not what I studied in graduate school. Anyway, on that Friday November 22, 2019, I opened that aforementioned file and, as I was browsing through it, I came across the following sketch (Fig. 152) that I made 6 years earlier.

Fig. 152: Movement of a projectile in the air

In fact, at 2:04 AM on November 19, 2013, I was inspired that the movement of planets in the Solar System is like that of projectiles launched into space. In those days, I was inspired to use this analogy to define and map the structure of space. The above figure came to my mind with some of the cinematic equations associated with it:

$$y = ax + bx^2 + y_0$$

$$v = v_0 + at$$

While I was looking at that graph again on November 22, 2019, meaning 6 years and 3 days later, I remembered the Physics lessons I had in high school about the cinematics and dynamics of projectiles. I was so good in Physics in High School that after exams, my Physics teacher, Mr. Henry Oreochan, in Benin Republic (West Africa), used to ask me to explain to the entire class the exam using my exam papers. After school, I also tutored many classmates and some of them whose parents are wealthy paid me for it. In the last decade of the 20th century, one of them used to pay me about 1$ US per month, which was a big deal for me in those days, for with 25 cents, I could get a decent breakfast on my way to school, which most students could not afford. In those days, although I was very shy, I delighted in helping my classmates to understand physics and mathematics, which I loved so much, and which even caused me to transfer from "Biology and Physics" major to "Mathematics and Physics" when I was passing from 10th grade to 11th grade. That was why I obtained my first High School Diploma in Math and Physics by self-teaching. Then, the next year, I obtained the High School Diploma in Biology and Physics by self-teaching also.

Anyway, as I was looking at that graph, it instantly came to my attention that when a projectile is launched into space from the earth, its speed decreases as it goes against gravity until it reaches its highest point and then it starts descending. I remember that as a projectile is going higher and higher, one of the things that reduces its speed is gravity. In the case of the celestial bodies in the Solar System, the deceleration of their speed in the direction of the shear flow of their precursors can be estimated using their orbital speed, rotational speed, and semi major as I already did in the chapter on derivatives. Therefore, I sensed that the first derivative of the equation linking speed to semi major axis may explain the deceleration of speed in the direction of the shear flow of the precursors of the bodies. It appeared to me that, as the precursors of the satellites escaped the precursor of their planetary system, they started moving away as fluid masses in which vorticial structures were formed and gained shape, while their speed decreased until they reached their aphelia (a point whereby the precursor of a body could no longer gain more semi major axis due to the inability of its fluids to shear any longer); at that point, the daughter bodies started "falling" back, just as a projectile launched with a certain speed reaches an apogee or peak and then has to "fall". Here, the bodies were not falling back because of gravity, but because the position of their precursors in space was changing and they could no longer get any farther from the primary body in their system. From the aphelia, the orbital speed could have started increasing until it reached its maximum at the perihelia. Because of the expansion of the universe and the systems of bodies it contains, the current perihelia and aphelia of celestial bodies could be located outward of where they had been for the first time when the celestial bodies were formed and placed in orbit.

I also felt like the location of the satellites with the highest axial tilt could have been affected by some events in the shear flow. Again, recalling my high school course, when a projectile launched from the ground is going up, its speed decreases until it reaches the highest point, then, its speed increases until it reaches the ground. Usually, the position of the highest point is defined using the first derivative of the equation of its trajectory. At that point, the acceleration is null and the sign of that derivative also changes afterward. When the shear rate reached a certain level, the orientation of the orbit or rotational axis was a response to the forces which pulled or pushed the precursors of the bodies in the fluid flow. Therefore, as I was tiredly reclining in my house on that November 22, 2019, I felt that I should carefully investigate the trends of the:

- derivative of the innermost retrograde satellite, that of the satellite upstream of it, and that of the satellite downstream of it.
- derivative between the satellite with the highest axial tilt, that of the satellite immediately upstream of it, and that of the satellite downstream

of it.

Without hesitation, I felt like, if well explored, the ideas that just crossed my mind on November 22, 2019 would contain or lead to a key secret to unlock the code of the formation of the universe. Therefore, throughout the entire week of November 22, 2019, I tried hard to understand why the rotational axis is tilted for some bodies and why the orbital inclination is higher for some more than others. I was thinking that the way the vortices or the moving fluids of the precursors of bodies were pushed and/or pulled from one another contributed to tilting the rotational axis and inclining the plane of the orbit of some bodies. I was also feeling like the derivatives (of the orbital speed and rotational angular speed with respect to the semi major axis) of some key positions may hold a secret. Although some things I said in this section on the search for the connection between eccentricity, orbital inclination, axial tilt, and size of the celestial bodies may be incorrect or partial, they armed me with a well of ideas I tackled to see the connection I have been looking for.

28.2. Key facts about orbital inclination, eccentricity, axial tilt, and rotational angular speed

When I first looked at a picture of the axial tilt of the planets (see chapter on axial tilt) in 2013, I immediately felt like something pushed, pulled, tilted, or caused some bodies to flip over or roll over. As I explained in the chapter on axial tilt, only one satellite in each satellite system usually has a very high axial tilt, while for the other satellites, the axial tilt is zero, meaning the rotational axis of most bodies was not tilted. Similarly, the orbital inclination of more than 99% of the satellites located in Zones 1, 2, and 3 is very small, while for the satellites in Zones 4 and 5, it is higher. This means that the orbital plane of most satellites in Zones 1, 2, and 3 is not tilted or inclined. Moreover, the eccentricity of most satellites in Zones 1, 2, and 3 is much smaller than that of the satellites in Zones 4 and 5, meaning that the orbit of most satellites in Zones 1, 2, and 3 is almost circular, while that of most bodies in Zones 4 and 5 is very elongated (see chapter on eccentricity).

As for the planets, although their orbital inclination is small, the axial tilt of 3 of them is very high; and these highest axial tilts of the planets were observed with both an inner planet (e.g. Venus) and outer planets (e.g. Uranus and Pluto), signifying that the axial tilt cannot be explained just by the distance separating the bodies from their primary body but by something else. Else, a planet too close to the Sun like Venus and a planet far away from the Sun like Uranus and Pluto could not both have a very high axial tilt.

If the axial tilt was connected to the speed of the fluid flow of the precursors of the bodies, either the innermost planets would all be tilted or the outermost planets would all be tilted or vice versa. But that is not the case. Venus (the second innermost planet in the Solar System) is even more tilted than Pluto (the outermost planet) and Uranus (a giant planet). Although located between Uranus and Pluto, Neptune is not tilted. This means that the axial tilt of the planets is not connected to their semi major or to their orbital speed. Consequently, the speed of the fluid layers of the precursors did not define the axial tilt of the bodies. As I looked at the satellites, I noticed that, the satellites with the highest axial tilt are neither the innermost nor the outermost ones, but they are located around the end of Zone 3 and the beginning of Zone 4.

In each satellite system, no more than 2 satellites have a significant tilt. For instance, only one Saturnian satellite and one Neptunian satellite have an axial tilt higher than 90°. The rotational axis of no Uranian satellite is tilted. Similarly, except Europa and Ganymede which axial tilt is 0.1° and 0.16° respectively, no other Jovian satellite is tilted. Finally, only 2 Plutonian satellites have a significant axial tilt: Styx (82°) and Nix (132°).

In contrast, of the 9 planets in the Solar System, 3 have an axial tilt higher than 90°. Among all the variables I studied, axial tilt is the only one which shows this kind of trend according to which just a few celestial bodies have a significant value and the others do not. The satellites which rotational axis is the most tilted in their system, are not consistently the smallest ones nor the biggest ones. But the highest axial tilt is usually found around the biggest satellites, usually not inward of them or upstream of them, but sometimes downstream of them. For example, the highest axial tilt recorded on the Jovian and Neptunian satellites was recorded with the biggest satellite in their planetary system. In contrast, for the Saturnian and Plutonian satellites, the highest axial tilt was found downstream of the biggest satellite in their system. This implies that the axial tilt of the bodies was not defined by just their size.

When I tried to figure out whether the planets were pushed inward or outward of the direction of the flow of fluids of the precursors of the bodies orbiting the Sun, I realized that the scientific community does not even agree on what is the north and south pole of some planets like Uranus, which like Venus and Pluto, have a tilt higher than 90°. If I considered the planets as born from a flow running from the inner part of the Solar System toward its outskirt, most literatures show Venus as having fallen inward, meaning backward while Uranus and Pluto are shown as having fallen outward or forward. Therefore, although some people may not agree with these directions or senses or the so called "fall" or "knockout" (and I think the scientific community may even change those directions very soon), I considered them as they seem to be believed:

- Venus' tilt as oriented backward as if Venus has "fallen" inward to the Solar System
- Uranus' tilt as oriented outward/forward in the Solar System or Uranus as having been pushed outward/forward in the Solar System and
- Pluto's tilt as oriented outward/forward as well.

Considering what I was sensing from the data I was studying and what I read in the literature, I felt like to properly explain the root cause of the trends of the orbital inclination, axial tilt, eccentricity, and rotational angular speed of the celestial bodies, I better off first review the literature concerning advection, movement and interaction between vortices in an advective fluid layer or stratified fluid in the perspective of shearing, deformation, stretching, separation, tilting, and elongation of vortices in a fluid flow. That is what I addressed in the next section.

28.3. Movement and interactions between consecutive bodies (e.g. vortices) in layers of an advective flow or in a stratified fluid

While advection is the transport of fluid properties at a definite rate and direction which is that of the fluid velocity; fluid parcel trajectories are the characteristics of the advection equation (Price, 2006). According to the same author, every parcel of fluids that participates in a fluid flow is believed to be literally pushed and pulled by all of the surrounding fluid parcels by shear stress and normal stress. Consequently, it is believed to be impossible to predict the motion of a given fluid parcel in isolation from its surroundings, rather the motion of the surrounding fluid parcels must be predicted as well. When fluid layers of a precursor are considered, adjacent upstream or top layer and adjacent downstream or bottom layer are the immediate surrounding layers. In other words, each fluid layer could have been impacted by the fluid layer located before or on top of it and the fluid layer located after or at the bottom of it. Consequently, I felt like, to understand how fluid layers of the precursors of the bodies could have affected one another during the genesis of the universe, I also needed to study the bodies immediately inward and downward of each body. In the case of the satellites for instance, this strategy can be observed by studying the characteristics of the bodies upstream and downstream of each satellite.

In general, the instantaneous motion of a fluid has been resolved or decomposed into 3 components using what is called the "Cauchy-Stokes Decomposition Theorem":

- a translation,
- a rigid body rotation, and
- a linear deformation rate along two mutually perpendicular directions.

In most fluid flows, the rotation rate and the deformation rate are believed to vary spatially and in time, right along with the velocity itself, and each of them have important roles in theories that attempting to predict the evolution of fluid flows; and the deformation rate is believed to be proportional to the rate at which adjacent fluid parcels slide past one another (Price, 2006). Therefore, it appeared to me that the way the fluid layers of the precursors of the bodies passed one another could have affected the deformation rate of their daughter bodies. The changes in the orbital inclination, axil tilt, and eccentricity of the fluids can be

385
Science180: Source of Unconventional Wisdom and Knowledge on the Origin
of the Universe

explained by the deformation of the fluids in their precursors as the fluid layers flowed passed one another, or as they were being collected to form their daughter bodies. The fluid layers in Zones 1, 2, and 3 could have significantly affected the deformation of the bodies in the fluid layers in Zones 4 and 5. It appeared to me that the interaction of the fluid layers of the precursors of the outermost bodies in Zone 3 and that of the innermost bodies in Zone 4 could also hold a key secret about the dynamic of fluids in turbulence.

According to the same author, "when a column of fluid is squashed, it slows its rotation rate, and the reverse must happen if a column is stretched" (Price, 2006). This made me to believe that the squashing or stacking of the fluid layers in Zones 1, 2, and 3 may also contribute to explaining why their rotational angular speed decreased as the semi major axis increased. In contrast, once the weight or pressure of the satellites in Zones 1, 2, and 3 was removed from the precursor of the satellites in Zones 4 and 5, the fluid column or fluid layers of the latter could have stretched more. In other words, the fluid layers in Zones 1, 2, and 3 could have squashed the fluid layers in Zones 4 and 5; but as soon as the fluid layers in Zones 1, 2, and 3 were removed or split from the layers of Zones 4 and 5, the vortices in the layers in Zones 4 and 5 could have been more stretched and consequently, their rotation rate increased and I explained later, their orbital inclination also was affected.

The instability of a fluid can affect its turbulence. For instance, when the stratification of a fluid is sufficiently strong, a zigzag instability was found (Billant and Chomaz, 2000). In some stratified fluids, an "instability characterized by a rim region of strong vertical mixing" was found (Cariteau, 2005; Cariteau and Flor, 2003). It was proposed that this instability was probably due to a misalignment of the vortex axis with the direction of stratification (Boulanger et al., 2007). According to the previous authors, when a "*vortex axis is tilted with respect to the direction of stratification, the azimuthal symmetry of the vortex is broken and non-axisymmetric corrections of azimuthal wavenumber are created. For very turbulent fluids, these corrections are characterized by strong density variations and intense axial flow near a particular radial location associated with the critical-point singularity where the angular velocity of the vortex is equal to what is called the Brunt–Vaisala frequency of the fluid*". Critical for shear instabilities (Maslowe, 1986) and for internal wave propagation (Booker and Bretherton, 1967), these singularities were found in the deformation of a vortex by a rotating strain field (Le Dizes, 2000) for which the "*critical-point singularity occurs at a location where the angular velocity of the vortex is equal to the angular frequency of the strain*". These statements suggest to me that the tilting of a vortex can change its morphology in certain directions, and the location of the tilt can depend on the strain in the fluid. Applying this logic to celestial bodies, I think that the location of some tilted bodies may be explained by the strain in the fluid of their precursors and that of the precursors downstream or upstream of them. The stretching of some vortices could have increased the strain and shearing of their fluids. At one point, the forces (e.g. shearing) which acted on some fluids was strong enough to contribute to their overthrowing or tilting.

On September 22, 2020, as I was sitting outside watching my children run around in the backyard as they asked me to do so, and I had to stop working at my desk to satisfy their need, I got an idea about why the rotational axis and orbital plane of some satellites could have been highly tilted or inclined. It appeared to me that the speed of the vortices of the precursors of some celestial bodies could have been divided into 3 components:

- The X component which can be considered parallel to the direction of the shear flow and almost perpendicular to the direction of the revolution or orbital movement
- The Y component which can be assimilated to the direction of the revolution or orbital movement, meaning having a speed in the order of magnitude of the orbital speed, and
- The Z component which is a direction almost perpendicular to both X and Y, meaning in a direction almost perpendicular to the orbital plane or the equator of their primary body in some conditions.

The change (e.g. increase or decrease) of the speed in any of these components could have affected that of the other components. Therefore, as the precursors of the celestial bodies were being slowed down in their X and Y axis, meaning as their orbital speed and shearing rate was decreasing (due to the factors that defined the relationship between orbital speed and semi major axis and their derivatives), the Z component of the speed (e.g. associated with the vorticity) could have increased and could have been translated into the tilting of the orbital plane and/or rotational axis of some bodies impacted by the pulling and pushing. In other words, the axial tilt, eccentricity, and orbital inclination of the vortices of the precursors of the celestial bodies could have related to the magnitude of the vertical and horizontal components of their vorticity. Put another way, the orbital inclination of celestial bodies is connected to the nature of the vertical, horizontal, and axial movement of their precursors.

28.4. Shearing, deformation, stretching, separation, tilting, and elongation of vortices of fluid as a root cause of the diversity of the axial tilt, eccentricity, orbital inclination, and the increase of rotational angular speed of the celestial bodies

When addressing fluid dynamics, the equation of Navier-Stokes has been a reference for more than 100 years. Although the Navier-Stokes equation governing fluid dynamics is not perfect, and efforts are still being made to improve it, it is said to contain a component that points to the stretching and tilting of a vortex. Modern flow visualization and simulations allowed the detection of vortices, but the details of their dynamics are hard to grasp due to the huge number of points or data to resolve before properly having the big picture of the microscopic events affecting a turbulent fluid flow. Nevertheless, turbulence researchers are increasingly agreeing that turbulence is a nest of vorticial structures. According to Prof William K. George, *"a long vortex which is being stretched along its axis clearly spins up, but the rate at which it does so is determined by the large scales that are doing the stretching"* (George, 2013a). These statements suggest that stretching can affect the rotation of a vortex. As the fluid of the precursors was flowing, vortices and vortex tubes could have been formed in some layers of the precursors and carried by the current. As the fluid layers were flowing, they also moved and entrained the vortices in them.

Advection by a nonuniform velocity can cause linear and shear deformation and rotate a fluid parcel; and the rotation rate of the fluid, often called vorticity, is believed to follow a particularly simple and useful conservation law (Price, 2006). Fluids respond differently to shear stress than solids. The previous source mentioned that ordinary fluids (e.g. air and water) have no intrinsic configuration, and they do not develop a restoring force capable of providing a static balance to a shear stress. Consequently, no volume change is associated with a pure shear deformation of a fluid but, to respond to a shear stress, fluids move or 'flow', and they are believed to continue flowing so long as a shear stress is present. The stress applied to the vortices of the precursors and the strain which result from it contributed to shaping some daughter bodies. In other words, the variation of shape, form, and orientation of some bodies could have been caused by the deformation of the fluids of their precursors and how they collected themselves together after breaking from other precursors. The rate of change of the angles or the sides of the vortices could have defined the scale of deformation and the change of shape of their daughter bodies. Some daughter bodies were flipped over, while others were rotated by the forces acting on the fluids as they were gathered together. As the fluid layers were flowing, vortices were formed in them according to the proprieties of the fluids and the intensity of their turbulence. Because turbulence favored the formation of big bodies, I think that in each stack of fluid layers, larger celestial bodies could have been formed at the locations where the turbulence was more developed and where the fluid layers were also larger. Smaller bodies could have been formed in less turbulent fluid layers. For instance, in the fluid layers of the precursor of a satellite system, bigger satellites were formed where the turbulence was stronger. Then, inside a set of fluid layers of the precursor of satellites and rings, a satellite could have been formed where the turbulence was stronger, while the particles and rocks constituting the rings were formed where the turbulence was weaker. I revisited this notion in the chapter on rings.

It has been shown that when a vortex is stretched, its vorticity amplifies, its diameter decreases, and its rotation increases (Petitjeans and Bottausci, 2020). Therefore, I felt like the stretching of the vortex of the precursors of bodies could have also contributed to decreasing the radius and increasing the rotational angular speed of some bodies. For instance, the small radius of some satellites in Zones 4 and 5 could also be caused by fluid threads which stretched a lot before breaking. Furthermore, the stretching of some vortices could have accelerated some components (e.g. axial component) of their vorticity. After separating from vortices in Zone 3, the vortices of the satellites in Zones 4 and 5 could have been destabilized and broken up into small satellites with a higher rotational angular speed. The tilting of some satellites could have been due to the tilting of the vortices of their precursors. The precursors of most satellites in Zones 1, 2, and 3 could have been shaped on almost the same plane, hence their small and similar orbital inclination. Triton (the only satellite in Zones 1, 2, and 3 which axial tilt and orbital plane is highly inclined) could have been an exception; and in other segments below, I explained what happened to the precursor of Triton. By the time the bodies acquired their shape, the tilted vortices transferred their axial tilt to their daughter bodies. Here, when I used the term vortex, I do not mean a tiny filament of fluid, but a gigantic cluster of fluid layers which was curled as how filament of fluids can be curled into a vortex. Similarly, the axial tilt of most celestial bodies can be explained by a shift or change of direction of some components of their vorticity. The intensity of the entrainment and the characteristics of the vortices that were entrained could also have impacted the tilt and orbital inclination of some bodies.

Vorticity can also be generated through the interaction of density and pressure field. For instance, a term of the derivative equation of vorticity is believed to suggest that pressure and density interaction can affect vorticity. Around the region where vortices were tilted, the fluids of the precursors of bodies could have been turned into a direction or sense different from that of "non-tilted" fluids. In the end, the daughter bodies of tilted vortices were tilted. According to the variation of the speed of the vorticity in the downstream fluids and how their daughter bodies split-gathered, other downstream vortices could have been turned into other directions. Hence prograde and retrograde orbits can be found in Zones 4 and 5 according to what happened to the precursors of fluids. The pressure that the precursors of the outermost satellites in Zone 3 imposed on the precursors of the satellites downstream of them affected the vorticity of the latter, leading to the tilting of their orbit and rotational axis and the increase of their stretching, which can be noticed from the huge semi major increment of some bodies in Zones 4 and 5.

During the breakup of fluids in a laboratory setting, some droplets were overturned. For instance, droplet overturning was observed for low-viscosity liquids as well as more viscous liquids (Chen et al., 2002). According to the same authors, droplet overturning is suppressed when the liquid viscosity is sufficiently large such that viscous shear stress is efficiently dissipated into the liquid. These authors also found that the shape of the droplet at pinch-off is closer to its most favorable state – a perfect sphere – hence it is less subject to shape oscillations. The overturning of fluid layers in the boundary can explain the high inclination of some celestial bodies such as the outer satellites (e.g. in Zones 4 and 5). In other words, some bodies could have been overturned because their precursors were overturned when they were being collected together. The small viscosity of the precursors of the bodies in Zones 4 and 5 could have affected the orbital inclination of their daughter bodies. However, because some prograde bodies are also found in Zones 4 and 5, I understood that other factors may have also affected the orbital inclination of the bodies.

In a study on the wrap, tilt, and stretch of vorticity lines around a strong thin straight vortex tube in a simple shear flow (Kawahara et al., 1997), it was shown that a *"simple shear vorticity can wrap and stretch around a vortex tube by a swirling motion, forming double spiral vortex layers of high azimuthal vorticity of alternating sign. The spirals induce axial flows of the same spiral shape with alternate sign in adjacent spirals which in turn tilt the simple shear vorticity toward the axial direction. Consequently, the vorticity lines wind helically around the vortex tube accompanied by conversion of vorticity of the simple shear to the axial direction. The axial vorticity increases and their direction is sometimes opposite to that of the vortex tube when the vorticity magnitude is strongest. A viscous cancellation takes place in tightly wrapped vorticity of alternate sign, leading to the disappearance of the vorticity normal to the vortex tube. Only the axial component of the simple shear vorticity is left there, which is stretched by the simple shear flow itself. Consequently, the vortex tube inclined toward the direction of the simple shear vorticity (a cyclonic vortex) is intensified, while the one oriented in the opposite direction (an anticyclonic vortex) is weakened. The growth rate of vorticity due to this effect attains a maximum (or minimum) value when the vortex tube is oriented in a certain way. The present asymptotic solutions are expected to be closely related to the flow structures around intense vortex tubes observed in various kinds of turbulence such as helical winding of vorticity lines around a vortex tube, the dominance of cyclonic vortex tubes, the appearance of opposite signed vorticity around streamwise vortices and a zig-zag arrangement of streamwise vortices in homogeneous isotropic turbulence, homogeneous shear turbulence and near-wall turbulence"*.

Although the content of this quote is filled with technical jargons and I do not expect you to understand it all, at least, it points to some factors involved in the stretching, wrapping, and tilting of some vortices. I wondered if the zig zag shape mentioned above can help explain the zig zag of the orbital inclination (see chapter on orbital inclination) observed particularly for satellites in Zone 4 (a transition zone out of turbulence), where the stretching of vortices could have contributed to orienting the rotational axis and the orbital planes of celestial bodies born from the fluids in that region and downward. Because the precursors of some bodies were pushed or pulled too strongly, the axial component of their vorticity could have been unable to change much, unless their rotational angular speed or the rotational angular speed of the vortex above or below them hit a certain minimum value, and/or unless the orbital speed in the flow hits a certain threshold. Some bodies with the highest axial tilt are like an elbow through which the precursors of the bodies downstream of them were rotated, turned around, curved, curled, or released so that the configuration of the system they belong to could have been completed. Likewise, it appeared to me that the more the eddies or vortices in a top layer were unable to rotate much, the more the vortices in the layers below it could have tended to be tilted. And when the vortices below another one which is not rotating much are relatively big, the bottom vortex can flip the top one during their separation. Later in this chapter, I explained how this could explain how Triton was highly tilted.

The existence of at least one inflexion point in the profile of a parallel shear flow of an inviscid fluid is known to be a necessary condition for (linearized) instability of the flow (Moffatt, 2011). Because of the inflexion of rotational angular speed at the location of the outermost satellite in Zone 3, I felt like a special instability could have existed around the end of Zone 3 and at the beginning of Zone 4. That instability could have been one of the factors that affected the orbital inclination, eccentricity, rotational angular speed, and axial tilt usually around that zone. To put it in a more global perspective, the mega "vortices" in the fluids of the precursors of the bodies which are downstream of Zone 3 could have felt a sort of release once the precursors of the bodies in Zone 3 separated from them, and as a response, not only did their vorticity change in the direction of the shearing but also in the direction of their axial component (meaning the direction perpendicular to the shearing and the revolution), hence the unusual tilting of the rotational axis with the bodies I called the "hinge bodies, or elbow bodies, or knee bodies" and the increase of the orbital inclination for the bodies involved. As all of that was happening, the shearing of the flow in the precursors which were not separated yet contributed to stretching the mega "vortices" in those fluids until the break up occurred somewhere usually in Zone 4. I used the term "mega" vortices because the size of these fluid structures was astronomical, in the order of many kilometers, not just a few centimeters vortices observed on laboratory settings.

The stretching of vortices in a fluid flow of the precursors could have changed some components of their vorticity in the direction of the stretching and others. The slowing down of the rotational angular speed of some bodies (e.g. in Zone 3) could have increased the stretching or inclination of the axial component of the vorticity of the structures formed in their layers and/or the layers downstream of them. This could have ultimately affected the axial tilt, orbital inclination, and eccentricity of the daughter bodies formed in the affected fluid layers. The abundance of retrograde satellites in Zones 4 and 5 (see the chapter on orbital inclination) suggests that the axial vorticity of the vortex of their precursors could have been strong to a point that it could have been flipped over. It is as if there was a kind of balance between the orbital speed, rotational angular speed, orbital inclination, and the axial tilt of the celestial bodies. The decrease of the rotational angular speed in Zones 1, 2, and 3 could have increased the speed of other components of the vorticity (e.g. Y and Z axis), which in Zones 4 and 5 could have led to the tilting or inclination of the vortices and their daughter bodies. The change in the rotational angular speed of the vortices of the precursors of the satellites around the inflexion point in Zone 3 could have contributed to tilting and inclining the orbit of some satellites in Zones 4 and 5. Some vortices could have merged, making smaller ones bigger, while others could not have merged, but packed close to one another. Some fluid layers of the precursors could have formed sheets which rolled up in sequences matching the movement imposed on them by the turbulence in their environment. While some vortices could have been unstable enough and broke up into smaller ones, others could have been amalgamated to form bigger ones. The way some vortices were organized into tubes or sheets could have affected the stratification of their daughter bodies.

The "inclination of a vortex approximated the ratio of its vertical vorticity to its horizontal vorticity magnitude" (Canals et al., 2009). The same authors also found that, the "ability of tilted eddies to overturn the ambient fluid depends on their inclination, intensity, and the strength of the background stratification". A study on the vertical mixing of coastal ocean showed that "tilted vorticity can lead to turbulence and intense vertical mixing" (Farmer et al., 2002). These findings suggested to me that, on the scale of celestial bodies, the tilting of some precursors could have affected the orbital inclination of others. The tilting of some bodies could have affected the orbital inclination of the bodies located downstream of them. The ability of some strongly tilted eddies to overturn the ambient fluid of their precursors could also explain why the orbit plane of some retrograde satellites were highly inclined. Furthermore, the presence or alternation of prograde and retrograde satellites in Zones 4 and 5 suggests that some initially turned vortices could have been again overturned as they split from their neighbors. Although the speed of the fluid flow of the precursors dominated the rotational speed of the vorticial structures formed, the orbital plane and the rotational axis of the vorticial structures were still tilted, flipped, and inclined.

The fact that most of the retrograde satellites downstream of the one having the highest axial tilt do not have a high axial tilt can be explained by their precursors having adjusted its orientation to align with the flow of the fluids carrying them. It is also possible that the rotational axis of the satellites downstream of the one which has the highest axial tilt were not inclined because the location of the precursors of the body with the highest axial tilt was a place where the plane of the downstream fluid layers was bent as an arm flexes and extends from the elbow. Once the precursors of the satellites downstream of the one with the highest axial tilt flexed, they could have continued flowing as an extension movement toward their respective orbit. Once the highest axial tilt was reached, no significant additional tilt of satellites occurred to change the direction of the flow in any major way, but the impact of the stress and strain in the remaining flow was felt on the orbital inclination instead of the axial tilt. In other words, no instability downstream of the precursor of the satellite with the highest axial tilt was able to significantly tilt again the rotational axis of another satellite. The thinness of the fluid layers of the precursors of the satellite downstream of the satellite with the highest axial tilt could have also contributed to preventing them from forming larger bodies which could pull another one upstream of it. However, in the case of the planets, the large size of their precursors and the way they were distributed could have made it possible to have many planets highly tilted.

CHAPTER 28: VORTEX STRETCHING, TILTING, INCLINATION, ELONGATION, ROTATION, DENSITY, AND SIZE

The way the orbit of some satellites was elongated or stretched could have affected their orbital inclination. For instance, as of 2020, the outermost prograde Saturnian satellite found in Zone 5 interestingly has the smallest eccentricity (0.049) found on any of the satellites in Zones 4 and 5. This suggests that the lack of elongation of the orbit of the precursor of that satellite (and/or the stretching and tilting of its vortex into another direction after breaking from its neighbor) could have favored its prograde motion. How the precursor of that outermost prograde Saturnian satellite was launched could have also worked together to set its orbit from being retrograde as its neighboring satellites to being prograde. The intermittence or alternation of retrograde satellites by prograde ones can also be related to how the orbit of their precursors were flipped over or tilted according to their position with respect to the breakup point of the ligament of their precursor and how they may have been pulled and/or pushed into different directions during their formation. This can also be why prograde satellites with small eccentricity sandwiched between retrograde satellites could have just been caused by precursors pushed inward or downward or flipped over during their stretching and breakup from neighboring precursors, therefore decreasing their eccentricity and flipping their orbit as compared to the retrograde near them. That can explain why the highest eccentricities are found in Zones 4 and 5.

The stretching and tilting of vortex can affect their inclination. For instance, while investigating vortex stretching in a homogeneous isotropic turbulence, Hirota et al. (2017) found that:

"The directions of larger vortices near the segments of the fast-stretching vortices tend to be orthogonal to the direction of the stretching segments, and the locations of the larger vortices that contribute most to the stretching of smaller vortex segments are likely to be found in the direction with the relative angle of 45° from the axes of the stretching vortex segments. And, the vortices with the second highest contributions tend to be in the directions 45° from the stretching segments' axes and orthogonal to the directions of the highest contributing vortices."

Although this research was done in conditions different than those prevailing during the formation of the universe, it points to possible links between vortex stretching, inclination, and tilting. It should not be surprising that around the region covering the outermost satellites in Zone 3 and the satellites coinciding with the innermost peak of semi major increment (meaning most of the inner satellites in Zone 4), retrograde and prograde satellites are usually mixed or alternate, and the distance between the satellites grows as the semi major increases, therefore pointing to a significant stretching which could have occurred in that zone during the formation of the bodies. For more details, please revisit the chapter on semi major increment and that on orbital inclination.

In a study on a stratified layer with horizontal shear (Basak and Sarkar, 2006), it was found that, as a fluid was flowing, some incoherent structures led to the formation of interspersed braids, some of which progressively became more distinct, then thickened by amalgamation while others braid regions thinned by exuding vorticity. The same authors reported that the core of some vortices corotated, paired, and as some vortices evolved to get a more defined shape, the vertical vorticity of structures near their edge were stretched, resulting in the thinning of some nearby braids as other vortices interacted, including merging to form bigger vortices. In the meantime, the vorticity lump of some structures stretched and turned as they interacted with neighboring vorticities, which led to the thinning of some nearby braids. At the same time, the vorticity of some regions amalgamated to form thick vortex cores.

Similarly, considering the characteristics of the celestial bodies, I think that the split-gathering of the fluids in the precursors could have involved some of the steps or processes mentioned above: 1) formation, thickening, thinning, and amalgamation of braids, and 2) co-rotation, stretching, merging, interaction of vortices and braid cores with others as the flow carrying them was being pushed or pulled by the stress or forces communicated to it including that by its mother precursor. As some vortex filaments were forming, they could have weaved or interweaved together with those near them as they were moving, therefore forming braids, some of which could have been amalgamated, wrapped, or collected into bigger ones, while others could have been shattered or scattered, leading to the formation of smaller clusters of particles (e.g. rocks, compounds, or particles) which could have ended up in rings or dusty particles embedding the trajectory or orbit of some bodies. Some braids could have also been formed by the interaction of vortices formed by fluids in tinny adjacent layers, which interacted as their precursors were being moved by the turbulence around them. The movement of the fluid layers as a whole could have wrapped, curled, amalgamated, or collected most of their constituents with a speed at least in the same order of magnitude of the orbital speed of their bodies. Although I used the word "braids" to address structures in the precursors of the celestial bodies, it is important to emphasize that at the astronomical scale of the celestial bodies, structures much bigger than microscopic braids found in a laboratory setting were formed. In other words, I could have used or invented another word to explain how some "mega braids" could have been formed on a larger scale during the formation of the celestial bodies or could have explained the mixing of constituents of celestial bodies. Like, vortices in the order of magnitude of hundreds of even thousands of kilometers were formed.

28.5. Explanation of the axial tilt and eccentricity of the satellites and planets in the Solar System

Armed with what I explained in the previous sections of this chapter, my perception on the cause of the axial tilt, orbital inclination, eccentricity, and rotation of the celestial bodies has improved. Therefore, I felt like I needed to specifically explain the variation of these variables for the planets in the Solar System. I could have engaged in some modern mathematical modeling here, but I felt like it is better to explain the machinery behind the trends of the aforementioned variables. Later I will use that insight to model things.

28.5.1. Why Triton has a very high axial tilt despite its big size and why Nereid located downstream of it has a chaotic motion

Because I suspected that the characteristics of the satellites downstream of the outermost satellites in Zone 3 could help explain the retrograde motion of Triton, I decided to explore them more carefully. I first explored the ratio between the radius of the innermost satellite in Zone 4 and that of the outermost satellite in Zone 3. I found that in the planetary systems of the Solar System, that ratio varies between 0.166% and 12.56% and the highest value was observed between the satellites downstream of Triton and Triton itself (Fig. 153). In other words, the radius of Nereid (the innermost Neptunian satellite in Zone 4) is 12.56% that of Triton (the outermost Neptunian satellite in Zone 3 and the largest Neptunian satellite).

Fig. 153: Radius of satellite *100 / Radius of the satellite upstream of it (%)

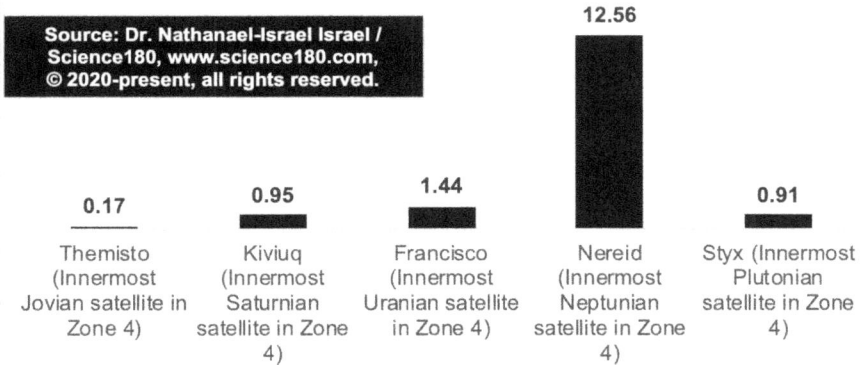

Themisto (Innermost Jovian satellite in Zone 4)	Kiviuq (Innermost Saturnian satellite in Zone 4)	Francisco (Innermost Uranian satellite in Zone 4)	Nereid (Innermost Neptunian satellite in Zone 4)	Styx (Innermost Plutonian satellite in Zone 4)
0.17	0.95	1.44	12.56	0.91

I also explored the ratio of the total kinetic energy of the innermost satellite in Zone 4 with respect to that of the outermost satellite in Zone 3. I found that the total kinetic energy of the innermost satellites in Zone 4 varies between 1.7E-07% and 0.0104% that of the outermost satellite in Zone 3 (Fig. 154). The highest ratio was again observed with Nereid. To put it another way, the ratio between the total kinetic energy of Nereid and Triton is higher than that between the innermost satellite in Zone 4 and the outermost satellite in Zone 3 of any other planetary system in the Solar System.

395
Science180: Source of Unconventional Wisdom and Knowledge on the Origin
of the Universe

Fig. 154: Total kinetic energy of satellite *100 / Total kinetic energy of the satellite upstream of it (%)

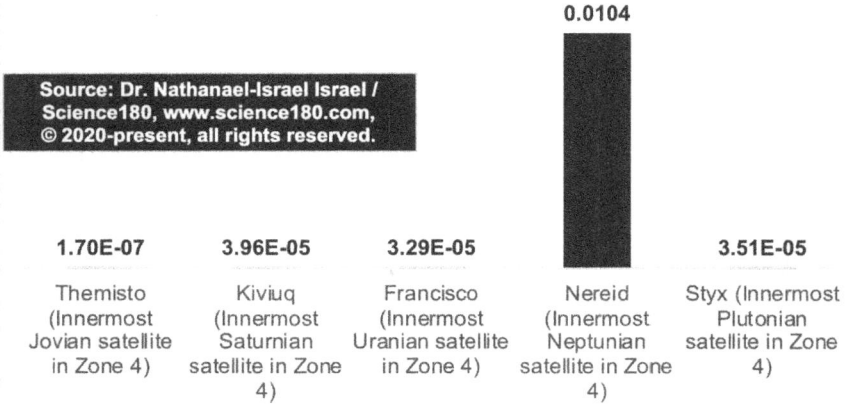

			0.0104	
1.70E-07	3.96E-05	3.29E-05		3.51E-05
Themisto (Innermost Jovian satellite in Zone 4)	Kiviuq (Innermost Saturnian satellite in Zone 4)	Francisco (Innermost Uranian satellite in Zone 4)	Nereid (Innermost Neptunian satellite in Zone 4)	Styx (Innermost Plutonian satellite in Zone 4)

Considering the other data I analyzed, it appeared to me that the relatively high radius and kinetic energy of the precursor of Nereid with respect to that of Triton implies that the precursor of Nereid could have been able to pull the precursor of Triton, and consequently, Triton was flipped over, which in the end, explains its high axial tilt. Furthermore, because the rotational angular speed of Triton was very small (being the outermost Neptunian satellite in Zone 3), the axial component of the vortex of the precursor of Triton could have increased its speed, which consequently could have helped to flip the rotational axis. At the same time, the vortex of the precursor of Nereid could have pulled that of Triton as they were separating. I also suspected that the force associated with the shearing of the fluids downstream of Triton could have had an effect.

The fact that part of the kinetic energy of the precursor of Nereid was used to pull the precursor of Triton can also explain why, unlike the innermost satellite in Zone 4 of the other planetary systems, the highest semi major increment of the Neptunian satellites was not found with Nereid (the innermost Neptunian satellite in Zone 4), for the energy which could have been used to stretch the semi major of Nereid could have been used to pull the precursor of Triton. Indeed, the semi major increment of the Neptunian satellites continued to increase until reaching a first peak with Halimede (the Neptunian satellite downstream of Nereid) after which a major breakup of a ligament could have happened. The pulling of the precursor of Triton by the precursor of Nereid could also explain why Nereid has a chaotic movement. Considering what I already discussed in previous segments about how the precursor of Nereid could have affected that of Triton, I think that the way the vortex of the precursor of Nereid was separated from that of Triton and from that of Halimede (the satellite downstream of Nereid) could have also affected the motion of Nereid. This also implies that, after the precursor of Nereid could have pulled that of Triton, its movement could have been perturbated.

The data also showed that, the rotational speed of Nereid (2.57E-02 km/s) is higher than that of any other innermost satellite in Zone 4 of the planetary systems in the Solar System. It is even higher than the rotational speed of:

- Triton (the Neptunian satellite): 1.67E-02 km/s
- Titan (the largest Saturnian satellite): 1.17E-02 km/s
- Titania (the largest Uranian satellite): 6.59E-03 km/s
- Charon (the largest Plutonian satellite): 6.90E-03 km/s
- Moon (the largest and only satellite of the Earth): 4.61E-03 km/s
- Phobos (the largest Martian satellite): 2.57E-03 km/s

Only 3 Jovian satellites in Zone 3 (Ganymede, Europa, and Io) and one Uranian satellite (Sycorax) have a rotational speed higher than that of Nereid. It then appeared to me that, the vortex of the precursor of Nereid could have been spun around its rotational axis in a special way, hence Nereid has a very high rotational speed. The high rotation of the precursor of Nereid and the conditions surrounding how the shearing of the flow could have tried to flip it, and how its axial speed could have tried to increase due to the slow rotational angular speed of the precursor of Triton, could also help explain why Nereid ended up having a chaotic motion.

However, it is important to mention that Nereid is not the only satellite having a chaotic motion and which is located downstream of a large satellite. For instance, Hyperion, a Saturnian satellite located immediately downstream of Titan (the largest Saturnian satellite) has a chaotic motion. Nix (the second innermost Plutonian satellite in Zone 4) and Hydra (the outermost Plutonian satellite) both located downstream of Charon (the largest Plutonian satellite) have a chaotic motion. Because, unlike Nereid and Hyperion (which are both located immediately downstream of the largest satellite in their planetary system), neither Nix nor Hydra are located immediately of Charon (the biggest satellite in their planetary system), I deduced that the chaotic motion of satellites is not just caused by their proximity to a large satellite. Else, Styx (the innermost Plutonian satellite in Zone 4) and Calisto (the Jovian satellite immediately downstream of Ganymede (the largest Jovian satellite) could have a chaotic motion, but they do not.

The size of the 2 satellites downstream of Triton suggested that, as the precursor of Triton was "sitting" on top of the precursors of these 2 satellites and was not rotating much, the vortices in the precursors of these 2 satellites or at least the one immediately below Triton, which is that of Nereid, was "released" and as it was about to relax, stretch more and tilt to adjust its altitude or the axial component of its "vorticity", it could have tapped the precursor of Triton and flipped it or pulled it over and/or the precursor of Triton was ejected or fell back or forward according to the direction or sense of the axial tilt that the scientific community would like to adopt. In contrast, because the precursors of the innermost satellite in Zone 4 of the other planetary systems were not big enough, their vortices could have not been able to tap, flip, or tilt the vortex of the outermost satellite in Zone 3. Hence it was only with Triton that such a tilt of a big satellite occurred. The shear rate or deformation rate (of both the rotational angular speed and orbital speed with respect to semi major) of the fluid around the precursor of Triton could have predisposed the precursor of Triton to be affected by its flipping over or what happened to the fluids in it and/or beneath it.

28.5.2. Axial tilt of Venus and Earth

Based on what I explained above for the satellites, I think that the high axial tilt of some planets can be explained by how their precursors were separated from the precursors of the planets adjacent to them. In addition to what I already said, the precursor of Venus could have been flipped over because it was unable to rotate much and its vortex was flipped over to increase the "vorticity" component associated with the rotational axis (e.g. the Z component or direction). Here, I used the term vorticity (instead of others like circulation) because I did not find any better way to explain the 3D motion of the precursor of Venus as it was being collected as a planet. The precursor of Venus could have separated from that of Mercury (located upstream) and that of the Earth-Moon system (located downstream) so strongly that it was flipped over.

Considering the full picture of the connection between the variables, I felt like the way the precursors of some bodies split from one another and the characteristics of the precursors downstream and upstream of them could have affected the axial tilt. It appeared to me that, when the precursor of Venus was splitting from the precursor of the planet downstream of it (meaning the precursor of the Earth), it may also have pulled the precursor of the Earth. By the time the precursor of Venus collected itself together and finalized its shape, its rotational axis was tilted. The precursor of Mercury also pulled the precursor of Venus, but I felt like the pulling of the precursor of Earth could have also tilted or flipped over Venus. The slow rotation of the precursor of Mercury on top of the precursor of Venus could have affected the vorticity the precursor of Venus and in the end, the response was the tilting of Venus.

When I looked at images of axial tilt in the literature, I felt like the rotational axis of the Earth is tilted downstream of the direction of the shear flow of the precursor of the bodies orbiting the Sun (meaning away from the Sun), while the rotational axis of Venus seems tilted inward, meaning toward the Sun or against the shear flow of the precursor of the bodies orbiting the Sun. This can be due to the fact that the precursor of Venus and the precursor of the Earth-Moon system were sent into two different directions after they separated from one another just as 2 persons holding and pulling a rope in 2 different directions could fall if the rope is cut. Because of the high speed of the precursor of Venus, it could have been projected inward (meaning against the shear flow carrying the precursors of the bodies downstream of Venus) too much and consequently

Venus was flipped over, hence its high axil tilt. In other words, at one point, when the precursor of the Earth-Moon system was separating from that of Venus, there could have been a time when they were connected by a ligament of fluid. That ligament could have been stretched until it broke, sending Venus backward and Earth forward, meaning that after the breakup of the ligament linking the precursor of the Earth-Moon system from the precursor of Venus, the precursor of the Earth-Moon system could have been propelled or pushed forward or toward the direction of the shearing of the fluid of the secondary bodies orbiting the Sun.

It is also possible that the small rotation of the vortex of the precursor of Venus could have also contributed to accelerating the Z component of the "vorticity" of that vortex, hence the precursor of Venus gained speed in the axial direction, leading to a higher rotational axis. However, although the precursor of Venus could have had a small rotation, it could not have caused the vortex of the precursor of the Earth to gain speed in the axial direction, but to gain speed in the X and Y direction, hence the Earth has a very high rotational speed and a relatively small axial tilt.

28.5.3. Axial tilt of Mercury

Although its rotational speed was small, the precursor of Mercury was not flipped over but was pushed, or pulled, or stretched more, which elongated its orbit, giving Mercury the highest eccentricity recorded on any planet in the Solar System. In other words, the precursor of Mercury could have increased the component of its vorticity in the direction of the shear flow, hence its higher eccentricity. The rotational axis of Mercury was not increased despite the small rotational speed of Mercury because no other significant precursor was sitting on top of the precursor of Mercury. Consequently, because the vortex of Mercury could not be accelerated in the Z or axial direction, its orbit was elongated in the direction of the shear flow of the precursors of the bodies orbiting the Sun, hence Mercury ended up having the highest eccentricity recorded on the planets in the Solar System. In other words, although Mercury also has a low rotational speed, the effect of the shearing or pulling by the fluid flow downstream of it could have been felt just by its eccentricity.

28.5.4. Axial tilt of Mars

Mars could not have been tilted much because the precursor of the Martian planetary system could have been influenced by the precursor of the Main asteroid belt, which is not much heavy or energetic. Additionally, the small energy of the Martian planetary system could have reduced its tilting.

28.5.5. Axial tilt of Jupiter and the Jovian satellites

Jupiter could have resisted the impact of the stretching and/or pulling by the precursor of the Saturnian planetary system because of its huge mass and energy. As they were being formed, some bodies could have resisted the tilting, maybe due to the viscosity of the fluids of their precursors or their position in the shear flow. The precursor of Ganymede (the largest Jovian satellite) may have resisted much of the pushing of the satellites upstream or the pulling of the satellites downstream of it, but because it could have been pulled by a very energetic vortex and fluid flow downstream, it was tilted just a little bit. Therefore, despite being the largest Jovian satellite, Ganymede was the most tilted of all of the Jovian satellites. Europa (the Jovian satellite upstream of Ganymede) was a little bit tilted because the vortex of the precursor of Ganymede was strong enough to pull it. No other Jovian satellite was tilted because the precursor of the satellites downstream of them was not strong or energetic enough to pull it. The viscosity of the Jovian satellites could have reduced the extent to which the rotational axis of the Jovian satellites was tilted. In the next chapter, I also explained how viscosity also played a role in the formation of the bodies.

28.5.6. Axial tilt of Saturn and the Saturnian satellites including Phoebe

Similarly, the precursor of Saturn could have been pulled by that of Uranus and the tilt of Saturn could been affected by the characteristics of Uranus. Because the precursor of Saturn was not as heavy or energetic as that of Jupiter, Saturn was more tilted than Jupiter.

The precursor of Titan (the largest Saturnian satellite) was pushed much farther away as compared to that of any Saturnian satellite located upstream of it, causing it to have a high semi major increment. Titan is not tilted because the vortex of the precursor of the satellite downstream of it was not strong or energetic enough to pull it. But a tilt later happened downstream with the rotational axis of the precursor of Phoebe (the largest Saturnian satellite in Zone 4). Because the precursor of Phoebe could not resist the tilting of its rotational axis, it was flipped over. The nature of the shearing at the vicinity of the precursor of Phoebe could have also affected the movement of its axis and orbit. The data on semi major increment also suggested that some precursors were stretched and pushed farther away rather than being flipped or tilted over.

I cannot finish talking about Saturn and its satellites without tackling why Phoebe has the highest axial tilt among all of the Saturnian satellites. Considering what I explained above for Triton and Nereid, I think that Kiviuq (the innermost Saturnian satellite in Zone 4) was not able to tilt Iapetus (the outermost Saturnian satellite in Zone 3) because the size and energy of its precursor was too small. Among the Saturnian satellites in Zones 4 and 5, Phoebe is the biggest. Knowing that no Saturnian satellite in Zones 1, 2, and 3 has an eccentricity and an orbital inclination as high as that of Kiviuq (eccentricity = 0.334 and orbital inclination = 48.7°), I think that something happened to the fluid of the precursor of Kiviuq to elongate, or stretch, and incline its orbit. The energy which was used to stretch or push away the precursor of Kiviuq from that of the outermost satellite in Zone 3 and to tilt its orbit could have prevented its rotational axis from being tilted. Similarly, Ijiraq (the second innermost Saturnian satellite in Zone 4) has a very high eccentricity (0.3163), but a relatively small orbital inclination (49.1°). The energy used to elongate its orbit to give it such a high eccentricity could have prevented it from tilting the rotational axis. However, the eccentricity (0.164) of Phoebe (the 3rd innermost Saturnian satellite in Zone 4) is smaller than that of the 2 innermost Saturnian satellites in Zone 4 (Kiviuq and Ijiraq) that I just talked about, but its axial tilt is the highest (152.14°). I think that the precursor of Phoebe instead was not pulled or pushed too much by the shear flow (hence a small eccentricity), but its rotational axis was tilted, while its orbital plane was flipped over. Hence Phoebe not only has the highest axial tilt, but also one of the highest orbital inclinations (174.8°) recorded on the Saturnian satellites. This underlined how eccentricity, orbital inclination, and axial tilt of the satellites connect to one another. The relatively high semi major increment of Phoebe as compared to that of the satellites in its vicinity suggested to me that the distance over which its precursor flowed before being collected could have also played a role in its characteristics. Indeed, the semi major axis increment of Ijiraq (the satellite upstream of Phoebe) is 10,000 km, while the semi major axis increment of Phoebe is 1,824,000 km, meaning that the semi major axis increment of Phoebe is 182.4 times that of Ijiraq. In other words, after splitting from the precursor of the satellite (Ijiraq) upstream of it, the precursor of Phoebe traveled a distance equivalent to 182.4 times that travelled by the precursor of that satellite (Ijiraq) after it (Ijiraq) split from the precursor of the satellite (Kiviuq) located upstream of it (Ijiraq). A secret is hidden here!

28.5.7. Axial tilt of Uranus and the Uranian satellites

Considering the axial tilt of Uranus, it sounds to me that the precursor of the Uranian planetary system was flipped over after splitting from the precursor of the Saturnian planetary system (located inward or upstream) and the precursor of the Neptunian planetary system (located downward or downstream). It is possible that the precursor of the Neptunian planetary system could have pulled the precursor of the Uranian planetary system as they were separating. Although it is possible that the precursor of the Uranian planetary system were tilted after breaking away from the precursor of the Saturnian planetary system, I felt like the pulling of the precursor of the Uranian planetary system by the precursor of the Neptunian planetary system better explains the tilting of the Uranian planetary system. I felt like the precursor of the Uranian planetary system could have been tilted before its bodies (primary planet, satellites, and rings) were formed. Another way of better saying this is that, as the bodies in the Uranian planetary system were being formed, the precursor of that planetary system was already tilted or being tilted and therefore, it transferred that tilt to its daughter bodies. Hence, Uranus and the Uranian satellites orbit the Sun like a rolling ball. The orbital plane of the Uranian satellites could have been tilted due to the impact of the pulling of their precursor by the precursor of the Neptunian planetary system. The precursor of the Uranian planetary system could have been tilted before its planets and satellites were born; hence they mostly moved in a rolling manner. In other words, people may think that the Uranian planetary system may seem as knocked out on its side but as for me, it appeared that its precursor was pulled over by the energetic precursor of the Neptunian planetary system, which was located downstream of it. The precursor of the Saturnian planetary system which was located on top of the precursor of the Uranian planetary system could have also played a role, but nothing called knockout or collision ever played a role in the tilting of Uranus!

Finally, the rotational axis of the Uranian satellites were not tilted because, among other things, the entire Uranian planetary system was already flipped over and Uranus could have absorbed what could have tilted its satellites.

At the time of Voyager 2's flyby in 1986, Uranus' south pole was said to have been pointing toward the Sun. Its north's pole was said to have been under the elliptic as if Uranus was pushed or pulled outward away from the Sun.

28.5. 8. Axial tilt of Neptune

Neptune was not as tilted as Venus, Uranus, or Pluto because the precursor of the Plutonian planetary system (located downstream of it) was not able to pull the precursor of the Neptunian planetary system much and/or the precursor of the Neptunian planetary system resisted the pull due to its high energy.

28.5.9. Axial tilt of Pluto and the Plutonian satellites

The splitting from the precursor of the Neptunian planetary system could have pushed the precursor of the Plutonian planetary system outward, hence it was pushed forward. Although it is also possible that the precursor of some asteroids such as Haumea (radius = 743.67 km and semi major axis = 43.22 AU) located at 3.74 AU downstream of Pluto could have pulled the precursor of the Plutonian planetary system, hence the rotational axis of Pluto was very tilted in the end, I personally think the fact that the vortex of the precursor of Pluto was unable to rotate much contributed to the tilting of its rotational axis and its orbit, hence Pluto has a high axial tilt. Seeing another perspective, the precursor of the Plutonian planetary system could be seen as pulled instead of pushed!

28.6. Clustering of vortex to form celestial bodies and explanation of the cyclones encircling Jupiter's poles and of Saturn's polar hexagon

After the precursors of the celestial bodies in the fluid layers separated, they could have continued moving independently until their final shape and position of orbit was "completed". After breaking from their mother and the ligament which could have connected them to neighboring precursors, the precursors of some bodies could have continued their movement (including stretching) before breaking up later. The rotating eddies in the precursors could have gathered together and reached their "final" shape, while their constituents were mixed with others. At one point, the configuration of the materials in their eddies could no longer change, and as the temperature on the surface of some of these bodies dropped due to environmental factors, some bodies could have solidified at their surface, while deep inside, toward their center, some fluids could still exist, which can explain the presence of lava or magma inside some bodies. That was how some celestial bodies broke from their precursors, solidifying at the surface, but having fluids (e.g. lava and other fluid-like materials) at their center. Big bodies which overcame the cold of their environment and could not be cooled down kept their fire or heat such as stars. Hence stars are not solid but are usually big and less dense. In the chapter on galaxies and stars, I also detailed how the physics of vortex also helped me to understand the formation of the diverse types of galaxies, stars, and globular clusters.

Although I tried to explain how adjacent vortices in layers or threads of fluid could have interacted, it is important to notice that the precursors of some celestial bodies consisted of the amalgamation, wrap, or collection together of many vortices. Some larger or bigger vortices could have contained smaller ones, which at their turn, could have contained other smaller vortices and so one and so forth until the smallest vortices were reached. This hierarchy or inclusion or fragmentation of the precursors can also explain the presence of smaller bodies or systems of bodies inside bigger and more complex ones. That is why for instance, a stellar system can contain planetary systems, whereas a planetary system can contain a planet and satellites, while a planet can contain many vortices. While some people may use the term "nesting" to address what I just said, I chose to stick with my simple vocabulary, hoping it can help more people to get my point. In my book on the origin of life, I also addressed how things I handled in this chapter and elsewhere in this book also apply to the organization, growth, branching, and many other characteristics of living things, but with specific scaling factors, constraints, opportunities, and turbulence rules.

On the scale of particles or fluid elements, vorticity can explain the curling or 3D movement, but on the astronomical scale or that of an entire layer of the precursors of the celestial bodies, the layers were wound up, wrapped up, "folded" or collected to form well defined and structured celestial bodies of different sizes. Just as a vortex is a compact region of vorticities, the precursor of a celestial body was a compact region of one or more vortices of various sizes. The magnitude of the squeezing, compaction, or compression of the vortices of the precursors of the celestial bodies affected the density and many other parameters of the celestial bodies. Celestial bodies which were strongly or tightly squeezed are denser than those which are loosely squeezed. The precursors of some celestial bodies were very turbulent, birthing large bodies, but because their vortices were not highly squeezed, the density of their bodies is generally small. However, because the correlation between density and radius of the bodies is not perfect, other factors besides radius contribute to explaining the density of the bodies.

While some vortices could have merged, others did not. Under some circumstances, the matter in some vortices solidified, forming rocks or clusters of minerals. In other conditions, some vortices stayed unchanged while others were squeezed even until today. Some of those vortices can be found in some celestial bodies which surface is not solid. For instance, the Jovian poles are invisible from Earth (using a telescope), and most of the Jovian spatial missions have focused on the equatorial plane instead of the poles as Juno spacecraft did. Indeed, thanks to the Juno spacecraft which has been orbiting Jupiter since 2016, observations at Jupiter's poles by 2018 provided some insights into what may be lying underneath its surface (Adriani et al., 2018). These authors reported that the northernmost and southernmost regions of Jupiter have large cyclones. According to that reference, visible and infrared images taken from above each pole by the Juno spacecraft during its orbits of Jupiter revealed persistent polygonal patterns of large cyclones arranged in a "*strange pattern*":

- At the north pole, eight circumpolar cyclones are observed about a single polar cyclone. In other words, eight cyclones surround one large one in the north while
- At the south pole, one polar cyclone is encircled by five circumpolar cyclones, meaning that five cyclones surround a central cyclone in the south.

How these cyclones were formed, evolved, and persisted without merging until today is unknown. Considering what I already explained about the formation of the universe, I think that the origin of these cyclones or vortices is related to the turbulence during the genesis of the universe, and particular during the formation of these planets. Those cyclones are the consequence of the vortices which were formed in the precursor of Jupiter and/or could be new vortices formed after the formation of the universe. In other words, these cyclones could have been present for many years, even centuries, or even since the formation of Jupiter. The fact that the configuration of the cyclones at the north pole are different than that at the south pole of Jupiter suggests that the turbulence at these poles has not been the same. These vortices could have influenced one another, and the waves they generated could have contributed to pushing them and locking them into their current positions. The flow of their precursors can explain why the position of these cyclones can change with time. If certain conditions were met, the constituents of these vortexes could have solidified and formed a crust. However, because the conditions did not allow the formation of a solid body, these vortexes stayed like a fluid in a kind of "dynamics equilibrium" with one another and with their environment. Finally, the encirclement of one cyclone by many other circumpolar cyclones seems to imitate the organization of systems of bodies into a primary body orbited by secondary bodies.

Vorticial clouds were also found at the north pole of Saturn. The vortex at the north pole of Saturn formed a hexagon which is referred to as Saturn's hexagon. You can visualize the hexagonal vortex at Saturn's north pole by searching for them online. The origin of this mysterious six-sided weather-like pattern has puzzled the scientific community since its discovery. The difference of the intensity of the turbulence in the precursors of Jupiter and of Saturn can also explain the difference in the configuration of their polar vortex. The vortex of Saturn's pole may not contain many visible cyclones because the turbulence in the precursor of Jupiter could have been stronger than that in the precursor of Saturn. Consequently, the vortices in Jupiter and its precursor could have been bigger and stronger than those in Saturn and its precursor. The difference in the density and viscosity of the precursors of these planets could have also played a role in the characteristics of their vortices.

Because the celestial bodies were formed and shaped inside a fluid flow, as their precursors were going through "metamorphic" changes, some fluids in the flow which embedded or surrounded them birthed smaller particles, rocks, and dusts which in the end appear as trails of dusts or rings or grainy particles surrounded by some celestial bodies or in the path of their orbit. That is why even planets like Venus and Mercury, which have no satellite, are also embedded in trails of "dusts" or particles appearing as a ring of the trajectory of their orbit.

I wish I could expound on this a little longer now, but the day I was getting this inspiration, I had to go to my living room to talk to my kids because, I was writing this section on my birthday, and I heard them loudly saying in their bedroom that "*today is daddy's birthday*". I naturally do not celebrate my birthday but, they have been slowly converting me into a "birthday celebrator". Therefore, to please them, I stopped the flow of ideas in my mind to go show them some love and enthusiasm. As I later realized, the flow of ideas about this topic never came back to me again yet. The day my mind was first enlightened by what I said in this chapter, I felt like the data I was handling was begging me to listen to them and, as I opened up myself, I could hear them as if they have waited for thousands of years to be heard by a human being, but no one ever really paid attention to them as I did. Since that day, I had my own mysterious "proof" that nonliving things can speak, but the problem is that we, human beings, do not want to listen and we think that we are the only entities which are smart or intelligent. In the next chapter, I also explained how viscosity also played a role in the formation of the celestial bodies and in some of the things I handled in this chapter.

28.7. Take home message

If you have properly read this chapter until this point, you would have really understood why, after my keynote speeches (www.Science180.com/speaking), people always rush to me to thank me for wholeheartedly sharing with them (and now with you) the colossal amount of priceless secrets I revealed in this chapter, and which we are going to use to break so many myths about the universe-origin in the next chapters.

Because it tackled questions connected to many variables, this chapter contains codes to decrypt the meaning and significance of the variation among certain variables (e.g. eccentricity, axial tilt, rotation, orbital inclination, semi major axis, and radius) fundamental to the understanding of the origin of the universe with a turbulence perspective. I showed that the way the fluid layers of the precursors of the bodies sheared over one another could have affected the deformation rate of their daughter bodies. The squashing of the fluid layers of some precursors can explain why the rotational angular speed of some bodies decreased as the semi major axis increased. The strain and stress in the fluid of the precursors of bodies can explain the location of some tilted bodies. The stretching of some vortices could have increased the strain and shearing, overthrowing, or tilting of some fluid layers. The tilting of some bodies could have been due to the tilting of the vortices of their precursors. Around the region where the vortices were tilted, the fluids of the precursors of bodies could have been turned into a direction different from that of "non-tilted" fluids. According to the variation of the speed of the vorticity in the downstream fluids and how their daughter bodies broke up, other vortices downstream could have been turned again in other directions. Hence prograde and retrograde orbits can be mixed with one another according to what happened to the precursors of their fluids. The way the precursors of some bodies squashed the precursors of others affected the vorticity, tilting of orbit and rotational axis, and the increase of the stretching of some downstream bodies. I also explained why the precursors of some celestial bodies solidified at their surface, while others did not, but ended up forming stars or giant gas, which are less dense than terrestrial planets. The precursors of some celestial bodies birthed large bodies, but because their vortices were not highly squeezed, the density of most of these bodies is generally small.

CHAPTER 29

DID ALBERT EINSTEIN AND ISAAC NEWTON REALLY MISS THIS INCREDIBLE LINK? CHECK OUT WHAT THESE CRITICAL VARIABLES ARE SAYING ABOUT HOW TO CRACK THE UNIVERSE-ORIGIN?

- Is the "war" between mathematics, physics, and faith drifting science in wrong directions?
- How can mathematicians and scientists enhance their methodology of research for the benefit of humankind?
- How does Science180 cosmological framework profit mathematicians, scientists, and even philosophers?
- What is the one thing scientists and mathematicians need to understand about their theories and how they approach reality?

In this chapter, you will discover how 5 little-known variables or concepts (i.e. ligament, necking, pinching off, breakup, and viscosity of fluids) worked together to help me crack a major portion of the code of the universe formation. People I trained at Science180 Academy (www.Science180Academy.com) usually thank me for revealing the secrets in this chapter with them. I am here to help you decode those essential secrets too.

Indeed, during my investigation of the origin of the universe, as I explored the trends and distribution of the radius and semi major increment of celestial bodies, I felt like the way that their precursors were split-gathered follows some patterns found with the breakup of fluids in laboratory settings. In this chapter, I will use what is known about the physics of fluid breakup to shockingly explain how the fluids in the precursors of the bodies in the universe were broken up and how some of the processes involved could have affected their distribution and the distance separating them.

29.1. Necking and pinching off of the precursor of the bodies

A fluid ligament is a "more or less columnar object (a jet) attached by its foot to the liquid bulk from which it has been stripped" (Joseph et al., 1999. A neck of fluid is a thread like area connecting two bodies of fluid which are about to break up. As I looked at the bodies in the Solar System, I felt like when their precursors were splitting from one another, fluid ligaments were formed between some of them. The separation of the precursor of secondary bodies from that of their primary bodies involved processes such as necking, pinching off, stretching, etc. For instance, in each system of bodies, as the fluids of the precursor of the secondary bodies were leaving their mother precursor to form a ligament of the precursors of the secondary bodies, a neck was formed to connect the ligament to the secondary bodies to the ligament of the primary body. Another way to say this is that as the mother precursor of a system of bodies was splitting, the ligament of the precursors of the secondary bodies was connected to the ligament or mass of the precursor of the primary bodies by a neck. Just as the neck of a human being connects the head to the rest of the body, a neck of fluids connected 2 bodies of fluids. That neck would have stayed in place and allowed the fluid of the mother precursor to flow into the ligament of the secondary bodies until a moment was reached for the ligament of the secondary body to pinch off from the precursor of the primary body and for the neck to break. Fig. 55 presents a ligament of fluid showing a neck connecting 2 bodies of fluid.

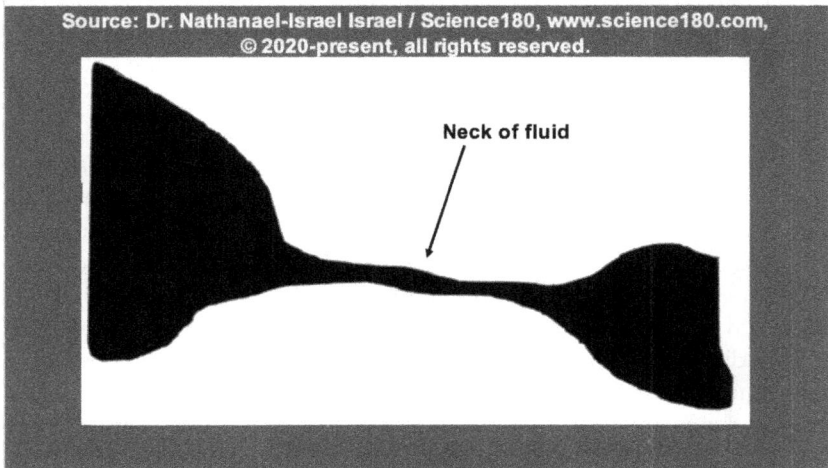

Fig. 155: Ligament of fluid showing a neck connecting 2 fluid bodies

Neck of fluid

Nathanael-Israel Israel: Has had the Honor to be Acknowledged the First Human Being that Scientifically Reconciled Science and Biblical Creation

Another way to illustrate this is that, just as water leaving a faucet or a nozzle forms a filament (or drip of water) attached to that faucet or nozzle and then forms a neck, so also when fluids were leaving the fluid mass of a mother precursor to go into its daughter bodies, a neck was formed to connect the precursor of the primary body to that which would become its secondary bodies. Although a fluid neck can sometimes rapidly retract to a faucet and prevent breakup, in other cases, it cannot do so but will destabilize, detach, and disintegrate into several small satellite drops. Similarly, after the neck connecting the precursor of the secondary bodies detached from the precursor of the primary body, the precursor of the secondary bodies went through processes which split-gathered it into daughter bodies along the current of the flow.

In experiments where water is released from a nozzle, the minimum distance from the nozzle over which the liquid jet is still connected is usually called liquid intact length. Although the movement of the fluid in a mother precursor toward the precursor of the secondary bodies was a completely different scenario, if a comparison could be made, the equivalent of the liquid intact length for the secondary bodies in each system could have been smaller than the semi major axis of the innermost secondary body.

As the neck connecting the precursor of the fluids destabilized under turbulence, it could have broken up and freed the fluid ligament connecting it to the source of the flow. At the time a fluid neck broke up, its radius could have gone to zero, for if the radius did not near zero, a breakup could not happen. As the thread of the neck was thinning, it could have broken at many places under some conditions, leading to the formation of many smaller bodies instead of breaking neatly at only one point. Similarly, as the thin thread connecting bigger vortices or precursors or clusters of fluids broke up, it could have also led to the formation of additional small bodies between the bigger one. These processes could have contributed to the intermittence of bodies in the universe. Like I explained later in this chapter, the size of the bodies formed around the neck or the breaking point contributed to the intermittence of the bodies in the systems of bodies in the universe.

After the ligament of the precursor of the secondary bodies broke up from the ligament of the precursor of their primary body, it continued its journey until all its bodies were fully formed, released, and put in orbit. To start flowing from a mother precursor toward the daughter bodies and form a continuous fluid jet, the fluid had a speed which was sufficiently large enough to allow its kinetic energy to overcome the surface energy and the precursor of the surface energy. For if the fluid did not have or was not propelled or ejected with enough energy, it could not have been able to "hand over" the energy holding it to its other precursors and consequently, could not have broken up from it. After breaking or pinching off from the larger fluid mass of its primary body, the fluid in the precursor of the secondary bodies could have stayed as a single mass or break up into many bodies according to many factors including the recoil disturbance imposed on it by the separation of the main droplet or other environmental factors.

For instance, as the precursor of the Solar System was split-gathering into the precursor of the Sun and the precursor of the bodies orbiting the Sun, a neck was formed and connected these 2 types of precursors. After the fluid of the mother flowed into the ligament of the precursor of the bodies orbiting the Sun, a pinch off occurred so that the precursor of the Sun and that of the bodies orbiting could be separated. Similarly, during the split-gathering of the precursor of a planetary system, a neck was formed and connected the precursor of the primary planet and the precursor of its satellites. After the ligament of the precursor of the secondary bodies escaped the precursor of the primary body, it later went through other stages until all the secondary bodies were formed. In other words, by the time the precursor of the satellites pinched off from the precursor of their primary planet, all that was left for the mother precursor was the precursor of the planet. Another way of saying this is that after the precursor of the satellites pinched off to be molded into its distinct satellites, the precursor of the primary planet was released and had to go through some stage of shaping to become a primary planet. By the time that all of the precursors of the bodies were separated, the flow stopped. In some conditions, some precursors of the secondary bodies could have separated from one another before the neck connecting the ligament of the precursor of the secondary bodies broke up from the precursor of the primary body. For instance, some satellites could have been formed and separated from one another before the neck connecting the primary planet to the satellites broke up. This also implies that some outer bodies could have been formed and split before some inner ones could be separated.

Just as artificial satellites launched from Earth have to be escorted into orbit following specific procedures or maneuvers, so also from the time they were separated from their mother precursor and later from other daughter bodies (of their mother) adjacent to them (e.g. located upstream and downstream of them), celestial bodies went through changes and processes which "escorted" them into their orbit. After their separation from the ligament of fluid connecting them to others, the blobs and vortices of the precursors could have continued their movement until their configuration or shape was "settled" when the degree of freedom of some of their deformations could have been reduced enough for them to relatively "stop" changing. By that time, the bodies could have been positioned in their orbit or trajectory path and a sort of equilibrium could have been found between their characteristics and environmental conditions. Another way to say this is that, from the time the daughter precursors departed their mother and/or split from their mother precursor to the time they "arrived" at their "final" location or destination in orbit, their constituents and form could have been changing. By "final" location or destination, I did not necessarily mean that the position of bodies in the universe is static, but I mostly wanted to signify that the location on orbit from which their characteristics could have been almost similar to what they are today. For, although bodies in the universe have been changing their position since the early moment of their genesis, there was a time or position in the past at which their current identify or characteristics were nearly reached. Similar processes could also have happened for subatomic particles, atoms, and other types of systems of bodies on smaller length scales. While on a laboratory scale, blobs can be seen in ligaments of fluids, on the scale of the celestial bodies, bigger mass of bodies or vortices could have been present in the ligaments.

To sum it up, necks were not formed only between the precursors of primary bodies and the ligament of the precursor of their secondary bodies, but also between the precursors of consecutive secondary bodies as they were breaking from one another. In other words, detachment or pinching off at the base of the ligaments of the precursors could have occurred not only between the fluid bulk connecting a mother precursor to the bulk of the precursors of its daughter bodies but also between daughter bodies of the same mother. The time it could have taken to complete the pinching off or the breakup of the neck connecting the ligament could have depended on the intensity of their turbulence, the speed, size, volume, viscosity of their fluids, and other factors involved in their stretching and breakup including forces involved (e.g. "capillary" forces). Below, I provide additional details about how the precursors of the satellites could have broken up.

29.2. Formation of conical tips between the precursor of some bodies during their breakup from one another

The following graph (Fig. 156) shows a conical shape in a ligament of fluids before and after a breakup.

On a laboratory scale, surface tension was shown to play a crucial role in the pinching of a drop in a laboratory setting. For instance, *"for large capillary numbers, the effect of surface tension is initially small and the drop stretches following the simple shear flow, elongates, until eventually, the cross section of the drop evolves to a circular shape, length scales are reached at which surface tension becomes important, and pinching begins"* (Cristini and Renardy, 2006). The pinch off involves many stages. The last stages of pinching are very fast, far below the time resolution of the eye; as pinching progresses and the radius of the jet gets smaller, the time scale becomes shorter and pinching precipitates to form a drop in finite time (Eggers, 1997). The same author reported that at the pinch point, the radius of curvature goes to zero, and the small amount of fluid left in the pinched region is driven by increasingly strong forces. Experimental observations showed that, regardless of initial conditions, free surface shapes of fluids are similar near the pinching point (Eggers, 1997). Furthermore, that same author reported that at the pinching point of water jets, a conical tip was attached to a sharp front. For high viscosities, threads are formed. After breakup, the way the end of the neck recedes, recoils, or retracts, can affect whether or not and how it will break to lead to smaller drops between bigger ones. In other words, the destiny of the fluids near the breaking point is not always the same and can depend on many factors approached below. To the question of the intermittence of the size of the bodies, some studies suggested that big bodies alternate generally with small ones probably because of "*an asymmetrical evolution of the interface of the precursors near the pinch-off region into cones which would have had different angles during the final stages of pinch-off*" (Lister and Stone, 1998).

Fig. 156: A ligament of fluids showing masses of fluids before and after a breakup showing the conical shape formed during the breakup of a neck of fluid

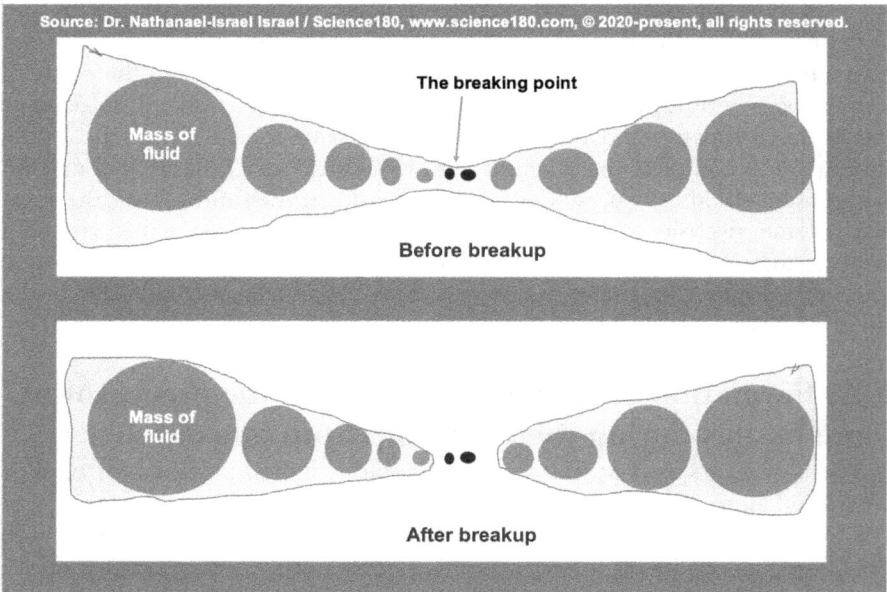

The study of the distribution of the semi major increment and radius of satellites with perspective to the cones that are said to usually form around breaking points gave me an insight into the existence and location of some breaking points. Indeed, I noticed that the highest semi major axis increments are generally near the bodies with the smallest radius. Between the biggest bodies and the smaller ones, the shape of the radius points to cones that could have been formed by the precursors of the bodies during their breakup. In other words, the trends of the radius and semi major axis increment of the bodies suggested to me that the breakup of some precursors of bodies could have occurred around the location of some bodies with the smallest radius whereas the coalescence of the fluids in the ligament of some precursors after breakup could have occurred

Nathanael-Israel Israel: Has had the Honor to be Acknowledged the First Human Being that Scientifically Reconciled Science and Biblical Creation

around the biggest bodies situated between the smaller ones, particularly in Zones 4 and 5. In Zone 3, the biggest bodies can be explained by the size of the fluid layers and the accumulation of the fluid of the precursor fluids due to how the orbital speed and rotational angular speed evolved in Zones 1, 2, and 3. I also elaborated on that in the chapter on rotational period and rotational speed. The notion of fluids in adjacent precursors can coalesce and merge suggested to me that not all consecutive bodies came from different layer of fluids, but some could have just merged or split from neighboring vortices or fluid ligaments.

The repetitive conical shape of the radius with respect to the semi major increments in Zones 4 and 5 points to the possible breakup points and coalescence of fluids in the ligaments of the precursors in those zones. The data in Zones 4 and 5 gave some clues as to how after their breakups, some precursors were spaced and others coalesced according to their position. Indeed, the peaks of the semi major increments are usually between a crescendo of increments upstream and a decrescendo of increments downstream. I also observed similar trends with the radius and like I explained in the chapter related to radius, bodies with smaller radius are usually found around those with higher semi major increments, whereas the biggest radius are around the smallest semi major increments. More pronounced in Zones 4 and 5, this trend suggests that some smaller bodies would have been displaced (pushed or pulled) farther away than some bigger bodies. As I carefully compared the trends of the semi major increments and radius, it appeared to me that the breakup points of the fluid ligaments of the precursors in Zones 4 and 5 would have been near the location of the smaller bodies around which the semi major increments peaks were observed. In contrast, some biggest bodies in Zones 4 and 5 would have been due to the coalescence of fluids after their ligaments broke up, ejecting smaller bodies closer to their breakup points.

For a ligament to break, its jet or fluid column must thin to a point whereby it has to break. In other words, the ligament near the breaking points is smaller so that the breaking could occur. After the breakup, the bodies on either side of the breaking points could have been displaced farther because of the momentum given to some of them during the breakup. In some cases, it could have been like 2 people holding hands and pulling each other in different directions, when their hands are released to symbolize a break up, their momentum can affect their movement in 2 different directions and even cause them to fall depending on their mass and how they were pulling against each other. To better illustrate this, imagine 2 persons pulling each other, one person pulling toward the north and the other person pulling toward the south. As their hands are released from one another, according to their weight or energy, one may be pulled toward the north and the other toward the south. In the case of the precursors of celestial bodies, at their breaking point, it could have been as if one was pulling toward the downstream of the flow and the other pulling toward the upstream of the flow.

413

Hence, after the breakup, the bodies downstream of the breakup can increase its semi major increment while some bodies upstream of the breakup could coalesce, leading to a relatively bigger size and a reduced semi major increment. After the break up, the precursors of some vortices of fluids which are farther away from the breaking point could have coalesced, meaning they could have been gathered together to form relatively bigger bodies as compared to those in their neighborhood. In the case of the precursors of some satellites, before they broke up, a cone could have been formed as their ligaments connecting to adjacent precursor of the satellites were thinning. At the breaking time, the cone which was formed could have affected the size and semi major increment distribution of the resulting bodies. Some breakups could have occurred around the zone where the semi major increments are high. The crescendo and decrescendo trends observed with the radius and the semi major axis increment particularly in Zones 4 and 5 could be due to how the breakup and the coalescence of the fluids in the ligaments of the precursors occurred.

In Zones 4 and 5, the breakup of fluid ligaments could have occurred around some bodies with smaller radius whereas coalescence could have happened around the biggest bodies situated between the smaller ones. In Zone 3 in contrast, the position of the biggest bodies can be explained by the accumulation of fluids in the precursor due to many factors including the change in speed, angular speed, the shear of the fluids, the size of the fluid layers, etc.

It was found that during the breakup of fluids, the axial motion of a drop pushes the fluid toward the center of the drop, and therefore prevents further breakup while at the same time, the radial motion can expand and/or contract the interface of the fluid (Tjahjadi et al., 1992). According to the same authors, when the axial motion does not balance the radial motion, the drop can undulate and fragmentate. Similarly, I think that the extent to which the axial motion overcomes the radius motion could have contributed to determining whether the precursor of some celestial bodies could form spherical or very oblate daughter bodies. Celestial bodies which are very oblate could have had an axial motion unable to overcome the radial motion. In contrast, for bodies that are very spherical and which equatorial and polar radius are similar, the radius and axial motion could have balanced.

After a pinch off, a drop which can appear as having a shape with a pointed end can quickly retract and smooth as the drop relaxes and gets a spherical shape (Tjahjadi et al., 1992). Likewise, the breakup events of the precursors of some celestial bodies could have generated shapes which have pointed ends, while the end of the jet of the ligament at other breaking points could have been quickly rounded up. When the fluids surrounding the pointed conical are made to go through another breakup, it could have formed much smaller bodies. Under some circumstances, instead, a "tube" of fluid can connect vortices before their ligament broke up to yield additional satellites, particles, and rocks found in rings or in the atmosphere of some bodies.

Nathanael-Israel Israel: Has had the Honor to be Acknowledged the First
Human Being that Scientifically Reconciled Science and Biblical Creation

Fig. 157: Necking, stretching, and breakup of a ligament of fluids and intermittency of daughter bodies

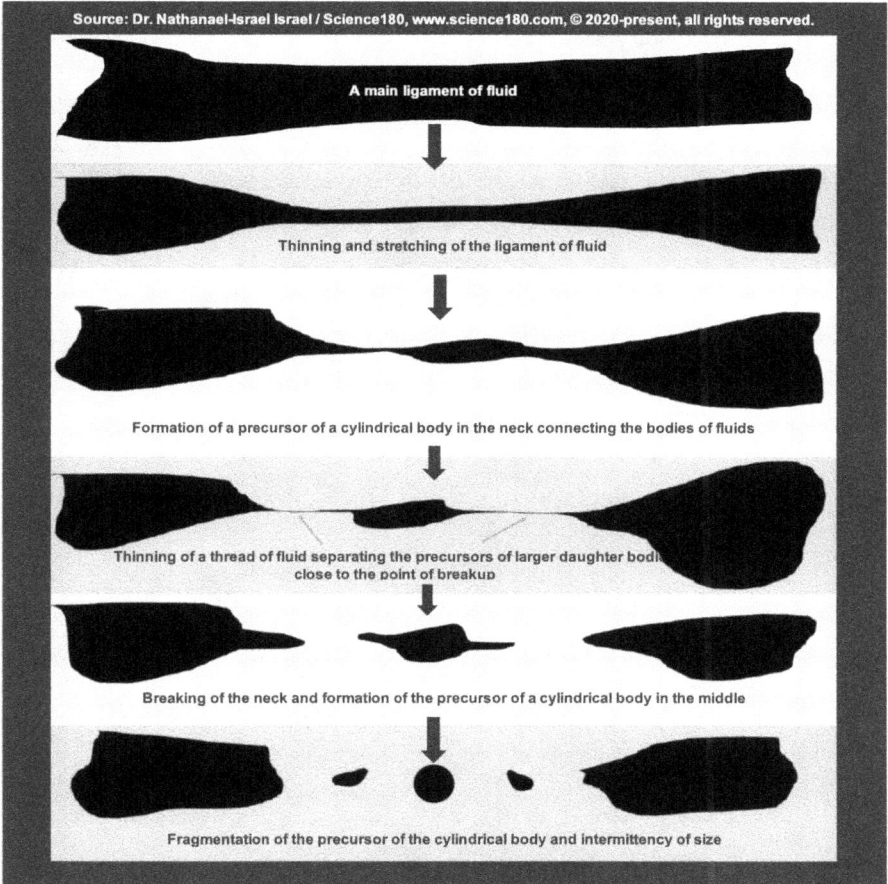

A main ligament of fluid

Thinning and stretching of the ligament of fluid

Formation of a precursor of a cylindrical body in the neck connecting the bodies of fluids

Thinning of a thread of fluid separating the precursors of larger daughter bodies close to the point of breakup

Breaking of the neck and formation of the precursor of a cylindrical body in the middle

Fragmentation of the precursor of the cylindrical body and intermittency of size

During a breakup, a drop can change its form from being ellipsoidal to a dumbbell with a central portion continuing to stretch and thin (Cristini and Renardy, 2006). According to the same authors, at one point, the bulb of drop can detach from the mother drop to form a daughter drop, and when the mother drop is large enough, surface tension can cause *"the ends of its remaining portion to retract slightly and then bulb up and end-pinching repeats. However, if the capillary number is high enough, end-pinching leaves behind a long cylindrical thread, where capillary wave breakups are observed, leading to the formation of large satellite droplets interspersed with small droplets"* (Cristini and Renardy, 2006). This can imply that the level of intermittency of the size of the bodies descending from a same mother precursor can depend also on the surface, capillary forces, and the size of the mother precursors. Fig. 157

illustrates the necking, stretching, and breakup of a fluid ligament and the intermittency of its daughter bodies.

In Fig. 158, I sketched how a mass of fluid connecting a spherical main body to a neck or ligament can break up to form small daughter bodies or can recoil around the primary body near them.

Fig. 158: Destiny of a mass of fluid connecting a spherical main body to a neck or ligament it broke from

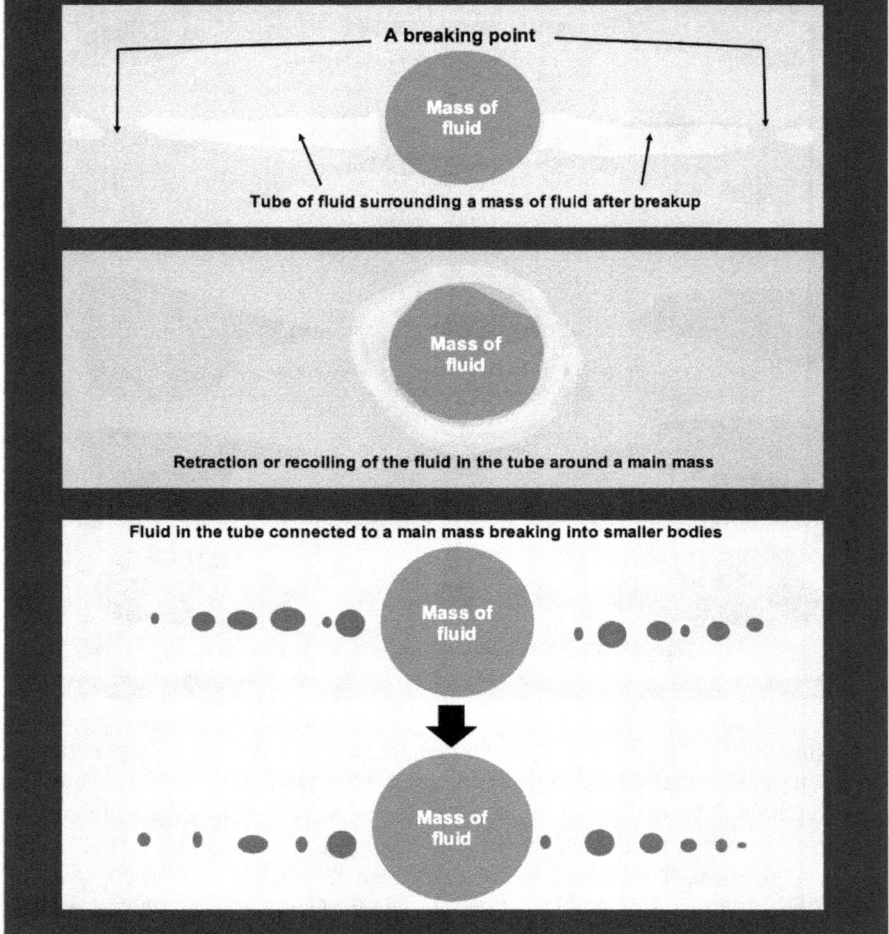

A breaking point

Mass of fluid

Tube of fluid surrounding a mass of fluid after breakup

Mass of fluid

Retraction or recoiling of the fluid in the tube around a main mass

Fluid in the tube connected to a main mass breaking into smaller bodies

Mass of fluid

Mass of fluid

29.3. Undulation and breaking of precursors into daughter bodies

Instead of being layers of fluids stacked on top of one another, the precursors of some bodies (e.g. Zones 4 and 5) could have been in a same thread that broke up and released its daughters. Considering the trends of the semi major increment of the satellites and their relationship with their radius, it appeared to me that, the distribution of the radius and semi major increment of the satellites can be explained using the way the precursors were broken up and how their fluid layers interacted. During the formation of the bodies, many fluid ligaments were formed, and at one point, some of them were covered with fluid masses of various sizes corresponding to the local thickness of the ligaments. At one point, the precursor of the secondary bodies was connected to the precursor of its primary body by a neck, which in some cases was later broken during the pinch off. After their pinch off, release, or ejection from their mother precursors, the fluids inside the ligaments that became the secondary bodies (e.g. satellites) continued to flow for a while. At one point, that ligament was broken up into pieces. The top layers separated first and the bottom layers separated last. After the precursor of the bodies in Zones 4 and 5 pinched off from the precursor of the bodies in Zones 1, 2, and 3 (which by then separated), it could have been propelled very far, hence the highest semi major increment is usually observed with the innermost satellite in Zone 4. In the previous chapter, I explained how the stretching of the vortices could have affected the propulsion or increase of the semi major increment. The way the fluid layers slid over one other also played a crucial role in the spacing between the bodies.

After the initial pinch off of the precursors of the bodies in Zones 4 and 5, additional breakups occurred as the segmented fluids in the ligament divided into different portions. The precursor of the bodies in Zones 4 and 5 split into different portions which then were undulated and after pinching off at many other points, could have led to the formation of strings of satellites which were either pulled apart or coalesced together according to their position from the breaking points. Some precursors which were farther away from the breaking points coalesced to yield relatively bigger bodies. In contrast, precursors that were closer to the breaking point or pinch off points differentiated into individual bodies but with smaller size. The viscosity of the precursors could have affected the formation of the satellites and also the pinch off from their mother precursors. For instance, because the precursors of the Jovian satellites in Zone 3 could have been very viscous, no smaller satellite was formed between them during their breakup. Hence, the Jovian satellites in Zone 3 are very big. In contrast, the precursor of the Saturnian satellites in Zone 3 could have been less viscous than that of the Jovian satellites in the similar zone. Therefore, the ligament of the Saturnian satellites in Zone 3 could have undulated a lot and then pinched off into many more structures which were broken up into many satellites. Hence the Saturnian satellites in Zone 3 are more than the Jovian satellites in Zone 3. Similarly, because it could have been very viscous, the precursor of the Moon did not undulate too much to be broken into many satellites, but it kept itself in one body which was propelled far away from the Earth. The higher viscosity of the precursors of Mercury and Venus could have also prevented them from breaking into a primary planet and secondary satellites, but to just keep its consistency and form a planet with no satellite.

Some previous studies found that, as a shear flow extends and stops, the fluid filament can undergo end pinching before decaying into drops (Stone et al., 1986). These authors explained that the end pinching was manifested by the formation of bulbous heads at the end of the stretched filament and then, their separation while the rest of the filament remains intact before later breaking into many small drops. In the next graph (Fig. 159), I sketched the end pinching of a ligament of the precursor of secondary bodies and its breakup into daughter bodies.

Fig. 159: End pinching of a ligament of the precursor of secondary bodies and its breakup into daughter bodies

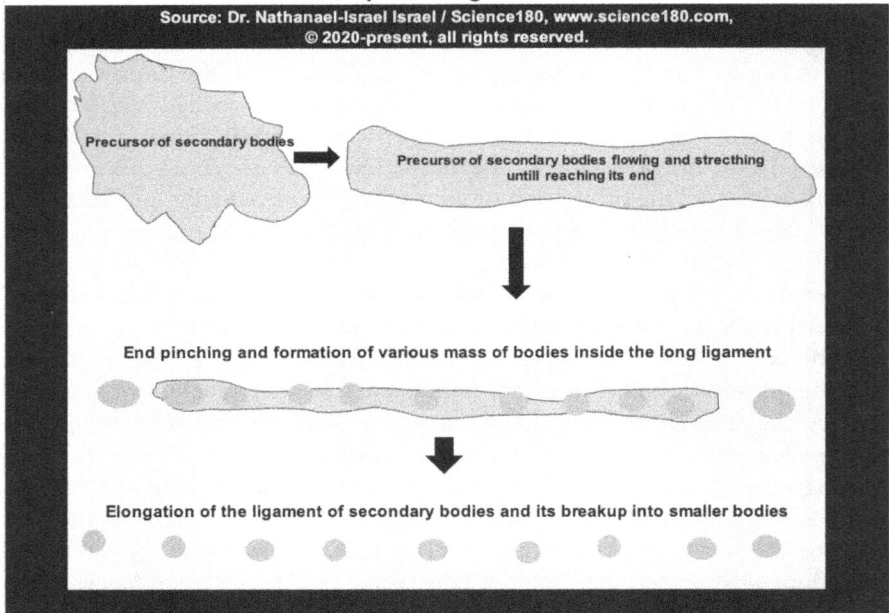

Precursor of secondary bodies

Precursor of secondary bodies flowing and strecthing untill reaching its end

End pinching and formation of various mass of bodies inside the long ligament

Elongation of the ligament of secondary bodies and its breakup into smaller bodies

As illustrated in Fig. 159, the fluids of the precursors of some bodies (e.g. Zones 4 and 5) could have been extended until the flow stopped or until it could no longer stretch or slide over other fluid layers. Then, the tube or fluid ligament could have pinched off at the end in some conditions, ejecting the outermost body, and then decaying or breaking into satellites differently spaced according to the additional breakup and coalescence which some of them had to undergo. The early pinching of the precursor of the outermost satellite could explain why the semi major increment of the outermost satellite is usually much higher than that of the other satellites in Zone 5 for some satellite systems. For instance, the

Nathanael-Israel Israel: Has had the Honor to be Acknowledged the First Human Being that Scientifically Reconciled Science and Biblical Creation

highest semi major increment of the Jovian, Saturnian, and Uranian satellites in Zone 5 were observed with the outermost satellite in those planetary systems. The fact that the outermost Neptunian and Plutonian satellite does not have the highest semi major increment in Zone 5 suggested to me that other factors can also limit how pinching can also affect the semi major increment of the outermost satellite.

Under some conditions, the fluids in Zones 1, 2, and 3 could have kept flowing and stayed connected to the mother precursors until they reached a point where they could not flow anywhere. On other conditions, the fluid of all the bodies in a system could have been connected until an instability in their ligament shook or undulated them before they separated into different daughter bodies. Fig. 160 illustrates a sequence of some events which could have occurred during the breakup of the ligament of a precursor of secondary bodies.

Fig. 160: Sequence of events during the breakup of the ligament of a precursor of secondary bodies

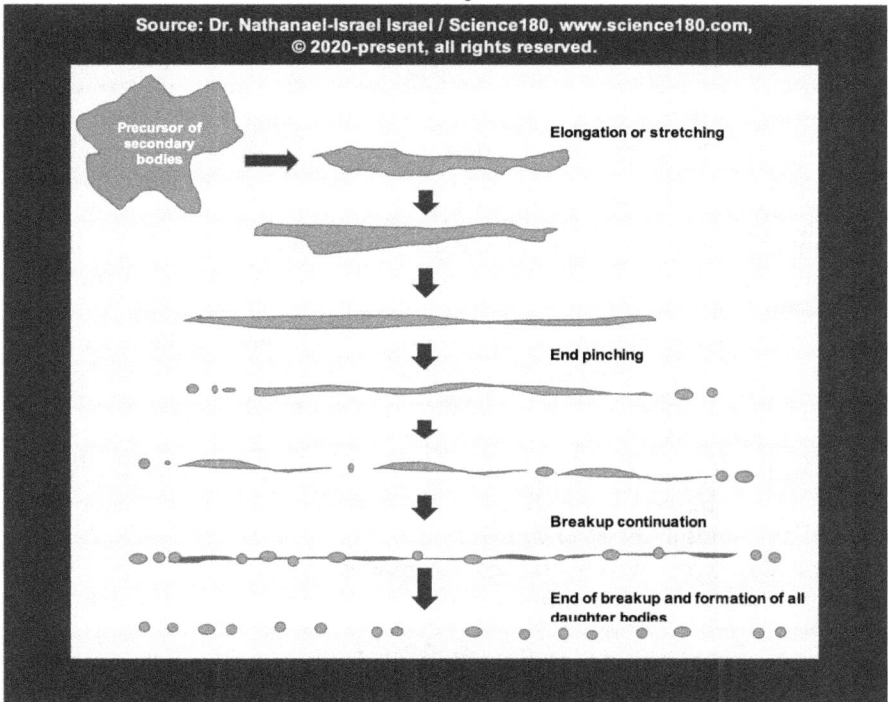

Science180: The Premier Organization that Scientifically Decoded the Universe-origin Accurately

29.4. Impact of the roughness and smoothness of the ligament of vortex (blobs and conglomerate) of fluids on the number and size distribution of their daughter bodies

One of the questions I had asked for years before getting an answer was what can explain the distribution and the number of bodies in each turbulence zone. As I reviewed the literature on fluid breakup and turbulence, I perceived that the nature of turbulence in the fluid flow of the precursors could have impacted:

- the smoothness or roughness of some fluid ligaments,
- the stability and destabilization of the neck connecting the bodies,
- the detachment of vortices and blobs (according to the scales) and consequently
- the number and size distribution of the daughter bodies.

For instance, "an initially smooth ligament of liquid drops forming in a gaseous environment produced a set of close to mono-disperse drops, while a rough ligament broke into drops with a broad collection of sizes" (Eggers and Villermaux. 2008). In other words, the number of bodies is higher in smooth ligaments, whereas it is smaller in rough ligaments. Additionally, the previous authors reported in their review of literature on the physics of liquid jets that the size of the bodies was smaller in smooth ligaments than in rough ligaments. To put it another way, for smooth ligaments, the distribution of the size is narrow whereas rough ligaments are strongly corrugated and have a broader size distribution. The previous authors also reported that "ligaments which break into one big drop, plus a few smaller ones, produce the broadest possible size distribution". Other sources testified that "thinner, but nevertheless corrugated ligaments produce not only finer drops, but also narrower distributions" (Hennequin et al., 2006).

These findings suggested that the intensity of turbulence in the precursors of the bodies could have affected the smoothness or roughness of their ligaments and defined the number, diversity, and distribution of their daughter bodies after breakup. For instance, in each planetary system, the size, number, and distribution of the satellites were affected by the roughness and smoothness of the ligament of fluid which birthed them. Because the turbulence of their precursor was different, the fluids in the ligaments of the precursors of the celestial bodies could have had different smoothness and roughness. Therefore, it should not be surprising that, during the formation of the universe, smooth fluid ligaments could have produced more mono-disperse daughter bodies, while rough ligaments could have produced daughter bodies with a broader collection of size. Because the highest intensity of turbulence is usually in Zone 3, I perceived that the fluid ligaments in Zone 3 could have been very rough, while those in the other zones could have been smoother. As explained in the chapter on radius, the highest standard deviations of radius of the satellites were recorded in Zone 3, whereas the smallest values were found in Zones 1 and 5, confirming the huge size variation in Zone 3 and a smaller radius variation in Zones 1 and 5. Consequently, the ligaments of fluids of the precursors in Zone 3 could have been rougher than that of the fluids in the other turbulence zones, whereas the smoothest ligament could have been in Zones 1 and 5. To be more specific, the smoothest ligaments could have been in Zone 5 followed by Zone 1.

The fact that the smallest standard deviation of radius in Zone 5 was recorded with the Saturnian satellites (see the chapter on radius) suggested that the ligament of the Saturnian satellites in Zone 5 could have been smoother than that in Zone 5 of the other satellite systems in the Solar System. The smoothness and roughness of the ligament in Zones 2 and 4 are intermediary. In other words, the roughness of the ligament of fluids in Zones 2 and 4 could have been intermediary because they were the transition zone into turbulence (Zone 2) and the transition zone out of turbulence (Zone 4). The near-uniformity of the size of satellites in Zone 5 could be explained by the smoothness of their ligaments before breaking up. What I said for the ligament of the precursors of the celestial bodies could also have been true for microscopic particles at the atomic and subatomic length scales. In the next graph (Fig. 161), I illustrated the breakup of smooth and rough ligaments into daughter bodies of various sizes.

Fig. 161: Breakup of smooth and rough ligaments into daughter bodies of various sizes

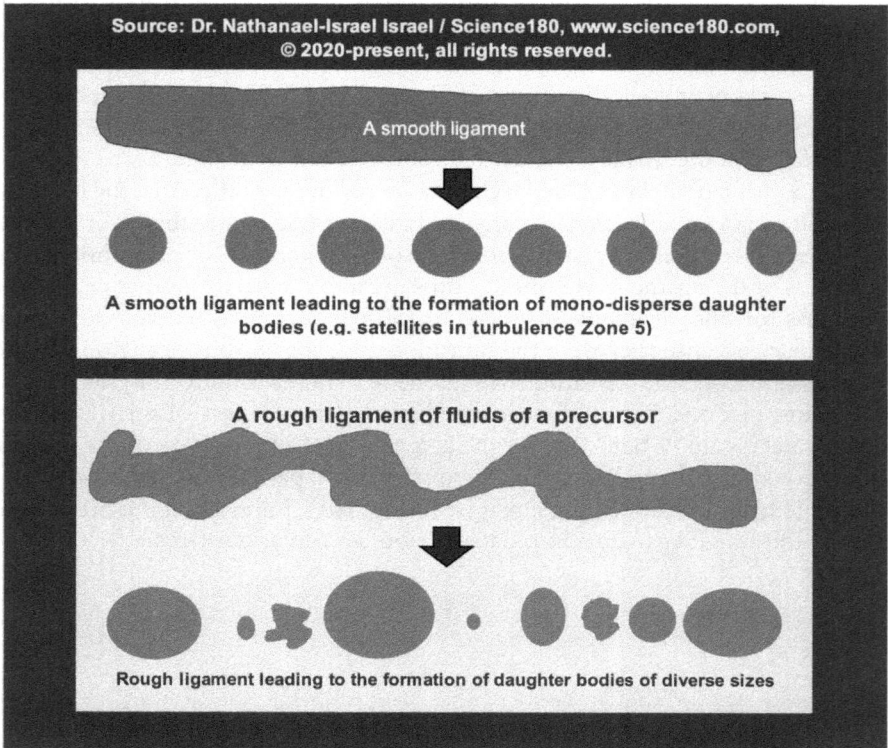

A smooth ligament

A smooth ligament leading to the formation of mono-disperse daughter bodies (e.g. satellites in turbulence Zone 5)

A rough ligament of fluids of a precursor

Rough ligament leading to the formation of daughter bodies of diverse sizes

Reading this document before anyone else, someone commented: "*this graph really helped me to understand what you* [I] *were saying*". This comment put a smile on my face, for I really did not expect that person to comprehend most of the complex and highly encrypted things I have been trying to decode. I hope by now, you are also piecing together what I have been trying to explain. Anyways, let's continue in the same direction.

The smoothness of the ligaments of the precursors of bodies in Zone 5 may also explain the narrow distribution of the radius of the satellites in that zone (see the chapter on radius). In other words, the fact that more bodies with near-similar radius are found in Zone 5 corroborates with the fact that the fluid ligament in that zone was smoother than that in other zones. The subdivision of the fluid into layers could also have affected the number of bodies. Hence the smallest number of bodies is not always found in Zone 3. To put it in another way, while many bodies in Zone 5 could have likely descended from a same ligament that stretched and then broke up, in Zone 3 for instance, the precursors of the bodies could have been more organized into layers. Hence the number of bodies in the inner zones could have been defined by the factors which divided the fluids into

Nathanael-Israel Israel: Has had the Honor to be Acknowledged the First Human Being that Scientifically Reconciled Science and Biblical Creation

a certain number of layers. Some of the small satellites found between larger ones in Zone 3 could also have been due to the way the fluids in the precursors of larger ones split and collected, causing an intermittence.

Rain atomization is one of the phenomena where size distribution of daughter bodies was studied. It has been proven that the heaviness of a rain can affect the size of the rain drop. For instance, drop sizes are more broadly distributed in heavy storms than in fine mists (Laws and Parsons, 1943; Marshall and Palmer, 1948; Bentley, 1904). Finally, for rain atomization, "bag breakup" (Pilch and Erdman, 1987) is a process which illustrates the three stages shared by all atomization processes (Eggers and Villermaux. 2008):

- Topological change of an initial object during which a big drop flattens into a pancake shape as it decelerates downwards.
- Formation of ligaments during which a toroidal rim of the bag collects most of the initial drop volume.
- Broad distribution of fragment sizes during which the rim is highly corrugated and breaks into many small drops, and a few larger drops.

Knowing that it was shown that a "liquid drop falling in a counter-ascending air current can first deform, then destabilize and finally break up into stable fragments (Pilch and Erdman, 1987), it is possible that during the formation of the universe, some drops or vortices of the precursors of the bodies could have been broken into fragments leading to the formation of small bodies (e.g. rocks and dust particles). Some small rocks and particles found in rings could have descended from the fragmentation of a mother precursor upon its contact with the air in space. In other words, it is possible that the air through which some precursors of bodies passed through could have affected their breakup and the size of their daughter bodies.

I would like to mention that the size of the daughter bodies can also relate to that of their mothers. Indeed, on small scales like those in a laboratory (e.g. experiments of water drop breakup), blobs of drops can be formed in a fluid ligament indeed, but on the scale of the precursors of celestial bodies, instead of blobs, large vortices (of various astronomical sizes) were formed inside the fluid flow of a ligament. Just as drop sizes were found to be proportional to the initial thread diameter (Rayleigh, 1879), so also do I think that the size of some celestial bodies could have been defined by the dimension of the layers of fluids or fluid "thread" of their precursors. In other words, the size of some celestial bodies can be a good indicator of the size of the fluid layers that birthed them. However, because some precursors could have coalesced, while others could have split into many more daughter bodies after stretching and thinning, the radius of some celestial bodies could be different than that of their precursors. In Zone 5 where most bodies have a similar size, the size of some satellites could reflect that of the thread of their precursors. Other studies also made me believe that, as they detached, retracted, and relaxed, some blobs and vortices (according to the scale) which made up the ligament of some precursors could have coalesced or amalgamated, therefore forming bigger and bigger blobs (Oliveira and McKinley, 2005). Some bodies could have broken up into pieces forming many smaller bodies because of some size restriction in their environment.

As I was trying to explain the intermittency of celestial bodies following the breakup of their precursors, I felt like I should report some findings about the formation of satellite droplets in laboratory settings where fluid breakup is usually analyzed using water drops and droplets. Indeed, also known as secondary droplets, satellite drops are produced during the thread breakup in addition to large main droplets. The control and prediction of the precise size of satellite droplets have been very difficult. While the production of satellite droplets can be predicted based on fluid properties, their precise location and volume cannot be predicted (Henderson et al., 1997). Because of how drops are produced in laboratory settings, satellites drops obtained in the lab do not orbit the primary ones as is the case of celestial bodies. Satellite drop formation in a lab leads to a bimodal drop size distribution and the size of the main drop is set by the jet radius and thus the size of the orifice (Eggers and Villermaux, 2008). The same authors reported that the asymmetry of the pinch-off profile implies that satellite drops form at each breakup event, therefore leading to a bimodal distribution of drops. Thus, I think that, on the scale of the celestial bodies (e.g. bodies in Zones 4 and 5), it is possible that some smaller satellites which are upstream and downstream of some relatively bigger ones could have formed when bigger ones detached and retracted after breakup just as how satellite droplets can form after the primary drops. The primary droplet is usually bigger, while the satellite droplets are smaller and formed from the breakup of the thin thread between the jet and the primary droplet. In some conditions, the "satellite droplet pinched off from the primary droplet (at $t \approx 0\mu s$) before it pinched off from the jet (at $t \approx 4\mu s$), resulting in a deceleration of the satellite droplet" (Pimbley and Lee, 1977). On the scale of the celestial bodies, if such a phenomenon could have been observed, the speed of the bodies could have been decelerated because of their early pinching off from the main thread or ligament of fluid.

29.5. How the viscosity of the fluids of the precursors could have affected the formation of their daughter bodies

Viscosity measures the resistance of a fluid to shear or to angular deformation. Honey, lava (of volcanos), and motor oils are examples of viscous fluids, and consequently they feel "sticky". In many studies related to fluid dynamics, viscosity is one of the parameters that significantly affects the equations. Considering the data that I have analyzed, and particularly the processes involved in split-gathering, I felt like the viscosity of the fluids of the precursors of the bodies was changing as they were being shaped into their daughter bodies. I suspected that the way that the precursor of the bodies broke from one another was affected by the viscosity of their fluids. Here, I did not seek to know or estimate the viscosity of the fluids of the precursors, but to explain how some characteristics of the celestial bodies in the universe as of today point to the role that viscosity of their precursors could have played during their formation. As I tackled viscosity, I focused on turbulence Zone 5 because it is one of the zones where turbulence is minimum and where the impact of viscosity could have been higher or more easily visible.

An increased liquid viscosity can slow down capillary breakup (Tucker and Moldenaers, 2002; Mabile et al., 2003), therefore causing the ligament to be stretched longer (Bremond and Villermaux, 2006). Stretching can also prevent a ligament from emptying and hinder its destabilization (Eggers and Villermaux, 2008). High-viscosity liquids tend to produce very long and thin threads (Eggers,1997). For instance, because it is very viscous, honey threads usually resist some perturbations that can quickly break some less viscous fluids (like water), therefore leading to the formation of long filaments of honey rather than quickly breaking into individual droplets. When a cylinder of fluid was suspended in another fluid or outer fluid (having a much larger viscosity), recoil of fluid necks was slow and the necks broke multiple times (Tjahjadi et al., 1992). These findings suggested to me that viscosity could have affected the size and the distance between the daughter bodies. It could have affected the necking, breakup, recoiling, and the number of daughter bodies after fluid breakup.

Because the viscosity of the Jovian satellites could have been very high, the recoiling of the fluids of their precursors in Zone 5 could have been slow, and consequently, the neck connecting them could have broken up multiple times, yielding many Jovian satellites in Zone 5. Hence no other satellite system has as many satellites in Zone 5 as the Jovian satellite system. When the turbulence of the fluid of the precursors was very weak, viscosity could have affected their behavior and dampened disturbances which could have otherwise deformed and broke the fluid thread. In other words, a high viscosity can delay the time breakup of the fluid jet of the precursors of the bodies. In the chapters on semi major axis and semi major axis increment, I also showed how viscosity could have affected the sliding fluid layers and the stretching of the fluid ligament of the precursors.

Viscosity of the precursors of the bodies could have also affected the size of the bodies in Zone 5, where turbulence is minimum as compared to Zone 3. The small radius of the Jovian satellites in Zone 5 as compared to the satellites in the same zone of the other planetary systems could also have been caused by a high viscosity of the Jovian satellites, which could have caused the fluid layers of their precursors to slid or be stretched over a long distance, therefore thinning more before breaking up. Viscosity could also have affected the split of the precursors of the planetary systems. It appeared to me that the high viscosity of the precursor of Mercury and Venus could have been one of the factors that prevented the formation of a satellite around those planets. I also think that the viscosity of the precursor of the Earth-Moon system was also very high and that could have contributed to preventing it from forming many satellites around the Earth. Nevertheless, I believe that as more thorough investigations will be done on the satellites of the Earth, more satellites of the Earth located far beyond the Moon could exist and be orbiting the Earth and they could have not been found yet because they are too small and located very far away. Some of those satellites could be farther away from the Moon than the Moon is from the Earth. For when I considered the satellites is Zone 4 of the satellite system of the other planets, they are usually located far away from the biggest satellite in their satellite system (see the chapter on semi major increment). Knowing that the Moon is a satellite in Zone 3, I expect other satellites of Earth to be found beyond the Moon to behave like those in Zone 4: higher orbital inclination, higher rotational angular speed, higher eccentricity, smaller size, etc.

I also personally think that the viscosity of the lava of the celestial bodies could depend on their semi major axis. Although it may be easy to think that the viscosity of the bodies which are closer to their primary bodies may be higher, it is important to keep in mind that the impact of temperature and pression on viscosity can also cause the precursors of some inner bodies to be less viscous than those of the outer bodies. The difference in the composition and characteristics (e.g. density) of the crust of the terrestrial bodies could have been also affected by the viscosity of their precursors.

In a study related to the breakup of a liquid drop, a high viscosity was found to delay the time of breakup and if the delay is great, some drops accelerated before breaking (Joseph et al., 1999). Because breakup time can increase with viscosity, precursors (in Zones 4 and 5 for instance) that were very viscous would likely have had more difficulties breaking up into daughter bodies. As a reminder, I also found a negative correlation between the number of satellites and the density of the planets in the Solar System ($r2=0.7$, $p=0.004$), meaning that in certain conditions, a less dense planet tends to have more satellites. A positive correlation could also exist between the viscosity of some precursors and the density of their daughter bodies. For instance, a positive correlation could have existed between the viscosity of the precursors of the planetary systems and the density of the planet and satellites that they birthed. This suggests that some precursors which birthed the densest bodies could have been unable to break up into many pieces that could have led to many satellites. In other words, the number of satellites of a planet can also be connected to the viscosity of its precursor. That may be why the densest precursors of planets and satellites yielded the smallest number of satellites.

The impact of viscosity on the separation of turbulence was a subject of a heated debate during the 2011 Turbulence Conference in Marseille, France, 50 years after the 1961 Marseilles conference, which was dominated by the

Nathanael-Israel Israel: Has had the Honor to be Acknowledged the First
Human Being that Scientifically Reconciled Science and Biblical Creation

Kolmogorov's theory of cascade breakup. During that debate in 2011, Prof George William (an expert in turbulence) pointed that *"viscosity plays a very important role in separating the dissipation range from the energy range, … Viscosity very much affects the second-order moments."* (Turbulence Conference, 2011). One way of viscosity to inhibit turbulence can be its ability to "absorb" kinetic energy, therefore reducing movement of fluids. This may also be why some people think that viscosity can slow down and even bring turbulence to a stop at the molecular level.

Some studies found that the "shortest eddy scale is set by viscosity. The shorter the eddy scale, the more important is the fluid viscosity and consequently, the shortest eddy scale can be defined as the length scale at which viscosity is dominant" (Cushman-Roisin, 2015). This finding suggests that when the viscosity of the fluids of the precursors is too high, the size of the bodies in Zone 5 can be smaller, therefore supporting my idea that the precursor of the satellites in Zone 5 could have had a viscosity negatively correlated with the semi major axis of their primary satellites. That is why the size of the satellites in Zone 5 increased as the semi major of their primary planet increased (see the chapter on radius). At the scale of the satellites in Zone 5, the diffusion and transfer of energy and momentum could have been significantly impacted by viscosity, but in turbulence zones where the turbulence was very high (e.g. Zone 3), the impact of viscosity could have been dominated by turbulence. According to the previous author, "the greater the turbulence, the shorter the finest eddy scale", suggesting that in addition to and/or in cooperation with viscosity, the intensity of the turbulence could have also contributed to defining the size of the bodies in Zone 5.

In experiments done in a laboratory setting, the minimum diameter of the eddies at the dissipation scale is said to occur in the order of or shorter than a few millimeters (Cushman-Roisin, 2015). However, at the scale of the precursors of satellites, because the size of satellites in Zone 5 (the least turbulent zone) is in order of 0.3 km – 30 km (see the chapter on radius), it appeared to me that the turbulence during the formation of the satellites was much more than what could be experienced in a laboratory setting. Nevertheless, because smaller particles and dusts are usually found between celestial bodies, it is also understandable to say that turbulence was not dissipated at the scale of the satellites in Zone 5. In my book *"Turbulent Origin of Chemical Particles"*, I provided more details.

As I close this section on viscosity, I would like to notice that most of the flows studied in physics are usually considered incompressible (implying the density stays the "same") and their viscosity is also usually assumed "constant". However, considering the trends of the celestial bodies (e.g. density, semi major increment, size), it seemed to me that neither the density nor the viscosity of the turbulent flows (at least of the precursors of the bodies during the formation of the universe) was constant. Although assuming that density and viscosity are constants might have helped scientists to simplify some of their equations related to fluid dynamics, it could have also hidden some keys that should have helped explain turbulence. In other words, even in fluids labeled as incompressible, density cannot always be the same everywhere. Although it may not be perceived and sometimes its variation can be very small, even a small variation in density and pression of the fluids can affect the organization of its molecules and the physics of their interaction. As I have explained, the changes of the configuration of the turbulent prima materia contributed to forming diverse subatomic particles and atoms in the early universe. Although new molecules or atoms may not be formed in some turbulent fluids today, the impact of density, pression, viscosity changes in them (e.g. during turbulence) can significantly affect and bias some equations related to fluid dynamics of turbulent flow. Considering the data that I have studied, the variation of the size and other characteristics of the vortices in turbulent flows suggested to me that their density cannot be the same everywhere!

I understand that some physicists celebrate some of the achievements of the French physicist and mathematician, Joseph Valentin Boussinesq (1842-1929)—considered as one of the most important persons in the history of natural convection and turbulence—for his work on the "Boussinesq approximation" that allegedly "simplified" the problems connected with density variations in a flow. However, I personally think that this kind of simplification of the complex problems of fluids dynamics prevents the apprehension of the problems and its solution as a whole. It seemed to me that the Boussinesq approximation made on density can be a huge problem. For example, the density of the celestial bodies in the Solar System varies according to many factors and it can be misleading to average it or minimize its effect on other variables. The approximation of Boussinesq may have helped scientists to get rid of a big problem, but it may have taken away the main solution of their problem. For, avoiding a problem or finding a way to avoid it is not the best way to find its solution. It would have been better to keep complicated formulas in models and expect someone to find the best model someday than removing or replacing the difficult variables by simple ones, which gives people the illusion that they solved their problem. From one simplification to another, and throughout generations, complex problems have been left with simplistic prejudices and heavy formulas or equations that some people cling to as the universal solution, not knowing the many unverified assumptions which led them there. Hence, some people tend to accept lies instead of the truth. For instance, the inability of current scientific methods to detect configurational and relational changes in some atoms and molecules in turbulence or even in laminar fluid flows, and the many scientific assumptions aiming at simplifying some equations have been causing physicists to miss certain facts crucial for the real understanding of natural laws. I revisited this problem in other books.

Finally, by the time I reached this stage of my writing, it appeared to me that, the theories of turbulence which assume that turbulence dissipated at the molecular level where/when viscosity is said to control the dissipation of energy, are partial and neglect to consider the turbulence which occurred and is still

Nathanael-Israel Israel: Has had the Honor to be Acknowledged the First
Human Being that Scientifically Reconciled Science and Biblical Creation

occurring at the scale of galaxies, atoms, and subatomic particles and even on scales smaller than theirs. I have reviewed the literature and found that no comprehensive study has been done on the turbulent origin of the chemical elements. Hence, I devoted a completely different book (*"Turbulent Origin of Chemical Particles"*) to the origin of chemical particles with perspective of the mother of all turbulences. Talking about the dissipation of turbulence, I have also realized that, once the kinetic energy in the early turbulent world was converted into other forms of energy (including the energy stored in celestial bodies and particles in the universe), the turbulence in the precursors of bodies could have rapidly "dissipated" on some scales, while it could have continued on other scales, leaving behind daughter bodies, some of which are still turbulent. For instance, while turbulence is not visible at the surface of the terrestrial planets (e.g. Earth), it is still present on the surface and in the atmosphere of some giant planets (e.g. Jupiter and Saturn) where large cyclones are found. Turbulence still exists at the center of the Earth and of many celestial bodies. The surface of the Sun is still very turbulent and intense solar activities are still recorded until today. In other words, while turbulence has ended in some precursors of the bodies, in others, it is still going on although not with the same characteristics as in the beginning of the universe. As I was working of the physics of vortex, I realized that the formation and evolution of a vortex bear some facts, which could help me to explain the formation of celestial bodies such as planetary systems, stellar systems, and galaxies. In the chapter on galaxies, I detailed how their precursors were shaped and how everything in the universe points to a "turbulent organization" (e.g. "nesting, branching, spiraling, etc.") under the influence of the split-gathering I have explained in this book. In my book on the origin of life, I also expounded on these topics at the biological level.

As I close this chapter on the breakup of fluids, I would like to emphasize that the physics of fluid breakup also points to the precision of the turbulence at the beginning of the universe. I also understood that, if the initial events at the origin of the universe were not precisely calibrated, a different universe could have been formed. For instance, throughout this book, I have shown that the systems of bodies and matter in the universe were molded after a precise program of split-gathering of particles and bodies. Some studies have shown that the breakup mode of a fluid depends very sensitively on the amplitude of the initial perturbation (Pimbley and Lee, 1977; Hilbing and Heister, 1996), as well as on the presence of higher harmonics (Eggers, 1997; Chaudhary and Maxworthy, 1980; Xing at al., 1996; Barbet et al., 1997). Consequently, a small detuning at the beginning could have led to a different distribution and characteristics of bodies in the universe. Another way of saying this is that, if the turbulence at the beginning of the universe was different, the universe could also have been different. For the order and harmonics of and between the bodies in the universe are highly organized.

29.6. Take home message

As precursors of bodies were splitting from one another, fluid ligaments were formed between some of them. For instance, in each system of bodies, as the fluids of the precursor of the secondary bodies were escaping the precursor of their primary body, a fluid ligament was formed so the fluids could flow through it. The neck of that ligament could have stayed in place and allowed fluids to flow into the ligament of the secondary bodies until a moment came for the precursor of the secondary bodies to pinch off from the precursor of the primary body. Afterwards, the ligament of the secondary bodies continued its journey until all its daughter bodies were fully formed and put into orbit. Fluid necks were also formed between the precursors of consecutive secondary bodies as they were breaking from one another. The time it could have taken to complete the pinching off or the break-up of the neck connecting ligaments of fluids could have depended on the intensity of their turbulence, the speed, size, volume, viscosity of their fluids, and other factors involved in their stretching and break-up including forces involved (e.g. "capillary" forces). I showed that the intensity of turbulence in the precursors of the bodies could have affected the smoothness or roughness of their ligaments and defined the number, diversity, and distribution of their daughter bodies after breakup. Smooth ligaments of fluids could have produced more mono-disperse daughter bodies while rough ligaments could have produced daughter bodies with a broader collection of size. Because the most developed turbulence is usually in Zone 3, the fluid ligament in Zone 3 could have been rougher than that of the fluids in the other turbulence zones whereas the smoothest ligament could have been in Zones 1 and 5. The smoothness of the ligaments of the precursors of bodies in Zone 5 may also explain the narrow distribution of the radius of the satellites in that zone. The viscosity of the fluids of the precursors of the bodies could have been changing as the latter were being shaped into their daughter bodies. Likewise, the viscosity of the precursor of the fluids could have affected how they broke up, the number of their daughter bodies as well as their size and the distance between the latter. When the viscosity of the precursors of bodies was very high, the recoiling of their fluids was slow, and consequently, the neck connecting them could have broken up multiple times, yielding many bodies. When the turbulence of the fluids of the precursors was very weak, viscosity could have dampened disturbances which could have otherwise deformed and broke the fluid thread. In other words, a high viscosity can delay the time of break-up of the jet of fluids of the precursors of the bodies. The high viscosity of the precursor of Mercury and Venus could have been one of the factors that prevented the formation of a satellite around those planets. Similarly, the viscosity of the precursor of the Earth-Moon system was also very high and that could have contributed to preventing it from forming many satellites around the Earth.

'Science180 Academy' Success Strategy
HOW TO RAISE RATIONAL CHILDREN IN OUR MODERN WORLD

In our modern secular world, and with the many things that kids are taught at school and over which parents have little control once the kids head to public school, parents have a lot to worry about. But it does not have to be that way. Universe-origin and life-origin scientist Dr. Nathanael-Israel Israel has discovered that, more than ever, parents have a crucial responsibility to rationally prepare their kids to have a strong worldview that properly embraces both science and faith, so their kids are not pulled on one side by the secular education and on the other side by religious belief. But how can parents and their children achieve that common goal?

Listen to this Beninese-American scientist and mathematician Dr. Nathanael-Israel Israel to figure it out. Nathanael-Israel is the author of the acclaimed book *"How Baby Universe was Born"*, an easy to understand scientific book primarily written for children age 7-12 years old to help them properly crack the code of the formation of the universe in a language they completely enjoy, and that prepares them to fight any secular or religious theory that may try to rationally drift them away from the reality of everything!

Sample questions that will get answered include the following and many more:

- How can parents use the latest breakthrough about the universe-origin to rationally raise their kids?
- How can parents prepare their children from being victims of the danger of wrong theories and dogmas on the origin of life and the universe?
- What can parents do to shield their children from the influence of religious and scientific beliefs that try to enslave them in the name of reason or faith?
- Why is wrong science not the only danger of raising rational children, but wrong belief as well?
- How can we help children to positively navigate the intersection of science and faith?

Learn more at Science180.com/children

Science180: The Premier Organization that Scientifically Decoded the Universe-origin Accurately

CHAPTER 30

SO YOU THINK YOU REALLY KNOW THE ORIGIN OF THE RING SYSTEMS OF THE PLANETS INCLUDING THE BEAUTIFUL RINGS OF SATURN? NOT SO FAST MY FRIEND, WAIT UNTIL YOU READ THIS FIRST!

30.1. Generalities on rings

If you want to accurately know how the ring systems of planets and asteroids were formed, don't miss this?

Consisting of solid materials (e.g. dusts, moonlets, and other small objects), rings are traditionally defined as a disc orbiting an astronomical object. Although for a long time, rings were originally thought to be present only around the 4 giant planets in the Solar System, recent evidences showed that rings are around other celestial bodies including satellites, asteroids, terrestrial planets (e.g. Mercury, Venus, and the Earth). For instance, NASA has observed "grainy rings", or "fine haze of cosmic dust" overlying the orbits" of Earth, Venus, and Mercury; and some of these "rings that can be seen with telescopes on Earth" (NASA, 2019). While Mercury's ring was estimated at 9.3 million miles wide, that around Venus is said to be about 16 million miles from top to bottom and 6 million miles wide; and if all the dust around Venus could be packed together, they are believed to form an asteroid with a radius of 1 mile (NASA, 2019; Pokorny and Kuchner, 2019). Although it was believed that Pluto may have its own rings, observations made by New Horizons did not detect any such ring system.

Rings are also observed around asteroids. For instance, although having a radius of 125 km, 10199 Chariklo, a minor planet or an asteroid orbiting between Saturn and Uranus has 2 rings that were discovered in 2014 (Braga-Ribas et al., 2014). Some satellites also have their own rings. For instance, Triton (the largest moon of Neptune) has its own rings. Triton having a ring made me believe that its precursor was not able to gather together all of the material around it, hence some were left over to form that ring. In most pictures of the Solar System where all the planets are shown, the only planet that is usually shown with rings is obviously Saturn. However, many other planets besides Saturn have a set of rings around them. All of the giant planets (Jupiter, Saturn, Uranus, and Neptune) have rings. As of 2020, at least 48 rings have been discovered around the 4 giant planets in the Solar System and by order of increasing distance from their primary planet, they are:

Jovian rings:
- Halo
- Main Ring
- Amalthea Ring
- Thebe Ring
- Thebe Extension

Saturnian rings:
- D Ring
- C Ring
- Titan ringlet
- Maxwell gap/ringlet
- 1.470 Rs Ringlet
- 1.495 Rs Ringlet
- B Ring
- Cassini division
- Huygens Ringlet
- A Ring
- Encke gap
- Keeler gap
- F Ring
- Janus/Epimetheus Ring
- G Ring
- E Ring
- Methone Ring
- Anthe Ring
- Pallene Ring
- Phoebe Ring

Uranian rings:
- Six

- Five
- Four
- Alpha
- Beta
- Eta
- Gamma
- Delta
- Lambda
- Epsilon
- Nu (R/2003 U 2)
- Mu (R/2003 U 1)

Neptunian rings:
- Galle (1989N3R)
- LeVerrier (1989N2R)
- Lassell (1989N4R*)
- Arago (1989N4R*)
- Unnamed (indistinct)
- Adams (1989 N1R)
- Courage (Arcs in Adams Ring)
- Liberté (Arcs in Adams Ring)
- Egalité 1 (Arcs in Adams Ring)
- Egalité 2 (Arcs in Adams Ring)
- Fraternité (Arcs in Adams Ring)

Like I explained in my book on the delimitation of the turbulence zones, I finished dealing with the boundaries of the turbulence zones of the satellites before realizing, while studying the rings, that some rings are located between the primary planets and their innermost satellite. In other words, according to the delimitation I did for the satellites, I did not address the turbulence zones of the bodies that could be located between the primary planets and their innermost satellite. As I came across rings being located in that region, I felt like, instead of extending the limit of the 5 turbulence zones I already defined for satellites, planets, and asteroids, I better define another zone which can increase the precision to the location of the bodies found between the primary bodies and their innermost secondary body. I called that zone, turbulence Zone 0. Therefore, although you will not find any mention of turbulence Zone 0 in the previous chapters on satellites, planets, and asteroids, you will find a lot written about them in this chapter on rings. Below, I studied some of the characteristics of rings.

30.2. How I collected data on rings

The raw data I used to make the graph and tables presented in this chapter came from the ring's fact sheets on NASA's website (NASA, 2018). Unlike other types of bodies such as satellites, asteroids, planets, and stars, very little information exists on rings in the Solar System. Besides color and brightness, most of the data available on rings in the literatures are about their:

- Inner boundary radius,
- Outer boundary radius,
- Middle boundary radius,
- Width,
- Optical depth,
- Albedo,
- Surface density,
- Eccentricity, and
- Thickness range.

Unfortunately, all of those variables are not even reported for all rings. In some cases, they are reported as ranges, not as a value which can be easily used in statistical analysis. Therefore, to make the data analyzable, I had to do some mathematics on the missing values reported as range. For instance, the width of the rings is calculated as:

Width of ring = Outer boundary radius - Inner boundary radius

Middle boundary radius = Inner boundary radius + 0.5*(Width of ring)

The Inner boundary radius and the Outer boundary radius were not reported for some rings (e.g. Uranian rings and Neptunian rings). Therefore, to calculate them, I used the value of the Middle boundary radius and the Width of the ring as:

Outer boundary radius = Middle boundary radius + Width of ring

Inner boundary radius = Middle boundary radius - Width of ring

For some rings, the width was reported as a range. To estimate the average width, I calculated the average of the reported range. Similarly, because most of the following variables were reported as range, the value I used for them in the quantitative analysis was the average of the reported range: Optical depth, albedo, surface density, and thickness.

30.3. Albedo of rings

Measured for 24 rings, the albedo of the rings varies between 0.015 and

10,000,000 (Fig. 162). The smallest albedo was recorded on Main Ring, a Jovian ring located upstream of the innermost Jovian satellite, meaning it is located in the region between Jupiter and its innermost satellite. In contrast, the highest value was recorded on E Ring, a Saturnian ring located in turbulence Zone 2, meaning in the transition zone to turbulence. The second highest albedo (100, 000) was recorded on G Ring, a Saturnian ring located in turbulence Zone 2. All other reported albedos are not higher than 0.6, which was recorded on F Ring, a Saturnian ring in Zone 1. In general, the 9 smallest albedos were recorded with Uranian rings while the 7 highest ones were with Saturnian rings. The albedos of Neptunian rings are intermediate between those of Uranian rings and those of Saturnian rings.

Fig. 162: Albedo of the rings according to their types and turbulence zones

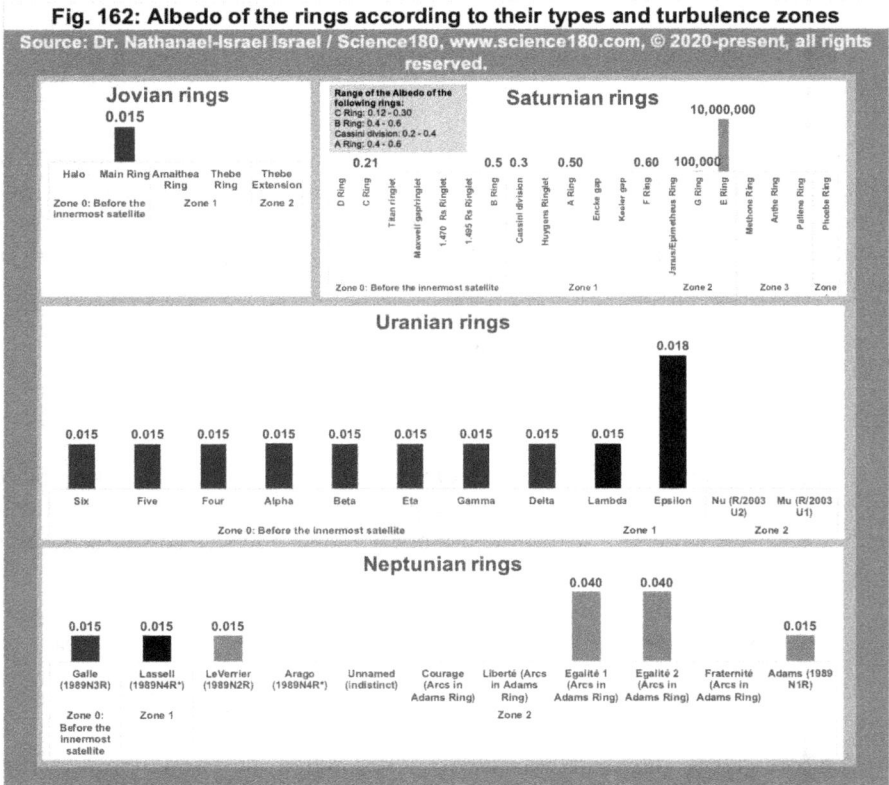

30.4. Inner boundary radius of rings

Measured on 47 rings, the inner boundary radius of the rings varies between 40,900 km and 7,722,240 km (Fig. 163). The smallest inner boundary radius was recorded on Galle (1989N3R), a Neptunian ring found in the region inward or upstream of the innermost Neptunian satellite, meaning located before the innermost Neptunian satellite. This means that no ring is closer to its primary planet as the Neptunian ring mentioned above. The next 10 smallest inner boundary radii of rings were recorded on Uranian satellites. In contrast, the highest inner boundary radius was recorded on Phoebe Ring, a Saturnian ring located in turbulence Zone 4. The second highest inner boundary radius of rings was recorded on Thebe Extension, a Jovian ring located in Zone 2, at a distance of 221,900 km from Jupiter. In other words, the distance separating the inner boundary of the Saturnian outermost ring from Saturn is 34.8 times that separating the outermost Jovian ring from Jupiter.

The repartition of the rings according to the turbulence zones is:
- 18 rings in the region located upstream of Zone 1,
- 11 rings in Zone 1,
- 15 rings in Zone 2,
- 3 rings in Zone 3, and
- 1 in Zone 4.

The 18 rings located inward of the innermost satellite of their primary planets are:
- 2 Jovian rings,
- 7 Saturnian rings,
- 8 Uranian rings, and
- 1 Neptunian.

All of the rings in Zones 3 and 4 are Saturnian rings. In other words, beside Saturn, no other planet known as of 2020 has a ring in Zones 3 and 4. No planet has rings in Zone 5, suggesting that the precursors of the bodies in Zone 5 may have not met the requirement for ring formation. The size, amount of material, and the speed of the precursors in Zone 5 may have played a role.

Fig. 163: Inner Boundaries Radius (km) of the rings according to their types and turbulence zones

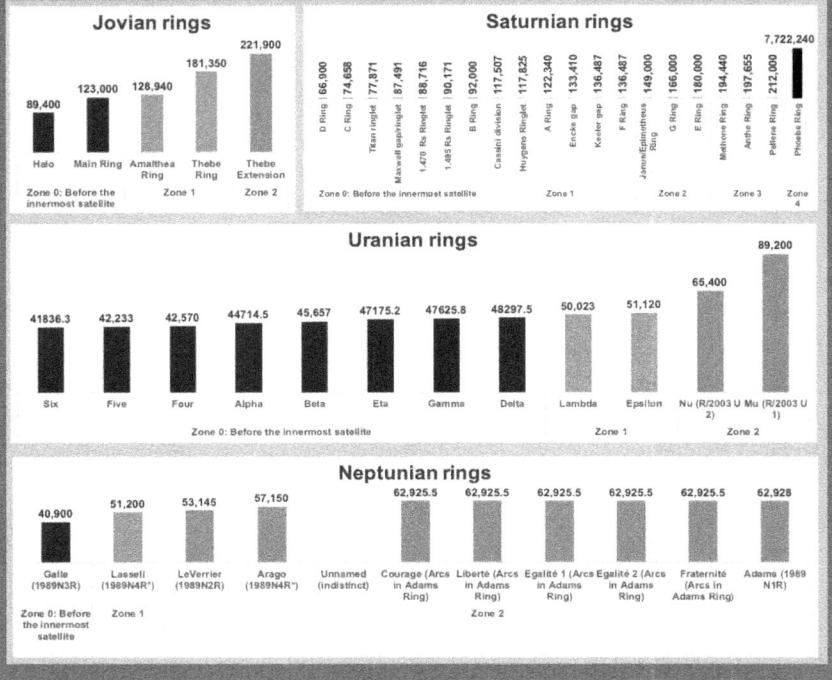

The minimum inner boundary radius of the rings is negatively correlated with the semi major of the primary planets (r2=0.99) (Fig. 164). In other words, the farther away a primary planet is from the Sun, the smaller the minimum inner boundary radius of its rings. To put it another way, as the precursors of the planetary systems were positioned farther away from the precursors of the Sun, the closer their rings were formed from their primary planets.

Nathanael-Israel Israel: Known as the #1 Universe-Origin, Life-Origin, and Chemicals-Origin Scientist & Mathematician

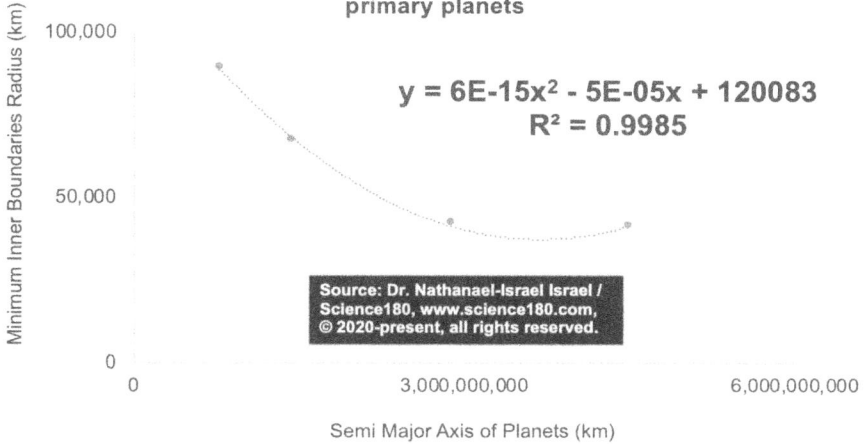

Fig. 164: Relationship between the inner boundaries radius of the innermost rings and the semi major axis of their primary planets

$$y = 6E\text{-}15x^2 - 5E\text{-}05x + 120083$$
$$R^2 = 0.9985$$

Before moving to the next segment, I need to say a few things about how the inner boundary radius of the rings has helped me to better investigate the formation of rings. Indeed, when I began working on the origin of rings, I wrongly thought at one moment that they could have been made of particles that escaped the largest satellites as the later were being formed just as if you violently open a packet of powder and pour it out, some powder can escape into the atmosphere. However, after I calculated the inner boundary radius and the outer boundary radius of the rings, and then mixed them with the semi major axis of the satellites, I was shocked to see that many rings were close to their planets than the innermost satellites. Moreover, most prominent rings are not even found near the largest satellites. I quickly understood that the origin of the rings cannot just be connected to the formation of the largest satellites. This early insight gave me a solid ground on which I confidently stood to quickly rule out certain thoughts that could have drifted me away from the real origin of the rings. Before presenting my understanding of the formation of the rings, I will present some of the facts which led me there.

30.5. Middle boundary radius of rings

Studied for 45 rings, the middle boundary radius of the rings varies between 41,837 km and 10,105,275 km (Fig. 165). The smallest middle boundary radius was found with Six, a Uranian ring located in Zone 0. The highest middle boundary radius was found with the Phoebe Ring, the only Saturnian ring in turbulence Zone 4. The second highest middle boundary radius (330000 km) was recorded with E Ring, a Saturnian ring in Zone 2. Most of the smallest middle boundary radius were found with Uranian rings.

The middle boundary radius of the rings is negatively correlated with the semi major of the primary planet (r2=0.99).

Fig. 165: Middle Boundaries Radius (km) of rings according to their types and turbulence zones

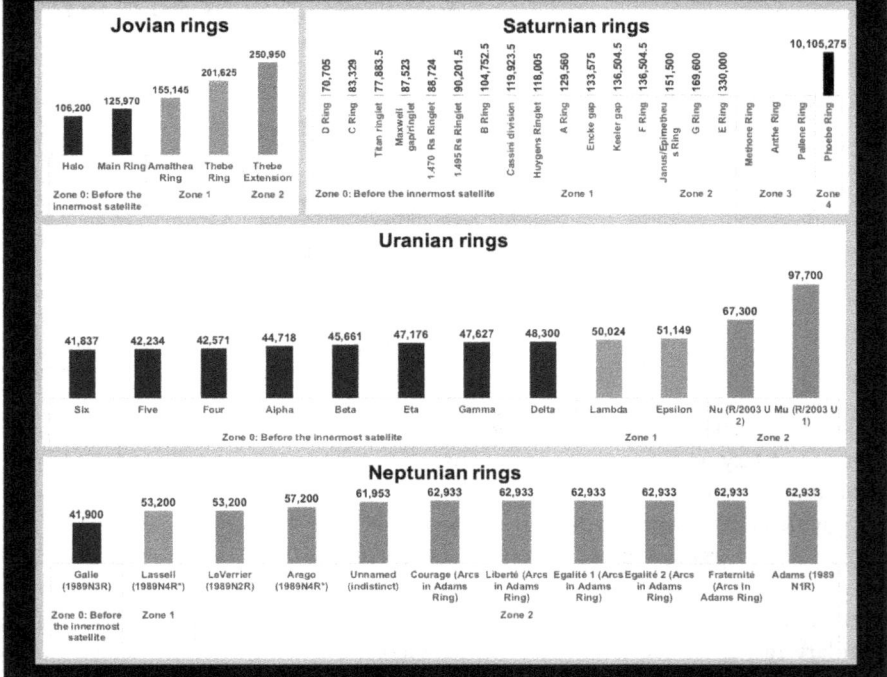

30.6. Outer boundary radius of rings

The outer boundary radius of the rings in the Solar System varies between 41,837.8 km and 12,488,310 km (Fig. 166). The highest outer boundary radius was found with the Phoebe Ring, a Saturnian ring in turbulence Zone 4. The second highest outer boundary radius was also recorded with another Saturnian ring called E Ring, located in turbulence Zone 2. Then, the 3rd, 4th, and 5th highest outer boundary radius were recorded with the Jovian rings:

- Thebe Extension (located in Zone 2): 280,000 km
- Thebe Ring (located in Zone 1): 221,900 km
- Amalthea Ring (located in Zone 1): 181,350 km

In contrast, the smallest outer boundary radius was recorded with Six, a Uranian ring located in the space between Uranus and its innermost satellite. The 3 smallest outer boundary radii were recorded on the Uranian rings. Then, the 4th smallest outer boundary radius is for a Neptunian ring.

The outer boundary of 18 rings is located upstream of the innermost satellite of their primary planets. In other words, 18 rings are entirely located between the primary planets and their innermost satellite:

- 2 Jovian rings,
- 7 Saturnian rings,
- 8 Uranian rings, and
- 1 Neptunian ring.

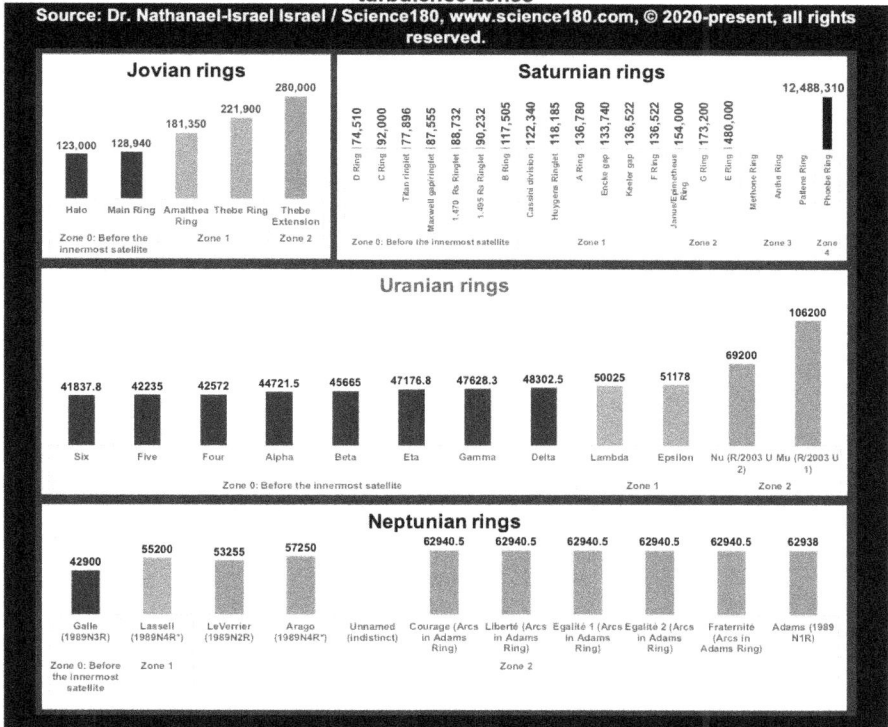

Fig. 166: Outer Boundaries Radius (km) of rings according to their types and turbulence zones

The outer boundary radius of about 91.67% of the rings in the Solar System are located upstream of Zone 3. To put it another way, of the 48 rings I studied in the Solar System, 44 are located before or inward of Zone 3 (the zone where the largest satellites are found). To explain myself better, I am saying that 91.67% of the rings found around the 4 giant planets are located upstream or before the zone of the biggest satellites.

No Jovian, Uranian, or Neptunian ring is located beyond turbulence Zone 2. Saturn is the only planet which has rings in Zones 3 and 4. Rings are not found in turbulence Zone 5. This suggests that in general, conditions in the innermost turbulence zones could have favored the formation of rings than in the outermost turbulence zones.

The outer boundary radius of the outermost Saturnian ring (Phoebe Ring) is 44.6 times that of the outermost Jovian ring (Thebe Extension). The large area covered by the Saturnian rings and their thickness may explain why they are the most visible from Earth.

30.7. Eccentricity of rings

Reported for just 14 rings (3 Saturnian and 11 Uranian), the eccentricity of the rings varies between 0 and 0.0079 (Fig. 167). The most eccentric ring is Epsilon, a Uranian ring in turbulence Zone 1. The second highest eccentricity (0.0026) was recorded with F Ring, a Saturnian ring in Zone 1. In contrast, the least eccentric rings are the Uranian. Compared to the eccentricity of satellites, planets, and asteroids, the eccentricity of rings is too small.

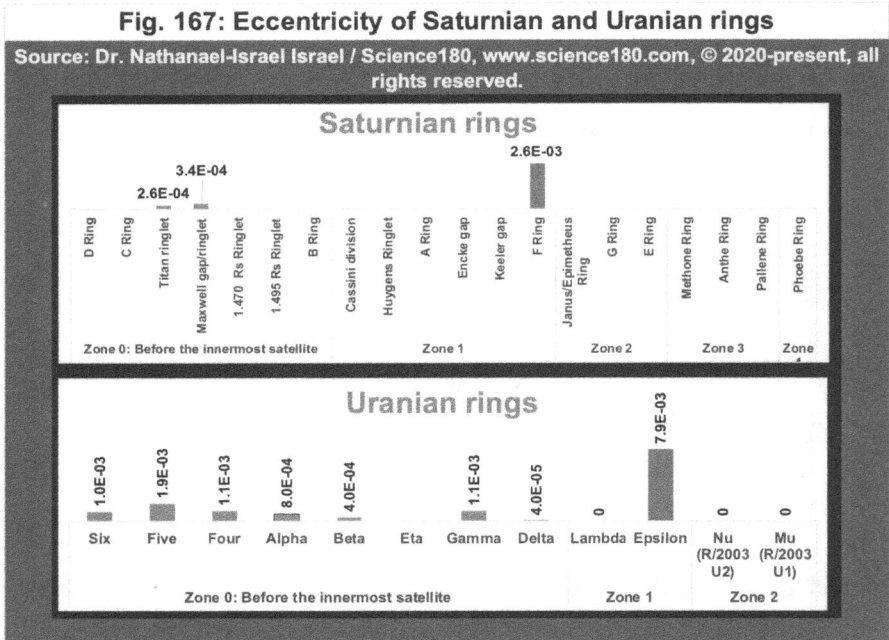

Fig. 167: Eccentricity of Saturnian and Uranian rings

30.8. Optical depth of rings

The optical depth of the rings in the Solar System ranges from 0 to 5 (Fig. 168 and Fig. 169). The 3 smallest values were recorded with the Saturnian rings, while the highest value was with Six, a Uranian ring located in turbulence Zone 0, meaning the turbulence zone positioned between Uranus and its innermost satellite. The optical depth of the Jovian rings is among the smallest ones.

Fig. 168: Maximum optical depth of rings according the turbulence zones

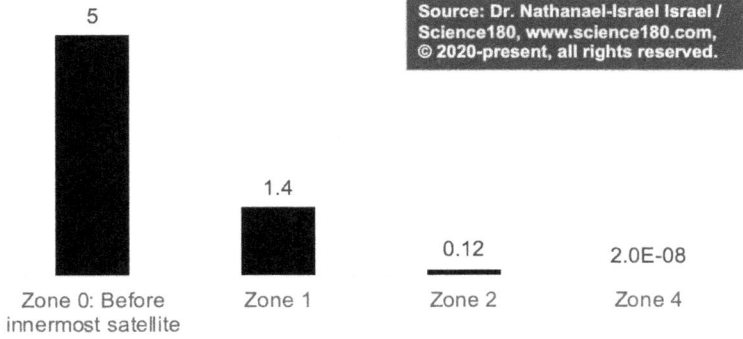

Fig. 169: Average Optical Depth of rings according to their types and turbulence zones

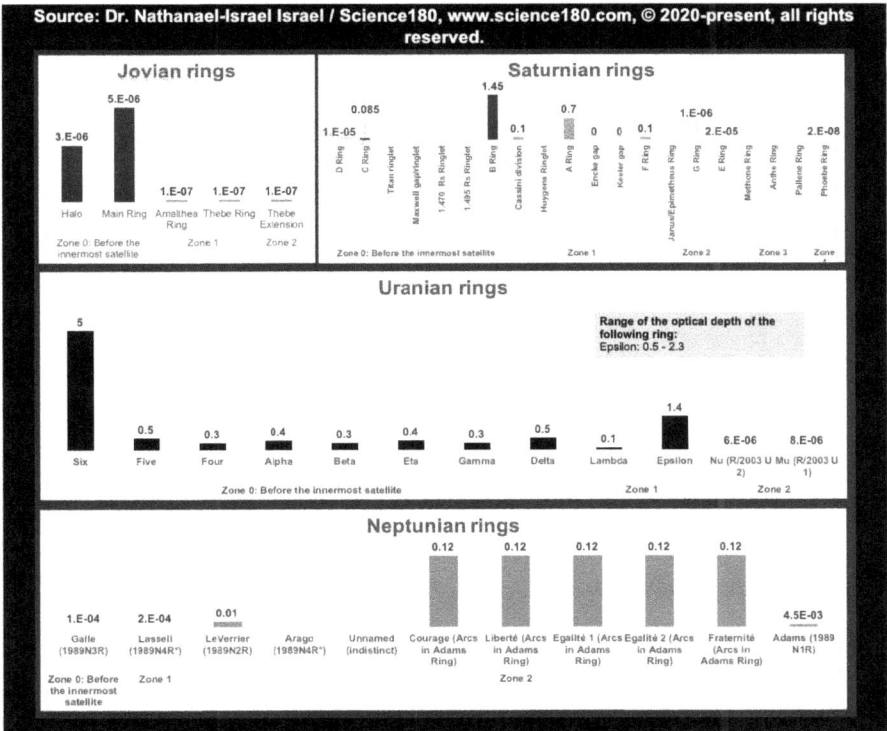

The 11 highest optical depths are recorded on either a ring in Zone 0 or in Zone 1 whereas none of the 7 smallest optical depths were found in Zone 0. Some of the 7 smallest optical depths are in Zones 1 and 2 or 4. This suggested

that the optical depth of some inner rings tend to be higher than that of the outer rings.

As shown in the next graph, the maximum optical depth of the rings decreases from the innermost rings to the outermost ones.

The highest optical depth of the Jovian and Saturnian rings is recorded on the ring immediately upstream of the innermost satellite. In contrast, for the Uranian rings, the highest optical depth was recorded with the innermost ring, which is farther from the innermost satellite. For the Neptunian rings, the highest optical depth was recorded on those in Zone 2.

30.9. Surface density of rings

Data on surface density was reported for only 7 rings: one Jovian ring and 6 Saturnian rings. The surface density ranges from 5.E-06 g/cm^2 to 60 g/cm^2 (Fig. 170). The smallest value was obtained with Main ring, a Jovian ring. The highest surface density was found with B ring, a Saturnian ring located in the outer portion of Zone 0 just before the innermost Jovian satellite. In other words, the B ring which got the highest surface density is positioned between Saturn and its innermost satellite. Similarly, the surface density reported for the Jovian rings was also obtained with the ring located just inward of the innermost Jovian satellite.

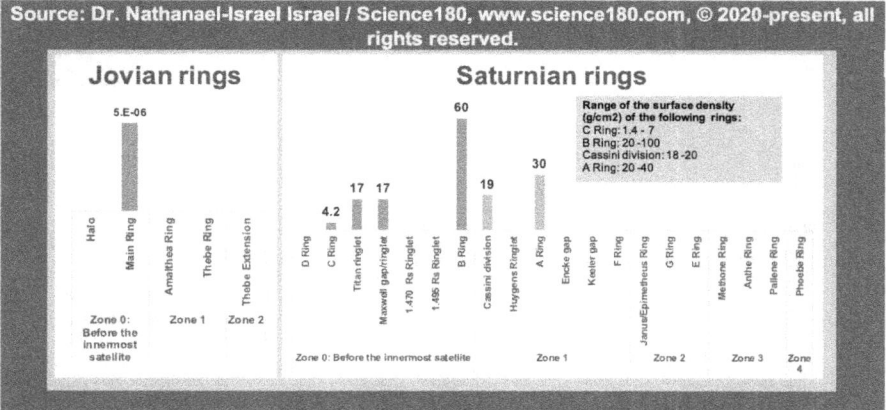

Fig. 170: Surface Density (g/cm2) of Jovian and Saturnian rings according to their types and turbulence zones

30.10. Thickness of rings

Thickness is reported for only 9 rings (5 Jovian and 4 Saturnian). It varies between 5 meters and 10,000 km (Fig. 171). The 4 smallest were recorded on the Saturnian rings while the 5 highest were on the Jovian rings. The thickest ring is Halo, the innermost Jovian ring located in turbulence Zone 0, which is the zone between Jupiter and its innermost satellite. The least thick Jovian ring, which is the Main ring (thickness = 100 km) is 5,000 times to 20,000 times thicker than the Saturnian rings.

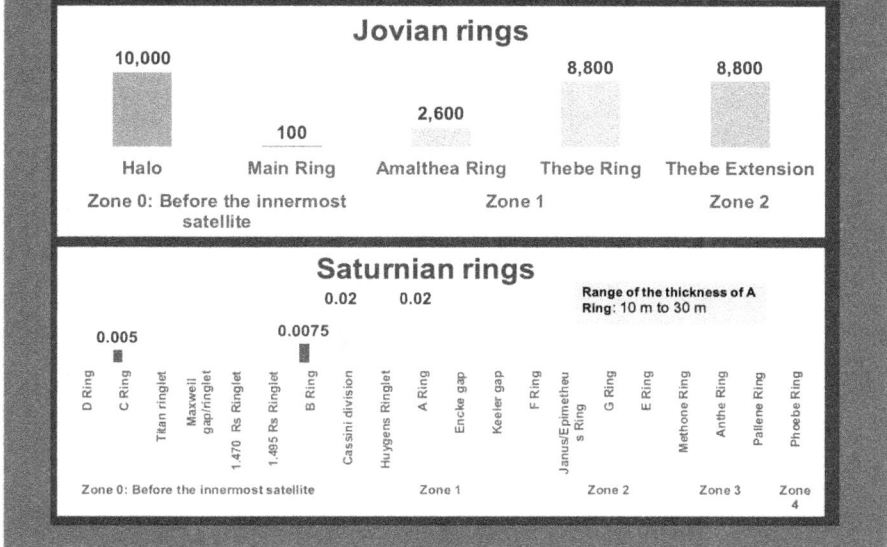

Fig. 171: Thickness (km) of Jovian and Saturnian rings

30.11. Width of rings

The width of the rings varies between 1.5 km and 4,766,070 km (Fig. 172). The smallest width was recorded with Six, a Uranian ring located in turbulence Zone 0. Varying between 1.5 km and 8 km, the 9 smallest widths of the rings were recorded on the Uranian satellites. Eight of those 9 rings are located in Zone 0 of Uranian satellites, meaning between Uranus and its innermost satellite. The highest width was recorded with Phoebe Ring (a Saturnian ring in Zone 4). The second highest width (300,000 km) was found with E Ring, a Saturnian ring in Zone 2.

The innermost Jovian rings are not as wide as the outer ones. For Saturnian rings, wider rings are separated by many thinner ones. Except for the Neptunian rings where the widest is in Zone 1, the widest rings for the other planetary systems are generally the outermost ones.

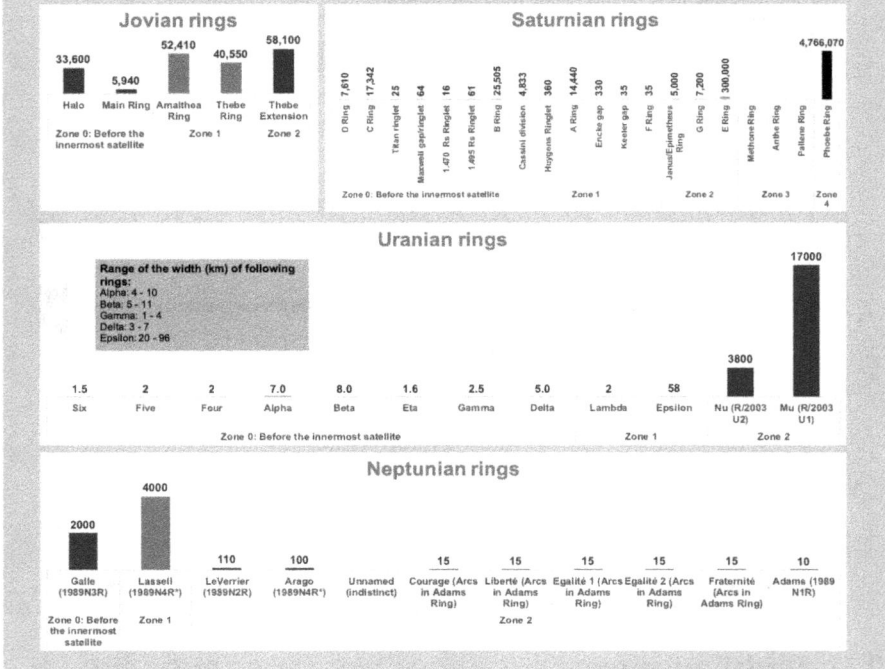

Fig. 172: Width (km) of rings according to their types and turbulence zones

30.12. Size of the particles in rings

The size, nature, and chemical composition of the particles and rocks found in rings are not the same, but they vary according to the types of rings. Because a lot of uncertainty still exists about the chemical composition of some rings, I did not focus on that here but I focused on the size of the particles. Indeed, unlike the rings of the other 3 giant planetary systems, the Uranian rings are said to consist of large bodies 20 cm to 20 m in diameter (Ockert et al., 1987). The bigger size of the particles in the Uranian rings may be connected to the nature of the turbulence of their precursors. In contrast, the Jovian rings are mostly made of dusts (Smith et al., 1979; Showalter et al., 1987). While the size of particles in the Halo ring is estimated to be in the order of submicrometers, particles in the Jovian rings are estimated to have a radius of about 15 μm (Throop et al., 2004). In contrast, the Saturnian rings are said to consist mostly of water ice and trace amounts of rock, and the size of their particles varies between micrometers to meters (Porco, 2012). Saturnian rings are also believed to be very rich in water ice (Esposito, 2002). Finally, the Neptunian rings are believed to be very dusty and the densest regions are said to be like the less-dense Saturnian rings. Both Uranian and Neptunian rings are very dark and are alleged to contain organic compounds

(Smith et al., 1989). As long as size and faintness are concerned, Neptunian and Jovian rings are said to look similar, both consisting of faint, narrow, dusty ringlets, and even fainter broad dusty rings (Burns et al., 2001). Finally, unlike what many people think, *"rings are not a series of tiny ringlets, but are more of a disk with varying density"* (Tiscareno, 2013). Below, I explained the origin, formation, and characteristics of the rings.

30.13. Origin and formation of the rings

At this point, based on the data I presented above on rings and also the data I studied about the planetary systems and asteroids in the Solar System, I would like to present how I think the rings were formed. Indeed, to better understand the origin of the rings, I had to analyze their features in perspective with the characteristics of the satellites imbedded in them or located near them. I also considered the characteristics of the turbulence zones where those rings are found.

To address the formation of the rings, I felt like I needed to add some details to the formation of the satellites first. In previous chapters, I showed that the precursor of the planetary systems split-gathered to birth the precursor of the satellites, which at its turn split-gathered into individual satellites after going through some events. Because I did not want to put the horse before the cart, I did not introduce the precursor of the rings when I was addressing the precursors of the satellites. To be precise, when the precursors of the planetary systems split-gathered, they birthed:

- the precursors of the primary planet and dusts surrounding them, and
- the precursors of the satellites and rings.

To ease the understanding, I called the precursor of the satellites and the precursor of rings the precursor of the secondary bodies in the planetary systems. Then, I called the precursor of a primary planet and that of the dusts around them the precursor of the primary bodies. In other words, the precursor of the planetary systems split-gathered into the:

- precursor of the primary bodies, and the
- precursor of the secondary bodies.

The trail of dusts in which some planets orbit the Sun could have come from some "left over" of the precursor of their planetary system that was not gathered together with neither their planet nor their satellites. As the fluid layers in the precursor of the secondary bodies in the precursor of a planetary system started going through turbulence, they were not always able to gather together all of their particles into a single body and form a single satellite that has no trail of dust around it. Instead, some particles were gathered together into rocks and other particles of various sizes. The particles which were gathered together into larger bodies became the satellites, while those which were not incorporated into the precursor of the satellites become the particles or clusters of matter which together formed the rings. In other words, because the particles in some fluid layers of precursors of secondary bodies could not be gathered together into satellites, they were collected and scattered into large particles and/or rocks which became rings. Hence some rings do not contain satellites while others do. To say this another way, not all layers were gathered together into satellites. Those which were gathered into satellites were not able to collect all their particles. The layers which were not gathered together at all and the left-over particles from the layers which were gathered formed the particles and rocks present in the rings. Another way of explaining this is that during the formation of the rings and satellites, there was a point, where the precursors of the rings and satellites were precursors of particles which were split-gathered into the precursor of the rings and the precursors of the satellites as their turbulence was underway (Fig. 173). As turbulence was developing and fluids were moving, mixing, rotating, spiraling, and packing, some fluids were packed into bigger clusters which yielded the satellites while other fluids did not rotate enough to be amass into bigger bodies, but smaller particles spread over the space occupied by their mother precursors. The daughter bodies which were not amassed into satellites became the particles constituting the rings. In contrast, the gathering together of the precursors of particles into denser bodies led to the formation of the satellites. Hence some satellites are embedded inside rings. Additionally, the size of the particles in the rings depends on the type of planetary system.

At this point, I felt like I still needed to explain why all particles in the precursor of the secondary bodies in a planetary system were not incorporated into the satellites. Indeed, when fluids are flowing in a current, not all the fluids collect into vortices. While some fluid filaments can swirl and form vortices, other fluids just flow along the current. Moreover, all vortices in a fluid flow do not rotate the same way. Likewise, in the fluid layers of the precursors of the secondary bodies in a planetary system, not all the fluids were collected into vortical structures, nor all vortices were big. Some vortices were piled together or collected together to form larger bodies including satellites, while the vortices of the fluids that could not be integrated into the bodies of the satellites formed rocks which surround those satellites; and those rocks all together form the rings. Also, some fluid which did not swirl to form vortical structures were frozen in some cases and became also ring components. Remember, I showed in previous chapters that, during the turbulence of the precursors, vortices in fluid layers were amalgamated, meaning smaller vortices were swallowed by big ones, which at their turn were incorporated in even larger ones and so on and so forth until the largest bodies producible by those fluid layers were reached. Likewise, in the fluid layers of the precursors of the secondary bodies in the planetary systems, not all vortices which were formed were incorporated into the precursor of the satellites. In the end, the vortical structures that did not land into the precursor of satellites formed rocks and other types of particles constituting the rings. Unlike what some theories said, the rocks and other particles in rings did not migrate from other places to their current location, but they were formed at the same time as the satellites they surround.

By the time the primary planet, the satellites and the rings finished their formation, they were organized into a planetary system as follows (Fig. 174):

Fig. 173: Layers of fluids in the precursor of the secondary bodies split-gathered into rings and satellites

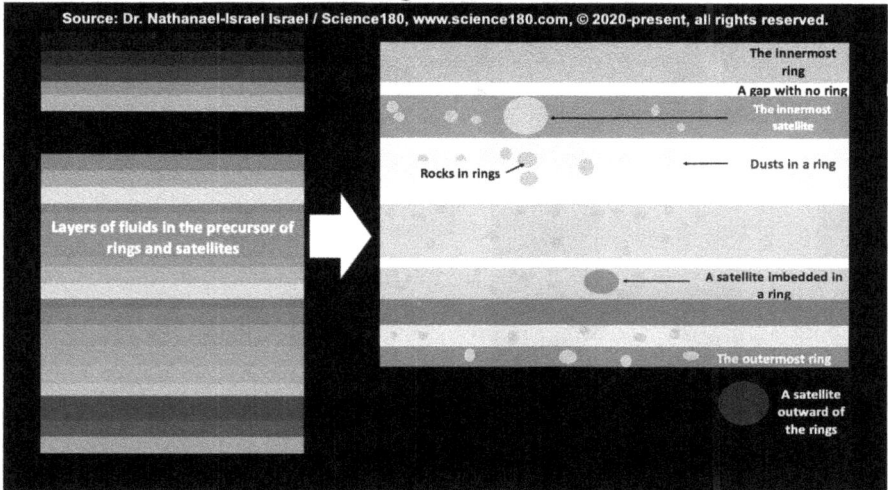

To get a very beautiful colored version of this graph, visit http://www.science180.com/RingFluidLayers

For instance, no satellite was found inside the Halo ring, the innermost Jovian ring. While the Halo ring is very thick, the other Jovian rings are faint. The fact that the Halo ring is the vertically thickest Jovian ring suggested to me that its precursor was unable to gather into a satellite but its particles were packed into a thick torus. In other words, the Halo ring is a good example of a layer of fluids of a secondary body in a planetary system which was not -gathered into a satellite. The other Jovian rings are faint because the precursors of some particles present in their neighborhood were gathered together into some of the satellites embedded into those rings and/or the layers of fluids which birthed those rings could not be gathered together into a satellite. Some dust and other smaller particles in the precursors of the secondary body diffused after being unable to form a satellite.

The same trend is observed for Saturnian rings. At least 7 rings are found upstream of the innermost Saturnian satellite and they are:

- D Ring
- C Ring
- Titan ringlet
- Maxwell gap/ringlet
- 1.470 Rs Ringlet

- 1.495 Rs Ringlet
- B Ring

Fig. 174: Layers of fluids in the precursor of the primary planet, satellites, and rings split-gathered to form the planetary system

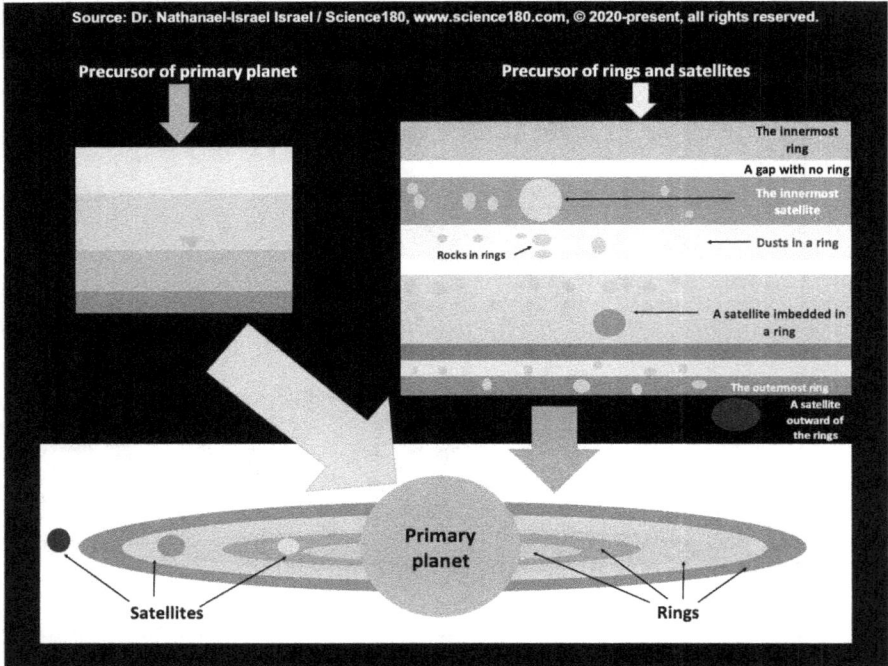

Get a pretty colorful version of this graph at
www.Science180.com/RingFormation

Of these inner Saturnian rings, B Ring (located just upstream of the S/2009 S1, the innermost Saturnian satellite) is the opaquest of Saturn's ring. The opacity of B Ring is not by chance, but can be explained by the amount of material gathered into its precursor. The precursor of B Ring could have been dense, but its gathering together did not yield a satellite. However, downstream of the precursor of B Ring, conditions were met and allowed the gathering together of the precursors of the secondary bodies into S/2009 S1, the innermost Saturnian satellite known as of 2020. Together with the C Ring (which is upstream or inward of B Ring), the A and B Ring (which are separated by the Cassini Division), are generally considered as the main Saturnian rings. The main rings are the densest rings and they contain larger particles, suggesting that their precursors failed to amass into a single satellite but were gathered into particles constituting the rings. Some of the precursors of the secondary bodies in the vicinity of the main rings which were "favorable" were gathered together into the innermost Saturnian satellite (S/2009 S1), located inside the main ring. Part of the particles or rocks that were located in the gap between the A ring and the B ring could have also been gathered together to form the innermost Saturnian satellite. Some of those gaps can also explain the space created after fluid layers were separated just as "empty" voids exist between bodies in the universe. In contrast, because the precursors of the bodies downstream of the main rings were able to split-gather together into satellites, their remaining particles that were not incorporated into those satellites became the rings. Hence those rings are usually dusty, narrow, very faint, and are composed of much smaller rocks and particles.

As for the Uranian rings, the main Uranian rings are also located inward of the innermost Uranian satellite. This also corroborates the location of the main Jovian rings and the main Saturnian rings: inward of the innermost satellite of their primary planet. Outward of the innermost Uranian satellite, the rings are faint and consists of smaller particles and generally lack dust, suggesting that outward of the innermost Uranian satellite, more precursors of the particles or rocks which could have ended up in the rings were mostly incorporated into the satellites.

Some of the empty spaces or gaps between rings are due to demarcation of the space between the layers of fluids in the precursors of the secondary bodies (rings and satellites) and/or due to the incorporation of the precursors of those particles into satellites. For instance, the Saturnian satellite called Pan is located inside the Encke gap, while Daphnis (another Saturnian satellite) is located inside the Keeler gap. Some of the materials used to form these satellites could have been particles which otherwise would have filled these gaps within the rings. In other words, the more the particles in the precursors of the secondary bodies were gathered together into satellites, the less materials were left to form the rings, hence a gap or a faint ring. As the fluid layers of the precursors of the secondary bodies were separated, distances were also born between them. The density and brightness of the rings could have been determined by the number of particles or amount of matter in the precursors of bodies that were not gathered together into the satellites, but which were left out to become part of the rings. The fewer the particles in the rings, the brighter it may look, while the more particles in the rings, the denser it would be. The variation of the size of the rocks and particles constituting the rings could depend on the size of the vorticial or spiraling clusters of matter in their precursors. In other words, as the precursors of the rings were being shaped into their daughter bodies, different amounts of matter were gathered along the way according to the nature of the turbulence they went through and according to the factors that split-gathered them. The vortices and other spiraling structures which occurred inside the precursors of the rings and satellites can explain why some spiral patterns and spiral waves are still observed inside some rings until today. Because most of the precursors of the secondary bodies in Zone 3 were gathered together into the satellites in that zone, rings are not usually found in or outward of Zone 3 and when they are found (like in the case of some Saturnian rings ranging from Methone ring to Phoebe ring), they are usually faint because they consist of tiny, smaller particles or rocks, and just dusts. This also suggests that most of the rocks, particles, and dust in rings are not just escapees from the big satellites or from any other satellites. In contrast, rocks have been falling from the rings onto neighboring satellites or onto their primary planets.

Also called in the literature as a herder moon, watcher moon, a shepherd moon is defined by Wikipedia (2020a) as a small natural satellite that "clears a gap in planetary-ring material or keeps particles within a ring contained". I think that the gaps around "shepherd moons" are cleared because their precursors were gathered together into those moons. Some of those shepherd moons are located at the edge of the rings because what is called edge or gap is in fact caused by the removal of what could have been rings and its incorporation into the satellites. The process by which shepherd moons were formed was not a simple pick up or deflection of particles from their original orbits as some people think. That is why I am not comfortable with the usage of the term "shepherd" for it is misleading.

Beyond Phoebe, no visible ring is reported because the layers of fluids of the precursors of the secondary bodies were so thin that most of them were gathered together into satellites. Although some satellites downstream of Zone 4 may be embedded in tiny trails of particles, these particles are not as many as needed to form visible rings. In other words, although some particles or dust in rings may have escaped from the precursor of the satellites near them as they were being wrapped, spiraled, and packed together, the trends of the data I analyzed suggested that dust escape is not the major driver of the formation of rings, else Zone 3 could have been the richest in rings; for satellites in Zone 3 are bigger and their precursors could have been very turbulent and could have released a lot of dust if that was the case. While some satellites protrude above the rings, others are embedded inside them and others still are located at the edge of the rings. Some rings could have been moved outward because of the shearing of their precursors and the expansion of their system from inward to outward. Because the precursor of the satellites and the precursors of the rings imbedding them descended from neighboring mother precursors, there must be a link between the characteristics of the rings and those of the satellites embedded inside of them. This can also explain why the Phoebe ring tilted more than the rings inward of it. I felt like the precursor of the Phoebe ring could not have been the same as that of Phoebe but near the precursor of some of the inner satellites of Zone 4 located inward of Phoebe. Hence that ring does not have a tilt or inclination as high as that of Phoebe, but instead an inclination that better matches the inclination of the satellites of Zone 4 located inward of Phoebe.

The thing that explains why Saturn is the only planet which has rings in Zones 3 and 4 can also explain why the Saturnian rings are the most massive and prominent in the Solar System. Indeed, from Earth, the rings of Jupiter are not as clear and well defined as Saturnian rings. In fact, for most pictures of the Solar System, Jovian rings are not usually mentioned. It is only the Saturnian rings that are mentioned because they are the largest and most visible. To properly explain the size of the Saturnian rings, I needed to make a connection with the radius, number, and semi major increment of the Saturnian satellites with respect to the same variables of the other satellites. Indeed, like I addressed in the chapter on radius of the celestial bodies, the smallest satellite in the Solar System (S/2009 S1) is the innermost Saturnian Satellite and many other Saturnian satellites are small. For instance, in Zones 3 and 4 (the zones where no planet other than Saturn has rings), the mean radius of the Saturnian satellites is the smallest as compared to the mean radius of the satellites of the other planets (Fig. 175 and Fig. 176).

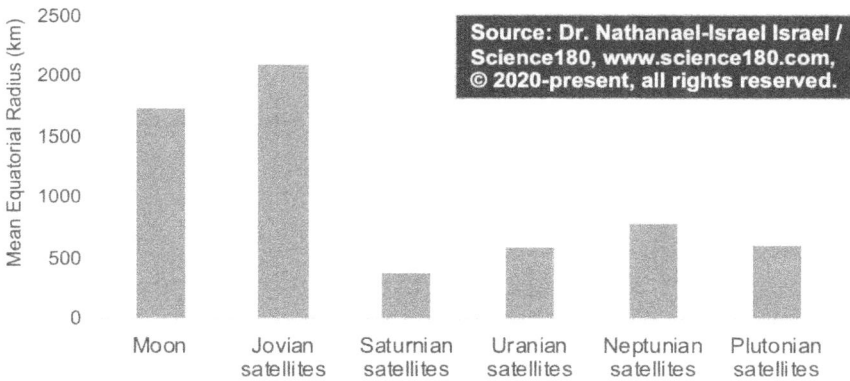

Fig. 175: Mean Equatorial **Radius** (km) of Satellites in Turbulence **Zone 3**

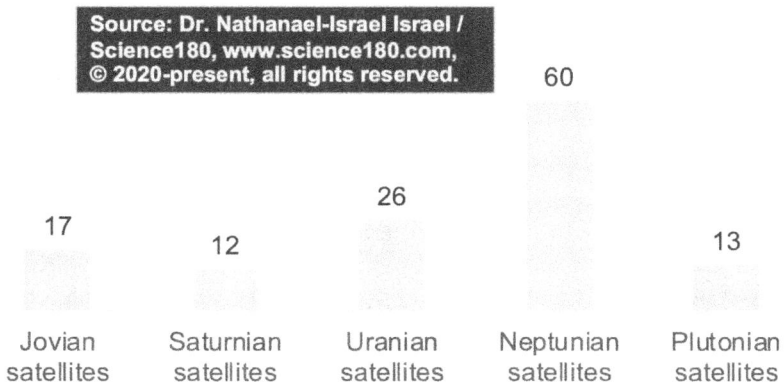

Fig. 176: Mean Equatorial **Radius** (km) of Satellites in Turbulence **Zone 4**

Moreover, when the radius of the satellites in Zones 1, 2, and 3 were compared across the planetary systems, zone by zone, Saturnian satellites recorded the smallest radius in each of the zones. In Zones 1, 2, and 3, Saturnian satellites are not only relatively smaller than the satellites of the other planets, but also, they are many more. For instance, in Zone 3 alone, Saturn has 15 satellites while no any other planet has more than 5 satellites in Zone 3. In Zone 4, Saturn has 16 satellites while none of the other planets has more than 9 (Table 14). In other words, the precursor of the Saturnian satellites in Zones 3 and 4 was broken up into many small satellites more than the precursors of the satellites in the same zones of the other planets.

Table 14: Number of satellites according to the types of satellites and turbulence zones

Turbulence zones	Moon	Martian satellites	Jovian satellites	Saturnian satellites	Uranian satellites	Neptunian satellites	Plutonian satellites
Zone 1	1	2	2	4	5	2	
Zone 2			2	5	8	4	
Zone 3			4	15	5	2	
Zone 4			9	16	6	4	1
Zone 5			62	42	3	2	4

As far as the semi major of the biggest satellites in each of the planetary systems is concerned, Titan (the biggest Saturnian satellite) recorded the highest value of them all, suggesting that the precursors of the Saturnian satellites are carried over a longer distance before forming its biggest satellite. The inability of the precursor of the Saturnian secondary bodies to be gathered into big satellites (beside Titan) was also manifested on the microscopic and macroscopic levels, making more of these precursors to be gathered into smaller satellites and also leaving out many of them which were instead collected into small rocks and dusts visible as rings. In other words, the precursors of the materials in the Saturnian rings which could have been used to form bigger Saturnian satellites or to increase the size of the Saturnian satellites could have been instead spread out to form the prominent Saturnian rings. It is possible that the viscosity of the precursor of the Saturnian planetary system could have been smaller than that of the precursor of the Jovian planetary system. In the chapters on the semi major increment, vortex, and fluid ligament, I also explained why the viscosity of the bodies in the Jovian planetary system could have been higher than that of the bodies in the other giant planetary systems. Consequently, the materials constituting the Saturnian rings could have been looser than those of the Jovian rings. This could have also contributed to the higher dispersion or diffusion of the precursor of the Saturnian rings. All this could have been facilitated by the ease that the precursor of Titan had to slide over the precursor of the bodies located beneath it. As far as density is concerned, the bodies in the Saturnian planetary system are usually the least dense. Saturn itself is the least dense planet in the Solar System. Because the precursors of the Saturnian satellites could have been less dense than that of the Jovian satellites, it is also possible that the precursors of the Saturnian rings were less dense than that of the Jovian rings. Therefore, the precursors of the Saturnian rings could have diffused more and/or were unable to amass into bigger rocks and bigger satellites. In contrast, because the precursors of the Jovian satellites in Zone 3 could have been viscous, they gathered together their constituents into daughter satellites and did not leave out much particles to form rings. Hence no ring was formed in Zone 3 of the Jovian satellites. In contrast, most of the rings are inward or upstream of Zone 3 because the precursors of the bodies in that zone were not able to amass all their particles into bigger bodies. That is why most of the rings in Zone 3 are thin and faint, if any at all. Moreover, in the zone inward of Zone 1 where no satellite is found, the precursors of the bodies were scattered into rocks and particles forming rings.

To summarize, most of the particles found in rings were not transported to their current location by other celestial bodies (as some theories allege) but they were born from the precursor of the secondary bodies in their planetary system. The split-gathering that defined the formation and size of matter on different scales, allowed that in the regions where rings are found, small rocks or clusters of particles were formed, incapable to be amassed into larger celestial bodies such as satellites or incorporated into the bodies of their neighboring satellites. As the precursors of the rings could not gather together into bigger bodies, they were spread all over the domain near the trajectory of their orbit, that could have been cleared or that looked like a gap in these precursors incorporated into satellites. The color, density, brightness, thickness, and other characteristics of the rings depend on their precursors and how their environmental conditions shaped them. The inclination of the rings is a product of the inclination of their precursors. The variation of the density, faintness, or brightness of the rings can be explained by the variation of the proportion of the precursors of the secondary bodies gathered together into satellites or left over to be incorporated into rings and how this left over was scattered throughout their surrounding space.

Although the precursors of Venus, Mercury, and Earth could have been dense, all of their particles were not gathered together without leaving out some which formed co-orbital trails of dusts or rings. For as they were being gathered together, all particles or matters in the precursors of the celestial bodies rarely collected themselves without leaving some out. Just as a gas torus exists around some satellites, for all the matter in the precursor of the secondary bodies in their neighborhood were not gathered together into the satellite but some were left to form the torus, so also, all of the particles in the precursors of the planets or planetary systems were not gathered together, but some of them were collected into small asteroids (including some trojans), while other left over particles became trails of dusts or rings surrounding some planets or in the path of their orbit. In other words, all the particles in the layers of the precursors of the planetary systems were not incorporated into the planetary systems. Some leading trojans could have descended from the leading portion of the precursors of a planetary system which was not gathered together with the planetary system. In contrast, trailing trojans could have been the portion of the precursors of the planetary systems that were lagging behind the portion which became the planetary system. In other words, some of those asteroids which were formed directly in front of the planets became some of the leading trojans while those which were formed behind the planets are some of the trailing trojans. Some trojans which, with respect to the position of the planet to the Sun, are inward of the planet and those located outward of the planet, could have been formed when the precursor of the entire planetary system was separating from the precursor of the asteroids or planetary system located before and after it. I already elaborated on the origin of the trojans in the chapter on precursors of the celestial bodies.

Therefore, I felt like, to some extent, the main belt asteroids are for the Solar System what the rings are for the planetary systems. For what explains the formation of the main belt asteroids also can explain the formation of rocks in the rings. To put it another way, what explains why the main belt asteroids did not stick together to form a planet or a unique and bigger asteroid can also to some extent explain why the particles and rocks in the rings were not gathered together to form other satellites or to increase the size of the satellites they surround. Just as most of the densest planets are inward of the main belt asteroids and the largest planets are outward of the main belt asteroids, so also the densest, thickest, and most opaque rings are inward of the innermost satellites and to a larger extent, the rings systems in the planetary systems are generally located inward or upstream of the largest satellites. All of this is the same law of split-gathering affecting the outcome of the precursors according to their location. Even on the scale of atoms and subatomic particles, the same law exists and could also explain why electrons around a nucleus could be organized into clouds instead of being very dense and in orbit.

30.14. Take home message

I showed that when the fluids in the precursor of the secondary bodies in the precursor of a planetary system started going through turbulence, they were not always able to gather together all of their particles into satellites that have no trail of dust around them. Instead, some particles were gathered together into rocks and other particles of various sizes. The particles which were gathered together into larger bodies became the satellites, while those which were not incorporated into the precursor of the satellites became those which all together formed the rings. Because the particles in some fluid layers of precursors of secondary bodies could not be gathered together into satellites, they were collected and scattered into large particles and/or rocks which became rings. Hence, some rings do not contain satellites, while others do. The fluid layers that were not gathered together at all and the left-over particles from the layers which were gathered formed the particles and rocks present in the rings. Another way of explaining this is that during the formation of the rings and satellites, there was a point, where the precursors of the rings and satellites were precursors of particles which were split-gathered into the precursor of the rings and the precursors of the satellites as their turbulence was underway. As turbulence was developing and fluids were moving, mixing, rotating, and spiraling, some fluids were packed into bigger clusters, which yielded the satellites, while other fluids were not amass into bigger bodies, but smaller particles spread over the space occupied by their mother precursors. The daughter bodies which were not amassed into satellites became the particles constituting the rings. In contrast, the gathering together of the precursors of particles into larger bodies led to the formation of the satellites. Hence some satellites are embedded inside rings. All this depended also on the size of the fluid layers of the precursors. Some of the empty spaces or gaps between rings are due to the demarcation of the space between the layers of fluids in the precursors of the secondary bodies (rings and satellites) and/or due to the incorporation of the precursors of those particles into satellites. Some of the materials used to form these satellites could have been particles which otherwise would have filled these gaps within the rings. The size of the vortices or clusters of spiraling matters in the precursors of the rings defined the size of the rocks and other particles constituting the rings. I showed that most of the rocks, particles, and dust in rings are not just escapees from the big satellites or from any other satellites. These particles found in rings were not transported to their current location by other celestial bodies (as some theories allege) but were born from the precursor of the secondary bodies in their planetary system. As the precursors of the rings could not have been gathered together into bigger bodies, they were spread all over the domain near the trajectory of their orbit, which could have been cleared or which looked like a gap in some rings. The color, density, brightness, thickness, and other characteristics of the rings depend on their precursors and how their environmental conditions shaped them. The variation of the density, faintness, or brightness of the rings in a ring system can be explained by the variation of the proportion of the precursors of the secondary bodies gathered together into satellites or left over to be incorporated into rings and how this left over was scattered throughout.

Another Book by Nathanael-Israel Israel:
TURBULENT ORIGIN OF CHEMICAL PARTICLES

FIND ALL THE RELIABLE, CONVINCING, SCIENTIFIC ANSWERS YOU NEED TO SUCCESSFULLY DECODE THE ORIGIN OF CHEMICAL PARTICLES SAFELY

Where did all elementary particles and composite particles including atoms, molecules, minerals, and rocks come from? What are the fundamental factors, the machinery, and the generic processes that defined their formation and proprieties? What was the nature of their precursors at the beginning of the universe and what underlying processes shaped or molded them into the chemicals we know today? What was the primary cause of the abundance and diversity of chemicals in the celestial bodies in the universe? What is the accurate link between the formation of chemical particles and the formation of galaxies, stars, planets, asteroids, and satellites? What light can the origin of chemicals shed on the real cause and meaning of gravity and the other so-called fundamental forces in nature? How does the formation of the chemical particles fit into the big picture of the formation of the universe?

After studying these questions for more than 12 years, Dr. Nathanael-Israel Israel discovered that the proper understanding of the origin of chemical particles is a very challenging but profitable task that requires original, scientific, mathematic, and philosophic efforts beyond the current state of modern science—until recently.

The solution for all of these puzzling problems: *"Turbulent Origin of Chemical Particles"*, the straightforward and trustworthy book that will help you to quickly, cheaply, easily, and efficiently navigate everything you need to know to finally solve the hard problems about the origin, the formation, and the functioning of all chemical particles. Whether you are a chemist, a biochemist, any other scientist, an engineer, as long as you have a reasonable background in chemistry but ignore how to scientifically demonstrate the origin of all chemical particles, this marvelous book is for you!

Amazingly packed with eye-popping analysis, fantastic graphs, tables, and the historic formula that broke the universe-origin code, *"Turbulent Origin of Chemical Particles"* will:

- Make it easier than ever for you to properly understand, decrypt, and articulate the real origin of natural chemical particles in the universe, therefore freeing you from false and boring explanations of the origin of all matters, and embrace the proven theory that opens doors to unparallel opportunities

- Professionally teach you how to transform the true knowledge of the origin of chemical particles into insights that significantly add value to your life in less time, and successfully establish you as a symbol of freedom, power, creativity, and originality in your field of expertise
- Fire you up to become the best version of you, and to cause positive changes to your initiatives that will profit you nonstop
- Discover thrilling illustrations and unconventional explanations of the formation of all matter in the universe, written in a simple language that brings humankind much closer to the complete deciphering of the mysteries at the very heart of chemistry, and open the way to a future of technology, innovation, discoveries, and breakthroughs
- Equip you to bypass technical knowledge that restricts non-experts from accessing the origin-related secrets contained in the massive scientific data, and get to the bottom of origin-related mysteries regardless of your background so you can empower yourself to leave unforgettable marks in your field of expertise. Learn more at Science180.com/chemical

With "*Turbulent Origin of Chemical Particles*", the accurate decrypting and understanding of the formation of chemicals has never been profitable and easy. Hence this great book is THE ultimate how-to guide for great people wanting to correctly decode the origin of the chemicals and positively transform their lives.

Get this celebrated book today. Don't wait!

Known as the nonconformist, rule-breaker, and accurate demonstrator of the universe-origin, Dr. Nathanael-Israel Israel is the founder of Science180, the one-stop for answering the most crucial universe and life's origin questions. He has had the honor to be acknowledged as the fearless universe-origin decryption trailblazer. Learn more at Israel120.com.

CHAPTER 31

DON'T LISTEN TO ANYONE TALKING ABOUT THE ORIGIN OF GALAXIES, GLOBULAR CLUSTERS, AND STARS IN THE UNIVERSE BEFORE YOU READ THIS CHAPTER

31.1. Constellations

The sky at night is filled with bright objects, some of which are stars, others are planets and asteroids, etc. Perplexed by the diversity and immensity of these celestial bodies, human beings for thousands of years have tried to map, name, and study them. Before modern astronomy, efforts were made to divide stars into groups and map them using terminologies such as constellations defined as imaginary outlines or patterns of the celestial bodies in the sky symbolizing animals, mythological entities (e.g. individuals, gods, or idols) and other things. For Oxford Dictionary, "a constellation is a group of stars considered to form imaginary outlines or meaningful patterns in the celestial sphere, typically representing animals, mythological people or gods, mythological creatures, or manufactured devices".

Different countries and nations have different boundaries for their constellations, therefore causing some confusions in the classification of stars into constellations. By the 20th century, the boundaries of constellations were changed and based on more accurate outliners defined by the International Astronomical Union (IAU). Some of the main characteristics of constellations are called "right ascension" and "declination", which defined the angles and distance of the constellations in the celestial sphere according to celestial references.

In the days of the Greco-Roman astronomer Ptolemy in the second century, 48 constellations were identified. However, as navigation increased and the southern hemisphere was better explored, more constellations were discovered.

As of 2020, 88 constellations have been acknowledged by the International Astronomical Union (Wikipedia 2017b):

- 36 constellations are predominant in the northern sky and the remaining
- 52 are predominant in the southern sky.

These 88 constellations have been recognized by the International Astronomical Union since 1922, while their boundaries were clearly defined and mapped in 1930. Some of these constellations are Andromeda (predominant in the Northern hemisphere), Hydra (one of the largest constellations), Crux (one of the smallest constellations). Some constellations are even mentioned in the Bible: Orion, Bear, Pleiades (e.g. Job 9:9 and Job 38)—Pleiades is also known as the Seven Sisters—proving that human exploration of the night sky is thousands of years old. For instance, the Bible recounts that, at the birth of Jesus, wisemen from the East followed a star to find Him. However, with the depth of the information available on the celestial bodies, other more advanced terminologies (e.g. galaxies and their clusters) are used to define stars and their association in the universe. Therefore, I did not use the constellations to study the stars, but rather galaxies and their clusters, and globular clusters.

31.2. Stars

Stars are the most visible things in the sky to the naked eye. During the day, the Sun can be clearly seen but, in the night, stars are observable in the sky. Some stars can also be seen in the sky during Solar eclipses. Astronomers believe that, besides the Sun, the nearest star to Earth is Proxima Centauri, which they said is 39.9 trillion km, or 4.2 light-years away; a light-year (designated "ly") being the distance traversed by light in one year, at a speed of about 300,000 km/s.

Some of the variables used to characterize stars include their chemical composition, classification, diameter, distribution, kinematics, magnetic field, mass, radiation, rotation, and temperature. Stars have different sizes. Some stars are believed to have a diameter as small as 20 km, while others like Betelgeuse are said to have a diameter 950 times that of the Sun (Graham et al., 2008). That star is said to have a density much lower than that of the Sun (Davis, 2000). Although the estimation of these sizes may contain errors, they point to the fact that bigger stars would likely be less dense than the Sun and consequently brighter. However, it is important to keep in mind that size alone cannot be sufficient to estimate the density of stars, for some stars can be bigger than the Sun and yet denser than the Sun. For the relationship between size and density of the bodies boils down to the type of turbulence of their precursors and the environment they were formed.

Research conducted in 2010 suggested that about 300 sextillion (3×10^{23}) stars exist in the observable universe (Borenstein, 2010). Although stars have been thought to only exist within galaxies, some intergalactic stars have been discovered between galaxies (Hubble News Desk, 1997). The intergalactic stars can be the consequence of intermittency of the precursors of the stars on a larger scale (e.g. astronomical scale). For example, when the precursors of the galaxies were splitting from one another, some of the material between them could not have been big enough to form a galaxy, but just some stars, which could not be collected into galaxies but into individual stars or globular stars, which I explained in another section toward the end of this chapter. On a much smaller scale, a similar phenomenon is found in the Solar System where smaller asteroids are found between planets because, as the precursors of the planetary systems were splitting, precursors of smaller bodies were formed between them. In other words, while some stars in the universe are isolated, others are organized into clusters like open clusters, globular clusters, while others are arranged into galaxies and clusters of galaxies. I already discussed this in the chapter on intermittence. Below, I will explore these associations and also provided more specific data on the stars in the galaxies.

Stars are known to produce, release, or radiate energy produced into space in the form of electromagnetic radiation and particle radiation, which manifests as the stellar wind (Koppes, 2003), which is said to stream as electrically charged protons, alpha particles, beta particles, neutrinos, etc. Stellar electromagnetic radiations are said to span the entire electromagnetic spectrum, from the longest wavelengths of radio waves through infrared, visible light, ultraviolet, to the shortest X-rays, and gamma rays. Particles in the stellar wind are believed to travel until they interfered with the interstellar medium. The area in space where the solar wind extends is said to be a bubble-shaped region called the heliosphere (Burlaga et al., 2005).

Some characteristics of the Sun include:

- Rotation period at the equator: ~ 27 days
- Rotation period at the poles: ~ 36 days
- Equatorial radius: 695,700 km (which is about 109 times that of the Earth)
- Equatorial speed: 1.994 km/s.

I already expounded on these characteristics in the chapters corresponding to them and I also contrasted and compared them with those of the planets, asteroids, and satellites in the Solar System. Before I finish this section on stars, I would like to mention that the characteristics of the limit of the Solar System is still not well known. For instance, based on some data collected from spacecrafts sent into the deep space, a lot is known about the Solar System, while many questions still remain unanswered. For instance, Voyager 1 is a space probe which was launched by NASA on September 5, 1977, to study the outer Solar System. By November 20, 1980, that probe finished its flybys of Jupiter, Saturn, and Titan (the largest Saturnian satellite), therefore allowing the scientific community to have a first close glimpse at the structure of the giant planets in the Solar System.

In June 2010, as Voyager 1 was at approximately 116 AU from the Sun, the scientists that were following it alleged that the Solar wind at that distance turns sideways (Wikipedia, 2016m). They interpreted the slanting of the Solar wind as a consequence of interstellar wind pushing against the heliosphere. When I came across that data, I realized that, regardless of the interpretation of the sideways of the Solar winds, it can be deduced that toward the edge of the Solar System, even smaller and lighter things like the Solar wind can be tilted, a fact which corroborates the tilting of the rotational axis and of the orbital plane observed with some satellites in Zone 4.

On December 5, 2011, NASA said that *"Voyager 1 had entered a new region referred to as a "cosmic purgatory"*, a stagnation region where charged particles streaming from the Sun allegedly slow and turn inward, and the Solar System's magnetic field is postulated to double in strength" (CNN, 2011). NASA's scientists thought that the increase of the magnetic field was caused by the "interstellar space which is applying pressure" on the Solar System (Wikipedia, 2016m). If that spacecraft was really near the edge of the Solar System (what can be doubted), I think that this increase in magnetic field may be caused by:

- the consequence of air or gases or waves being pushed from the top and bottom of the Solar System because of its rotation and/or the rotation of the outer bodies in the Solar System, or because
- the spacecraft was passing through a region dominated by retrograde asteroids which revolution around the Sun is opposite to the conventional direction and consequently things appeared as though the spacecraft was pushed by an inward force.

According to NASA, *"energetic particles originating in the Solar System declined by nearly half, while the detection of high-energy electrons from the outside increase 100-fold in the cosmic purgatory ... The inner edge of the stagnation region is at about 113 AU from the Sun"* (CNN, 2011). However, when I considered the ratio of the semi major axis of the outermost satellites with respect to that of the biggest bodies in their planetary system, and then used that trend to explore how extended the Solar System can be (knowing the semi major axis of the planets), I felt like Voyager 1 is not out of the Solar System yet. The Solar System, which is often defined as the region of space occupied by the bodies orbiting the Sun, should not be confounded with the heliosphere. Voyager 1 is still in the Solar System and, as of 2016, it was less than one seventh the distance to the aphelion (936 AU) of Sedna, one of the remote Trans Neptunian Objects in the Solar System (Wikipedia, 2016m). The increase of cosmic ray particles suspected toward the edge of the Solar System also supports what is found in the planetary systems, where several small bodies are found toward the outskirts. On April 9, 2013, scientists started noticing an 80-fold increase in electron density based on the frequency of plasma oscillations (Gurnett et al., 2013). According to NASA, the mission of Voyager 1 is supposed to end around 2025 when its radioisotope thermoelectric generators would be theoretically out of energy and subsequently, none of its 11 scientific instruments can function (Wikipedia, 2016m).

31.3. Generalities on the galaxies

According to NASA, a "*galaxy is a gravitationally bound system of stars, stellar remnants, interstellar gas, dust, and dark matter*". In other words, a galaxy is a term used to design a cluster of stars that scientists think is allegedly bound together because of their physical distribution. Usually gravity is what some scientists think binds the stars in a galaxy together, but as I explained in the chapter on gravity and the other fundamental forces in nature, gravity is not the reason of the association of stars into clusters, including galaxies.

In 2005, it was believed that the observable universe contains 200 billion galaxies (Gott et al., 2005). However, a 2016 review of that number suggested that the observable universe contains more than 2 trillion (2×10^{12}) galaxies consisting of more than 10^{24} stars, meaning there are more stars in the universe than all the grains of sand on planet Earth (Mackie, 2002; Staff, 2019). While dwarf galaxies are believed to contain less than 10 million stars, giant galaxies are believed to contain billions of stars. Because they are very far from Earth, some galaxies have not been clearly seen using modern telescopes, and it is likely impossible for human beings to know the limits of the universe and map all of its constituents. As more evidences and details on the universe are found, these numbers will be reviewed.

While some galaxies are "well" catalogued, others are not. Some of the "well-catalogued" include the:

- Milky Way, (the local galaxy to which the Solar System belongs),
- Andromeda Galaxy, (the northern sky's brightest galaxy believed to be the nearest large galaxy from the Milky Way),
- Magellanic Clouds, (the nearest external galaxies visible in the skies of the Southern Hemisphere),
- Whirlpool Galaxy, and the
- Triangulum Galaxy.

Although I believe that errors exist with the measurement of the dimensions of the galaxies, I felt like it may be useful to use them to illustrate what is currently known about the number, dimensions, and distributions of galaxies and their stars in the universe. For instance, according to Wikipedia (2019c), as of 2019, the diameter of most galaxies is estimated at 1,000 to 100,000 parsecs (equivalent to about 3,000 to 300,000 light years). Additionally, galaxies are said to be separated by distance equivalent to millions of parsecs. According to the same source, the Milky Way has a diameter higher than 30,000 parsecs (100,000 ly). Andromeda Galaxy and the Milky Way are believed to be separated by 780,000 parsecs (2.5 million light years).

While some galaxies called solitary galaxies, isolated galaxies or field galaxies are isolated and seem not to be interacting with other galaxies, most galaxies are organized into structures called groups, clusters, superclusters, and clusters of superclusters. A group of galaxies is the smallest and the most common type of galaxy association. Clusters of galaxies are said to be made of hundreds to thousands of galaxies bound together (ESA/Hubble Press Release, 2015). Clusters of galaxies are said to be habitually dominated by a giant elliptical galaxy, called the brightest cluster galaxy (Dubinski, 1998). This biggest galaxy in the cluster may be what could have been the primary body or primary galaxy if I can compare clusters of galaxies to the planetary systems in the Solar System. Superclusters of galaxies are believed to have tens of thousands of galaxies, and on the supercluster scale, "galaxies are organized into sheets and filaments surrounding vast empty voids" (Bahcall, 1988). Recent studies suggested that some superclusters can also be organized into even larger superclusters called cluster of superclusters (Gibney, 2014). For instance, the Milky Way belongs to the Local Group, which diameter is said to be about one megaparsec (van den Bergh, 2000). According to Encyclopedia Britannica (Hodge, 2019), the Local Group contains about 50 galaxies which are dominated in brightness by the Milky Way and Andromeda Galaxy respectively the brightest galaxy and the second brightest galaxy within the Local Group. For the rest of this chapter, whenever I talk about Encyclopedia Britannica, I am referring to its article written by Paul W. Hodge in 2019 as cited above (Hodge, 2019). According to Encyclopedia Britannica, the Andromeda Galaxy, also known as the Andromeda Nebula or M31 is located at 2.48 million light-years from Earth. The Local Group is part of the Virgo Supercluster, which is said to belong to the Pisces-Cetus Supercluster Complex, a giant galaxy filament. Because of the size of the galaxies, errors can still exist in their measurements. For instance, according to Encyclopedia Britannica, less than a 100 years ago, the renowned American astronomer Edwin Hubble (after whom the Hubble telescope was named) estimated the mass of the Andromeda Galaxy to be about 3.5 billion times that of the Sun. Today, astronomers say that the real mass of that galaxy must be at least 100 times greater than Hubble's value. In the future, this estimation will likely change. The high error in the measurement on galaxies and on larger scale systems of bodies is one of the reasons I don't trust most distance and size measurements of bodies beyond the Solar System, where significant errors are also still made despite the proximity of the Solar System objects from the Earth. Even in the Solar System, significant errors are made.

31.4. Structure of galaxies

The structure of the galaxies is defined by 3 components:

- Spheroidal component,
- Disk component, and
- Spiral arms.

Below, I explored each of these 3 components. Beforehand, I would like to mention that as a whole, galaxies are also characterized by their plane. For instance, the Milky Way galactic plane is said to be inclined by about 60° to the elliptic (the plane of Earth's orbit). Because of the positioning of their galactic planes, some galaxies could not be properly seen from the Earth. The position of the galactic planes could have been defined by the position of the precursor of the galaxies landed during its formation. Some of them could have been tilted or rolled over.

31.4.1. Spheroidal component

Most galaxies have a spheroidal component of stars although it may be less remarkable or even absent in some irregular galaxies. According to Encyclopedia Britannica, while some galaxies have an "oblate spheroid that rotates around their short axis", other galaxies have a "prolate spheroid that rotates around their long axis". Considering the trends that I found for the celestial bodies in the Solar System, the precursors of some galaxies could have been rolled over, tilted, or inclined during their formation and consequently, the rotational axis of their daughter galaxies changed. Some galaxies that have a prolate spheroid could have been rolled over just as some celestial bodies having a higher axial tilt or higher orbital inclination in the Solar System seemed rolled over. It should not be surprising that some galaxies lack a spheroidal component. If I can use the primary and secondary bodies terminology I applied to the celestial bodies in the Solar System, I can say that the spheroidal component is like the primary system of the galaxies, while the arms and disk are like a system of secondary bodies. Just as some celestial bodies in the Solar System do not have a system of primary and secondary bodies, some galaxies also may lack some components.

I would also like to mention that just as some bodies orbit others in the Solar System, so also within some galaxies, stars are said to "orbit" the bulge (also called galactic center). Bulges are found with some spiral galaxies and some elliptical ones. The bulge of the Milky Way Galaxy is said to encompass stars in the innermost 10,000 light-years radius (Grant and Lin, 2000). The galactic center of the Milky Way is said to be an intense radio source known as Sagittarius A*. Moreover, the luminosity of the compact region at the center of some galaxies is much higher than normal luminosity and these regions are termed "active galactic nucleus" (Wikipedia, 2019c). The brightest areas of these active galactic nucleus are called quasars and they are believed to be a region where new stars are formed. Talking about star formation, I would like to add that, some galaxies (e.g. Starburst galaxy) are believed to be experiencing an extremely high rate of star formation, suggesting that it is possible that somewhere in the universe, conditions can still be favorable for stars to be formed "just as" it happened at the beginning of the universe. I don't believe in those theories, and very soon, I will explain how stars were really formed.

Unlike planets and satellites which are separated from their primary bodies by a distance generally considered as the semi major, stars in the galaxies are separated from the center of their galaxies by a distance termed the radial distance from the galactic center. Sometimes it is referred to as the distance from the galactic bulge. For instance, believed to be located on the inner edge of the Orion Arm (a spiral-shaped concentration of gas and dust), the Solar System is said to be positioned at about 26,490 light-years (25,000–28,000 light-years) from the Galactic Center (Grant and Lin, 2000).

31.4.2. Disk component

Found in most spiral galaxies and irregular galaxies, the disk component is a flat structure of stars that astronomers believe emits most of the brightness of these galaxies. Elliptical galaxies are not believed to have disks. The fact that the disk emits most of the brightness of the galaxies makes me believe that it does so because, as being the outer bodies in those galaxies, the stars in the disk were not as compressed, nor as highly wound up, nor as condensed during their formation as those located in or near the spheroidal component. The small density or compaction of the stars in the disk implies that they could have less dense particles in them and consequently, they are brighter and give most of the brightness for the galaxies that have disk. The nature of the turbulence on the astronomical scale of galaxies can explain these variations.

31.4.3. Spiral arms

While some spiral galaxies have unclear spiral arms with small pitch angles, others have larger pitch angles which are reflected by their more-open arms Hodge, 2019). While some arms extend around the center for more than 2 complete rotations, others have what Encyclopedia Britannica called a *"chaotic arm structure made up of many short fragments that extend only 20° or 30° around the center"*. When I looked at the spiral arms, it seemed to me as if their precursors were ejected from the spherical component as the precursors of the entire galaxy they belonged to were being split-gathered to birth all their daughter bodies embedded in the components of the galaxy. Unlike systems of bodies in the Solar System where a primary body is clearly separated from its secondary bodies, on the galactic level, the gathering together of the precursor of some galaxies could not have been strong enough to release or separate the precursor of the arms farther away from the spheroidal component. The size and the degree to which the precursors of the spiral galaxies were able to push away, eject, or release the precursor of their secondary stars away from the primary stars (which constitute the spheroidal component) could have determined the properties of the arms such as the size of the pitch angles, the angle of the extension around the center, the closeness or openness, and the number of "complete rotations" around the center of the arms. The precursors of spiral galaxies, which seem to have long arms extended around the center many times, could have flown over a long distance before its arms were wrapped or spiraled around the center of their galaxies and/or their precursor was huge and, in the end, yielded spirals. In contrast, the precursors of the galaxies which do not have long arms or which are more spherical could have not been spread over the long distance before being spiraled into its "final" shape. Having spent a lot of time studying the celestial bodies in the Solar System made it easy for me to quickly figure out the genesis of the galaxies.

31.5. Morphological classification of galaxies

Scientists have made significant efforts to characterize galaxies including measuring their size, speed, temperature, chemical composition, radiation, rotation, and distance between them, etc. However, because of the distance separating them from Earth, inaccuracy and a lot of errors can be found with these measurements. Therefore, I chose not to focus on most of those data, but mostly on the morphology of these galaxies. For the morphology of some galaxies can be seen from Earth even with the naked eye. Galaxies come with different sizes. Those which contain a few hundred million stars are called dwarf galaxies while those containing a hundred trillion stars are termed giant galaxies. According to Encyclopedia Britannica, the size of galaxies can range from 100 light-years across (for dwarf galaxies) to more than 3,000,000 light-years (e.g. for giant radio galaxies).

31.5.1. Elliptical galaxies

Elliptical galaxies (annotated E) range from nearly spherical or perfectly circular (E0) to highly elongated (E7). According to Encyclopedia Britannica, elliptical galaxies are red in color and are flatter than $b/a = 0.3$ (with b and a being the minor and major axes of the elliptical image, respectively). They are said not to have many open clusters. The largest galaxies are believed to be giant ellipticals which are often found near the core of large galaxy clusters (Cornell University, 2005). According to Encyclopedia Britannica, the images of many elliptical galaxies are "*more circular in the central regions than in the outer parts*". According to the same source, the major axes of elliptical galaxies do not line up and their position angles vary in the outer parts. These data suggested to me that the elliptical galaxies may have been inclined and those located toward the central region may have been more squeezed or compressed during their formation than the outer elliptical galaxies, which "orbit" may have been tilted or elongated more than the innermost ones. These trends match what I found for the celestial bodies in the Solar System where the orbit of the outermost satellites and planets are usually more tilted or elongated than that of the innermost ones. The elliptical galaxies are darker or less bright than the other types of galaxies because their stars may have been more compressed and/or contained denser particles, hence a lower brightness. Elliptical galaxies could be for clusters of galaxies what terrestrial planets are for the Solar System. For because their individual stars are not spread in spirals, elliptical galaxies could have a higher density. In my book *"Turbulent Origin of Chemical Particles"*, I explained how chemical composition can affect the brightness of stars.

31.5.2. Spiral galaxies

According to Wikipedia (2019c), "spiral galaxies consist of a rotating disk of stars and interstellar medium, along with a central bulge". Bright arms of stars spread outward from the bulge of the spiral galaxies. Encyclopedia Britannica described spiral galaxies as having a *"bright nucleus surrounded by a thin outer disk, and a superimposed spiral structure"*. I believe that their galactic outer disk is thin because it consists of stars smaller than some of those near or at the center of the galaxy. Just as the celestial bodies in the outer turbulence zones (e.g. Zone 5) are generally smaller than those in Zone 3, I think that the outermost stars in the galaxies are also smaller, hence their least brightness. Spiral galaxies are categorized into 2 classes:

- normal spirals (designated S) and
- barred spirals (designated SB)

According to Encyclopedia Britannica, normal spirals have "arms that emanate from the nucleus, while barred spirals have a bright linear feature called a bar that straddles the nucleus, with the arms unwinding from the ends of the bar". I think that the precursors of normal spiral galaxies could have had less "difficulty" ejecting and split-gathering their secondary stars than the precursors of barred spirals. For the stars that are in the arm are secondary, while those in the nucleus can be perceived as primary stars. The precursors of barred spiral galaxies could not have been able to eject their secondary stars away directly from the primary stars, but they could have flown for a while before spiraling. At one point, the precursor of the barred region of barred spirals could have been almost in the middle of the arms organized as if one is leading and the other trailing. As the precursor of the barred spirals was winding up, because it could not wind up all of its stars to form an elliptical galaxy, the leading or the trailing arms were tilted or wound up according to the direction or sense of the movement of the precursor. The central part of the stars could not have been compressed a lot or wound up much because of its thickness, resistance to winding up, and the lack of energy that could have been required to compress it or spiral it.

Encyclopedia Britannica described the nucleus of spiral galaxies as a *"sharp-peaked area of smooth texture, which sometimes make up the majority of the galaxy"*. I think that such spiral galaxies which do not have much stars in their arms are like the planets which have fewer satellites. In contrast, those spiral galaxies which have a long arm of stars, are like planets that have many satellites. Encyclopedia Britannica reported that the *"arms and the disk of a spiral galaxies are blue in color, whereas their central areas are red like an elliptical galaxy"*. This observation made me think that the stars in the disk and arms of spiral galaxies are not as dense as those in the central areas, hence they give a blue color implying they shine more or release more energy than the central ones which are red, implying that they radiate less energy which I link to their denser composition or the presence of denser particles in them which consequently reduced their brightness. In other words, the central part of spiral galaxies may be closer in composition or structure to the elliptical galaxies, hence they share the same red color. The same trend was observed with the celestial bodies in the Solar System where the innermost bodies are generally denser than the outermost bodies. For instance, innermost planets in the Solar System were generally denser than those outward. Just as Pluto is dense although located at the outskirt of the planets, some stars not located toward the galactic core can also be dense. Hence elliptical galaxies are not only found at the center of galaxy clusters.

Barred spiral galaxies have a *"linear, bar-shaped band of stars that extends outward to either side of the core, then merges into the spiral arm structure"* (Eskridge and Frogel, 1999). For instance, the Milky Way Galaxy is believed to be a large disk-shaped barred-spiral galaxy (Alard, 2001). Claimed to have a diameter of approximatively 30 kiloparsecs and a thickness of one kiloparsec, the Milky Way galaxy is believed to contain about 200 billion stars (Sanders, 2006) and a total mass of about 600 billion times the mass of the Sun (Bell and Levine, 1997). Although I personally believe that many of these estimations may be inaccurate or false, they, at least give a glimpse at what the scientific community believes about the scope of the galaxies in the universe or at least the visible universe seen from Earth.

Unlike barred spiral galaxies, super-luminous spirals are believed to generate a substantial amount of ultraviolet and mid-infrared light. Some of these galaxies may be closer to us than those which have less light, for their ultraviolet and mid-infrared light does not necessarily mean that they have more energy, but that they may also be closer to us. Similarly, I think that some stars which are blue can be closer to us than those which are not. The size of the galaxies can also play a role in their luminosity.

The bulge of spiral galaxies is believed to be a large, firmly packed group of stars. While the bulge of some spiral galaxies is thick, that of others is thin and dense. According to Encyclopedia Britannica, the arm of most spiral galaxies is shaped like logarithmic spirals. Regardless of their classes, spiral galaxies are classified (into a, b, and c types) according to the degree of tightness or coiling of their spiral arms and the size of their nucleus (central bulge). For instance, Sa galaxies have *"tightly wound, poorly defined arms and a relatively large (sometimes small nucleus) core region or large amorphous bulge in the center"* whereas Sc galaxy have open, well-defined arms and a small core region (Smith, 2000). According to Encyclopedia Britannica, Sc galaxies have a very small nucleus and several open spiral arms with fairly large pitch angles. An example of Sc galaxy is the Whirlpool Galaxy (M51). Finally, Sb galaxies are intermediate between Sa and Sc and the size of their nucleus is medium. Andromeda Galaxy is classified as a Sb galaxy. A huge variation was observed with the shape of Sb galaxies. I believe these variations are related to the way the precursors of the stars in their arm were split-gathered and ejected from the nucleus.

According to Encyclopedia Britannica, "a protogalaxy with a huge angular

Nathanael- Nathanael-Israel Israel: Author of "Reconciling Science and Creation Accurately"

momentum can become a spiral galaxy, whereas one with a small angular momentum can develop into an elliptical galaxy", suggesting to me that the way the precursors of the galaxies may have rotated or tried to rotate could have affected the structure of morphology of the component of their components.

According to Encyclopedia Britannica, the diameter of normal large spiral galaxies, varies between 100,000 to 500,000 light-years. The limit of the outer edge is not always very clear. For instance, the limit of the edge of the Milky Way Galaxy is not sharply set in such a way that beyond it, no star can be found. Instead, the concentration of stars declines with the radial distance galactic center. For instance, beyond a distance of about 40,000 light years from the center of the Milky Way, the number of stars per a certain volume of space is said to drop much faster with radius (Sale et al., 2010).

31.5.3. Irregular galaxies

Irregular galaxies are neither elliptical nor spiral. They do not have a clear symmetry or clear central nucleus. I think that some of the irregular galaxies are just remote galaxies that current tools are failing to properly resolve. Others were born from precursors which failed to properly curl or wrap around. They come with many shapes, some of which are lenticular galaxies, ring galaxies, and interacting galaxies. Even among asteroids, some are well shaped and others are not. It all boiled down to how their precursors were gathered together. In other words, the irregularities in the shape of galaxies were caused by how their precursors were split-gathered.

31.5.3.1. Lenticular galaxies

Lenticular galaxies (also called Hubble type S0 galaxies) are intermediate between elliptical galaxies and spiral galaxies for they have properties of both elliptical and spiral galaxies. Harvard-Smithsonian Center for Astrophysics (2006) defined them as having "ill-defined spiral arms with an elliptical halo of stars". In other words, S0 galaxies are a type of peculiar irregular galaxies that have the disk shape characteristic of spiral galaxies but have no spiral arm. S0 galaxies may seem as lacking arms, but it is possible that some stars in those arms are too small or too distant to be resolved from current telescopes on Earth. In other words, some S0 galaxies may have stars in thinner arms that are invisible from Earth. Just as some planets do not have a satellite, while others have many satellites differently organized, so also galaxies do not have the same components. Some galaxies have more components than others.

31.5.3.2. Ring galaxies

Ring galaxies (e.g. Fig. 177) are galaxies having a circle-like appearance or a ring-like structure of stars surrounding a "bare" core (Nemiroff and Bonnell, 2002). A famous example of ring galaxies is Hoag's Object, which was discovered in 1950. As scientific measurements of galaxies improve, the core of some galaxies including the "bare" core of the center of ring galaxies will be better resolved into stars, even if those stars may be few and small. For I think that the core of some galaxies may not show much stars or any star at all because their brightest may be too faint to be visible from Earth using current technology. I also believe that some stars at the center or the core could be less bright, very dense, and consequently they could not be resolved. For just as some planets around stars are not easily resolved due to the low brightness, some stars that are less bright can evade detection. In other words, the lack of stars at the center of some ring galaxies is not a guarantee that those kinds of galaxies lack stars at their core or at their center of mass. I also believe that as the layers of the precursors of some ring galaxies were being split-gathered, they could not yield a primary star or system of stars, and consequently, the precursors of the stars in some layers could have been spread over in a "circle". It would have been possible that the precursors of the ring galaxies "exploded" and freed their stars which then, arranged like a ring accordingly, just as the main belt asteroids are organized around the Sun and how some planets in the Solar System could have been arranged in rings if their precursors were not gathered together into unique bodies. The precursors of some ring galaxies were initially fluid layers which split and gathered together.

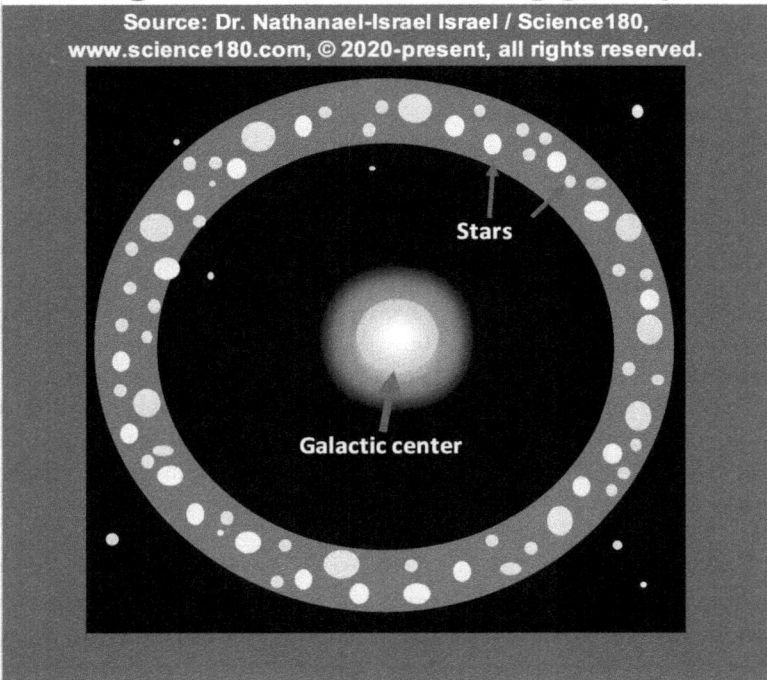

Fig. 177: Sketch of a ring galaxy

Nathanael- Nathanael-Israel Israel: Author of "Reconciling Science and Creation Accurately"

31.5.3.3. Interacting galaxies

These galaxies look as if they are interacting with one another or being scattered. Some people think that they are merging.

31.5.3.4. Others types of galaxies

Some irregular galaxies have a double bulge around a central nucleus while others have an almost central bar structure dominating what Encyclopedia Britannica called a *"chaotic arrangement of material"*. I think that the diversity of the turbulence which dominated the universe where some galaxies were formed explains the various nature of the irregular galaxies. The Large Magellanic Cloud is mentioned as an example of such a galaxy with chaotic arrangement.

31.6. Orbital speed of stars and galaxies, and problems of the galaxy rotation curve

Also termed velocity curve, the rotation curve of a disc galaxy describes the relationship between the orbital speeds of visible stars or gases and their radial distance from that of the galaxy's center. According to Wikipedia (2017a), the rotation curves of each side of the spiral galaxies is not the same and is said to be asymmetric. According to the same source, although in some conditions the speed of outer stars in galaxies is said to be smaller than that of the inner ones, other works suggest that the orbital speed of most stars in spiral galaxies is roughly the same. I think this can be due to the fact that the fluid layers in the precursors of the spirals were moving at different speeds, hence their daughter stars of gas inherited different speeds according to their location in spiral galaxies. The way that the layers of the precursors of galaxies wound up or spiraled could have positioned the top layers (which could have had a higher speed) in such a way in the end, the daughter stars of the leading and trailing parts of the top layers could be positioned toward the outskirts of the spirals, and consequently some outer stars in spirals have a higher speed than some inner ones. In other words, outer stars in some galaxies have a higher orbital speed than the inner stars because of how the top layers of the fluids of the precursors were positioned or sandwiched into the other layers as the stars were being formed. In contrast, some of the fluids which were not on top or which were toward what could have been called Zone 3 could have ended up in the spheroidal component of the bulge of the spiral galaxies. In the chapters on fluid layers and orbital speed, I already amply explained how the position of the fluid layers of the precursors of bodies affected the speed of their daughter bodies.

Stars in galaxies are usually believed to orbit the center of mass of their galaxy. For instance, the Solar System is said to travel through the Milky Way galaxy in a circular orbit roughly 30,000 light years from the galactic center (the center of the galaxy) at a speed of about 220 km/s (Leong, 2002). The orbital speed of other stars and gases at a wide range of distances from the Milky Way Galactic Center is also approximately 220 km/s. According to the Median Statistics Estimate of the Galactic Rotational Velocity, away from the central bulge or outer rim, the typical stellar orbital speed in the Milky Way is between 210 ± 10 km/s. Other measurements found some stars at the outer edge of the Milky Way orbiting the center of the galaxy at 254 km/s (Finley and Aguilar, 2009). As a whole, the Milky Way is believed to be moving at approximately 630 km/s with respect to the local co-moving frame of reference or the extragalactic frame of reference (Mark et al., 2004).

Although the vast majority of galaxies in the universe are said to be moving away from the Milky Way, the Andromeda Galaxy, is said to be heading toward it at about 120 km/s (Fraser, 2007). The globular clusters can follow rosette orbits about the Milky Way, in contrast to the elliptical orbit of a planet around a star (Gnedin et al., 1999). The motion of some stars is prograde while that of others is retrograde. For instance, several globular clusters in the Milky Way Galaxy are said to have a retrograde orbit around the Galaxy (Kravtsov, 2001). About 40% of the Milky Way's clusters are believed to be on retrograde orbits, which means they move in the opposite direction from the Milky Way rotation (Dauphole et al., 1996). The motion of the stars suggested that, what happened to some bodies and caused them to be retrograde in the Solar System could have also happened to some stars, and even to some galaxies or clusters of galaxies, to cause some to be prograde and others to be retrograde according to their position in the universe. The processes I used to explain the prograde and retrograde motion of celestial bodies in the Solar System can also explain (on the astronomical scale) the prograde and retrograde motion of galaxies. It is all about how, according to their position, the fluid layers of their precursors were split-gathered, and moved away from the precursors of neighboring galaxies.

Fig. 178: Sketch of the Milky Way Galaxy

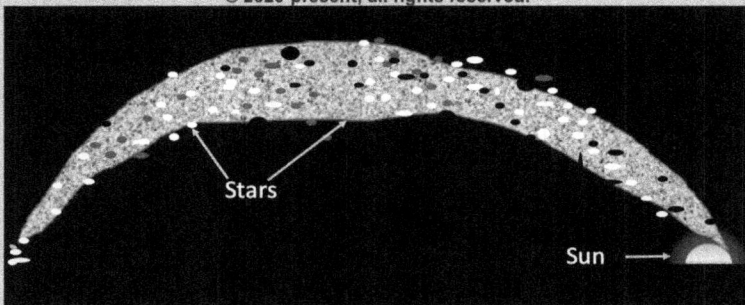

Furthermore, it was said that the Solar System would take about 240 million years to complete one orbit of the Milky Way (a galactic year also called cosmic year) (Sparke et al., 2007). The Milky Way is said to be moving in the general direction of the Great Attractor and other galaxy clusters, including the Shapley supercluster, behind it (Kocevski and Ebeling, 2006). Most satellite galaxies of the Milky Way lie in a very large disk and orbit in the same direction (Kivivali, 2014). As I finish this section on clusters of galaxies, I would like to mention that, just as some bodies orbit others in the Solar System, so also in galactic systems, some galaxies called satellite galaxies are said to orbit larger ones too. Likewise, the largest satellite galaxy of the Milky Way is believed to be the Large Magellanic Cloud (CFA, 2017). The movement of the satellite galaxies in the same direction reminded me of the movement of the planets around the Sun and of the satellites around their primary planet in a same direction. This suggested to me that the precursors of some satellite galaxies could have been launched in the same direction and even same sense, therefore making it harder to rule out the existence of a potential common precursor of the Milky Way Galaxy (Fig. 178) and its satellite galaxies.

The fact that the shape of the Milky Way Galaxy is like an arc suggested to me that its precursor was something which was trying to wind up or rotate or was thrown and behaved like a kind of projectile.

31.7. Globular clusters and open clusters

Stars are usually organized into clusters. Typically found in the halo of galaxies, or orbiting galaxies, globular clusters are spherical gatherings of stars that are said to orbit a galactic core, as a satellite (Wikipedia, 2019d). They are believed to be very tightly bound or bound altogether. Their centers are believed to have higher stellar densities. In contrast, open clusters are structures or collections of stars inside a galaxy, usually in the disk of a galaxy. Open clusters are said not to be found in elliptical galaxies, but in spirals and in irregular galaxies only (Wikipedia, 2019d). The Pleiades (also known as the Seven Sisters) exemplify an open cluster of stars.

Most of the globular clusters found in the Milky Way Galaxy are said to be in the halo around the galactic core (Shapley, 1918). I think that the precursor of some globular clusters may have been ejected from the precursors of the galaxies they belong to (or near them) and others may have been born by the way the precursors of the galaxies around which they are found were split from one another, causing a phenomenon of intermittence of stars between galaxies of clusters of stars. I already dealt with this in the chapter on intermittence. In some circumstances, the precursors of globular clusters could not have spun or spiraled enough to cause their daughter stars to orbit the primary star or primary clusters of stars as some stars go around the center of some galaxies. Some precursors of globular clusters could have exploded without giving their daughter stars enough ability to form a well-defined system of secondary stars orbiting a well-defined primary core. Another way of saying this is that the precursors of globular clusters may have also lacked the ability to gather their daughter stars into a unique and condensed cluster, but instead released or scattered them without making a primary star to "dominate" them. All this can be boiled down to the way the precursors of the clusters of stars were split-gathered into their daughter bodies.

31.8. Clusters of galaxies

The structure of the galaxy clusters in the universe is not always the same. According to Encyclopedia Britannica, clusters of galaxies "are not evenly spaced in the sky" and "superclusters of galaxies consist of 3 to 10 clusters" extended over about 200 million light-years. While some clusters of galaxies are spherical, others are spiral and others are mixed. According to Encyclopedia Britannica, irregular clusters of galaxies are *"large loosely structured assemblages of mixed galaxy types* (mostly spirals and ellipticals)", while spherical clusters of galaxies are *"dense and consist almost exclusively of elliptical and S0 galaxies"*. According to the same source, spherical clusters of galaxies are *"massive and their diameter can reach to 50,000,000 light-years and can contain 10,000 galaxies concentrated toward the center of the cluster of galaxies"*. These characteristics of the clusters of galaxies suggested to me that, during their formation, the precursors of spherical clusters of galaxies could have been more tightly wound, spiraled, or compressed, hence their daughter galaxies are denser than those of the irregular or spiral clusters. Considering the distribution of bodies in the systems of bodies in the Solar System, the distribution of clusters of galaxies can yield a clue on the genesis and initial location of their mother precursors. For some denser spherical galaxies could likely be located toward the center of the clusters of spherical galaxies because such a location could have favored the spiraling of their precursors into dense and more compacted galaxies. Clusters of spherical galaxies are said to exclusively contain elliptical and S0 galaxies because the conditions of the environment of their precursors could have favored the formation of galaxies which are dense and spherical. In other words, spiral galaxies are absent in spherical clusters of galaxies because their precursors could have been transformed into spherical or S0 galaxies for the conditions in those environments could not have tolerated any precursors to remain spiraled or irregular.

As the precursors of galaxies were split-gathering, stars were being formed until it came to a point when the split of precursor of galaxies could no longer occur or continue to occur due to the size of the precursors and the environmental conditions. At this moment, the formation of new stars could have stopped. Each of the precursors of the stars were wound up to yield its individual star. According to their environments, some stars were tightly wound up, while others were loose. The degree of the tightness and the turbulence of the precursors of the stars could have defined the internal composition and consequently the intensity of the brightness of their daughter stars. Consequently, some stars that contain denser constituents ended up being redder than those which constituents were less dense particles and which appear bluer.

CHAPTER 31: CONSTELLATIONS, GALAXIES, GLOBULAR CLUSTERS, AND STARS IN THE UNIVERSE

On scales larger than that of superclusters, it is believed that proprieties of the universe are the same in all directions observed from the Earth. This assumption of a homogeneous and isotropic universe is known as the cosmological principle. On the largest scale, meaning the scale of the whole visible universe, clusters of superclusters, superclusters, and cluster of groups of galaxies are usually arranged into sheets and filaments surrounded by enormous voids (University of Cambridge, 2007). When the filaments of the network in the human brain were compared to the filaments of galaxies in the universe, no apparent difference can be easily detected, yet, the brain and the universe are different. I believe that the apparent similarities in the physical appearance of things in the universe caused people to mistakenly believe that the universe is really the same everywhere and therefore, they mistakenly misunderstood its formation. The aforementioned similarity between the filaments can be due to the fact that both the brain and the universe are products of a turbulence, and the presence of filaments can be a manifestation of turbulence when many daughter bodies of that turbulence are viewed. A picture or scan of a brain may show similar filament networks, but the brain is wired differently and each portion of a brain has a specific function. Likewise, the universe as a whole may look like a uniform filament, but the law governing each location may be different. Based on my study, the assumption that the laws of nature are the same everywhere in the universe is one of the things that have fouled some scientists and caused them to design and believe in wrong theories that have drifted them away from the truth.

31.9. Voids in the universe

As I finished this chapter on galaxies, stars, and clusters of bodies, I felt like it will also be useful to remark that the universe is dominated by apparent empty space that is alleged to contain "nothing". Indeed, looking at the sky with the naked eye suggests that the universe has massive "empty" regions. The largest known of the empty spaces was estimated to be 1.8 billion light-years across (Devlin, 2015). Some of the voids seen in space are the atmosphere of some bodies, but most of them are the space between the systems of bodies.

Why is the universe mostly void and why is there a huge distance between the celestial bodies? The physics of jets has been used to try to explain the large-scale structure of the universe and the support of the galaxy clusters. Based on a study done at the Smithsonian Astrophysical Observatory in 1993 on the "Filamentarization' of the galaxies in the universe", the universe would consist mostly of voids (90%), with filaments and sheets of galaxies comprising the rest (Eggers and Villermaux, 2008).

Although it can be impossible to know exactly the proportion of the universe which is void or not, a look at the night sky suggests that a lot of void spaces exist between the bodies in the sky. The "empty" spaces as seen from Earth are not necessarily void, but they contain things that most people are unable to see with the naked eye and even with advanced telescopes. However, no matter the size of the content of space, there are more voids than material things in space. In other words, matter is not present everywhere in the universe. Something happened to allow matter to be present in some locations and absent in others.

Why was matter differently distributed throughout the universe and it is not present everywhere? I think that during the formation of the universe, in addition to the forces deployed to fragment or split-gather the precursors including the original mysterious scattering (see the chapter on "In the beginning"), the light present in the precursors of the celestial bodies being formed might have also contributed to increasing the distance between the bodies. For light travels at a high speed and has the ability to move things. Therefore, light emanating from a precursor of the bodies could have also contributed to pushing away some bodies in its vicinity. For in the beginning, when celestial bodies were being formed, they were initially full with energy that can also take the form of light. In addition to the force that scattered the precursors of the matter across the universe, the radiation emanating from those bodies could have increased the distance between them until an equilibrium was reached from which point, the energy of those lights could not be able to significantly affect the location of the neighboring bodies much.

Even on the atomic level, a distance is found between the subatomic particles. A distance exists between electrons, protons, and neutrons. On scales smaller than the subatomic ones, a distance exists between everything although current scientific equipment fails to measure them. This is because, during the formation of the matter in the universe, things were pushed away, moved around and the initial energy inside the precursors of all known matter contributed to distancing the types of matter being formed. Even at the microscopic level, distance exists. Some scientists have tried to explain the laws governing the distance between the celestial bodies. Unfortunately, they thought that celestial bodies were attracting one another and, ignoring the laws related to turbulence scaling on different scales, they thought that the force of this so called "attraction" is correlated to their radius and the distance between them. Hence, they also struggled to reconcile gravity with the other so called fundamental forces in nature. During the formation of the universe, when fluid layers of precursors of bodies were moving away from the precursors of primary bodies and from the precursors of neighboring bodies, spaces were created and amplified.

31.10. Sketch of the formation of galaxies

In the previous segments of this chapter and in previous chapters, I addressed many processes related to the formation of galaxies. Here, I will generalize how galaxies were formed. Indeed, as I was studying vortices, I realized that the processes that explained the formation and evolution of a vortex can also explain the formation of galaxies. Indeed, looking at some pictures of galaxies and what I have said concerning vortices in a fluid flow, I noticed that the configuration of galaxies has similarities with vortices. Indeed, when I first looked at the picture of some galaxies, I saw some similarities with pictures of hurricanes. By the time I arrived at this level of my investigation and writing, it appeared to me that, just as a fluid filament can curl to form a vortex, so also the precursors of galaxies could have initially been an astronomical fluid layers, which, due to a turbulent instability, were "curled" or "wrapped" into an astronomical 3D body. The same thing happened to many celestial bodies, meaning that their precursors were fluid layers, which were "curled" to yield them. The difference lays in how the size and other characteristics of the fluid layers of the precursors responded to the forces that acted on them to swirl, squeeze, or compact them to form (at their best) nearly spherical bodies. Below, I will explain what I meant.

Indeed, the shape or the configuration of the celestial bodies gave me a clue as to how the curling, or wrapping, or gathering together of their fluid layers could have been. Just as on the microscopic scales, some vortices can be spherical and very dense, so also on the astronomical scales of the galaxies, the precursors of some galaxies and their clusters were tightly collected into spherical bodies forming for instance elliptical galaxies. In other cases, the fluid layers of the precursors of some galaxies were unable to compact into elliptical galaxies, but instead into spiral galaxies. In other words, the precursors of spiral galaxies were unable to tightly wrap all their fluid layers around their central or spherical bodies. The same is true for barred spiral galaxies, but here, the wrapping around or the spiraling occurred around a longer central body, implying that the plane of a significant portion of the fluid layer of the precursor was not bent. Hence the presence of a bar in the "middle" or center of the barred spirals. The precursors of the arms of some spiraled galaxies were fluid layers that were loosely wound around a primary portion of their precursor. The central part or the core, or the spheroidal component of some spiral galaxies was the place where the fluid layers of their precursors were most stuck during the processes which gathered them together.

As the gigantic fluid layers of the precursors of the galaxies were being gathered together, stellar systems were formed inside of them. As I already explained, as those stellar systems were being formed, their planetary systems and asteroid systems were also formed. The formation of planetary systems also implies the formation of their planets and satellites. In other words, as the precursors of galaxies were being formed, on different scales and levels of hierarchy, stars, planets, satellites, asteroids, and microscopic particles (e.g. atoms, subatomic particles) were also being formed in the universe as components of the bodies found in and between those galaxies. On a scale larger than galaxies, larger fluid layers were split-gathered so that galaxies could be formed and so on and so forth until the highest level of clusters of galaxies were formed.

Just as stars are orbited by their planets and asteroids, so also the core of the spherical components of galaxies are "orbited" by stars and stellar systems. Just as rings are found orbiting some planets, so also on the galactic scale, the galactic disc contains many stars and planets imbedded in them. Just as in a fluid flow, fluid elements or vortices are pulled and pushed by others in the flow, so also, in the fluid layers of the precursors of gigantic galaxies, some precursors of galaxies and of stars were pushed or pulled by those adjacent to them. Just as I explained in the chapters on vortex and fluid ligaments (where I linked orbital inclination, axial tilt, eccentricity, radius, and semi major increment), some stars could have been titled or inclined in a certain direction and angle because of the pulling or pushing of their precursors by the precursors of adjacent stars. Likewise, the orientation of the plane of some galaxies could have been caused by the pulling or pushing of the precursors of the galaxies adjacent to them and/or caused by other events associated with their breakup up from others.

The galaxies in the universe are positioned in different planes and some of them cannot be seen from telescopes placed on Earth because of the position that their precursors landed during their formation after breaking up from others. That is why while some galaxies can be clearly seen from Earth, only a tiny line of light is seen for others, which likely lay in the same plane as the Milky Way, the galaxy that the Sun belongs to. The characteristics of the fluid layers or their stratification could have affected their turbulence and how their fluids were mixed. As the fluids were mixing, some from different layers could have been mixed and the vortices could have been differently flipped over, inclined, and even overturned for in the vicinity of the interface of two adjacent layers. The shearing of the fluid flow due to the fluid layers moving at different speeds could have also contributed to inclining, flipping over, and even overturning some stars, stellar systems, and even galaxies (depending on the scale). Sometimes, layers of clouds in the atmosphere lead to the formation of billows, some of which have structures resembling spiral galaxies (but on a smaller scale).

I was not surprised to learn that since *"modern flow visualization and simulations (e.g. Direct Numerical Simulations: DNS) in which instantaneous motions are numerically produced in a computer using equations governing fluids allowed the detection of coherent structures, turbulence researchers are increasingly believing that turbulence is a nest of vortical structures"* (George, 2013a). In other words, wherever there is turbulence, vortical structures should be expected to exist and even nest. Hence, on all scales of turbulence, the size of vortical structures present in the precursors of the celestial bodies affected the size of the celestial bodies and their clusters. As I addressed it in the chapter on vortex, just as a long vortex which is stretched can spin up at a rate which can depend on the size of the larger bodies controlling the stretching, so also the size, rotational speed, and rotational period of galaxies, stars, planets, satellites, and even microscopic particles could have been affected by the characteristics of their vortices and the size of the fluid layers of their precursors.

In Fig. 179, I sketched the developmental stages that some spiraled and elliptical bodies (e.g. vortices and galaxies) born from different configurations of the fluid layers of their precursors could have gone through.

CHAPTER 31: CONSTELLATIONS, GALAXIES, GLOBULAR CLUSTERS, AND STARS IN THE UNIVERSE

Fig. 179: Developmental stages of some spiraled and elliptical bodies (e.g. vortices, galaxies) born from different configurations of the fluid layers of their precursors

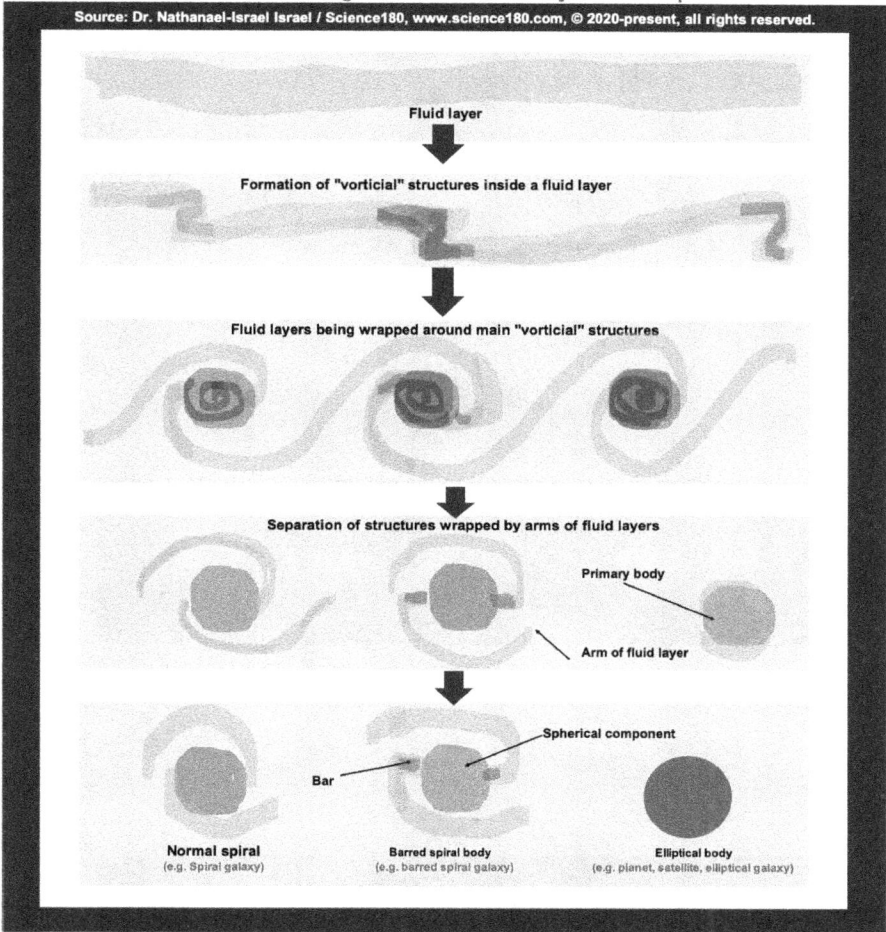

Next (Fig. 180), I sketched some of the configurational changes in which the precursor of the elliptical, normal spiral, and barred spiral galaxies could have gone through.

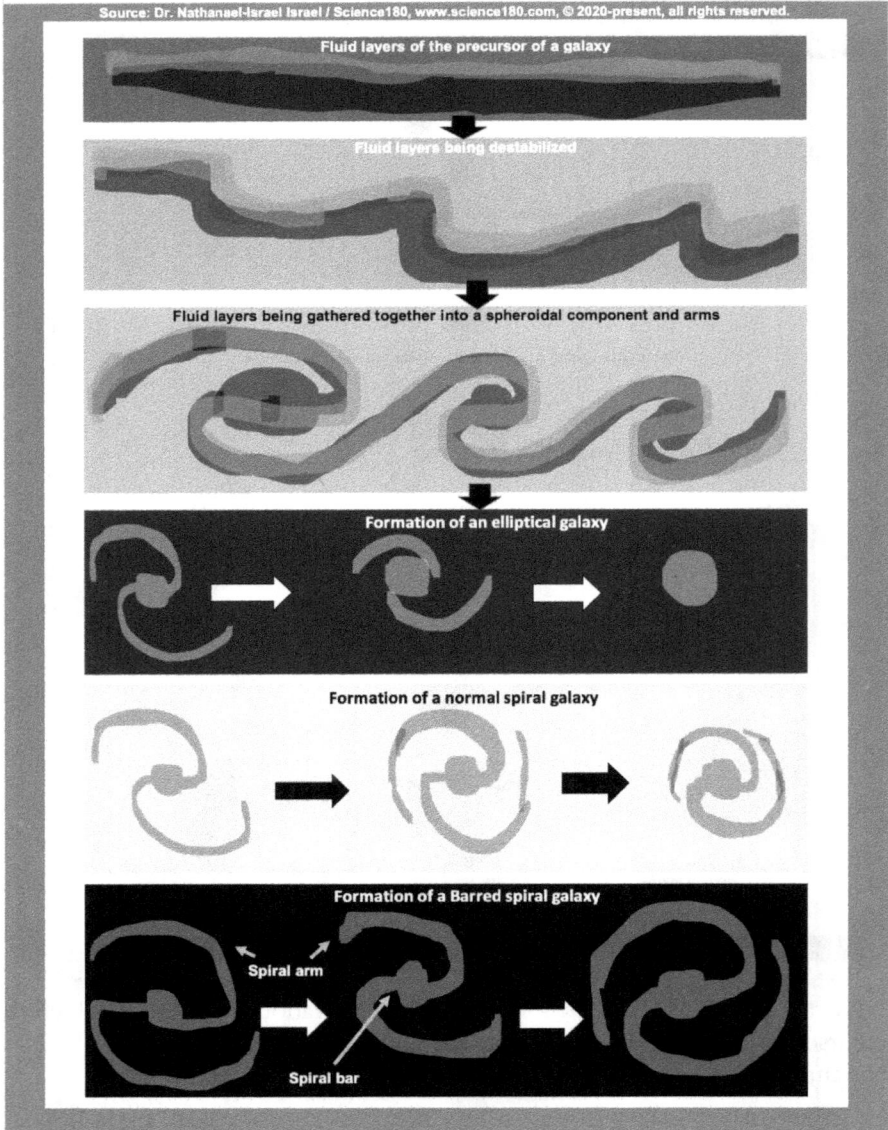

Fig. 180: Formation of elliptical, normal spiral and barred spiral galaxies

Visit http://www.science180.com/GalaxyFormation to get a very beautiful colored copy of this graph.

Fig. 181 shows the formation of the arms of some spiral galaxies. Some spiral arms are not a continuum but seem broken because indeed, their precursors were broken and therefore separating the layers of their fluids. Some arms are longer than others because the size of the fluid layers of their precursors were large or

Nathanael- Nathanael-Israel Israel: Author of "Reconciling Science and Creation Accurately"

deep. The speed of the stars in the arms could depend on the speed of the fluid in the layers of their precursors and how they were wound up or spiraled.

Fig. 181: Fluid layers gathered together to form a galaxy under the influence, instability, and turbulence of their precursors

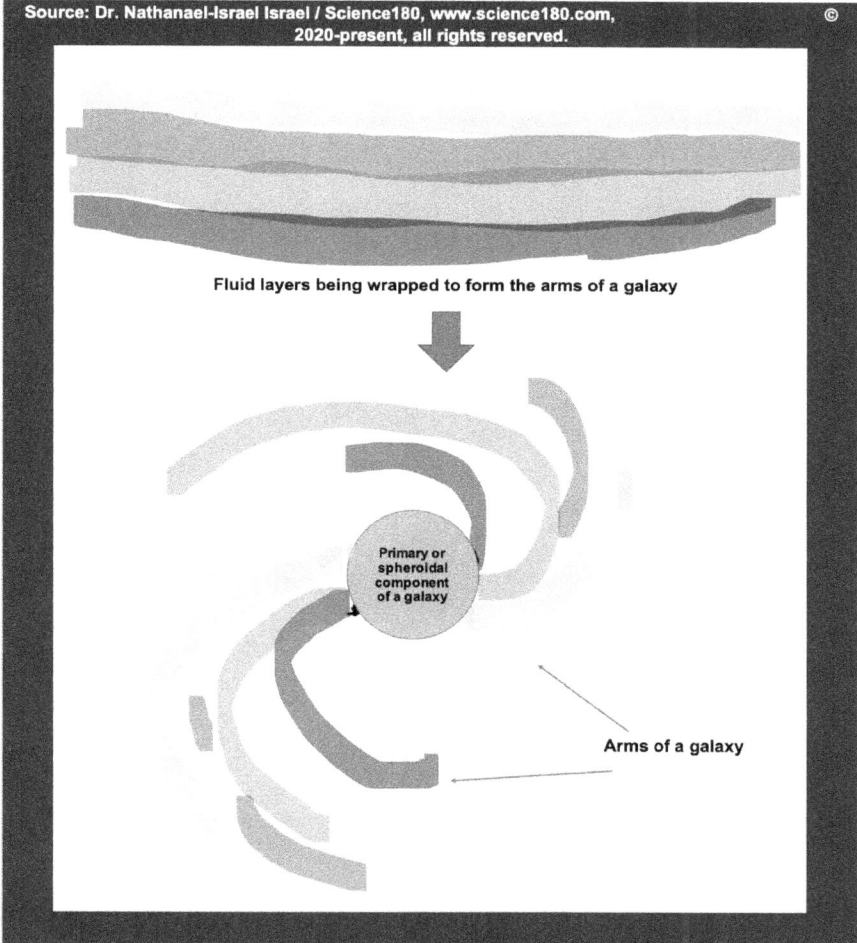

Fluid layers being wrapped to form the arms of a galaxy

Primary or spheroidal component of a galaxy

Arms of a galaxy

In Fig. 181, I did not show the stars because I did not want to overload it. However, as explained next (Fig. 182), the arms and disc and even the spheroidal components of a galaxy are filled with stars. As sketched in the above graph, some fluid layers which were on top of the layers of fluids of the precursors of a galaxy can end up in the outer portion of the arm as well as in the inner portion. Considering what I have already explained in the chapter on orbital speed, the orbital speed of the stars in the top layer of the precursor of each galaxy can have a higher speed than that of the stars in the bottom layer. Likewise, the orbital speed of the stars in the galactic arms could depend on the position of the fluid layers out of which they were gathered together. The way the fluid layers were mixed and split between the central or spheroidal component, and the arms could define the speed of their stars. Hence the speed of stars should not be expected to always depend on their position in an arm or disc in a galaxy. Because galaxies are oriented differently, and some are flipped over, the speed of the stars would also depend on those orientations.

Fig. 182: Stars spread throughout the arms and disc of a galaxy

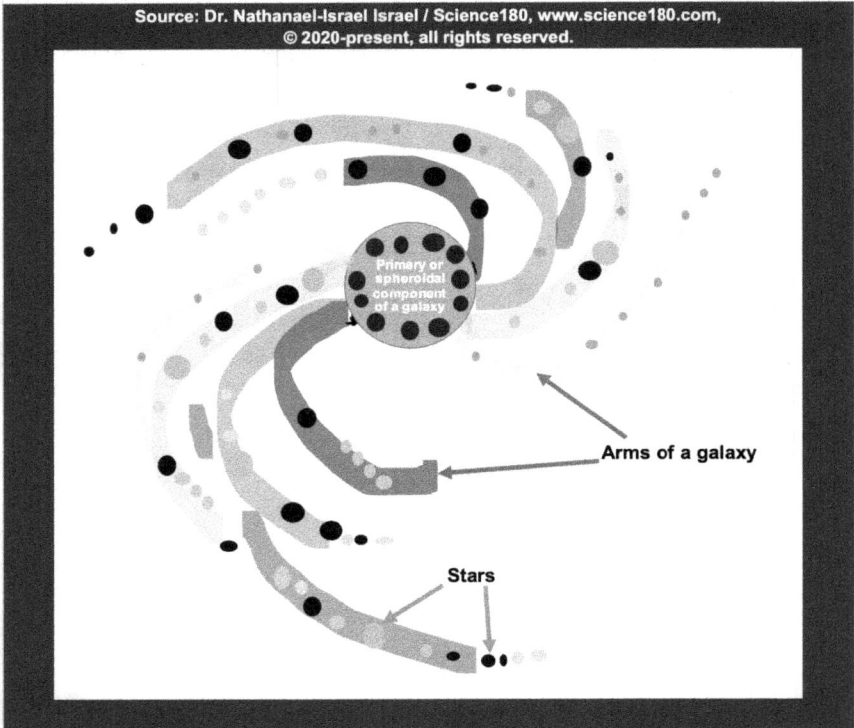

31.11. Take home message

After carefully reviewing the literature concerning the diversity, classification, and characteristics of the constellations, galaxies, globular clusters, and stars in the universe, I elucidated the processes which formed them. I showed that galaxies, galaxy clusters, and star clusters descended from gigantic fluid layers. When the precursors of the galaxies were splitting from one another, some of the material between them could not have been big enough to form a galaxy but just some stars, which could not be collected into galaxies but into individual stars or star clusters. The precursors of the stars were wound up to yield individual stars. During the process, huge voids separated the celestial bodies and their various clusters. Hence, the universe mostly seems void (when seen with the naked eye) and huge distances exist between the celestial bodies. I showed how the processes which explain the formation and development of a vortex can also explain the formation of galaxies. Just as a fluid filament can curl to form a vortex, so also the precursors of galaxies were initially astronomical fluid layers, which, due to a turbulence they went through, were "curled" or "wrapped" into astronomical 3D bodies. Just as on the microscopic scales, some vortices can be spherical and very dense, so also on the astronomical scales of the galaxies, the precursors of some galaxies and their clusters were tightly collected into spherical bodies forming for instance elliptical galaxies. In other cases, the fluid layers of the precursors of some galaxies were unable to compact into elliptical galaxies, but instead into spiral galaxies. In other words, the precursors of spiral galaxies were unable to tightly wrap all their fluid layers around their central or spherical bodies. The same is true for barred spiral galaxies, but here, the wrapping around or the spiraling occurred around a longer central body, implying that the plane of a significant portion of the fluid layer of the precursor was not bent or curled. Hence the presence of a bar in the "middle" or center of barred spirals. The precursors of the arms of spiraled galaxies could have just been layers of fluid which were loosely wound around a primary portion of their precursor. The central part or the core, or the spheroidal component of spiral galaxies could have been the place where the fluid layers of their precursors were most stuck during the processes which gathered them together. As the gigantic fluid layers of the precursors of the galaxies were being gathered together, stellar systems were formed inside of them. On a scale larger than galaxies, larger fluid layers could have been split-gathered so that galaxies could be formed and so on and so forth until the highest level of clusters of galaxies were formed. Just as in a fluid flow, fluid elements or vortices are pulled and pushed by others in the flow, so also, in the fluid layers of a gigantic precursor of galaxies, some precursors of galaxies and of stars could have been pushed or pulled by those adjacent to them. Some stars could have been titled or inclined in a certain direction or angle because of the pulling or pushing of their precursors by the precursors of adjacent stars. Likewise, the orientation of the plane of some galaxies could have been caused by the pulling or pushing of the precursors of the galaxies adjacent to them and/or caused by other events associated with their break-up up from others. Just as a long vortex that is stretched can spin up at a rate which can depend on the size of the larger bodies controlling the stretching, so also the size, rotational speed, and rotational period of galaxies, stars, planets, satellites, and even microscopic particles could have been affected by the characteristics of their vortices and the size of the fluid layers of their precursors. I showed the developmental stages that spiraled and elliptical bodies (e.g. vortices and galaxies) born from different configurations of the fluid layers of their precursors could have gone through. I also presented the configurational changes that the precursors of the elliptical, normal spiral, and barred spiral galaxies went through. Moreover, I showed how the orbital speed of the stars in the galactic arms depend on the position of the fluid layers out of which they were gathered together and also on the way the fluid layers were mixed and split between the central or spheroidal component and the arms.

At Science180 Academy (www.Science180Academy.com) that I founded, I have many programs that can help you learn more about the formation of galaxies and their content. One of them is "Science180 Academy of Cosmology", designed for all scientists who want to scientifically study cosmology, and dig into details such as those I handled in this chapter. If you are not interested in that, you can also register for "Science180 Academy of Turbulence" to learn more about the turbulence that shaped everything in the universe.

CHAPTER 32

YOU THOUGHT THAT ALBERT EINSTEIN AND ISAAC NEWTON WERE RIGHT ABOUT GRAVITY? READ THIS SHOCKING STORY TO GET THE LATEST BREAKTHROUGH ABOUT THE MEANING OF GRAVITY

32.1. Generalities

What if Albert Einstein's and Isaac Newton's explanations of gravity were all wrong? Read this important chapter to find out!

Before the days of Science180 (that I, Nathanael-Israel Israel, pioneered), when the fundamental forces in nature are addressed, existing theories usually point at 4 phenomena:

- the strong force also called the strong nuclear interaction,
- the electromagnetic force,
- the weak force also called the weak nuclear interaction, and
- gravity.

These co-called fundamental forces made me ask questions like:

- Are these interactions the only (fundamental) forces in the universe?
- Are they even worth calling forces?
- Do they really explain reality as seen in nature?
- What is their real explanation?

I could have addressed these issues a long time ago, but because they relate to many other chapters, I had to place them here to ensure the concepts I would use are understood first. In this chapter, I not only addressed some weaknesses of the existing theories of the fundamental forces in the universe, but I also explained why they do not properly illustrate how the universe was formed. In perspective of Science180 explanation of the formation of clusters or systems of matter in the universe, I tackled the real meaning of the so-called fundamental forces in nature, beginning with gravity.

NASA defines gravity (m/s^2) as the "gravitational acceleration on the surface at the equator, including the effects of rotation. The gravity of the gas giant planets is given at the 1 bar pressure level in the atmosphere". The surface acceleration is defined as the "effective equatorial gravitational acceleration at the surface of the body or the 1 bar level, including the effects of rotation". In contrast, the surface gravity is defined as the "equatorial gravitational acceleration at the surface of the body or the 1 bar level, not including the effects of rotation".

The understanding of gravity has been one of the main mysteries that scientists have been trying to unravel for centuries. Isaac Newton was one of the scientists that tried to explain gravity. The gravitation theory that Isaac Newton wrote to explain gravity and gravitation has been a main finding in science for centuries. After the theory of Albert Einstein on relativity, the theory of Isaac Newton was "rejected" although it still explains many things on the small scales. Nevertheless, the way that Albert Einstein explained gravity has not addressed and cannot address most of the scientific data available today, yet many people view the work of Albert Einstein as the ultimate explanation of the universe. Therefore, many people have not dared to challenge that theory. The current explanation that most scientists give to gravity is based on that of Albert Einstein from a century ago. Similarly, certain controversial ramifications of this theory (e.g. theory of black holes) are also based on the work of Albert Einstein. Seeing the importance of gravity in most interpretations of the origin of the universe, the late Stephen Hawking erroneously said that "*God did not create the universe, gravity did*". What an aberration and a blasphemy, some believers may say! If gravity created the universe, where did gravity come from?

Nonetheless, many critical questions remain unanswered about gravity, implying that a demonstration of any mistake in the gravitational theory of Albert Einstein would have a huge impact on theories based on his work. Most well-informed scientists know that gravity is still not explained yet. The observation that astronomers are making regarding the movement of galaxies "convinced" some of them that the theory of gravity elaborated by Albert Einstein has some problems. Beside the theories of Isaac Newton and Albert Einstein, many other theories of gravity have been made, but none of them has fully explained most of the scientific data available. Although the General Theory of Relativity tried to set the geometry of gravity, many other scientists stated that it does not describe the underlying physical mechanics of gravity. Addressing the mechanics of gravity, Richard Feynman (a laureate of the Nobel Prize in Physics) said:

> "*What about the machinery of gravity? All we have done is to describe how the Earth moves around the Sun, but we have not said what makes it go. Newton made no hypotheses about this; he was satisfied to find what it did without getting into the machinery of it. No one has since given any machinery.*" (Feynman, 2006).

Nathanael-Israel Israel: Member of the American Society of Agricultural and Biological Engineers

Because I was dissatisfied with the explanations given to gravity in the literature, I decided to investigate gravity using the insight I got from the turbulence that took place during the formation of the universe. Before I explained gravity in perspective of the turbulence which shaped the universe, I will first present below, the trends of the gravity of celestial bodies according to existing data. Then, I will explain what I think is the root cause of gravity.

32.2. Description of the gravity of the celestial bodies in the Solar System

Gravity data is available for only 123 celestial bodies in the Solar System. Most of these values are theoretical, therefore implying that if the theories on which they are based are wrong, they will also be wrong. It would have been great if scientists could have experimentally collected data on gravity. Because of the potential bias associated with the theories that were used to calculate the gravity available in the literature, I was very cautious in drawing conclusions while using the data on gravity.

The gravity of the objects I studied in the Solar System varied from 1.7E-05 m/s² to 274 m/s². Among the data I studied, the smallest gravity was recorded with the Aten asteroid called 2002 AA29, whereas the highest gravity in the Solar System was obtained with the Sun. About 87.9% of the gravities I studied were less than 1 m/s². The gravity of the planets is higher than 1 m/s² except that of Pluto, which is 0.62 m/s². Among the bodies which gravity is higher than 1 m/s² are 4 Jovian satellites, one Saturnian satellite, and the Moon:

- 1.235 m/s²: Callisto (the outermost Jovian satellite in Zone 3)
- 1.314 m/s²: Europa (the second innermost Jovian satellite in Zone 3)
- 1.352 m/s²: Titan (the biggest Saturnian satellite)
- 1.428 m/s²: Ganymede (the biggest Jovian satellite)
- 1.62 m/s²: Moon (Earth's only satellite)
- 1.796 m/s²: Io (the innermost Jovian satellite in Zone 3)

32.2.1. Gravity of the planets

The planet that has the highest gravity (24.79 m/s²) in the Solar System is Jupiter, whereas the one with the lowest gravity (0.62 m/s²) is Pluto (Fig. 183). The gravity of the Earth is 9.798 m/s². Of the 4 giant planets (all of which are bigger than the Earth), Uranus is the only one which gravity (8.87 m/s²) is smaller than that of the Earth. Although the radius of Uranus is 9.96 times that of Venus, they both have the same gravity (8.87 m/s²). Similarly, although the radius of Saturn is 9.44 times that of the Earth, the gravity of the Earth is just 93.85% that of Saturn. Finally, although the radius of Jupiter is 1.18 times that of Saturn, the gravity of Jupiter is 2.4 times that of Saturn. These data suggested that gravity may not just be due by to the radius of the planets.

According to Encyclopedia Britannica (Faller et al., 2020), the gravity of the Earth varies between 9.78 m/s² at the Equator and approximately 9.83 m/s² at the poles. Gravity is usually higher at the surface of the oceans and lower at the top of mountains. According to the previous source, soon after the days of Newton, "*it was found that the gravity on top of large mountains is less than expected on the basis of their visible mass while on ocean surfaces, gravity is unexpectedly high*". Instead of acknowledging that the theory of gravity has flaws, its proponents defend that discrepancy by postulating the "*low-density rock 30 to 100 km underground the mountains*" versus the "*dense rock 10 to 30 km beneath the ocean bottom*".

Fig. 183: Gravity (m/s²) of **Planets** in the Solar System

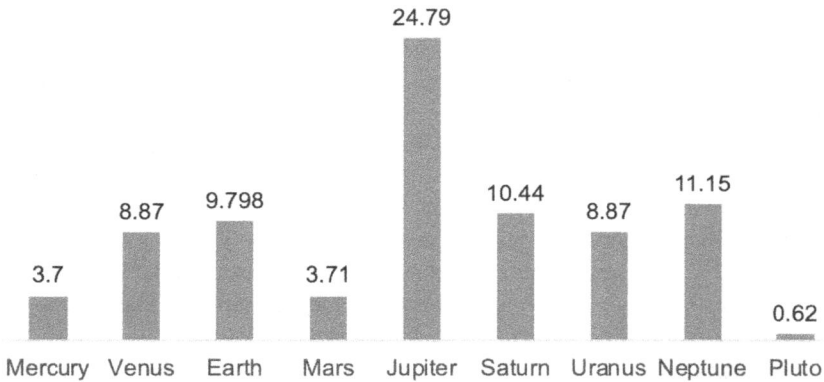

32.2.2. Description of the gravity of the satellites in the Solar System

The gravity of the satellites in the Solar System depends on their types and turbulence zones (p<0.000). Out of the 210 satellites known in the Solar System as of 2020, gravity is available for only 61 and it ranges from 2.5E-04 m/s² to 1.796 m/s² (Fig. 184). The smallest gravity was found with Daphnis, the third innermost Saturnian satellite in Zone 1. In contrast, the highest gravity was recorded with Io, the innermost Jovian satellite in Zone 3. The second highest gravity recorded on the satellites was with the Moon (1.62 m/s²), which is higher than that (1.428 m/s²) of Ganymede (the biggest Jovian satellite), and that (1.352 m/s²) of Titan, the biggest Saturnian satellite. To sum it up, the gravity of the biggest satellite in each planetary system is:

- 1.62 m/s²: Moon (Earth's only satellite)
- 0.0057 m/s²: Phobos (the biggest Martian satellite)
- 1.428 m/s²: Ganymede (the biggest Jovian satellite)
- 1.352 m/s²: Titan (the biggest Saturnian satellite)

Nathanael-Israel Israel: Member of the American Society of Agricultural and Biological Engineers

- 0.379 m/s^2: Titania (the biggest Uranian satellite)
- 0.779 m/s^2: Triton (the biggest Neptunian satellite)
- 0.288 m/s^2: Charon (the biggest Plutonian satellite)

In general, 90.16% of the gravity of the satellites is inferior to 1 m/s^2. The highest gravities were recorded in Zone 3, but not always with the biggest satellite.

Fig. 184: Gravity of the satellites in the Solar System according to their types and semi major axis

Fig. 185 is made from the same data as that above except that its style allowed me to spread out the clusters which are at the origin of the Fig. 184. It permitted me to pinpoint that the highest gravities are not with the innermost or outermost satellites, but somewhere in between. In fact, 3 years before I could analyze most data with the perspective of turbulence, Fig. 185 helped me to know that the highest gravities are not recorded on the innermost satellites nor on the outermost satellites, but somewhere between the innermost and the outermost satellites. During those days, by contrasting the gravity data with the radius data, I knew that the satellites with the highest gravity are the biggest ones. As I explained below, when I finally tested the impact of the turbulence zones on the gravity of the satellites, the highest gravities were observed in Zone 3 (Fig. 186), suggesting that gravity may have something to do with turbulence and/or the size of the bodies.

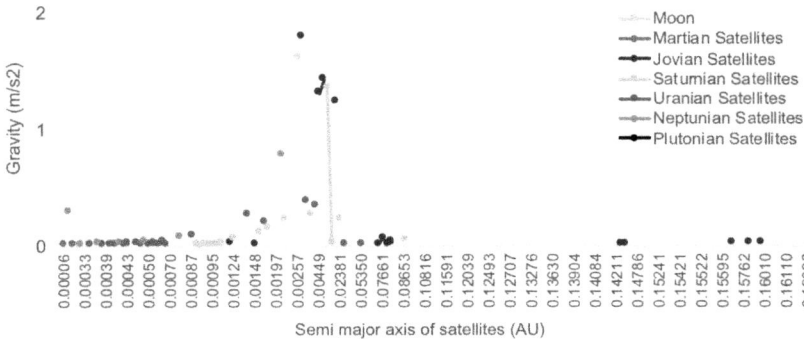

Fig. 185: Gravity (m/s^2) of the **satellites** in the Solar System according to their types and semi major axis

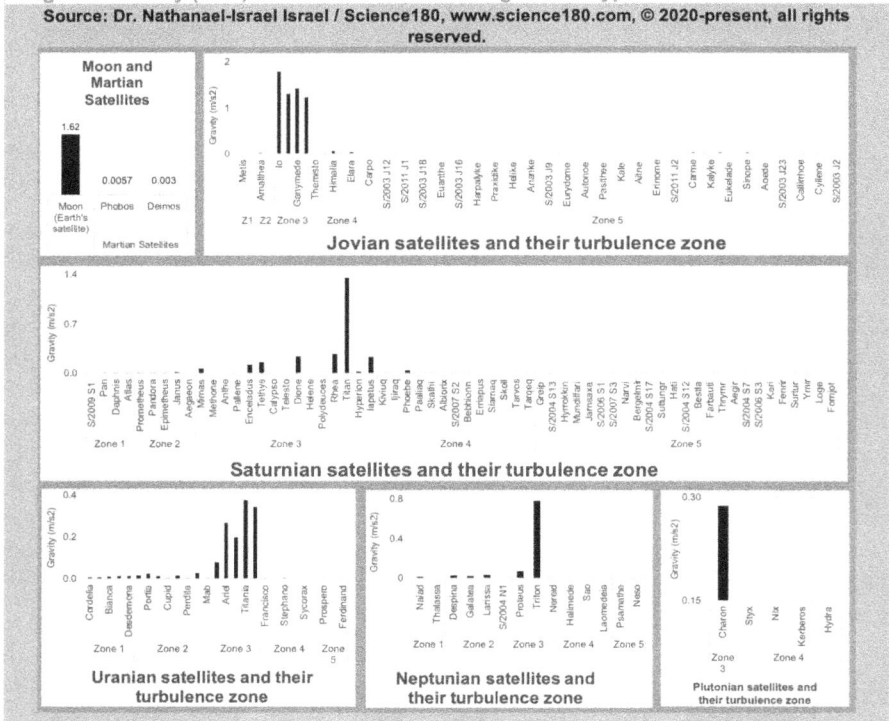

Fig. 186: Gravity (m/s^2) of the satellites according to their types and turbulence zones

Jovian satellites and their turbulence zone

Saturnian satellites and their turbulence zone

Uranian satellites and their turbulence zone

Neptunian satellites and their turbulence zone

Plutonian satellites and their turbulence zone

32.3. What existing theories say about gravity

Conventionally, when fundamental forces in nature are considered, the one most people are familiar with is gravity. In *"Turbulent Origin of Chemical Particles"*, I also explained the Standard Model theory of particles, which is alleged to connect most of the known particles together.

Most human beings can easily relate to experiences of things falling to the ground after they are thrown into the air. Gravity changes with the location or with the position on the surface of a celestial body. On Earth, the force of gravity is said to be the weakest at the equator but the strongest at the poles. For instance, according to Encyclopedia Britannica, the *"acceleration gravity [of the Earth] varies from about 9.78 meters per second per second at the Equator to about 9.83 meters per second per second at the poles. In addition to this broad-scale variation, local variations are* [believed to be] *caused by variations in the density of Earth's crust as well as height above sea level"* (Faller et al., 2020).

Theories have been built to try to explain gravity. As of 2025, the year I published the first edition of this book, the two main schools of thoughts about the theory of gravity are:

- The theory of Isaac Newton based on a hypothetical "action at a distance" and
- The theory of Albert Einstein based on a hypothetical "curvature of space"

Indeed, Isaac Newton perceived gravity as a "force that causes any two bodies to be attracted to each other, with the force believed to be proportional to the product of their masses and inversely proportional to the square of the distance between them." In other words, *"Newton's theory of gravitation postulated that the force between two bodies is proportional to the product of their masses and the inverse square of their separation and that the force depends on nothing else"* (Faller et al., 2020). In contrast, Albert Einstein in his general theory of relativity labeled gravity "not as a force (as viewed in Newton's law of universal gravitation), but as a consequence of the curvature of spacetime that he believed to be caused by the uneven distribution of mass". According to Encyclopedia Britannica (Faller et al., 2020), the field theory of Einstein's general relativity assumed that the *"acceleration due to gravity is a purely geometric consequence of the properties of space-time in the neighborhood of attracting masses"*.

Since the time of Isaac Newton, a lot of efforts were made to classically explain gravity, but they all failed. Early in the 20th century, most attempts to explain gravity using classical physics have been halted or abandoned too soon because the scientific community thought and still thinks that the theory of relativity of Albert Einstein has solved the old problem of the cause of gravity. In other words, although advances in subatomic physics and the forces that govern it are relatively recent, a lot of attention has been given to gravity for many centuries. For instance, Newton developed and published a classical theory to try to explain gravity in 1686, but his attempt to link gravity to the attraction between the celestial bodies or the action at a distance has been rejected, particularly after the work of Albert Einstein in 1915. Until today, Albert Einstein's theory of relativity is mostly used by scientists to try to explain celestial bodies. However, on the scale of the Earth, the classical theory of Isaac Newton is mostly accepted.

Although efforts have been deployed to try to design a theory explaining the interaction between subatomic particles, no quantum theory has successfully explained gravity. In others words, scientists have been having serious difficulties to construct a quantum model of gravity. Scientists believe that one of the obstacles in designing a quantum theory of gravity is that general relativity is incompatible with some of the fundamental assumptions of quantum theory. One of the current quantum theories being worked out and on which some scientists are putting their hope to explain gravity is called string theory.

However, after years of impartial analysis of the immense data collected on celestial bodies and microscopic particles in the universe, I realized that, the gravity theories advocated by Isaac Newton and Albert Einstein are wrong although part of them applies to certain things in nature. For, when I considered the amount of data that I used to address the origin of the universe (as detailed in this book and others I wrote), I realized that nothing called "action at a distance" (as postulated by Isaac Newton) or "curvature of space" (as assumed by Albert Einstein) applied to the physics that prevailed during and after the formation of the universe. For instance, although the orbit of most celestial bodies is curved, space itself is not curved nor does it define gravity. The curvature of space advocated by Albert Einstein is an abstract mathematic hypothesis or attempt to geometrically explain a physical phenomenon without physical proofs. Similarly, although some of the equations of the speed, radius, movement, and spacing characterizing the celestial bodies can be twisted to deduct some ideas of attraction between some bodies, in reality, the celestial bodies are not attracting one another, although the field around them can sometimes act as attracting smaller things in their vicinity, but not on the astronomical scale of celestial bodies. Below I provided details.

32.4. Problems with gravity and flyby anomaly based on gravity assist suggested that existing gravity models are probably wrong

A lot of problems surround the theories on gravity and the dark matter associated with them. In my book *"Turbulent Origin of Chemical Particles"*, I explained how the theory of dark matter was forged to try to patch the misunderstanding surrounding gravity. On June 29, 2018, I realized that although efforts were made to try to reconcile the fundamental forces in atoms, gravity is like a plank in the eyes of some theorists who are looking for the speck of sawdust in the eyes of others. Used in many equations in physics, gravity is the foundation of many scientific theories, implying that no one can properly explain other forces in nature if they cannot first understand gravity. Unfortunately, instead of acknowledging that their interpretation of natural "forces" is false and rethink everything from a fresh perspective, most scientists usually try to patch old errors with new ones. For instance, scientists have thought that galaxies are held together by their mutual gravitational interactions. However, although they also found that galaxies move so fast that gravity cannot be able to hold them together, they never audaciously rejected their own gravity theories related to galaxies. In other words, scientific data suggested that the observed matter contained in galaxies are not able to generate enough gravity to keep the stars locked in their stately orbits; yet they kept addressing gravity as the cause of the orbit of all celestial bodies including galaxies and even clusters of galaxies.

The mismatch of some aspects of the gravity theory and the reality of physical data collected on galaxies and other bodies in the universe led certain scientists to realize that the current understanding of gravity is inaccurate and needs to be modified. Instead of fixing the scientific interpretation of gravity, some scientists invented the notion of "dark matter", which they have been holding onto to defend the weakness of their theory of gravitation and of matter in the universe. Other scientists even think that dark matter is the mysterious force accelerating the expansion of the universe. Yet after several years of investigation, no trace of dark matter has ever been found in the universe. Similarly, many theoretical particles (including graviton) postulated to explain gravity have not been found and will likely never be found, for they just do not exist. The belief in dark matter and the misinterpretation of gravity may explain why secular or traditional science has been unable to properly elucidate the formation of the universe using the current massive scientific data, which some people think is not enough yet. In other words, if scientists had understood that their interpretation of gravity was wrong, they would not have forced certain theories to create virtual particles, which no empirical experiment has ever confirmed. Although the dark energy, dark matter or antiparticles associated to gravity has ever been seen, they still constitute a "foundation" for many theories which are trying to explain the universe. I think that most theories of dark energy, dark matter, or the classical and relativistic interpretation of gravity and gravitation are false because they do not consider the secrets hidden in the turbulence that took place at the beginning of the universe.

Besides the Earth where gravity was measured by studying the movement of falling objects, for other celestial bodies in the Solar System, gravity is theoretically estimated using models, most of which are based on theories that I came to realize are incompatible with the true origin of the universe as demonstrated in this book. Therefore, I think that the gravity estimated for most celestial bodies is biased. If scientific expeditions devoted to experimentally measure gravity at least for all the planets can be done to compare the theoretical gravity data with reality, errors will be found.

Although flyby and gravity assist data suggest that existing gravity models are wrong, little or no effort is deployed to fix our conception of gravity quickly. For instance, most flyby missions proved that the gravity of the planets are different from what was projected using scientific theoretical models. While flying by planets to boost their speed, many artificial satellites gained more speed than the predicted value. In other words, according to the "flyby anomaly", many spacecrafts have experienced greater acceleration than expected during gravity assist maneuvers (Wikipedia 2020b). This gain of speed could have been caused by the underestimation of gravity. For most (if not all) existing theories about the origin of the universe are anchored on and/or are strongly connected to theories of gravity or gravitational force. For instance, gravitational "attraction", gravitational pull, or gravitational force are some of the variables that some scientists refer to when they try to explain galaxies and stars. Some scientists stick so much to gravitation as if its modeling was already 100% correct. Yet, most of the experiments in the Solar System to test the theory of gravity failed and this failure was also seen in "flyby anomaly" data. If gravity was correctly estimated, no significant anomaly should have been observed. Yet, the scientists who designed those theories still cling to them.

Indeed, scientists have observed that when spacecrafts fly by the Earth, their energy and speed increased. Scientists have been talking of this phenomenon and using artificial spacecrafts to fly by natural celestial bodies (e.g. planets) in order to increase their speed. Also called gravitational slingshot, gravitational assist, gravity assist maneuver, or swing-by, gravity assist was reported by Wikipedia (2020c) as a technique that uses the movement and gravity of a celestial body (e.g. planet) to modify the path and speed (e.g. accelerate or decelerate) of a spacecraft and to save propellant, time, and other expenses. For instance, when a spacecraft gets near the planet from which it will get a gravity assist, the planet is believed to pull the spacecraft and consequently, increase its speed. This technique has been used to increase the speed of several spacecrafts that were used to explore deep into space. Such a process was used to increase the speed of the two Voyager probes (Voyager 1 and Voyager 2) by flying them by Jupiter and Saturn on their way to the deep space. When I looked at the speed data (Wikipedia, 2020c) of some artificial satellites launched with very high speed, I felt like they tend to normalize their speed with that of the space they go through. If their initial speed is too high, it decreased until it nears or matches that of the celestial bodies orbiting in the same neighborhood. This suggests that space might have been wired in a way that the speed of the celestial bodies does not depend on their mass, but on their spatial position. The forces that control the speed of the bodies in space may be strong enough to influence the speed of the artificial space crafts.

During flyby maneuvers, the geometry of the trajectory, the position and velocity of the spacecraft are continually tracked. Some spacecrafts have experienced an increase of speed just by flying by the Earth. That was the case for the Near Earth Asteroid Rendezvous (NEAR) spacecraft which velocity increased "anomalously" by 13.46 mm/s after encountering the Earth (ESA, 2009; Anderson et al., 2007). According to the same report, other spacecrafts that gained speed include:

- Cassini–Huygens extended its speed by 0.11 mm/s in 1999
- Rosetta speed increased by 1.82 mm/s in 2005.

Nathanael-Israel Israel: Member of the American Society of Agricultural and Biological Engineers

These discrepancies between theoretical estimation and real data suggested that significant mistakes still exist in the understanding of gravity or gravitation. As for me, the errors surrounding the misinterpretation of gravity and its origin contribute to why the so called "gravitational constant G" is an erroneous constant, yet it is used to many theories.

32.5. What gravity is NOT

Before handling what gravity really is, I explained below what it is NOT based on many processes I explained in this book and which do not require gravity to hold.

32.5.1 Gravity is not responsible for the orbit of celestial bodies and does not control their trajectory

According to the mechanics of Newtown, gravity is alleged to provide the centripetal force which is responsible for the orbit of the celestial bodies. However, according to my findings, gravity is NOT responsible for and does not control the orbit or the trajectories of the celestial bodies in the Solar System or in any other stellar system in any galactic system. Gravity does not control the orbit of the stars in a galaxy nor does it define the orbit or galaxies in clusters of galaxies. Indeed, the precursors of celestial bodies were already launched and were on their way to their orbit before gravity was completely established. It is the way that the bodies in the universe were gathered together (including the spiraling of the precursors of the bodies and the formation of vorticial structures according to the scale of the turbulence) that progressively birthed gravity and the other so called fundamental forces. For, as the fluid layers of the precursors of bodies were spiraling, curling, or swirling, their rotational angular speed changed and, they were compressed or squeezed, progressively creating a field, which effect can also be felt as gravity. Experiences with artificial satellites for instance should have been enough to help some theorists to correct their mistakes on gravity. For instance, the trajectory and orbit of the artificial satellites launched from Earth are not controlled by the gravity of the celestial bodies, but by the processes which launched those artificial satellites (including the launch speed) and progressively put them in orbit. Similarly, the orbit of the celestial bodies is not controlled by gravity. It is because previous theories failed to explain how the universe was formed that they failed to properly explain gravity. Just as a product of something cannot be the source of that thing, so also gravity which is a product or a consequence of how the celestial bodies and system of bodies were formed cannot be the main source or the cause of their formation.

32.5.2. Gravity is not the process that causes secondary bodies to orbit their primary bodies and it is not a force of attraction acting between all matters

Unlike what some theories say, gravity is not the process that causes secondary bodies to orbit their primary bodies. For instance, gravity is not the process which causes satellites to orbit their primary planets. It is not the process which causes planets and asteroids to orbit their primary star. In the Solar System, gravity is not responsible for the revolution or the movement of planets and asteroids around the Sun. Similarly, gravity is not responsible for the movement of the Sun or the Solar System around the center of the Milky Way Galaxy. Gravity is neither responsible for the movement of the Milky Way Galaxy around the center of the galaxy cluster it belongs to. Gravity is not responsible for the large-scale structures in the Universe, but the way the bulk of the turbulent prima materia (the precursor of all matters in the universe) was split-gathered is.

Unlike what is thought in existing theories, gravity (also called gravitation in mechanics) is not a force of attraction acting between all matters. For instance, the Sun is not attracting the planets or the celestial bodies in the Solar System (as some people think or some theories alleged) and vice versa. When they are very close to one another, some celestial bodies can influence one another by the magnitude of their gravity, but the structures of the celestial bodies in the universe are not governed by gravity. For instance, I explained how the proximity of some celestial bodies can affect their motion and cause them to move chaotically. Similarly, when the positions of the orbit of two celestial bodies which are on distinct orbits happen to near one another beyond certain limits determined by their size and speed, they can temporally affect the motion of one another before sometimes getting back to their normal motion. These temporary affects should not be viewed as a reason to evoke gravity as the factor affecting the motion or orbit of all of the celestial bodies. The fact that most human beings leave their home, go to work, do some errands, and then return home afterwards on a daily basis does not mean that their home is attracting them like a physical magnet. Likewise, the revolution of celestial bodies around others should not be interpreted through the glance of some bodies being attracted by others.

32.5.3. Gravity is not responsible for the mass, structures, and development of the universe

Unlike what some theories alleged, gravity is not responsible for the structures and the development of the universe and the bodies it contains. For internal constituents of the celestial bodies were being fully established before the gravity (known as today) of the big bodies they belong to were completed. Unlike what the conventional definition says (Comins et al., 2008) and based on the trends in the data I analyzed, I don't think that gravity is the natural phenomenon by which all things with mass or energy—including celestial bodies (planets, satellites, stars, and galaxies) are brought toward or gravitate toward one another. The process that split-gathered the precursors of the bodies in the universe is not just a matter of gravity. Gravity is just a tiny component of the story of the formation of the universe, yet a lot of attention is given to it. As you can see in this book, many variables I explored are crucial to properly address the origin of the universe, but unfortunately no or very little attention has been given to them while the focus seems to be mainly on gravity!

The mass of particles and celestial bodies cannot be properly explained without an intimate understanding of the process that split-gathered the precursors of the bodies in the universe. For instance, as particles were formed, some were clustered together to an extent that their clusters were significant enough to generate mass. The mass of a particle (small or big) was defined by the amount of the turbulent prima materia which was amassed and squeezed to yield it. Some masses are significant enough to be felt or measured, while in the case of some very tiny particles (e.g. some subatomic particles and particles smaller than them), mass cannot always be measured. For the sensitivity of current scientific materials could not allow its detection. In other words, some particles are classified as massless (meaning having no mass) not because they really have no mass, but because current scientific equipment or theoretical models could not properly measure them.

The extent to which the original particles were gathered together into matter defined the density of the matter and system of matter. In other words, if the turbulent prima materia was not split-gathered into bodies under a turbulence process, nothing could have gained a mass and if the precursors of the matter and system of matter were not squeezed or compressed differently, the density of everything in the universe could have been the same. But because the bulk of the turbulent prima materia could not have stayed stable without being scattered, the series of split-gathering occurred and mass was differently gained by the clusters of particles on many scales. The size of the fluid of the precursor of the bodies in the universe defined their mass. Larger precursors produced more massive bodies then smaller precursors.

32.5.4. Gravity is not responsible for the configuration and constitution of atoms and subatomic particles

As I extensively explained in *"Turbulent Origin of Chemical Particles"*, the conditions which affected the formation of atoms and subatomic particles were not defined by gravity. Celestial bodies with the highest gravity are not those which are dominated by the heaviest atoms. In fact, the greater the gravity, the lighter or smaller the dominant atoms. For instance, the Sun has the greatest gravity in the Solar System, yet, it is not abundant in heavy elements as some terrestrial planets are. Similarly, although Jupiter is the planet with the highest gravity in the Solar System, it is not dominated by heavy elements. In contrast, most terrestrial planets which abound with heavy elements do not have the greatest gravity in the Solar System. More details about these trends can be found in my book *"Turbulent Origin of Chemical Particles"*. The diversity and variation of the characteristics of atoms and subatomic particles and even of celestial bodies were defined by the impact of the process their precursor went through during and after their split gathering from their mother precursors and from adjacent daughter bodies of their mothers. Gravity is not the force which confers a certain configuration to microscopic particles.

Existing theories viewed gravity as weaker than the other fundamental forces or interactions of physics, and "about 10^38 times weaker than the strong interaction, 10^36 times weaker than the electromagnetic force and 10^29 times weaker than the weak interaction" (Wikipedia, 2019b). Gravity is not perceived as having a significant influence at the level of subatomic particles (Krebs, 1999). Gravity may even end up not being the weakest known force in nature. According to the scale of interest, (e.g. planets, stars, galaxies, cluster of galaxies, clusters of clusters of galaxies, etc.), other forces may exist and which contribute to "maintaining" the force of the central body, or primary systems of bodies or the core of systems of bodies in the universe with an intensity smaller than what can be attributed to gravity. In other words, gravity may not be the weakest known force in nature. The misunderstanding of gravity has caused some people to think that it played the most significant role in the internal properties of matter.

Because the gravity of a body decreases as the distance away from that body increases up to a point where no significant gravity can be felt, it is possible that the precursors of some celestial bodies could have been extended to the upper limits where a significant gravity can be felt in their atmosphere. And it was the processes which squeezed or compressed the precursors of the bodies which could have reduced their size to their current one. For if gravity is able to pull things down, it is common sense to accept that some precursors of bodies which were present in the precursors of the gravitational field of the precursors of the celestial bodies could have been pulled to the surface or the atmosphere of these bodies. At the same time, it is important to keep in mind that some bodies or particles present in the atmosphere of the celestial bodies could have been ejected or leaked from the interior or the surface of these celestial bodies. Similarly, the size of the precursors of some celestial bodies (e.g. the giant gases) may be increased due to the expansion of diffusion of their constituents, meaning that in certain cases, some particles could have been released to the atmosphere from the surface of the celestial bodies more than others pulled to the surface by gravity and its precursor. I am trying to bring all these scenarios up because the processes which formed the universe were very complex in their dynamics and careful attention needs to be given to them so some errors can be avoided in their modeling.

32.6. What gravity really is

Gravity relates to the vortical structures (on small and large scales) in the fluid flow during the formation of the celestial bodies. To put it another way, gravity could be the consequence of the vortices of the precursors of the celestial bodies during the turbulence event which formed the universe. Then, after the formation of the universe, the impact was left as a field of gravity around each celestial body.

CHAPTER 32: GRAVITY AND OTHER FUNDAMENTAL FORCES IN NATURE

Removing gravity from a celestial body would require reversing the processes involved in the formation of the universe. In other words, if the processes that formed the universe could be reversed and the celestial bodies unwound, the universe can be brought back to a state near that of its origin just as I explained in the chapter titled "In the beginning …". For the bodies in the universe to be formed, a process was needed to split and pull together precursors of the bodies in their nearby environment into different "compartments".

The way fluid layers and the vorticial bodies formed in them were moved, sheared, revolved, rotated, tilted, and elongated had collected some precursors into bigger bodies and others into smaller ones according to the split-gathering laws I extensively explained in this book. For some bodies to get their size, fluids in the surroundings of their precursors were accumulated as they were moving until they reached a position where their volume could no longer allow them to be sheared any farther by the shear flow. Therefore, bodies of fluids were kind of "stationed" at semi majors which agreed with their shear rate, orbital speed, rotational speed, and other characteristics. In other words, the force that sheared the fluids was not able to move their bodies farther away beyond some limits defined by their energy, location, and other factors. Therefore, each body was positioned at a location matching its size, movement, constitution, and the split-gathering history of its precursor. Just as the wind can easily move a leaf but not a heavy house made of concrete, so also when a big body was formed in the way of the fluid flow, the ability of the flow to displace or position it depended on many factors including the energy, the stress in the flow, and the energy of the body itself, etc. But if the body contains more energy, it may not be moved by the flow beyond a certain distance. Hence the largest bodies are not usually very far from their mother precursors. Similarly, the energy that could have caused the shear flow could have had a power that can pull or push things in its way or into its surroundings. The viscosity of the precursors could have also affected their stretching. As all this was happening, bodies were being rotated (on various scales) and gravity was progressively formed according to the rotation and size of the precursors of the bodies. As of today, artificial gravity can be "created" using rotation of bodies, therefore suggesting that natural gravity itself could as well have been the consequence of the rotation of the precursors of the celestial bodies.

Considering all the variables I explored in this book and the processes that explain the formation of the universe, I came to realize that gravity is a consequence of the process which split-gathered the precursors of celestial bodies into their daughter bodies. Indeed, without a mechanism, process, or force to combine together or assemble the precursors into unified bodies, most daughter bodies could have not been able to form bodies after their split from their mothers. The process of gathering together the matter in the precursors of bodies as they started splitting from their mother was not initially controlled by gravity. It was later that gravity itself was born or formed. Another way of explaining this

Science180: Complex Universe-Origin Questions Solved Accurately in a Simple Language is that, before what is known today as gravity was established, forces and interactions were already acting on the precursors of bodies to help gather together their matters into unified bodies and systems of bodies. Gravity was like the matured or last state of expression of some of the forces that acted to collect dispersed matters in the precursors of bodies.

Under the influence of turbulence, vorticial structures present in the fluid flow and/or fluid layers were progressively squeezed, amassed, and wound up into bigger bodies as fluid layers were pulled and pushed while movements were being born. Some precursors of bodies started rotating more than others. Gravity was progressively established as the fluid layers of the precursors were amassed together under the influence of their movement (e.g. rotation, or rotational-like, or spiraling motion). The forces involved (e.g. centrifugal and centripetal forces) to shear the fluids and to compress them acted on the precursors according to the intensity of their rotation and their size and consequently, affected the magnitude of the gravity of these bodies. In other words, gravity and its precursor allowed the precursors of some bodies not to be dispersed but to form a unique and unified body. Another way of explaining this complex process is that as the precursors of the celestial bodies were being split and were moving, a squeezing force started compressing and shaping them at the same time that other internal processes were also gathering together the clusters of rocks, minerals, atoms, subatomic particles, and other smaller matters inside of them. Because gravity itself depends on the radius of the celestial bodies and their rotational speed, its value as it is today could have not been established before the bodies got their current shape. In other words, as the precursors of the bodies were getting their shape, size, and speed (including rotational speed), a process of gathering was collecting them together and established gravity. Just as all bodies today have had their own precursors, so also all forces and interactions in nature had their precursors. Similarly, gravity as of today had its precursor, which was embedded into how turbulence allowed the precursors of bodies to be gathered together on astronomical scales as they were being split-gathered as I extensively explained in this book. By the time that the size and movement were set and the celestial bodies were put on their "final trajectory" or orbit, gravity was established.

Without a mechanism to match gravity with the size, movement, and composition of the bodies, gravity and all the systems of gathering together at all scales of the precursors could have highly compressed some bodies while others could have been scattered or dispersed. As they were moving inside the fluid layers of the precursors of the bodies, the vorticial structures could have been able to pull some bodies just as, even as of today, when a vortex is rotating, it can pull some matter toward itself. By the time the celestial bodies were formed, a zone of a seemingly "attractive" field was created around them and most things which are in that zone can be pulled toward that body. On a small scale also, all bodies have a field that was formed to exert a force upon matter to define and maintain the form of these bodies. Hence, even at the atomic and subatomic level, there is a field around particles. It is this field which allowed these particles to exist, meaning that without it, matter would have never existed.

Gravity can also be explained by the air, gases, and waves coming from the poles due to the rotation of the celestial bodies. In other words, the rotation of the celestial bodies and even any other body can cause the formation of a field around it, which intensity can depend on the size of the body, the rotational speed, and other factors. For instance, when fluid in a bowl is rotated, a force can be born which can cause a conical structure at the surface of the fluid and also attracts and pulls some air from the outside toward the inside of the fluid. A similar effect is found with hurricanes and many other moving things. Gravity is higher at the poles because the force of the air being pulled toward the poles directly hits them, while it is lower at the equator because the equator is farther from the poles. The force or the air descending at the poles is accompanied by winds, which explain why the poles are colder than the equator although some of them are in more contact with the Sun that the equator. Both the mass and the rotational angular speed play a role, consequently, the way the fluid layers of the precursors were spiraled played a big role. For instance, Saturn is more ablate because of the small density of its precursor had predisposed it to be highly squeezed or compressed at the poles. Hence its sphere was flattened more than that of Jupiter.

Gravity is like the leftover of the force which defined the existence and organization of the celestial bodies. To explain this concept, I drew a comparison from living things. Indeed, every living organism and nonliving thing seeks to protect its domain. Human beings, wild animals, and planets do the same. On the national level, countries and states try to protect their limits and neighborhood. On the individual level, human beings like to hide themselves inside fences. In each nation, borders exist to delimitate states, cities, counties, etc. Additionally, human beings protect themselves with clothes to fight against adverse weather. Without those strategies, human beings can be destroyed by perilous things. From the herbivores to the carnivores, wild animals have their ways to protect themselves according to their ecological niche. Although carnivores try to dominate herbivores while bigger animals try to dominate smaller ones, each wild species manages to have a terror to enjoy. Because they cannot move from one place to another by themselves, plants adapt to their environment. They also know how to defend their ecological niche and even compete with other plants in their environment. Plants of the same species seem to enjoy living together than with others from a different species. When a stranger enters the property of someone else, he or she can be attacked and charged with trespassing. Some immigrants are sometimes caught, detained, and deported from their new country of residence if they do not have their immigration papers properly set. When a stranger enters into a new environment or niche of a different species, the later can oppose, attack, destroy or try to dominate the former. This is not true just for human beings but also for wild animals and plants. These behaviors hide a code or law of defense not only of ecological niches but also fundamental laws of existence that sustain life in the midst of the challenges in the environments and to maintain some global laws at the root of the formation of the universe. This was made known to me in 2013, the year I was supernaturally inspired about the fundamentals about this book and many others.

As far as nonliving things are concerned, matter is equipped with an ability to defend its environment. Atoms know how to maintain their integrity and not "voluntary" engage themselves into reactions or interactions that will destroy them. And when a "stranger" enters into their environment, "frustrations" are expressed in the form of reactions including radioactive reactions, which can hurt some living organisms near them. Likewise, on the scale of the celestial bodies (e.g. satellites, planets, and stars), the bodies know how to defend their environment or propriety. When a stranger enters their environment, celestial bodies can attract them and pull them toward their surface using gravity. In other words, gravity is one of the mechanisms the celestial bodies "use" to maintain and "defend" their environment or property in the universe. The impartation of energy, motion, and other characteristics of anything that exists in nature was not meant to be lost easily, nor is the force to reverse these laws cheap!

Gravity is not the force that explains the gravitation or revolution of the celestial bodies or their so-called attraction toward one another, but it is a "force" that proceeds from the way the precursors of the celestial bodies were amassed and which ensure the existence and "cohesion" of the celestial bodies. It was a major error that some scientists made by thinking that planets attract their satellites or one another or even attract the Sun, and vice versa. The misunderstanding of gravity is one of the reasons of the main scientific drifting such as those in astrophysics and cosmology, which in many ways seem to become a religion where things are claimed and accepted without proofs but defended by some confusing statistical and other mathematical analysis in the name of physical things we cannot see. On the scale of subatomic particles and atoms, the process which formed them and defined their limits also locked them into compartments in such a way that some theorists can mistakenly think that those particles are attracting one another. In other words, just as some theories on gravity make people believe in an "action at a distance" (e.g. hypothetical attraction between a primary body and its secondary bodies), some people mistakenly think that nucleons and electrons are attracting one another. In *"Turbulent Origin of Chemical Particles"*, I extensively provided more proofs.

By the time I came to this point of my investigation, I realized that gravity will never be sufficient to explain the movement and the functioning of celestial bodies and their clusters. If gravity can be reversed for all bodies in the universe, matter may not hold together freely and celestial bodies cannot maintain their state, and in the end, clusters of matter will be spread out in the universe into smaller particles similar to the first particle in the early universe indeed. In other words, anything that can reverse the direction of the rotation and revolution of the celestial bodies, meaning causing celestial bodies to rotate and orbit in a direction and sense contrary to what they are as of today, could also reverse gravity. For the processes that progressively established gravity were also connected to the ones which established the rotation, revolution, and many other properties of the celestial bodies and particles in the universe. But because human beings and the level of their technology are not and will not be able to reverse the course of the movement of the celestial bodies, they can neither change the gravity of the celestial bodies nor change the design, functioning, and the course of the history, story, and destiny of the universe. What is that destiny? To learn more about that and how I also expounded on my thought on the fundamental forces in nature, read the following books:

- *"Reconciling Science and Creation Accurately"*
- *"Origin of the Spiritual World"*
- *"Turbulent Origin of Life"*
- *"How God Created Baby Universe"*
- *"Science180 Accurate Scientific Proof of God"*

32.7. Establishment of the fundamental forces in nature (e.g. nuclear forces, electromagnetism, gravity)

If scientists cannot properly explain gravity that can be easily measured with tools that can be seen, how can we even trust measurements and theories done on smaller particles like atoms and subatomic particles that nobody has ever seen? The data I analyzed made me to conclude that, although many models including the Standard Model (see my book *"Turbulent Origin of Chemical Particles"*) claim to explain the so-called fundamental forces in nature, I personally think that they miss the point. For if they had properly understood how things in the universe interact, they could have also understood how everything was formed. The radical and turbulent explanation of the origin of the universe also calls for a new perspective on the fundamental forces in the universe and how the universe functions as a whole.

Indeed, as the turbulent prima materia was pushed into movement, destabilized, and entered turbulence, different fields could have been birthed and formed across the compartments of matter which were undergoing changes. These fields could have been the precursors of some interactions that were twisted by theorists to invent the so-called fundamental forces. For instance, the processes which compressed or squeezed the subatomic particles and even the atoms may give an impression of forces between them. But in reality, no subatomic particle carries any force just as the presumed graviton (which is never and will never be found) is not responsible for gravity. In other words, although it can be ok to talk about forces inside atoms, I am not convinced that they are mediated by particles, but by the processes that determined the way matter and its clusters were formed. This way of thinking agrees with the explanation of gravity that does not either mean that celestial bodies are attracting one another. For instance, if I apply my hands to compress or squeeze something in my hands, that does not mean that the constituents of that thing in my hands are attracting one another. The interactions between human beings through love and sometimes hate are "strong" and may be perceived by some people as a force of love or force of hatred, yet they are not mediated by a particle even if the settings of the human heart may play a role.

Similarly, by the end of the formation of the universe and its bodies, natural "forces" may seem present between the bodies, but they are not the root cause of the formation of the universe. Some of the "forces" which acted on the precursors of the bodies and which are still acting on their daughter bodies today have been a product of their movement including rotation and revolution. For instance, as bodies rotate, they can attract some air toward their poles. Consequently, the rotation of the celestial bodies can be responsible for the influx of cold air toward their poles. The forces that accompanied the incoming air at the pole can also explain why gravity can be higher at the poles and while the poles are flatter than the equator and while the equators are oblate.

Unlike what some people think, the so-called fundamental forces were not forged in one step or instantly. However, as the structures and characteristics of the precursors of the bodies (on different scales) in the early universe were changing, so also did the intensity and nature of their interactions between them, some of which would later be called fundamental forces. The characteristics of these forces and the things that mediate them evolved until a kind of equilibrium was reached and most of the constituents of matter (small and big) were locked into systems which dynamism is much smaller than what prevailed at the beginning of the world. In other words, the changes that occurred inside bodies or systems of bodies in the universe today is much smaller than those which occurred during the formation of the universe. Some of the current modifications occurring inside the bodies and systems of bodies are not very significant and could not even be perceived of course, but many things in the universe are changing. The magnitude of those changes may depend on the size of the bodies bearing them and their localization.

Concerning the electromagnetic force, for more than a century, it has been proven that the movement of electrons in a medium can cause electricity. Electric fields can induce or cause a magnetic field perpendicular to it. This suggests that the movement of electrons around the atomic nucleus can generate an electric field that can induce a magnetic field. Consequently, an electromagnetic force can be formed with a tendency of "attaching" the electrons to the atomic nucleus they belong to. In other words, the electromagnetic force found in atoms can be the manifestation of physical forces arising from the movement of electrons and which tend to "bind" the electrons to or around the nucleus. In contrast, to what people think, it is not the protons, nor the neutrons, nor the electrons which are

Nathanael-Israel Israel: Member of the American Society of Agricultural and Biological Engineers

mediating the electromagnetic force in atoms.

As I carefully think about the interactions between the subatomic particles, I realized that, bosons may not even be responsible for a fundamental force. Indeed, also called strong interaction, strong force, strong nuclear force, nuclear strong force or color force, the nuclear force is reported by Wikipedia to be about 100 times stronger than electromagnetism, 10^6 times as great as the weak force, and about 10^{39} times that of gravitation. Although these estimations may bear some errors, they at least suggest that electromagnetism is stronger than the weak force, which at its turn is stronger than gravitation. Although some scientists think that bosons are responsible for or are the carriers of the nuclear force, I do not believe that. As for me, the fundamental forces were caused by the ways the precursors of particles (of various sizes) were compressed or squeezed by the turbulence that took place during the genesis of all matters. Without the impact of turbulence on the precursor of matters, bosons and other particles could not have been compressed to form dense matter. Bosons found in the nucleus are just particles sandwiched between others by the processes which split-gathered particles. For instance, if the Solar System can be squeezed tightly, some innermost planets and asteroids may be like a trace of matter. Similarly, some subatomic particles found in the nucleus are just sandwiched there by the process that compressed or squeezed the precursor of the nucleus. In *"Turbulent Origin of Chemical Particles"*, I extensively presented the origin of the chemical elements and the interpretations of the interactions between them.

If the interactions between particles and celestial bodies can be properly inventoried, more than 4 fundamental forces would be found in nature. As the gravity of a planet in the Solar System is smaller than that of the Sun, so also the gravity of the Sun itself may be smaller than that of the core of the Milky Way Galaxy. But a problem of this kind of comparison is that although the Sun is one body, the core or center of the Milky Way Galaxy is a system of bodies having its own "gravity" without all of them as a whole being bound together by "gravity". Therefore, the spheroid component of a spiral galaxy like the Milky Way may not be held together by what is generally known as "gravity". Similarly, on the scale of clusters of galaxies, other interactions different from gravity can explain the organization of their structures. In other words, while on astronomical scales, gravity may be used to explain the field around a single body, on the scale of system of stars, galaxies and their clusters, gravity is inappropriate. In other words, to be accurate, a new concept different from gravity must be used to explain the mechanism holding together of clusters of bodies such as:

- A stellar system,
- A galaxy,
- A cluster of galaxies,
- A cluster of clusters of galaxies, etc., and
- All the clusters of galaxies in the universe.

In the chapter on galaxies, I also elaborated on how large scales of vorticial structures could have been aggregated to form astronomical systems of bodies. Spiritual forces are also highly involved in almost every natural thing. But due to the nature of this book, I cannot deal with that subject here. In my book dealing with the universal laws of nature, I also expounded on my thought on other aspects of the fundamental forces in nature. The way I explained gravity will change the perspective of turbulence and will open a new era in science as well as in all domains where gravity is used.

32.8. Take home message

This chapter is one of the keys needing to be properly understood before grasping the turbulent origin of the universe. I showed that gravity and the other so-called fundamental forces in nature were progressively birthed by the way the precursors of bodies were split-gathered. The misunderstanding of gravity is one of the reasons for the main scientific drift in some scientific fields, which in many ways seem to have become a religion where things are claimed and accepted without proofs but defended by some confusing statistical and other mathematical analysis in the name of physical things we cannot see. Before addressing what gravity is, I first said what it is not. Unlike what some theories postulated, gravity is not responsible for the orbit of the celestial bodies nor does it control the trajectories of the bodies. Gravity does not control the orbit of the stars in a galaxy nor does it define the orbit of galaxies in clusters of galaxies. Indeed, the precursors of the bodies were already launched and were on their way to their orbit before gravity was later completely established. As fluid layers of the precursors were spiraling, they were compressed or squeezed, progressively creating a field, which effect can also be felt as gravity.

Gravity is not the process that causes secondary bodies to orbit their primary bodies and it is not a force of attraction acting between all matters. For instance, gravity is not the process which causes satellites to orbit their primary planets. It is not the process which causes planets and asteroids to orbit their primary star. In the Solar System, gravity is not responsible for the revolution or the movement of planets and asteroids around the Sun. Similarly, gravity is not responsible for the movement of the Sun or the Solar System as it "orbits" the center of the Milky Way Galaxy. Gravity is neither responsible for the movement of the Milky Way Galaxy around the center of the cluster of galaxies it belongs to. Gravity is not responsible for the large-scale structures in the Universe, but the way the bulk of the turbulent prima materia (the precursor of all matters in the universe) was split-gathered is. Gravity is not a force of attraction acting between all matters.

Gravity is not responsible for the mass, structures, and development of the universe. The process which split-gathered the precursors of the bodies in the universe is not just a matter of gravity. Gravity is just a tiny component of the story of the formation of the universe, yet a lot of attention is given to it. The mass of particles and celestial bodies cannot be properly explained without an intimate understanding of the process which split-gathered the precursors of the bodies in the universe. For instance, as particles were formed, some were clustered together to an extent that their clusters were significant enough to generate mass. The mass of a particle (small or big) was defined by the amount of the turbulent prima materia which was amassed and squeezed to yield it. Some masses are significant enough to be felt or measured, while others (e.g. the case of some very tiny particles) cannot be measured. The extent to which the original particles were gathered together into matter defined the density of the matter and system of matter. Gravity is not responsible for the configuration and constitution of atoms and subatomic particles.

Gravity can be traced back to the vorticial structures in the fluid flow during the formation of the celestial bodies. Removing gravity from a celestial body would require reversing the processes involved in the formation of the universe. In other words, if the processes which formed the universe could be reversed and the celestial bodies unwound, the universe can be brought back to a state near that of its origin. Without a mechanism, process, or force to combine together or assemble the precursors into unified bodies, most daughter bodies could not have been able to form bodies after their split from their mothers. The process of gathering together the matter into the precursors of the bodies as they started splitting from their mother was not initially controlled by gravity. It was later that gravity itself was born or formed. Before what is today known as gravity was established, forces and interactions were already acting on the precursors of bodies to help gather together their matters into unified bodies and systems of bodies. Gravity and its precursor allowed the precursors of some bodies not to be dispersed but to form a unique and unified body. Another way of explaining this complex process is that as the precursors of the celestial bodies were being split and were moving, a squeezing force started compressing and shaping them at the same time that other internal processes were also gathering together the clusters of rocks, minerals, atoms, subatomic particles, and other smaller matters inside of them. Because gravity itself depends on the radius of the celestial bodies and their rotational speed, its value as it is today could have not been established before the bodies got their current shape. By the time that the size and movement were set and the celestial bodies were put on their "final trajectory" or orbit, gravity was established.

If the interactions between particles and celestial bodies can be properly inventoried, more than 4 fundamental forces will be found in nature. As the gravity of a planet in the Solar System is smaller than that of the Sun, so also the gravity of the Sun itself may be smaller than that of the core of the Milky Way Galaxy. Similarly, on the scale of clusters of galaxies, other interactions different than gravity can also explain the organization of their structures. To put it another way, while on the astronomical scale gravity may be used to explain the field around a single body, on the scale of systems of bodies such as stars, galaxies, and their clusters, gravity is inappropriate. A new concept different from gravity must be used to explain the mechanism holding together the clusters of bodies in a planetary system, a stellar system, a galaxy, a cluster of galaxies, and all the clusters of galaxies in the universe.

'Science180 Academy' Success Strategy:
SCIENCE180 SERVICES AND PRODUCTS YOU WILL LOVE

Because you are reading this book, you are probably very interested in answering your questions about the origin of the universe, of life, and of chemicals. Imagine you want to be trained by Dr. Nathanael-Israel Israel and his team so you can benefit from their outstanding expertise to empower yourself or your team. Or you want him to give a keynote speech, a seminar, or any other kind of talk or conference at your organization. Or you want him to mentor you or some people or team at your organization. Maybe you have critical origin-related questions that you need his help to accurately answer. You want a true expert to talk with you about the customized program or game plan that fits your needs. You want him to tailor his advice, expert feedback, and proven shortcuts to the stage of life you are in and help you get to where you want to be in your desire to properly understand the origin of the universe, life, and chemicals and harness the benefits that come with it. Perhaps you don't know how to properly get any of these important tasks done according to your specific needs or the needs and demands of your organization.

That is what Science180 Academy is all about. Visit Science180.com/services for more details about how to benefit from the services that Science180 provides.

Maybe you are a leader that wants to hire Dr. Nathanael-Israel Israel and his team to train some departments at your organization. Or you want to refer them to other companies like a good dish passed around the dinner table, and you want to explore how Nathanael-Israel Israel can pay you something for that referral. Maybe you attended Nathanael-Israel Israel's speaking program, for which, without going into details, he accurately raised your awareness about how the universe, life, and chemicals were formed. Or maybe you attended his training, in which he detailed and showed you how he decoded the scientific data using various tools and certain thinking strategies that helped him and which transferred some skills to you; and now, you are interested in a long term one-on-one consulting, or mentoring program with him, so that, he delves into more details about how to use proven techniques to decode the universe (strategies for data collection, data analysis, data presentation, writing, and even tips for future research) and change your behavior on a long term basis. If you related to any of the points mentioned above, Science180 Academy is the right fit for you!

Other customizable services that Science180 provides include:

- Assessments
- Books and other products (e.g. booklets, online courses, posters, how-to-guides, study guides, and field guides)
- Book publishing (Yes! Science180 can publish your books!)
- Conferences
- Consulting
- Executive mastermind groups
- Face-to-face visits
- Licensing programs including corporate <u>licensing</u> programs
- Master classes
- Online courses
- Podcasting
- Retreats
- Seminars
- Speaking engagements (offline and onsite–e.g. seminar, keynotes, and workshops)
- Survey and research tools
- Training
- Video programs
- Virtual presentations

Here are other reasons why you should choose to work with or hire Nathanael-Israel Israel and the team at Science180:

- A simple universe-origin theory that made no assumption
- All the products and services you need to accurately and easily decode the universe-origin
- All the universe-origin solutions you love
- Bringing people together through the power of the accurate decoding and understanding of the universe-origin
- Complex universe-origin questions solved accurately in a simple language
- Discover the key variables needed to decode the universe-origin

- Efficient, trustworthy, and cost-effective company to add to your strategic journey toward your best tomorrow
- First stop for your universe-origin, life-origin, and chemicals-origin needs
- Irrefutable scientific demonstration of the universe-origin
- Nonconformist, rule-breaker, and accurate demonstrator of the universe-origin
- One-stop for answering your universe-origin questions
- Personalized universe-origin decoding package
- Properly understand turbulence. Only on Science180
- Receive an exceptional universe-origin decoding perspective
- Source of unconventional wisdom & knowledge on universe-origin
- State-of-the-art decoding experience of the universe-origin
- The all-in-one proven & uncomplicated universe-origin formula
- The best universe-origin decoding experience. Only on Science180
- The go-to source for valuable universe-origin information
- The new physics that will revolutionize science forever
- The only one formula accurate enough to explain the formation of the universe
- The place where scientists get all their universe-origin questions properly answered
- The place where the accurate interpretation of universe-origin data matters & Universe-origin formula accurately is made easy
- Theory that helps you avoid dangerous dogma and irrational thinking and that helps you fight wasteful programs
- Think universe-origin differently and unconventional approach
- Understand the origin of the universe. Increase your glory and peace of mind

Nathanael-Israel Israel: Member of the American Society of Agricultural and Biological Engineers

CHAPTER 33

CAN THERE BE AN ACCURATE ROAD MAP TO PROPERLY CALCULATE THE TIMELINE OF THE UNIVERSE FORMATION?

33.1. Is it possible to properly calculate the timeline of the universe formation?

Indeed, throughout the ages, human beings have raised questions about the age of the universe and the duration of its formation. The "answers" to these questions vary mainly according to the religious belief systems and the theories used (many of which are not scientific at all, although they claim to be). In the secular literatures for instance, some people have postulated and maintained that the duration of some processes connected to their theories of the formation and of the age of the universe is in the order of magnitude of "billions of years". Several religions also point to various myths and dogmas of the creation of the universe as a matter of millions of years of processes. Each religion seems to have its own story and some of these religious beliefs seem to be framed after some recent scientific theories, in which I found flaws as you will see in the chapters to come.

Recorded by Guinness World Records, as the best-selling book of all times, the Bible (the oldest book known to humankind, also the most printed and most distributed book of all times more than any other best seller book ever written by humankind) points to the creation of the universe as a matter of 6 days. The Books of Enoch, which is also mentioned in the Bible, but which I will not address in this book, but in another one I titled *"Origin of the Spiritual World"*, is one of the oldest books in the world, also talk about 6 days of creation and many other details found nowhere in any other book. Some people believe in the Biblical story but reject the scientific theories. Others believe in the scientific theories but reject the Biblical account. Some people believe in both the scientific theories of billions of years and also in the Biblical story of 6 days of creation. Some people say they believe in the Biblical story of creation, but they also believe that creation was not done in 6 days as the Bible says, but in millions or billions of years. Some people said that those who believe in both evolutionism and the Bible story of creation are confused. Many believers reject the scientific theories of billions of years, but they are not able to scientifically demonstrate how things in the universe were created in the 6 days of creation story highlighted in the Bible; they just simply have faith in the account Moses recorded and don't question how it really happened. Here, I am not saying that some believers are unable to provide proofs that back their 6 days of creation belief, but I am saying that no scientist has ever proven how things were created in 6 days. To give a specific example, no scientist has ever demonstrated how and why in Genesis 1 (the first book in the Bible), it is said that waters were separated from waters by the second day, the Earth was created by the 3rd day, the Sun and the Moon were formed by the 4th day, etc. Some people think that using natural processes to track the creation story back to God is bad, impossible, and removes the need for God. Therefore, many believers do not even attempt at looking into a scientific demonstration of the creation timeline. Some believers accept the creation as a miracle which timeline can never be demonstrated using a scientific method. Anyways, some people believe in none of what I have said so far. Others say that things are relative, yet, they hold onto their relativity as an absolute fact. This attitude is similar to how some people hold onto their faith.

- But who is right among all of these people, theories, and religious beliefs?
- How can we know for sure the duration of the creation and the age of the universe?
- How can we predict the fate of the universe?
- Is it possible to use scientific evidences based on concrete data (not just theories) to reconcile the story of the formation of the universe with what is known in any religion?
- How can we know for sure if the universe was really created in 6 days like the Bible says or by billions of years process as the Big Bang theory said?
- Can a human being ever be able to really demonstrate the story of creation of the universe?
- Is it possible to test and know which God created the universe if that was the case?

Therefore, I cannot answer questions surrounding the origin and fate of the universe without addressing some of those critical issues related to the duration of the formation of the universe and its age. In the following chapters, I tackled the aforementioned challenging questions and many more. The goal of this chapter is to lay a foundation for the method I designed to handle the timeline of the formation of the universe. Then, in the next 5 chapters, I detailed each of the things I mentioned in this chapter as being crucial in properly defining the timeline of the formation of the universe. Then, I will prove whether science disproved or proved God and if and which God created the universe. As you continue reading this chapter, you will figure out how a simple formula alone helped me crack the code of the universe and answer all these questions easily.

33.2. Semi major axis holds a secret of the distance travelled by the precursor of the bodies

A comprehensive timeline of the formation of the universe requires an intimate understanding and timing of the processes which took place. Throughout this book, I have already established and extensively detailed that, for the bodies present in the universe to be born, precursors had to be split-gathered by processes involving for instance the formation of precursors of primary and secondary bodies, separation of fluid layers and their gathering into the current shape known for the bodies in the universe. In other words, the duration of time it took for the celestial bodies to form must include the time that their precursors "traveled" from the position of their mother precursor to the current orbit of their daughter bodies and the time these precursors took to wrap, collect, or gather themselves together into the shape known for the bodies as of today. It is one thing for the bodies to be formed but it is another thing for them to be placed in orbit. Therefore, in addition to the duration of time it took for the bodies to collect or get their shape, I was interested in knowing how long it took to position the bodies into their orbit. Wouldn't you have liked to have seen this happen? I imagine it as the best fireworks show in history.

In the rest of this book, I adopted the term timescale (also written time scale) to refer to the duration, quantity of time, or the order of magnitude of the time during which a certain event took place during the formation of the bodies. The difficult task was to know how to time the events related to the formation of the universe. To address that issue, I focused on a few processes in the Solar System and then designed a method to approach not only the distance that the precursors of the bodies could have "travelled" before they occurred, but also the speed of such "travel". Let us not forget that things were moving as the precursors of the bodies were being shaped into their daughter bodies.

Typically, distance and speed are habitually used to calculate the duration of events in the physical realm. In other words, knowing the distance travelled by something and the speed of that travel, it is possible to calculate the duration of that travel by dividing the distance by the speed. In this chapter and the others related to timescale, I used a similar approach by focusing on few key distances and speeds related to the formation of the universe. The distance separating the celestial bodies and the speed of some of their movements hold secrets related to the duration of their formation. In the case of the celestial bodies, I had to estimate the time it took for some precursors to split into the precursors of their primary bodies and secondary bodies. I also needed to estimate the time it took for the precursor of the primary bodies to gather together. I also needed to estimate the time it took for the fluid layers of the precursors of the secondary bodies to split once they escaped the precursor of their primary body. Because the fluid layers split from one another, each at its turn, I also needed to know the increment of time during which those splits occurred from one fluid layer to the next one. To achieve that goal, I focused on the split-gathering of the precursor of the Solar System. The estimation of the duration of the time it took for the precursors of secondary bodies to leave the precursor of their primary bodies (after splitting from them) until they reached their orbit required an intimate understanding of the distance travelled and the speed. This distance for the secondary bodies can be estimated using the semi major axis, the aphelion (the farthest distance from the primary body), or the perihelion (the shortest or closest distance from the primary body). Because semi major axis is based on aphelion and perihelion, I decided to use it as the mean distance travelled. In other words, I will be using the semi major axis to handle the distance travelled by the bodies. Like you will see later, I sometimes use semi major increment to estimate the increment of distance travelled between two bodies. Then, a big challenge was the estimation of the speed of the travel. In the rest of this chapter, my goal is to present the timeline of each of the events I narrated above and which I have spent most of the previous chapters to handle.

33.3. Timescale of the split of the precursor of a system of bodies into the precursor of its primary bodies and the precursor of its secondary bodies

In some of the previous chapters, I already explained how the precursors of a system of bodies split into the precursor of a primary body and the precursor of its secondary bodies. The precursor of a primary body would have needed the precursor of its secondary bodies to move out, or to separate, or to escape before it can collect itself together. In other words, from the moment the precursor of a system of bodies started splitting into the precursor of its primary body and the precursor of its secondary bodies, a certain amount of time elapsed before their split could be completed so that the precursor of the primary body could start collecting its body and also so that the precursor of the secondary bodies could continue its journey and split into its daughter bodies according to their position in the fluid layers as I already extensively explained in previous chapters. The duration of the split of the precursor of the secondary bodies from the precursor of the primary body depended on the speed with which the fluid flowed from the precursor of the mother of the system into or through the ligament connecting the precursor of the primary body and the precursor of the secondary bodies. It was as if the mother of a system of bodies emptied part of its content into the precursor of the secondary bodies and what was left over became the precursor of the primary body of that system. For example, after the precursor of the bodies orbiting the Sun escaped the precursor of the Solar System, what was left was the precursor of the Sun. Although not perfectly expressed, it can also be said that the precursor of the bodies orbiting the Sun escaped the precursor of the Sun.

CHAPTER 33: ROADMAP OR METHODOLOGY OF THE CALCULATION OF THE TIMELINE OF THE UNIVERSE FORMATION

For it was the precursor of the Solar System which emptied itself to yield 2 main daughter bodies: the precursor of the bodies orbiting the Sun and the precursor of the Sun. It took a certain amount of time for the precursor of the bodies orbiting the Sun to escape the precursor of the Sun before the precursor of the Sun could have been clearly visible and could be worth calling the precursor of the Sun. That separation time could have elapsed before the precursor of the Sun could fully wrap itself and also before the precursor of the bodies orbiting the Sun could start their own journey completely away from the precursor of the Sun. Put another way, as much as there was a time when the precursor of the Sun was "connected" to the precursor of the bodies orbiting the Sun, there was also a time when all the fluids that should go into the precursor of the bodies orbiting the Sun were cut off from the precursor of the Sun. Like I explained in the chapters on vortex and ligament, as a fluid thread breaks, a ligament is formed so the fluid can pass through. Before the breakup, the fluid neck (connecting fluid bodies) thinned until reaching a zero radius and then, it breaks. Likewise, before the breakup of the precursor of the Solar System, a fluid ligament or neck was formed between the precursor of the Sun and the precursor of the bodies orbiting the Sun. The moment that the neck connecting both fluid bodies broke was the end of the precursor of the Solar System and the birth of the precursor of the Sun and the precursor of the bodies orbiting the Sun. If an observer was there, the lifespan of the precursor of the Solar System could have ended just at the breaking time of its 2 main daughters. That time could have passed before the precursor of the Sun and that of the bodies orbiting it could have started their own gathering or split-gathering. However, because I did not have a way to accurately know the time that had elapsed before the aforementioned breakup occurred, I had to estimate it another way. The example I mentioned above for the precursor of the Solar System can be applied to the breaking of the precursor of the planetary systems (into the precursor of their primary planets and the precursor of their satellites). The same can also apply for the breaking of the precursor of the asteroid systems (into the precursor of primary asteroids and the precursor of their satellites). Even at the microscopic level, the same principle can be applied to the breaking of the precursor of the atoms (into the precursor of the nucleus and the precursor of the particles such as electrons moving around them).

To reiterate, in each system of bodies, because the precursor of the primary body did not travel much away from the position of its mother precursor before being collected, it was as if the fluid of the bodies orbiting the primary body was stripped away before the primary body collected itself together. Put another way, the precursor of the primary body was the leftover of the precursor of the entire system after the precursor of their secondary bodies was "stripped away" or split from it.

33.4. The distance separating a primary body and its innermost secondary body is the maximum distance travelled by the precursor of the secondary bodies before splitting from the precursor of the primary body

To know the time elapsed before the precursor of the primary body started collecting itself, it is important to consider the time it could have taken for the precursor of the bodies orbiting that primary body to be released, or cut off, or moved away from the precursor of the primary body. After considering all the factors that could have been involved in the split-gathering, I concluded that the semi major axis of the innermost secondary body (i.e. the distance separating the innermost secondary body from the primary body) can give a clue into the maximum distance travelled by the precursor of the secondary body before it started splitting into the secondary bodies. I said "maximum" distance because the distance separating the primary body and its innermost secondary body include also the distance travelled by the innermost body before reaching its orbit. Likewise, the duration of time which separated the primary body and its innermost secondary body include the time it took for the precursor of the primary body to be split from the precursor of the bodies orbiting it.

Knowing the speed with which the precursor of the secondary bodies was "ejected" from the precursor of their primary body, it is possible to estimate the maximum amount of time it took for a body to travel the distance separating the primary body from the innermost secondary body. That duration of time is in the order of magnitude of the maximum time it could have taken for the precursor of the secondary bodies to split from the precursor of the primary body. This time does not include the time that the precursor of the secondary bodies would later take to split-gather into their constitutive bodies according to the size and number of its layers. For by the time the top secondary layer separated from the stack of the layers of the secondary bodies, all of the fluid layers of the secondary bodies could have been separated from the fluid of the precursor of the primary bodies. To recap, the duration of the split of the precursor of the primary body from the precursor of their secondary bodies was in the order of magnitude of the semi major axis of the innermost secondary body divided by the speed of the separation of the stack of fluid layers of the precursors of the secondary bodies. For example, the time it took for the precursor of the bodies orbiting the Sun to split from the precursor of the Sun could not have been longer than the time it took for the precursor of the stack of fluid layers of the bodies orbiting the Sun to move from the position of the Sun to the position of Mercury (the innermost body in the Solar System). Because, as of the time that I published the first edition of this book, Mercury is the most renowned innermost body in the Solar System, the semi major axis of Mercury is the maximum distance travelled by the precursor of the bodies orbiting the Sun before they started splitting.

Once I know the speed of the separation or the speed of the escape of this precursor, I can also estimate the amount of time elapsed beforehand. In other

CHAPTER 33: ROADMAP OR METHODOLOGY OF THE CALCULATION
OF THE TIMELINE OF THE UNIVERSE FORMATION

words, "Semi major of Mercury / Speed of the escape of the precursor of the bodies orbiting the Sun" can give a clue into how long it took for the precursor of the bodies orbiting the Sun to separate from the precursor of the Sun. This amount of time is also the maximum time which elapsed since the beginning of the formation of the Solar System until the precursor of the Sun could have started gathering itself together. Later in this chapter, I explain what the escape of the precursor of the bodies orbiting the Sun actually is.

All the fluid layers in a stack of precursors of bodies did not separate at the same time. The fluid layers at the top separated first. Then, those beneath it separated according to their position until the bottom layer separated last. Once a fluid layer separated from the stack of layers beneath it, the remaining stack of layers continued its split-gathering until the last body was formed. Once a fluid layer separated, it was then gathered together, meaning that after the separation, a certain amount of time passed before the gathering of the layer was completed. As the fluids in a separated layer were collecting themselves into a body, the remaining stack of fluids which were initially at its bottom continued its journey, until the following layer separated and so on and so forth until the bottom layer separated last. While the maximum distance travelled by a stack of fluid layer before its top layer separated was in order of magnitude of the semi major axis of the innermost body born from that stack of fluids, the layer which is beneath that of the innermost body traveled a distance in the order of magnitude of the semi major increment of the body born from it before the layer of its precursors separated from the stack of fluid beneath it. On average, the distance travelled by a stack of fluid before a layer separated is about the semi major increment of the bodies born on the layer. As a general rule, the distance travelled by the precursor of each body before its fluid layer(s) separated from the fluid layers at its bottom is about its semi major increment. Due to the expansion of the universe, the distance that the stack of fluid layers could have travelled before the first fluid layer separated could have been less than the semi major axis of the innermost body born from the stack of fluid.

Applied to the Earth-Moon system for instance, the distance travelled by the precursor of the Moon after it separated from the precursor of the Earth is less than the semi major axis of the Moon. For after the precursor of the Moon split from that of the Earth, it also took some additional time before the precursor of the Moon reached its position on orbit. Furthermore, since the end of the formation of the Moon, due to the expansion of the universe, the Moon has also slightly receded from the Earth, meaning that the current semi major of the Moon was not how it was soon after its formation long ago. Although I will deal with the distance separating the primary bodies from their innermost secondary body in a separate chapter later, Fig. 187 is a glimpse at the distance travelled by the stack of fluid layers of the precursor of the secondary bodies before their split began. As shown in Fig. 187, while on the scale of the Solar System, the precursor of the bodies orbiting the Sun travelled about 57,910,000 km before the first fluid layer (that of Mercury) separated from the rest, on the scale of the planetary system, the stack of fluid of the precursor of the satellites travelled between 9,378 km (recorded on Phobos, the innermost satellite of Mars) and 384,400 km (recorded on the Moon, the only satellite of the Earth).

Fig. 187: Name and semi major axis (km) of the innermost body in the system of bodies

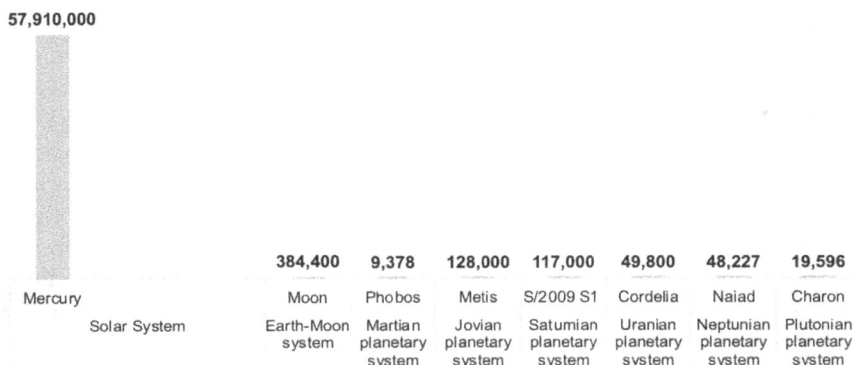

57,910,000								
		384,400	9,378	128,000	117,000	49,800	48,227	19,596
Mercury		Moon	Phobos	Metis	S/2009 S1	Cordelia	Naiad	Charon
	Solar System	Earth-Moon system	Martian planetary system	Jovian planetary system	Satumian planetary system	Uranian planetary system	Neptunian planetary system	Plutonian planetary system

33.5. Escape speed of the precursor of secondary bodies

Considering everything I said so far, I came to realize that the precursor of the secondary bodies orbiting a primary body in a system of bodies escaped the precursor of their primary body with a speed in the order of magnitude of the escape velocity of that primary body. The escape speed or escape velocity of the primary body in a system of bodies is about the speed with which the fluid of the precursor of the secondary bodies in that system escaped or were ejected from the precursor of the primary body. For instance, in a stellar system, the precursor of the bodies orbiting a star escaped the precursor of the star at about the escape velocity of that star. In other words, the fluid layers of the bodies orbiting a star were ejected or escaped the precursor of the star with a speed in the order of magnitude of the escape velocity of that star. Hence, the precursor of the bodies orbiting the Sun could have escaped the precursor of the Sun with a speed in the order of magnitude of the escape velocity of the Sun. In the same manner, the precursor of the satellites in a planetary system could have escaped the precursor of their primary planet at about the escape velocity of that primary planet.

To generalize, based on my investigation, the speed of the separation of the precursor of the secondary bodies from the precursor of their primary body was about the escape velocity of the precursor of the primary body. At the time of the separation of the precursor of the secondary bodies from the precursor of the primary body, no primary body or secondary body was fully formed yet, of course, but the potential of the precursors carried moving abilities (in the form of energy) that were transferred into their daughter bodies. Hence, it is reasonable to say that the speed with which the stack of fluids of the precursor of the secondary body was launched could have been about the escape velocity of their primary body. For instance, the speed with which the precursor of the bodies orbiting the Sun separated from the precursor of the Sun was about the escape velocity of the Sun. Likewise, the speed with which the fluid layers of the precursor of the satellites in a planetary system was separated from the fluid of the precursor of the primary planet was about the escape velocity of the primary body. However, let us not forget that the speed of the bodies formed inside the fluid layers in a system was not the same from one layer to the other. As I already explained in the chapter on orbital speed, the bodies formed in the top layers moved faster than the bodies formed in the bottom layers. The regression between orbital speed and semi major axis also suggested that as the precursors of the bodies were moving away from their precursor of their primary body, their orbital speed decreased. Yet, the entire stack of fluid of the secondary bodies could have been moved at the same speed for a while, or until the bottom body was freed and completely formed. I meant that the speed with which the fluid flow carried the precursor of bodies was much higher than the orbital speed gained by the celestial bodies formed in those moving fluids. For instance, as the fluid layers were separating and distancing themselves from the position of their primary body, the speed of the bodies formed inside of them decreased. For instance, as the fluid layers of the precursor of the satellites were distancing themselves from the fluid layers of the precursor of their primary planet, their orbital speed decreased. In the chapter on orbital speed, I expounded on those evidences. In general, the speed of the flow that carried the fluid layers was higher than the speed of the bodies born from them. For instance, the speed with which the fluid layers of the precursor of the satellites escaped the precursor of their primary planet was higher than the orbital speed of the satellites formed in those fluid layers. In the same manner, the speed with which the precursor bodies (asteroid systems and planetary systems) escaped the precursor of the Sun could have been higher than the orbital speed and the asteroids and planets known in the Solar System as of today. However, the semi major axis of the secondary bodies today can be higher than what it was at the end of the formation of the bodies. In summary, the time elapsed between the separation of 2 consecutive layers was in the order of magnitude of the semi major increment of the bottom layer divided by the escape velocity of the primary body of the bodies formed from the same layer. After the fluid layers separated, it also took some time before they could collect themselves. Later I elaborated on that.

It is important not to confuse the speed that carried the fluid layers with the speed of the bodies and structures formed inside those fluid layers. The speed of the flow that carried the layers of the bodies orbiting the Sun was in the order of magnitude of the escape velocity of the Sun. But the speed of the bodies formed inside or from those layers was in the order of magnitude of the orbital speed of the daughter bodies. For instance, as of today, when a current of water (e.g. river) is moving, structures can be formed in that water and these structures are carried by the water current. As they are being carried, some of those structures can be turning around themselves or making other movements inside the dominant movement of the current carrying them. Likewise, because the Sun contains more than 99% of the 9 system-additive variables (mass, orbital angular momentum, orbital moment of inertia, rotational angular momentum, rotational kinetic energy, rotational moment of inertia, total kinetic energy, translational kinetic energy, and volume) in the Solar System, it is appropriate to consider the escape velocity of the Sun to be in the order of magnitude of the escape velocity of the precursor of the Solar System. In fact, the bodies that orbit the Sun did not escape from the Sun, nor did they escape the Solar System. The separation or split of their precursor from the precursor of the Sun is a kind of escape of the precursor of the Sun. Hence, I felt very confident considering the escape velocity of the Sun as the speed with which the precursor of the bodies orbiting the Sun was pushed way, or "ejected" or escaped from the precursor of the Sun. Because of some constraints that the fluid layers of the precursor of the celestial bodies met during the journey that positioned them on their orbit after they split from the stack of fluids carrying their peers or neighbors, it is possible that as the fluid layers moved farther away from their primary body, their speed could have been less than the escape velocity of their primary bodies. In other words, it could have been possible that in some systems of bodies, some bottom fluid layers travelled at a speed smaller than the escape velocity of their primary body.

Now, I will briefly talk about the escape velocity of the celestial bodies in the Solar System. Indeed, because of gravity, anything on Earth is bound to Earth unless a sufficient force is used to propel it out of the atmosphere. Similarly, most celestial bodies that are big enough tend to hold anything that is on them or in their atmosphere... The gravitational pull is defined as the force that causes some celestial bodies to attract something toward them. To be able to pull something out of a celestial body zone where the gravity rules, an energy must be applied to allow that thing to move out. According to NASA, the escape velocity is the *"initial or minimum velocity that is required at the surface (or at the 1 bar pressure level for the gas giants) to escape a body's gravitational pull, while ignoring atmospheric drag"*. Once the escape velocity is applied to any object, that object is supposed to escape (without propulsion) the gravitational pull of the celestial body where it was without the gravity of that later body being ever able to manage to pull it back. For instance, if you throw something in the air, it will likely fall back on Earth. However, if the escape velocity is applied to that thing, it will escape the Earth, meaning that it will leave the Earth and go into space and never return to Earth because of Earth's gravity.

The escape velocity does not depend on the mass of the body which it is going to escape but on the mass of the body to be escaped. The data on escape velocity played a key role in my estimation of the timeline and duration of the formation of the celestial bodies in the universe. In fact, according to what I already established, the escape velocity of a primary body is about the speed with which the precursor of the secondary body of that primary body escaped the precursor of that primary body.

Escape velocity is reported for just 127 of the bodies that I studied in the Solar System. In this segment, I will present what is available. Indeed, available for about a third of the celestial bodies I studied in the Solar System, the escape velocity ranges from 0 to 617.6 km/s. The Sun recorded the highest escape velocity. The escape velocity of the planets varies between 1.2 and 59.5 km/s. As the biggest planet in the Solar System, Jupiter recorded the highest escape velocity (59.5 km/s) among the planets. The smallest value among the planets was recorded on Pluto. The escape velocity of the Earth (11.19 km/s) is higher than that of Mercury, Venus, Mars, and Pluto. Because escape velocity depends on the radius of the body to be escaped, it should not be surprising that the bigger satellites have the highest escape velocity. Based on the data available, the highest escape velocity recorded among the satellites are found with those orbiting Jupiter and Saturn. As a take home message, I would like to say that the escape velocity of the celestial bodies played a key role in my estimation of the timeline and duration of the formation of the universe. I showed that the escape velocity of a primary body is about the speed with which the precursor of their secondary body escaped the precursor of that primary body. The escape velocity of the Sun and the planets in the Solar System is as presented in Fig. 188.

Fig. 188: Escape velocity (km/s) of some celestial bodies in the Solar System

The precursor of the fluids of the secondary bodies of each of these primary bodies escaped the later with a speed in the order of magnitude of the escape velocity of their primary body.

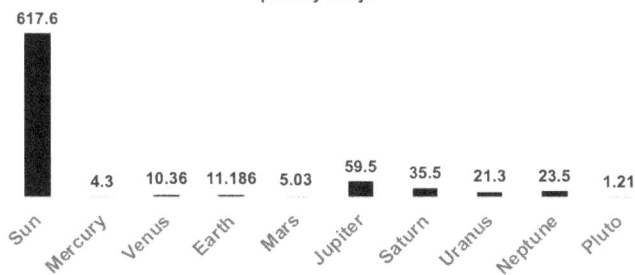

33.6. Lifespan of the precursor of the systems of bodies

By lifespan of the precursor of a system of bodies, I mean the length of time that such a precursor could have existed as one body before it was split into the precursor of a primary body and the precursor of secondary bodies (which would later split into individual bodies). In other words, by lifespan, I do not mean the duration of time after which all the matter in a precursor stop to exist or disappear, but instead the maximum duration of time after which that precursor could have been split into the precursor of its daughter bodies (e.g. primary body and/or secondary bodies). I understand that people usually think of lifespan from birth to death; but to some extent, the description I provided here for lifespan is like the reproductive years in which women can conceive and give birth and then die themselves but leaving behind their offspring.

Based on what I already established, the duration of the formation of the primary bodies includes the time the precursor of their system was split into the precursor of a primary body and the precursor of its secondary bodies. That duration of time could have elapsed before the precursor of the primary body could start wrapping itself into its final shape. Unlike the precursor of secondary bodies which needed to be moved to a specific location before their daughter bodies could be born or gathered together, the precursor of the primary body did not need to move with respect to their system but needed to "wait" until the precursor of their secondary bodies was cleared or ejected, or escaped before it can gather it together.

The lifespan of the precursor of the Solar System for instance is about the time that lapsed before it was split into the precursor of the Sun and the precursor of the bodies orbiting the Sun. Based on what I mentioned above, that time is not higher than the "Semi major of Mercury" divided by the "Escape velocity of the Sun". Likewise, the lifespan of the precursor of each planetary system is not higher than the "Semi major of its innermost satellite" divided by the "Escape velocity of the primary planet".

Lifespan of the precursor of a system of bodies
$$\leq \frac{\textbf{Semi major axis of the innermost secondary body}}{\textbf{Escape velocity of the primary body}}$$

Applied to the Solar System and the planetary systems, I got the following formulas:

Lifespan of the precursor of the Solar System
$$\leq \frac{\textbf{Semi major axis of the innermost planet}}{\textbf{Escape velocity of the Sun}}$$

Lifespan of the precursor of a planetary system
$$\leq \frac{\textbf{Semi major axis of the the innermost satellite}}{\textbf{Escape velocity of the primary planet}}$$

Based on the above formula, Fig. 189 illustrates the maximum lifespan of the precursor of the Solar System and of the precursor of the planetary systems.

Fig. 189: Maximum lifespan (hours) of the precursor of the system of bodies

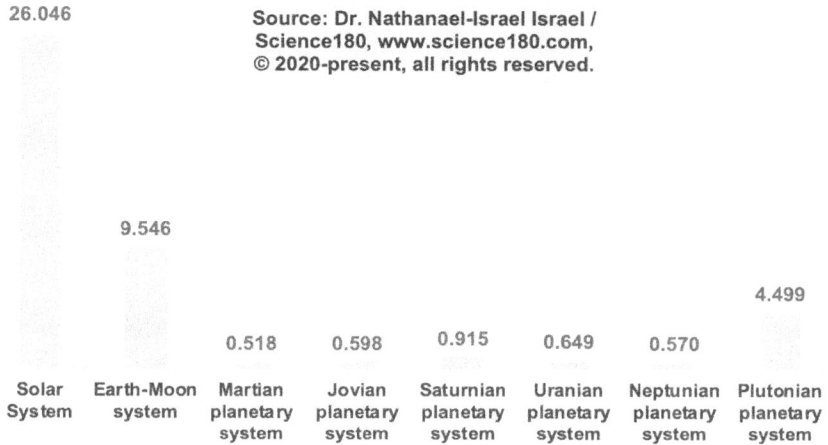

Solar System	Earth-Moon system	Martian planetary system	Jovian planetary system	Saturnian planetary system	Uranian planetary system	Neptunian planetary system	Plutonian planetary system
26.046	9.546	0.518	0.598	0.915	0.649	0.570	4.499

As shown in Fig. 189, except for the Earth-Moon system, the lifespan of the precursor of the other planetary systems in the Solar System is less than 4.5 hours. For the precursors of the giant planets, the lifespan is even less than 1 hour. Despite the slow speed of the precursor of the Plutonian planetary system, its lifespan is less than 4.5 hours. Because the Moon is in Zone 3, the time its precursor took to split from the precursor of the Earth must have been less the 9.5 hours maximum estimation of its lifespan. Therefore, to calculate the birthdate of the primary planets, I did not consider the time it took for the precursor of their satellites to escape first. For that time is negligeable, as it is less than 2 hours on average and is insignificant before the days that the fluid layers took before splitting from the stack of fluids of the precursor of the bodies orbiting the Sun.

In each system of bodies, before the precursor of the primary body was born, a certain amount of time passed to allow the precursor of the secondary bodies to move out. I coined the term "escape time" to designate the duration of that time. In other words, the escape time of the precursor of a primary body was the time needed for the precursor of the bodies orbiting that primary body to escape the precursor of that primary body. Because the bodies orbiting the primary body escaped as fluid layers and moved at about the escape velocity of their primary body before later splitting and birthing the innermost secondary body followed by other secondary bodies, the escape time of the precursor of the primary body cannot be higher that the semi major axis of the innermost secondary body divided by the escape velocity. In other words, the escape time of the precursor of the Sun is not higher than the semi major axis of Mercury (the innermost body in the Solar System) divided by the escape velocity of the Sun. Likewise, the escape time of the precursor of the planets is not higher than the semi major axis of the innermost satellite divided by the escape velocity of the primary planet. For the sake of simplifying the formula, I will consider the escape time of the precursor of the primary body as the semi major axis of the innermost secondary body divided by the escape velocity of the primary body.

33.7. Fluid flow of the precursor of the secondary bodies as it escaped the precursor of the primary body

The exact timescale of the separation of the precursor of a system of bodies into the precursor of its primary body and secondary body was about the time needed for the volume of all the fluid of the precursor of the secondary bodies to fly away or pass by a certain location just outside the boundary of the precursor of the primary body. This time would have depended on the flow rate of the escape of the fluid of the precursor of the secondary bodies. By the way, a flow rate is the volume of fluid passing by a location through an area during a period of time. This means that if the flow rate is slow (meaning the fluid is flowing slowly), the time needed to evacuate or to escape a certain amount of fluid can be higher. For example, if you take a 1-liter bucket of water and make a 5 cm hole in the bottom of the bucket, the water will flow out faster than if you make a 2 cm hole in the bottom of another 1-liter bucket of water. Likewise, if the volume of the fluid is larger, the time needed to flow all of it will be also higher. An example, would be if you took a 5 liters bucket of water and you put a 2 cm opening in the bottom of the bucket, the amount of time for it to empty is going to take longer than if you took a 1-liter bucket of water and you put a 2 cm opening in the bottom of the bucket. In other words, the time elapsed before the separation of the precursor of the primary body from the precursor of the secondary bodies would have depended on the flow rate and the volume of the fluid of the precursor of the secondary bodies. Like I explained in the chapters on vortex and ligament, the viscosity of the fluid might have played a role in some cases. In another book, I expounded on this concept of fluid flow of precursors. The follow up of the fluid flow of the precursors led me to what I called leaf-like structures of the precursors of the bodies and I delved into that in my book on the origin of life.

As the fluids flowed, they then broke up into daughter bodies. For instance, considering details I already gave in the chapters on vortex and fluid ligaments, by the time the fluid layer(s) of the innermost secondary body in each system split, a certain ligament or neck was formed and elongated. Then, the neck could have thinned and then broke. Afterward, the fluid bodies located on each side of the breaking point could have retracted somehow. In other words, upstream of the breaking point was the precursor of the primary body and downstream of the breaking point was the precursor of the bodies orbiting that primary body. For instance, during the breaking or escape of the precursor of the bodies orbiting the Sun, there was a time when the precursor of the Sun was located upstream of the breaking point and the precursor of the bodies orbiting the Sun were located downstream of the breaking point. Likewise, during the split of the precursor of a planetary system, there came a time when the precursor of the primary planet was located upstream of the breaking point and the precursor of the satellites and other bodies orbiting that planet were located downstream of the breaking point.

33.8. Branching of the precursors of the bodies

Before I delve into the notion of the branching of the fluid layers of the precursors, I would like to recall something that I dealt with in the chapter on "Inclusion level of primary and secondary bodies". Indeed, in that chapter, I showed that, on the scale of the Solar System, the Sun is the first level primary body. Because they directly orbit the Sun, planets and asteroids in the Solar System are first level secondary bodies of the Sun or of the Solar System. In contrast, satellites of planets or satellites of asteroids are level 2 secondary bodies in the Solar System, and this order would also be true for other stellar systems. In other words, the fluids that flowed away from the precursor of the Sun branched to yield the precursor of the planetary systems, and then branched again before yielding the precursor of the satellites. Hence, satellites do not directly orbit the Sun, but orbit their primary planets, which at their turn, orbit the Sun. I will show you a graphical illustration of this process very soon.

Using that classification of primary and secondary bodies, I talked about 2 types of timescales based on semi major:

- timescale at the branching level 1 = timescale of branching of the precursor of the planetary systems, asteroids systems, and of the planets (e.g. Mercury and Venus) and asteroids which have no satellite;
- timescale at the branching level 2 = timescale of branching of the precursors of the satellites from the precursor of the planetary systems they belong to.

Before I close this segment, I would like to mention that the formation of layers in a fluid is a form of separation or a form of splitting of fluids, but the branching site is the point or position where a fluid layer leaves, splits away, or is removed from a stack of fluid layers so it can continue its own journey of gathering without any layer (of upstream or downstream bodies) directly above or below it. As a layer branched out, the remaining of the main stack that it had escaped continued its journey and released others layers with time according to their position in the stack of fluid, the top layer first and the bottom layer last. The branching site or branching point is where the splitting of a layer is consumed or completed. From that point until the completion of the formation of the body, the gathering of the layer at the scale of its size dominated the rest of its split-gathering. For from this point, no more split is needed, but smaller layers located inside the major ones can be found inside the bodies being gathered. Hence the presence of strata or layers of materials in some celestial bodies. An example is the layer of some rocks (which are not sedimentary) found in the Earth's crust. Unfortunately, some people think that all layers in the crust are deposited by sedimentation.

For instance, when I considered the stack of fluid layers of the precursors of the planets, the branching site of a precursor is the site at which each of them splits from the rest of the stack of the precursor of the bodies orbiting the Sun. More specifically, the site at which the precursor of the Earth-Moon system split from the rest of the fluid layers of the precursor of bodies beneath it is its branching site. The position of that branching site is about 1 AU from the precursor of the Sun. Put another way, the branching site of the precursor of a planetary system is about the semi major of their primary planet. The distance between 2 consecutive branching sites is the semi major increment separating them. Likewise, the fluid layer of the precursor of the satellites split from the stack of the fluid layers of the satellites of their system at different branching sites. The distance separating two branching sites of satellites is the semi major increment of these satellites. Considering what I said above, the "Semi major axis timescale" is another term to express the distance separating 2 branching sites divided by the speed of such a separation.

The "Semi major axis timescale" of the bodies is the time right at the beginning of the branching of the fluid layer of the precursor of each body. For the precursor of the planetary systems, it is the time that they took before splitting from the precursor of all bodies orbiting the Sun. For satellites, that time includes the time that the precursors of the planetary system took before splitting (as mentioned in the previous sentence), but also the time that the precursor of the satellites took before splitting from the stack of fluid layers of the precursor of the satellites in their system. To better understand what I am talking about, I made a graph for those who are visual learners and I will show it to you in a few moments. In the incoming chapters, I consecrated a special chapter to the "Semi major axis timescale" of the celestial bodies.

In Fig. 190, I illustrated the branching of a stack of 7 fluid layers of the precursor of a celestial body. That stack of fluid layers moved and split into its individual layers. After a layer splits, the rest moved for a certain distance before another one split, and so on and so forth until all the layers are separated. Each of them is collected into a special body after splitting from the rest.

Fig. 190: Branching of a stack of fluid layers of the precursor of celestial bodies

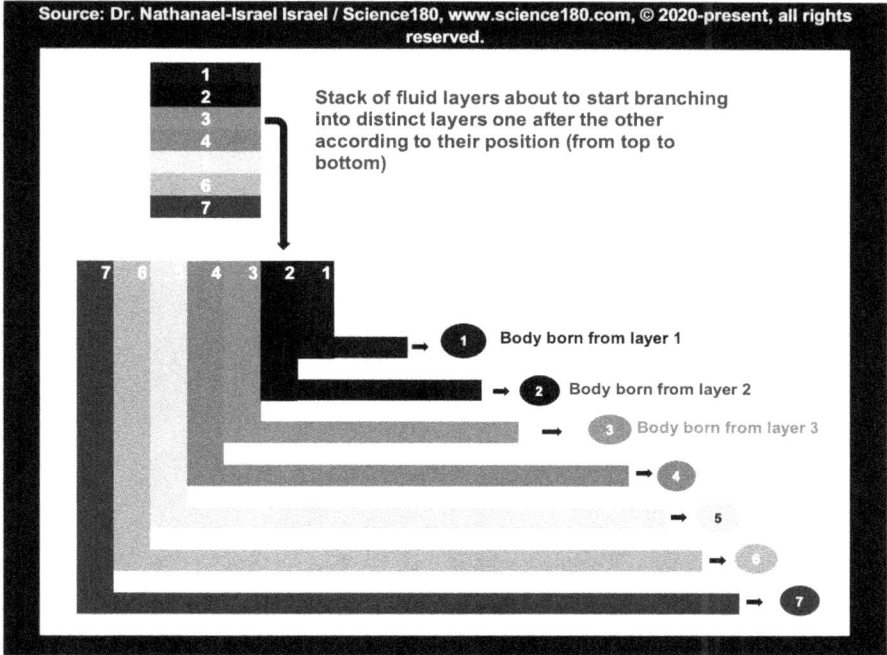

In Fig. 191, I showed a generic graph which presents with more details the split-gathering processes of a stack of 4 fluid layers of the precursors of bodies.

Fig. 191: Split-gathering processes of a stack of 4 fluid layers of the precursors of bodies

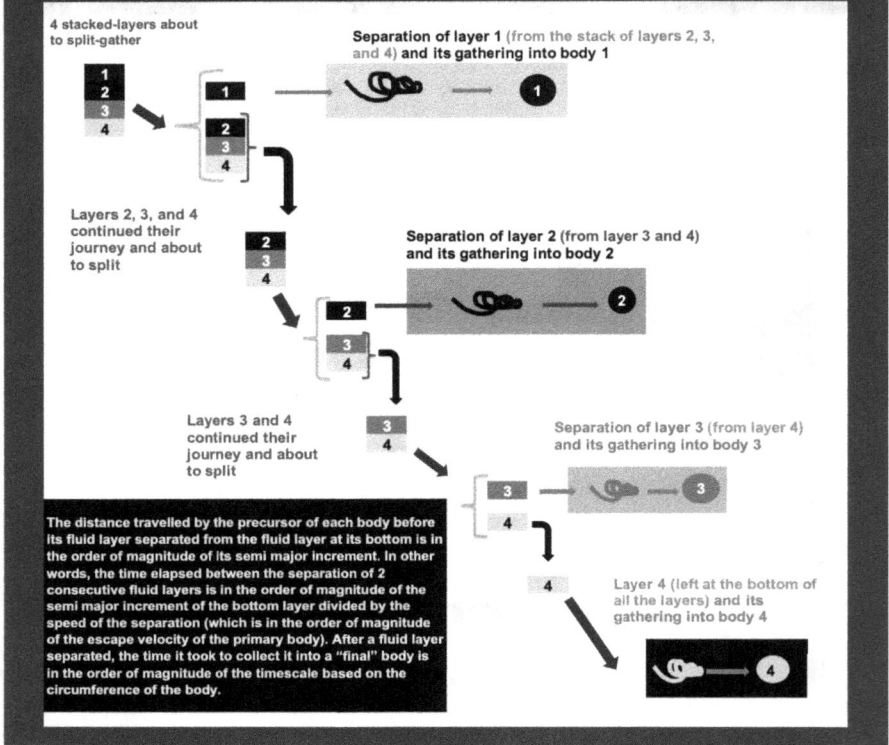

If you want this graph is a colored format, visit www.Science180.com/SplitGathering4Layers

Fig. 192 illustrates the split-gathering processes of a stack of fluid layers of the precursors of systems of bodies into a primary body and secondary bodies. First, the precursor of the system of bodies split into the precursor of a primary body and the precursor of secondary bodies. Each of them goes through some processes before yielding its daughter bodies.

Fig. 192: Split-gathering processes of a stack of fluid layers of the precursor of a system of bodies into a primary body and secondary bodies

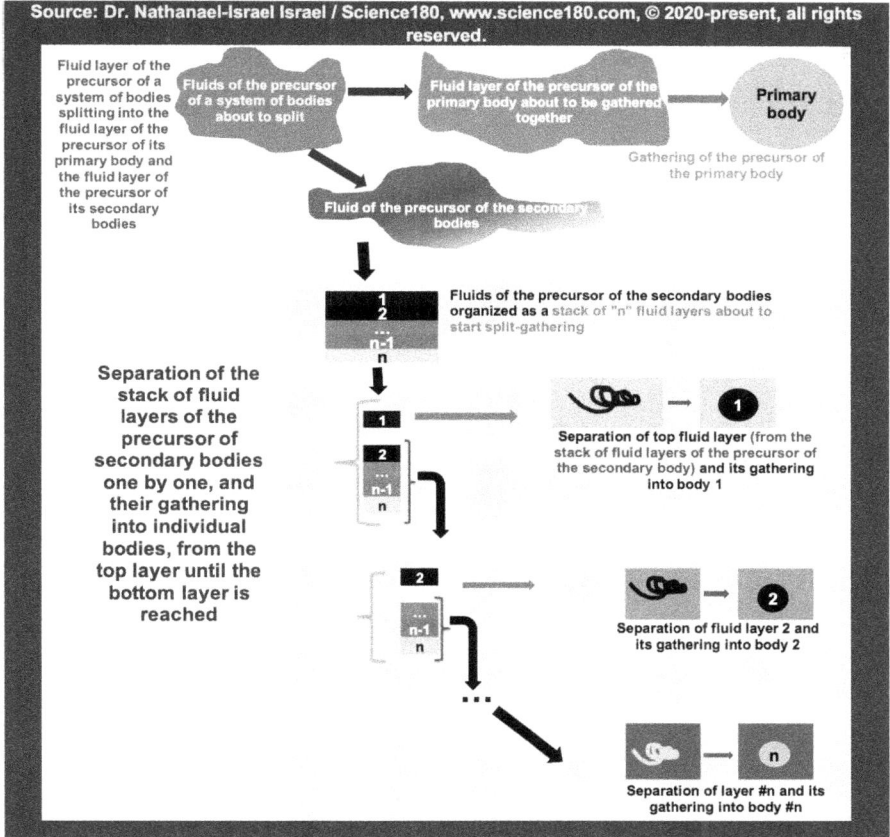

I did Fig. 193 to show the stack of fluid layers of the precursor of all the bodies orbiting the Sun at the moment of the split of the fluid layer of the precursor of Mercury.

Fig. 193: Stack of fluid layers of the precursor of all the bodies orbiting the Sun at the moment of the split of the fluid layer of the precursor of Mercury

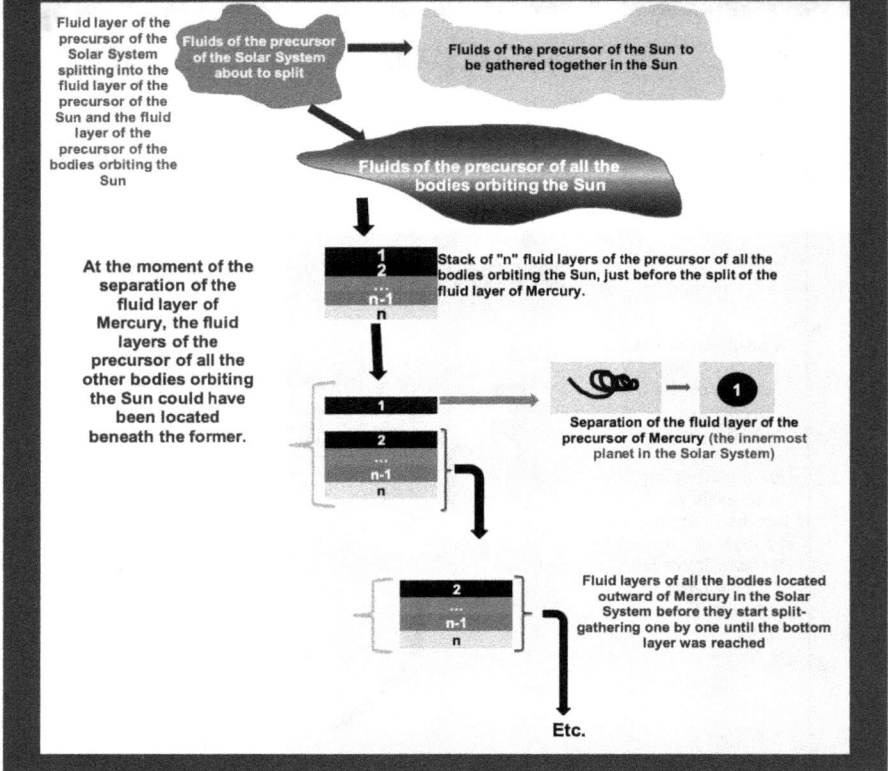

In some of the incoming chapters, I detailed the amount of time each of the processes enunciated in all the graphs I showed in this chapter took.

33.9. Circumference timescale

So far, I spent a lot of time on how to estimate the timescale based on the semi major axis and the escape velocity of the celestial bodies. However, the story of the birth of the celestial bodies did not stop just after their fluid layers split and took a path or trajectory different from that of the precursor of the bodies which were beneath them before the split. Indeed, after the split, the fluid layers must be gathered together. Some fluid layers were collected into spherical bodies, while others did not get a clear shape. Using the data on the dimensions of the bodies, I was able to calculate the average radius and consequently, the circumference of the celestial bodies that I studied. The Circumference timescale is about the time it took for a split fluid layer (which could not shear anymore) to be gathered together into a well-defined celestial body. Just as the timescale based on semi major axis (that I already explained in this chapter), in addition to a distance (here referred to as a circumference), a speed is needed to calculate the Circumference timescale of the bodies. In one of the incoming chapters, I detailed the Circumference timescale of many celestial bodies in the Solar System. In that chapter, I explained why the orbital speed of the celestial bodies is about the speed with which the fluid layer of the precursor was gathered together. Hence, after the gathering together of the precursor, the orbital speed was acquired by the daughter bodies. In other words, the Circumference timescale of a body is calculated by dividing the circumference of that body by its orbital speed. To sum up, it is the time it took to wrap or collect a fluid layer after it split or separated from the stack of the other fluids.

33.10. Birthdate of the celestial bodies

By birthdate of a body, I meant the time that elapsed from the beginning of the formation of the universe until the body was formed. I devoted a specific chapter to this topic. As I have been explaining, and as you will see in the aforementioned chapter, the birthdate of the celestial bodies is calculated by adding the time it took for their precursor before they started gathering to the time it took for them to be gathered together. The time it took before the precursor started gathering is related to the semi major axis timescale. For the Sun, the time elapsed before the beginning of the gathering is related to the time it took for the precursor of the bodies orbiting the Sun to travel from the position of the Sun to the orbit of Mercury (the innermost body orbiting the Sun). For the primary planets, the time which passed before the gathering of their precursors started was not higher than the time that elapsed before the precursor of the innermost satellite split. But, as I explained in the next chapters, the escape time of the precursor of the planets is negligeable. Therefore the birthdate of planets will be about the their semi major axis timescale and their circumference timescale. One of the incoming chapters is solely devoted to the birthdates of the celestial bodies.

33.11. Many characteristics of the celestial bodies were defined by the history of their precursors

By the time the precursors of the bodies were born, some of the characteristics were affected by the processes they went through. For instance, the orbit of some bodies is elongated (leading to a higher eccentricity), while that of others is tilted (leading to higher orbital inclination). In some cases, the rotational axis is tilted, leading to a higher axial tilt, while for some bodies, the axial tilt is null. Due to the processes that their precursors went through, some bodies rotate faster than others. Some bodies ended up being denser than others.

Like I explained in some previous chapters, at the breaking of a fluid layer, some fluid layers were elongated, others tilted as they retracted or were ejected.

The retraction of the precursor of the bodies orbiting the Sun after they escaped the precursor of the Sun may have affected the top layers (which were the fluid layers of Mercury) and in the end that retraction may have contributed to elongating the orbit of Mercury. Hence Mercury has a huge eccentricity, having the second highest eccentricity of the planets in the Solar System just after the record held by Pluto. According to the size of the fluid layers, some daughter bodies could not be squeezed enough to yield a solid body. Hence some celestial bodies are larger but less dense, while others are very dense. Like I mentioned in the chapter on galaxies and stars, the Sun is huge and consist of plasma (close to a fluid) because among many things, its fluid layers were too large and could not be collected and squeezed enough to yield a solid body. The same thing happened to the precursors of some giant planets (Jupiter and Saturn) which ended up being large but filled with gas.

In the end, the strategic organization of these timescales and the events surrounding them allowed me to establish the timeline of the formation of the universe. In the following chapters, you will read more details about each of the following topics that I introduced above:

- Semi major axis increment timescale
- Semi major axis timescale
- Circumferential gathering start date
- Circumference timescale
- Birthdate of the celestial bodies

33.12. Take home message

Considering the secrets that I have unearthed, I decided to calculate the timeline of the formation of the universe. Such a task required an intimate understanding of the timing of the processes that shaped the universe. I had to craft a unique methodology. I established that, for the celestial bodies to be formed, precursors were split-gathered into precursors of primary and secondary bodies, which were then separated and gathered together. Therefore, the duration of time it took to form the celestial bodies must include (1) the time that their precursors "traveled" from the position of their mother precursor to the current orbit of their daughter bodies and (2) the time these precursors took to wrap, collect, or gather themselves together into their final shape known for them today. The strategy I used took advantage of secrets hidden in the size (e.g. radius), orbital speed, escape velocity, and distance separating the celestial bodies. For instance, the semi major axis of the celestial bodies holds a secret of the distance travelled by their precursors. I showed that, the precursor of the secondary bodies must have moved out or escaped the precursor of their primary body before the latter could fully collect itself into a unique body. I demonstrated that the distance separating a primary body and its innermost secondary body is the maximum distance travelled by the precursor of the secondary bodies before the precursor of the primary body was fully formed and ready to swirl to form itself. Knowing the speed with which the precursor of the secondary bodies escaped the precursor of its primary body, I showed that it is possible to estimate the maximum amount of time it could have taken for a body to travel the distance separating a primary body from its innermost secondary body. I found that the duration of the split of the precursor of a primary body from the precursor of its secondary bodies was about the semi major axis of the innermost secondary body divided by the escape velocity of the primary body. I showed that the escape velocity of the primary body is about the speed with which the stack of fluids of the precursor of the secondary bodies separated from the precursor of their primary body.

I showed that the fluid layers at the top separated first. Then, those beneath it did the same according to their position until the bottom layer separated last. Once a fluid layer separated, a certain amount of time passed before its gathering was completed. I showed that the escape speed of the precursor of the secondary bodies from the precursor of its primary body is in the order of magnitude of the escape velocity of the primary body. For instance, the precursor of the bodies orbiting the Sun could have escaped the precursor of the Sun at the escape velocity of the Sun. Likewise, the precursor of satellites could have escaped the precursor of their primary planets at the escape velocity of their primary planet. I showed that the lifespan of the precursor of a system of bodies is inferior to or equal to the ratio "Semi major axis of the innermost secondary body" / "Escape velocity of the primary body". I showed that the time it took for the precursor of a secondary body to travel from the position of its primary body to the position it was fully formed was about the "Semi major axis of that secondary body divided by the escape velocity of the primary body".

I proved that the story of the birth of the celestial bodies did not stop just after their fluid layers split and took a path or trajectory different from that of the precursor of the bodies which were beneath them before the split, but, after the split, the fluid layers were gathered together. I calculated the Circumference timescale of the bodies which is the time it could have taken for the bodies to wrap their layers once that fluid layer split from a stack of fluid layers and could no longer shear. The orbital speed of the celestial bodies is about the speed with which the fluid layer of the precursor was gathered together. Hence, after the gathering together of the precursor, the orbital speed was acquired by the daughter bodies. Hence, the Circumference timescale of a body is calculated by dividing the circumference of that body by its orbital speed. Defining the birthdate of the celestial bodies as the time that elapsed from the beginning of the formation of the universe until the body was formed, I showed that the birthdate of the celestial bodies can be calculated by adding the time it took for their precursor before they started gathering to the time it took for them to be gathered together. For the Sun, the time elapsed before the beginning of the gathering is related to the time it took for the precursor of the bodies orbiting the Sun to travel from the position of the precursor of the Sun to the orbit of Mercury (which as of 2025 is the most known innermost body orbiting the Sun). For the primary planets, the time which passed before their gathering started was not higher than the time that elapsed before the precursor of the innermost satellite split, which is negligeable. In the next chapters, I will detail each of the components of the birthdate I introduced above. Then, I will review some creation stories before my conclusive remark.

When I do seminars or one-on-on-one consulting with my clients or give a keynote speech (see details at www.Science180.com/speaking), I like to offer this advice to the audience: "Do not ignore what I share in this chapter if you want to decode the date of birth of celestial bodies". To get the same result, you should do the same, for I want you to succeed.

Another Book by Nathanael-Israel Israel:
FROM SCIENCE TO BIBLE'S CONCLUSIONS

THE # 1 UNIVERSE-ORIGIN MASTERPIECE OF ALL TIME ... AND THE MOST ACCURATE SCIENTIFIC FORMULA THAT STOOD AND WILL STAND THE TEST OF TIME AND OF MATHEMATICS

The real reason scientists have been struggling to accurately understand the universe-formation is because they have spent centuries collecting expensive, complicated, and massive amounts of data, but learned very little, if not nothing, about how to unconventionally step back to properly analyze it to decode the universe. Consequently, people learned to collect all kinds of data everywhere to build models and imaginary concepts that betray their discernment, but they never learned to unlearn wrong theories, nor learned how to stop trashing great raw data hidden in theories they dislike or misunderstand, never knew where to find and how to properly combine the fundamental variables without which it is impossible to ever clear the way so their data can properly work for and precisely lead them to the real origin of the universe. How can people abandon the dangerous theories they think are correct because they don't know any better ones?

Lucky you, that is where Dr. Nathanael-Israel Israel, the founder of Science180 (Science180.com) came in to properly reanalyze and put under control these costly, underrated data to provide the accurate and simple solution people have been looking for throughout the ages, but that they have ignored.

In *"From Science to Bible's Conclusions"*, you will:

- Get a world class explanation of the 4 fundamental variables without which it is unquestionably impossible to ever decode the universe-formation scientifically

- Save time and money, and enjoy a life filled with the wonderful peace that the accurate understanding of the universe-origin can create
- Discover the errors in the scientific theories and religious belief systems about the universe-formation that are putting you at risk, and learn how to take control over cosmological threats lurking at the edge of your rational mind, faith, disbelief, or doubt
- Unlock the accurate scientific formula to rationally test the existence of God in a historic way that uncompromisingly satisfies both believers and skeptics (*Science180.com*/public)
- Get all you need to become a knowledgeable person who will never again need anybody else to explain to you the origin of the universe, for, you will fully understand and articulate it yourself and rationally know whether science is really at war with religion
- Receive deep insights that even those who went to university for years were not able to decrypt by themselves, so you can equip yourself to eliminate all forms of scientific and religious universe-origin prejudices
- Discover whether the scientific data finally confirms that the formation of the Earth was completed on the 3rd day, while that of the Moon and the Sun was on the 4th day of creation like the Bible says, or whether the data proves that it took billions of years to progressively form the universe
- Understand the celebrated scientific formula that rationally puts to rest all debates about the relationship between science, faith, and all theories about the universe-origin so you can properly develop yourself, expand your network, and shape your future

Quickly grab and read this scientifically verifiable, bestselling book to finally get the accurate, jaw-dropping answer that has been rationally shaking both believers, skeptics, and all freethinkers. Don't wait!

Dr. Nathanael-Israel Israel has had the honor to be acknowledged as the #1 universe-origin, life-origin, and chemicals-origin expert. He is the author of *"Turbulent Origin of the Universe"*, *"Reconciling Science and Creation Accurately"*, *"Turbulent Origin of Chemical Particles"*, *"Turbulent Origin of Life"*, *"How Baby Universe Was Born"*, *"Science180 Accurate Scientific Proof of God"*. Visit Israel120.com to learn more about this world's most trusted expert that helps scientists and laypeople to properly decode the origin and formation of the universe, life, and chemicals so people can live more effectively nonstop.

CHAPTER 34

MISS THIS SCIENTIFIC FORMULA ABOUT THE SEMI MAJOR AXIS INCREMENT TIMESCALE AND YOU WILL ALWAYS GET THE UNIVERSE-ORIGIN WRONG

34.1. Formula of the Semi major axis increment timescale

The "Semi major axis increment timescale" is a term I coined in 2025 to express the duration of time the fluid layers of the precursor of a celestial body travelled before splitting from the fluid layer below it once the fluid layer above it split. In other words, because the fluid layers were stacked one on top of the other, when one layer split from the stack, the layer originally below it also traveled with the remainder of the stack of fluid layers for a certain amount of time before splitting at its turn. The "Semi major axis increment timescale" is about the maximum time it could have taken for a layer to split after the one above had done the same. The split of the fluid layers of the precursors of bodies is like the branching at different branching sites of a growing tree. When one branch merges, the trunk continues growing until another branch emerges again and so on and so forth until the tree reaches its maximum height and eventually dies. Each branch continues growing at its own rate which can be different from that of the trunk. I better elaborated on this analogy in my book on the origin of life.

In the rest of this chapter and in others to come, whenever I mention "Semi major axis increment timescale", please remember that it is not calculated just by using the semi major only, but also by using the escape velocity of the primary body of the concerned bodies. I already established why the escape velocity is (in the order of magnitude of) the speed with which the precursor of the secondary bodies escaped the precursor of their primary body in each system of bodies. For instance, I showed that the escape velocity of the Sun is almost the speed with which the precursor of the bodies orbiting the Sun (planetary systems and

Nathanael-Israel Israel: Acknowledged as Undisputable Specialist of All
Questions at the Intersection of Science and Biblical Creation

asteroid systems) escaped the precursor of the Sun. In the same manner, in each planetary system, the escape velocity of the planet is about the speed with which the precursor of the satellites in that planetary system escaped the precursor of their primary planet. By the way, when I say "semi major", I mean "semi major axis". Therefore, the formula that I invented for the "Semi major axis increment timescale" is:

"Semi major axis increment timescale" of a body = Semi major increment of that body / Escape velocity of the primary body of that body

For the innermost bodies in each system, I considered their semi major increment to be their semi major, which is the average distance from their primary body. To express the above formula differently using the term of primary body and secondary bodies, the time it took for the precursor of a body to travel from the position of the precursor of the body upstream of it to its current position was in the order of magnitude of:

Semi major axis increment of the body / Escape velocity of the primary body of that body

Considering what I said above, in each system, the time it took for the precursor of the innermost secondary body to travel from the position of the precursor of its primary body to the current position of that innermost secondary body is about:

Semi major axis of the innermost secondary body / Escape velocity of the primary body

Because in the previous chapters, I already devoted a chapter to the semi major increment, and another one to escape velocity, I will not dwell much on their calculation or meaning in this chapter. If needed, you can refer to those chapters to refresh your mind. Because the primary body of the planets and asteroids in the Solar System is the Sun, the above formula applied to asteroids and planets in the Solar System becomes:

"Semi major axis increment timescale" of a planet = "Semi major axis increment" of the planet" / "Escape velocity of the Sun"

"Semi major axis increment timescale" of an asteroid = "Semi major axis increment of the asteroid" / "Escape velocity of the Sun"

Likewise, because the primary body of the satellites in a planetary system is their primary planet, the same formula applied to satellites as:

"Semi major axis increment timescale" of a satellite = "Semi major increment of the satellite" / "Escape velocity of the primary planet"

Because the semi major of the Sun is 0 when estimated with respect to the Solar System, I did not calculate the "Semi major axis increment timescale" of the Sun. Likewise, with respect to the planetary system, the semi major of the planets is zero. Hence, as you will see below, the "Semi major axis increment timescale" of the planets is calculated with respect to the Sun not with respect to the planetary system.

34.2. Semi major axis increment timescale of the celestial bodies according to their types

Calculated for 461 celestial bodies, the "Semi major axis increment timescale" of the bodies that I studied in the Solar System varies between 0 and 2,102,643.9 hours (which is equivalent to 240.03 years). The highest value was recorded with C/1999 F1, a comet allegedly said to be located at 4.98161E+12 km (meaning 33,300 AU, which is 33,300 times the distance separating the Sun and the Earth). Some of the bodies which "Semi major axis increment timescale" is 0 are:

- 76P West-Kohoutek-Ikemura (a Comet)
- 87 Sylvia (a Main belt asteroid)
- Isonoe (a Jovian satellite) and
- 5 Saturnian satellites (i.e. Calypso, Telesto, Helene, Polydeuces and S/2007 S3).

For the Main belt asteroids, the "Semi major axis increment timescale" varies between 0 and 8.68 hours. That value for Ceres (the largest main belt asteroid) is 0.17 hours, meaning 10 minutes and 6 seconds.

Calculated by considering the distance between the planets only, the timescale increment of the planets varies between 18.6 and 729.8 hours (equivalent to 30.4 days) (Fig. 194). The shortest timescale was recorded with the Earth while the longest timescale was found with Neptune. This means that, traveling at a speed equivalent to the escape velocity of the Sun (i.e. 617.6 km/s), it took about 18.6 hours for the precursor of the Earth-Moon system to move from the position of Venus to its position which is 1 AU (astronomical unit) from the Sun. Likewise, it took about 30.4 days for the precursor of the Neptunian planetary system to move from the position of the precursor of Uranus to arrive at its position. The "Semi major axis increment timescale" of Mercury (the innermost planet in the Solar System) is 26.05 hours (equivalent to 1.09 days). This means that the precursor of Mercury was separated from the precursor of all the bodies orbiting the Sun on the second day since the beginning of the split of the precursor of the bodies in the Solar System. As of 2021, no significant celestial body in the Solar

System is known to be closer to the Sun than Mercury. Because the fluid layers of the precursor of Mercury were on top of all the fluid layers of the precursor of the bodies orbiting the Sun, the maximum timescale it could have taken for the precursor of the bodies orbiting the Sun to completely split from the precursor of the Sun was the "Semi major axis increment timescale" of Mercury, which, as explained above, is 26.05 hours. In other words, the separation of the fluid layers of the bodies orbiting the Sun from one another started on the second day after the beginning of the formation of the Solar System. Before I close this segment of the "Semi major axis increment timescale" of the planets, let me emphasize that in reality, the precursor of the planetary system split from the precursor of the asteroids located near them according to their position. In other words, as you view the graph of the "Semi major axis increment timescale" of the planets, keep in mind that between the precursor of the planetary system from which descended the planets, there were also precursors of asteroids. Hence, in the main table I did for the "Semi major axis increment timescale" of all the bodies in the Solar System, I calculated the "Semi major axis increment timescale" of the planets by considering them as having split from the precursor of the asteroids, thus their value is smaller than when I calculated it by "ignoring" the precursor of the asteroids. If I did not use this strategy, some people may not realize the real time which elapsed between the separation of the precursors of the planetary systems from the remainder of the fluid layers of the bodies orbiting the Sun. Hence, I calculated the "Semi major axis increment timescale" of the planets in 2 different ways: with respect to the planets only and also with respect to the asteroids near them.

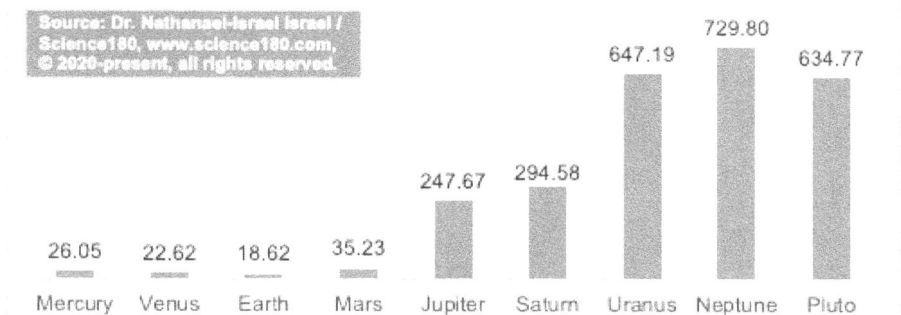

Fig. 194: "Semi Major Axis Increment Timescale" (Hours) of the Planets in the Solar System

The "Semi major axis increment timescale" of the planets is significantly affected by the semi major of the primary planet ($r^2=0.99$) (Fig. 195).

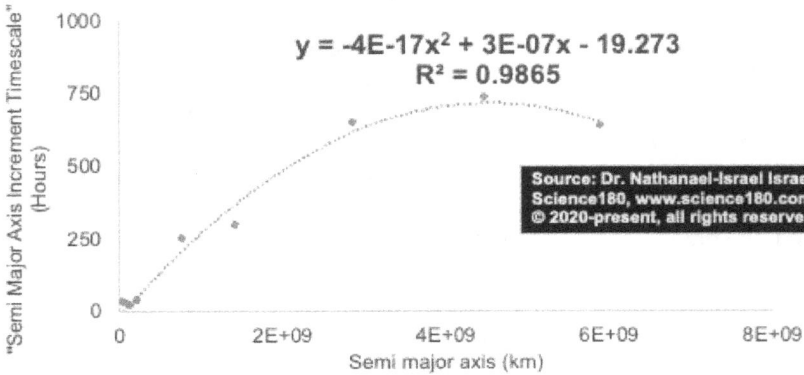

Fig. 195: Regression between "Semi Major Axis" and "Semi Major Axis Increment Timescale" of the Planets in the Solar System

$$y = -4E{-}17x^2 + 3E{-}07x - 19.273$$
$$R^2 = 0.9865$$

34.3. Semi major axis increment timescale of the satellites according to their types and turbulence zones

Like I explained earlier, I calculated the "Semi major axis increment timescale" of the satellites as "Semi major increment of the satellites" / "Escape velocity of their primary planet". For the innermost satellites, I considered their semi major increment to be their semi major. The "Semi major axis increment timescale" of the satellites here is with respect to their primary planets, not with respect to the precursor of the Solar System.

Investigated for 210 satellites in the Solar System, the "Semi major axis increment timescale" of the satellites varies between 0 and 273.4 hours (equivalent to 11.4 days) (Fig. 196). In other words, the time between the separation of 2 consecutive fluid layers of satellites in the Solar System is no longer than 11.4 days. The highest "Semi major axis increment timescale" was obtained with Psamathe (the 13th innermost Neptunian satellite), meaning the Neptunian satellite located just inward of the outermost Neptunian satellite known as of 2021. As a reminder, as of 2025, Neptune has 14 satellites and some of them hold the record of the longest semi major axis recorded on satellites in the Solar System. The relatively high semi major of some Neptunian satellites contributed to the higher value of the "Semi major axis increment timescale".

CHAPTER 35: SEMI MAJOR AXIS INCREMENT TIMESCALE

Fig. 196: "Semi Major Axis Increment Timescale" of the Satellites according to their Types and Semi Major Axis

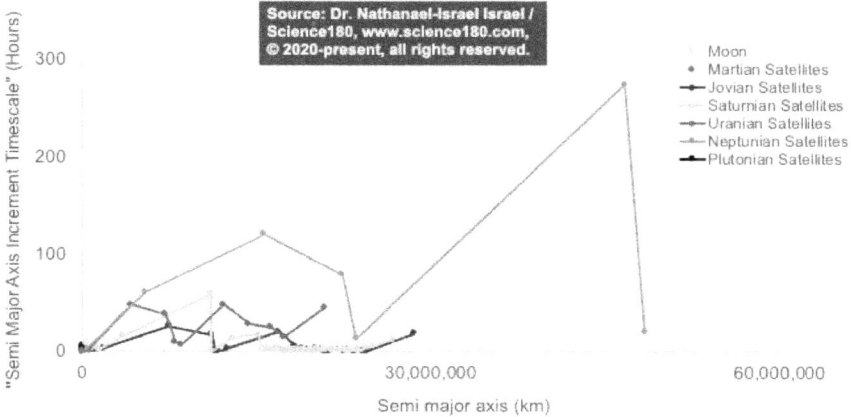

In each planetary system, the "Semi major axis increment timescale" of the innermost satellite is not the smallest, nor the highest. The "Semi major axis increment timescale" for the Moon (Earth's satellite) is 9.55 hours (i.e. 9 hours 32 minutes and 44.38 seconds). In other words, with a speed of 11.186 km/s (i.e. the escape velocity of the Earth), the precursor of the Moon could have reached the current position of the Moon in 9.55 hours after splitting from the precursor of the Earth-Moon system.

In contrast, the "Semi major axis increment timescale" of the Martian satellites is much smaller:

- 31.07 minutes for Phobos (the innermost and largest Martian satellite) and
- 46.66 minutes for Deimos (the outermost Martian satellite).

This data for the Martian satellites also means that the lifespan of the precursor of the Martian planetary system after it split from the remainder of the precursor of the bodies orbiting the Sun located beneath or outward of the precursor of the Martian planetary system could not have been higher than 31.07 minutes (the "Semi major axis increment timescale" of the innermost Martian satellite).

The "Semi major axis increment timescale" of the innermost Jovian satellite (Metis) is 35.85 minutes. Considering what I already said about the innermost satellite in each planetary system, the maximum time it could have taken for the precursor of the Jovian planetary system to split into the precursor of Jupiter and the precursor of the Jovian satellites could have not been longer than 35.85 minutes (which I just showed is the "Semi major axis increment timescale" of the innermost Jovian satellite). After that time, the fluid layers of the precursor of the other Jovian satellites split from one another according to their rank in the stack of their fluids. In other words, as the precursor of the Jovian satellites continued

Science180: One-Stop for Answering Your Universe-origin Questions its journey, the fluid layer of the precursor of the Jovian satellites split or "branched out" one after the other according to their position. The "Semi major axis increment timescale" of the Jovian satellites varies from 0 to 26.26 hours. The smallest timescale was recorded with Isonoe (a Jovian satellite in Zone 5). In contrast, the highest timescale for the Jovian satellites was found with Themisto (the innermost satellite in Zone 4). In other words, after the precursor of Callisto (the outermost Jovian satellite in Zone 3) split from the precursor of all the Jovian satellites located below it, it took about 26.26 hours before the precursor of Themisto (the innermost satellite in Zone 4) split from the bulk of the rest of the precursors of the Jovian satellites located beneath it. The length of that timescale of Themisto is due to the long distance separating Callisto and Themisto. In the chapters on "semi major increment", "vortex", and "fluid ligament", I already explained the reason behind the high semi major increment of some innermost bodies in Zone 4. For instance, I explained that the presence and the slow rotational angular speed of the precursors of the large satellites in Zone 3 on top of the precursor of the bodies in Zones 4 and 5 could have changed the vorticity of the fluid layers located beneath them, and affected other things, hence the fluid layers of the precursor of some satellites in Zone 4 changed the direction of their vorticity and as they were ejected, their semi major increment increased. The second highest "Semi major axis increment timescale" (20.68 hours) by the Jovian satellites was recorded with Carpo (the outermost Jovian satellite in Zone 4). Finally, the third "Semi major axis increment timescale" (18.81 hours) of the Jovian satellite was found with S/2003 J2 (the outermost Jovian satellite in Zone 5). I showed in some of the previous chapters that the end pinching of the outermost satellites could explain their high semi major increment, hence their high "Semi major axis increment timescale".

The "Semi major axis increment timescale" of the innermost Saturnian satellite (S/2009 S1) is 54.93 minutes, meaning that in less than 1 hour after the precursor of the Saturnian satellites escaped the precursor of Saturn, it started splitting into the precursor of individual satellites. Considering the formula that I developed for the lifespan of the precursor of the planetary systems, it took no longer than 54.93 min ("Semi major axis increment timescale" of the innermost Saturnian satellite) before the precursor of the Saturnian planetary system split into the precursor of Neptune and the precursor of the Saturnian satellites. Once the precursor of the innermost Saturnian satellite split by taking its own path, the precursor of the other Saturnian satellites did the same according to their position in the rest of the stack of fluid layers of the precursor of Saturnian satellites. As the precursor of the Saturnian satellites continued its journey, the fluid layers of the precursor of the satellites split or "branched out" from the rest of the main stack of fluid according to their position. The "Semi major axis increment timescale" of the Saturnian satellites varies between 0 and 59.07 hours. Four Saturnian satellites (Calypso, Telesto, Helene, and Polydeuces) out of the 5 Saturnian satellites which "Semi major axis increment timescale" is 0 are found

544
Nathanael-Israel Israel: Acknowledged as Undisputable Specialist of All
Questions at the Intersection of Science and Biblical Creation

in Zone 3. The highest "Semi major axis increment timescale" of the Saturnian satellites was recorded with Kiviuq (the innermost Saturnian satellite in Zone 4).

Because the "Semi major axis increment timescale" of the innermost Uranian satellite (Cordelia) is 38.97 minutes, I deduced that the maximum lifespan of the precursor of the Uranian planetary system after it split from the rest of the fluid layers of the precursor of the bodies orbiting the Sun located beneath or beyond it was no higher than 38.97 minutes. Once the innermost Uranian satellite split, the rest of the fluid layers of the precursor of the Uranian satellites separated at their turn one after the other according to their rank. The next precursor of Uranian satellite to split after Cordelia was Ophelia and it split no longer than 3.13 minutes after the former. In general, the "Semi major axis increment timescale" of Uranian satellites varies between 42.25 seconds (recorded with Desdemona, the outermost Uranian satellite in Zone 1) and 48.15 hours (recorded with Francisco, the innermost Uranian satellite in Zone 4).

The "Semi major axis increment timescale" of the innermost Neptunian satellite (Naiad) being 34.2 minutes, I determined as explained above that the maximum lifespan of the precursor of the Neptunian planetary system could have been smaller than 34.2 minutes. The "Semi major axis increment timescale" of the Uranian satellites varies between 1.31 minutes and 273.4 hours (meaning 11.39 days). The smallest timescale increment was recorded with Thalassa (the outermost Neptunian satellite in Zone 1) whereas the highest value was with Psamathe (the innermost Neptunian satellite in Zone 5, also located just upstream of the outermost Neptunian satellite). After the split of the precursor of Triton (the largest Neptunian satellite), it took about 60.98 hours (meaning 2 days 12 hours 58 mins and 36.6 seconds) before the fluid layers of the precursor of Nereid (located just beneath the precursor of Triton) could split from the fluid layers of the precursor of the bodies located outward of Nereid. Then, it took another 120.76 hours (meaning 5 days and 45 minutes before the precursor of Halimede (located downstream of Nereid) could split.

The "Semi major axis increment timescale" of Plutonian satellites varies between 1.44 hours (recorded on Nix, the second innermost Plutonian satellite in Zone 4) and 5.24 hours (recorded on Styx, the innermost Plutonian satellite in Zone 4). The "Semi major axis increment timescale" of Charon (the innermost and the largest Plutonian satellite) is 4.5 hours, suggesting that the lifespan of the Plutonian planetary system could have been less than 4.5 hours.

The "Semi major axis increment timescale" of the largest satellites in each planetary system ranges between 31.07 minutes and 9.55 hours (Fig. 197). The smallest value was recorded with Phobos (a Martian satellite), while the highest value was with the Moon (Earth's satellite). The highest value being recorded with the Moon can be explained by the escape velocity of the Earth (11.2 km/s) and the semi major of the Moon (384,400 km) as compared with the semi major increment of the largest satellite in each planetary system. For instance, the semi major axis of the Moon is higher than the semi major increment travelled by the biggest satellites of the other planets, and also the largest satellite of the giant planets escaped the precursor of their planet with a speed higher than that of the escape velocity of the Earth. Indeed, the escape velocity of the Earth (11.2 km/s) is smaller than that of the giant planets:

- 59.5 km/s for Jupiter
- 35.5 km/s for Saturn
- 21.3 km/s for Uranus
- 23.5 km/s for Neptune

Also, the semi major increment of some of the largest planets is smaller than or in the same order of magnitude of that of the Moon although they escaped their primary with a higher speed. Indeed, the semi major increment of the largest satellite of the other planet is:

- Phobos: 9,378 km (i.e. 2.44% the semi major of the Moon)
- Ganymede: 399,300 km (1.04 times the semi major of the Moon)
- Titan: 694,790 km (1.81 times the semi major of the Moon))
- Titania: 170,300 km (i.e. 44.3% the semi major of the Moon)
- Triton: 237,113 km (i.e. 61.68% the semi major of the Moon)
- Charon: 19,596 km (i.e. 5.1% the semi major of the Moon)

In each planetary system, the "Semi major axis increment timescale" of the largest satellite (Fig. 197) is higher than that of any satellite located between them and their primary planet.

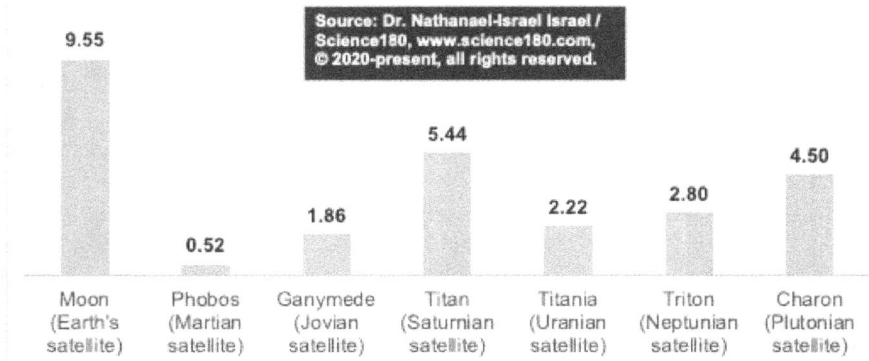

Fig. 197: "Semi Major Axis Increment Timescale" (Hours) of the Largest Satellite in each Planetary System

Source: Dr. Nathanael-Israel Israel / Science180, www.science180.com, © 2020-present, all rights reserved.

In general, the highest "timescale increments based on semi major" of the satellites were found in Zones 4 and 5, meaning that the separation of the bodies in those turbulence zones took longer to occur than it was the case for the bodies in Zone 1, 2, and 3.

Fig. 198 presents the "Semi major axis increment timescale" of the satellites according to their types and semi major.

Fig. 198: Relationship between "Semi Major Axis Increment Timescale" and "Semi Major Axis" of the Satellites according to their Types

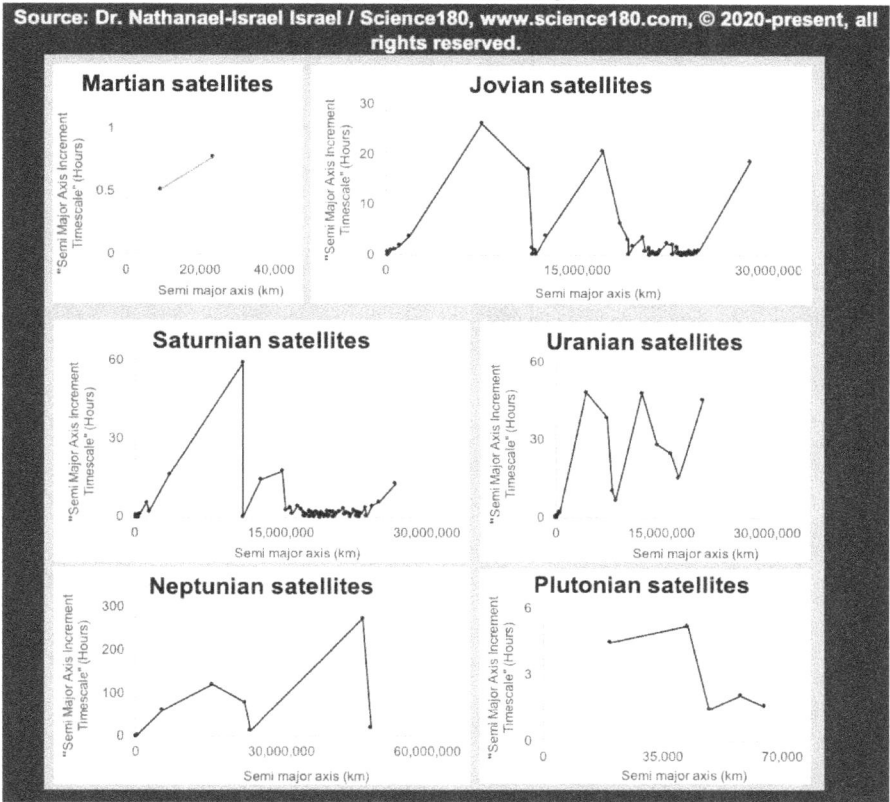

In Zones 4 and 5 of the Jovian and Saturnian satellites, there is a region where the satellites are clustered and where the "Semi major axis increment timescale" is very small, suggesting that the precursors of the satellites in that region split from one another much faster than was the case for the precursors in other regions.

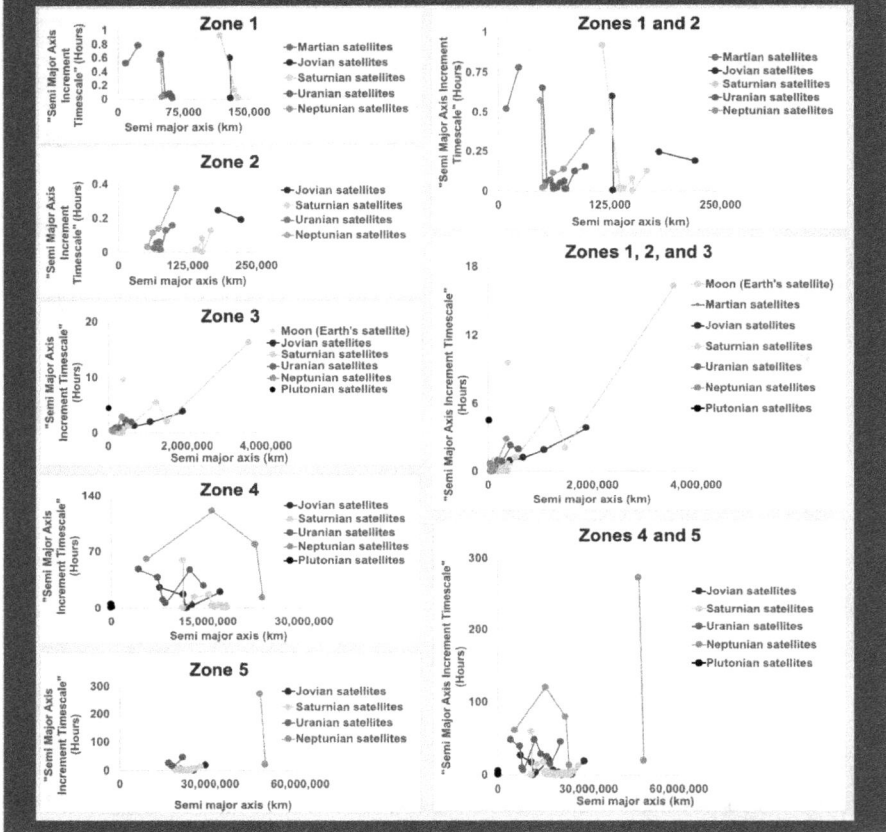

Fig. 199: "Semi Major Axis Increment Timescale" (Hours) of the Satellites according to their Types and Turbulence Zones

Next, I will present the sum of the "Semi major axis increment timescale" of the satellites according to their types and turbulence zones (Fig. 199). Based on the formula I designed to calculate the "Semi major axis increment timescale" of the satellites, the sum of the "Semi major axis increment timescale" of the satellites represents the total duration of time of the split of the fluid layers according to the types of satellites or turbulence zones of interest. For instance, the sum of the "Semi major axis increment timescale" of the satellites in a satellite system represents the duration of time from the moment the precursor of all the satellites started moving away from the precursor of their primary body until the time the last or bottom fluid layer (representing the precursor of the outermost satellite) just separated from all of the fluid layers which were above it and started gathering together to form the outermost satellite. To be precise, because there is a certain distance between the outermost satellite and the satellite located just

Nathanael-Israel Israel: Acknowledged as Undisputable Specialist of All
Questions at the Intersection of Science and Biblical Creation

upstream of that satellite, it is important to keep in mind that it took some time for the bottom fluid layer (meaning the fluid layer of the outermost satellite) in each planetary system to move away and position itself at a distance equal to its semi major axis before its gathering together started and/or was completed. Therefore, the sum of "Semi major axis increment timescale" of the satellites in each planetary system includes the maximum time that the precursor of the outermost satellite travelled away from the precursor of the satellites upstream of it before it started gathering together. When I talked about the gathering of fluid layers, I did not mean that nothing inside the fluid layers was not being gathered together before the fluid layers split from their peers or neighbors. Some particles and compounds could have been formed (up to some point) inside all of the fluid layers before their split from one another, but the gathering of the fluid layers as a whole occurred after they split. Some particles were formed after the fluid layers of the celestial bodies they belong to were split from their peers. Else, the chemical composition of the celestial bodies could not be so different (for instance across the planets and satellites in the Solar System). In my book *"Turbulent Origin of Chemical Particles"*, I provided ample details about the formation of the chemical elements. This also means that no celestial body was fully formed inside the fluid layers of its precursor before those fluid layers were split from the stack of fluid layers they belonged to.

As explained in the following graph (Fig. 200), the duration of time from the moment the precursor of the satellites started moving away from the precursor of their primary planet until the moment that the bottom fluid layer separated from the rest and started gathering (into the outermost satellite in that system) varies between 1.3 hours (equivalent to 1 hour 17 minutes and 43.81 seconds) and 571.99 hours (equivalent 23 days 19 hours 59 minutes and 8.94 seconds). The smallest duration was recorded on the Martian satellites, whereas the longest duration was obtained with the Neptunian satellites. Putting this another way, if someone could start counting time from the moment the fluid layers of the precursor of the Neptunian satellites started moving away from the precursor of Neptune until the moment the bottom fluid layers of the precursor of the Neptunian satellites separated from the rest and started gathering into the outermost Neptunian satellite, that person could have counted about 571.99 hours (which is equivalent to 23.83 days). The distance travelled by the precursor of the Neptunian satellites militated in favor of them holding the record of the longest duration of split of fluid layers of precursors of satellites in the Solar System. By the time the bottom fluid layers of the precursor of the Neptunian satellites started gathering into the outermost Neptunian satellite, they could have travelled a distance equal to or near the semi major axis of the outermost Neptunian satellite. As a general rule, at the time that all of the fluid layers above the bottom fluid layer split and the later was about to start gathering into its final shape, the bottom fluid layer would have travelled away from its primary body a distance no higher than the semi major axis of the outermost body known in its planetary system.

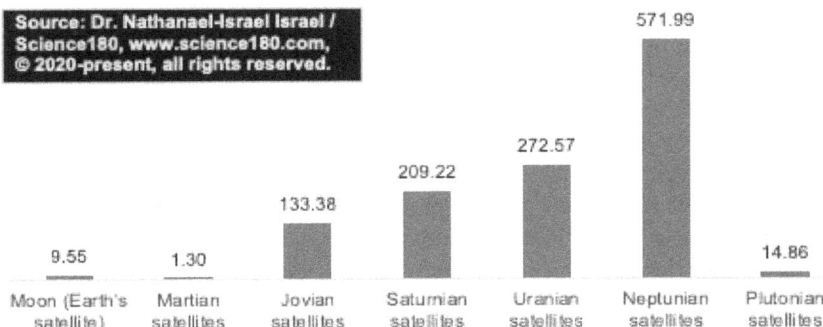

Fig. 200: Sum of the "Semi Major Axis Increment Timescale" (Hours) of the Satellites according to their Types

The sum of the "Semi major axis increment timescale" is the highest in Zone 5, followed by Zone 4 (Fig. 201). This means that it took longer for all of the bodies in turbulence Zone 5 to split from one another. In contrast, the smallest sum was recorded in Zone 2, meaning that the duration of time of the split of consecutive fluid layers was smaller in Zone 2. By the way, when I sometimes said that the duration of split was quicker, I was not referring to the speed of the split, but to the duration of time separating the split of 2 consecutive fluid layers. For instance, in each system of satellites (meaning all the satellites surrounding a planet), the speed with which the fluid layers of the precursor of the satellites escaped the precursor of their primary planet is the order of magnitude of the escape velocity of their primary planet. Yet, due to the semi major increment (representing the distance between two consecutive splitting sites) of the fluid layers of the precursors of the satellites, the time elapsed between some consecutive splits occurred faster than others.

Fig. 201: Sum of the "Semi Major Axis Increment Timescale" (Hours) of the Satellites according to their Turbulence Zones and Types

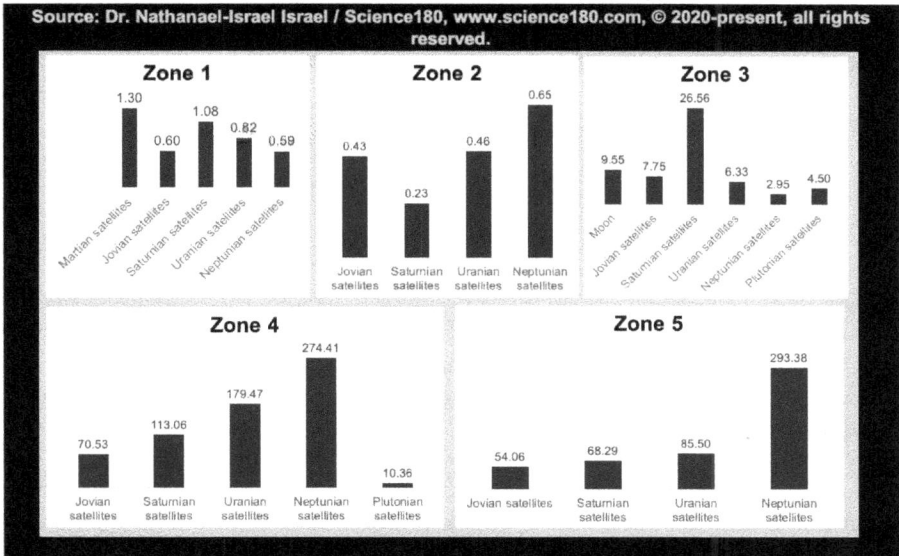

In Zone 1 for instance, the duration of time of the split of all of the fluid layers varied between 0.59 hours (equal to 35.51 minutes, recorded among the Neptunian satellites) and 1.3 hours (equivalent to 77.73 minutes, recorded for the Martian satellites).

In Zone 2, the total duration of time of the split of the fluid layers is not higher than 0.65 hours (equivalent to 39.17 mins), suggesting that the transition into a developed turbulence was very quick. Let's remember that Zone 2 is the zone of transition into the major turbulence in each system of satellites. Because the total duration of events occurring in Zone 2 is the quickest as compared to events in the other turbulence zones, it means that the transition toward the most developed turbulence zone (which would occur in Zone 3) is very quick.

Considering what I said above and illustrated in the graph of the sum of the "Semi major axis increment timescale" of the satellites according to the zones, the sum of the duration of time it took for the satellites in Zones 1 and 2 combined to split is:

- 1.30 hours for Martian satellites
- 1.04 hours for Jovian satellites
- 1.31 hours for Saturnian satellites
- 1.27 hours for Uranian satellites
- 1.24 hours for Neptunian satellites

While all the Neptunian satellites in Zone 3 split within 2.95 hours, it took 26.56 hours for the fluid layers of all the Saturnian satellites in Zone 3 to split. This means that it took longer for the fluid layers of all the satellites in Zone 3 to split from one another than it took for all of the satellites in Zones 1 and 2 combined. Put another way, the duration of time it took for the fluid layers of the precursors of all the satellites in Zone 3 to split from one another is 2.37 to 20.26 times that it took to split all fluid layers in Zones 1 and 2 combined. In fact, the ratio between the total time of the split of the fluid layers in Zone 3 divided by that in Zones 1 and 2 combined is:

- 7.48 for Jovian satellites
- 20.26 for Saturnian satellites
- 4.97 for Uranian satellites
- 2.37 for Neptunian satellites

The size of the bodies in Zone 3 and the distance that the fluid layers of their precursors had to travel before splitting from one another explains why their split took longer than that of the fluid layers in Zones 1 and 2 combined.

Except for the Neptunian satellites, it took longer to split the fluid layers in Zone 4 than it did for any other turbulence zone. The duration of the split in Zone 4 ranges between 10.36 hours (recorded on Plutonian satellites) and 274.41 hours (meaning 11.43 days), recorded on Neptunian satellites. The huge semi major increment and the number of the satellites in Zone 4 contributed to the high amount of time it took to split the fluid layers of their precursors.

It took longer to split all the fluid layers of the Neptunian satellites in Zone 5 (duration of split = 293.38 hours) than it took to split the fluid layers in all the satellites in any other zone of the satellite systems in the Solar System. When I talk about satellites systems here, I am not talking about satellites around asteroids, but satellites around planets only. The duration of the split of the Jovian, Saturnian, and Uranian satellites in Zone 5 is smaller than that in Zone 4. In fact, it took longer to split all the fluid layers of the Neptunian satellites in Zone 4 than it took to split the fluid layers of all the Jovian, Saturnian, and Uranian satellites in Zone 5 (total duration of such a split = 207.85 hours). The huge distance (i.e. semi major axis) that the fluid layers of the precursor of the Neptunian satellites had to travel before splitting from one another and the relatively small escape velocity of Neptune (with which such a travel of the fluid layers of the Neptunian satellites occurred) as compared to the semi major of the Jovian, Saturnian, and Uranian satellites and the relatively higher escape velocity of Jupiter and Saturn explain the highest duration of the split of the fluid layers of the Neptunian satellites in Zone 5.

34.4. Take home message

The "Semi major axis increment timescale" is a short term I invented to express the timescale increment based on the escape velocity of the primary body and the semi major of the celestial bodies. I calculated and established that the "Semi major axis increment timescale" properly estimates the duration of time a fluid layer of the precursor of a celestial body travelled before splitting from the fluid

552
Nathanael-Israel Israel: Acknowledged as Undisputable Specialist of All
Questions at the Intersection of Science and Biblical Creation

layer below it once the fluid layer above it split. Because the fluid layers were stacked one on top of the other, when one layer split from the stack, the layer originally below it also traveled with the remainder of the stack of fluid layers for a while before splitting at its turn. The split of the fluid layers of the precursors of the bodies is like the branching at different branching sites of a growing tree. When one branch merges, the trunk continues growing until another branch emerges again and so on and so forth until the tree reaches its maximum height and eventually dies. For instance, using the distance between the planets, the "Semi major axis increment timescale" of the Earth was 18.6 hours, meaning that it took about 18.6 hours for the precursor of the Earth-Moon system to move from the position of Venus to its position, which is 1 AU from the Sun. I also showed that the precursor of the Moon reached the current position of the Moon in about 9.55 hours after splitting from the precursor of the Earth. I proved that, after splitting from the precursor of the Sun, the precursor of Mercury spent no more than 26.05 hours (equivalent to 1.09 days) before reaching its position. In other words, the precursor of Mercury was separated from the stack of fluid layers of the precursor of all the bodies orbiting the Sun on the second day since the beginning of the split of the precursors of the bodies in the Solar System. The lifespan of the precursor of the Jovian planetary system was 35.85 minutes.

CHAPTER 35

EXPERT REVEALS ONE OF THE 4 SURPRISING ESSENTIAL VARIABLES FOR SMART PEOPLE TO DISCOVER THE UNIVERSE-ORIGIN FORMULA

35.1. Definition and formula of the types of "Semi major axis timescale" of celestial bodies

The "Semi major axis timescale" of a body is the "Semi major axis timescale of that body (with respect to its primary body). Because the precursor of the secondary bodies in a system of bodies escaped the precursor of their primary body at about the escape velocity of the primary body, the formula of the "Semi major axis timescale of a body (with respect to its primary body)" is:

"Semi major axis timescale of a body (with respect to its primary body)" = "Semi major axis of the body" / "Escape velocity of the primary body of that body"

Because the primary body of the planets and asteroids in a stellar system is the star they orbit, the "Semi major axis timescale" of planets and asteroids is defined with respect to the star they orbit. In other words, the "Semi major axis timescale" of a planet is the "Semi major axis timescale of that planet (with respect to the star it orbits". Likewise, the "Semi major axis timescale" of an asteroid is the "Semi major axis timescale of that asteroid (with respect to the star it orbits)".

Planets and asteroids have one type of "Semi major axis timescale", which is their "Semi major axis timescale (with respect to the star)" that they are orbiting. In short, I called this the "Semi major axis timescale". In other words, wherever I say "Semi major axis timescale" of a planet, I mean the "Semi major axis timescale of that planet (with respect to the star it orbits)".

The "Semi major axis timescale of a planet (with respect to the star) is the

Nathanael-Israel Israel: Acknowledged as Undisputable Specialist of All Questions at the Intersection of Science and Biblical Creation

"Semi major axis" of that planet divided by the "escape velocity" of the star that the said planet orbits.

Semi major axis timescale of a planet
$$= \frac{\textbf{Semi major axis of the planet}}{\textbf{Escape velocity of the primary star of that planet}}$$

Likewise, the "Semi major axis timescale of an asteroid (with respect to the star) is the "Semi major axis" of that asteroid divided by the "escape velocity" of the star that the said asteroid orbits.

Semi major axis timescale of an asteroid
$$= \frac{\textbf{Semi major axis of the asteroid}}{\textbf{Escape velocity of the primary star of that asteroid}}$$

Now, I will talk about the "Semi major axis timescale" of satellites. Indeed, because planets are the primary bodies which satellites orbit, the "Semi major axis timescale" of a satellite is primarily defined with respect to the primary planet of that satellite. In that case, the "Semi major axis timescale" of a satellite is usually the "Semi major axis timescale of that satellite (with respect to the primary planet)" of that satellite. However, because planets orbit their primary star, the "Semi major axis timescale" of a satellite can also be calculated with respect to the star that the primary planet of that satellite is orbiting. In that case, it is termed "Semi major axis timescale of the satellite (with respect to the star)". Therefore, satellites can have 2 types of "Semi major axis timescale":
- "Semi major axis timescale (with respect to the primary planet)"
- "Semi major axis timescale (with respect to the star)"

The "Semi major axis timescale of a satellite (with respect to its primary planet)" is the "Semi major axis timescale of that satellite" divided by the "Escape velocity" of the primary planet of that satellite. To shorten this term, wherever I say "Semi major axis timescale of a satellite" I mean the "Semi major axis timescale of that satellite (with respect to its primary planet)".

Let's now talk a little more about the "Semi major axis timescale of a satellite (with respect to the star)". Indeed, the "Semi major axis timescale of a satellite (with respect to the star)" considers the time that elapsed since the moment that the precursors of the bodies orbiting the precursor of a star escaped the latter. Because the precursor of a satellite was born after it escaped the precursor of its primary planet, the "Semi major axis timescale of a satellite (with respect to the star)" is the sum of:
- the "Semi major axis timescale of the satellite (with respect to the primary planet) and

Science180: One-Stop for Answering Your Universe-origin Questions

- the "Semi major axis timescale of the planet (with respect to its star)"

"Semi major axis timescale of a satellite (with respect to the star)" = "Semi major axis timescale of the satellite (with respect to the primary planet)" + "Semi major axis timescale of the primary planet (with respect to the star)

In short,

"Semi major axis timescale of a satellite (with respect to the star)" = "Semi major axis timescale of the satellite" + "Semi major axis timescale of the primary planet of the satellite"

Therefore, unless specified, the "Semi major axis timescale of the satellite" is the "Semi major axis timescale of the satellite (with respect to the primary planet)", which is the "Semi major axis of that satellite" divided by the "Escape velocity of the primary planet of the satellite".

"Semi major axis timescale of a satellite (with respect to the primary planet)" = "Semi major axis of the satellite" / "Escape velocity of the primary planet of the satellite"

But when the words "(with respect to the star)" is used to qualify the "Semi major axis timescale of a satellite", then the "Semi major axis timescale of the primary planet" of that satellite must be added to the "Semi major axis timescale of the satellite (with respect to the primary planet)" to get the complete amount of time that elapsed since the split gathering of the precursor of the stellar system that they belong to. For the precursor of the "Planetary system" (containing what would become the precursor of the primary planet and the precursor of its satellite(s)) once traveled a certain distance (which is about the semi major of the primary planet) before split-gathering into the precursor of the primary planet and the precursor of the satellite(s). Afterward, according to its position in the stack of fluid layers of the precursor of the satellites orbiting a planet, the precursor of a satellite traveled a certain distance before reaching its semi major axis. Therefore,

"Semi major axis timescale of a satellite (with respect to the star)" = "Semi major axis timescale of the primary planet" of that satellite" + "Semi major axis timescale of the satellite (with respect to the primary planet)"

In other stellar systems, the above formulas can be applied, but a special care needs to be taken to sync the data with the reference I set in the Solar System accordingly.

At this point, I will apply the definition and formula of the "Semi major axis timescale" to the planets, asteroids, and satellites in the Solar System. In fact, the "Semi major axis timescale of a celestial body (with respect to its primary body)" is the total amount of time that elapsed from the moment the precursor of that body started moving away from the precursor of its primary body until the moment that the precursor of the body reached about the position of the semi major axis, ready to be swirled to get its spherical or near spherical form. It was about this time that the precursor of some bodies that have irregular forms were ready to move around to get their shape.

As a reminder, on the scale of the Solar System, the Sun is the primary body, while the planets and asteroids are the secondary bodies of the Sun. Consequently, the "Semi major axis timescale of a planet or an asteroid" is the time that elapsed since the precursor of all the bodies orbiting the Sun escaped the precursor of the Sun until they reached the position of their semi major axis, ready to swirl and get their final shape, which for most large celestial bodies is spherical or near-spherical.

In contrast, on the scale of the planetary systems, the primary planets are the primary bodies, while the satellites are the secondary bodies. Subsequently, the "Semi major axis timescale of a satellite (with respect to its primary body)" is the time that elapsed since the precursor of the satellite escaped the precursor of the primary planet until the time that the precursor of the satellite reached the semi major axis of that satellite, ready to swirl and get its final shape. Here, the "Semi major axis timescale of a satellite (with respect to its primary body" is not with respect to the precursor of the Sun, but with respect to the precursor of the primary planet.

To summarize, the formula of the "Semi major axis timescale of a planet or an asteroid (with respect to the Sun) is:

"Semi major axis timescale of a planet or an asteroid" in the Solar System = "Semi major of the planet or of the asteroid" / "Escape velocity of the Sun"

"Semi major axis timescale of a satellite (with respect to the primary planet)" in the Solar System = "Semi major axis of the satellite" / "Escape velocity of the primary planet of the satellite"

"Semi major axis timescale of a satellite (with respect to the Sun)" = "Semi major of the primary planet of that satellite" / "Escape velocity of the Sun"

+ "Semi major axis of the satellite" / "Escape velocity of the primary planet of the satellite"

For now, in this chapter, I am interested in studying the "Semi major axis timescale of the planets, asteroids, and satellites (with respect to their primary body)". This means that the "Semi major axis timescale of the planets and asteroids" that I am dealing with in this chapter is with respect to the Sun, while that of the satellites is with respect to their primary planet. In the next chapter, I will calculate the "Semi major axis timescale of the satellites (with respect to the Sun)", and also elaborate on its significance.

35.2. Semi major axis timescale (with respect to the primary body) of celestial bodies

The "Semi major axis timescale (with respect to the primary body)" of the celestial bodies in the Solar System varies from 0.518 hours (equivalent of 31.07 minutes) to 2240577.356 hours (equivalent to 255.8 years) according to their types and their primary bodies. The highest value was recorded on C/1999 F1, a comet which semi major axis is reported to be 33,300 AU (a value I don't think it true). In contrast, the smallest value was found with Phobos (the innermost and largest Martian satellite). In other words, the fluid layer of the precursor of no celestial body in the Solar System split from the stack of fluid layers of the precursors of the secondary bodies in its system faster than the fluid layers of the precursor of Phobos.

About 30.95% (meaning 65 out of the 210) of the satellites known in the Solar System as of 2021 have a "Semi major axis timescale (with respect to the primary body)" smaller than the "Semi major axis timescale" of Mercury, the innermost body orbiting the Sun. This means that with respect to their primary bodies, the fluid layers of the precursor of 65 satellites in the Solar System took less time to split from the stack of precursors of satellites in their planetary system than the fluid layers of the precursor of Mercury took to split from the stack of fluid layers of the bodies orbiting the Sun. On top of these 65 satellites, the fluid layers of the precursors of 11 other satellites took less time to split than the precursor of the Earth-Moon system took to split from the remainder stack of fluid layers of the precursor of the bodies located beyond the semi major of the Earth. In other words, 76 satellites in the Solar System have a "Semi major axis timescale (with respect to the primary planet)" less than 67.29 hours, which is the "Semi major axis timescale" of the precursor of the Earth. In other words, after the precursor of the bodies orbiting the Sun escaped the precursor of the Sun, it took about 67.29 hours before the precursor of the Earth-Moon system split from the rest of the stack of fluid layers of the precursor of the bodies orbiting the Sun. By that time, the fluid layers of the precursors of many bodies orbiting the Sun had already split. These include the precursor of Mercury, Venus, and that of all the asteroids inward of the Earth.

Out of the 210 satellites known in the Solar System by 2021, only 2 (both of which are Neptunian satellites) have a "Semi major axis timescale (with respect to the primary body)" higher than that of Jupiter (350.18 hours):

- Psamathe: 552.01 hours (equivalent to 23 days) and
- Neso: 571.99 hours (equivalent to 23.83 days)

In other words, the fluid layers of the precursors of 99.5% of the satellites in the Solar System took less time to split after escaping the precursor of their primary planets than the fluid layers of the precursor of Jupiter (the largest planet in the Solar System) took before splitting from the remainder of the fluid layers of the precursor of the bodies orbiting the Sun since the latter escaped the precursor of the Sun. Finally, the fluid layers of the precursor of no satellite in the Solar System took more time to split from their peers than the fluid layers of the precursor of Saturn took (644.76 hours, i.e. 26.86 days) to split from the rest of the stack of fluid layers beneath of it.

Of the 62 satellites present in Zones 1, 2, and 3, 61 of them (meaning 98.39% of them) have a "Semi major axis timescale (with respect to the primary body)" inferior to the "Semi major axis timescale of Mercury (with respect to the Sun)". The only satellite which "Semi major axis timescale (with respect to the primary body)" (27.87 hours) is higher than the "Semi major axis timescale of Mercury (with respect to the Sun" is Iapetus (the outermost Saturnian satellite in Zone 3).

Of the 148 satellites in Zones 4 and 5, only 4 (all of which are the Plutonian satellites in Zone 4), meaning 97.3% of these satellites, have a "Semi major axis timescale (with respect to the primary body)" less than the "Semi major axis timescale of Mercury" (with respect to the Sun).

The "Semi major axis timescale (with respect to the Sun)" of the main belt asteroids varies between 145.87 hours (i.e. 6.08 days) to 234.82 hours (i.e. 9.78 days). This also means that 6.08 days after the beginning of the formation of the bodies in the Solar System, the fluid layers of the precursors of the main belt asteroids started splitting from one another and within 88.95 hours (i.e. 3.71 days) all of them split from one another.

The "Semi major axis timescale (with respect to the Sun)" for the planets in the Solar System ranges from 26.05 hours to 2656.51 hours (i.e. 110.69 days) (Fig. 202). The shortest duration of split was recorded with Mercury, while the longest duration was with Pluto. In other words, about 26.05 hours after the fluid layers of the bodies orbiting the Sun escaped the precursor of the Sun, the precursor of Mercury split. The fluid layers of the precursor of Pluto could have split about 110.69 days since the start of the journey of the fluid layers of the bodies orbiting the Sun away from the precursor of the Sun. Putting this another way, in about 110.69 days since the beginning of the formation of the bodies in the Solar System, the fluid layers of the precursor of the outermost planetary system (which is the Plutonian planetary system) split from the stack of the remaining fluid layers of the precursor of the bodies orbiting the Sun. The same value for the Earth is

67.29 hours (equivalent to 2.8 days), meaning that by the 3rd day since the formation of the Solar System, the precursor of the Earth-Moon system split from the rest of the stack of fluids of the precursor of the bodies orbiting the Sun.

Fig. 202: "Semi Major Axis Timescale" of the Planets in the Solar System

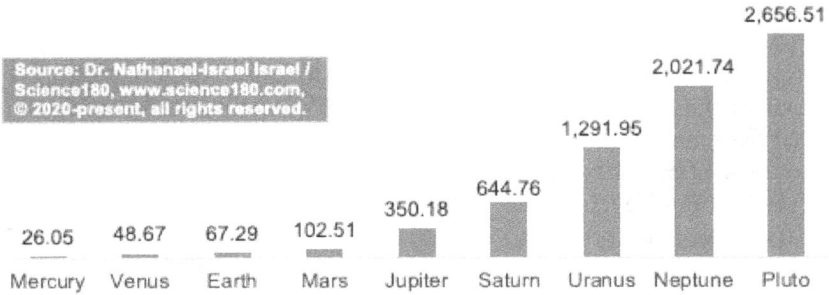

									2,656.51
								2,021.74	
							1,291.95		
					644.76				
				350.18					
26.05	48.67	67.29	102.51						
Mercury	Venus	Earth	Mars	Jupiter	Saturn	Uranus	Neptune	Pluto	

The "Semi major axis timescale (with respect to the Sun)" of the planets is positively correlated with their semi major axis ($r^2=1$) (Fig. 203).

Fig. 203: Regression between "Semi Major Axis" and "Semi Major Axis Timescale" of the Planets in the Solar System

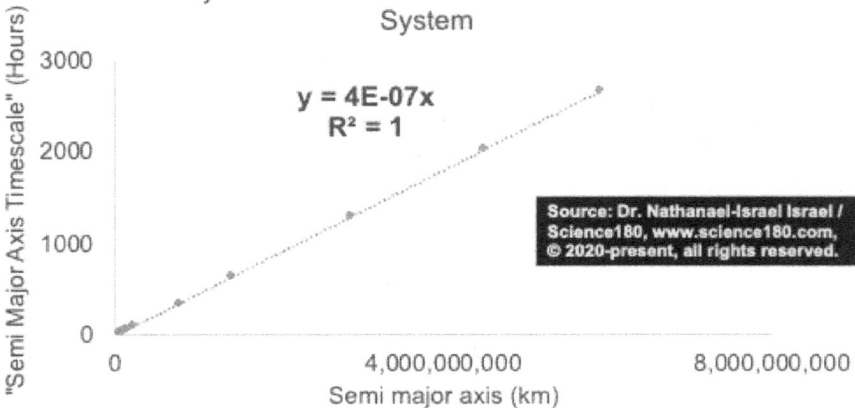

$$y = 4E\text{-}07x$$
$$R^2 = 1$$

"Semi Major Axis Timescale" (Hours)

Semi major axis (km)

Nathanael-Israel Israel: Acknowledged as Undisputable Specialist of All Questions at the Intersection of Science and Biblical Creation

35.3. "Semi major axis timescale of the satellites (with respect to their primary planet) according to their types and turbulence zones

As a reminder, the formula I designed for the "Semi major axis timescale (with respect to the primary planet) is:

Semi major axis timescale of the satellite (with respect to the primary planet) = Semi major axis of the satellite / Escape velocity of its primary planet

Again, the reference for this timescale is the start of the escape of the precursor of all the satellites in each planetary system. The shortest split of fluid layers of precursors of satellites (31.07 minutes) was recorded on Phobos (the innermost Martian satellite), whereas the longest (571.99 hours i.e. 23.83 days) was recorded on Neso (the outermost Neptunian satellite) (Fig. 204 and Fig. 205).

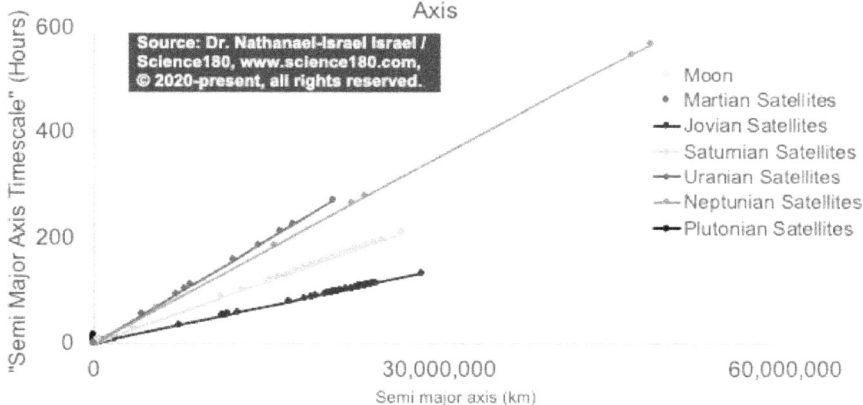

Fig. 204: "Semi Major Axis Timescale (with respect to the Primary Planet)" of the Satellites according to their Types and Semi Major Axis

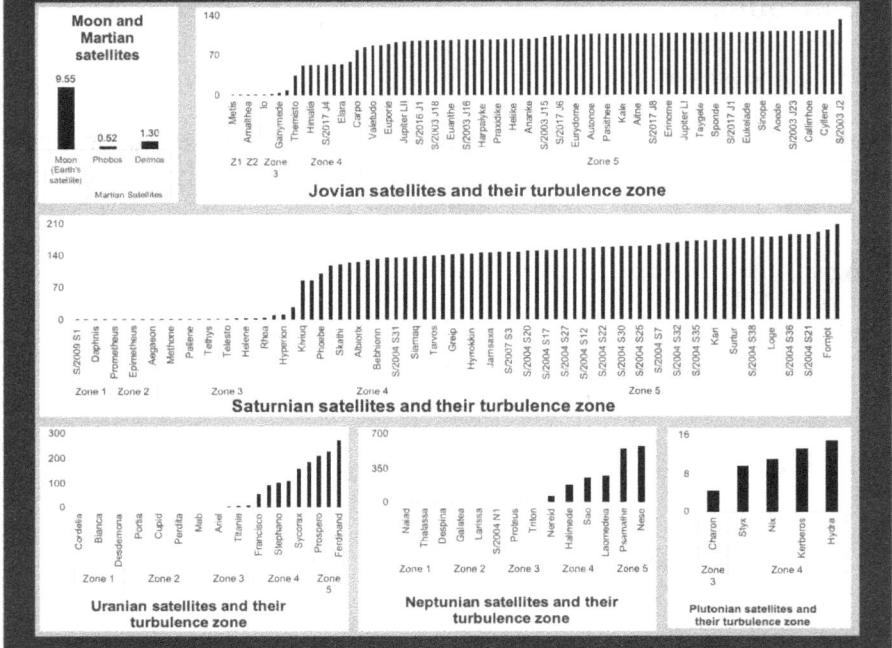

Fig. 205: "Semi Major Axis Timescale (with respect to the Primary Planet)" (Hours) of the Satellites according to their Types and Turbulence Zones

The date (with respect to the primary planet) of the split of the fluid layers of the precursor of the satellites is positively correlated with their semi major ($r^2=1$) (Fig. 206).

Fig. 207 illustrates the "Semi major axis timescale of the satellites (with respect to the primary planet)" according to the turbulence zones, types, and semi major axis. In general, the highest "Semi major axis timescale of the satellites (with respect to its primary planet)" recorded on satellites in Zones 1, 2, and 3 is 27.87 hours (obtained with Iapetus, the outermost Saturnian satellite in Zone 3). This means that with reference to each primary planet, the fluid layers of no satellite in Zones 1, 2, and 3 took more than 27.87 hours before splitting from the fluid layers of the precursor of their peers. As a reminder, about 9.55 hours after the fluid layers of the precursor of the Moon split from the fluid layers of the precursor of the Earth, they were about to start gathering together into the Moon known as of today.

Besides the 4 Plutonian satellites in Zone 4 which "Semi major axis timescale (with respect to Pluto)" ranges between 9.74 hours and 14.86 hours, no satellite in Zones 4 or 5 has a "Semi major axis timescale (with respect to their primary planet)" inferior to 35.05 hours (recorded on Themisto, the innermost Jovian satellite in Zone 4).

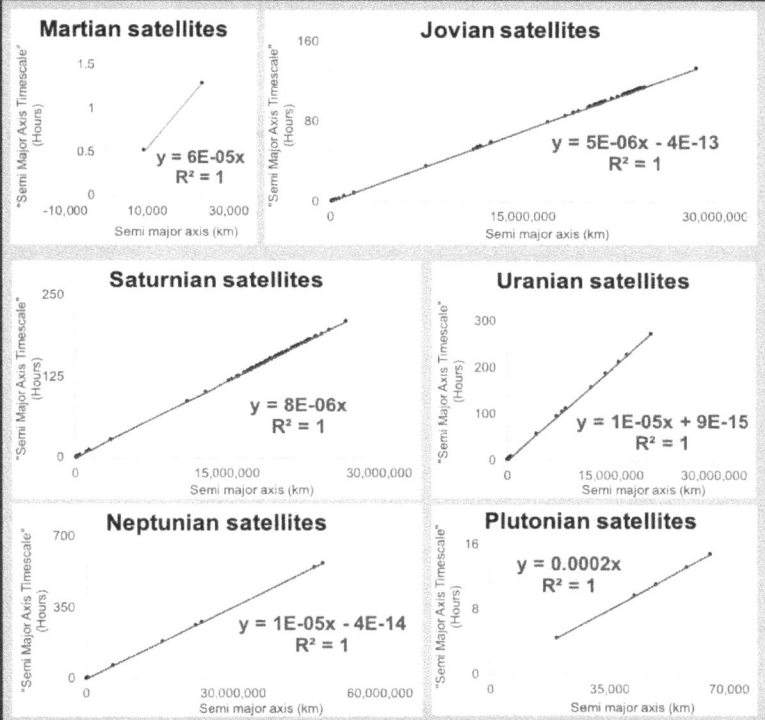

Fig. 206: Relationship between "Semi Major Axis" and "Semi Major Axis Timescale (with respect to the Primary Planet)" of the Satellites according to their Types

At this time, I would like to review some statistical parameters (minimum, maximum, and range) of the "Semi major axis timescale (with respect to the primary planet)" of the satellites.

Fig. 208 illustrates the minimum "Semi major axis timescale (with respect to the primary planet)" of the satellites. The fluid layers of the precursor of the satellites of the giant planets took less than an hour to start splitting from one another. "Semi major axis timescale (with respect to the primary planet)" of the satellites. The fluid layers of the precursors of the satellites which took longer before starting to split were those of the Saturnian satellites (0.915 hours, i.e. 54.93 minutes).

Fig. 207: "Semi Major Axis Timescale (with respect to the Primary Planet)" (Hours) of the Satellites according to their Types and Turbulence Zones

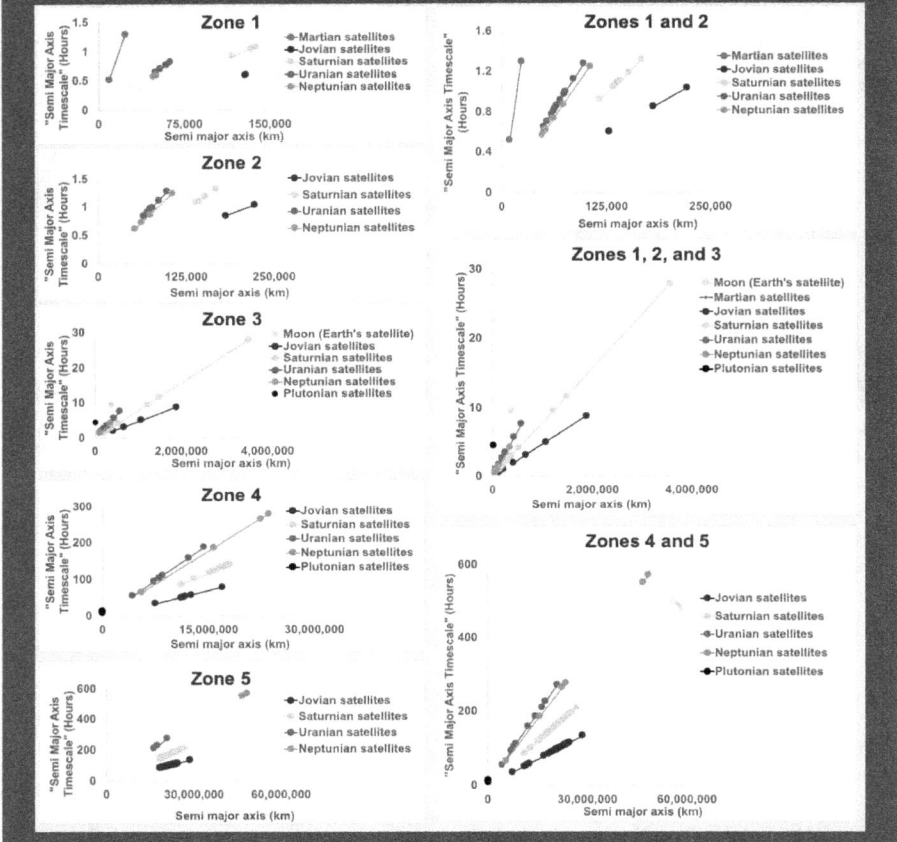

Fig. 209 is about the maximum time which elapsed from the moment the precursors of the satellites escaped the precursor of their primary planet until the moment all their fluid layers split according to their types and turbulence zones. For instance, while all of the fluid layers of the precursors of Martian satellites started splitting from one another about 1.296 hours from the time they escaped the precursor of Mars, it took 571.99 hours (i.e. 23.83 days) before all the fluid layers of the precursor of the Neptunian satellites were all split from one another, leaving the bottom Saturnian fluid layer ready to start gathering together. By 14.86 hours, all the fluid layers of the precursor of the Plutonian satellites were split and ready to gather themselves together into the Plutonian satellites. Despite the gigantic size of some Jovian satellites, about 133.38 hours (i.e. 5.56 days) after the precursor of the Jovian satellites escaped the precursor of Jupiter, all of their fluid layers were split from one another.

Fig. 208: Minimum "Semi Major Axis Timescale (with respect to the Primary Planet)" (Hours) of the Satellites according to their Types

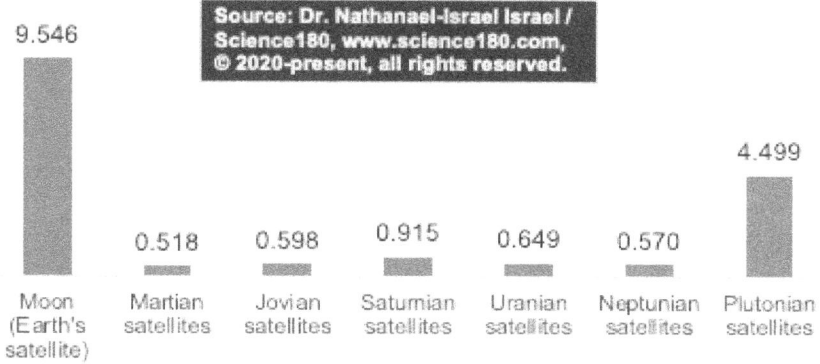

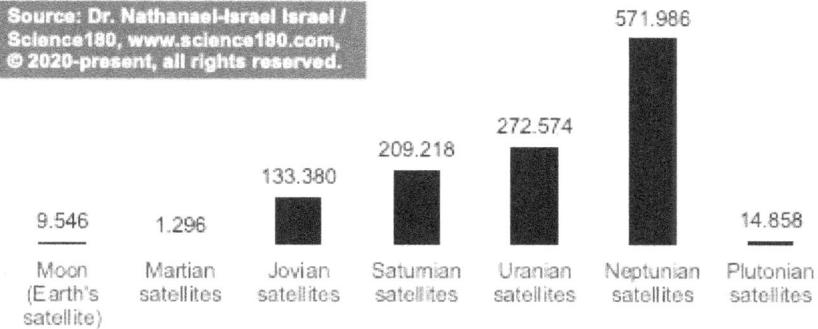

9.546						
			0.915			4.499
	0.518	0.598		0.649	0.570	
Moon (Earth's satellite)	Martian satellites	Jovian satellites	Saturnian satellites	Uranian satellites	Neptunian satellites	Plutonian satellites

Fig. 209: Maximum "Semi Major Axis Timescale (with respect to the Primary Planet)" (Hours) of the Satellites according to their Types

					571.986	
				272.574		
			209.218			
		133.380				
9.546	1.296					14.858
Moon (Earth's satellite)	Martian satellites	Jovian satellites	Saturnian satellites	Uranian satellites	Neptunian satellites	Plutonian satellites

About 1.31 hours (i.e. 1 hour 18 minutes and 38.3 seconds) from the time the precursor of the satellites escaped the precursors of their primary planets, all the fluid layers in Zones 1 and 2 were split. By about 27.87 hours (Fig. 210) since the precursor of the satellites escaped the precursor of their primary bodies, all the fluid layers of the satellites in Zone 3 split from one another, each according to its position in the stack of fluid in which they "branched out" from.

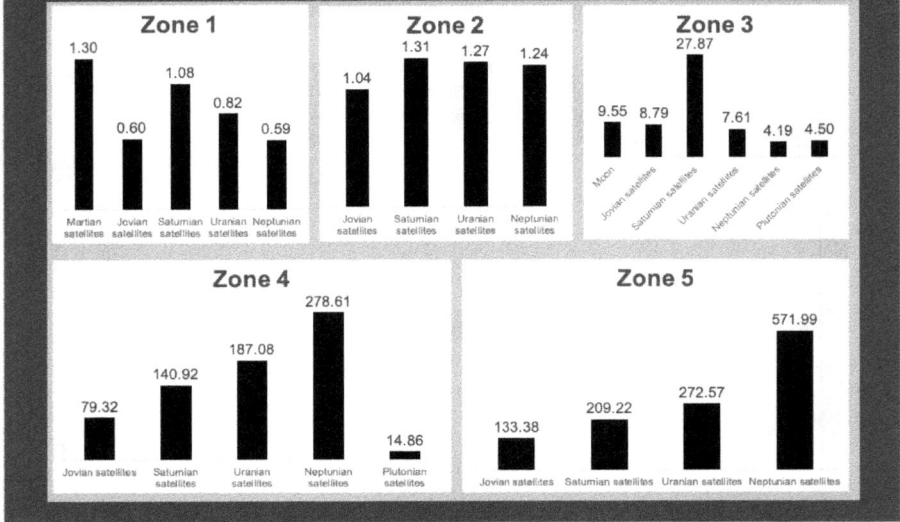

Fig. 210: Maximum "Semi Major Axis Timescale (with respect to the Primary Planet)" (Hours) of the Satellites according to their Turbulence Zones and Types

The range of the "Semi major axis timescale (with respect to the primary planet)" for the satellites gives a glimpse at the time that elapsed before the first split and the last split according to the types of satellites. For example, between the split of the top and the bottom fluid layer of the precursors of Jovian satellites, about 132.78 hours had elapsed while for Plutonian satellites, about 10.36 hours had elapsed (Fig. 211 and Fig. 212)).

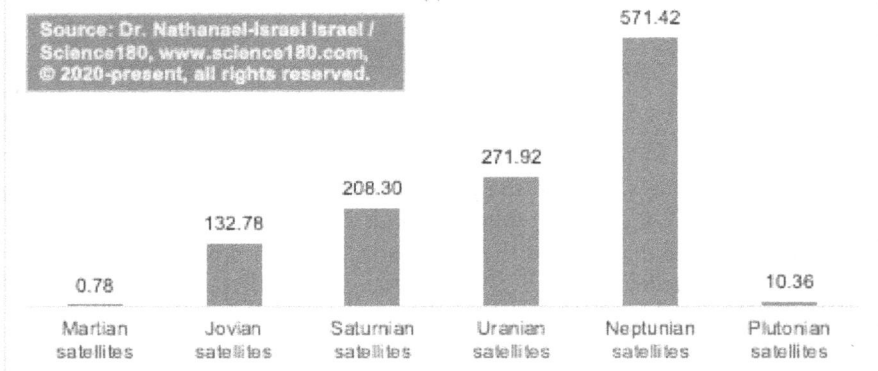

Fig. 211: Range of the "Semi Major Axis Timescale (with respect to the Primary Planet)" (Hours) of the Satellites according to their Types

Nathanael-Israel Israel: Acknowledged as Undisputable Specialist of All Questions at the Intersection of Science and Biblical Creation

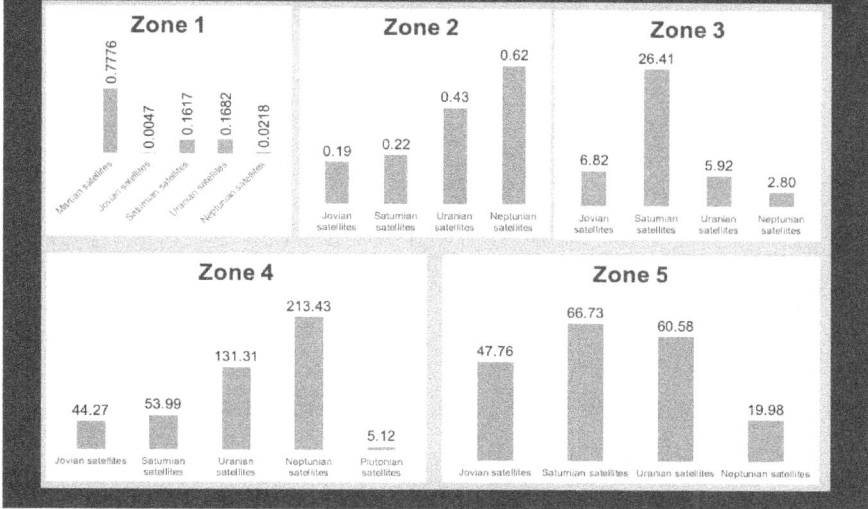

Fig. 212: Range of the "Semi Major Axis Timescale (with respect to the Primary Planet)" (Hours) of the Satellites according to their Turbulence Zones and Types

35.4. Take home message

The "Semi major axis timescale of a body (with respect to its primary body)" is a concept I invented to express the total amount of time that elapsed since the precursor of the secondary bodies in the system of bodies started moving away from the precursor of their primary body. For instance, about 26.05 hours after the fluid layers of the bodies orbiting the Sun escaped the precursor of the Sun, the precursor of Mercury split. The "Semi major axis timescale" of the Earth is 67.29 hours (equivalent to 2.8 days), meaning that by the 3rd day since the formation of the Solar System, the precursor of the Earth-Moon system split from the rest of the stack of fluids of the precursor of the bodies orbiting the Sun.

CHAPTER 36

CAN THIS FORMULA END ALL DOUBTS OR STRUGGLES ABOUT THE EXACT DATE THAT CELESTIAL BODIES STARTED GATHERING THEIR CONTENTS?

36.1. Formula of the "Circumferential gathering start date"

In the previous chapter, I explained the amount of time (with respect to the primary bodies) the fluid layers of the precursors of the celestial bodies took before splitting from one another according to their position in the stack of fluid layers. Around the time, the split of a fluid layer occurred, that fluid layer was gathered together according to its size. I did not mean that nothing inside the fluid layer of a precursor was not being collected or has not been collected before that fluid layer split. However, to allow people to have a better understanding of the timeline, I considered the fluid layers of the celestial bodies to have gathered together into their corresponding celestial bodies after splitting from other fluid layers. The "Circumferential gathering start date" of a body is the amount of time that separates the beginning of the formation of the Solar System from the moment the fluid layers of that body started gathering together into the current form or shape of its daughter bodies.

Although the Sun has no semi major axis with reference to no celestial body in the Solar System, the precursor of the Sun did not start gathering itself immediately at the beginning of the formation of the Solar System. In fact, from the beginning of the formation of the bodies in the Solar System until the time the precursor of the Sun started gathering into the Sun, a certain amount of time had passed during which some events took place, including the clearing or escape of the precursor of all the bodies orbiting the Sun. In other words, the precursor

Nathanael Nathanael-Israel Israel: Author of "Science180 Accurate Scientific Proof of God"

of the bodies orbiting the Sun had to be removed from the surface of the precursor of the Sun before the later could start gathering together. Another way to express this is that the split-gathering of the precursor of the Solar System yielded the precursor of the Sun and caused the precursor of all the bodies orbiting the Sun to escape. Therefore, the time that elapsed before the precursor of the Sun could fully start gathering itself together is about the time at which the escape of the fluid layers of the precursor of the bodies the Sun was completed. As I explained in previous chapters, I established that the "Circumferential gathering start date" of the precursor of the Sun is not higher than the date that the precursor of Mercury (the innermost body orbiting the Sun) split from the rest of the precursor of the bodies orbiting the Sun. Although the precursor of the Solar System had to split into the precursor of the Sun and the precursor of the bodies orbiting the Sun before the precursor of Mercury could later split from the stack of the fluid layers of the bodies orbiting the Sun, it is not possible to know exactly the time it took before such a split of the precursor of the Solar System occurred. But I considered it to have occurred before the split-gathering together of the precursor of Mercury.

For the asteroids and planets, the "Circumferential gathering start date" is the same thing as their "Semi major axis timescale (with respect to the Sun)". According to the formula I developed and explained in the previous chapter:

"Circumferential gathering start date" of a planet in the Solar System = Semi major axis of the planet / Escape velocity of the Sun

"Circumferential gathering start date" of an asteroid in the Solar System = Semi major axis of the asteroid / Escape velocity of the Sun

For the satellites, the formula is a little different, for it considers the time elapsed before the precursor of their planetary system split from the precursor of the bodies orbiting the Sun and the time elapsed from the moment the precursor of the satellites escaped the precursor of their primary planet until they reached the position at which they split from the stack of the fluid layers of the precursor of the satellites in their planetary system. Because the data I analyzed suggested that the lifespan of the precursor of the planetary systems varied between 31.07 minutes and 9.55 hours, the date of the split of the planetary systems into the precursor of a planet and the precursor of all their satellites is about the "Semi major axis timescale of their primary planet". As a reminder, as I demonstrated in Chapter 33, the lifespan of the precursor of a planetary system is not higher than the "Semi major axis timescale (with respect to the primary planet)" of the innermost satellite in that planetary system, which is equal to:

Lifespan of the precursor of a planetary system ≤ "Semi major axis of the innermost satellite in the planetary system / Escape velocity of the primary planet"

In fact, the "Semi major axis timescale (with respect to the primary planet)" of the innermost satellites ranges from 31.07 minutes (recorded on the innermost Martian satellite with respect to Mars) to 9.55 hours (recorded on the Moon (Earth's satellite). Therefore, the date that the precursor of a satellite started gathering together is almost the sum of:

- the "Semi major axis timescale of the primary planet" of that satellite and
- the "Semi major axis timescale (with respect to the primary planet)" of the satellite:

"Circumferential gathering start date" of a satellite = "Semi major axis of the primary planet of that satellite / Escape velocity of the Sun" + "Semi major axis of the satellite / Escape velocity of the primary planet of that satellite"

In other words,

Circumferential gathering start date of a satellite = Semi major axis timescale (with respect to the Sun) of the satellite = Semi major axis timescale of the planet + Semi major axis timescale of the satellite (with respect to its primary planet)

Although sometimes I considered and wrote the "fluid layer" of a celestial body as singular, it is important to note that the main fluid layer of a precursor as a whole consisted of many smaller fluid layers, but which altogether is the fluid layer of the body. In other words, even if I write "fluid layer" of a celestial body as if I am talking about a single layer, please keep in mind that it was actually a stack of many fluid layers, which were "bound together" by a kind of "destiny" that led to their gathering into a single celestial body. The presence of the many layers in the precursors also explains why celestial bodies have many layers or strata. For instance, here on Earth the crust of the soil has many layers and those strata can be seen while viewing cliffs and canyons or some bluffs along the highways in which those who constructed the road, blasted the rocks (Fig. 213).

Fig. 213: Bluffs showing rock layers along a highway

This can also explain the stratification of the atmosphere of some bodies (more details on this are in my book *"Turbulent Origin of Chemical Particles"*). However, sometimes, all the fluid layers in a body were not collected into the celestial bodies, but some were left out, leading to the formation of smaller chunks of matter such as rocks and chemical compounds of various sizes as seen in rings embedded in or encircling some celestial bodies (e.g. rings of satellites). In the chapter on rings, I elaborated on how the processes of the gathering together of the fluid layers led to the formation of rings and trails of dusts along the orbit or trajectory of some planets, asteroids, and their satellites.

Before I provide any more data in this chapter, I would like to emphasize that the gathering together of the fluid layers of the celestial bodies did not happen the same way, nor with the same intensity, or force, hence it led to the formation of various bodies having different compositions. In the chapter on "Constellations, galaxies, globular clusters, and stars in the universe", where I also talked about the generic process of the gathering of the precursor of various bodies, I explained why the gathering of some precursors led to the formation of galaxies, stars, planets (some of which are solid, others gas and some made up of ice), asteroids, and satellites according to their size and position and many other factors I explored throughout this book. In some cases, the gathering together led to the formation of spirals, in other cases, the fluid layers were so tightly gathered that no spiral could be seen in their daughter bodies. Hence some gigantic systems of celestial bodies like galaxies appeared as spirals, while other bodies like terrestrial planets are very compacted and dense, while other planets like the giant planets in the Solar System are like gas or ice. In my book *"Turbulent Origin of Chemical Particles"*, I also explained why and how, on the microscopic scale, the same thing happened and led to the formation of chemical particles (subatomic particles, atoms, molecules, minerals, rocks, and various chemical compounds), having various densities and other proprieties that still puzzle some scientists who think they know how they were formed, but who fooled themselves by believing in myths and pseudoscience, while rejecting the truth. However, in this chapter, I will focus on the date that the gathering of the fluid layers of the precursors of the Sun and the other bodies in the Solar System occurred.

36.2. Circumferential gathering start date of the celestial bodies

The "Circumferential gathering start date" of the celestial bodies varies between 26.046 hours all the way to 93,357.39 days (which is about 255.77 years. The smallest value was recorded with the Sun and Mercury while the highest value was recorded with C/1999 F1, a comet allegedly located at 33,300 AU from the Sun. In other words, even years after the formation of the Sun, the precursor of some bodies could have still been going through gathering. It is also possible that those bodies were formed earlier, but were pushed away from the Sun because of the expansion of the Solar System after its formation. Furthermore, because of the intense cold in the outer space, the fluid layers of some precursors of the outer celestial bodies could have been frozen, for it may be impossible for a tinny fluid layer to keep traveling in a very cold environment without being frozen. This may explain why many asteroids are known to be rich in icy materials. In other words, due to their distance from the Sun, the fluid layers of some remote celestial bodies may not have been gathered together as it would have been the case in a warmer environment, but could have been frozen while still on their way to their orbit. This reminds me of how, when people are moving through a blizzard, the snow can be so thick that, as it sticks to people, they can begin to freeze; and if left outside too long, these people can freeze to death and their body will stiffen.

The "Circumferential gathering start date" of the planets ranges from 26.05 hours to 2,656.51 hours (i.e. 110.69 days). Pluto recorded the highest value while Mercury has the smallest one. Fig. 214 presents the data. The "Circumferential gathering start date" of the precursor of the Earth-Moon system is 2.8 days.

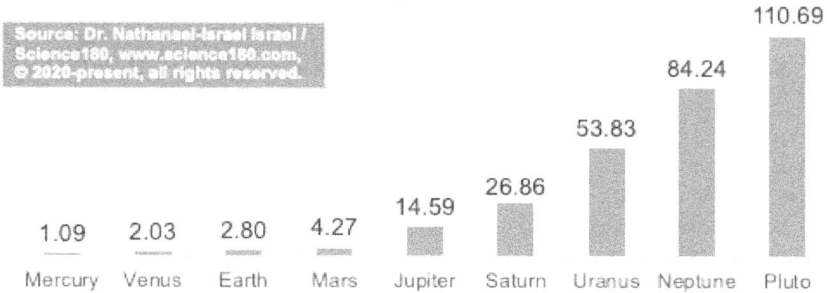

Fig. 214: "Circumferential Gathering Start Date"
(Days) of the Precursors of the Planets in the Solar
System

36.3. Circumferential gathering start date of satellites

As I already explained, the "Circumferential gathering start date" of the satellites is the time elapsed from the beginning of the formation of the Solar System until their fluid layers started gathering together into satellites. The "Circumferential gathering start date" of the satellites varies between 3.2 and 111.3 days since the beginning of the formation of the Solar System. The earliest date was recorded with the Moon (Earth's satellite), meaning that early on the 4th day of the formation of the Solar System meaning that by the 76.83th hour since the beginning, the fluid layers of the precursor of the Moon started gathering into the Moon as it is known today. However, the last satellite in the Solar System to gather its fluid layers was Hydra (the outermost Plutonian satellite). Fig. 215 presents the data according to the types and turbulence zones of the satellites according to their types and semi major axis.

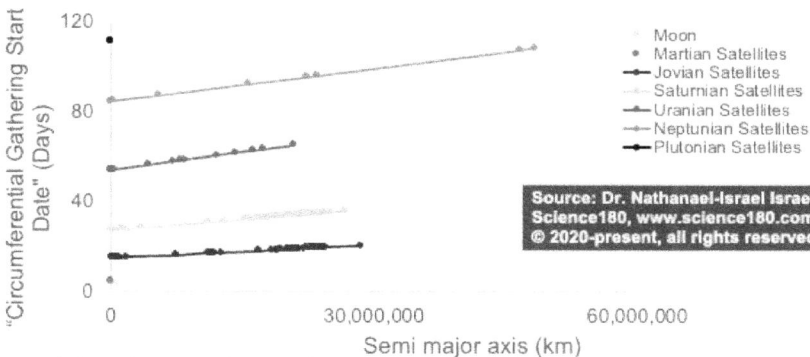

Fig. 215: "Circumferential Gathering Start Date" of the precursors of Satellites according to their Types and "Semi Major Axis"

Fig. 216 presents the "Circumferential gathering start date" of the satellites according to their types and turbulence zones. After the Moon started gathering together, it took 26.2 hours before the precursor of the innermost Martian satellite started gathering. After the outermost Martian satellites started gathering, it took another 10.29 days before the innermost Jovian satellite started gathering. Likewise, it was 6.76 days after the start of the gathering of the outermost Jovian satellite that the innermost Saturnian satellites started going the same. From the start of the gathering of the outermost Saturnian satellite to the start of the gathering of the innermost Uranian satellite, 18.28 days have passed. Similarly, from the time the outermost Uranian satellite started gathering to the time the innermost Neptunian started doing the same, it took 19.1 days. However, the innermost Plutonian satellite started gathering just 2.8 days after the start of the gathering of the outermost Neptunian satellite.

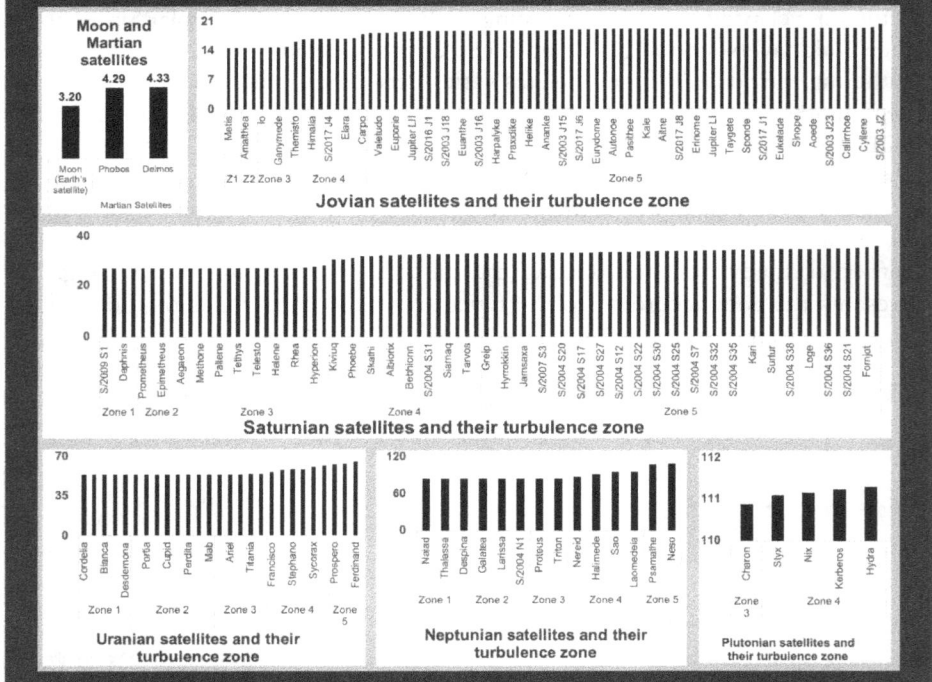

Fig. 216: "Circumferential Gathering Start Date" (Days) of the Precursors of the Satellites according to their Types and Turbulence Zones

The time that elapsed between the start of the gathering of the outermost satellite in a planetary system and the innermost satellite of the next planetary system is partially due to the time that the precursor of the planetary systems took to split from one another before the precursors of their satellites could split each

Nathanael Nathanael-Israel Israel: Author of "Science180 Accurate Scientific Proof of God"

at their turn according to their position in the stack of fluid layers.

The "Circumferential gathering start date" of the satellites is positively correlated with their semi major axis ($r^2=1$) (Fig. 217).

Fig. 217: Relationship between "Circumferential Gathering Start Date" and "Semi Major Axis" of the Satellites according to their Types

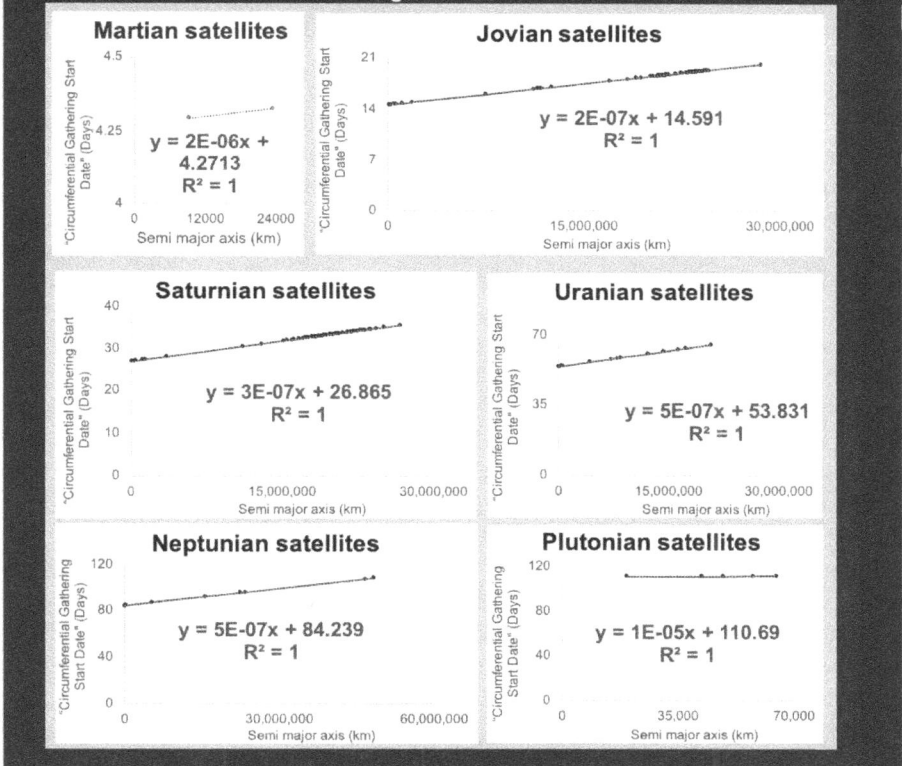

Martian satellites
$y = 2E\text{-}06x + 4.2713$
$R^2 = 1$

Jovian satellites
$y = 2E\text{-}07x + 14.591$
$R^2 = 1$

Saturnian satellites
$y = 3E\text{-}07x + 26.865$
$R^2 = 1$

Uranian satellites
$y = 5E\text{-}07x + 53.831$
$R^2 = 1$

Neptunian satellites
$y = 5E\text{-}07x + 84.239$
$R^2 = 1$

Plutonian satellites
$y = 1E\text{-}05x + 110.69$
$R^2 = 1$

36.4. Why did the fluid layers of the precursors split in the first place and why some layers birthed large bodies?

A question some people may still be asking could be, why did the layers in the fluid flow start splitting in the first place? As I have already explained throughout this book, when the precursors of the secondary bodies escaped the precursor of their primary body at about the escape velocity of the primary body, as a whole, they "flowed" with the same speed. During their movement, fluid structures started forming inside of them and these structures moved at different speeds. Quickly, the bulk of the fluid layers stratified, and because of their inability to move altogether as a solid could do, the fluid structures being formed inside some fluid layers moved faster than others. Those on top moved faster and those at the bottom which moved more slowly. The difference in their speed contributed to causing the top layer to split from the rest until the bottom layer is left alone and ready to gather together its fluids as those on top if it was done by then. The variation of the speed of the structures in the fluid layers also contributed to amalgamating them and causing some of them to collect into bigger and bigger structures. For instance, when runners are running a race and they are all packed together and everybody is moving at the same speed (no matter if it is a fast speed or a slow speed) traffic jams occur less or not all. But when some people start moving much faster or slower than others, traffic jams occur as some try to pass slower runners who will not give them the way. Similarly, when cars are moving at the same speed or a near speed, no matter if it is a fast or slow speed, they do not cause a traffic jam as when some which are moving much faster than others they are trying to pass. In the same manner, fluids which are on top (meaning moving fast) and those at the bottom (meaning moving slow) of a fluid flow do not usually cause as much of a traffic jam as the fluids which are between them and around which consists of some moving very fast and others very slow. Using the classification I did for fluid layers according to their turbulence zones, it is in Zone 3 that the most traffic jams occurred, for in that zone, the variation of the speed of the fluid structures in the layers is so big that a kind of traffic jam occur. The "concentration" or gathering of the structures in the fluid layers in Zone 3 contributed to the formation of some larger fluid structures in that zone, but not to an increase of their speed. Hence, in Zone 3, larger bodies can be found but their speed is not higher than that of the bodies located in Zones 1 and 2. But compared with the bodies in Zones 4 and 5, the speed in Zone 3 is higher. In short, the speed of the fluid layers is not controlled by the concentration or gathering of the fluid layers, but by other factors related to the way the fluids are launched into motion and also to their proprieties according to their environment. As they were being split-gathered, some precursors could have looked like tongues of fiery materials (sometimes just like of fiery lava flying out of a volcano), while others could have been covered with darkness.

36.5. Take home message

While the asteroids and planets started gathering together as soon as the fluid layers of their precursors split from the remainder of the fluid layers of the bodies orbiting the Sun, for the precursor of the satellites, they had to travel a distance close to the semi major of the satellites born from them before they could start gathering together. In other words, the "Circumferential gathering start date" of the satellites considered:

- the time elapsed before the precursor of their planetary system split from

Nathanael Nathanael-Israel Israel: Author of "Science180 Accurate Scientific Proof of God"

the precursor of the bodies orbiting the Sun, and

- the time elapsed from the moment the precursor of the satellites escaped the precursor of their primary bodies until they reached the position at which they split from the stack of the fluid layers of the precursor of the satellites in their planetary system.

I established that about 26.046 hours after the beginning of the formation of the Solar System, the precursor of the Sun started gathering itself together. The precursor of the Earth-Moon system started gathering its bodies together 2.8 days since the beginning. The Moon was the satellite which gathering together occurred first before that of any other satellite. In other words, the gathering together of the Moon occurred on the 4th day since the beginning. The last satellite in the Solar System to be gathered together was Hydra (the outermost Plutonian satellite). I found that the time which elapsed between the start of the gathering of the outermost satellite in a planetary system and the gathering of the innermost satellite of the next planetary system is partially due to the time that the precursor of the planetary systems took to split from one another before the precursor of their satellites could split, each at its turn according to their position in the stack of fluid layers. I also explained why the fluid layers of the precursors split in the first place and how the size of the fluid layers which split defined the size of the celestial bodies born from them.

'Science180 Academy' Success Strategy:
SCIENCE180 MODELS OF THE ORIGIN OF THE UNIVERSE AND ITS CONTENT

Science180 Models consist of all the theories elaborated by Nathanael-Israel Israel regarding his ground breaking discovery on the origin of the universe and its content including all forms of life and chemical particles. These theories are detailed in various books written by Dr. Nathanael-Israel Israel encompass the following:

1. *SCIENCE180 MODEL OF COSMOLOGY*, also called Science180 Cosmology, Science180 Model of Cosmology, Science180 Cosmological Model, a scientific theory that explains Science180 to the scientists. Discover the details of this model in Nathanael-Israel Israel's book titled *"Turbulent Origin of the Universe"*. In that book, you will also unearth the new physics that will revolutionize science forever and land you into a zone of original ideas that improve lives nonstop regardless of your expertise.

2. *SCIENCE180 CREATIONISM*, also called Science180 Model of the Creation of the Universe and Life by God, a scientific theory that presents the origin of the universe in a biblical language. If you want to learn more about how to scientifically prove the Biblical account of the creation of the universe and the existence of God in a way that makes the head of God deniers to spin faster than a DJ's turntable, then get Nathanael-Israel Israel's book titled *"Reconciling Science and Creation Accurately"*.

3. *SCIENCE180 MODEL OF THE ORIGIN OF CHEMICAL PARTICLES*, a scientific theory that explains the origin of chemical particles with the perspective of Science180 Turbulence. If you want to professionally learn how to transform the true knowledge of the origin of chemical particles into insights that significantly add value to your life in less time, successfully establish you as a symbol of freedom, power, creativity, and originality in your field of expertise, get Nathanael-Israel Israel's book *"Turbulent Origin of Chemical Particles"*, THE ultimate how-to guide for great people wanting to correctly decode the origin of the chemicals and positively transform their lives. Get this celebrated book today. Don't wait!

4. *SCIENCE180 MODEL FOR THE GENERAL PUBLIC* (which explains the origin of the universe and life to the general public in a language that laypeople can understand). Find out more in Nathanael-Israel Israel's book called *"From Science to Bible's Conclusions"*, a scientifically verifiable, bestselling book to finally get the accurate, jaw-dropping answer that has been rationally shaking believers, skeptics, and freethinkers. Get this very popular book today.

5. *SCIENCE180 MODEL OF LIFE-ORIGIN*, or Science180 Model of the Origin of Life, a scientific theory that explains the origin of all forms of life using turbulence. To unlock the step-by-step pathway to decode the origin of life and get the power, freedom, and boldness to detect, correct, and remove all misinformation, ambiguity, and misleading claims and theories surrounding the life-origin and take advantage of the opportunities that an accurate understanding of the life-origin creates, get Nathanael-Israel Israel's book titled *"Turbulent Origin of Life"*.

6. *SCIENCE180 MODEL FOR CHILDREN*, a children's version of the theory of the origin of the universe and life in a language that 7-12 years old children can properly understand. To know the proven formula that helps children to easily answer their huge universe-origin and life-origin questions with confidence, humor, and joy, get *"How Baby Universe Was Born"*, the pragmatic book that has been causing children to belly laugh and thank those who offered it to them.

7. *SCIENCE180 MODEL OF PSEUDEPIGRAPHA*, a deep explanation of the secrets of the origin of the universe and life revealed a long time ago, but hidden from the general public. To discover how the only one ancient blueprint has the reliable power to help you to accurately decrypt the spiritual origin and history of everything in the universe, get Nathanael-Israel Israel's book called *"Origin of the Spiritual World"*. In it, you will discover deep rejected secrets that have prevented humankind from unearthing the beginning of the universe and know how to properly use the lost and rejected scriptures to articulate the process by which the universe was formed, so you can use that insight to improve your understanding of the Bible, innovate in your domain of interest, and improve your life perpetually.

8. *SCIENCE180 MODEL OF THE PROOF OF THE EXISTENCE OF GOD*, a theory that ties together most of Nathanael-Israel Israel's discoveries that scientifically prove the existence of God. With Nathanael-Israel Israel's book *"Science180 Accurate Scientific Proof of God"*, you will surely know the only way to scientifically know if God exist, and if so, which of the thousands of beings worshipped across the globe is the true God. In that book, you will also discover the errors in the scientific and religious theories (about the origin of the universe, life, and chemicals) that are putting you at a high risk you will never recover from if you don't quickly and confidently learn how to rationally take control over threats lurking at the edge of your efforts to understand the universe and life today.

9. SCIENCE180 THEORY OF EVERYTHING, (also called the theory of all theories), ties together everything in the universe into a single theory. Checkout Science180.com to learn more about the incoming book that covers this extremely important topic. Go to www.Science180Academy.com to register today.

CHAPTER 37

WOULDN'T IT BE GREAT IF WE CAN KNOW WITHOUT A DOUBT THE TIME IT TOOK FOR THE PRECURSORS OF CELESTIAL BODIES TO PASS FROM "LINEAR" FLUID LAYERS TO "SPHERICAL" CELESTIAL BODIES? IF SO, TAKE THE CIRCUMFERENCE TIMESCALE OF CELESTIAL BODIES SERIOUSLY

37.1. Definition and formula of the Circumference timescale

Also called "eddy turn-around time", or "eddy turnover time", or "intrinsic timescale" in some literatures, timescale is a key variable that can give a clue into the time it could have taken for the celestial bodies to be gathered together. Two formulas are generally used to calculate timescale:

- Timescale = 3.14 * diameter / orbital speed
- Timescale = diameter / orbital speed

In the above formula 3.14 is a short expression π (pi). The first timescale is what I termed Circumference timescale and the second one is the timescale based on diameter (or timescale based on the characteristic length). In most literatures, the characteristic length of the bodies is used instead of their diameter, meaning that in that case, the diameter of the bodies is perceived as the characteristic length.

The timescale is considered as the time that a particle (assumed to be spherical) takes to cover its circumference (3.14 * diameter) as it moves at its orbital velocity (Cushman-Roisin, 2015). Some people consider it as a timescale

Nathanael-Israel Israel: Member of the American Society of Agricultural and Biological Engineers

of energy transfer. Others consider timescale as the typical timescale for an eddy to undergo a significant distortion (Frisch, 1995). Because distortion is believed to be a mechanism for energy transfer, timescale is also viewed as the typical time for the transfer of energy from a scale similar to the length of the eddy to even smaller scales.

Considering the impact of spiraling on the formation of the bodies in the Solar System, I do not personally think that the "Timescale based on diameter" is a good indicator of the time it took for the bodies to be formed. In contrast, the Circumference timescale seems to me to be more relevant to address the duration of the gathering together of the bodies. For the fluid layers of the precursors of celestial bodies had to pass from being "linear" to curling or swirling themselves to yield spherical bodies. Considering the above formula and the processes I explained in this book, I established that, once the fluid layers of the precursors of the celestial bodies reached their orbit, their length were about the circumference of the bodies they would birth, and these fluid layers swirled at about the orbital speed of their daughter bodies to form the later. In other words, the Circumference timescale of the celestial bodies is about the time it took for the fluid layers of the precursor of the bodies to pass from being "linear" fluid layers to being well-shaped (e.g. spherical) bodies know today. Put another way, the Circumference timescale of the bodies is about the duration of time needed to roll or curl the fluid layers of the precursors of the bodies into the spherical or near-spherical bodies as of today. Below I presented the Circumference timescale of the celestial bodies.

37.2. Circumference timescale of the bodies in the Solar System

Studied for 408 bodies in the Solar System, the Circumference timescale is affected by the types of bodies in the Solar System (p<0.000). In the Solar System, it varies between 0.0005 seconds and 2.61 days. The smallest Circumference timescale was recorded with 2003 SQ222 (an Apollo asteroid). In contrast, the highest timescale was with the Sun. About 23.77% of the bodies that I studied in the Solar System have a <u>circumference timescale </u>less than 1 second. Of these bodies which timescale is less than 1 second, about 95% are asteroids. Of the 408 bodies which timescale I studied, 98.28% have a timescale less than 1 hour. Only 7 bodies have a <u>circumference timescale </u>higher than 1 hour and they include 2 satellites (the Moon and Charon), the 4 giant planets, and the Sun:

2.96 hours: Moon (Earth's only satellite)
4.74 hours: Charon (the biggest Plutonian satellite)
6.56 hours: Uranus (third largest planet in the Solar System)
7.96 hours: Neptune (fourth largest planet in the Solar System)
9.55 hours: Jupiter (largest planet in the Solar System)
10.86 hours: Saturn (second largest planet in the Solar System)
62.58 hours (i.e. 2.61 days): Sun (a star)

The data above suggested that besides the Sun, all the bodies in the Solar

System could have gotten their shape or gathered together in less than 12 hours after their precursors split from the stack of fluid layers of the precursors that they "descended" from. Although their shape could have been formed quickly, their internal composition could have taken a little longer. It could also have taken some additional time to position the bodies on their orbit. In the next chapter, I handled that aspect.

The asteroid with the highest timescale (39.34 minutes) is Sedna, a Trans-Neptunian object allegedly located at about 506.2 AU from the Sun. For the main belt asteroids, the Circumference timescale varies between 0.63 seconds and 2.74 minutes (Fig. 218). The highest value was observed with Ceres, the biggest asteroid in the main belt.

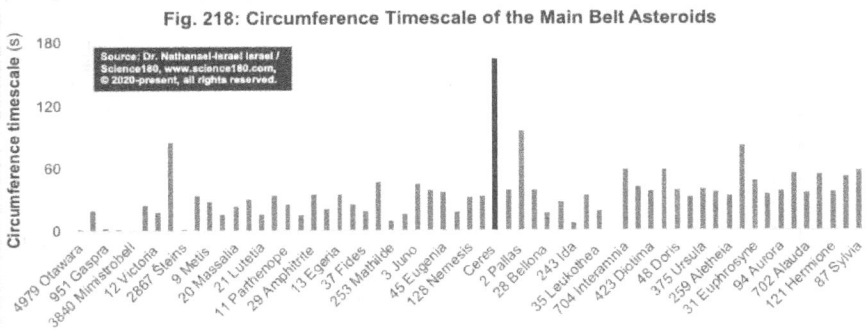

Fig. 218: Circumference Timescale of the Main Belt Asteroids

The Circumference timescale of the planets varies between 5.39 minutes and 10.86 hours (Fig. 219). The highest value was recorded with Saturn, while the smallest value was found with Mercury:

- Mercury: 5.39 minutes
- Venus: 18.09 minutes
- Earth: 22.42 minutes
- Mars: 14.77 minutes
- Jupiter: 9.55 hours
- Saturn: 10.86 hours
- Uranus: 6.56 hours
- Neptune: 7.96 hours
- Pluto: 26.6 minutes

Nathanael-Israel Israel: Member of the American Society of Agricultural and Biological Engineers

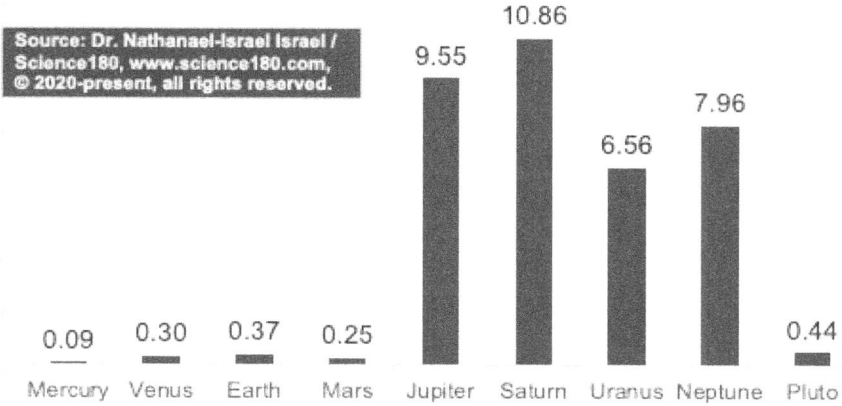

Fig. 219: Circumference Timescale (hours) of the Planets in the Solar System

Investigated for 178 satellites in the Solar System, the Circumference timescale of the satellites depends on their types and turbulence zones (p<0.000) and varies between 0.1 second and 4.74 hours (Fig. 220). The smallest Circumference timescale of the satellites was recorded with S/2009 S1 (the innermost Saturnian satellite), while the highest was with Charon (the biggest Plutonian satellite). The second highest timescale recorded among the satellites was with the Moon (i.e. 2.96 hours). Besides the Moon and Charon, no satellite has a Circumference timescale higher than 1 hour. Furthermore, 74.72% of the 178 satellites which timescale I studied have a timescale of less than 1 minute, meaning that the gathering of most of the satellites could have been completed extremely fast.

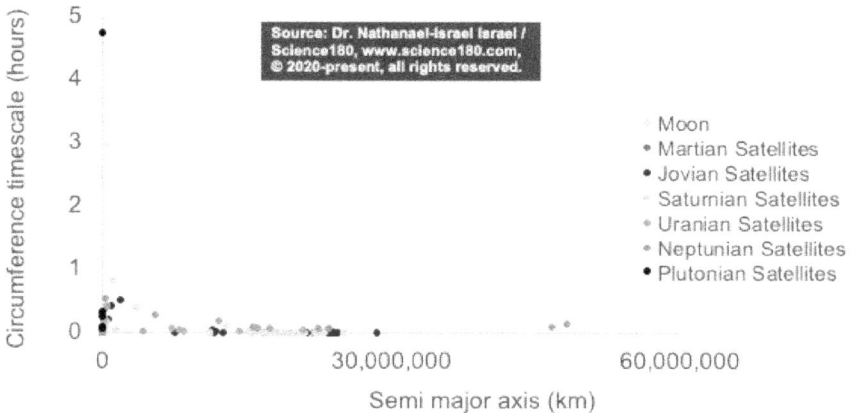

Fig. 220: Circumference Timescale (hours) of the satellites according to their types and semi major axis

Science180: Improve Your Understanding of the Universe-Origin with New, Accurate Products and Services

Most of the highest Circumference timescales were found with the biggest satellites except for the Jovian and Uranian satellites (Fig. 221):

- Moon: 2.96 hours
- Callisto (the outermost Jovian satellite in Zone 3): 30.75 minutes while the timescale of the biggest Jovian satellite (Ganymede) is 25.34 mins
- Titan (the biggest Saturnian satellite): 48.37 mins
- Oberon (the outermost Uranian satellite in Zone 3): 25.29 minutes while the timescale of the biggest Uranian satellite (Titania) is 22.66 mins
- Triton (the biggest Neptunian satellite): 32.27 mins
- Charon (the biggest Plutonian satellite): 4.74 hours

Fig. 221: Circumference Timescale (hours) of the Satellites according to their Types and Turbulence Zones

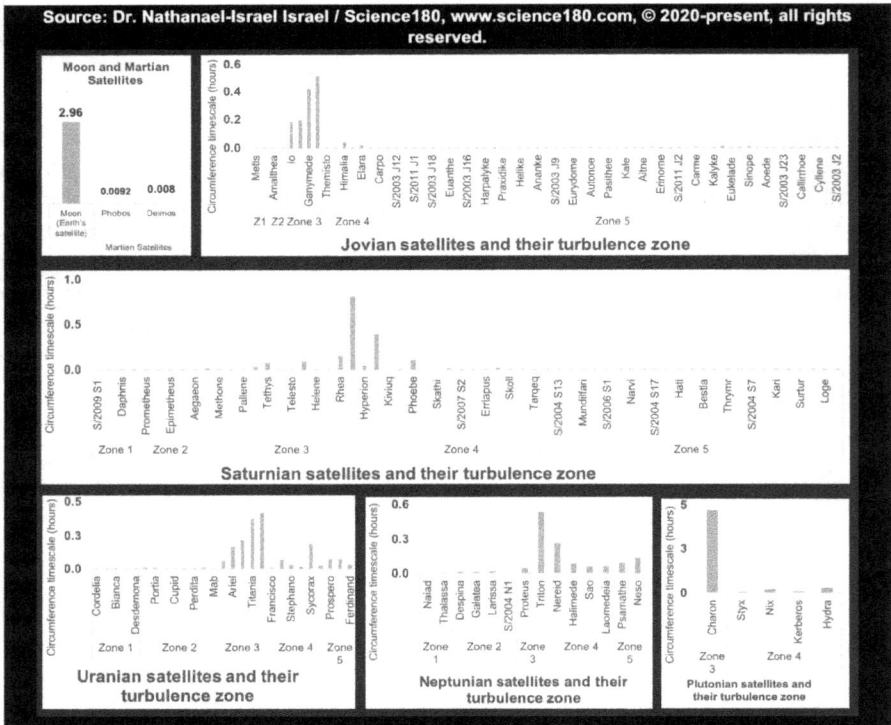

The highest Circumference timescales were obtained in Zone 3 (Fig. 221), suggesting that some bodies in Zone 3 took more time to complete their gathering together than the bodies in the other turbulence zones. As you will see in the next chapter, the Circumference timescale is not the whole story of the time it took to form the bodies. Because it took some time for the fluid layers to separate from one another, some upstream bodies finished gaining their shape before the fluid layers of the bodies downstream of them were separated.

While the Circumference timescale of the Moon is about 2.96 hours, that of the Martian satellites is less than a minute: 33.09 seconds for Phobos and 28.81 seconds for Deimos (Fig. 222). The Circumference timescale of the Moon (2.96 hours) is longer than that of the Earth (22.41 mins) which means that, after splitting from their common precursor, the precursor of the Moon took a little longer before wrapping its fluid into one body after reaching its "final position".

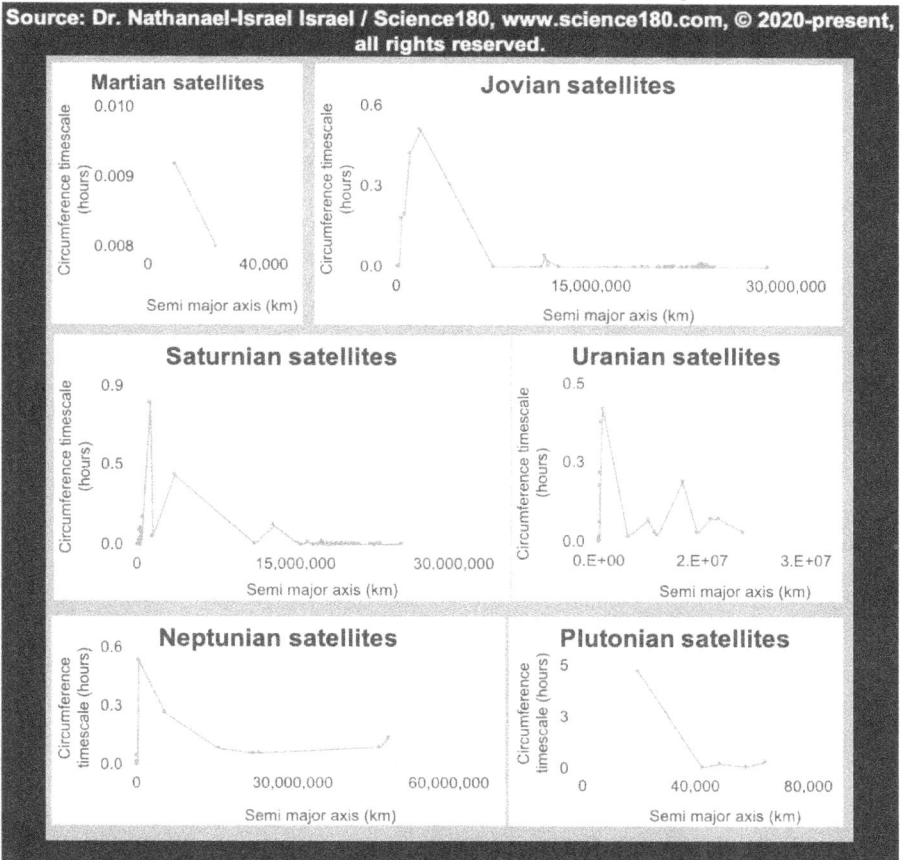

Fig. 222: Circumference Timescale of the Satellites according to their Types and Semi Major Axis

The Circumference timescale of the Jovian satellites varies between 1.66 seconds and 30.75 mins; the smallest value was with the second innermost Jovian satellite, Adrastea, while the highest timescale was with the outermost satellite in Zone 3.

The smallest Circumference timescale recorded on any satellite (0.1 second) was with S/2009 S1, the innermost Saturnian satellite. In contrast, the highest timescale (48.37 mins) was with Titan, the biggest Saturnian satellite.

For the Uranian satellites, the smallest Circumference timescale was with Cupid (3.59 seconds) while the highest (25.29 minutes) was with Oberon, the outermost Uranian satellite in Zone 3.

For the Neptunian satellites, S/2004 N1 (the outermost Neptunian satellite in Zone 2) recorded the smallest Circumference timescale (14.02 seconds) while the highest value (32.27 minutes) was with Triton, the biggest Neptunian satellite.

For the Plutonian satellites, the Circumference timescale ranges between 3.76 minutes and 4.74 hours. The smallest value was with Styx (the innermost satellite in Zone 4), whereas the highest timescale was with Charon (the biggest Plutonian satellite).

Differences were observed according to the turbulence zones (Fig. 223). For instance, the Circumference timescale in Zone 1 varies between 0.1 second and 33.09 seconds. The smaller timescale was recorded by S/2009 S1 (the innermost Saturnian satellite) while the highest value in Zone 1 was recorded by Phobos (the innermost Martian satellite).

In Zone 2, the smallest timescale (0.21 seconds) was recorded by Aegaeon (the outermost Saturnian satellite in Zone 2), while the highest value (63.82 seconds) was found with Larissa (a Neptunian satellite).

In Zone 3, the Circumference timescale of the satellites ranges from 0.45 seconds to 4.74 hours. The 10 smallest timescales in Zone 3 were recorded by the Saturnian satellites with the smallest value being found with Anthe. The highest Circumference timescale in Zone 3 was recorded by Charon (the biggest Plutonian satellite). The second highest timescale in Zone 3 was obtained with the Moon (2.96 hours). Except for Charon and the Moon, all of the timescales in Zone 3 are less than 1 hour.

In Zone 4, the timescale varies between 3.95 seconds (recorded on Dia, a Jovian satellite) and 19.11 minutes (recorded on Hydra, the outermost Plutonian satellite).

Finally, in Zone 5, the timescale varies between 2.42 seconds and 8.36 minutes. The 43 smallest timescales were recorded with the Jovian satellites. The 2 highest timescales were recorded with the 2 Neptunian satellites in Zone 5:

- 8.36 minutes for Neso (the outermost Neptunian satellite) and
- 5.62 minutes for Psamathe (the satellite downstream of Neso).

The most striking fact in this chapter is that out of all the bodies I studied in the Solar System, only 7 bodies have a circumference timescale higher than 1 hour and they include 2 satellites (the Moon and Charon), the 4 giant planets, and the Sun. Besides the Sun, all of the bodies I studied in the Solar System could have gotten get their shape or gathered together in less than 12 hours from the time their precursors started to be spiraled or wrapped around. The small values of the timescale, suggest that most of the celestial bodies could have gotten their shape shortly after reaching their semi major axis.

In hour(s), the circumference timescale looks as follows (Fig. 223).

Fig. 223: Circumference Timescale (hours) of Satellites according to their Types and Turbulence Zones

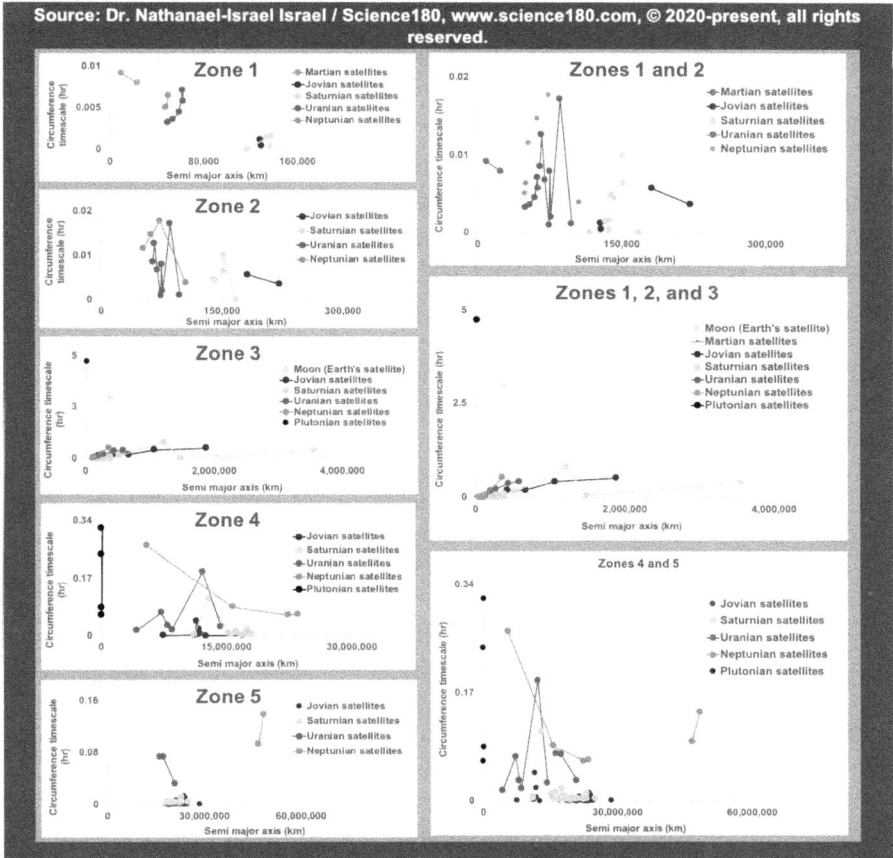

37.3. Take home message

The Circumference timescale gives a clue into the time it could have taken for the celestial bodies to be gathered together after they reached their semi major axis or their orbit. It ranged between less than a second to 2.61 days, with the highest timescale recorded with the Sun. About 23.77% of the bodies that I studied have a Circumference timescale less than 1 second. Of these bodies which timescale is less than 1 second, about 95% are asteroids. Of the 408 bodies which timescale I studied, 98.28% have a timescale less than 1 hour. Only 7 bodies have a timescale higher than 1 hour and they are 2 satellites (the Moon and Charon), the 4 giant planets, and the Sun. The Circumference timescale of the planets varies between 5.39 minutes and 10.86 hours, the highest value being recorded with Saturn, while the smallest value was with Mercury. As for the Earth, the Circumference timescale was 22.42 minutes. For the satellites in the Solar System, the timescale ranges between 0.1 second and 4.74 hours. The highest values were obtained in Zone 3. The Circumference timescale of the Moon is about 2.96 hours. I showed that, besides the Sun, all of the bodies in the Solar System could have been able to get their shape or gathered together in less than 12 hours from the time their precursors started to be spiraled or wrapped around once they reached their semi major axis.

'Science180 Academy' Success Strategy:
SCIENCE180 INTERVIEW REPORT (AKA SCIENCE180 INTERNET-TV-RADIO INTERVIEW REPORT)

Science180 Interview Report is the newsletter to read for guests and unconventional show ideas at the intersection of science and faith. Indeed, many hot questions are still unanswered on the road leading to the correct understanding of the origin of the universe, of life, and of chemicals. But most people don't know where to find the accurate answers to those challenging questions. What if with one simple call you can accurately answer all of those questions. You need to get in touch with or interview Dr. Nathanael-Israel Israel on your show, radio, tv, podcast, and even website, or invite him for a live presentation at your organization if your audience can benefit from any of the following show, talk, speaking, or interview ideas:

- Can we explain the formation of the universe through natural processes without evoking evolution and Big Bang?
- Is it a waste of time to attempt to prove the Biblical creation using science or historical investigation?
- Can anyone scientifically prove the Bible to be true?
- Is it possible to scientifically demonstrate without opening or quoting the Bible that God created the universe?
- Can a single variable play a crucial role in cracking the universe?
- Can the distance separating celestial bodies give a clue to the universe's birthdate?
- What is the master key to crack the universe-origin?
- How to deal with the fear of not knowing the origin of the universe and life?

Nathanael-Israel Israel: Member of the American Society of Agricultural and Biological Engineers

- Can you judge a pastor based on the creation lies he is spewing from the pulpit and that the people like and praise God for because they don't know?
- Can you be really free from doubt about the universe-origin and God's existence?
- Can mathematics and science collide to accurately explain the formation of the universe, of life, and of chemical particles?
- Can we mathematically prove that the formation of the Earth was completed on the 3rd day of creation like the Bible says?
- Can we scientifically demonstrate without a doubt that the Moon and the Sun were really formed on the 4th day of creation like the Bible says?
- Can mathematics and science rescue Christians in their efforts to rationally prove the existence of God, the Creator?
- Did the Quran and any other religious book make any gigantic error about the universe creation that any scientific formula proves the Bible got right?
- Why the secularist world doesn't care much if Christians and their leaders believe in Evolutionism, but they actually care much if they don't believe in the billions of years process?
- How to scientifically crack the code of turbulence
- How can people benefit from understanding the mother of all turbulences?
- What has prevented scientists from understanding turbulence?
- What is one thing experts in turbulence need to do from now forth if they want to reap the fruits of the turbulence mysteries you have decoded?
- Is science making you doubt your faith?
- Does the Bible scientifically teach anything about the universe-origin that most scientists and Christians ignore?
- Must Christians apologize to atheists, rationalists, and all freethinkers for the proofs creationist scholars and preachers have used to argue creation?
- Can anyone really be scientifically 100% sure and prove that God created the universe?

- Is the war between mathematics and physics drifting science in the wrong direction?
- Can we explain the formation of the universe through natural processes without evoking evolution and Big Bang?
- Is it a waste of time to attempt to prove the Biblical creation using science or historical investigation?
- Is there any need to prove the Biblical creation to be true?
- Why are most nations wasting millions of dollars on universe-origin and life-origin researches they don't need—or do they?
- Can anyone scientifically solve the most-asked questions on the universe creation after 3500 years of rational demands?

I know you may be tempted to answer these questions by yourselves, but avoid landing yourself on wrong paths that caused some people to lose contact with reality, it is better to get the accurate answer from the know-how expert, Dr. Nathanael-Israel Israel, the author of many books on the origin of the universe, of life, and of chemicals, and the standout expert who accurately decoded the scientific formula that forces science to bow to the truth. If you would like to register to Science180 Interview Report so we can periodically send you show ideas and interview opportunities related to the origin of the universe, of life, and of chemicals particles, please visit Science180Interviews.com for more details. You can also visit Science180.com to contact Nathanael-Israel Israel about how to get him to answer any of these questions.

CHAPTER 38

THE ONLY FORMULA THAT ACCURATELY TELLS YOU THE DATE OF BIRTH OF ANY CELESTIAL BODY IN THE UNIVERSE

38.1. Birthdate formula of celestial bodies

If you think human beings are the only ones that have and celebrate birthdays, wait until you discover what the scientific formula of the birthdate of celestial bodies is saying!

When I talk about the birthdate of celestial bodies, I do not mean the date that they spontaneously appeared as if there was no trace of their precursor. Indeed, like I introduced in the previous chapters, the birthdate of a celestial body is the approximated time that elapsed from the beginning of the formation of the universe until the formation of that body was completed. Beforehand, the precursor of each body went through changes, including:

- change of position until it reached the appropriate semi major,
- changes of configurations (e.g. internal, external, structural),
- elongation of their orbit, sometimes causing the tilting of the rotational axis, compression or compacting from a higher volume to a smaller one, etc.

In addition to what I already said, I also later explained why I considered the beginning of the formation of the Solar System and the beginning of the formation of the universe to be nearly the same. Although in this chapter and also in this book, I focused on the formation of celestial bodies, I would like to remind you that I devoted a completely different book (*"Turbulent Origin of Chemical Particles"*) to the origin and formation of chemical particles (e.g. subatomic particles, atoms, molecules, minerals, rocks, and other chemical compounds). I also consecrated a different book to the origin and formation of living things. In those books, I addressed the timeline related to events surrounding the formation of chemical particles and life. In other words, just because you will not hear much about the timeline of chemical elements and living things in this chapter does not mean that I have not handled them anywhere else.

Before I enunciate the formula of the birthdate of the celestial bodies, I would like to first recall a few things about the variables involved in it. In fact, in some of the previous chapters related to timescales, I explained that the calculation of the birthdate or date of birth of the celestial bodies requires an understanding of some key concepts:

- Lifespan of the precursor of celestial bodies
- Relationship between the escape velocity of primary bodies and the speed of the separation of the precursor of their secondary bodies
- Escape time of the precursors of primary bodies
- Branching out of the precursors of bodies according to their position in the stack of fluid layers of their system
- "Semi major axis increment timescale" of secondary bodies
- Circumference timescale

In the previous chapter, I explained the formula of these parameters. Now, I will explain the birthdate formula:

Birthdate of a celestial body = Circumferential gathering start date + Circumference timescale of the body

Considering the demonstrations that I already did in the previous chapters, I deduced the following birthdate formulas of the celestial bodies in the Solar System:

Birthdate of a planet = Semi major axis of the planet / Escape velocity of the Sun + 2 * 3.14 * (Radius of the planet) / Orbital speed of the planet

Birthdate of an asteroid = Semi major axis of the asteroid / Escape velocity of the Sun + 2 * 3.14 * (Radius of the asteroid) / Orbital speed of the asteroid

Birthdate of a satellite = (Semi major axis of the primary planet of that satellite / Escape velocity of the Sun) + (Semi major of the satellite / Escape velocity of its primary planet) + 2 * 3.14 * (Radius of the satellite) / Orbital speed of the satellite

Birthdate of the Sun = (Semi major axis of Mercury / Escape velocity of the Sun) + 2 * 3.14 * (Radius of the Sun) / Orbital speed of the Sun

In the above formula 3.14 is a short expression π (pi). In Figures 224-226, I illustrated the formula I enunciated above.

Fig. 224: Nathanael-Israel Israel's Formula of the Birthdate of a Star

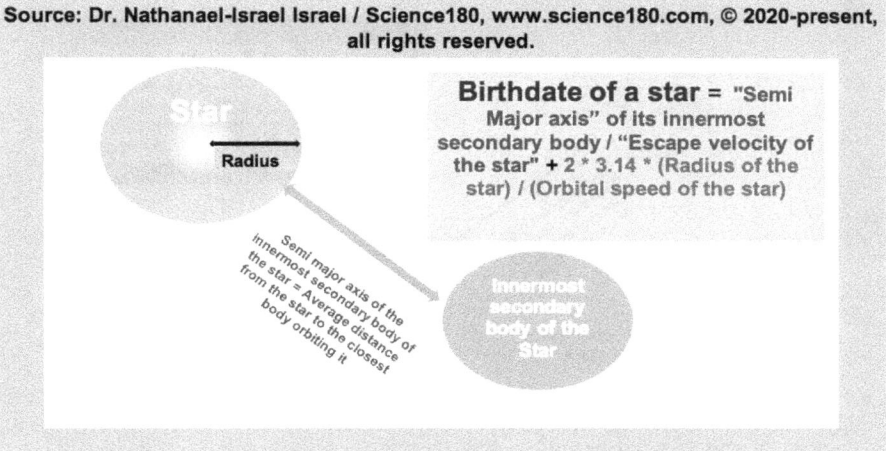

Go to www.Science180.com/BirthdateFormulaStar to get a prettier and color version of Fig. 224.

Fig. 225: Nathanael-Israel Israel's Formula of the Birthdate of Planets and Asteroids Orbiting a Primary Star

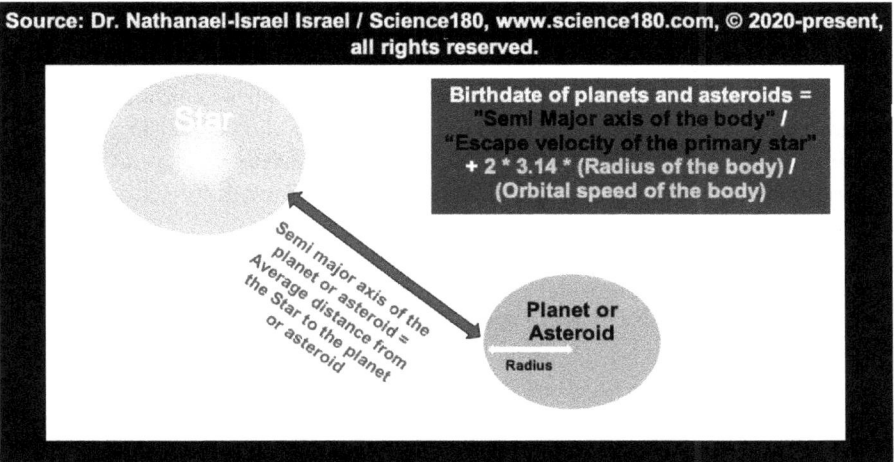

If you want to get a prettier and colorful version of Fig. 225, visit the following link: www.Science180.com/BirthdateFormulaPlanetAsteroid

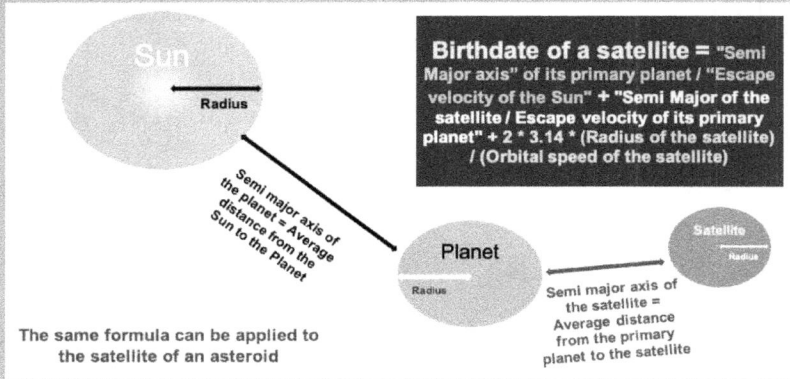

Fig. 226: Nathanael-Israel Israel's Formula of the Birthdate of the Satellite of a Planet or the Satellite of an Asteroid

Do you want a prettier and colored version of Fig. 226? If yes, then get it at: www.Science180.com/BirthdateFormulaSatellite

38.2. Birthdate of the celestial bodies according to their types

Investigated for 440 bodies in the Solar System, the birthdate varies between 1.089 days (or 26.14 hours) and 3266.11 hours (i.e. 8.95 years). The first body to be born was Mercury, the innermost body orbiting the Sun. Based on the available data, the birthdate is not calculated for all the celestial bodies I studied in the Solar System because either their semi major axis or their radius is missing. While the Sun was born 3.693 days after the beginning, the Earth was about 2.82 days (meaning 67.66 hours) (Fig. 227) after the beginning, while the Moon (Earth's satellite) was born 12.14 hours later, meaning 79.79 hours (or 3.32 days) after the beginning of the formation of the world. In addition to Mercury and Venus, a lot of asteroids (e.g. Aten asteroids) were formed before the Earth. In addition to the Aten asteroids, Mercury, Venus, and the Earth, several Apollo asteroids were born before the Moon.

By the time the Sun was born, no other satellite of the planets (known as of 2022) was formed besides the Moon. Of the 464 celestial bodies I studied in the Solar System, only 34 were formed before the Sun, meaning that more than 92.9%

were formed after the completion of the formation of the Sun. Some celestial bodies in the Solar System were formed before the Sun because the precursor of the Sun was very large and, after the precursor of the bodies orbiting the Sun escaped the precursor of the Sun, it took 2.61 days (i.e. 62.58 hours) alone for the precursor of the Sun to gather together into the Sun. Because the precursors of some inner bodies in the Solar System were not too large, they were quickly gathered together before the Sun was formed. Hence, many inner bodies were formed before the Sun.

After the formation of the Sun was completed, many other Apollo asteroids, and Mars trojans were formed before Mars. Indeed, Mars was formed about 14.13 hours after the formation of the Sun was finished, meaning that Mars was formed 102.76 hours or 4.28 days after the beginning. Within 1.057 hours meaning within 1 hour and 3.4 minutes, the formation of the Martian planetary system was completed. Any celestial body which semi major axis is higher than that of Mars was formed after Mars. Knowing the plethora of celestial bodies located beyond the Mars, it is fair to say that, except for Mercury, Venus, Earth, the Moon, some Aten asteroids and Apollo asteroids, most celestial bodies in the Solar System were formed after the formation of the Sun was completed.

The 2 Martian satellites were the next satellites to be formed after the Moon. Next, many Amor asteroids and Apollo asteroids were formed before the main belt asteroids, which formation was completed 6.078 to 9.785 days after the beginning of the formation of the Solar System. As the main belt asteroids were being born, several comets, Apollo asteroids, and Amor asteroids were also formed. Some of the comets formed when the main belt asteroids were being born were:

- 2P/Encke (formed 6.218 days since the beginning)
- 9P/Tempel (formed 8.76 days since the beginning)

Although it took 6.56 to 10.86 hours for the precursors of the giant planets to gather their fluids together (i.e. 6.56 hours for Uranus, 7.96 hours for Neptune, 9.55 hours for Jupiter, and 10.86 hours for Saturn) once they separated from the rest of the stack of fluid layers of the precursors of the bodies orbiting the Sun, because of the position of their precursors in the aforementioned stack of fluid layers, it took days before their turn to separate and gather themselves came. In fact, it was 14.99 days after the beginning that Jupiter was formed. Beforehand, many Jupiter trojans and all of the Jovian satellites in Zones 1, 2, and 3 were already formed. In other words, the formation of many Jovian satellites was completed before the formation of Jupiter was finalized. Furthermore, Jupiter was born 4.13 hours after Ganymede (the largest Jovian satellite). All of the Jovian satellites in Zones 4 and 5 were formed at least a day after Jupiter. Indeed, while Jupiter was formed 14.99 hours (i.e. 359.73 hours) after the beginning of the universe, Themisto (the innermost Jovian satellite in Zone 4) was formed 6.05 days (meaning 385.23 hours) from the beginning. All of the other Jovian satellites in Zones 4 and 5 came afterwards. The birthdate of the outermost Jovian satellite

(S/2003 J2) was 20.15 days (meaning 483.56 hours) after the beginning.

It took 6.75 days (i.e. 162.12 hours) after the formation of the outermost Jovian satellite before the innermost Saturnian satellite (S/2009 S1) was born. Those 6.75 days include the time it took from the moment the precursor of the Jovian planetary system split from the stack of fluid layers of the precursor of the bodies outward or beneath its layer until the moment the precursor of the Saturnian planetary system started its split-gathering. After the birth of the innermost Saturnian satellite 26.9 days after the beginning, 21 other Saturnian satellites, meaning all of the Saturnian satellites ranging from the innermost one all the way to Titan (the largest Saturnian satellite) were formed before the formation of Saturn was completed. In other words, the last Saturnian satellite to be born just before Saturn was formed was Titan. Saturn was born 27.31 days after the beginning, which was just 29.65 minutes after that of Titan, its biggest satellite. The only Saturnian satellites in Zone 3 which were born after Saturn were Hyperion and Iapetus: the former was born 46.56 minutes after Saturn and the later was born 17.4 hours after Saturn. This also implies that all the Saturnian satellites in Zones 4 and 5 were born after Saturn, with the outermost (S/2004 S26) of the 82 Saturnian satellites known by 2021 born 853.98 hours (i.e. 35.58 days) after the beginning.

For the next 18.28 days after the formation of the outermost Saturnian satellite, many centaurs were born before the innermost Uranian satellite was born 53.86 days after the beginning. Some of the centaurs born between Saturn and Uranus were:

- 95P Chiron (located at 13.64 AU) and born on day 38.23 since the beginning,
- 10199 Chariklo (located at 15.79 AU) born 44.27 days since the beginning, and
- 1P/Halley (Halley's Comet), located at 17.8 AU and born 49.9 days since the beginning

All of the innermost 17 Uranian satellites, meaning Uranian satellites ranging from Cordelia to Titania (the largest Uranian satellite) were born before Uranus. In fact Titania was born 54.08 days after the beginning, while Uranus was born just 29.36 minutes later, meaning 54.1 days since the beginning. Of the Uranian satellites in Zone 3, only one (Oberon) was born after Uranus and its birth occurred 1.47 hours after that of Uranus. This also implies that all the Uranian satellites in Zones 4 and 5 were born after their primary planet, Uranus. After the birth of Uranus, 11.09 days passed before the outermost Uranian satellite (Ferdinand), which is also the 27th Uranian satellite, was born on 65.19 days after the beginning. Afterwards, many centaurs (e.g. Amycus), comets (e.g. 109P/Swift–Tuttle), and Neptune trojans (e.g. 2008 LC18) were born before the first Neptunian satellite (Naiad) came into existence.

In fact, between the formation of the outermost Uranian satellite and that of the innermost Neptunian satellite, 19.07 days had passed. Like I already explained for the other planetary systems, the 19.07 days that I just mentioned is part of the days that passed before the precursor of the Neptunian planetary system could have moved and positioned itself to start split-gathering. In fact, like I presented in the chapter on "Semi major axis increment timescale", it took about 729.8 hours (meaning 30.4 days) before the precursor of the Neptunian planetary system moved from the position where the precursor of the Uranian planetary system split from the stack of fluid of the bodies orbiting the Sun to the position where the former did the same. Triton (the largest Neptunian satellite) was born 84.44 days after the beginning. Then, Neptune was born 3.22 hours later. Nereid (the innermost Neptunian satellite in Zone 4) was born 2.39 days after Neptune, meaning 2.53 days after Triton. As has been the case for all the giant planetary systems, all the Neptunian satellites in Zones 4 and 5 were formed or born after their primary planet.

As the Neptunian satellites were being formed, many asteroids including Neptune trojans and Trans-Neptunian objects, were doing the same. From the formation of the innermost Neptunian satellite and the outermost Neptunian satellite, about 23.8 days passed. Then, between the formation of the outermost Neptunian satellite and the innermost Plutonian satellite (Charon), almost 3 days passed. As it was taking days for all of the Neptunian satellites to form, the fluid layers of the precursor of the Plutonian planetary system had also progressed a lot on its "journey" to reach its appropriate position before splitting from the rest of the fluid layers of the precursor of the bodies orbiting the Sun. Hence, by the time the outermost Neptunian satellite was formed, the precursor of the Plutonian planetary system just had a few more days to go before starting its split-gathering. In the end, Charon was formed 111.07 days (i.e. 2665.75 hours) after the beginning, while Pluto (its primary planet) was form 8.79 hours earlier. Pluto is not the only planet which was formed before its largest satellite. Indeed, as a reminder, the Earth and Mars were born before their satellites. However, as I previously showed, all of the 4 giant planets were born after their largest satellites, and usually after most of their satellites in Zones 1, 2, and 3, but before any of their satellites in Zones 4 and 5. This suggests that the size of the precursor of some planetary systems may have affected the timing of the birth of the largest satellites. By this time of the formation of the bodies in the Solar System, several Trans-Neptunian objects and Kuiper belt objects (e.g. 90482 Orcus) were formed. The latest body to be formed in the Plutonian planetary system was Hydra (the outermost Plutonian satellite) born 14.73 hours after Pluto, meaning 111.32 days after the beginning of the formation of the universe. After the end of the formation of the last body found in each planetary system in the Solar System, many other asteroids were formed as the remainder of the stack of fluid layers of the precursor of the bodies orbiting the Sun continued its journey:

- Trans-Neptunian objects (e.g. Haumea located at 43.2 AU and formed on day 121.17; Makemake located at 45.72 AU and formed on day 128.2; Eris located at 67.78 AU and formed on day 190.05)

- Kuiper belt objects (e.g. 120347 Salacia located at 41.9 AU and formed on day 117.36; 50000 Quaoar located at 43.37 AU and formed about 121.6 days after the beginning).

By the way, all the birthdates I mentioned in this chapter are with reference to the beginning of the formation of the Solar System. Toward the end of this chapter, I will provide more details on the formation of comets. In the next segments, I will illustrate the birthdate of the celestial bodies with figures.

As I announced earlier, the birthdate of the planets in the Solar System varied between 1.09 days (recorded on Mercury) and 110.7 days (recorded on Pluto). Fig. 227 illustrates the birthdate of the planets in the Solar System. According to the birthdate of the Sun (i.e. 88.63 hours after the beginning of the formation of the Solar System), and that of the planets, the Sun is:

- 2.6 days younger than Mercury
- 1.65 days younger than Venus
- 20.97 hours younger than the Earth
- 8.84 hours younger than the Moon
- 14.13 hours **older than** Mars
- 11.3 days **older than** Jupiter
- 23.62 days **older than** Saturn
- 50.41 days **older than** Uranus
- 80.88 days **older than** Neptune
- 107.01 days **older than** Pluto

Fig. 227: Birthdate (Days) of the Planets in the Solar System

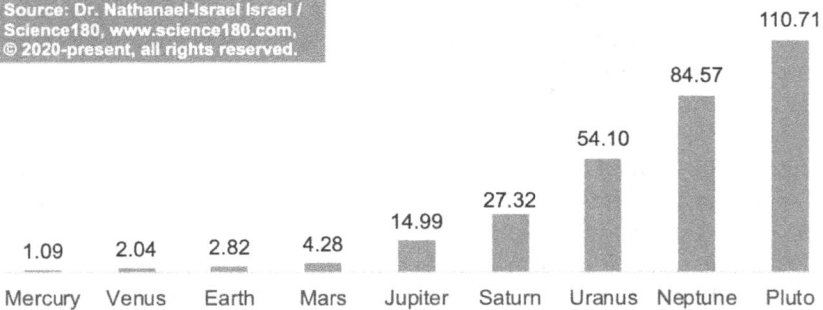

Mercury	Venus	Earth	Mars	Jupiter	Saturn	Uranus	Neptune	Pluto
1.09	2.04	2.82	4.28	14.99	27.32	54.10	84.57	110.71

A positive relationship exists between semi major and birthdate of the planets (r2=1) (Fig. 228).

Fig. 228: Regression between Semi major and Birthdate of the Planets in the Solar System

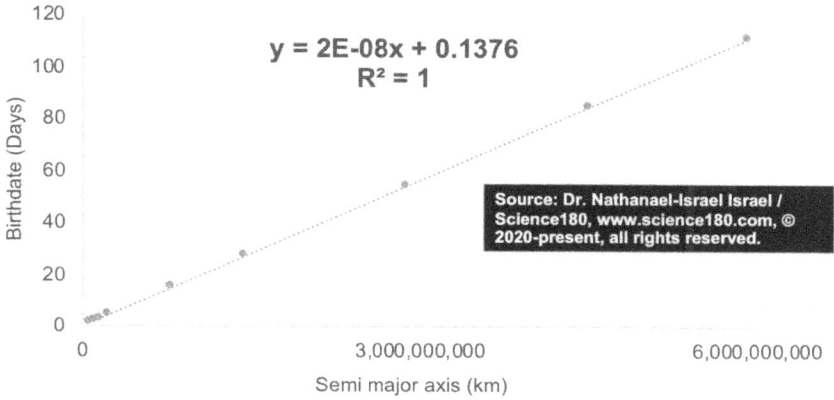

$$y = 2E\text{-}08x + 0.1376$$
$$R^2 = 1$$

Even when I combined the planets and all the asteroids, a strong regression was found between their semi major and birthdate (r2=1) (Fig. 229).

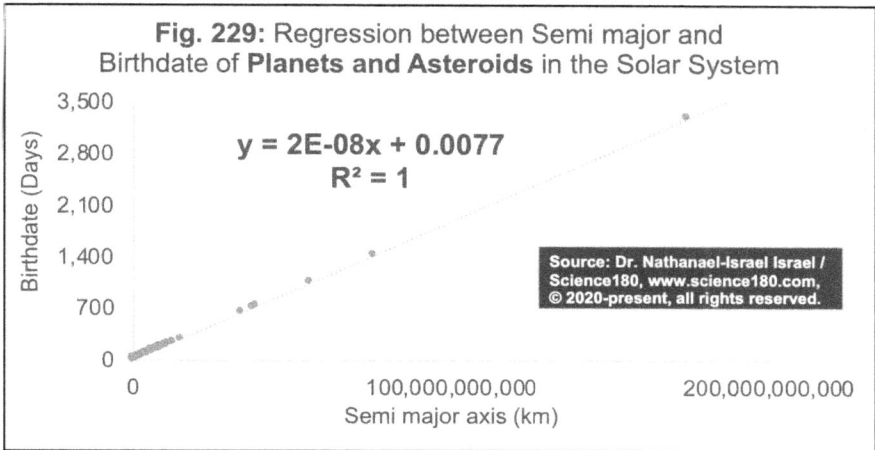

Fig. 229: Regression between Semi major and Birthdate of **Planets and Asteroids** in the Solar System

$$y = 2E\text{-}08x + 0.0077$$
$$R^2 = 1$$

Fig. 230 shows the birthdate of the main belt asteroids.

Fig. 230: Birthdate (Days) of the Main Belt Asteroids

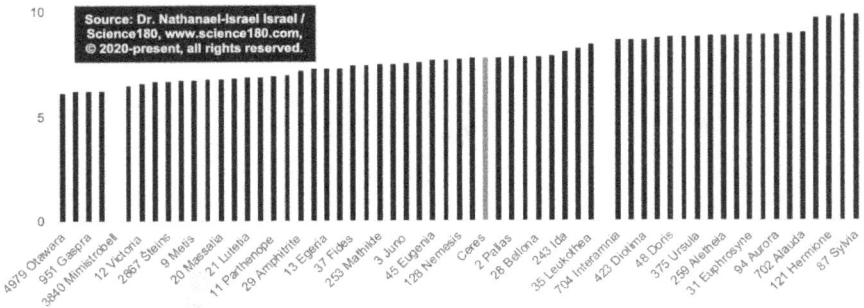

A positive regression was also found between semi major axis and birthdate of the main belt asteroids.

38.3. Birthdate of the satellites

Like I already said, the birthdate of the satellites is the time that elapsed from the beginning of the formation of the Solar System until their formation was completed. It ranges between 3.325 days (recorded on the Moon) and 111.32 days (recorded on Hydra, the outermost Plutonian satellite) (Fig. 231). From the birth of the first satellite to the birth of the last satellite in the Solar System, about 108 days elapsed. If the oldest satellite means the one born first and the youngest satellite means the one born last, the oldest satellite of a planet in the Solar System is therefore the Moon, whereas the youngest is Hydra, the outermost Plutonian satellite. Based on what I said above, the oldest satellite in the Solar System is older than the youngest by about 108 days.

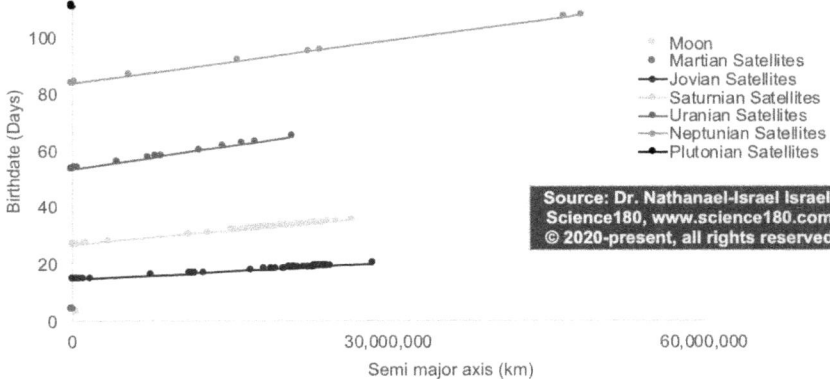

Fig. 231: Birthdate (Days) of the satellites according to their types and semi major axis

The satellites were born not only according to their semi major axis with respect to their primary planet, but also according to the position of their primary planet with respect to the Sun. All the satellites of each planetary system were born before the birth of the first satellite of the planetary system downstream occurred. Considering what I already said about the birthdate of the planets, all the bodies in a planetary system were born before the first body in the downstream planetary system was born (Fig. 232). For instance,

- The innermost Martian satellite was born 23.24 hours after the formation of the Moon (Earth's satellite)
- The innermost Jovian satellite was born 10.29 days after the formation of the outermost Martian satellite
- The innermost Saturnian satellite was born 6.75 days after the formation of the outermost Jovian satellite
- The innermost Uranian satellite was born 18.28 days after the formation of the outermost Saturnian satellite
- The innermost Neptunian satellite was born 19.07 days after the formation of the outermost Uranian satellite
- The innermost Plutonian satellite was born 3.00 days after the formation of the outermost Neptunian satellite

Fig. 232: Birthdate (Days) of the satellites according to their types and turbulence zones

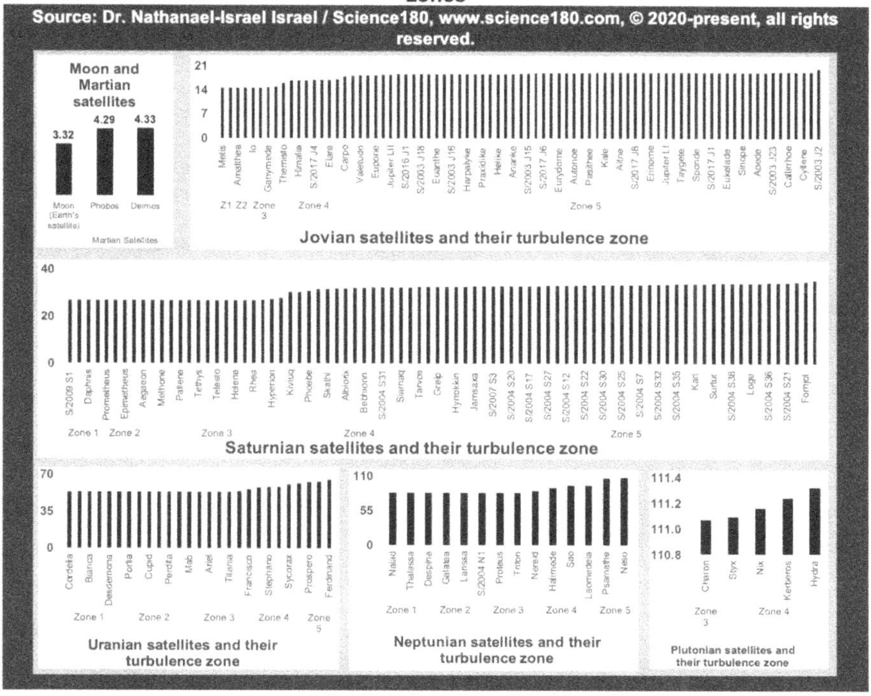

A strong positive regression (r2=1) was found between birthdate and semi major axis of the satellites according to their types (Fig. 233).

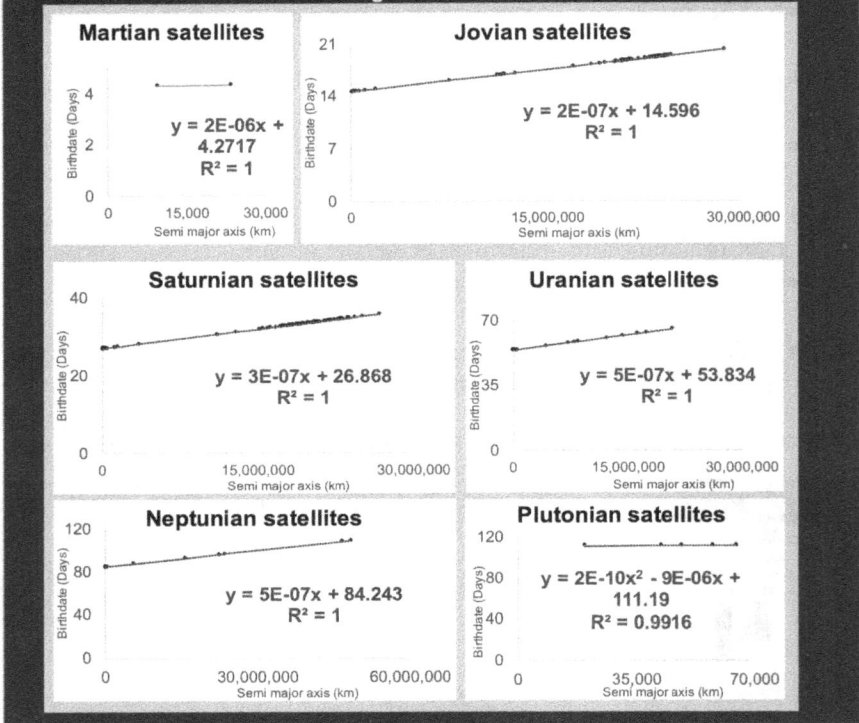

Fig. 233: Relationship between birthdate and semi major axis of the satellites according to their types

Fig. 234 illustrates the birthdate based on the turbulence zones, types, and semi major of the satellites.

Fig. 234: Birthdate (Days) of the satellites according to their types, semi major axis, and turbulence zones

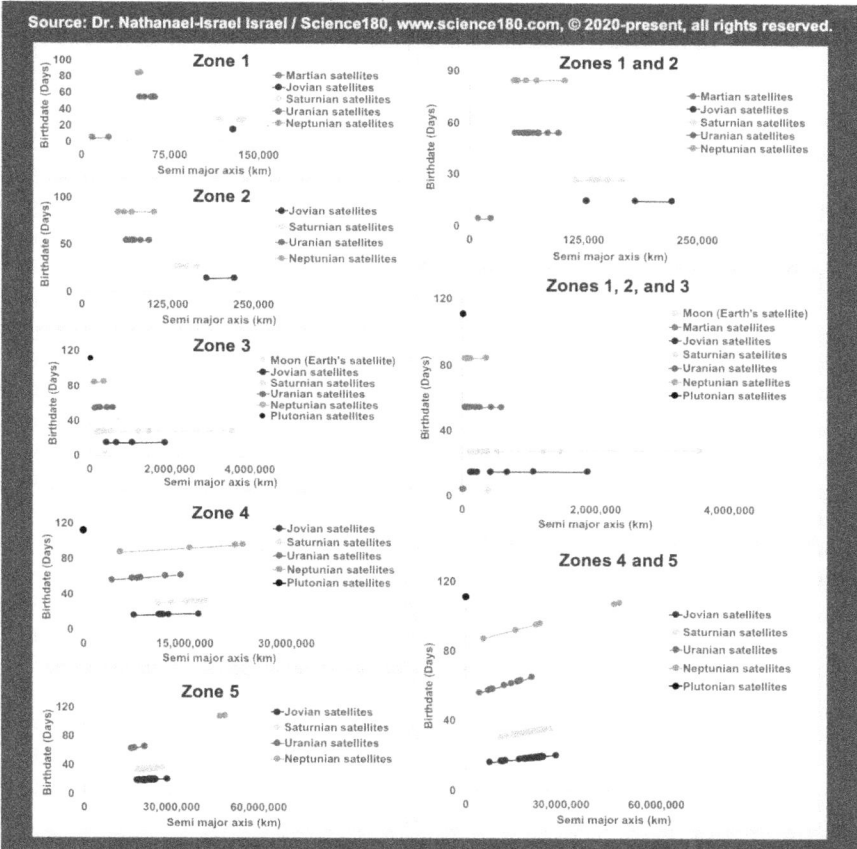

Visit www.Science180.com/BirthdateSatellites to download a beautiful colored version of this graph.

38.4. Formation of comets and of the "turbulence clustered belt of Zones 4 and 5" of the Solar System

According to the data I collected on the semi major of the celestial bodies, some bodies could be located thousands of AU from the Sun, meaning that even if their precursors travelled at a speed equal to the escape velocity of the Sun, it could have taken years before they could have reached such remote positions. Considering the cold in the outer region of the Solar System, I don't think the precursor of some bodies orbiting the Sun could have travelled for years and still kept its original state. A simple look at how lava turns into solid after erupting from their hot location to Earth's atmosphere (which is cooler than the interior of the Earth) suggested to me that, as the precursors of the outer bodies were moving into deep space, they (or at least their surface) could have been frozen as their layers were thinning (as some were splitting from the rest of the fluid layers or branching out), while confronting or passing through colder and colder portions of space, and therefore forming comets and other types of asteroids filled with ice. Hence, many (if not the majority of the) remote objects in the Solar Systems are comets known for their icy tails, which "emit" (or appear to emit) light as they pass through space. In other words, during their orbit, as they come near the Sun, some comets lose material into space and consequently give off a 'tail' called a coma. Those comas are beautiful to see of course, and most comets are known for them. In general, beyond the position of the Earth and particularly beyond Mars and the giant planets, as the temperature in space dramatically dropped, some tinny fluid layers could have been frozen and yielded comets. Depending on the distance at which the precursor of the comets split-gathered, some yielded short-period comets, which are said to have an orbital period usually less than 200 AU. But fluid layers that split-gathered far beyond 200 AU birthed what is known as long-period comets. Depending on the conditions of their split (e.g. precursors of celestial bodies that may have squashed the fluid layers of the precursors of the comets), the orbit of some comets is elongated, therefore having a high eccentricity. Just as with most innermost bodies, some short-period comets will likely have a lower orbital inclination than most long-period comets. In other words, the precursors of comets originated from the precursor of the bodies orbiting the Sun just as the precursor of the planets and other asteroids orbiting the Sun did. Unlike what some scientific theories claim, comets are not the origin of the water on Earth. In my book *"Turbulent Origin of Chemical Particles"*, I talked about the origin of water in more details. The way the precursors of comets and other asteroids were tilted led to them orbiting backwards. In the chapters orbital inclination, axial tilt, vortex, and fluid ligaments, I already detailed the cause behind the prograde and retrograde motion of the celestial bodies. The same logic applies to comets as well.

Furthermore, considering the distribution of Jovian and Saturnian satellites in Zones 4 and 5, according to which some of them are clustered around a small region around the end of Zone 4 and the beginning of Zone 5 (see the chapter of semi major axis and semi major axis increment), I think that there is a region toward the edge of the Solar System where many asteroids would be clustered, most of which would be retrograde. I called that zone the "turbulence clustered belt of Zones 4 and 5". Just as the clustered satellites around Jupiter and Saturn are small (in the order of 1 km), I think the same would be true for some outer asteroids in the Solar System. Some of those asteroids could be comets. Although I think that the theories behind the Kuiper belt and the Oort cloud are wrong (for they are based on a wrong explanation of the origin on the universe), I think that there is indeed a zone toward the edge of the Solar System where many asteroids are clustered. In the planetary systems for instance, there are more bodies in the turbulence Zones 4 and 5 than in Zones 1, 2, and 3. Most of the bodies in Zones 4 and 5 are clustered around a region. For more details, you can revisit the distribution of the semi major axis, radius, and orbital inclination of the satellites in Zones 4 and 5.

Considering the semi major axis of the satellites in the Solar System, I tried to estimate the position of the "turbulence clustered belt of Zones 4 and 5" in the Solar System. Indeed, I already showed in the chapter on semi major axis that, for the Jovian satellites, the semi major axis of the outermost satellites in Zone 3 is 1,882,700 km and the semi major of the outermost satellites in Zone 4 is 16,990,000. In contrast, the semi major of the outermost Saturnian satellite in Zone 3 is 3,561,300 whereas that of the outermost Saturnian satellite in Zone 4 is 18,010,000 km. In other words, the ratio of the semi major of the outermost satellite in Zone 4 and Zone 3 is 9 for the Jovian satellites and 5.1 for the Saturnian satellite. Put another way, the distance separating the position of the "turbulence clustered belt of Zones 4 and 5" for the Jovian satellites is at least 9 times the semi major of the outermost bodies in Zone 3, the zone of the largest satellites. Likewise, that ratio is at least 5 for Saturnian satellites. Knowing that the energy with which the precursor of the bodies orbiting the Sun escaped the precursor of the Sun was high (see escape velocity of the Sun versus that of the planets), it is more likely that it would have travelled much farther before the "turbulence clustered belt of Zones 4 and 5" could have been formed. Furthermore, knowing that Neptune (which is the outermost largest planet orbiting the Sun) can be positioned in Zones 3 or 4, I think that the position of the "turbulence clustered belt of Zones 4 and 5" of the Solar System can be at least 5 to 9 times the semi major of Neptune. Knowing that the semi major of Neptune is about 30 AU, I think that the "turbulence clustered belt of Zones 4 and 5" of the Solar System can be at least 150 to 270 AU from the Sun. In other words, at a position at least 150 AU from the Sun, a very huge belt of mostly retrograde and small asteroids will be found orbiting the Sun.

38.5. Birthdate of galaxies and clusters of galaxies

I cannot complete the story of the formation of the bodies in the Solar System without addressing the birthdate of the galaxies and their clusters in the universe. In a previous chapter, I already addressed the formation of galaxies and stars. In this segment, I will explain why the speed of the separation of the precursors of stars, galaxies, and galaxy clusters could have been extremely higher than that which human equipment could measure.

Just as I considered the time it took for the precursors of the bodies in the Solar System to escape the precursor of their mother and reach their orbit, so also it took some time for the precursor of the Solar System to move from a certain initial position inside the precursor of the Milky Way to its current position inside the Milky Way. Likewise, it could have taken some time for the precursor of the Milky Way Galaxy to move from a certain position with respect to what can be perceived as the precursor of the galaxy cluster that the Milky Way belongs to. In the same manner, it could have taken some time for the precursors of that galaxy cluster and other clusters of galaxies to move from a certain position with respect to their precursors and so on and so forth until the largest organization of clustering in the universe could be reached.

As I showed in the segment on escape velocity, the larger a body, the higher its escape velocity. For instance, the escape velocity of the Moon is 2.38 km/s, that of the Earth is 11.2 km/s, while that of the Sun is 617.6 km/s. As you can see, the escape velocity increases as the level of clustering increases. Subsequently, as one considers higher and higher levels of clustering of bodies (e.g. from stars to galaxies, and from galaxies to galaxy clusters, and still from galaxy clusters to clusters of clusters of galaxies, etc.), the escape velocity must increase significantly as well.

However, as explained in previous chapters, the fluid layers of the precursors of the systems of bodies (e.g. galaxy clusters, galaxies, stellar systems, planetary systems, atomic and subatomic systems) were split-gathered to form their components at the same time that their daughter bodies and components were also going through changes. To put it another way, precursors of galaxies and stars did not have to move to a certain place first before what would become the precursors of their daughter bodies could start organizing themselves toward their formation. As the precursors of the galaxies were being moved, the precursors of their stars (which were their components) were also doing the same, while what became the precursors of the planetary systems and asteroids were also moving. Because of the size of the fluid layers according to the scale of the systems of celestial bodies, the precursors of all the stars in galaxies could not have travelled from a position closer to the galactical center, but could have just travelled from a certain position in the fluid layers of the precursor of their galaxy. This means that, on the scale of galaxies, it can be misleading to consider the distance separating a star from the center of the galaxy it belongs to and the "orbital speed" of that star to estimate the time it could have taken for the precursor of that star to be put into orbit. Furthermore, it is also important to keep in mind that the precursors of the bodies in a system were not collected in a tiny point like some theories have claimed. Knowing that the universe is expanding, secondary bodies in the universe are much more distant from their primary body than they were just at the end of their formation. Knowing that a lot of errors and bias exist regarding the measurement of the distance separating galaxies and galaxy clusters, I did not entertain the thought to estimate the time it could have taken for them to split-gather. In other words, because of the inaccuracy surrounding the distance separating galaxies and also their speed, I decided to rather focus on the bodies in the Solar System, which data has been followed for years. Even so, it is with time that some errors in the Solar System are corrected, confirming that measurement errors exist with the celestial bodies outside of the Solar System, but scientists have no means yet to properly detect and correct them.

For instance, on the scale of the planetary systems, the time it took to form a satellite include not only the timescale based on its semi major axis and the semi major of their planet, but also the Circumference timescale of that satellite. For instance, when I considered the Moon, the time it took for it to be formed accounted for the duration of the travel of the precursor of the Earth-Moon system with respect to the precursor of the Sun, but also the distance travelled by the precursor of the Moon after it escaped the precursor of the Earth-Moon system, and also the Circumference timescale of the Moon. A similar logic can be applied to the precursors of stars with respect to the position of the "center" of the galaxy they belong to, and also the position of galaxies with respect to the position of the "center" of the galaxy clusters they belong to. But, the speed with which these gigantic astronomical bodies (galaxy cluster) were separated was so huge that even considering the vast distance separating them, the ratio of those distances and an immense speed would boil down to the same story of their daughter bodies having been formed within a few minutes, hours, or days. Put another way, what I showed in this book pertaining to the bodies in the Solar System can be applied to other stellar systems with diverse levels of recalibration or scaling. For instance, extrasolar planets, or 'exoplanets' are real, but they may function with laws differently scaled than those present in the Solar System. Nevertheless, the same logic I used to demonstrate the origin of the Solar System can be applied and come up with the same conclusion I found in this book: the formation of the celestial bodies was a matter of days, NOT millions or billions of years.

Furthermore, as I demonstrated in another book, it is important not to think that the law of physics as found in the Solar System must be exactly the same in all other stellar systems. For instance, depending on the conditions surrounding its split-gathering, the fluids of the precursor of a star can split into two bodies which will then be gathered into 2 different stars, forming a system of stars orbited by planets and asteroids as is the case for individual stars. In other words, it is possible to find planets orbiting a binary star system. Furthermore, planets and asteroids should not be expected to be found in galaxies only, for even in globular clusters and around isolated stars, planets and asteroids can be found. For, as some stars were being formed, some materials could have escaped their precursors and formed bodies orbiting those stars as was the case of the precursor of the Solar System. In fact, just as some clusters of nucleons are inside atoms, it is possible to find some planets and asteroids orbiting clusters of stars. In other words, a system of stars can be orbited by planets and asteroids. Put another way, some planets and asteroids can be found orbiting more than one star. It all depends on whether or not the fluid layers of the precursor of the star were gathered together into a single star or was split into different fluid layers forming different stars, or a single layer was cut into pieces forming different stars. To make a long story short, anything was and is possible.

TURBULENT ORIGIN OF THE UNIVERSE

Furthermore, based on the data on the birthdate of the celestial bodies in the Solar System, some secondary bodies are always born before their primary bodies or vice versa. For instance, before the formation of the Sun was completed, some planets and asteroids were already formed. Likewise, before the formation of some planets was completed, some of their satellites were fully formed. In the same manner, the shape of the Milky Way which is a cluster of stars could have been completed after the formation of the Sun. Similarly, the shape of some galaxy clusters could have been completed after the formation of some of their galaxies. The same thing applies for clusters of clusters of galaxies all the way to the highest level of superclusters. In other words, the formation of everything in the universe may not have been finished before the formation of the Sun. Some bodies or systems of bodies may even be still going through some processes of formation even as of today. For as things in the universe are moving, changes are happening in them. However, the destiny of every celestial body in the universe was set by the laws that characterized the mother of all turbulences, which produced the bodies in the Solar System and in the other systems in the universe. In other words, everything in the universe as a whole may not have been completed within days, yet, the course of their formation and fate was set by the end of the first few days after the beginning. I will revisit this concept in the next chapter.

As I pondered on the speed with which the precursors of some galaxy clusters could have been ejected or have escaped the precursor of their mothers, I did not have any reason not to accept that many precursors of bodies, or let's say many daughter bodies of the turbulent primary materia (the original matter in the universe), could have travelled at a speed higher than that of light, which some people erroneously think is the highest achievable speed by matter. As of today, photons which are the particles of light are considered to be elementary indeed, but deep inside me, I feel like it will be just a matter of time before photons and many other particles so called elementary today could be proven to be composite particles, meaning composed of other particles. Considering what I already elaborated on in previous chapters, the precursor of daughter bodies usually travelled or were ejected or escaped their own mother precursors more slowly than the later. In other words, the speed of the particles that were used to form the photons could have been much higher than the speed of the photons as of today. Applying the same logic, the speed of the precursor of stars could have escaped the precursor of the primary body of their galaxy at a speed slower than the speed that the precursor of their galaxy could have escaped the precursor of their galaxy cluster, etc. By climbing the "ladder" of the cascade of split-gathering (as I explained in the chapter on "Inclusion levels of primary and secondary bodies"), one can find that the speed at which the "turbulent prima materia" (the original matter in the universe) was scattered was very huge, even higher than the speed of light as I explained in the segment entitled *"original mysterious scattering"* of the chapter "In the beginning ...". Being a product of the "turbulent prima materia", light cannot travel faster than the former. Likewise, the precursor of certain bodies in the universe could have travelled faster than light. In light of this and considering the expansion of the universe (which has been increasing the original distance between bodies in the universe), the time it could have taken to separate the stars and galaxies in the entire universe, or at least to put them in the final course of their formation, could be boiled down to a few days as I showed for all the bodies in the Solar System. For no matter the distance separating the precursors of stars and other bodies, a speed or energy of their separation in the order of magnitude of infinite or extremely higher than what human beings can imagine can easily bring the ratio "distance / speed" to just a few days or years, instead of billions of years just as I already proved that the formation of the Earth, Moon, and the Sun was a matter of less than 4 days. Remember this: if the speed with which the precursor of the Moon escaped the Earth-Moon system is about 11.2 km/s (i.e. escape velocity of the Earth), and the speed with which the precursor of the bodies orbiting the Sun was about 617.6 km/s (i.e. escape velocity of the Sun), meaning about 55.14 times the escape velocity of the Earth, it cannot be ruled out that as the level of inclusion and the size of the system of bodies increases, their escape velocity will do the same, meaning that the speed with which their daughter bodies escaped or were ejected was extremely higher than what is known in the Solar System.

Furthermore, just as the formation of the Earth (the primary body of the Moon) was completed before that of the Moon (the secondary body), it is possible that the formation of the core or central part of some clusters of bodies could have been finished before that of some of the bodies or systems of bodies orbiting them. It was all about the size, position, energy, and the nature of the turbulence which birthed the bodies. The day that scientists may collect more accurate data on the distance, and speed of the galaxies and cluster of galaxies, by plugging the number into some of the formulas I invented in this book, they will realize that nothing in the universe was formed over billions of years as some theories have been claiming for so long.

Moreover, it is worth mentioning that, as the precursors of the celestial bodies were being mixed together to gain their mass, their speed decreased. For instance, although photons are known as the fastest particles in nature, when they were bound to one another to form bigger "molecules" of photons, the speed of the heavier cluster of photons decreased, suggesting that as precursors of bodies were getting bigger or were gathering their constituents together, their speed could have decreased. In the case of the precursors of the celestial bodies, as their constitutive particles or matters were being gathered together into more unified bodies having a defined shape (e.g. spherical), the speed of their daughter bodies could have decreased. This suggests that the time it took for the precursors of the galaxies and galaxy clusters to migrate from the position of their mothers to their current position could have been shorter than what their orbital speed and the distance separating them could suggest.

38.5. Take home message

The birthdate of a celestial body is the time that elapsed from the beginning of the formation of the universe until its formation was completed. To fully understand the date of birth of the celestial bodies, it is important that you first read the previous 5 chapters related to the timelines of the universe. The formula I invented for the calculation of the birthdate of the celestial bodies involves the timescale based on semi major axis and the Circumference timescale, which at their turn involved factors related to the:

- Lifespan of the precursor of bodies;
- Relationship between the escape velocity of the primary body and the speed at which the precursor of the secondary bodies escaped the precursor of their primary body;
- Escape time of the precursor of the primary body (i.e. timescale of the separation of the precursor of the primary body from the precursor of its secondary bodies);
- "Semi major axis timescale" of the secondary bodies;
- Circumference timescale of the bodies.

After 9 years of extensive research on the origin of the universe, I established the following formulas for the birthdate of celestial bodies with reference to the beginning of the formation of the stellar system they belong to:

Birthdate of a planet = (Semi major axis of the planet) / (Escape velocity of the star that the planet is orbiting) + 2π(Radius of the planet) / (Orbital speed of the planet)

Birthdate of an asteroid = (Semi major axis of the asteroid) / (Escape velocity of the star that the asteroid is orbiting) + 2π(Radius of the asteroid) / (Orbital speed of the asteroid)

Birthdate of a satellite = (Semi major axis of the primary planet of that satellite) / (Escape velocity of the star that the planet is orbiting) + (Semi major axis of the satellite) / (Escape velocity of its primary planet) + 2π(Radius of the satellite) / (Orbital speed of the satellite)

Birthdate of a star = (Semi major axis of the innermost body orbiting the star) / (Escape velocity of the star) + 2π(Radius of the star) / (Orbital speed of the star)

CHAPTER 38: BIRTHDATE OF THE CELESTIAL BODIES IN THE UNIVERSE

Here, the beginning or the reference of these birthdates is the beginning of the split-gathering of the precursor of the stellar system that these bodies belong to. In the Solar System, the reference that I used is the beginning of the split-gathering of the precursor of the Solar System. I demonstrated that (based on what was known as of 2021) the first major celestial body to be formed in the Solar System was Mercury, the innermost body orbiting the Sun. While the Earth was formed about 2.82 days (meaning 67.66 hours) after the beginning, and the Moon (Earth's satellite) was born 12.14 hours later, meaning 79.79 hours (or 3.32 days) after the beginning, the Sun was born 3.693 days after the beginning.

By the time the Sun was born, no other satellite of the planets (known as of 2021) was formed besides the Moon. Of the 464 celestial bodies that I studied in the Solar System, only 34 were formed before the Sun, for the precursor of the Sun was very large and it took 2.61 days (meaning 62.58 hours) for it to gather together into the Sun after it split from the precursor of the bodies orbiting the Sun. Some of the bodies that were formed before the Sun include Mercury, Venus, Earth, the Moon, some Aten asteroids, and Apollo asteroids. After the formation of the Sun, many other Apollo asteroids, and Mars trojans were formed before Mars. Mars was formed about 14.13 hours after the formation of the Sun was completed, which means 4.28 days since the beginning. Within 1.057 hours, the formation of the Martian planetary system was completed. Many Amor asteroids and Apollo asteroids were formed before the main belt asteroids, which formation was completed 6.078 to 9.785 days after the beginning of the formation of the Solar System. As the main belt asteroids were being born, several comets, Apollo asteroids, and Amor asteroids were also formed. Jupiter was formed 14.99 days after the beginning and after the formation of all the Jovian satellites in Zones 1, 2, and 3. All the Jovian satellites in Zones 4 and 5 were formed at least a day after Jupiter. The outermost Jovian satellite known as of 2021 was formed 20.15 days after the beginning. About 6.75 days after the formation of the outermost Jovian satellite, the innermost Saturnian satellite was born. After the birth of the innermost Saturnian satellite 26.9 days after the beginning, 21 other Saturnian satellites, meaning all of the Saturnian satellites ranging from the innermost one all the way to Titan (the largest Saturnian satellite) were formed before the formation of Saturn was completed 27.31 days after the beginning. By this time, the Saturnian rings were also formed. Saturn was born 29.65 minutes after Titan. All of the Saturnian satellites in Zones 4 and 5 were born after Saturn, with the outermost Saturnian satellite known by 2022 which was born 35.58 days after the beginning of the world.

Many centaurs were born before the innermost Uranian satellite was born 53.86 days from the beginning. The 17 innermost Uranian satellites (i.e. Uranian satellites ranging from Cordelia to Titania, the largest Uranian satellite) were born before Uranus. In fact, Titania was born 54.08 days after the beginning, while Uranus was born just 29.36 minutes later, meaning 54.1 days after the beginning. The Uranian satellites in Zones 4 and 5 were born after Uranus. After the birth of Uranus, 11.09 days passed before the outermost Uranian satellite was born on 65.19 days after the beginning. Afterwards, many centaurs, comets, and Neptune trojans were formed before the first Neptunian satellite (Naiad) came into existence. Between the formation of the outermost Uranian satellite and the formation of the innermost Neptunian satellite, 19.07 days had passed. Triton (the largest Neptunian satellite) was born 84.44 days since the beginning and Neptune was born 3.22 hours later. Nereid (the innermost Neptunian satellite in Zone 4) was born 2.39 days after Neptune. As the Neptunian satellites were being formed, many asteroids including the Neptune trojans and Trans-Neptunian objects were doing the same. From the formation of the innermost Neptunian satellite to the outermost Neptunian satellite (Neso), about 23.8 days passed. Then, between the formation of the outermost Neptunian satellite and the innermost Plutonian satellite (Charon), almost 3 days passed. Charon was formed 111.07 days after the beginning, while Pluto (its primary planet) was form 8.79 hours earlier. Pluto is not the only planet which was formed before its largest satellite. The last body to be formed in the Plutonian planetary systems was Hydra (the outermost Plutonian satellite) born 14.73 hours after Pluto, meaning 111.32 days after the beginning of the formation of the universe. Then, many other asteroids (e.g. Trans-Neptunian objects and Kuiper belt objects) were formed as the remainder of the stack of fluid layers of the precursor of the bodies orbiting the Sun continued its journey. I showed that the Sun is:

- 2.6 days younger than Mercury
- 1.65 days younger than Venus
- 20.97 hours younger than the Earth
- 8.84 hours younger than the Moon
- 14.13 hours older than Mars
- 11.3 days older than Jupiter
- 23.62 days older than Saturn
- 50.41 days older than Uranus
- 80.88 days older than Neptune
- 107.01 days older than Pluto

I also showed that, as the precursors of some outer bodies were moving into deep space, they could have been frozen as their layers were thinning while confronting or passing through colder and colder portions of space. Therefore, they formed comets and other types of asteroids filled with ice. Hence, many remote objects in the Solar System are comets known for their icy tails, which "emit" light as they pass through space. I also made some predictions about the distribution of the celestial bodies toward the outskirts of the Solar System. I invented a term called the "turbulence clustered belt of Zones 4 and 5" in the Solar System to reflect a very huge belt of mostly retrograde and small asteroids that I predict would be beyond 150-270 AU from the Sun.

Another Book by Nathanael-Israel Israel:
RECONCILING SCIENCE AND CREATION ACCURATELY

THERE IS ONLY ONE SIMPLE, COMPELLING, SOLUTION-DIRECTED SCIENTIFIC FORMULA ACCURATE ENOUGH TO RATIONALLY EXPLAIN HOW GOD CREATED THE UNIVERSE

"Reconciling Science and Creation Accurately" is a landmark book in universe-origin writing from a rare perspective by one of the most respected minds of our time. It scientifically explores the most challenging questions of all times that believers, nonbelievers, and all freethinkers are interested in: How can we rationally demonstrate, without checking our brain at the door in the name of faith, that God created the universe? How did the universe begin and what processes did God use to create it? Are these processes still operating in the universe or not? Can believers abandon wrong theories if they think it is impossible for science to literally prove the Genesis story, or if they think that science is evil and diametrically opposed to faith, or if they compromisingly embrace scientific theories that contradict the Biblical account of creation written before the scientific era? What can believers do to help the skeptics believe in the Biblical narrative of creation?

Lucky you, Dr. Nathanael-Israel Israel successfully navigated all those questions with an accuracy that both scientists and nonscientists have been applauding across the globe. After reading *"Reconciling Science and Creation Accurately"*, you will confidently:

- Scientifically prove the Biblical account of the creation of the universe and the existence of God in a way that makes the head of those who deny God to spin faster than a DJ's turntable
- Know how to rationally talk to anti-creationists, evolutionists, Big Bang proponents, atheists, skeptics, and other freethinkers about the universe-formation and they will beg you to know more about God, the Creator, that they mistakenly rejected
- Discover very accurate, rare, and factual truths about the universe-origin that will save you time and money, and get you much closer to the better and joyful life you want to live today and forever
- Improve your health and faith by knowing that the existence of God can be scientifically justified using Science180 Cosmology and particularly Science180 Creationism

- Enter a new area of freedom and power by crushing the head of and breaking free from the suffocating expectations of all wrong theories that have highjacked secular and religious education, and that have held the Biblical account of creation captive for almost 3500 years
- Break free from the suffocating expectations of some forms of creationism that have sequestered the mind of some believers for a long time
- Uncompromisingly, intelligently, and scientifically explode the myth of those who, instead of literally taking the Biblical days of creation as 24-hours consecutive days, think that they were millions of years, or were representative of long ages, or that millions of years existed before them or were positioned between them
- Understand the accurate standard to interpret the Biblical account of creation thanks to Science180's breakthrough that transformed science and laid a foundational bedrock for the inerrancy of Scripture

Now that Genesis (the oldest manuscript in the world, written before science and most religions were born) is scientifically proven to be correct (*Science180.com*/biblical), what unstoppable, jaw-dropping paradigm shift will the discovery of the perfect alignment between science and the Bible bring into the religious, rational, and secular world today? Get this thoughtful book now to figure out what happened at the beginning, what is coming up, and why it is time to urgently rethink everything you have been told about the universe-origin so you don't eventually regret! Don't say nobody told you!

Founder of Science180 Academy, **Dr. Nathanael-Israel Israel** is acknowledged worldwide as the discoverer of the all-in-one, proven, and simple scientific formula that accurately cracked the origin of the universe, of life, and of chemicals, and that scientifically unearthed the holy grail at the intersection of science and the Biblical account of creation. Learn more at Israel120.com.

CHAPTER 39

CAN THE SCIENTIFIC EVIDENCES OF THE UNIVERSE FORMATION CONFIRM ANY RELIGIOUS NARRATIVE OF CREATION AND UNVEIL THE SCIENTIFIC PROOF OF GOD'S EXISTENCE?

- Can science and faith meet on a hot topic such as the origin of the universe?
- Can any creation story be demystified in the 21st century during which churches in Europe and America are being emptied at the profit of secularism?
- Can freethinkers celebrate anything rational with creationists that they think are irrational?
- Does the Biblical account of creation have a scientific power?

39.1. Does any religious narrative scientifically crack the code of the universe formation?

During my investigation, I realized that my research on the origin of the universe would be incomplete if it ignored all of the creation perspectives found in the mythology and religions of the world. I detailed my findings in my books *"Reconciling Science and Creation Accurately"*, *"Origin of the Spiritual World"*, and *"Science180 Accurate Scientific Proof of God"*. Because of the objective and audience of this book, I will not detail those findings here, but I will highlight some key points.

Indeed, to account for all types of knowledge concerning the origin of the universe as perceived by human beings throughout the ages, and to compare what was said long before the scientific era to what scientific evidences point to today,

I reviewed the creation narratives found in the literature. Details on that can be found in my book called "*Science180 Accurate Scientific Proof of God*" and "*Reconciling Science and Creation Accurately*". In these books, I presented the creation stories across the world into 2 categories:

- the Judeo-Christian creation narratives (e.g. found in Judaism and Christianity) and
- the non-Judeo-Christian creation narratives (e.g. found in Animism, Buddhism, Confucianism, Hinduism, Islam, and Evolutionism).

I scrutinized the world's largest religions to see if any religious story of the formation of the universe matched the scientific story I presented in this book. I discovered that none of the animist creation stories matched any aspect of the story of the universe origin I presented in this book. I figured out that, in Buddhism, there is no creator and the belief in a creator God is even often rejected. As the most dominant philosophy or religion in China, Confucianism advocates that the universe creates itself out of a primary chaos of material energy. That religion did not elaborate on the mechanism of such a creation. In contrast, Hinduism does not provide a single undisputed and straightforward account of creation, but various theories of creation grounded on a concept of repeated cycles of creation and destruction. Furthermore, the Quran teaches that Allah created the Earth from the 1st day to the 2nd day of creation. The Quran also teaches that the stars, the Sun, the Moon, and the angels were created on the 6th day of creation. It is very important to understand that the creation story in the Quran is DIFFERENT from that in the Bible. In other words, although Islam supports a form of creationism, the details of the Islamic creation story is different from that of the Judeo-Christian narratives, which is based on the Bible's Book of Genesis found in the Hebrew Bible or the Old Testament of the Christian Bible.

Indeed, Christians and Jews share the same story, which is reported in Genesis, the first book of the Bible. In other words, Judeo-Christian religions (i.e. Judaism and Christianism) drew their creation narrative from the Book of Genesis found in the Tanakh or the Hebrew Bible or the Old Testament of the Christian Bible. In the Book of Genesis, Moses reported that:

Genesis 1: *1 In the beginning God created the heaven and the earth. 2 And the earth was without form, and void; and darkness was upon the face of the deep. And the Spirit of God moved upon the face of the waters. 3 And God said, Let there be light: and there was light. 4 And God saw the light, that it was good: and God divided the light from the darkness. 5 And God called the light Day, and the darkness he called Night. And the evening and the morning were the first day. 6 And God said, Let there be a firmament in the midst of the waters, and let it divide the waters from the waters. 7 And God made the firmament, and divided the waters which were under the firmament from the waters which were above the firmament: and it was so. 8 And God called the firmament Heaven. And the evening and the morning were the second day. 9 And God said, Let the waters under the heaven be gathered together unto one place, and let the dry land appear: and it was so.*

10 And God called the dry land Earth; and the gathering together of the waters called the Seas: and God saw that it was good. 11 … 12 … and God saw that it was good. 13 And the evening and the morning were the third day. 14 And God said, Let there be lights in the firmament of the heaven to divide the day from the night; and let them be for signs, and for seasons, and for days, and years: 15 And let them be for lights in the firmament of the heaven to give light upon the earth: and it was so. 16 And God made two great lights; the greater light to rule the day, and the lesser light to rule the night: he made the stars also. 17 And God set them in the firmament of the heaven to give light upon the earth, 18 And to rule over the day and over the night, and to divide the light from the darkness: and God saw that it was good. 19 And the evening and the morning were the fourth day. (King James Version of the Bible)

In other words, the Bible specifically mentioned the date that the Earth, the Moon, and the Sun were fully formed. How does this story compare with the scientific evidences I presented in this book? Before I address that question, I would like to quickly say that the various interpretations of the Biblical account of creation by people (including Christians themselves) throughout the generation birthed several creationist theories (most of which differ by the duration of the processes involved and the interpretation of some keywords). Some of these theories include what is known as:

- Gap creationism
- Intelligent design
- Neo-creationism
- Old Earth creationism
- Progressive creationism
- Theistic Evolution (Evolutionary creationism)
- Young Earth creationism
- Science180**Error! Bookmark not defined.** creationism and many more that I detailed in my books *"Science180 Accurate Scientific Proof of God"* and *"Reconciling Science and Creation Accurately"*. Science180 creationism is spearheaded by myself, Nathanael-Israel Israel.

Without arguing too much, let's briefly review what science actually says about the genesis of the universe, particularly the split-gathering of fluid layers, and the creation of the Earth, the Moon, and the Sun.

39.2. Summary of the split-gathering of fluid layers

My findings on the origin of the universe proved that, at one point in the beginning, fluid layers were really present and their impact can be felt even today by carefully studying the characteristics of celestial bodies including the trends of their orbital speed, eccentricity, rotation, and semi major axis. Under the influence of turbulence, these fluid layers were moved at specific speeds, and split-gathered into bodies which characteristics match the environment of their formation. I proved that for the Earth, the Moon, and the Sun to be formed, their precursors went through changes and traveled a certain distance at a certain speed.

Fig. 235: Split-gathering of the fluids of the precursor of the Solar System into the Sun and the bodies (e.g. planets) orbiting it

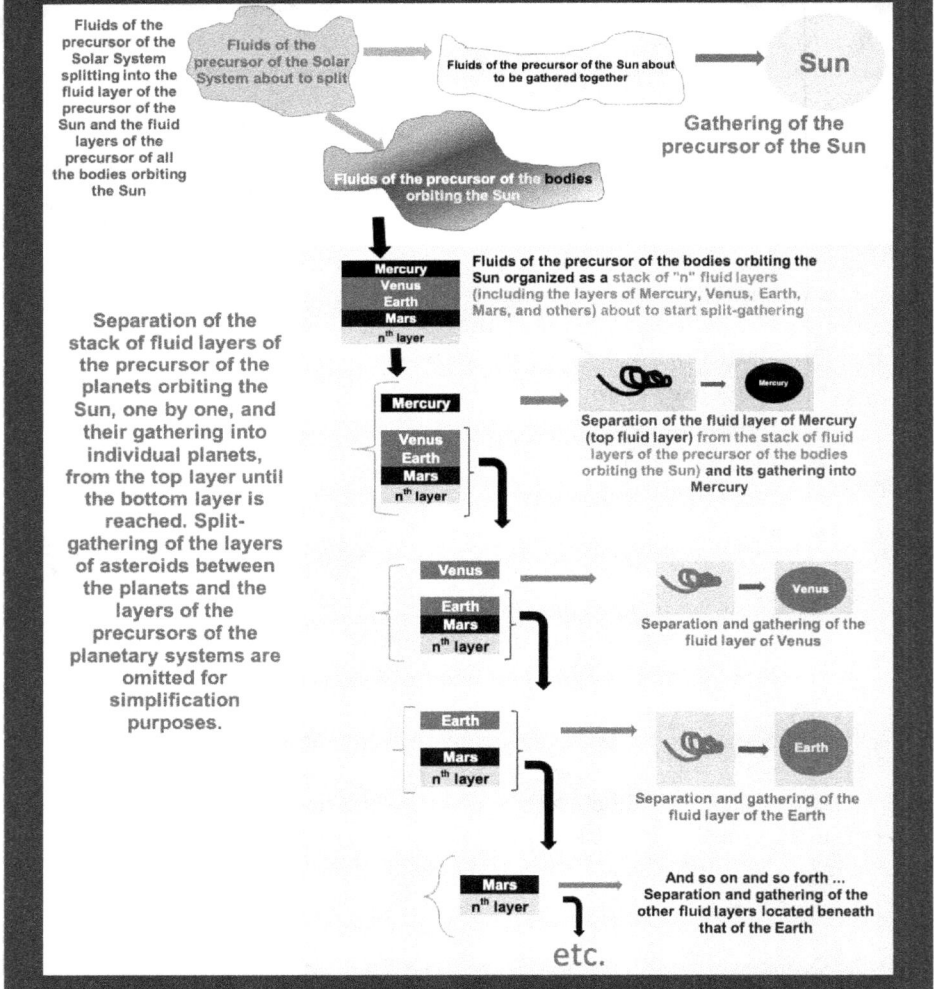

I proved that it took approximately 26.05 hours maximum for the precursor of the bodies orbiting the Sun to split from the precursor of the Sun and reach the position of Mercury, the innermost planet in the Solar System. These 26.05 hours is what I termed the escape time of the precursor of the Sun.

In Fig. 235, I exemplified how the fluid layers of the precursors of the planets and the Sun split-gathered according to their position in the stack of fluid layers of the precursor of the Solar System. For clarity purposes, and so that the graph is not overloaded, I focused on the precursor of the Sun and that of a few planets.

Of all the religious books in the world, only the Bible mentioned that, on the second day of creation, meaning after the first 24 hours from the beginning, waters started dividing from waters. Because no (major) celestial body is closer to the Sun than Mercury, the 26.05 hours means that waters were not yet separated from other waters (in the Solar System) before the beginning of the second day. This fact agrees with the Biblical account about the separation of fluids on the 2nd day. The Book of Genesis talks a lot about waters referred to as "Mayim" in Hebrew. In the days when Moses revealed the Biblical narrative of creation, very little was known about chemical elements, and much less about fluid dynamics. Water being a fluid, the Bible was right by saying that, on the 2nd day of creation, waters were really being divided or separated from waters just like the Bible said (Genesis 1:6-8).

39.3. Summary of the formation of the Earth

I also proved that the precursor of the bodies orbiting the Sun escaped the precursor of the Sun at about 617.6 km/s (i.e. escape velocity of the Sun), and was organized as a stack of fluid layers which embedded the precursor of the Earth-Moon system. I demonstrated that the precursor of the Earth-Moon system had to "wait" until all the fluid layers of all the bodies located above it (e.g. precursor of Mercury, of Venus, and of many asteroids located between the Sun and the Earth) were removed before it could (at its turn) split from that stack of fluid layers and gather itself into the precursor of the Earth and the precursor of the Moon. I specifically showed that the precursor of the Earth-Moon system traveled about 149,600,000 km (i.e. the semi major axis of the Earth) away from the precursor of the Sun before birthing the precursor of the Earth and the precursor of the Moon.

Dividing the semi major axis of the Earth by the escape velocity of the Sun, I proved that it took about 67.29 hours (i.e. 2.804 days) for the precursor of the Earth-Moon to travel from about the position of the precursor of the Sun to the location of the Earth.

$$149{,}600{,}000 \text{ km} / 617.6 \text{ km/s} = 67.29 \text{ hours}$$

I also showed that the fluid layers of the precursor of the Earth were wrapped around at about 29.78 km/s (the orbital speed of the Earth) to form the Earth, which equatorial radius is 6378.137 km. I also demonstrated that, when the precursor of the Earth reached about its semi major axis, the length of its fluid layers were about the circumference of the Earth (which I calculated using the radius of the Earth):

$$\text{Earth's circumference} = 2 * 3.14 * \text{Earth's radius} = 2 * 3.14 * 6378.137 \text{ km}$$
$$= 40054.7 \text{ km}$$

By dividing the final length of the fluid layers of the Earth's precursor by the orbital speed of the Earth (the speed at which they were wrapped around), I proved that the amount of time it took for these fluid layers to be collected into the spherical Earth is:

$$40054.7 \text{ km} / 29.78 \text{ km/s} = 1345.02 \text{ seconds} = 22.42 \text{ minutes}$$

Knowing that, as I explained above, it took 67.29 hours for the fluid layers of the precursor of the Earth-Moon system to move from about the precursor of the Sun to the orbit of the Earth, and knowing that it took 22.42 minutes for these fluid layers to be gathered together into the spherical Earth, I deduced that the amount of time that lapsed before the Earth was completely formed was:

$$67.29 \text{ hours} + 22.42 \text{ minutes} = 67.66 \text{ hours} = 2.82 \text{ days}$$

This proved that the formation of the Earth was really completed on the 3rd day. Fig. 236 recapitulated the formula of the birthdate of the Earth.

This time I calculated scientifically matched the Biblical story of creation according to which the formation of the Earth was completed on the 3rd day (Genesis 1:9-13):

> *9 And God said, Let the waters under the heaven be gathered together unto one place, and let the dry land appear: and it was so. 10 And God called the dry land Earth; and the gathering together of the waters called the Seas: and God saw that it was good. 11 ... 12 ... and God saw that it was good. 13* **And the evening and the morning were the third day.**

Fig. 236: Formula of the Birthdate of the Earth

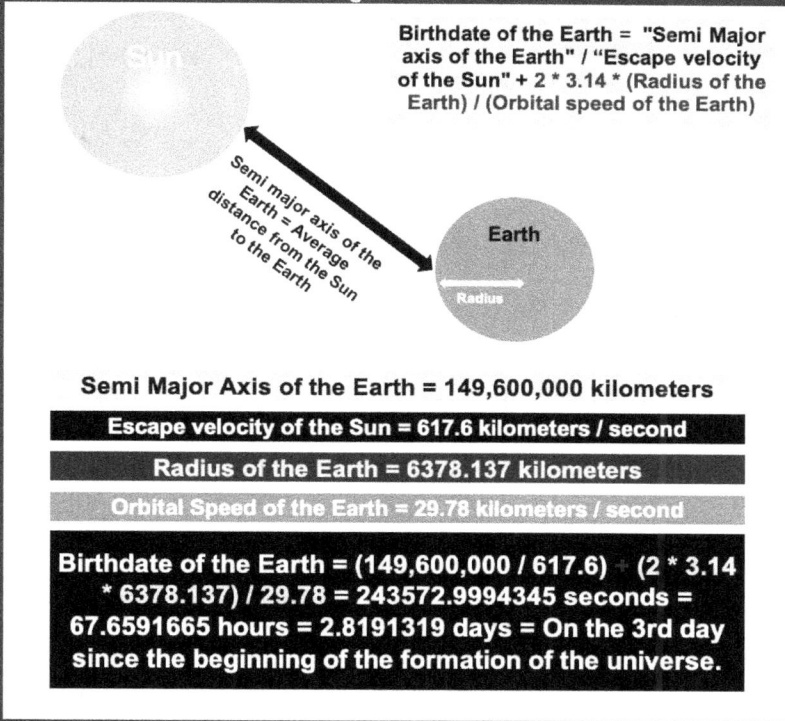

Birthdate of the Earth = "Semi Major axis of the Earth" / "Escape velocity of the Sun" + 2 * 3.14 * (Radius of the Earth) / (Orbital speed of the Earth)

Sun

Semi major axis of the Earth = Average distance from the Sun to the Earth

Earth

Radius

Semi Major Axis of the Earth = 149,600,000 kilometers

Escape velocity of the Sun = 617.6 kilometers / second

Radius of the Earth = 6378.137 kilometers

Orbital Speed of the Earth = 29.78 kilometers / second

Birthdate of the Earth = (149,600,000 / 617.6) + (2 * 3.14 * 6378.137) / 29.78 = 243572.9994345 seconds = 67.6591665 hours = 2.8191319 days = On the 3rd day since the beginning of the formation of the universe.

39.4. Summary of the formation of the Moon

In the previous segment, I demonstrated that it took about 67.29 hours before the precursor of the Earth-Moon system birthed the precursor of the Earth and the precursor of the Moon. In the previous chapters, I also evidenced that the precursor of the Moon escaped the precursor of the Earth at 11.186 km/s (i.e. the escape velocity of the Earth), and traveled about 384,400 km (i.e. semi major axis of the Moon) before reaching the position where its fluid layers were collected into the spherical Moon. By dividing the semi major axis of the Moon by the escape velocity of the Earth, I found that it took 9.54 hours for the precursor of the Moon to travel away from the precursor of the Earth to arrive at the position of the Moon:

384,400 km / 11.186 km/s = 34364.38 seconds = 9.54 hours

My calculation also proved that, once the precursor of the Moon arrived at the semi major axis of the Moon, it was as long as about the circumference of the Moon and was wrapped around at about 1.02316 km/s (i.e. the orbital speed of the Moon) before forming the spherical Moon, which radius is 1738.1 km. Knowing that radius, I proved that the length of the fluid layers of the precursor of the Moon just before they swirled around to form the Moon was:

$$2 * 3.14 * 1738.1 \text{ km} = 10,915.27 \text{ km}$$

By dividing that distance by the orbital speed of the Moon, I revealed that it took 2.96 hours for the precursor of the Moon to be gathered together after reaching its semi major axis:

$$10,915.27 \text{ km} / 1.02316 \text{ km/s} = 10,668.2 \text{ seconds} = 2.96 \text{ hours}$$

Because it took 9.54 hours for the precursor of the Moon to move from about the precursor of the Earth to the semi major axis of the Moon, and because once it arrived there it took 2.96 hours before the fluid layers of the precursor of the Moon could be collected into the Moon, I deduced that it took 12.5 hours (i.e. 9.54 hours + 2.96 hours) from the moment the precursor of the Moon escaped the precursor of the Earth to the moment the formation of the Moon was completed.

$$9.54 \text{ hours} + 2.96 \text{ hours} = 12.5 \text{ hours}$$

Because it took 67.286 hours before the precursor of the Moon was formed––(this amount of time elapsed before the precursor of the Earth-Moon system arrived at the position of the Earth after it moved away from the precursor of the Sun)—the total amount of time it took for the Moon to be completely formed with reference to the beginning of the formation of the Solar System was:

$$67.286 \text{ hours} + 12.5 \text{ hours} = 79.786 \text{ hours} = 3.324 \text{ days}$$

In other words, from the beginning of the formation of the Solar System, 3.324 days elapsed before the Moon was completely formed. In Fig. 238, I illustrated the formula I used to calculate the birthdate of the Moon as 3.32 days.

Fig. 238: Formula of the Birthdate of the Moon (Earth's Satellite)

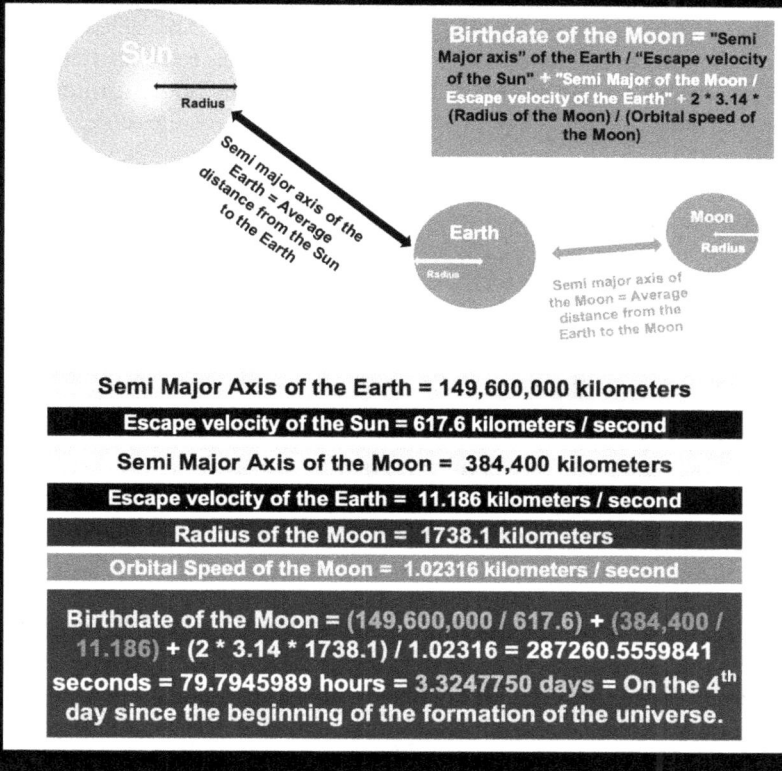

Birthdate of the Moon = "Semi Major axis" of the Earth / "Escape velocity of the Sun" + "Semi Major of the Moon / Escape velocity of the Earth" + 2 * 3.14 * (Radius of the Moon) / (Orbital speed of the Moon)

Semi major axis of the Earth = Average distance from the Sun to the Earth

Semi major axis of the Moon = Average distance from the Earth to the Moon

Semi Major Axis of the Earth = 149,600,000 kilometers

Escape velocity of the Sun = 617.6 kilometers / second

Semi Major Axis of the Moon = 384,400 kilometers

Escape velocity of the Earth = 11.186 kilometers / second

Radius of the Moon = 1738.1 kilometers

Orbital Speed of the Moon = 1.02316 kilometers / second

Birthdate of the Moon = (149,600,000 / 617.6) + (384,400 / 11.186) + (2 * 3.14 * 1738.1) / 1.02316 = 287260.5559841 seconds = 79.7945989 hours = 3.3247750 days = On the 4th day since the beginning of the formation of the universe.

Surprisingly, this timeline perfectly matches what the Biblical account of creation indicated more than 3500 years ago as the birthdate of the Moon. The Moon was indeed formed on the 4th day of creation (Genesis 1:14-19):

*14 And God said, let there be lights in the firmament of the heaven to divide the day from the night; and let them be for signs, and for seasons, and for days, and years: 15 And let them be for lights in the firmament of the heaven to give light upon the earth: and it was so. 16 And God made two great lights; the greater light [the Sun] to rule the day, and the lesser light [**the Moon**] to rule the night: he made the stars also. 17 And God set them in the firmament of the heaven to give light upon the earth, 18 And to rule over the day and over the night, and to divide the light from the darkness: and God saw that it was good. 19 **And the evening and the morning were the fourth day**.*

39.5. Summary of the formation of the Sun

As a reminder, I demonstrated that it took about 26.05 hours maximum for the precursor of all the bodies orbiting the Sun to escape the precursor of the Sun. In fact, the precursor of the bodies orbiting the Sun escaped the precursor of the Sun at about 617.6 km/s (i.e. escape velocity of the Sun). And before such a split was consumed, so that the precursor of the Sun could be ready to start gathering its fluid layers into a spherical body, the precursor of the bodies orbiting the Sun traveled a distance no higher than 57,910,000 km, the average distance between the Sun and Mercury (Mercury being the major closest body to the Sun). Considering the speed I mentioned earlier, that distance, which is known as the semi major axis of Mercury, was traveled in about 26.046 hours:

57,910,000 km / 617.6 km/s = 93766.2 seconds = 26.046 hours

This time was needed before the precursor of the Sun could be fully formed and ready to be wrapped around to form the spherical body called the Sun. I also proved that, just before the fluid layers of the precursor of the Sun were collected into the Sun, they were long as about 4,370,880 km, the circumference of the Sun that I calculated using the radius of the Sun (696,000 km):

Sun's circumference = 2 * 3.14 * 696,000 km = 4,370,880 km

I showed that, once the fluid layers of the precursor of the Sun reached that length at a position near the current location of the Sun within the Solar System, they were wrapped around at approximately 19.4 km/s (i.e. Sun's orbital speed). Therefore, by dividing the circumference of the Sun by the orbital speed of the Sun, I calculated that it took 62.58 hours for the Sun to be completely formed once all the fluid layers of the bodies orbiting it left the precursor of the Sun:

4,370,880 km / 19.4 km/s = 225,303 seconds = 62.58 hours

But because, like I said a few paragraphs earlier, about 26.046 hours passed before the precursor of the Sun was formed, I proved that the total amount of time it took for the Sun to be completely formed was:

26.046 hours + 62.58 hours = 88.63 hours = 3.693 days

In other words, the Sun was formed the 4th day after the beginning of the formation of the Solar System. In Fig. 237, I recapitulated the mathematics of the formation of the Sun in 3.69 days.

Fig. 237: Formula of the Birthdate of the Sun

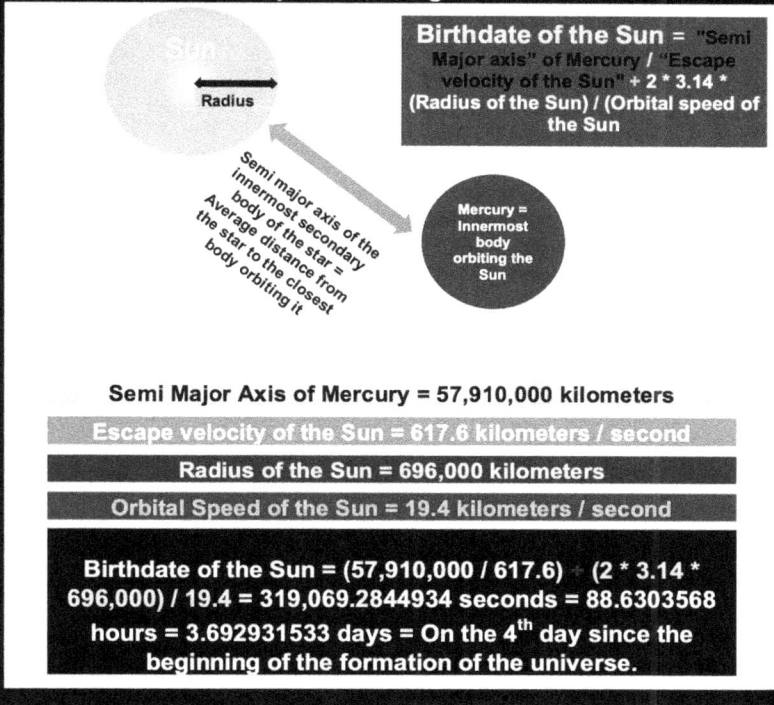

Birthdate of the Sun = "Semi Major axis" of Mercury / "Escape velocity of the Sun" + 2 * 3.14 * (Radius of the Sun) / (Orbital speed of the Sun

Radius

Semi major axis of the innermost secondary body of the star = Average distance from the star to the closest body orbiting it

Mercury = Innermost body orbiting the Sun

Semi Major Axis of Mercury = 57,910,000 kilometers

Escape velocity of the Sun = 617.6 kilometers / second

Radius of the Sun = 696,000 kilometers

Orbital Speed of the Sun = 19.4 kilometers / second

Birthdate of the Sun = (57,910,000 / 617.6) + (2 * 3.14 * 696,000) / 19.4 = 319,069.2844934 seconds = 88.6303568 hours = 3.692931533 days = On the 4th day since the beginning of the formation of the universe.

In other words, according to the scientific evidences that no human being has properly examined until my days, the days of Nathanael-Israel Israel, the Sun was really formed on the 4th day since the beginning of the formation of the Solar System. But guess what! One more time, that is another scientific fact the Biblical account of creation got right: the Sun was formed on the 4th day of creation (Genesis 1:14-19):

*14 And God said, let there be lights in the firmament of the heaven to divide the day from the night; and let them be for signs, and for seasons, and for days, and years: 15 And let them be for lights in the firmament of the heaven to give light upon the earth: and it was so. 16 And God made two great lights; the greater light [**the Sun**] to rule the day, and the lesser light [the Moon] to rule the night: he made the stars also. 17 And God set them in the firmament of the heaven to give light upon the earth, 18 And to rule over the day and over the night, and to divide the light from the darkness: and God saw that it was good. 19 And the **evening and the morning were the fourth day**.*

39.6. Take home message from the intersection of science and religion

The next 2 graphs summarized all of the timelines that I scientifically demonstrated for the separation of fluid layers during the formation of celestial bodies in the universe, and for the formation of the Earth, the Moon, and the Sun perfectly matched the Biblical account of creation.

Fig. 239: Birthdate of selected major events in the history of the formation of the universe (Days)

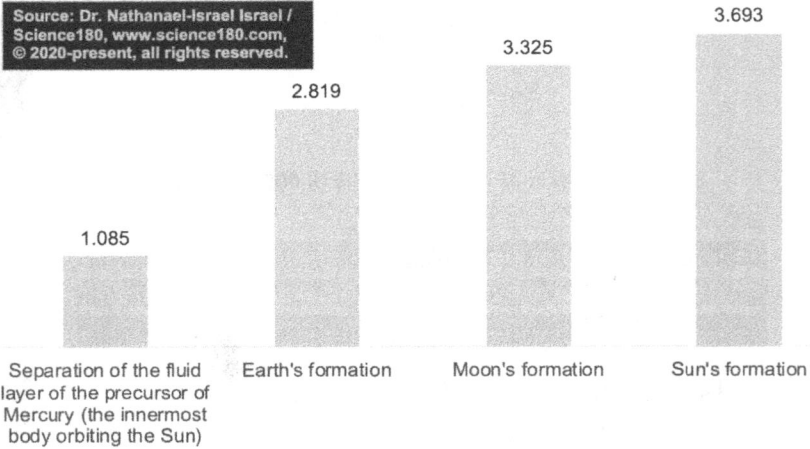

3.693

3.325

2.819

1.085

Separation of the fluid layer of the precursor of Mercury (the innermost body orbiting the Sun)

Earth's formation

Moon's formation

Sun's formation

Nathanael-Israel Israel: Beninese-Born American That Cracked the Universe-Origin Code

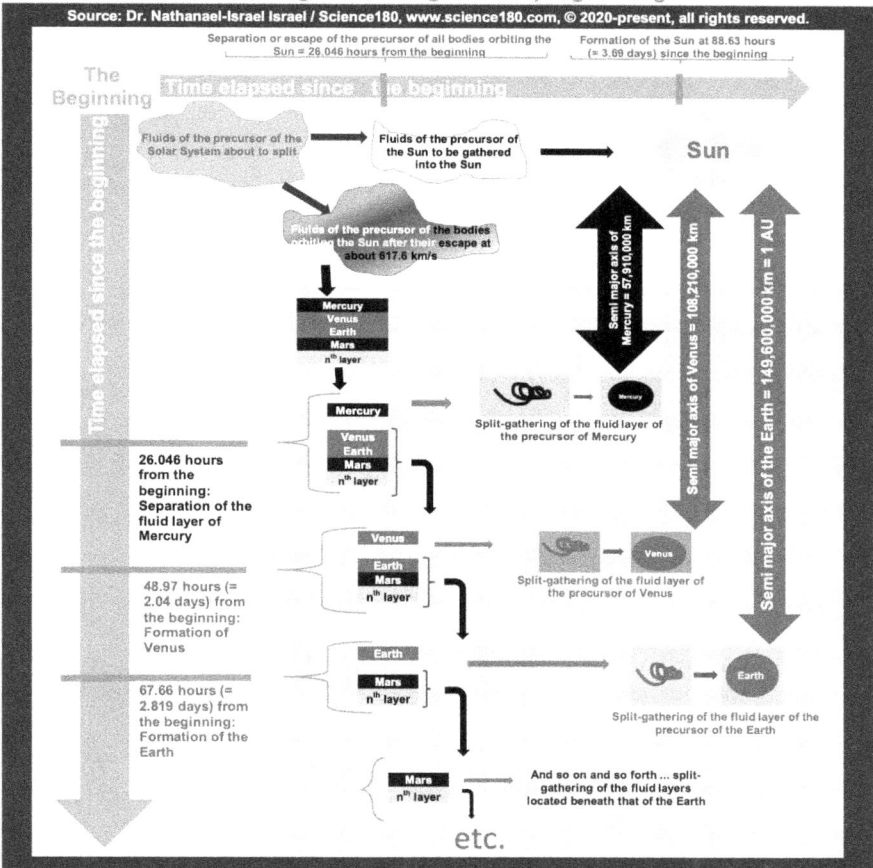

Fig. 240: Illustration of the semi major axis and birthdate of some planets and the Sun according to the timing of their split-gathering

Having taught the perfect match between science and the Biblical account of creation to countless people who registered and attended Science180 Academy (www.Science180Academy), which you can join as well, I understand that you still have many unanswered questions including:

- Are Biblical creationists really explaining creation as they say?
- Why did it take more than 3500 years before the Biblical account of creation is finally scientifically demonstrated? Why now? What is its significance and what is going to happen next?
- Why did scientists miss the formula that leads to the Biblical account of creation?
- How and why did Moses who lived thousands of years before the scientific age accurately decode and recount the formation of the universe in the Biblical creation narrative?

Science180: Find the Origin of the Universe at a Reasonable Price

- Why is the Islamic creation narrative different from the Biblical account of creation?
- What was there before the beginning of the created world?
- What and where was the beginning?
- Where was God before creating the world and during creation?
- What did God do before creation?
- Are the creation days really 24 hours each or millions of years long?
- Were all celestial bodies really formed completely by the end of the 6th day of creation
- Why did God choose to create the universe in 6 days?
- Is it a waste of time to attempt to prove the Bible true by means of science and historical investigations?
- How can mastering the demonstration of God's existence make you happy and improve your life and goals?
- What will you gain or save by reading Science180's creation books (see www.Israel120.com/books)?
- How can knowing more about God, the Creator, help you to do things better to get closer to the best version of yourself that you deserve?
- Can a God's existence book really help you to overcome, alleviate, or eliminate your pains?
- How does Science180 creationism best suit your needs and what makes it stand out from the other theories specialized in universe-origin and God's existence?
- How can Science180's books on God's existence actually give you a unique opportunity to examine religion using science and critical thinking?
- How is my discovery on creation and God's existence a comprehensive scientific resource that doesn't just criticize evolution or the Big Bang theory, but that actually uses modern science to rationally prove that God created the universe just as the Bible says?
- How can my writings help you enjoy reviewing your opinions about the Biblical account of creation based on reason and rational decision-making without feeling influenced by tradition, authority, established belief, or religious pressure?
- Do you need more explanation on how Science180's interpretation of the Biblical Genesis narrative of creation shockingly led to a perfect match between science and the Biblical account of creation, a milestone that until recently was thought impossible?
- Can Science180's discovery on creation satisfy your desire for freedom from beliefs and scientific theories about God's existence?
- Can you use Science180's breakthrough to eliminate the struggles that

Nathanael-Israel Israel: Beninese-Born American That Cracked the Universe-Origin Code you face when trying to decode the universe creation, therefore helping you to gain the respect, admiration, power, and confidence you deserve while addressing God's existence problems?

- Can this discovery remove your worries of not being sure of God's existence, therefore helping you sleep better at night?

- Can you eradicate the common mistakes people make while decrypting God's existence, and use it to challenge the status quo of existing theories on the origin of the universe, life, and chemicals?

- Can you gain the essential knowledge to interpret difficult scientific data and solve the most complex scientific problems related to God's existence?

- How can you test for yourself if what you have learned about creation or God is scientifically true or not?

- Is science really against the Biblical account of creation?

- Can you disrupt all religious and scientific chains of repetitive nonsenses about God's existence and turn your attention toward the original discoveries that lead to greater innovation and prosperity?

- How can you defy the cosmological status quo and embrace the real change that will break the cages that have been holding many people and that they unfortunately ignore?

- Is there any additional material that can help you to know for sure that God exists and that He really created the Earth, the Moon, and the Sun as the Bible says?

- Do you need more details about how modern science proves the inerrancy of the Biblical account of creation of the universe so you can put a full stop to all scientific and religious efforts threatening to undermine faith or to oppose it with science?

- If you open your mind to the truth in the Biblical account of creation, can that benefit you or humankind and really make you a resourceful person for people seeking to understand the relation between science and the Biblical account of creation?

- Will Christians use Science180 creationism to sue evolutionists and Big Bang proponents?

- Can evolutionism be banned from public schools or will the introduction of God in public schools impede science?

- Will Christians use Science180's findings to avenge the secularists who have casted God out for the public systems for a long time?

- Will any political leader take executive orders to enforce the teaching of Science180 creationism in public schools to honor God and make the Bible famous?

- Must Christians apologize to freethinkers for the wrong proofs or pseudoscience that some creationists have used to argue biblical creation scientifically?

- Will school boards, policy makers, and administrators start allowing the teaching of Science180 creationism in public schools and embody it in doctrines and regulations, even political laws?
- Can we continue to ignore the intersection of science and the Bible while still expecting to build a holistic support to advance humankind without letting modern science lead us to a place we won't regret?
- Did some Christians complicate the logical trail that some freethinkers could use to scientifically understand God and believe in Him, or will freethinkers persevere in their direction and brace for the impact they could not see yet?

Because of the scientific objectives of this book, I don't want to overload you with religious things here, even when I proved that the Biblical account of creation is correct, but in other books I detailed the answers to many questions I know you are still having concerning creation and God's existence. I discussed these hot topics in other books because, there, I felt more comfortable to address those interested in spiritual or religious things rather than alienating some people or diluting my message in this already-big book targeting all scientists including unbelievers. Those other books of mine include:

- *Science180 Accurate Scientific Proof of God*
- *From Science to Bible's Conclusions*
- *Reconciling Science and Creation Accurately*
- *Origin of the Spiritual World*

Mindful for the needs of children and their parents, I also wrote 2 children's versions of these books that all kids ages 7-12 and beyond can understand:

- *How God Created Baby Universe* (this book is excellent for children who or whose parents believe in God)
- *How Baby Universe Was Born* (this is very good for children who or whose parents don't believe in God, but in secular theories)

I also published 2 other books in 2025 handling the origin problems of life and of chemical particles:

- *Turbulent Origin of Life*
- *Turbulent Origin of Chemical Particles*

Although I knew that everybody will not agree with everything I said in this chapter, I felt like it would be unthoughtful, irresponsible, and a scientific mistake from my part not to tell you that science confirmed that the universe was really formed in a matter of days just like the Bible said a long time, even before science was established by human beings. I also want you to reap the benefits associated with correctly knowing the truth so you better prepare for what lays ahead of us that some people deny. Learn more at Science180Creationism.com

CHAPTER 40

DOES SCIENCE TEACH ANYTHING ABOUT CREATION AND GOD'S EXISTENCE THAT WE CAN DEMONSTRATE WITHOUT FALLING INTO THE TRAP OF TAKING SIDES BETWEEN RATIONALITY AND FAITH?

For the first time in history, I scientifically explained the underlying cause of the formation of the universe, and I settled the disagreements between the scientific explanation of the origin of the universe and the creation narratives. I proved that some of the key drivers of the formation of the universe are the turbulence that took place at the beginning of the universe and the split-gathering process associated with it. I coined the term "Science180 Cosmology", "Turbulence 1", or the "Mother of all turbulences" to designate the theory of the formation of the universe that I demonstrated in this book. Not only did this theory summarize many naturals laws, but it also settled the longstanding contradiction or disagreement between the scientific data and the creation stories, particularly the Biblical creation according to which the Earth was born on the 3rd day, the Moon, and the Sun on the 4th day of creation, each creation day being literally 24 hours long. Because of the scope of the data that I handled and the bridges I built across many scientific fields, I know that this theory has opened the doors to new areas and methodology of investigation, hopefully for the advancement of humankind, if the scientists who will use my findings as the foundation of their research will lay proper hypotheses.

> I coined the term "Science180 Cosmology", … or the "Mother of all turbulences" to designate the theory of the formation of the universe that I demonstrated in this book

A careful study of the characteristics of the celestial bodies in the universe and a strategic thinking in a reverse-engineering fashion allowed me to pinpoint the footprint of the turbulence that birthed everything in the universe. For instance, instead of focusing just on mathematical equations to capture the essential mechanisms behind the formation and functioning of the universe, I chose to properly have an intimate understanding of the mechanisms first, and then to explore which mathematics can unearth the relationship between the key players of the story. For, instead of first chasing mathematical equations or rules governing the universe, I proved that the bodies in the universe were formed and maintained by a complex and diversified program that all scientists before me have failed to decode. Hence, scientists have been having a hard time unraveling or modeling it using the conventional mathematics of complex systems. But once I understood the program of split-gathering of the precursors of bodies, I was able to explain the equations governing the universe formation. In other words, after years of investigation, I realized that the program that embodies the fundamental model of the universe is hidden in the turbulent split-gathering processes that I extensively explained throughout this book. Below, I will highlight some of the most salient points.

Indeed, I showed that, in the beginning of the formation of the universe, an initial matter that I termed the "turbulent prima materia" (with characteristics different from those of any current matter) mysteriously appeared in the universe out of nothing, and through turbulent processes, it was progressively fashioned into all matters currently in the universe. That initial matter was formless and occupied a very vast and deep portion of space. The state of the "turbulent prima materia" was different from the main four states of matter known today (e.g. solid, liquid, gas, and plasma). As it went through changes, the turbulent prima materia birthed the current states of matter. Soon after its appearance, the turbulent prima materia was broken open into giant and small pieces by a violent event like an "explosion" that initiated what I termed the "original mysterious scattering" which sent into specific directions blocks or clusters of the bulk initial matter.

As the clusters of matters started moving, time and energy as we know today were born. At some point, the matter in some precursors reached a fluid state. Cascades of turbulent "explosions" occurred until a moment when the clusters could no longer explosively split. Meanwhile, turbulent changes were already occurring in the bulk of matter that were breaking apart, leading to the formation of precursors of various systems of bodies (clusters of galaxies, galaxies, and all other celestial bodies and particles found in the universe). All these precursors went through processes including fluid instability, fluid breakup, transfer of energy, initiation of movement (e.g. revolution and rotation), fluid flow, formation of fluid layers, development of turbulence, fluid mixing and breakup, stretching and tilting of precursors, gathering together of fluid layers, spiraling of fluid layers, squeezing of fluid layers, etc.

As the precursors of the celestial bodies were splitting, their internal components were also being gathered together. To express the joint process of splitting and gathering together of the precursors of bodies, I coined the term "split-gathering together" or "split-gathering" in short. As the precursors of matters started flowing, fluid layers were formed, and stacked on top of others according to the system they belonged to. In each stack of fluids, the fluid layers moved at specific speeds which contributed to producing turbulence on different scales, with different intensities, at different positions, and led to the formation of various bodies on various scales in the universe. From the invisible or the smallest microscopic and even submicroscopic scales to the largest astronomical ones, processes similar to what I just explained occurred and led to the formation of subatomic particles, atoms, molecules, minerals, rocks, celestial bodies (e.g. stars, planets, asteroids, and satellites), planetary systems, stellar systems, globular clusters, galaxies, galaxy clusters (and higher orders of galaxy clusters), and everything else in the whole universe. I showed that the fluids in the precursors were characterized by:

- Their formation and shearing
- Their intermittence and mixing
- The top layers of the precursors of secondary bodies led to the formation of the innermost secondary bodies whereas the bottom ones birthed the outermost secondary bodies
- The stratification of the fluid layers of the precursors is responsible for the stratification of the crust and atmosphere of the celestial bodies
- Speed of the fluid layers was later "translated" into orbital and rotational speed of the daughter bodies according to the turbulence their precursor went through
- Direction of the flow of the fluid layers can explain the direction or sense of the revolution and rotation of some bodies
- Topological and structural changes of the fluids of the precursors of bodies account for many of the variations in the composition of the bodies

After an extensive study of the variables of the celestial bodies, I realized that, to draw meaningful information from them, I needed to creatively organize them into groups so I could better analyze them not only statistically, but holistically. Toward that end, to facilitate the study of the fluid layers of the precursors of the bodies, I divided them into 7 relevant regions that I called the turbulence zones, ranging from turbulence zone 0 to turbulence zone 6. For instance, turbulence Zone 1 of the planetary systems is where the innermost and fastest satellites are usually found. Turbulence Zone 3 is where the highest turbulence leading to the largest secondary bodies in a fluid flow (of the precursor of the secondary bodies) are usually found. A key characteristic of Zone 3 of the satellite systems is that, the rotation period increased from the innermost satellite in Zone 1 until the outermost satellite in Zone 3, meaning that the outermost satellite in Zone 3 usually has the highest rotation period among all of the bodies in Zones 1, 2, and 3, and after it the rotation period dropped. Turbulence Zone 5 is the outermost zone, where the remote and the slowest celestial bodies in each system are found. Zone 6 is located outer of Zone 5 and is a transition zone toward adjacent downstream or upstream systems. I devoted a specific book to the turbulence zones and if you are interested in it, please contact me at Israel120.com/contact.

I showed that during the cascade of breakup of the turbulent prima materia, precursors of galaxy clusters and globular clusters were formed, and birthed precursors of galaxies, which at their turn birthed precursors of stellar systems, which birthed precursors of primary stars and the precursors of the bodies orbiting them, which at their turn birthed the precursors of planetary systems and asteroid systems, which birthed the precursors of planets, asteroids, and satellites, in which precursors of minerals, mineraloids, and rocks were formed, containing precursors of chemical compounds, molecules, atoms, which birthed the precursors of subatomic particles, and all other small particles that may never be scientifically discovered.

During the split-gathering of the aforementioned precursors, smaller precursors were born between them through what I called turbulent intermittence. Due to the cascade of split-gathering, the universe appears as a nest of clusters of bodies.

One of the key elements of the process without which it would have been very difficult for me to explain the formation of the universe was that, very early in the beginning, the precursors of the bodies were organized as layers, some of which became fluids at one point. The interaction of the fluid layers of the precursors of bodies helped me understand the relation between the orbital speed and the semi major axis of the bodies.

To easily compare some characteristics of the split-gathering, I invented the term 9 system-additive variables to refer to the following: mass, orbital angular momentum, orbital moment of inertia, rotational angular momentum, rotational kinetic energy, rotational moment of inertia, total kinetic energy, translational kinetic energy, and volume of the bodies I studied. I demonstrated that, during the split-gathering of the precursor of a system of bodies, more than 99% of the 9 system-additive variables was pushed into the precursor of the primary body and less than 1% was pushed into the precursors of the systems of secondary bodies. I also proved that during the split-gathering of the precursors of the secondary bodies, more than 99% of the 9 system-additive variables were accumulated in their most turbulent zone, usually Zone 3.

I showed that, after the precursor of the bodies orbiting the Sun escaped the precursor of the Sun at about 617.6 km/s (the escape velocity of the Sun), it was organized as a stack of fluid layers, which moved at different speeds according to their position. In that stack of fluids, the precursor of the Earth-Moon system was embedded somewhere and had to "wait" until the fluid layers of the bodies located above it (e.g. precursor of Mercury, of Venus, and of asteroids located between the Sun and the Earth) split and were removed from above the precursor of the Earth-Moon before the latter could (at its turn) split-gather into the precursor of the Earth and the precursor of the Moon.

The distance traveled by the precursor of the Earth-Moon system before splitting from the fluid layers below it was about the semi major axis of the Earth (149,600,000 km =1 AU). In other words, after the precursor of the bodies orbiting the Sun escaped the precursor of the Sun, the distance traveled by the precursor of the Earth-Moon system before it split from the fluid layers below it was about the semi major axis of the Earth (149,600,000 km).

After the fluid layers of the precursor of the Earth-Moon system split from the stack and became "independent" from all of the other fluid layers that were above and below it, it did not take too long before it split into the precursor of the Earth and the precursor of the Moon. By this time, the precursor of the Earth started moving at about 29.78 km/s (i.e. the mean orbital speed of the Earth), and was collected together to form a spherical planet called Earth, which equatorial radius is 6378.137 km. Dividing the semi major axis of the Earth (distance separating the Sun and the Earth: 149,600,000 km) by the escape velocity of the Sun (617.6 km/s), the time it took for the precursor of the Earth-Moon to reach its position is about 67.29 hours (i.e. 2.804 days).

$$149{,}600{,}000 \text{ km} / 617.6 \text{ km/s} = 67.29 \text{ hours} = 2.804 \text{ days}$$

I then showed that, about 22.42 minutes (circumference of the Earth / orbital speed of the Earth) after reaching the semi major axis of the Earth, the fluid layers of the precursor of the Earth were gathered together by swirling or curling to form a spherical body almost just as a fluid "filament" long as the circumference of the Earth would curl to form a vortical structure, but in this case a compacted, spherical planet.

Adding together the time it took for the fluid layers of the precursor of the bodies orbiting the Sun (which embedded the precursor of the Earth-Moon system) to move from about the precursor of the Sun to the orbit of the Earth (67.29 hours) and the time it took for the precursor of the Earth to be gathered together around its circumference (22.42 minutes), the duration of the formation of the Earth is:

Nathanael-Israel Israel Earth's birthdate = 67.29 hours + 22.42 minutes
= 67.66 hours = 2.82 days

In other words, the Earth was formed on the 3rd day since the formation of the Solar System.

I showed that the precursor of the Moon escaped the precursor of the Earth at about 11.186 km/s (i.e. the escape velocity of the Earth) and traveled about 384,400 km (i.e. the semi major axis of the Moon) before reaching a point where it was in a position to be collected together into a satellite. By dividing the semi major axis of the Moon by the escape velocity of the Earth (i.e. speed of the travel of the precursor of the Moon), I showed that 9.54 hours after escaping the precursor of the Earth, the precursor of the Moon was in a position to collect itself into the Moon. The time it took for the fluid layers of the precursor of the Moon to gather themselves into the spherical Moon (radius = 1738.7 km) was about the circumference of the Moon divided by the orbital speed of the Moon:

$$2 * 3.14 * 1738.1 \text{ km} / 1.02316 \text{ km/s} = 10{,}668.2 \text{ seconds} = 2.96 \text{ hours}$$

With respect to the precursor of the Earth-Moon system, the time elapsed before the Moon was fully formed is the sum of the time it took for the precursor of the Moon to escape the precursor of the Earth-Moon system and reach the orbit of the Moon (9.54 hours) and the time it took for the fluid layers of the precursor of the Moon to be wrapped around to form a spherical body having a radius like that of the Moon (2.96 hours) = 9.54 hours + 2.96 hours = 12.5 hours.

$$9.54 \text{ hours} + 2.96 \text{ hours} = 12.5 \text{ hours}$$

In other words, the Moon was formed 12.5 hours after the split of the precursor of the Earth-Moon system into the precursor of the Earth and that of the Moon.

Because it took 67.286 hours for the precursor of the Earth-Moon system to move from the precursor of the Sun before splitting into the precursor of the Earth and the precursor of the Moon, the total amount of time that elapsed since the beginning of the formation of the Solar System before the Moon was formed was:

$$\text{Nathanael-Israel Israel Moon's birthdate} = 67.286 \text{ hours} + 12.5 \text{ hours} =$$
$$79.786 \text{ hours} = 3.324 \text{ days}$$

This means that the Moon was formed on the 4th day after the beginning of the formation of the Solar System.

I showed that, before the precursor of the Sun could start gathering its fluids into a spherical body, the precursor of the bodies orbiting the Sun traveled a distance no higher than the average distance between the Sun and Mercury (Mercury being the closest body to the Sun). That distance, which is also called the semi major axis of Mercury, is equal to 57,910,000 km. Considering that distance and the speed of that journey, the maximum time it took for the fluid layers of the bodies orbiting the Sun to escape the precursor of the Sun is 26.046 hours.

$$57{,}910{,}000 \text{ km} / 617.6 \text{ km/s} = 26.046 \text{ hours}$$

I coined the term escape time of the precursor of the Sun to designate that duration of time. Once the precursor of the bodies orbiting the Sun escaped the precursor of the Sun, the fluid layers of the precursor of the Sun moved at about

Nathanael-Israel Israel: Author of "From Science to Bible's Conclusions"

the orbital speed of the Sun (19.4 km/s) and was gathered together to form the Sun. Knowing the radius of the Sun (696,000 km), I estimated its circumference, 2 * 3.14 * 696,000 km = 4,370,880 km, which is in the order of magnitude of the distance a fluid parcel could have traveled to complete one turn around the Sun. Dividing the circumference of the Sun by its orbital speed, I showed that the time it took for the precursor of the Sun to gather together around all its fluids to form the spherical body like the Sun is about 62.58 hours.

$$2 * 3.14 * 696,000 \text{ km} / 19.4 \text{ km/s} = 62.58 \text{ hours}$$

This duration of time is what I originally called the circumference timescale of the Sun. The total amount of time it took for the Sun to form is the sum of the time it took for all of the fluids of the precursor of the bodies orbiting it to clear the way or to escape the precursor of the Sun (26.046 hours = escape time of the precursor of the Sun) and the time it could have taken for the fluid layers of the precursor of the Sun to be wrapped around to form a spherical body which radius is equal to that of the Sun (62.58 hours: Circumference timescale of the Sun). That amount of time is 3.693 days:

Nathanael-Israel Israel Sun's birthdate = 26.046 hours + 62.58 hours = 88.63 hours = 3.693 days

In other words, the formation of the Sun was completed on the 4th day after the beginning of the formation of the Solar System.

At this point, I decided to explore whether any religious story of creation matches the scientific evidences I just summarized. While studying the world's religions, I shockingly discovered that these timelines perfectly match the Biblical creation story revealed in the Bible's Book of Genesis, which mentioned that, on the first day of creation, God created the (precursor of the) heaven and Earth, but on that day, the Earth was still not fully formed yet but was fully formed on the 3rd day. In other words, the Bible said that the formation of the Earth was completed on the 3rd day, while the formation of the Moon and the Sun was completed on the 4th day of creation. Undoubtably, the scientific evidences confirmed that the creation days were 24 hours each, NOT millions of years as some people think.

The Bible also talks about waters (meaning fluids) started separating from waters on the second day of creation, therefore confirming the scientific evidences according to which fluid layers of the precursor of the bodies orbiting the Sun started separating by the 2nd day of the beginning of the formation of the Solar System.

In the Biblical story of creation, Moses did not say that the universe was formed by chance, but he clearly stated that God created it. Because the story that

Moses told more than 3500 years ago is scientifically found to be true today, we must also accept the fact that Moses said that the God of Israel is the Creator. Furthermore, in addition to the Book of Genesis, Moses also authored 4 other books, in which he revealed that there is only one God, the God of Israel, referred to in the Biblical creation story as Elohim, but known to the Jews by many other names: Adonai, Yahweh, Jehovah, the God of Abraham, Isaac, and Jacob, etc. In other words, although people across the globe and throughout history labeled their idols as "gods", there is only one true God, the God of Israel (whom I am proud to call the God of Nathanael-Israel Israel), the God who created the heavens and Earth and everything within and between them. Any other being called a "god" or even "God" besides Him is a false god or an idol. Moreover, it is a huge mistake to think or pretend to unite people across the globe by claiming that there are many gods or that the being called "god" in all the religions of the world are the same. Not only was I able to decode how the world was formed, but I also calculated the date of creation and the age of the universe and the date that the world will end. I addressed the age of the universe in another book. Contact me to learn more about that.

Before my discovery, all of the scientific models that attempted to unify the theories on the formation of the universe and its content have failed because, among many things, the people working on them did not take the time to properly think (without major prejudices) through the machinery of the main processes involved before dumping their data into their so-called "sophisticated" software or mind, as if the power of their computational modeling is sufficient to properly resolve the complex problems associated with their data and the way they are interpreted by people who have no clue of what they are doing. With time, I understood that what people think is disorder or chaos is just a plurality of orders that they failed to grasp. In other words, it is because some people's mind is unable to comprehend the orders coming from many angles or directions that they labeled turbulence and other orders in the universe as disorder or chaos. In fact, having gone deep into the study of turbulence, which some people labeled as chaos or lack of order, I have apprehended that chaos (at least natural ones) has a maximum level of order, but which can be decoded only with the proper code. To put it in other words, it is the unintelligence, or lack of wisdom and other useful aptitudes of higher levels that caused some people to think that so-called chaos (in nature) is a lack of order. Put another way again, those who allude to chaos as a lack of order are just trying to superpose their own sense of order to the multiple dimensional orders offered by (so-called) chaotic events or opportunities in nature. Therefore, if we can start looking at chaotic things with more tolerance, impartiality, an open mind, and diligence by "listening" to or considering the "need", "voice", or contribution of each of the things or beings involved in the so-called "chaos" or "disorder", we would realize how much information or wealth of information we have lost by neglecting to deeply analyze nature using unconventional means that go beyond modern science.

Moreover, the variations observed among the variables I studied in the

Nathanael-Israel Israel: Author of "From Science to Bible's Conclusions"

universe suggested to me that the calibration of the law beneath the formation and functioning of the systems of bodies in the Solar System is not the same and should not be expected to be the same across all the stellar systems in the universe. Therefore, it will be impossible for scientists to properly explain the variables available on the celestial bodies in the universe by just focusing on laws discovered on Earth or in the Solar System using a mind that denies the impossible. Consequently, unless someone has a mysterious or divine revelation, it may be very useless and dangerous to try to explain everything in the universe based on the few laws found or forged on Earth. Therefore, we need to be modest and humble in how we extrapolate laws we find or create in nature. Likewise, I would caution you to be careful in how you may extrapolate the findings I presented in this book and the 8 others I published in 2025 (see www.Israel120.com/books) as you may try to explore other systems. For each system is calibrated according to specific rules, and applying to the same rules to decode everything will not work, but lead to more problems.

Furthermore, the same mathematics that allowed me to calculate the birthdate of the Earth, the Moon, and the Sun (just as the Bible says) also proved that some celestial bodies were born several days, even months after the beginning. In other words, although my findings proved that the Biblical story of the creation of celestial bodies is correct, it is important to warn people not to think that the 6 days of creation mentioned in the Bible are referring to the formation of all the celestial bodies in the universe today. In other words, although the precursors of all celestial bodies were born and going through changes before the end of the 6th day of creation, it took many more days and even months before the precursor of some celestial bodies could reach their orbit and be fully formed. To make a long story short, the creation by God did not stop on the 6th day after the beginning, but since then, new things and beings were formed through processes or codes already established in the precursors of the bodies by the end of the 6th day. Hence, it is imperative we obey God's law so we can live in harmony with all created things. For everything in nature has a mission encoded in them or their precursor since the beginning.

I understand that while some people believe in God as the Creator, others credit an intellectual designer (which they do not always view as God) of the great order in the universe, while others yet see chaos and randomness, and others completely deny God. As for me, I did not find any other source or origin of the universe than the omniscient and omnipotent God, specifically the God of Israel revealed in the Bible. I understand that some people will not like the fact that I mentioned the name of God here, but as I was finishing this book, I needed to, for some people who are reading this book may never read the other books in which I expounded on how God created the universe, and because I don't want to be held accountable for withholding the truth, I needed to clarify my conclusions on God's existence. And also inform you that I devoted a particular book to that:

TURBULENT ORIGIN OF THE UNIVERSE

Science180 Accurate Scientific Proof of God: Can We Scientifically Explain the Formation of the Universe Through Natural Processes Without Evoking Evolution and Big Bang?

Without a proper guidance and compass, no one could be able to imagine the ramifications of the events that surrounded the origin of the world. But by properly setting my mind and learning from my own life journey, it was "easy" for me to reject certain pathways of thinking that have blinded and complicated the understanding of the functioning and origin of the universe by most of the people who lived before me. Although this complicated task could have been impossible some centuries ago, if people did not blindly accept some scientific methods, which have been claiming to not be seeking for truth, they may have laid the foundation for a better theory a long time ago. Moreover, based on my journey, I understood that some of the details I unraveled in this book can never be discovered using experiments in a laboratory and others may have not been properly understood without some advancement in fluid dynamics and an intimate understanding of many scientific disciplines. Put another way, a detailed insight into the origin of the universe requires a more comprehensive approach than a mere linear laboratory experiment can allow. The small size of lab equipment also suggests that it may be impossible for scientists to conduct large scale experiments that can allow them to clearly see and study some turbulence details. Even if scientists can design larger scale experiments than what they can currently handle, one thing is to collect data, and another is to know how to analyze them. Moreover, the cost, collection, and analysis of bigger experimental data is another challenge facing turbulence researchers, without forgetting the mindset of the lobbyists who are financing most scientific researches, and drifting them away from the truth that science (let's say the scientific methodology) claimed it is not seeking! This also means that human beings are limited in the replication of the large-scale turbulence found in nature. Doing bigger scale experiments is like simulating the formation of the universe with the limited human power. Consequently, the accuracy of scientific modeling is limited as many details are missing.

There are many aspects of the origin of the universe that I cannot handle in the secular version of this book, but that I addressed in other books you can get at www.Israel120.com/books.

To you, please receive my sincere gratitude for setting aside time to read this book. I know that by the time you reached this point, you have learned a lot of new things that have improved your understanding of the universe. If you have any questions, suggestions, or comments, please feel free to contact me at www.Israel120.com. I will be waiting to hear from you. If you like the book, inform others about it. One more time, thank you for your consideration and time. To God be the glory!

Nathanael-Israel Israel: Author of "From Science to Bible's Conclusions"

NEXT STEPS OF THE JOURNEY

Get free resources on Science180.com

If you have finished reading this book and would like to learn more about my discoveries and how they can help you, you are at the right place. Indeed, I am really committed to helping you address any questions that you may still have concerning the origin, functioning, and fate of the universe, and how you can partner or collaborate with me for greater results.

To get free resources that will help you understand other aspects of the universe formation not covered in this book, visit Science180.com and my personal website Israel120.com. On those sites, I will be sharing guides and strategies to get the most out of my initiatives. I will also be sharing my favorite references, tips, next-steps readings, and other important things in the pipeline that will help you regardless of your field of expertise, interest, and needs.

Subscribe to "Science180 Newsletter": The only accurate universe-origin, life-origin, and chemicals-origin newsletter in the whole world!

Be a part of decoding the universe-origin, life-origin, and chemicals-origin! Get origin-related news, information, discoveries, updates, announcements, news, reviews, articles, educational materials, and opportunities, from a holistic perspective not available anywhere else so you can participate in and enjoy decoding the origin, current state, and fate of the universe and its content. You will also receive priceless tips about how Nathanael-Israel thinks, what are his secrets and initiatives, what he has accomplished, and what he recommends. Without any delay, sign up for Science180 Newsletter today at Science180.com/newsletter. It is free!

Speaking engagement

In addition to writing groundbreaking books and engaging in other business endeavors, Nathanael-Israel Israel is a renowned speaker, who you can invite to speak at your organization.

Values that Dr. Nathanael-Israel Israel can add to your life include:

- Rare expertise and tips that will increase your abilities
- Usefulness that will advance your impact regardless of your field of expertise
- Understanding of the world that will sharpen your perspective
- Critical information that will positively change your life
- Experiences turned into insight that will motivate and guide you

- Accurate demonstration of the historic formula that reconciled science and faith
- Enlightenment that will help people to start using their brain instead of just praying and expecting God to do everything for them

For speaking inquiries, including how you can get Dr. Nathanael-Israel Israel to speak to your organization or at an event, visit Science180.com/speaking for more details.

As the standout scientific authority who accurately decoded the universe, Nathanael-Israel Israel has been helping countless people across the globe to discover and understand the complex origin of the universe without leaving out the challenging questions that people of all ages have been struggling to answer for thousands of years! As the true go-to expert when it comes to the formation of the universe and of life, Nathanael-Israel believes that, regardless of age, background, culture, religion, profession, everyone deserves to understand how the universe and life were formed and how they can leverage on that knowledge to improve lives nonstop. Therefore, his groundbreaking discoveries of the formation of the universe, life, and chemicals have been broken down into books tailored to scientists (including physicists, chemists, biologists, mathematicians), laypeople or general public, believers and freethinkers, philosophers, children, etc., therefore maximizing the benefits to humanity. These historic, internationally-acclaimed origin books include:

- "Turbulent Origin of the Universe"
- "Reconciling Science and Creation Accurately"
- "Turbulent Origin of Chemical Particles"
- "From Science to Bible's Conclusions"
- "Turbulent Origin of Life"
- "Origin of the Spiritual World"
- "How Baby Universe Was Born"
- "How God Created Baby Universe"
- "Science180 Accurate Scientific Proof of God"

When you hire Nathanael-Israel Israel to speak at your organization, you will:

- get specific in-depth knowledge, up-to-the-minute information, ideas, and insights about the universe-origin, life-origin, and chemicals-origin so that you expand your market, cut useless costs, stop wasting time on inadequate projects, and start focusing on the profitable solutions
- get relevant universe-origin stories that are specific to your field of expertise

Nathanael-Israel Israel: Acknowledged as the World's Most Trusted Expert that
Properly Decoded the Formation of the Universe, Life, and Chemicals

- learn from a cooperative, flexible, and an easy to work with expert who will respond to your universe formation needs and position you to stay on top of your competitors
- interact with a renowned expert that will not just lecture you, but that will help you sort out your origin-related questions using strategies to tap into deep secrets you ignore
- listen to an experienced expert who discovered outstanding secrets about the origin of all there is
- learn authentic information not from someone who just reads you a PowerPoint, but from the true go-to expert (when it comes to critical cosmological problems) who will share with you both his mistakes and successes that will help you get much closer to the better life you want to live
- revolutionize every origin-related domain with your accurate understanding of the universe-origin
- hear Dr. Nathanael-Israel Israel's personal selection and teaching of key topics that will help you break the code of the universe formation and functioning, and strategically enlighten you, guide you to navigate and filter the massive data collected on the universe and its content so you know how to answer the world's most challenging origin questions, remove any scientific and philosophical cataracts that may be blocking you, and help bring you many steps closer to your best life today and forever
- hear the greatest scientific and philosophic lessons of some top scientists, philosophers, thinkers, and public figures who have realized historic mistakes they made in life (concerning the origin of the universe, life, and chemicals), and that they corrected thanks to the discoveries of Nathanael-Israel Israel, who founded Science180, and who is acknowledged as the scientist that truly decrypted the universe-origin for the first time
- Get world key lessons successful people have learned in life and how people can learn from their experiences to improve lives instead of repeating their mistakes that many people still ignore at their own perils

To book Dr. Nathanael-Israel Israel for a speaking engagement purpose, visit Science180.com/speaking.

How you can make money by joining the affiliate program to sell Nathanael-Israel Israel's books

Greetings,

Do you want to make easy money by selling the #1 universe-origin, life-origin, and chemicals-origin books on your website, newsletter, and by mail? You can start making big money as you help sell Science180 Books including this one on your website and network. Indeed, by now, you know that I operate a website called Science180.com, specialized in helping people across the globe to scientifically decode and understand the formation of the universe, life, and chemicals.

Your contacts, site, blog, forum, podcast, and newsletter may be admired among my target audience. Some of my products and services may be of interest to your audience. My books are the first in history to scientifically demonstrate the universe formation in a way that satisfies both believers and nonbelievers, a historic achievement and discovery that is revolutionizing our view of the origin of the universe, life, and chemicals for the benefit of humankind.

Imagine you have a website where you can talk to people about my books and services and get a great percentage of every purchase they do on my site? Imagine you send a certain link about my books to your friends or network and, when any of your contacts buy a copy of my books, you get a percentage or a certain amount of what they pay on my sites. Imagine you can email your friends and spread the good news about my books and when anyone uses that link to buy my books, I give you something. Well! This is what the affiliate program is about. Apply today or learn more about it at Science180.com/affiliate. Likewise, if you own a website, you can apply for Science180's affiliate program, and I will send you a specific affiliate link that you will place on your website and newsletter, and if people click on it to buy my books, they will be led to my page and after they buy, I will pay you a certain amount, sharing the profit with you instead of just verbally saying thank you.

Would you be interested in reviewing some of my products and services with the aim of becoming an affiliate? We have a wonderful affiliate program and commissions are paid quickly and accurately.

If you are satisfied by the quality of our products and services, I am convinced you will also be impressed by our affiliate program.

I look forward to hearing from you

Nathanael-Israel Israel, PhD

Collaborate or partner with Nathanael-Israel Israel

If you have any lawful idea, initiative, or suggestion for a genuine partnership with Dr. Nathanael-Israel Israel or Science180, please visit Science180.com/partner to inform us.

How to be trained or mentored by or have a one-on-one consulting with Dr. Nathanael-Israel Israel

Hire Nathanael-Israel Israel to train you or your organization in the best ways to conduct yourself and your organization to align your initiatives with the real understanding of the origin of the universe, of life, and of chemical particles in a way that you will not hear anywhere else. Nathanael-Israel Israel offers training through the program called "Science180 University". For training purposes, please visit Science180Academy.com.

Visit Nathanael-Israel Israel's personal website to get for free great resources you won't find anywhere else

To stay in touch with, Dr. Nathanael-Israel Israel, and to get updates directly from him, please visit his website, Israel120.com, and sign up for his popular newsletter at Israel120.com/newsletter for free.

Ask for review

If you are a book reviewer or a professional wanting to review this book or others written by Nathanael-Israel Israel, please contact us at Science180.com/AskForReview

Donate and support Nathanael-Israel Israel's efforts and initiatives

To help humankind accurately understand the real origin of the universe and its content, like I have done in the groundbreaking books I published after 12 years of sacrifice, I need your financial support. If this book has been a blessing to you, please consider donating to me by visiting Israel120.com/donate to help me continue expounding on what I started. Your donation will be used to help me continue doing what I did to birth these books that you enjoyed and that you know will help many people across globe. No amount of money is too small or too big. Whatever you can give, please give.

Quantity discounts: Purchase Science180 books including this one in bulk at a special discount

To purchase Science180 books including this one in bulk at a special discount for sales promotion, corporate gifts, fund-raising, or educational purposes or to create special editions to specifications, contact specialsales@science180.com or visit Science180.com/discount.

Buy a copy of Nathanael-Israel Israel's books for your friends, family, or someone

If this book has been a blessing to you, and we know it has, please consider getting another copy and giving it to a friend, a family member, or someone you think it may help or challenge. If you want to get many copies, we can even give you a discount; just contact us as we previously explained.

Recommend Nathanael-Israel Israel's books to your organization

Because I know this book has been a blessing to you, I ask that you recommend it and others that I wrote to your organization, class, workplace, church, school, network, or clubs. Recommending this book will help others to tap into the blessing and opportunities that my books will open for them.

Share Nathanael-Israel Israel's groundbreaking discovery with others

To improve more lives, please share the findings of Nathanael-Israel Israel's books with others, for many people out there still do not understand how the universe was formed and sharing your experience of reading this book will help them. If you enjoy Nathanael-Israel Israel's books, please help other people find them by writing a book review on your blog or on online bookstores, or write it and share it with us. Likewise, share and mention this book on your social media platforms (e.g. Facebook, Twitter, YouTube, etc.).

Follow Nathanael-Israel Israel on social media

In our modern world, social medias have become a huge part of how messages spread across the globe today. To ensure more people hear about the good news revealed in my books, I need you to follow me and share my contents on your social medias and in your network. To know the full list of my social media accounts and follow me please visit Science180.com/socialmedia.

Share your feedback, critics, testimony, experience, adventures, story, or comment about this book with me

How has Nathanael-Israel Israel's books and services at Science180 improved your life? I would love to hear from you.

To help me know how I can better help you next and encourage others, I need to know and capture your testimony or critics. Please visit the feedback page, science180.com/feedback, to tell me:

- how this book impacted you or will impact you
- what you like or dislike or disagree with
- what you think, wish, or dream that I need to work on next
- what you wish to see in this book but that was absent
- what shocked you the most
- what got your heart pumping as you were reading this book
- what you found more insightful or thought-provoking
- what you want to do to be a part of my journey
- how my work changed your life or someone else's life

Nathanael-Israel Israel: Acknowledged as the World's Most Trusted Expert that Properly Decoded the Formation of the Universe, Life, and Chemicals

Message from the publisher of this book

Just like Nathanael-Israel Israel, you can publish your book(s) with us too. To get started and see how we may help you, please visit Science180Publishing.com today.

To contact Nathanael-Israel Israel or Science180

For any suggestions or questions, please visit Science180.com/contact and Nathanael-Israel Israel's personal website: Israel120.com. Feel free to ask me any questions you have about the universe formation, life formation, and chemicals formation.

Another Book by Nathanael-Israel Israel:
ORIGIN OF THE SPIRITUAL WORLD

ONLY ONE ANCIENT BLUEPRINT HAS THE RELIABLE POWER TO HELP YOU TO ACCURATELY DECRYPT THE SPIRITUAL ORIGIN AND HISTORY OF EVERYTHING IN THE UNIVERSE

Countless books talk about the origin of the universe and of life, but this amazing book is the first and the only one that has undeniably explained how the formation of the universe and everything in it was truly revealed in the rejected and hidden scriptures such as the Books of Enoch and others. In *"Origin of the Spiritual World"*, you will:

- Discover deep rejected secrets that have prevented humankind from unearthing the beginning of the universe
- Plainly see the scientific proof (hidden in scriptures) of the formation of the Earth, the Moon, and the Sun in a matter of days, a historic revelation that bizarrely and shockingly matches the scientific data as scientifically proved in *"From Science to Bible's Conclusions"*, a popular book written by Dr. Nathanael-Israel Israel
- Properly use the lost and rejected scriptures to articulate the process by which the universe was formed, and use that insight to improve your understanding of the Bible, innovate in your domain of interest, and improve your life perpetually
- Empower and align yourself with the historic breakthrough that has done what no other discovery has ever done: accurately unlock and decode mysteries concerning the origin of the cosmos and its content using scientific keys revealed in ancient scriptures that some elites have concealed (*Science180.com*/pseudepigraphic)

- - Discover and apprehend the complex formation of the universe and life without leaving out the challenging questions that people of all ages have been struggling to answer for thousands of years, while the answers were hidden
  - Find more joy in life through a clear interpretation of old and fresh revelations about the creation of the universe astonishingly backed by modern science, which some people wrongly think opposes the Bible
  - Make a difference and blaze new trails for those who depend on your leadership

If you believe in God, have some origin-related questions which answers you cannot find anywhere, not even in the Bible, and if you want to tap into historically neglected revelations to answer fundamental universe and life questions, then be sure to get a copy of *"Origin of the Spiritual World"* today.

Dr. Nathanael-Israel Israel happens to be the discoverer of the historic mathematical equations that scientifically demonstrated that the Earth was formed 2.82 days, the Moon 3.32 days, and the Sun 3.69 days after the beginning of the universe, therefore confirming the Biblical account of creation that revealed about 3500 years ago that the formation of the Earth was completed on the 3rd day, while that of the Moon and the Sun was completed on the 4th day of creation. Nathanael-Israel Israel is referred to as the "Undisputable Specialist of all Questions at the Intersection of Science and Biblical Creation". Learn more about this rare scientist at Israel120.com.

Nathanael-Israel Israel: Acknowledged as the World's Most Trusted Expert that Properly Decoded the Formation of the Universe, Life, and Chemicals

REFERENCES

During the course of the writing of this book, I read thousands of articles and gathered several references on them. Each of those references were cited according to the guidelines of their publishers. For the sake of time and efficiency, I did not force myself to rewrite all references to fit a single citation style. Instead, I alphabetically listed some of these references below just as cited in the source documents where I found them, hoping that those who are interested in them could follow up instead of me spending more time trying to standardize the reference citations and styles so they please grammarians! Some of these articles were key to my discoveries, and if you want to know about them, so you can focus your precious time on what matters the most, please consider joining "Science180 Academy" (www.Science180Academy.com), where you will find a program that best suits your needs, regardless of your educational background, and we start from there. "Science180 Academy" is designed to fit everybody, even laypeople who never attended college. As far as you can read, we can help you.

Adriani A. et al. (2018). Clusters of cyclones encircling Jupiter's poles. Nature volume 555, pages 216–219. doi:10.1038/nature25491. Visited on 8 March 2018 at https://www.nature.com/articles/nature25491.

Alard C. (2001). "Another bar in the Bulge". Astronomy and Astrophysics Letters. 379 (2): L44–L47. arXiv:astro-ph/0110491. Bibcode:2001A&A...379L..44A. doi:10.1051/0004-6361:20011487.

Anderson, John D.; James K. Campbell; Michael Martin Nieto (2007), "The energy transfer process in planetary flybys", New Astronomy 12 (5): 383–397, arXiv:astro-ph/0608087,Bibcode:2007NewA...12..383A, doi:10.1016/j.newast.2006.11.004.

Bahcall, N. A. (1988). "Large-scale structure in the Universe indicated by galaxy clusters". Annual Review of Astronomy and Astrophysics. 26 (1): 631–686. Bibcode:1988ARA&A..26..631B. doi:10.1146/annurev.aa.26.090188.003215.

Barbet B, Atten P and Soucemarianadin A (1997). J. Imaging Sci. Technol. 41 570.

Basak S. and S. Sarkar (2006). Dynamics of a stratified shear layer with horizontal shear. J. Fluid Mech. (2006), vol. 568, pp. 19–54.

Bell, G. R.; Levine, S. E. (1997). "Mass of the Milky Way and Dwarf Spheroidal Stream Membership". Bulletin of the American Astronomical Society. 29 (2): 1384. Bibcode:1997AAS...19110806B.

Bentley W A (1904). Mon. Weather Rev. 10 450–6.

Bhikku Bodhi (2007). "III.1, III.2, III.5". In Access To Insight
(http://www.accesstoinsight.org/tipitaka/dn/dn.01.0.bodh.html). The All
Embracing Net of Views: Brahmajala Sutta. Kandy, Sri Lanka: Buddhist
Publication Society.

Billant, P. & Chomaz, J.-M. (2000). Experimental evidence for a new instability of a
vertical columnar vortex pair in a strongly stratified fluid. J. Fluid Mech. 418,
167–188.

Booker, J. R. & Bretherton, F. P. (1967). The critical layer for internal gravity waves in a
shear flow. J. Fluid Mech. 27, 513–539.

Boulanger N., Meunier P. and S. Le Dizes (2007). Structure of a stratified tilted vortex.
J. Fluid Mech. (2007), vol. 583, pp. 443–458.
doi:10.1017/S0022112007006416.

Borenstein, Seth (2010). "Universe's Star Count Could Triple". CBS News. December 1,
2010. Retrieved 2011-07-14.

Braga-Ribas, F.; Sicardy, B.; Ortiz, J. L.; Snodgrass, C.; Roques, F.; Vieira-Martins, R.;
Camargo, J. I. B.; Assafin, M.; Duffard, R.; Jehin, E.; Pollock, J.; Leiva, R.;
Emilio, M.; Machado, D. I.; Colazo, C.; Lellouch, E.; Skottfelt, J.; Gillon, M.;
Ligier, N.; Maquet, L.; Benedetti-Rossi, G.; Gomes, A. R.; Kervella, P.;
Monteiro, H.; Sfair, R.; Moutamid, M. E.; Tancredi, G.; Spagnotto, J.; Maury,
A.; et al. (2014). "A ring system detected around the Centaur (10199)
Chariklo". Nature. 508 (7494): 72–75. arXiv:1409.7259.
Bibcode:2014Natur.508...72B. doi:10.1038/nature13155. PMID 24670644.

Bremond N and Villermaux E (2006). J. Fluid Mech. 549 273–306.

Burlaga, L. F.; et al. (2005). "Crossing the Termination Shock into the Heliosheath:
Magnetic Fields". Science. 309 (5743): 2027–2029.
Bibcode:2005Sci...309.2027B. PMID 16179471. doi:10.1126/science.1117542.

Burns, J.A.; Hamilton, D.P.; Showalter, M.R. (2001). "Dusty Rings and Circumplanetary
Dust: Observations and Simple Physics" (PDF). In Grun, E.; Gustafson, B. A.
S.; Dermott, S. T.; Fechtig H. (eds.). Interplanetary Dust. Berlin: Springer. pp.
641–725. Bibcode:2001indu.book..641B. ISBN 3-540-42067-3.

Canals M., Geno Pawlak, and Parker Maccready (2009). Tilted Baroclinic Tidal Vortices.
Journal of Physical Oceanography. Volume 39: 333-350.

Cariteau, B. (2005). Etude de la stabilité et de l'interaction de cyclones intenses en fluide
stratifié. PhD thesis, Université Joseph Fourier, Grenoble.

REFERENCES

Cariteau, B. & Flor, J.-B. (2003). Instability of a columnar vortex in stratified fluid. Bull. Am. Phys. Soc. 48 (10), 164.

CFA (2017). Satellite Galaxies. *https://www.cfa.harvard.edu/~lsales/SatelliteGalaxies.html. Retrieved 2017-08-20.*

Chaudhary K C and Maxworthy T (1980). J. Fluid Mech. p287.

Chen Alvin, Patrick Notz, and Osman A. Basaran. (2002). Computational and experimental analysis of pinch-off and scaling. Phys. Rev. Lett., 88(17): 174501.

CNN (2011). Spacecraft enters 'cosmic purgatory'". CNN. December 6, 2011. Retrieved December 7, 2011.

Comins, Neil F.; Kaufmann, William J. (2008). Discovering the Universe: From the Stars to the Planets. MacMillan. p. 347. Bibcode:2009dufs.book.....C. ISBN 978-1429230421.

Cornell University (2005). Galaxies. Cornell University. October 20, 2005. Archived from the original on June 29, 2014. Retrieved August 10, 2006.

Cristini V. and Y. Renardy (2006). Scalings for Droplet Sizes in Shear-Driven Breakup: Non-Microfluidic Ways to Monodisperse Emulsions. FDMP, vol. 2, no. 2, pp.77-93, 2006.

Cushman-Roisin Benoit (2015). Turbulence in Fluids: Kolmogorov's Theory of Inertial Turbulence. Lecture at Thayer School of Engineering Dartmouth College, New Hampshire, USA. www.dartmouth.edu/~cushman/courses.hmtl, visited on October 17, 2018.

Davis, Kate (2000). "Variable Star of the Month—December, 2000: Alpha Orionis". AAVSO. Archived from the original on 2006-07-12. Retrieved 2006-08-13.

Dauphole, B.; et al. (1996). "The kinematics of globular clusters, apocentric distances and a halo metallicity gradient". Astronomy and Astrophysics. 313: 119–128. Bibcode:1996A&A...313..119D.

Devlin, Hannah; Correspondent, Science (2015). "Astronomers discover largest known structure in the universe is ... a big hole". The Guardian. April 20, 2015.

Dubinski, J. (1998). "The Origin of the Brightest Cluster Galaxies". The Astrophysical Journal. 502 (2): 141–149. arXiv:astro-ph/9709102. Bibcode:1998ApJ...502..141D. doi:10.1086/305901. Archived from the original on May 14, 2011. Retrieved January 16, 2007.

Eggers Jens (1997). Nonlinear dynamics and breakup of free-surface flows. Reviews of Modern Physics, Vol. 69, No. 3, July 1997 0034-6861/97/69(3)/865(65).

Eggers Jens and Emmanuel Villermaux. (2008). Physics of liquid jets. Rep. Prog. Phys., 71(3):036601 (79pp). doi:10.1088/0034-4885/71/3/036601.

ESA (2009). Mystery remains: Rosetta fails to observe swing by anomaly". ESA. Archived from the original on 2009-12-23.

ESA/Hubble Press Release (2015). Hubble Pinpoints Furthest Protocluster of Galaxies Ever Seen". ESA/Hubble Press Release. Retrieved January 22, 2015 - http://www.spacetelescope.org/news/heic1201.

Eskridge, P. B.; Frogel, J. A. (1999). "What is the True Fraction of Barred Spiral Galaxies?". Astrophysics and Space Science. 269/270: 427–430. Bibcode:1999Ap&SS.269..427E. doi:10.1023/A:1017025820201.

Esposito, L. W. (2002). "Planetary rings". Reports on Progress in Physics. 65 (12): 1741–1783. Bibcode:2002RPPh...65.1741E. doi:10.1088/0034-4885/65/12/201.

Faller E. James et al. (2020). Gravity. Encyclopedia Britannica. https://www.britannica.com/science/gravity-physics. Visited on September 8, 2020.

Farmer, D., R. Pawlowicz, and R. Jiang (2002). Tilting separation flows: A mechanism for intense vertical mixing in the coastal ocean. Dyn. Atmos. Oceans, 36, 43–58.

Feynman Richard (2006). The Feynman Lectures on Physics, The Definitive Edition, California Institute of Technology, Pearson, Addison Wesley, Vol. 1, p. 7-9.

Finley, Dave and Aguilar, David (2009). "Milky Way a Swifter Spinner, More Massive, New Measurements Show"(Press release). National Radio Astronomy Observatory. January 5, 2009. Archived from the original on August 8, 2014. Retrieved January 20,2009.

Fraser Cain (2007). "When Our Galaxy Smashes into Andromeda, What Happens to the Sun?". Universe Today. Retrieved 2007-05-16.

Frisch Uriel (1995). Turbulence: The legacy of A.N. Kolmogorov. Cambridge University Press, 296 pp.

Gallup (2013). Evolution, Creationism, Intelligent Design. Retrieved 4 May 2013. http://www.gallup.com/poll/21814/Evolution-Creationism-Intelligent-Design.aspx.

George William K. (2013a). Lectures in Turbulence for the 21st Century. Department of Aeronautics, Imperial College of London, London, UK and Department of Applied Mechanics - Chalmers University of Technology Gothenburg, Sweden. Downloaded from www.turbulence-online.com on October 19, 2017.

REFERENCES

George William K. (2013b). An Introduction to Natural Convection Flows. Chalmers University of Technology Gothenburg, Sweden, 55 pages. Download from www.turbulence-online.com on October 19, 2017.

Georgis, Faris (2010). Alone in Unity: Torments of an Iraqi God-Seeker in North America. Dorrance Publishing. p. 62. ISBN 1-4349-0951-4.

Gibney, Elizabeth (2014). "Earth's new address: 'Solar System, Milky Way, Laniakea'". Nature. doi:10.1038/nature.2014.15819.

Gnedin, O. Y.; Lee, H. M.; Ostriker, J. P. (1999). "Effects of Tidal Shocks on the Evolution of Globular Clusters". The Astrophysical Journal. 522 (2): 935–949. arXiv:astro-ph/9806245. Bibcode:1999ApJ...522..935G. doi:10.1086/307659.

Gorokhovski M A and Saveliev V L (2003). Phys. Fluids 15 184–92.

Gott III, J. R.; et al. (2005). "A Map of the Universe". The Astrophysical Journal. 624 (2): 463–484. arXiv:astro-ph/0310571. Bibcode:2005ApJ...624..463G. doi:10.1086/428890.

Graham M. Harper; et al. (2008). "A New Vla-Hipparcos Distance To Betelgeuse And Its Implications". The Astronomical Journal. 135 (4): 1430–1440. Bibcode:2008AJ....135.1430H. doi:10.1088/0004-6256/135/4/1430.

Grant, J.; Lin, B. (2000). "The Stars of the Milky Way". Fairfax Public Access Corporation. Archived from the original on June 11, 2007. Retrieved May 9, 2007.

Gupta Ronak (2015). The 7 biggest unsolved mysteries in science. https://www.digit.in/features/general/7-greatest-unsolved-problems-in-science-26132.html. Published on May 26 2015.

Gurnett, D. A.; Kurth, W. S.; Burlaga, L. F.; Ness, N. F. (2013). "In Situ Observations of Interstellar Plasma with Voyager 1". Science 341: 1489–1492.Bibcode:2013Sci...341.1489G. doi:10.1126/science.1241681.

Harvard-Smithsonian Center for Astrophysics (2006). Spitzer Reveals What Edwin Hubble Missed". May 31, 2004. Archived from the original on September 7, 2006. Retrieved December 6, 2006. - http:/www.cfa.harvard.edu/press/pr0419.html.

Henderson, D; Pritchard W; Smolka Linda (1997). "On the pinch-off of a pendant drop of viscous fluid". Physics of Fluids. 9 (11): 3188. Bibcode: 1997PhFl....9.3188H. doi:10.1063/1.869435.

Hennequin Y, Aarts D G A L, van der Wiel J H, Wegdam G, Eggers J, Lekkerkerker H N W and Bonn D (2006). Phys. Rev. Lett. 97 244503.

Hilbing J H and Heister S D (1996). Phys. Fluids 8 1574.

Hirota M., Y Nishio, S Izawa and Y Fukunishi (2017). Vortex stretching in a homogeneous isotropic turbulence. Journal of Physics: Conf. Series 822 (2017) 012041   doi:10.1088/1742-6596/822/1/012041.

Hodge, Paul W. (2019). Galaxy. Encyclopedia Britannica. https://www.britannica.com/science/galaxy, consulted on December 3rd, 2019.

Hubble News Desk (1997). Hubble Finds Intergalactic Stars". Hubble News Desk. January 14, 1997. Retrieved 2006-11-06 - http://hubblesite.org/newscenter/archive/releases/1997/02/text/.

Islam for Today (2007). The Origin of Life: An Islamic perspective". Retrieved 2007-03-14. http://www.islamfortoday.com/emerick16.htm.

Israel Nathanael-Israel (2025a). Turbulent Origin of the Universe. Science180, Augusta, USA 683 pages.

Israel Nathanael-Israel (2025b). From Science to Bible's Conclusions. Science180, Augusta, USA 170 pages.

Israel Nathanael-Israel (2025c). Reconciling Science and Creation Accurately. Science180, Augusta, USA 299 pages.

Israel Nathanael-Israel (2025d). Turbulent Origin of Chemical Particles. Science180, Augusta, USA 397 pages.

Israel Nathanael-Israel (2025e). Turbulent Origin of Life. Science180, Augusta, USA 370 pages.

Israel Nathanael-Israel (2025f). Origin of the Spiritual World. Science180, Augusta, USA 151 pages.

Israel Nathanael-Israel (2025g). How Baby Universe Was Born. Science180, Augusta, USA 130 pages.

Israel Nathanael-Israel (2025h). How God Created Baby Universe. Science180, Augusta, USA 224 pages.

Israel Nathanael-Israel (2025i). Science180 Accurate Scientific Proof of God. Science180, Augusta, USA 214 pages.

Joseph D.D. et al. (1999). Breakup of a liquid drop suddenly exposed to a high-speed airstream. International Journal of Multiphase Flow 25, 1263±1303.

REFERENCES

Kawahara G., S. Kida, M. Tanaka and S. Yanase (1997). Wrap, tilt and stretch of vorticity lines around a strong thin straight vortex tube in a simple shear flow. J. Fluid Mech. (1997), vol. 353, pp. 115–162.

Kenneth, Kramer (1986), World scriptures: an introduction to comparative religions, p. 34, ISBN 978-0-8091-2781-8.

Kida Shigeo (2001). Life, Structure, and Dynamical Role of Vortical Motion in Turbulence (PDF). IUTAMim Symposium on Tubes, Sheets and Singularities in Fluid Dynamics. Zakopane, Poland.

Kivivali Lea (2014). "Nearby satellite galaxies challenge standard model of galaxy formation". Swinburne University of Technology. June 11, 2014. Archived from the original on March 16, 2015.

Kocevski, D. D.; Ebeling, H. (2006). "On the origin of the Local Group's peculiar velocity". The Astrophysical Journal. 645 (2): 1043–1053. arXiv:astro-ph/0510106. Bibcode:2006ApJ...645.1043K. doi:10.1086/503666.

Kolmogorov A N (1941). Dokl. Akad. Nauk SSSR 31 99–101.

Kolmogorov A N (1962). J. Fluid Mech. 13 82.

Konno M, Aoki M and Saito S (1983). J. Chem. Eng. Japan 16 313–19.

Koppes, Steve (2003). "University of Chicago physicist receives Kyoto Prize for lifetime achievements in science". June 20, 2003. The University of Chicago News Office. Retrieved 2012-06-15.

Kravtsov, V. V. (2001). "Globular Clusters and Dwarf Spheroidal Galaxies of the Outer Galactic Halo: on the Putative Scenario of their Formation" (PDF). Astronomical and Astrophysical Transactions. 20 (1): 89–92. Bibcode:2001A&AT...20...89K. doi:10.1080/10556790108208191. Retrieved 2010-03-02.

Krebs, Robert E. (1999). Scientific Development and Misconceptions Through the Ages: A Reference Guide (illustrated ed.). Greenwood Publishing Group. p. 133. ISBN 978-0-313-30226-8.

Laplace, P.S. (1805). Mechanique Celeste Supplement au X Libre. Paris: Courier.

Laws J O and Parsons D A (1943). Trans. Am. Geophys. Union 24 452–60.

Le Dizes S. (2000). 2000 Non-axisymmetric vortices in two-dimensional flows. J. Fluid Mech. 406, 175–198.

Leonard, Scott A; McClure, Michael (2004). Myth and Knowing (illustrated ed.). McGraw-Hill. ISBN 978-0-7674-1957-4.

Leong Stacy (2002). "Period of the Sun's Orbit around the Galaxy (Cosmic Year)". In Glenn Elert (ed.). The Physics Factbook (self-published). Retrieved 2008-06-26.

Liang et al. (2018). Observation of three-photon bound states in a quantum nonlinear medium. Science 16 Feb 2018, Vol. 359, Issue 6377, pp. 783-786. DOI: 10.1126/science.aao7293.

Lister, J. R.; Stone, H. A. (1998): Capillary breakup of a viscous thread surrounded by another viscous fluid. Phys. Fluids, vol. 10, pp. 2758–2764.

Mabile C, Leal-Calderon F, Bibette J and Schmitt V (2003). Europhys. Lett. 61 708–14.

Mackie, Glen (2002). "To see the Universe in a Grain of Taranaki Sand". Centre for Astrophysics and Supercomputing. February 1, 2002. Retrieved January 28, 2017.

Malvern (2016). Malvern Instruments White Paper - A Basic Introduction to Rheology, www.malvern.com, Worcestershire, UK, 20 pages.

Mark H. Jones; Robert J. Lambourne; David John Adams (2004). An Introduction to Galaxies and Cosmology. Cambridge University Press. p. 298. ISBN 978-0-521-54623-2.

Marshall J S and Palmer W M (1948). J. Meteorol. 5 165–6.

Martinez-Bazan C, Montanes J L and Lasheras J C (1999). J. Fluid Mech. 401 183–207.

Maslowe, S. A. (1986). Critical layers in shear flows. Annu. Rev. Fluid Mech. 18, 405–432.

Mayer J E and Mayer M G (1966). Statistical Mechanics (New York: Wiley).

McNamara Patrick and Wesley J. Wildman (2012). Science and the World's Religions [3 Volumes]. ABC-CLIO. pp. 180–. ISBN 978-0-313-38732-6. 19 July 2012. Retrieved 15 December 2012.

Moffatt H. K. (2011). A brief introduction to vortex dynamics and turbulence. WorldScientific_Abitvdat. Department of Applied Mathematics and Theoretical Physics University of Cambridge Wilberforce Road, Cambridge, UK.

Monin A S and Yaglom A M (1975). Statistical Fluid Mechanics: Mechanics of Turbulence vol 2 (Cambridge, MA: MIT Press).

REFERENCES

Mollo-Christensen Erik (1969). Flow Instability. The National Committee for Fluid Mechanics Film, under a grant from National Science Foundation. Massachusetts Institute of Technology. USA (https://www.youtube.com/watch?v=9lowHks4su8).

NASA (2014). Asteroid Fact Sheet. http://nssdc.gsfc.nasa.gov/planetary/factsheet/asteroidfact.html. 2014-8-24

NASA (2018). Planetary fact sheets. Fact sheets of the Sun, planets, satellites, rings and selected asteroids in the Solar System. Author/Curator: Dr. David R. Williams, NASA Goddard Space Flight Center, Greenbelt, MD, USA. http://nssdc.gsfc.nasa.gov/planetary/factsheet/. Visited on November 19, 2018.

NASA (2019). What Scientists Found After Sifting Through Dust in the Solar System. March 12, 2019. Retrieved on April 22, 2020 at https://www.nasa.gov/feature/goddard/2019/what-scientists-found-after-sifting-through-dust-in-the-solar-system.

Nemiroff, R.; Bonnell, J., eds. (September 9, 2002). "Hoag's Object: A Strange Ring Galaxy". https://apod.nasa.gov/apod/ap020909.html. Astronomy Picture of the Day. NASA. Retrieved March 31, 2012.

Novikov E A and Dommermuth D G (1997). Phys. Rev. E 56 5479–82.

Numbers, Ronald L. (2006) [Originally published 1992 as The Creationists: The Evolution of Scientific Creationism; New York: Alfred A. Knopf]. The Creationists: From Scientific Creationism to Intelligent Design (Expanded ed., 1st Harvard University Press pbk. ed.). Cambridge, MA: Harvard University Press. ISBN 0-674-02339-0. LCCN 2006043675.OCLC 69734583.

Ockert, M. E.; Cuzzi, J. N.; Porco, C. C.; Johnson, T. V. (1987). "Uranian ring photometry: Results from Voyager 2". Journal of Geophysical Research. 92(A13): 14, 969–78. Bibcode:1987JGR....9214969O. doi:10.1029/JA092iA13p14969.

Oliveira M S N and McKinley G H (2005). Phys. Fluids 17 071704.

Pavlov, Alexander A. (1999). "Irradiated interplanetary dust particles as a possible solution for the deuterium/hydrogen paradox of Earth's oceans". Journal of Geophysical Research: Planets. 104 (E12): 30725–28. Bibcode:1999JGR...10430725P. doi:10.1029/1999JE001120.

Petitjeans Philippe and Frédéric Bottausci (2020). Structures tourbillonnaires étirées: les filaments de vorticité. Laboratoire de Physique et de Mécanique des Milieux Hétérogènes (UMR CNRS 7636) Ecole Supérieure de Physique et de Chimie Industrielles 10, rue Vauquelin, 75005 Paris. 13 pages.

Science180: The Place Where Science Accurately Meets Biblical Creation

Pilch M and Erdman C A (1987). Int. J. Multiph. Flow 13 741–57.

Pimbley W. T. and H. C. Lee (1977). Satellite droplet formation in a liquid jet. IBM J. Res. Develop., 21:385–388.

Pokorny Petr and Kuchner Marc (2019). Co-orbital Asteroids as the Source of Venus's Zodiacal Dust Ring. The Astrophysical Journal Letters, 873: L16 (11pp), 2019 March 10.

Porco, Carolyn (2012). "Questions about Saturn's rings". CICLOPS web site. Archived from the original on 2012-10-03. Retrieved 2012-10-05.

Price F. James (2006). Lagrangian and Eulerian Representations of Fluid Flow: Kinematics and the Equations of Motion. Woods Hole Oceanographic Institution, Woods Hole, MA, 02543. http://www.whoi.edu/science/PO/people/jprice June 7, 2006.

Rayleigh Lord (1879). Proc. Lond. Math. Soc. 4 4–13.

Sale, S. E. et al. (2010). "The structure of the outer Galactic disc as revealed by IPHAS early A stars". Monthly Notices of the Royal Astronomical Society. 402 (2): 713–723. arXiv:0909.3857. Bibcode:2010MNRAS.402..713S. doi:10.1111/j.1365-2966.2009.15746.x.

Sanders, R. (2006). "Milky Way galaxy is warped and vibrating like a drum". UCBerkeley News. January 9, 2006. Retrieved May 24, 2006.

Seinfeld J H and Pandis S N (1998). Atmos. Chem. Phys. (New York: Wiley).

Shapley, Harlow (1918). "Globular clusters and the structure of the galactic system". Publications of the Astronomical Society of the Pacific. 30 (173): 42–54. Bibcode:1918PASP...30...42S. doi:10.1086/122686.

Sparke, Linda S.; Gallagher, John S. (2007). Galaxies in the Universe: An Introduction. p. 90. ISBN 9781139462389.

Staff (2019). "How Many Stars Are There In The Universe?". European Space Agency. Retrieved September 21, 2019.

Stern S A, Weaver H A, Steffl A J, Mutchler M J, Merline W J, Buie M W, Young E F, Young L A and Spencer J R (2006). Nature 439 946–8.

Showalter M. R., D. P. Hamilton (2015). "Resonant interactions and chaotic rotation of Pluto's small moons". Nature. 522: 45–49. Bibcode:2015Natur.522...45S. doi:10.1038/nature14469. PMID 26040889.

REFERENCES

Showalter, Mark R.; Burns, Joseph A.; Cuzzi, Jeffrey N.; Pollack, James B. (1987). "Jupiter's ring system: New results on structure and particle properties". Icarus. 69 (3): 458–498. Bibcode:1987Icar...69..458S. doi:10.1016/0019-1035(87)90018-2.

Smith, Bradford A.; Soderblom, Laurence A.; Johnson, Torrence V.; Ingersoll, Andrew P.; Collins, Stewart A.; Shoemaker, Eugene M.; Hunt, G. E.; Masursky, Harold; Carr, Michael H. (1979). "The Jupiter System Through the Eyes of Voyager 1". Science. 204 (4396): 951–972. Bibcode:1979Sci...204..951S. doi:10.1126/science.204.4396.951. ISSN 0036-8075. PMID 17800430.

Smith, B. A.; Soderblom, L. A.; Banfield, D.; Barnet, C; Basilevsky, A. T.; Beebe, R. F.; Bollinger, K.; Boyce, J. M.; Brahic, A. (1989). "Voyager 2 at Neptune: Imaging Science Results". Science. 246 (4936): 1422–1449. Bibcode:1989Sci...246.1422S. doi:10.1126/science.246.4936.1422. ISSN 0036-8075. PMID 17755997.

Smith, G. (2000). "Galaxies — The Spiral Nebulae". University of California, San Diego Center for Astrophysics & Space Sciences. March 6, 2000. Archived from the original on July 10, 2012. Retrieved November 30, 2006.

Stone, H. A., B. J. Bentley, and L. G. Leal (1986). J. Fluid Mech. 173, 131.

Stockmayer W H (1943). J. Chem. Phys. 11 45–55.

Ting, L. (1991). Viscous Vortical Flows. Lecture notes in physics. Springer-Verlag. ISBN 3-540-53713-9.

Tjahjadi M., H. A. Stone and J. M. Ottino. Satellite and subsatellite formation in capillary breakup. J. Fluid Mech. (1992), ad. 243, pp. 297-317.

Throop, H. B.; Porco, C. C.; West, R. A.; et al. (2004). "The Jovian Rings: New Results Derived from Cassini, Galileo, Voyager, and Earth-based Observations" (PDF). Icarus. 172 (1): 59–77. Bibcode:2004Icar..172...59T. doi:10.1016/j.icarus.2003.12.020.

Tiscareno, Matthew S. (2013). "Planetary Rings". In Oswalt, Terry D.; French, Linda M.; Kalas, Paul (eds.). Planets, Stars and Stellar Systems. Springer Netherlands. pp. 309–375. arXiv:1112.3305. doi:10.1007/978-94-007-5606-9_7. ISBN 9789400756052.

Torrance Robert M. (1999). Encompassing Nature: A Sourcebook. Counterpoint Press. pp. 121–122. ISBN 978-1-58243-009-6. Retrieved 15 December 2012.

Tucker C L and Moldenaers P (2002) Annu. Rev. Fluid Mech. 34 177–210.

TURBULENT ORIGIN OF THE UNIVERSE

Turbulence Conference (2011). Turbulence Conference in Marseille (http://turbulence.ens.fr), Session 2: Shear and Wake Flow Turbulence. General discussion report. 11 pages.

University of Cambridge (2007). Galaxy Clusters and Large-Scale Structure. University of Cambridge. Retrieved January 15, 2007 (http://www.damtp.cam.ac.uk/user/gr/public/gal_lss.html.

van den Bergh, S. (2000). "Updated Information on the Local Group". Publications of the Astronomical Society of the Pacific. 112 (770): 529–536. arXiv:astro-ph/0001040. Bibcode:2000PASP..112..529V. doi:10.1086/316548.

Wikipedia (2016a). Jupiter trojan. Retrieved on 2016-7-4, from https://en.wikipedia.org/wiki/Jupiter_trojan.

Wikipedia (2016b). Mars trojan. Retrieved on 2016-7-4, from https://en.wikipedia.org/wiki/Mars_trojan.

Wikipedia (2016c). Neptune trojan. Retrieved on 2016-7-4, from https://en.wikipedia.org/wiki/Neptune_trojan,

Wikipedia (2016d). Trans-Neptunian object. Retrieved on 2016-7-4, from https://en.wikipedia.org/wiki/Trans-Neptunian_object.

Wikipedia (2016e). Asteroid belt. Retrieved on 2016-7-4, from https://en.wikipedia.org/wiki/Asteroid_belt.

Wikipedia (2016f). Apollo asteroids. Retrieved on 2016-7-4, from https://en.wikipedia.org/wiki/List_of_Apollo_asteroids.

Wikipedia (2016g). Amor asteroid. Retrieved on 2016-7-4, from https://en.wikipedia.org/wiki/Amor_asteroid.

Wikipedia (2016h). Aten asteroid. Retrieved on 2016-7-4, from https://en.wikipedia.org/wiki/Aten_asteroid.

Wikipedia (2016i). Centaur (minor planet). Retrieved on 2016-7-4, from https://en.wikipedia.org/wiki/Centaur_(minor_planet).

Wikipedia (2016j). Comet. Retrieved on 2016-7-4, from https://en.wikipedia.org/wiki/Comet.

Wikipedia (2016k). Kuiper belt. Retrieved on 2016-7-4, from https://en.wikipedia.org/wiki/Kuiper_belt.

Wikipedia (2016l). Trans-Neptunian object. Retrieved on 2016-7-4, from https://en.wikipedia.org/wiki/Trans-Neptunian_object.

REFERENCES

Wikipedia (2016m). Voyager 1. Retrieved on July 7, 2016, from
https://en.wikipedia.org/wiki/Voyager_1.

Wikipedia (2017a). Galaxy rotation curve. Retrieved on October 10, 2017, from
https://en.wikipedia.org/wiki/Galaxy_rotation_curve.

Wikipedia (2017b). Constellation. Retrieved on October 10, 2017, from
https://en.wikipedia.org/wiki/Constellation.

Wikipedia (2019a). Fluid thread breakup. Retrieved on November 2019, from
https://en.wikipedia.org/wiki/Fluid_thread_breakup.

Wikipedia (2019b). Gravity. Retrieved on November 2019, from
www.en.wikipedia.org/wiki/Gravity.

Wikipedia (2019c). Galaxy. Retrieved on September 29, 2019, from
https://en.wikipedia.org/wiki/Galaxy.

Wikipedia (2019d). Globular cluster. Retrieved on on September 29, 2019, from
https://en.wikipedia.org/wiki/Globular_cluster.

Wikipedia (2020a). Shepherd moon. Retrieved on 22, 2020, from
https://en.wikipedia.org/wiki/Shepherd_moon.

Wikipedia (2020b). Flyby anomaly. Retrieved on September 8, 2020, from
wikipedia.org/wiki/Flyby_anomaly.

Wikipedia (2020c). Gravity assist. Retrieved on September 8, 2020, from
https://en.wikipedia.org/wiki/Gravity_assist.

Wikipedia (2023). Creationism. Retrieved on February 24, 2023, from
https://en.wikipedia.org/wiki/Creationism.

Xing J H, Boguslawsky A, Soucemarianadin A, Atten P and Attane P (1996). Exp.
Fluids 20 302.

Young, T. (1805). "An Essay on the Cohesion of Fluids". Philosophical Transactions of
the Royal Society of London. 95: 65–87. doi:10.1098/rstl.1805.0005.

Nathanael-Israel Israel: Historic Discoverer of the Universe Turbulent Origin
Formula™

INDEX

Nathanael-Israel Israel: Who is Told by People That He is the Universe-Origin, Life-Origin & Chemicals-Origin Accurate Decoder

Nathanael-Israel Israel: Who is Told by People That He is the Universe-Origin, Life-Origin & Chemicals-Origin Accurate Decoder

Nathanael-Israel Israel: Who is Told by People That He is the Universe-Origin, Life-Origin & Chemicals-Origin Accurate Decoder

Nathanael-Israel Israel: Who is Told by People That He is the Universe-Origin, Life-Origin & Chemicals-Origin Accurate Decoder

Nathanael-Israel Israel: Who is Told by People That He is the Universe-Origin, Life-Origin & Chemicals-Origin Accurate Decoder

Nathanael-Israel Israel: Who is Told by People That He is the Universe-Origin, Life-Origin & Chemicals-Origin Accurate Decoder

Nathanael-Israel Israel: Who is Told by People That He is the Universe-Origin,
Life-Origin & Chemicals-Origin Accurate Decoder

X

Science180: Universe-Origin Problems Final Bus Stop

DEDICATION

God Almighty! Creator of the Heavens and Earth, and of everything else I don't know! As I dare to start writing this book after so many years of thinking about it and deeply researching about the things You have created and formed, I feel blessed and honored to reveal in it my enlightened understanding of the universe and the laws that not only were at its beginning, but also at the core of its sustenance since then. As a human being aware of my limits, despite the ability I have with Your grace, I know that this book still contains errors, and is far from being the perfect expression and demonstration of what You have created, which I could not holistically understand nor apprehend in its totality. Yet, I dare to present this masterpiece as my effort to show the world that, despite the immense amount of scientific, philosophic, and vulgar data about the universe and the (usually wrong) theories written about them, there is still a way to review them scientifically and give the glory to You, God Almighty, who deserves to be praised instead of the creatures, which we sometimes pay more attention to. I hope this report will help someone believe in You, and bring some joy to You as a praise of a human being trying to tap into invisible things to explain part of the visible, that many long for understanding with their limited knowledge, which is usually filled with prejudices and unbelief, which block them. Therefore, I dedicate this work to You alone as a courageous sacrifice of a man who believes in You, and who is trying to honor You with the best scientific demonstration he could come up with, thanking You in advance that, regardless of the treatment human beings would give to this research and to me and to those who will believe in my efforts, You are a Merciful and Omniscient God, who forgives sins and who cleanses me from any imperfection in this work and in other things I have done in my life. Thank You for creating the beautiful universe and everything in it in 6 literal days of 24 hours each. You are good and everything You have done is indeed good. I can't wait to see You face to face very soon and know more about how You did it. Meanwhile, please accept my human account of how I think You created and formed the whole world and all that is in it! You are great and there is none like You, Oh Lord! Amen!

Nathanael-Israel Israel

June 18, 2020 at 9:14 PM

This was the first page I wrote when I started putting this book together and, because some people would have refused to finish reading the book if I had put this dedication at the beginning, I decided to place it at the end in an attempt to help many read what I have to say, hoping that by the time they finish the book, they could be more tolerant in their judgement or at least, they could have better understood something instead of stopping the reading because I dared to refer to God early!

ABOUT THE AUTHOR

From cracking the DNA of the universe-origin to unlocking the code of the formation of life and of chemical particles, this scientist and mathematician has done it all.

Dr. Nathanael-Israel Israel is the founder of Science180, the American company which mission is to improve the current and future state of human beings by accurately decoding and teaching them the real origin and formation of the universe, of life, and of chemicals, and meaningfully engaging business, nonprofit, political, academic, civil society leaders, and followers to properly shape local and global agendas that authentically value the truth. As the creator of the Universe Turbulent Origin Formula™. Dr. Nathanael-Israel Israel has revolutionized the way billions of people around the world think about the origin of the universe, of life, and of chemicals. Nobody understands and teaches the formation of everything in the universe (e.g. the Milky Way Galaxy, the Sun, the Earth, the Moon, and all other galaxies, stars, planets, satellites, and asteroids) better than Nathanael-Israel Israel. Individuals and organizations across the globe have been calling him so he helps them scientifically unlock the code of the universe-formation, helping veterans and rookies to have the real keys to decrypt the universe and turbulence (one of the top biggest unsolved mysteries in science) from the historic, unique, accurate, simple, easy-to-understand, nonconformist, trailblazing perspective that anybody can quickly learn at Science180 Academy (Science180Academy.com). Science180 Academy delivers outstanding value, insight, and lessons to assist people to accurately understand the true origin of the universe, chemicals, and life, so they can tap into that knowledge to improve lives perpetually. Nathanael-Israel's goal is to give you practicable and undeniable proofs of the formation of the universe so you can be fired up to become the best version of you, and to cause positive changes to your initiatives that will profit you today and forever. For Nathanael-Israel, accurately decoding and teaching the origin of the universe and everything in it is not a job, but his life mission, and helping others to fully understand that brings him closer to his assignment.

Dr. Israel earned his PhD in Plant, Insect, and Microbial Sciences in the USA, where he graduated first of his class of hundreds of PhD candidates. This Beninese-American is a member of the American Chemical Society, American Association for the Advancement of Science, American Society of Agricultural and Biological Engineers, American Society for Microbiology, American Society of Biochemistry and Molecular Biology, Ecological Society of America, American Society of Agronomy, Crop Science Society of America, and Soil Science Society of America. A scientist, a mathematician, a consultant, and the owner of Global Diaspora News, a news company in the USA, Dr. Israel is the author of the popular books:

- Turbulent Origin of Chemical Particles
- Turbulent Origin of Life
- From Science to Bible's Conclusions
- How Baby Universe Was Born
- How God Created Baby Universe
- Science180 Accurate Scientific Proof of God
- Turbulent Origin of the Universe
- Reconciling Science and Creation Accurately
- Origin of the Spiritual World

If you want to accurately understand the origin of anything, then be sure to get a copy of these amazing books. You cannot afford to ignore the greater, better, faster, simpler, cheaper, easier, and accurate formulas unlocked in these important books that successfully cracked the origin of the universe, of life, and of chemicals in a language that scientists, laypeople, adults, children, believers, skeptics, and anybody else can properly understand and enjoy.

Visit Israel120.com today to connect with this historic discoverer of the all-in-one proven and uncomplicated formula that accurately decoded the origin of the universe, of life, and of chemicals.